AN INTRODUCTION TO HYDRAULICS OF FINE SEDIMENT TRANSPORT

ADVANCED SERIES ON OCEAN ENGINEERING

Series Editor-in-Chief
Philip L- F Liu (*Cornell University*)

*For the complete list of titles in this series, please write to the Publisher.

Advanced Series on Ocean Engineering — Volume 38

AN INTRODUCTION TO HYDRAULICS OF FINE SEDIMENT TRANSPORT

Ashish J. Mehta

University of Florida, USA

World Scientific

NEW JERSEY · LONDON · SINGAPORE · BEIJING · SHANGHAI · HONG KONG · TAIPEI · CHENNAI

Published by

World Scientific Publishing Co. Pte. Ltd.

5 Toh Tuck Link, Singapore 596224

USA office: 27 Warren Street, Suite 401-402, Hackensack, NJ 07601

UK office: 57 Shelton Street, Covent Garden, London WC2H 9HE

Library of Congress Cataloging-in-Publication Data
Mehta, Ashish J.
  An introduction to hydraulics of fine sediment transport / Ashish J. Mehta.
  pages cm. -- (Advanced series on ocean engineering ; v. 38)
  Includes bibliographical references.
  ISBN 978-9814449489 (alkaline paper)
  1. Sediment transport. 2. Hydraulic engineering. 3. Marine engineering.
4. Coastal engineering. 5. Coastal sediments. 6. Estuarine sediments. 7. Marine sediments.
I. Title.
  TC175.2.M44 2013
  551.3'04--dc23

                                                              2013000653

British Library Cataloguing-in-Publication Data
A catalogue record for this book is available from the British Library.

Printed in Singapore by B & Jo Enterprise Pte

Dedication

Ishir, Aneesh and other seventh generation learned descendents of Bholanath Sarabhai Divetia (1822–1886)

Preface

I had the occasion to descend to the bottom in a current so swift as to require extraordinary means to sink the bell.... The sand was drifting like a dense snow storm at the bottom.... At sixty-five feet below the surface I found the bed of the river, for at least three feet in depth, a moving mass and so unstable that, in endeavoring to find a footing beneath my bell, my feet penetrated through it until I could feel, although standing erect, the sand rushing past my hands, driven by a current as rapid as that on the surface. I could discover the sand motion at least two feet below the surface of the bottom, and moving with a velocity diminishing in proportion to its depth.

–Description by James Eads of his descent to the bottom of the Mississippi River in mid-nineteenth century (from Barry [1998]).

Sediment transport is an observational science and will remain so; there is no substitute for the procurement and analysis of quality data from field and laboratory experiments. This work summarizes a variety of observations on the transport behavior of fine sediments, especially cohesive particles that aggregate in water. I have attempted to present these observations under process-related categories such as flocculation, erosion, deposition, and so on. The hydraulics-based approach reflects my understanding of transport phenomena in civil and chemical engineering during the past decades. Recent research suggests a shift towards the application of computational fluid dynamics to identify complex, non-linear causes and effects in fine sediment transport. This should lead to advanced simulations of sediment movement in a variety of settings, especially the marine environment. However, in typical coastal engineering projects' the need to carry out back-of-the-envelope calculations is unlikely to diminish, and to a fair extent this work is meant to support that need.

Inasmuch as cohesive sediment transport involves electrochemical as well as biochemical forces, coarse-grain transport is essentially a subset of particle dynamics in which these forces are ignored. I have included the main aspects of coarse-grain transport in Chapter 6, because understanding how coarse particles move in water under pressure, gravity, inertia and dissipative forces is in a sense a pre-requisite for dealing with the complexities of fine sediment movement. In fact, numerous developments in cohesive sediment transport are offshoots of theories and observations on sand movement.

This book is aimed at civil engineering seniors and graduate students who in the normal course of their curricula seldom come across this subject. The reader should have a basic understanding of the mechanics of fluid flow and open channel hydraulics. Refreshing the basics of chemistry can be quite useful. Admittedly, even though the subject as covered is "introductory," the twelve chapters taken together include far more material than what can be reasonably dealt with in a single semester. It will be necessary to be selective in choosing topics for a first course. Chapters 1 and 2 are merely preparatory. If entire chapters are chosen, Chapters 3, 4, 7, 9 and 12 could be selected. Geotechnical students are familiar with consolidation, which is only briefly covered in Chapter 8. Chapter 5, which includes a short review of mud rheology, is meant as the background for fluid mud transport (Chapter 10) and wave-mud interaction (Chapter 11). Students are encouraged to go through the exercises at the end of each chapter.

I am thankful to Cynthia Vey, who did almost all the typing for me for several years before personal computers became common. In the tedious task of text formatting, Candace Leggett provided excellent support at a critical juncture. Her meticulousness and rigor in handling word processing deserves a loud clap of hands. Exemplary professional help in the publication process was provided by Ms. Yun Hui En (Amanda) at World Scientific, Singapore.

For the technical content I have relied to a fair extent on studies by my students and research associates. In recent years some individuals have been directly connected with the production of this book. In particular I must acknowledge help provided by Bill McAnally (and his students), Han Winterwerp, Joe Letter, Mamta Jain, Dave Robillard, Yogesh Khare and Farzin Samsami. Earl Hayter, Andy Manning and Pravi Shrestha reviewed an early draft.

Credits for unpublished photographs not taken by me: Prof. Pradeep Talwani (Fig. 1.4a), Prof. David Robillard (Fig. 5.48), Dr. Robert Kirby (Fig. 9.1a), Prof. Lauro Calliari (Fig. 9.1b), Korea Ocean Research and Development Institute, Ansan (Fig. 9.60a), Profs. Dieter Muehe and Susana Vinzon (Fig. 12.47). The beach wave view in Fig. 11.29 was given by a knowledgeable and considerate graduate student whose name I am unable to recall.

Finally I must express my indebtedness to the professors who taught me how to think research and act engineering, in alphabetical order – Per Bruun, Bent Christensen, Robert Dean, Ray Krone, Morrough P. O'Brien, and last but not the least my thesis advisor Emmanuel Partheniades. Their imprint is all over these pages.

Ashish Mehta
University of Florida and
Nutech Consultants
Gainesville, Florida

Contents

Chapter 1

Introduction

1.1 Ancient Ports, Sea Level and Sedimentation

From archeological excavations and ancient records we know that humans began management of sediment shoaling at ports and harbors millennia before sedimentation became an engineering sub-discipline. Recent interpretations of historic developments in civil engineering technology suggest that methods to control sedimentation may have been widespread and many developed locally, as the rise of towns and ports followed farming 8,000 to 10,000 years Before Present (BP). Unfortunately, as research into past practices remains skewed with respect to its time-line and geographical distribution, we are left with sparse information on sedimentation engineering at pre-Columbian (*i.e.,* before year 1492 of Common Era) ports and harbors in numerous regions of the world [Graf, 1971; Albinia, 2008].

  We may surmise that in a broad sense the proliferation of ancient ports began when the rate of rise of post-glacial sea level[a] began to decrease around 6,000 yr BP. As seen in Fig. 1.1, a long-term mean rate of 0.67 m per century during 12,000 – 6,000 yr BP decreased to 0.05 m per century after about 6,000 yr BP. This order-of-magnitude drop in

---

[a] Sea level change of main interest to port engineers is associated with the relative level, which defines the local water depth. There are two categories of relative sea level change, eustatic and non-eustatic. Eustatic change is global. The two most important causes of global change are associated with the polar ice sheets, and steric change of near-surface ocean water referring to the specific volume of water which generally expands when heated or shrinks when cooled. Non-eustatic effects cause local changes. They mainly include tectonic lift or subsidence of the seafloor, and subsidence due to withdrawal of groundwater or hydrocarbons.

1

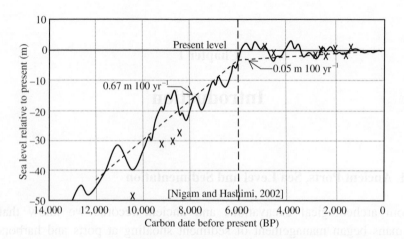

Figure 1.1. Sea level since ~13,000 yr BP and mean trends during pre- and post-6,000 yr BP periods (revised from Marine Board [1987]; based on Shepard [1963] with data points (×) for the western Indian continental margin from Nigam and Hashimi [2002]). Positive values of sea level indicate greater than present depth of water at a port.

the rate meant that shorelines began to remain stable over periods of decades rather than years, and enabled towns to flourish along shores.

When one considers the centennial time-scale over which some port structures may remain functional, a more complex relationship between coastal settlement and relative sea level emerges. There have been events of construction, relocation or abandonment, and even total destruction of ancient ports due to Holocene–Anthropocene transgressions and recessions of the sea, as the relative water level has risen and fallen. Figure 1.2 shows the historic sea level at Qalaat Al-Bahrain in the Persian Gulf. The greatest inundation of land occurred about 6,500 yr BP. The sea then receded as the level fell until a reversal around 4,700 yr BP, when the level rose. Around 4,300 yr BP the level began to fall again until about 3,500 yr BP. During this period of marine recession human occupation of the Bahrain seashore occurred. When the sea level began to rise once again the coastal part of the settlement had to be abandoned and was not reoccupied for about a millennium.

The coastal settlements of the people of the Sindhu–Sarasvati River Valley, also known as the Harappans, whose early habitations were along the banks of the Sindhu (Indus), followed nearly synchronous cycles of

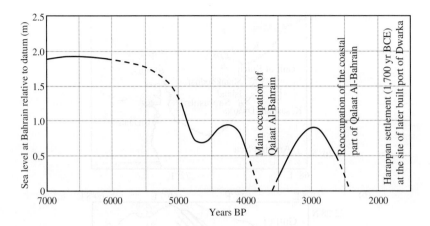

Figure 1.2. Variation of Holocene sea level at Qalaat Al-Bahrain (adapted from Rao [1994]).

growth and abandonment. Figure 1.3a marks some of the sites in the Kutch and Saurashtra regions of Gujarat (western India). At the site of the present-day city of Dwarka a settlement called Kusasthali was founded in the fifth millennium BP. It was destroyed when the sea level rose, and Dwarka was established in its place in the fourth millennium BP while the sea was receding. Later during that millennium Dwarka became submerged as the sea level rose and remained so until the city was reoccupied in the 3rd century Before Common Era (BCE). An extensive marine survey during the 1980's has revealed numerous rock structures of the still submerged portion of the port area (Fig. 1.3b).

Sedimentation was a critical concern and an engineering issue at many ancient ports. A case in point is an almost perfectly designed tidal basin in the Harappan settlement of Lothal (Fig. 1.3a), which pre-dated Dwarka. Excavations have revealed that the basin, measuring 214 m by 36 m in plan area, was built of masonry walls of burnt mud bricks (Figs. 1.4a, b). It was constructed around 2,350 yr BCE on the western side of a river reach, thought to be close to the confluence of Rivers Bhogao and Sabarmati. The depth of water in the basin was 2 m at low tide and 3.5 m at high tide. It is believed that coastal vessels could enter the basin at high tide through a 12 m wide entrance close to the northern wall of the basin. At the opposite end, a 1 m wide channel was constructed

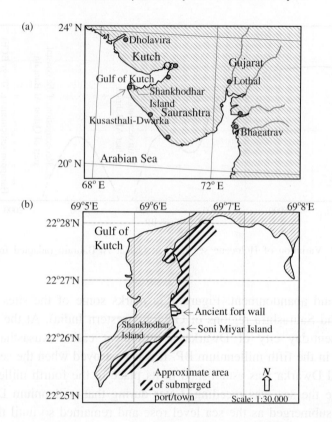

Figure 1.3. (a) Harappan coastal settlements (marked by dots) during c. 2,100–1,300 yr BCE; (b) submerged portion of Dwarka on Shankhodhar Island (adapted from Rao [1994]).

to allow excess water to escape the basin at high stages of tide. This channel served not only as a spillway but possibly also as a means to retain adequately deep water in the basin. The channel could be closed by lowering a wooden gate. The basin was in operation for about 350 years when a severe flood in the river changed its course by several kilometers away from the basin [Rao, 1973; Sankalia, 1987].

A tidal basin such as the one at Lothal with deeper water compared to the entrance is sometimes called a "half-tide" harbor (Fig. 1.5). Such a harbor allows vessels to enter when the water level is higher than the mean, and enables them to stay afloat while moored. Since river water

(a)

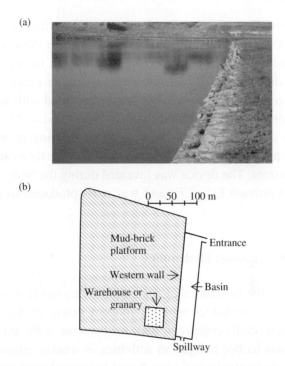

(b)

Figure 1.4. Tidal basin with a mud brick wall at Lothal: (a) Photograph taken in 2009 (Courtesy Pradeep Talwani); (b) Planview of the settlement mud brick base platform and tidal basin (modified from Schwartzberg [1992]).

at the outer end of the entrance is also deeper, the entrance "sill" prevents or reduces the influx of near-bed suspended matter into the basin. Part of the efficiency of access to the basin is exchanged for lower cost of dredging [Everts, 1980].

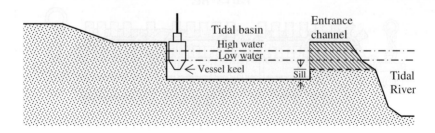

Figure 1.5. Schematic elevation view of a half-tide harbor.

Among later efforts to control sedimentation in navigation routes, one is reminded of the dredging implements mentioned by Needham [1974] in his monumental work on science and civilization in ancient China. One of these devices, called *hun jiang lung* (*"dragon which causes the river to become turbid"*), was a 2.4 m long roller fitted with iron blades for agitation-dredging of fluid-like mud in rivers and canals (Fig. 1.6). It was drawn along the bottom by a vessel proceeding upstream. The rolling blades resuspended soft bottom mud which was then carried away by the river current. The device was invented during the Sung dynasty in 11th century Common Era (CE) and was mass-produced to meet high demand.

1.2 Sediment Transport Hydraulics

Development of the hydraulics of fine sediment transport is an offshoot of sand transport mechanics with its modern origin in the 18th–19th centuries, when civil engineers had to contend with prolific port construction and harbor navigation activities in muddy estuaries. Since then valuable observations on fine sediment behavior have been made as, at times, corollaries to such construction projects. A recent example is the expansion of the port of Kochi (formerly Cochin) on the southwestern coast of India. As part of the project the chief engineer Robert Bristow created the present-day, 128 km^2 Willingdon Island by filling partly with mud found in the area. During his tenure with the erstwhile princely state of Travancore–Cochin, Bristow [1938] wrote two

混江龍

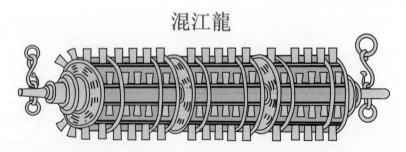

Figure 1.6. Agitation-dredger *hun jiang lung* invented during the Sung dynasty in China (redrawn based on Needham [1974]).

volumes of a work containing seminal observations on the dynamics of the famous mudbanks of Kerala coast known as *chakara*. These observations have provided the basis for much of the subsequent work in understanding the peculiar regularity with which mudbanks form during the monsoonal season and seemingly relocate thereafter [Narayana *et al.*, 2001].

In comparison with subaqueous sandy shoals, mudbanks are characterized by high biological productivity, an outcome of which is that they release living organisms at high rates. When *chakara* occurs at a location, water near the beach teems with fish and shrimp and a festival takes place. Figure 1.7 includes two views of the beach at Alappuzha (Alleppey) south of Kochi. Figure 1.7a shows traditional fishing boats during the *chakara* festival in nearly wave-free waters. In that area the beach is sandy, and a submerged mud layer begins near the waterline and extends seaward. Visually the presence of offshore mud is inferred only from the state of the sea surface. Comparison with Fig. 1.7b taken at about the same time approximately 2 km south of the mudbank shows a ~0.8 m high plunging breaker in an area where there was no mudbank and the bottom was sandy. This comparison illustrates the high capacity of bottom mud, no more than about 0.50 to 0.75 m in thickness, to absorb wave energy. At a pier on the beach at Alappuzha,

Figure 1.7. (a) Fishing boats during the *chakara* festival in the area of a mudbank at Alappuzha in the state of Kerala, southwestern India; (b) wave-breaking at an adjacent beach without a mudbank (from Mathew [1992]).

Mathew *et al.* [1995] reported as much as a 95% loss in monsoonal wave energy over a distance of about 1.1 km from the 5 m isobath to the shoreline. In the US there are only a handful of open-coast waters where wave damping due to mud is significant. A noteworthy region is the shallow inner shelf of the Gulf of Mexico off Atchafalaya Bay in Louisiana, where rapid decay of wave height has been recorded as waves travel over a bottom covered with mud whose source is the Mississippi and the Atchafalaya Rivers [Kineke *et al.*, 2007].

As mentioned, engineering treatment of the physics of marine sediment transport is of recent origin [*e.g.*, Stevenson, 1886]. As for fine-grained sediment, the development of its transport mechanics in the US must be credited to the pioneering work of Professor Hans Albert Einstein [*e.g.*, Einstein, 1941] and his students at the Berkeley campus of the University of California. Berkeley became a center of sediment transport research at least as far back as the work of Karl Gilbert [1914] of the U.S. Geological Survey. During the early part of the 20th century, Gilbert carried out experiments on sand transport in a flume located on the campus. These studies took place following large-scale sedimentation in the Sacramento–San Joaquin River delta area and the San Francisco Bay. The cause of sedimentation was hydraulic mining for gold beginning around mid-19th century in the hills and mountains of the Sierra Nevada range east of the city of Sacramento. Mining substantially increased the sediment load and reduced depths over the subaqueous delta. Overbank flooding became frequent, and brought riparian landowners in conflict with the miners. Eventually the State of California prohibited hydraulic mining, but as a result of sediment transported downstream by the rivers, the bottom of the northern part of San Francisco Bay known as San Pablo Bay became laden with layers of mud. Due to tidal currents and wave action, this mud has since spread over the entire San Pablo Bay area and southward into Central San Francisco Bay [Krone, 1979]. The sediment has a high degree of cohesion, with the flocculated particles having a complex structural hierarchy and erosion–deposition interaction with turbulent flow. In the US, much of the early experimental work and its interpretation with respect to the stochastic and phenomenological formulas for cohesive sediment erosion and deposition have been based on studies of Bay sediment by Ray Krone [1962] and Emmanuel Partheniades [1962] at Berkeley.

In the 1940's the Committee on Tidal Hydraulics of the U.S. Army Corps of Engineers began researching sedimentation in navigation channels due to the mobility of dense suspensions of fine sediment in San Francisco Bay, Savannah Harbor (GA) and other estuaries, often relying on the expertise of the Berkeley group [*e.g.*, Committee on Tidal Hydraulics, 1950; Committee on Tidal Hydraulics, 1960; Krone, 1962; Partheniades, 1962; Neiheisel, 1966]. During this period, equally important observations based on both laboratory studies and field work were made by Claude Migniot [1968] in France and Michael Owen [1977] at HR Wallingford, formerly the Hydraulics Research Station, in the UK. Owen's work followed that of Claude Inglis and others at the same institution [*e.g.*, Inglis and Allen, 1957]. At the Institute of Oceanographic Sciences (UK), Robert Kirby and W. Reginald Parker [1974] made the first set of comprehensive measurements describing the vertical fine-structure of dense sediment suspensions in the Severn estuary. As part of those studies, their lasting contribution to fine sediment transport is the term *lutocline* coined to denote the pycnocline in a stratified fine sediment suspension.

Early milestones of US developments in numerical modeling of fine sediment transport are also associated with northern California. A numerical code for a multi-dimensional hydrodynamic model was developed by Robert Shubinski and Gerald Orlob in the mid 1960's. It employed a quasi-two-dimensional link-node approach and was applied to San Francisco Bay [Shubinski *et al.*, 1965; Orlob *et al.*, 1967]. Subsequently a cohesive sediment transport model based on that work was proposed by Shubinski and Krone [1970]. Following this, as part of his doctoral thesis Ranjan Ariathurai [1974] developed the first "layered-bed" model for cohesive sediment transport prediction. This was a significant contribution to numerical modeling of fine sediments because soft cohesive beds are usually stratified. Bed layering in the model permits the assignment of different density and related properties to each layer. Ariathurai also applied a deterministic linear approximation of the non-linear stochastic equation for bed erosion flux proposed by Partheniades [1962]. Another milestone was the development of two-

and three-dimensional systems of hydrodynamic and sediment transport numerical models collectively known as TABS by Ian King, William Norton, Ranjan Ariathurai and the US Army Corps of Engineers Waterways Experiment Station (Vicksburg, MS) in the 1970's and 80's. Ariathurai's model was the prototype for cohesive sediment calculations in the TABS sediment code STUDH and the three-dimensional version of TABS called TABS-MDS [King *et al.*, 1973; Norton *et al.*, 1973; McAnally *et al.*, 1983; King, 1988].

Many scientists and engineers have contributed to topics that are closely associated with fine sediment transport hydraulics. We may note the fundamental contributions of Smoluchowski to the stochastic aspects of inter-particle collisions and Terzaghi to soil mechanics [Ulam, 1957; Goodman, 1999]. The work of Professor J. Th. G. (Theo) Overbeek [1952] on the stability of sols and flocculation kinetics of clays at the Agricultural University in Wageningen, the Netherlands, is just one of many early contributions without which it would be difficult to imagine the subsequent progress in understanding cohesive floc transport. With the establishment of the U.S. Environmental Protection Agency in 1970, interest in the role of flocs as carrier agents for nutrients and contaminants in bioaccumulation increased substantially. Turbidity, suspended solids and sedimentation are dominant polluting factors in rivers and streams, lakes, reservoirs, ponds, wetlands and open-coast waters. In 1998 approximately 40% of assessed river-miles in the US were reported to have sediment related problems [Berry *et al.*, 2003]. Such issues have spawned extensive research in observing and explaining the dynamic behavior of flocs [Gibbs, 1985; Lick *et al.*, 1992; Winterwerp, 1998; Arnot and Gobas, 2004].

1.3 Driving Factors and Sediment Response

Causal factors driving sedimentary particles in the marine environment tend to vary widely in frequency and energy. The frequency of gravity and ultragravity waves of interest ranges from 10^{-5} to 10^{1} Hz (Fig. 1.8). The lower frequency limit represents astronomical tide, and the upper limit roughly corresponds to the transition from gravity-dominated to

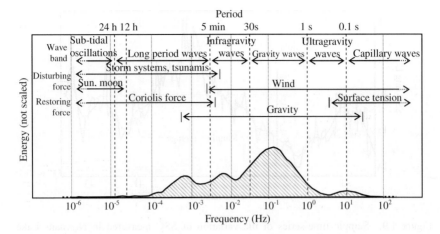

Figure 1.8. Schematic drawing of energy distribution among natural driving forces as a function of forcing frequency (adapted from Munk [1950]).

surface tension dominated wave regime. Frequencies lower than 10^{-5} Hz are associated with sub-tidal oscillations mainly due to quasi-periodic frontal passages (2×10^{-6} Hz). Frontal wave phenomena do not represent "conventional" wave motion. However, they do show periodicities linked to the marine climate and therefore to particle movement. Not included in the plot are frequencies associated with monthly and annual cycles (4×10^{-7} to 3×10^{-8} Hz). Their roles in modulating fine sediment transport can be significant because very fine particles in the bed tend to contribute to seasonal variability in the resuspension potential.

Lakes and reservoirs in which the water movement is mainly due to wind are examples of aquatic systems whose sediment dynamics can be difficult to characterize without considering the full spectrum of forcing frequencies and response. Newnans Lake in north-central Florida (with an area of 27 km^2 and a mean depth of 1.6 m) illustrates the complexity of fine sediment resuspension. Episodic coupling between wind-driven circulation, wind waves and rainfall-driven tributary discharge results in a variable and noisy signal of the suspended sediment concentration (SSC) spread over a wide frequency band. In Fig. 1.9, a segment of measured SSC time-series obtained in the middle of the lake illustrates this multi-frequency response. The running-average line representing

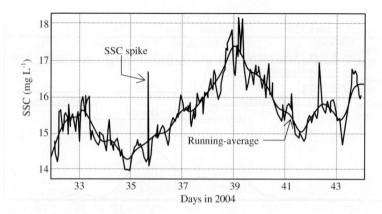

Figure 1.9. Sample time-series of the variation of SSC measured in Newnans Lake, Florida (from Jain *et al.* [2007]).

frontal oscillations on the order of days are modulated by higher frequency diurnal contributions. Figure 1.10 shows wind, air temperature and SSC spectra from a 90-day long (January–March 2004) time-series. The three noteworthy peaks seem to be coincident with the luni-solar tidal harmonics. These are the one-half and one-quarter lunar fortnightly periods of 6.8 and 3.4 solar days, respectively, and also 0.93 day, which is close to the solar day. A fourth, weaker peak corresponding to about 0.5 day could be the luni-solar semi-diurnal period (11.97 h). Sea breeze is known to respond to tidal effects, and its role as a causal factor would explain the correspondence between lunar motion and suspended sediment response [Richards, 2003]. A point to emphasize is that the SSC response illustrates difficulties one often faces in identifying a particular forcing with a given response. The unusually sharp spike in the SSC of unknown origin is illustrative in this regard.

Newnans Lake is an example of a very shallow body of water in which SSC can be characterized as low, with values almost never exceeding a few hundred milligrams per liter at mid-depth. In other bodies, concentrations can increase to tens of thousands of milligrams per liter (Table 1.1), and particle–particle interactions result in sedimentary regimes that can be far more complex than in Newnans Lake.

At the very extreme, we have the case of Yellow River in China, where concentrations as high as 800,000 mg L^{-1} (800 kg m^{-3}) flowing as

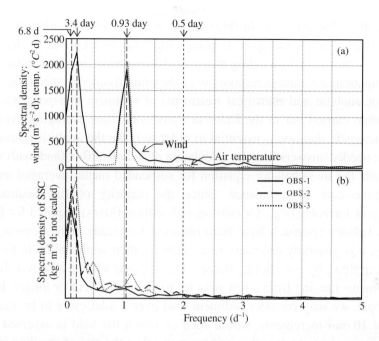

Figure 1.10. Wind, air temperature and suspended sediment concentration (SSC) spectra from Newnans Lake, Florida. SSC was measured with optical backscatter sensors (OBS) at three elevations above bottom (from Jain *et al.* [2007]).

a hyperconcentrated slurry have been recorded [Chien and Wan, 1999]. Notwithstanding the wide ranges of frequency over which forcing can occur, for convenience of data interpretation sediment response can be categorized within low and high frequency domains. Response at low frequencies arises from the hydrostatic pressure gradient. This includes transport by current in rivers and estuaries. At high frequencies the effect of the dynamic pressure gradient in water dominates. Wind waves

Table 1.1. Characteristic (nominal) ranges of SSC at about 1 m above bottom.

Location	Concentration	
	(mg L^{-1})	(kg m^{-3})
Newnans Lake, FL	5–300	0.005–0.3
San Francisco Bay, CA	30–1000	0.03–1
Amazon Estuary, Brazil	1,000–10,000	1–10
Yellow River, China	10,000–800,000	10–800

fall within that category. In this text we will conveniently identify and treat fine sediment transport as being either due to a current or due to waves with the understanding that a detailed analysis would require an examination of sediment response over the entire frequency spectrum.

For analytic and numerical treatments of sediment transport, which are often carried out in the Eulerian framework, time-averaging over a characteristic duration is essential in order to make the analysis relevant to the needed answers. For the calculation of sediment flux (and load) we will use the mass balance equation for suspended matter averaged over the time-scale of turbulence. Since the majority of fine sediment transport formulas (*e.g.*, for erosion and settling fluxes) required for the mass balance approach have been obtained for steady (zero-frequency) flows, it is generally necessary to account for their applicability to time-varying flows by considering transport to be piece-wise steady over time increments ranging from a few minutes to as much a year or longer. For example, we may consider sediment load over a tidal cycle to be made up of 10 min increments, within each of which the load is assumed to remain constant. At the other end we may select the rate of shoaling in a channel over a multi-year period at constant annual-mean rates. For high-frequency response the minimum time for averaging is usually the wave period. Further averaging over multiple wave periods may be required depending on the nature of application and the duration of data collection.

Developments in tracking the movement of fine sediment flocs in the Lagrangian framework require that the impact of fluid forces on the particles being followed be considered over the full range of active time-scales. These scales may vary from short durations associated with turbulence to the much longer sub-tidal periods. Calculations of sediment load can be quite sensitive to modeling assumptions, because at any given position and time floc properties depend on the stress-history with respect to the active fluid forces as well as inter-particle collisions. An illustrative case of large-scale model application is the tracking of particle-bound contaminants and bacteria in the receiving waters of coastal outfalls. There can be several critical steps in this type of simulation, including the selection of initial dilution of suspended sediment in the outfall, which influences the subsequent spreading of tagged particles. Predictions of long-term sediment and particulate zinc accumulation in the coastal benthos were carried out for tidal waters surrounding the city of Auckland (New Zealand) using a Lagrangian particle tracking model. The far-field output was found to be a strong function of near-field assumptions [Bogle *et al.*, 2007].

1.4 Particles and Flocs

The ability of clay particles to aggregate in water is similar to the behavior of aerosol particles, and understanding this behavior is basic to developments in the hydraulics of fine sediment transport. We will begin with a brief introduction to the subject of electrochemical flocculation, which is treated in greater depth in some of the subsequent chapters. In fine sediment transport it is essential to distinguish between the terms "particle" and "floc" for characterization of the sedimentary unit. We will consider the former to mean a primary particle, the basic mineral unit, and the latter to mean an agglomerate of thousands or more mineral particles that collectively behave as an identifiable sedimentary unit. Primary particles are often called dispersed particles because they are obtained by adding a chemical agent to a suspension of flocs, which attempts to disperse the flocs into their constituent mineral particles.

A dispersion of clay particles in water can take hours or days to settle out. During this process, as the larger particles deposit faster, the suspended matter becomes gradually finer. The ultimate deposit is densely packed and can be resuspended upon the slightest agitation of water. A suspension of particles smaller than about 1 μm is known as a sol. The momentum associated with water molecules due to their thermal energy may indefinitely retain sol particles in suspension. When a small amount of a salt, say about one-hundredth of a percent (by weight of water) of sodium chloride, is added to a sol the particles agglomerate and the resulting flocs tend to settle out. If the initially suspended particles are already large enough to settle slowly, addition of salt may accelerate the settling process. Flocs can capture and incorporate very small ($<<$ ~1 μm) and large (~4–10 μm) particles within their structure. Instead of a compact but easily-erodible deposit of dispersed particles, a loosely-packed and high-volume (but more resistant) deposit is formed — usually within minutes (Fig. 1.11). The type and degree of packing of the deposit depends on particle concentration, mineral composition as well as on the type of ions in water, temperature and time.

The process of particle agglomeration is known as flocculation or coagulation, and is of great importance because flocs occur in most environments due to sodium chloride or other salts and impurities in water. Even in supposedly pristine rivers or lakes flocs are found, either because small quantities of salts are present, or because biopolymeric (exopolymeric) substances bind sol particles to form aggregates that resemble those formed by salt. The role of organic substances in modulating, even controlling, floc properties can be critically important.

Flocs tend to be light-weight with densities close to water, and consist of interconnected, often card-house-like, matrices. As flocs move through water they may combine with other particles or flocs and grow, or they may break up due to high flow-induced shear. Consequently, floc density, size, shape and strength change with time. It is essential to account for these changes in calculations of the transport of flocs as sediment load.

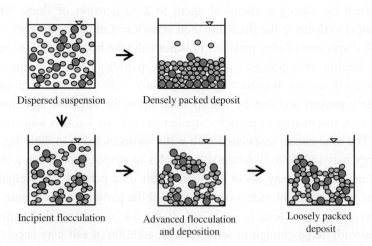

Figure 1.11.   Effect of addition of salt on suspended dispersed particles and deposit structure.

As we shall see in the subsequent chapters, by and large floc transport is modeled within a physics-based but semi-empirical framework. Beyond finding explanations for the observed features of floc transport, reducing empiricism remains a critical, but far from achieved, goal of research.

Whether particles are in the dispersed state or occur as flocs can make a great difference in problems of engineering interest. For instance, when the sedimenting matter is flocculated, shoaling, which is a volumetric measure of the thickness of the deposit, can occur at a high rate even when the mass of sediment contained in the loosely packed deposit is low.

Example 1.1: Calculate the thickness of a shoal formed in a year in the turning basin of a navigation channel by deposition of a flocculated sediment suspension having a concentration of 20 mg L^{-1}. The settling velocity of flocs is 3×10^{-5} m s^{-1} and the dry density (dry sediment mass per unit volume of deposit) of the deposited bed is 400 kg m^{-3}.

Since 20 mg L^{-1} = 0.02 kg m^{-3} and there are 3.15×10^{7} s in a year,

$$\text{Shoal thickness} = \frac{0.02\times3\times10^{-5}\times3.15\times10^{7}}{400} = 0.047 \text{ m} \approx 5 \text{ cm (per year)}$$

Shoaling of this order occurs at many harbors, channels and closed-end residential canals in, for example, southern Florida. In the northern part of the state, where there is a greater supply of fine sediment from rivers, the rate tends to be on the order of 10–15 cm per year [Hayter and Mehta, 1986; Lin and Mehta, 1997]. At some port channels in South Asia and Southeast Asia the annual deposition rate exceeds 1 m, and sediment removal requires near-continuous dredging. Such high rates occur where usually there is a riparian source of mud.

Example 1.2: Calculate the thickness of a shoal formed in a year by deposition of a dispersed (*i.e.,* unflocculated) sediment suspension having a concentration of 20 mg L^{-1}. The settling velocity of the particles is 3×10^{-6} m s^{-1} and the density of the deposit is 1,800 kg m^{-3}.

$$\text{Shoal thickness} = \frac{0.02\times3\times10^{-6}\times3.15\times10^{7}}{1800} = 0.0011 \text{ m} \approx 1 \text{ mm (per year)}$$

Although fully dispersed particles of inorganic minerals, say 1 μm or smaller, usually attach themselves to other particles, larger particles, for instance >10 μm may remain as individual units if these particles, which tend to be only weakly cohesive, are not bound into aggregates by, *e.g.,* biopolymers. When both small and large particles are present the settling process becomes complex. If 1 μm dispersed particles are weakly cohesive, their flocs and the dispersed particles will settle at different speeds. When 1 μm particles are highly cohesive their flocs may capture 10 μm particles during fall or while swirling in shear driven eddies [Partheniades *et al.*, 1966]. Binding by biopolymers can also result in a heterogeneous population of particles.

1.5 Natural Sediment: Definition of Mud

According to Webster's dictionary, sediment is: (a) matter that settles to the bottom of a liquid, or (b) material deposited by water, wind, or glaciers. This definition, based on the classical meaning of the word "sediment," emphasizes its significance as a deposit, and sedimentary particles as those that can form a deposit. Since we must account for sediment as muddy material in suspension as well as a deposit, the definition of mud must be explored further, especially in the context of natural sediments.[b]

Until about the mid-twentieth century, studies on fine sediment transport hydraulics dealt mainly with abiotic (inorganic) sediments. However, natural muds are rarely free of life forms and their products, which alter the properties of abiotic mud. For example, filter feeders and deposit feeders derive nourishment by passing fine material through their digestive tracks. Animal movement and sediment manipulation during feeding tend to change the size, surface texture, density and strength of flocs. The egested material consists typically of compact pellets containing mucoid biopolymers, *i.e.*, biofilms. Pellets at the bed surface change the bed roughness and boundary layer flow [Syvitski and Lewis, 1980; Wheatcroft *et al.*, 1989].

As mentioned earlier, the impact of biota on natural sediment can be considerable. This is so because, as suggested by data on the mud reworking rates by benthic macrofauna in Table 1.2, bed surface sediment may pass through the animal several times per year before the sediment is completely buried. Also present in very large numbers are

---

[b] Sediment can be thought of as referring to dry particles in large numbers, whereas mud is essentially a mixture of particles in water. Chapter 3 is concerned with sediment classification, and Chapter 5 deals with mud properties. With regard to transport, no well-defined distinction is made between "sediment" and "mud".

Table 1.2. Biological abundance and activity in marine muds.

Water depth (m)	Macrofauna		Bacterial density (# kg^{-1} dry sediment)	Carbon production (mg C m^{-2} d^{-1})
	Density (# m^{-2})	Reworking rate (L m^{-2} yr^{-1})		
< 10	10^1–10^5	10^{-2}–10^4	10^{11}–10^{15}	10^0–10^3
10–200	10^1–10^5	10^{-2}–10^{-4}	10^{11}–10^{13}	10^1–10^3
200–2000	10^{-2}–10^4	10^0–10^2	10^{10}–10^{13}	10^1–10^2
> 2000	10^{-4}–10^{-2}	10^{-4}–10^{-2}	10^{11}–10^{13}	10^{-2}–10^1

Source: Based on studies reported by Dade and Nowell [1991].

bacteria, resulting in high rates of carbon production, microalgae and their extracellular, or exopolymeric, secretions [Weaver, 1989; Decho, 1990]. There can be as many as 10^{15} micron sized particles per kilogram of dry sediment [Dade and Nowell, 1991]. In shallow (< ~10 m) waters the maximum bacterial number concentration may be of the same order of magnitude. This would mean that if the distribution of bacteria in sediment were uniform, there would be a bacterium on every particle. Experiments suggest that one bacterium can coat up to tens of thousands of particles with its secretion in a week.

Several definitions of mud are found in the literature, some of which appear to be interconnected possibly due to a common origin. Four noteworthy definitions are as follows.

Definition 1: Mud is an indurated mixture of clay and silt with water. It is slimy with consistency varying from that of a semi-fluid to that of soft and plastic sediment [Hunt and Groves, 1965].

Definition 2: Mud is a pelagic (*i.e.,* marine) or terrigenous detrital material consisting mostly of silt and clay-sized particles (less than 0.063 mm), but often containing varying amounts of sand and/or organic materials. It is a general term applied to any sticky fine-grained sediment whose exact size classification has not been determined [Tvler, 1979].

Definition 3: Mud is a fluid-plastic mixture of finely divided particles of solid material and water [Allen, 1972].

Definition 4: Mud is a dense mixture of water and particles of diameters that are predominatly smaller than 63 μm. Its rheological behavior varies with water content, composition and applied stress [modified from Mehta, 2002].

The first three definitions refer to the physical *state* of mud. However, the third definition makes no reference to composition, and therefore does not permit a distinction between abiotic fine sediment often used in laboratory experiments and natural fine sediment with biota. In the literature one finds that pure clays and clay mixtures have also been called mud. When making references to these studies the word mud will include such mixtures.

The fourth definition is based on the response of the bulk material to an applied (fluid) stress, and is therefore relevant to mud *transport* in water. Mud rheology (Chapter 5) concerns the relationship between applied stress and the material's response (strain, rate of strain) to stress. Since each mud has its own characteristic dynamic behavior, different responses to applied stress characterize different mud states. It follows that even if the difference between two muds is that one is abiotic and the other contains the same abiotic material impregnated with biotic matter, the rheological behaviors of the two muds will be different. In terms of understanding mud movement in water this difference can have considerable significance. For instance, to understand the damping of water-waves by energy loss in soft mud, knowledge of the viscous properties of mud is essential. With known viscosity, the amount of damping can be estimated without reference to the composition of mud. Since collisions between mud flocs, their growth and breakup in suspension depend on the dynamic response of flocs to fluid shear, rheological properties are also important in characterizing the transport behavior of flocs. We will further refer to this link between rheology and floc dynamics in subsequent chapters.

Fine sediment sizes of present interest range between about 0.1 and 62.5 μm. Particles smaller than 0.1 μm do not settle, while larger ones (> 62.5 μm) are considered "coarse". As we will see later, the fine/coarse boundary as defined is to a fair degree a convenient designation unrelated to particle dynamics in water. The boundary of interest between "course" and "fine" particles tends to be smaller than 62.5 μm. As mentioned in Chapter 3, this boundary can be defined in terms of a range of particle sizes over which a transition from cohesive to non-cohesive behavior occurs. The transition, which is never abrupt, is influenced by the sizes and types of particle minerals and the chemical composition of the pore fluid. Organic matter also plays a role.

In general, particles in the 0.1–62.5 μm size range can be present in the aquatic environment within one or more of the itemized categories in Table 1.3. The first three depend on the (inorganic) mineral, for which a distinction must be made between clay and non-clay minerals, and their dispersed particle size/shape distributions (Chapter 3). The next three categories include organic matter, whose composition can vary widely. The transport of organic aggregates is not well understood and is difficult to characterize in terms of usable bulk variables.

In summary, although the hydraulics of fine sediment transport includes the 0.1–62.5 μm size range, bound sol particles less than 0.1 μm may also be present. While only particles of clay minerals possess electrochemical cohesion, organic matter can effectively induce agglomeration by biochemical adhesion or polymeric binding which qualitatively mimics electrochemical cohesion. To this we must add effects from the presence of mixtures of inorganic and organic particles and substances, and chemicals in water.

1.6 Scope of Presentation

Detailed descriptions of wave–current–sediment interactions require non-linear systems of equations (see Exercise 1.3 at the end of the chapter). Large-scale analyses involving complicated geometries of natural

Table 1.3. Fine sediment categories in the aquatic environment.

Category	Description
1	Clay mineral particles as flocs
2	Non-clay mineral particles in dispersed state
3	Large clay mineral particles in dispersed state
4	Clay and non-clay mineral particle aggregates bound by biopolymers
5	Wholly organic aggregates
6	Unaggregated organic particles

marine systems, multi-frequency time-dependent flows and high suspended sediment concentrations can be so non-linear that they cannot be simplified even with severe approximations. As a result the application of numerical schemes to solve these equations becomes essential. In such applications it is critical to understand the underpinning sediment transport process relationships, which is the present subject matter.

The focus on hydraulics permits us to include estuaries, rivers and streams, lakes and reservoirs, and the shallow open-coast environment. Not treated is the cold region in which transportation by ice flows can be significant. Subjects such as terrestrial mud slides or aeolian transport of fine matter are not covered.

The absence of a comprehensive physical framework for the treatment of different processes poses a significant constraint. Unfortunately, when one views the present state of science from the perspective of hydraulics, this limitation is unavoidable, although there has been a gradual trend towards unification of concepts. In the present piecemeal approach, as the flow field changes between turbulent and laminar, and the volume fraction of solids in the fluid–plus–solids mixture changes as well, the physical framework changes between particle mechanics, Newtonian mechanics, non-Newtonian mechanics and solid (soil) mechanics (Fig. 1.12). In order to reduce any ensuing confusion, in Chapter 6 we will attempt to categorize relevant "unit" transport processes within definable sub-frameworks.

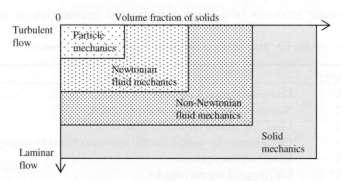

Figure 1.12. Physical sub-frameworks defined by the flow field and volume fraction of solids within which fine sediment hydraulics is conventionally investigated.

1.7 Exercises

1.1 From the literature, state the eustatic and non-eustatic components of the rate of sea level change for the past ~100 yr for: East Port, ME; Newport, RI; Charleston, SC; and Miami, FL.

1.2 (a) From the literature provide brief descriptions of two pre-Columbian ports other than those cited in the chapter as follows: (1) A port in which significant historic sedimentation is known to have occurred. Indicate the problem and the way in which it was solved. (2) A port that had to be abandoned due to reduced depths in the harbor. Cite your sources.

(b) From the literature provide brief descriptions of two ports other than those cited in the chapter and in Exercise 1.2a as follows: (1) A port that was submerged by rising sea level but was rebuilt at a higher elevation. (2) A port that was permanently abandoned due to submergence. Cite your sources.

1.3 Sediment transport equations tend to be highly nonlinear, with exponential-type relationships to fluid velocity and other independent parameters. The nonlinear growth equation E1.1 has been used as a simple predictor of the population of biotic communities. In this equation λ is the birth rate, χ is the normalized population count and n is the year counter. Thus χ_n is the population in year n divided by the maximum population that the natural system can support. Note that, for example, $\lambda = 2$ means that each pair of animals produces 2 offspring each year. Solve Eq. E1.1 for the following initial values χ_0 of χ_n at $n = 0$ and growth rates λ (12 cases in all): $\chi_0 = 0.1$, 0.4 and 0.5 and $\lambda = 1$, 2, 3 and 4. The variable n is an integer ranging from 0 to 100.

$$\chi_{n+1} = \lambda \chi_n (1 - \chi_n)$$
(E1.1)

As an example, given $\chi_0 = 0.1$ and $\lambda = 1$, we obtain

n	χ_n
0	0.1
1	0.09
2	0.0819
...	...

Plot χ_n against n. Do the results offer any insight into why sediment transport calculations can at times be quite wrong?

1.4 In text Examples 1.1 and 1.2 the bed dry mass densities are taken as 400 and 1,800 kg m^{-3}, respectively. On the other hand, the typical density of clay mineral particles in the deposit is 2,650 kg m^{-3}. Note that the bed is porous while particles are solid. Determine the porosity (Eq. 5.11) of each bed and suggest why two beds may have quite different dry mass densities.

1.5 From the scientific literature, try to find another definition of mud. How is it related to, or is different from, one or more of those described in the text? Cite the source of definition.

1.6 Commercial ports are defined by tonnage as in Table E1.1, and by water depth as in Table E1.2. In Table E1.3 bottom shoaling rates are classified as very high, high, moderate or low. Select three ports (two in US and one non-US), and try to find information that would enable you to define them by tonnage, depth and shoaling rate. Indicate the means employed to maintain channel depth in each port. Note: Channel dredging to maintain the required minimum depth in US ports is authorized by the U.S. Army Corps of Engineers [1981]. Estimation of shoaling is based on surveys of affected channel segments. Thus, the Corps of Engineers logs sediment volumes dredged from specified port channel segments. Dredged rates *per port* as a whole are not always available.

Table E1.1. Port definition by tonnage.

Cargo (metric tons yr^{-1})	Definition
<10^6	Small
10^6 to 10^7	Medium
>10^7	Large

Table E1.2. Port definition by depth.

Channel depth (m)	Definition
<7.5	Shallow
>7.5	Deep

Table E1.3. Port shoaling rates.

Shoaling rate (m yr^{-1})	Definition
>1	Very high
1–0.3	High
0.3–0.05	Moderate
<0.05	Low

Chapter 2

Topics in Fluid Flow and Wave Motion

2.1 Chapter Overview

In this chapter we will briefly review selected topics in two related subjects. These are boundary layer flow of incompressible fluids relevant to the hydraulics of fine sediment transport and water wave motion in inviscid fluid. Waves are assumed to be monochromatic and traveling over a rigid bed. Interactive effects of dissipative, compliant beds and fluid or fluid-like mud with waves are reviewed in Chapters 10 and 11.

Suspended sediment at a sufficiently high concentration can curtail the turbulent kinetic energy to such an extent as to render the flow entirely non-turbulent. In general, viscous fluid stresses cannot be ignored when the concentration is high, say greater than ~10 kg m^{-3}. Accordingly, we will review some aspects of viscous flow including the role of molecular viscosity and its contribution to the turbulent boundary layer velocity.

Reference is made to the significance of stratification due to flow layering by salinity or suspended matter. Flow through porous media is briefly introduced, followed by a summary of the wave theory and wave motion in coastal waters. The selection of topics and the succinctness of their treatment make them appear as snapshots, rather than as seamlessly connected conventional developments found in basic texts, some of which have been cited for further reading.

2.2 Time-Mean and Fluctuating Velocities

For the descriptive framework we will use the right-handed coordinate system (x, y, and z) with fluid velocity components u, v and w as functions of x, y, z and time t. Conventionally the origin 0 is at the bottom in open-channel flow (Fig. 2.1a) and at the water surface when wave motion is described (Fig. 2.1b). We will retain this convention because treatments dealing with currents and waves are given separately in order to reduce, although not completely eliminate, the awkwardness in shifting the origin of the vertical coordinate. In the majority of treatments the use of fully three-dimensional systems has been eschewed; one- and two-dimensional steady and unsteady flow problems are dealt with more commonly. In some cases reference has been made to zero-dimensional analysis, *i.e.*, one in which the system is time-dependent but is devoid of spatial variability.

For a description of turbulent flow in the conventional sense, each of the three fluid velocity components u, v and w is represented as the sum of a component averaged over the time-scale T_t for mean–value representation and a component to account for turbulent fluctuations. For the u velocity

$$u = \bar{u} + u'$$ (2.1)

where the overbar denotes a time-averaged quantity, and the prime

(a)   (b)

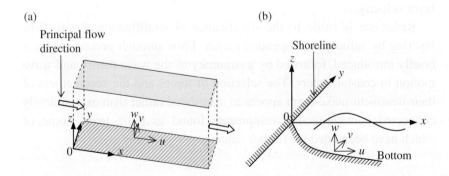

Figure 2.1. Right-handed spatial coordinates and fluid velocities for: (a) open-channel flow; (b) shoreline and waves.

represents a fluctuation (Fig. 2.2). Expressions for v and w are analogous.[a]

If the frequency of turbulence is 10^3 Hz, the time T_t of averaging can be, say, 1 s. Since turbulence manifests as intermittent bursting of flow energy that must be captured in the description, T_t is varied depending on the flow environment. Whereas over smooth beds in laboratory this scale may be on the order of 1 s, in the marine boundary layer 10 min may be required to include low-frequency, large energy bursts. However, increasing the averaging time also increases the likelihood that the effect of non-stationary processes such as tide could contaminate the bursting record. Thus, time-averaging over any record longer than 10 min may not be always suitable in the tidal environment [Soulsby, 1983; Dyer, 1986].

Due to the random nature of the u' fluctuations, by definition the following condition applies to the velocity u

$$\frac{1}{T_t}\int_0^{T_t} u\, dt = \frac{1}{T_t}\int_0^{T_t} \bar{u}\, dt + \frac{1}{T_t}\int_0^{T_t} u'\, dt = \bar{u} + 0 = \bar{u} \qquad (2.2)$$

with analogous conditions for v and w. Since random fluctuations have zero means, the representation of velocity magnitudes is characteristically done in terms of root–mean–square (*rms*) quantities,

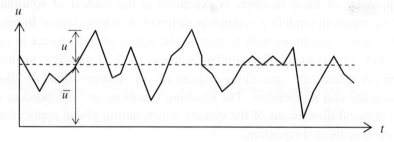

Figure 2.2. Time–mean and fluctuating components of water velocity u.

---

[a] For proper numerical simulations of fine sediment transport, compatibility is essential between the geometric scales of flocs in suspension and the turbulent eddy lengths [*e.g.*, Schweim *et al.*, 2000].

$$u_{rms} = \sqrt{\frac{1}{T_t} \int_0^{T_t} u'^2 dt} = \sqrt{\overline{u'^2}} \qquad (2.3)$$

e.g., with analogous expressions for v_{rms} and w_{rms}. They refer to the mean intensities of turbulence and an important quantity obtained from them is the turbulent kinetic energy per unit mass of fluid (TKE) associated with eddies. It is determined by time-averaging the mean–square values of the velocity fluctuations according to

$$TKE = \frac{1}{2}\left(\overline{u'^2} + \overline{v'^2} + \overline{w'^2}\right) \qquad (2.4)$$

The TKE is of essential significance in describing the transport of sediment suspensions. Note that, since in non-turbulent flows the fluctuating quantities do not exist, $u = \overline{u}$, and so on.

2.3 Characteristic Numbers

Since the state of fluid flow, *i.e.*, whether laminar (viscous) or turbulent, is critically important to sediment transport, it is necessary to introduce the significance of the Reynolds number and the Froude number in embodying viscous, gravity and inertia forces inherent in flows. The importance of these numbers is examined in the context of similitude requirements in modeling systems at different geometric scales. In brief, for dynamic similitude with respect to the action of fluid forces in any number of geometrically similar systems, it is necessary to non-dimensionalize the governing equations of conservation of flow continuity and momentum. The resulting equations are independent of the physical dimensions of the system, which permit global applications of the transformed equations.

Consider the *x*-component of the mean–flow momentum equation where H is the hydrostatic head, ρ is the (spatially invariant) fluid density, η is the (spatially invariant) molecular (or dynamic) viscosity and $\bar{p}$ is the mean pressure obtained by averaging the instantaneous pressure over the time-scale of turbulence[b]:

$$\frac{\partial \bar{u}}{\partial t} + \bar{u}\frac{\partial \bar{u}}{\partial x} + \bar{v}\frac{\partial \bar{u}}{\partial y} + \bar{w}\frac{\partial \bar{u}}{\partial z} = -g\frac{\partial H}{\partial x} - \frac{1}{\rho}\frac{\partial \bar{p}}{\partial x} +$$

$$\frac{\eta}{\rho}\left(\frac{\partial^2 \bar{u}}{\partial x^2} + \frac{\partial^2 \bar{u}}{\partial y^2} + \frac{\partial^2 \bar{u}}{\partial z^2}\right) \tag{2.5}$$

This equation can be made dimensionless by introducing the following non-dimensional quantities

$$\tilde{x} = \frac{x}{h_0}, \qquad \tilde{y} = \frac{y}{h_0}, \qquad \tilde{z} = \frac{z}{h_0}$$

$$\tilde{u} = \frac{\bar{u}}{U}, \qquad \tilde{v} = \frac{\bar{v}}{U}, \qquad \tilde{w} = \frac{\bar{w}}{U}$$

$$\tilde{t} = \frac{t}{h_0/U}, \qquad \tilde{p} = \frac{\bar{p}}{\rho U^2}, \qquad \tilde{H} = \frac{H}{h_0} \tag{2.6}$$

where h_0 and U are fluid depth and velocity, respectively, characteristic of the system. Substituting these quantities into Eq. 2.5 we obtain the following dimensionless equation of motion

$$\frac{\partial \tilde{u}}{\partial \tilde{t}} + \tilde{u}\frac{\partial \tilde{u}}{\partial \tilde{x}} + \tilde{v}\frac{\partial \tilde{u}}{\partial \tilde{y}} + \tilde{w}\frac{\partial \tilde{u}}{\partial \tilde{z}} = -\left(\frac{gh_0}{U^2}\right)\frac{\partial \tilde{H}}{\partial \tilde{x}} - \frac{\partial \tilde{p}}{\partial \tilde{x}} +$$

$$\left(\frac{\eta}{\rho U h_0}\right)\left(\frac{\partial^2 \tilde{u}}{\partial \tilde{x}^2} + \frac{\partial^2 \tilde{u}}{\partial \tilde{y}^2} + \frac{\partial^2 \tilde{u}}{\partial \tilde{z}^2}\right) \tag{2.7}$$

In this equation all except two terms have multiplier groups that are not equal to 1. The inverse square-root of the first group is the Froude number,

---

[b] For incompressible flows the equations of continuity and motion are given in Appendix B in Cartesian, cylindrical and spherical coordinates. Equation 2.5 is obtained from Eq. B.7 by changing the notation according to: $u_x = \bar{u}$, $u_y = \bar{v}$, $u_z = \bar{w}$, $g_x = g$, and expressing the pressure gradient as the sum of its dynamic and hydrostatic components.

$$F = \frac{U}{\sqrt{gh_0}} \equiv \sqrt{\frac{\text{Inertia force}}{\text{Gravity force}}} \tag{2.8}$$

which represents the ratio of inertia force to gravity force. The inverse of the second group is the Reynolds number

$$Re = \frac{\rho U h_0}{\eta} = \frac{U h_0}{\nu} \equiv \frac{\text{Inertia force}}{\text{Viscous force}} \tag{2.9}$$

where ν is the kinematic viscosity of the fluid. This number represents the ratio of inertia force to viscous force. Thus Eq. 2.7 can be written as

$$\frac{\partial \tilde{u}}{\partial \tilde{t}} + \tilde{u}\frac{\partial \tilde{u}}{\partial \tilde{x}} + \tilde{v}\frac{\partial \tilde{u}}{\partial \tilde{y}} + \tilde{w}\frac{\partial \tilde{u}}{\partial \tilde{z}} = -\frac{1}{F^2}\frac{\partial \tilde{H}}{\partial \tilde{x}} - \frac{\partial \tilde{p}}{\partial \tilde{x}} +$$

$$\frac{1}{Re}\left(\frac{\partial^2 \tilde{u}}{\partial \tilde{x}^2} + \frac{\partial^2 \tilde{u}}{\partial \tilde{y}^2} + \frac{\partial^2 \tilde{u}}{\partial \tilde{z}^2}\right) \tag{2.10}$$

The same dimensionless groups are obtained from the y- and z-components of the equation of motion. Finally the equation of flow continuity is

$$\frac{\partial \bar{u}}{\partial x} + \frac{\partial \bar{v}}{\partial y} + \frac{\partial \bar{w}}{\partial z} = 0 \tag{2.11}$$

which can be written in the dimensionless form as

$$\frac{\partial \tilde{u}}{\partial \tilde{x}} + \frac{\partial \tilde{v}}{\partial \tilde{y}} + \frac{\partial \tilde{w}}{\partial \tilde{z}} = 0 \tag{2.12}$$

This equation does not introduce any groups. Thus, dynamic similitude for any number of geometrically similar systems is fully satisfied if F and Re have the same values for a system at all geometric scales, *i.e.*, the equations of motion and continuity in their dimensionless forms are independent of the scale for that system.

A practical importance of similitude, *i.e.*, Reynolds number and Froude number equalities for geometrically similar systems, is in physically modeling prototype systems. In general both equalities must be satisfied for achieving full similitude. However, in laboratory models of estuaries Froude number equality is accepted but Reynolds number equality is ignored. This is necessary because, as shown easily, retention of both Reynolds number as well as Froude number equalities renders the viscosity of the model fluid dependent on the geometric scale between the model and the prototype. This makes it difficult, if not practically impossible, to construct models of large estuaries in the laboratory. The problem is minimized to a fair extent by maintaining fully turbulent flow in the model, as in this event bed friction representing energy loss in the model is independent of the Reynolds number. Since Reynolds number equality is not maintained, the use of water as the model fluid becomes permissible.

Example 2.1: It is desired to build a model of an ancient flow channel on Mars. The width of the channel is 78 m and in the model it is to be 0.7 m. From current velocity measurements in the model you are asked to estimate the corresponding velocity on Mars. Find the appropriate scaling relationship. The acceleration due to gravity on Mars is 3.73 m s^{-2}.

Let subscripts M and E represent Mars and the Earth, respectively. Froude number equality is

$$\frac{U_M}{\sqrt{g_M L_M}} = \frac{U_E}{\sqrt{g_E L_E}} \qquad (2.13)$$

Therefore

$$U_M = \left(\sqrt{\frac{g_M}{g_E}} \sqrt{\frac{L_M}{L_E}} \right) U_E \qquad (2.14)$$

Given $g_M = 3.73$ m s^{-2}, $g_E = 9.81$ m s^{-2}, $L_M = 78$ m and $L_E = 0.7$ m yields

$$U_M = 6.51 U_E \qquad (2.15)$$

If in the above example the sediment load in the Martian canal is to be estimated, modeling requirements would be additional to Eq. 2.15. Dimensionless numbers that depend on the mode of sediment transport would have to be introduced. For instance, as we shall see in Chapter 6 suspended sediment transport is dependent on the Rouse number.

Physical modeling of coarse sediment transport can be carried out by choosing smaller and, if further necessary, lighter particles to achieve similitude with respect to the prototype sediment [*e.g.*, Shemdin, 1970]. Unfortunately, since the electrochemical cohesive force or the biotic adhesive force cannot be scaled in this way, and in fact as their scaling laws are unknown, physical modeling of the transport of fine sediment flocs is not feasible for quantitative assessments. For a *qualitative* assessment of sediment shoaling in Brunswick Harbor in Georgia, granular gilsonite was used in a physical model as a surrogate for fine sediment in the estuary. Gilsonite is a natural, resinous hydrocarbon similar to hard petroleum asphalt. The specific gravity of the material was 1.04. It was ground to powder with particle size ranging between 0.4 mm and 0.6 mm. At selected points in the model the material was injected at constant rates as a slurry through tubes during portions of the tidal cycle. The volume of sediment accumulated in critical areas was measured by retrieving the material from these areas. Since gilsonite does not flocculate in water, results on shoaling pattern and rates required careful interpretation, but were shown to yield reasonable predictions [Letter and McAnally, 1981].

When the flow is stratified the effect of gravity is reduced due to buoyancy, and the Froude number becomes

$$F' = \frac{U}{\sqrt{g\frac{\Delta\rho}{\rho_m}h_0}} = \frac{U}{\sqrt{g'h_0}} \tag{2.16}$$

where g' is the reduced gravity, $\Delta\rho$ is the density difference between two adjacent fluid layers and ρ_m is the mean density of the two layers. The quantity F' is the densimetric Froude number. It is commonly used in engineering analysis such as the design of coastal outfalls, because they discharge fluids of density that is usually different from ambient water [Grace, 1978]. As we shall see later, F' is used in formulas related to saline wedges in estuaries. Suspended sediment wedges have also been examined on the basis of F' [Lin and Mehta, 1997]. In applications involving mixing and instability across haloclines and thermoclines, the use of Richardson number, equal to the inverse square of F', is common.

2.4 Viscosity

The dynamic viscosity η of mixtures of fine sediment and water, a measure of the capacity of the mixture to take out flow energy as heat, varies with mixture composition, density and temperature. The steady-state response of most fluids to an applied stress τ acting on the z = constant plane in the x-direction is expressed as

$$\tau = \eta\dot{\gamma} = \eta\frac{du}{dz} \tag{2.17}$$

where $\dot{\gamma}$ is the rate of strain. The bar over velocity u has been dropped because the flow is non-turbulent. The use of the symbol $\dot{\gamma}$ is meant to represent the shear rate in a more general sense than the velocity gradient du/dz, which is the shear rate in two-dimensional flows. The common SI unit of η is Pascal second (Pa.s). The Poise, which is an older unit, is equal to 0.1 Pa.s, and the centipoise (cp) is equal to one milli-Pascal second (mPa.s), *i.e.,* 0.001 Pa.s.

The relationship between shear stress and strain rate, also called a flow curve, can take a variety of forms. As we shall see in Chapter 5, all relationships represented in Fig. 2.3 are relevant to fine sediment transport. In a Newtonian fluid η is independent of $\dot{\gamma}$. However, both η and the kinematic viscosity ν (= η/ρ) vary with the fluid temperature (Table 2.1) as mentioned. In pseudoplastic (or shear-thinning) response, η decreases with increasing $\dot{\gamma}$, and in a dilatant (or shear-thickening) response η increases with increasing $\dot{\gamma}$. A Bingham "fluid" behaves as a solid until τ exceeds a threshold value, the yield stress, which is also the plastic yield strength of the material. At higher values of τ the material flows as a Newtonian fluid. The Bingham plastic (or viscoplastic) has two coefficients which depend on the material. Two, three as well as four-coefficient models have been developed for pseudoplastic and dilatant responses (Table 2.2). For most engineering studies on the transport of mud slurries, two-coefficient models have shown to have adequate accuracy, and are preferred over three or four-coefficient

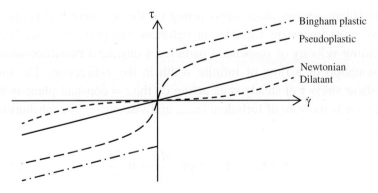

Figure 2.3. Relationships (or flow curves) between shear stress and rate of strain (shear rate in a fluid) at steady-state.

Table 2.1. Viscosities of freshwater and seawater at 35 psu salinity.

Temperature (°C)	Dynamic viscosity (Pa.s)		Kinematic viscosity (m² s⁻¹)	
	Seawater	Freshwater	Seawater	Freshwater
0	1.90×10^{-3}	1.80×10^{-3}	1.85×10^{-6}	1.80×10^{-6}
5	1.60×10^{-3}	1.50×10^{-3}	1.55×10^{-6}	1.50×10^{-6}
10	1.40×10^{-3}	1.30×10^{-3}	1.35×10^{-6}	1.30×10^{-6}
15	1.20×10^{-3}	1.15×10^{-3}	1.20×10^{-6}	1.15×10^{-6}
20	1.05×10^{-3}	1.00×10^{-3}	1.05×10^{-6}	1.00×10^{-6}
25	0.95×10^{-3}	0.90×10^{-3}	0.95×10^{-6}	0.90×10^{-6}
30	0.85×10^{-3}	0.80×10^{-3}	0.85×10^{-6}	0.80×10^{-6}
35	0.80×10^{-3}	0.70×10^{-3}	0.75×10^{-6}	0.70×10^{-6}
40	0.70×10^{-3}	0.65×10^{-3}	0.70×10^{-6}	0.65×10^{-6}

Source: Jumars *et al.* [1993].

models because commonly used rotational viscometers yield just two coefficients (viscosity and yield stress). Such devices consist of a cylinder containing the sample in which a metallic bob (a solid cylinder) or a cross-vane is spun. The stress is obtained from the torque applied to the bob/vane at a given angular speed, from which the shear rate is calculated. Complex rheological models require the use of more advanced application rheometers (Chapter 5).

2.5 Bed Shear Stress

The turbulence–mean shear stress acting on the sediment bed is one of the two most important variables in sediment transport, the other being the settling velocity of suspended particles. Consider a two-dimensional representation of a flow of infinite width in the y-direction. The total fluid shear stress τ at steady-state acting on the z = constant plane in the x-direction is the sum of turbulent (τ_{turb}) and viscous (τ_{visc}) contributions. Thus

$$\tau = \tau_{turb} + \tau_{visc} = \rho\varepsilon_m \frac{d\bar{u}}{dz} + \eta\frac{d\bar{u}}{dz} \qquad (2.18)$$

where ε_m is the kinematic turbulent viscosity, or eddy diffusivity, and $d\bar{u}/dz$ is the shear rate in which $\bar{u}$ is the turbulence–mean velocity.

Table 2.2. Non-Newtonian flow models.

No. of coefficients	Name	Source
2	Prandtl–Eyring	Bird *et al.* [1960]
2	Ostwald–de Waele	Bird *et al.* [1960]
2	Sisko	Sisko [1958]
2	Bingham	Barnes *et al.* [1989]
2	Herschel–Bulkley	Cousot and Piau [1994]
2	Casson	Casson [1959]
3	Ellis	Bird *et al.* [1960]
3	Reiner–Phillipoff	Bird *et al.* [1960]
4	Cross	Cross [1965]
4	Carreau	Carreau [1972]

This shear rate is the time–mean representation of shear production by turbulence [Tennekes and Lumley, 1972; Liggett, 1994].

Unlike η, which is a fluid property, the eddy diffusivity ε_m depends on flow and therefore varies with the intensity of turbulence. This dependence is conventionally described within a simple phenomenological framework. Referring to Fig. 2.4, consider the exchange of momentum due to turbulent fluctuations between fluid parcels in two thin layers separated by a small distance Δz. In this setup, mean flow exists in the horizontal direction only. Due to a fluctuation $+ w'$ in the z-direction, a lower parcel having a mean velocity $\bar{u}_A$ is transported to the upper layer. The mass flux associated with this transport is the product of fluid density and the fluctuation, *i.e.*, $\rho w'$.

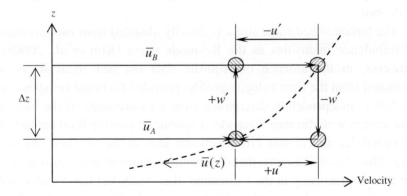

Figure 2.4. Exchange of momentum due to turbulence.

In the upper layer the velocity of the newly arrived parcel will be less than the mean velocity $\bar{u}_B$ of a parcel in the upper layer. That deficiency is $-u'$. Thus the exchange of instantaneous momentum is $-\rho w' \cdot u' = -\rho u'w'$. For conservation of water mass there must be synchronous transport of a fluid parcel from the upper layer to the lower one. In this case the mass flux will be $\rho u'$ and the velocity deficiency $-w'$. Therefore momentum exchange will again be $-\rho u'w'$.

The time–mean value of $-\rho u'w'$, i.e., $-\rho \overline{u'w'}$, which is the Reynolds stress representing drag resistance to motion, is the turbulent component of the local shear stress. Thus

$$\tau_{turb} = -\frac{1}{T_t}\int_0^{T_t}\rho u'w'\,dt = -\rho\overline{u'w'} \tag{2.19}$$

Therefore, Eq. 2.18 becomes

$$\tau = \eta\frac{d\bar{u}}{dz} - \rho\overline{u'w'} \tag{2.20}$$

An open channel with turbulent flow of a constant-density fluid is a convenient system to describe sediment transport. In such a channel the total shear stress τ increases linearly over depth h from nil at the surface (unless there is wind stress) to τ_b at the bed. The viscous contribution is largely confined to a near-bed viscous sublayer of notional thickness δ_l (Fig. 2.5). Its contribution to the total bed shear stress is often ignored, especially when δ_l is smaller than the height of the hydraulic roughness of the bed.

The turbulent bed shear stress is directly obtained from measurement of turbulence intensities as the Reynolds stress [Kim *et al.*, 2000b]. However, in the absence of requisite data the bed shear stress is estimated from the flow velocity profile, provided frictional resistance at the bed is independently determined from measurements of the loss of flow energy with distance. Consider a volume of flowing fluid (water) of elemental length Δx and cross-sectional area A_c in an open channel (Fig. 2.6a). We will assume that the channel bottom slope is small so that the flow is practically in the x-direction. It is steady but horizontally and vertically non-uniform with a depth–mean velocity $u_m(x)$. Therefore

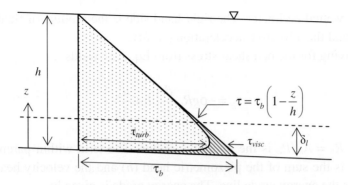

Figure 2.5. Variation of local shear stress with depth in an open channel.

$u_m(x)^2/2g$ is the local kinetic energy head. Due to friction at the bed and sides of the channel, the head loss over distance Δx is Δh, where h is the height of water above a datum. At steady-state the net fluid force in the x-direction must equal the product of water mass in the volume element multiplied by the acceleration of flow in the same direction, *i.e.,*

$$-\rho g A_c \Delta h - \tau_b P_w \Delta x = \rho A_c u_m \frac{du_m}{dx} \Delta x \qquad (2.21)$$

where P_w is the wetted perimeter, *i.e.,* the perimeter length of the cross-section wetted by water (Fig. 2.6b). The first term on the left hand side is the downstream gravity force due to weight of water in the elemental volume, and the second term is the resistance force due to friction acting

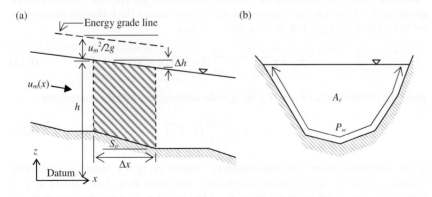

Figure 2.6. (a) Elemental flow volume of an open channel; (b) channel cross-section.

on the wetted perimeter. The right hand side is the product of fluid mass $\rho A \Delta x$ and the advective acceleration du_m/dx.

Solving for the bed shear stress from Eq. 2.21 yields

$$\tau_b = -\rho g R_h \frac{d}{dx}\left(h + \frac{u_m^2}{2g}\right) \qquad (2.22)$$

where $R_h = A_c/P_w$ is the hydraulic radius. The term within parentheses ($= H_t$) is the sum of the piezometric head (h) and the velocity head, and defines the energy grade line. The energy grade is given by

$$S_f = -\frac{d}{dx}\left(h + \frac{u_m^2}{2g}\right) = -\frac{dH_t}{dx} \qquad (2.23)$$

and therefore

$$\tau_b = \rho g R_h S_f = \gamma_w R_h S_f \qquad (2.24)$$

where $\gamma_w = \rho g$ is the unit weight of water. When the flow is uniform, u_m is independent of the x-coordinate and denotes the depth–mean (or boundary layer–mean) velocity. In this case, $S_f = -\Delta h/\Delta x = -\Delta z/\Delta x = S_o$, where $-\Delta h/\Delta x$ is the slope of the water surface and S_o is the bed slope.[c]

---

[c] When the flow is non-uniform, the effects of acceleration and pressure gradient must be accounted for. From the one-dimensional momentum balance the following equation for the bed shear stress τ_{bn} in non-uniform flows is found

$$\tau_{bn} = \gamma_w R_h\left[S_f - \frac{dh}{dx}(1 - F^2)\right] = \eta_n \tau_b \qquad (F2.1)$$

where the Froude number $F = u_m/\sqrt{gh}$, τ_b is the uniform-flow bed shear stress and

$$\eta_n = 1 - \frac{1}{S_f}\frac{dh}{dx}(1 - F^2) \qquad (F2.2)$$

is the correction factor for non-uniformity. Values of $\eta_n > 1$ indicate accelerating subcritical flow and $\eta_n < 1$ implies decelerating subcritical flow. From eleven tests in flume with current-driven flow, Lee *et al.* [2004] found η_n to range between 0.85 and 2.43 with a mean value of 1.50.

From dimensional analysis τ_b can be related to the velocity head $u_m^2/2g$ according to the relationship (quadratic law)

$$\tau_b = a_r \rho \frac{u_m^2}{2g} \tag{2.25}$$

where the coefficient a_r depends on bed resistance. The friction velocity is defined as

$$u_* = \sqrt{\frac{\tau_b}{\rho}} \tag{2.26}$$

Therefore

$$\tau_b = \rho u_*^2 = \rho g R_h S_f = a_r \rho \frac{u_m^2}{2g} \tag{2.27}$$

or

$$u_* = \sqrt{g R_h S_f} = \sqrt{\frac{a_r}{2g}}\, u_m \tag{2.28}$$

Several bed resistance coefficients related to a_r are in common use. The Darcy–Weisbach friction factor f_c is defined by the Darcy head loss formula

$$S_f = \frac{f_c}{4R_h}\frac{u_m^2}{2g} \tag{2.29}$$

Therefore

$$a_r = \frac{f_c g}{4} \tag{2.30}$$

Combining Eqs. 2.27 and 2.30 yields

$$\tau_b = \frac{f_c}{4}\rho \frac{u_m^2}{2} \tag{2.31}$$

In Fig. 2.7, u_m is plotted against τ_b obtained from flow velocity measurements over four tidal cycles at Johns Pass, an entrance channel on the Gulf of Mexico coast of Florida. The mean depth of water at the site was 7.6 m and the bed was composed of a mixture of sand and shells with the longest (of the three mutually perpendicular) dimension as much as 8 cm. The dashed line corresponds to Eq. 2.31 with $f_c = 0.027$. At the low and high velocity ends the data trends deviate from the equation. At velocities below ~0.3 m s^{-1} the flow was not in the hydraulically fully rough turbulent range. This range is characterized by the wall or roughness Reynolds number $Re_* = k_s u_* / v > 70$, where k_s is the Nikuradse bed roughness. In the transition range of flow ($5 < Re_* < 70$) the friction factor depends on Re_*. In the channel, f_c in this range was lower (flatter line slope) than the value predicted by Eq. 2.31. When the velocity exceeded about 1 m s^{-1} it was observed from underwater videos that large shells had begun to move. The result was that the additional kinetic energy loss for transportation of shells manifested as an increase in the friction factor (flatter line slope) relative to Eq. 2.31 [Mehta and Christensen, 1983].

The plot in Fig. 2.8 based on the laboratory flume data of Krone [1962] indicates close adherence to the quadratic law inasmuch as the exponent 1.94 is almost equal to 2.

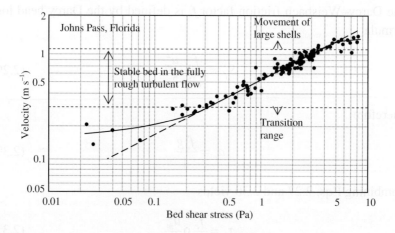

Figure 2.7. Boundary layer–mean flow velocity plotted against bed shear stress from Johns Pass in Florida (from Mehta [1978]).

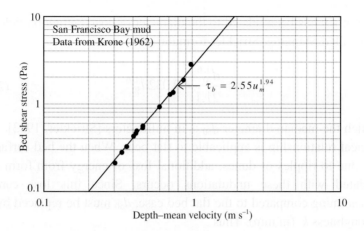

Figure 2.8. Bed shear stress versus depth–mean flow velocity based on flume data of Krone [1962] for a bed of San Francisco Bay mud.

The Chézy coefficient C_z is defined by the formula

$$u_m = C_z \sqrt{R_h S_f}$$ (2.32)

To be consistent with u_m as the depth–mean velocity, it must be taken as the unit discharge velocity, *i.e.*, discharge per unit width of the channel divided by the depth (h). From Eqs. 2.25 and 2.32 we obtain

$$a_r = \frac{2g^2}{C_z^2}$$ (2.33)

The bed resistance coefficient n is defined by the Manning formula

$$u_m = \frac{1}{n} R_h^{2/3} S_f^{1/2}$$ (2.34)

Therefore

$$a_r = \frac{2g^2 n^2}{R_h^{1/3}}$$ (2.35)

Manning's n has been related to the diameter of coarse-grain particles by the so-called Manning–Strickler formula

$$n = \frac{d_{50}^{1/6}}{21.1} = 0.032 d_{50}^{1/6} \qquad (2.36)$$

in which the median diameter d_{50} is in millimeters [Strickler, 1923]. This empirical relationship is applicable to flat beds. When the bed surface is wavy due to ripples or dunes, additional loss of energy from form drag associated with these undulations occurs. Since this loss can be overwhelming compared to the flat bed case, d_{50} must be replaced by the bed roughness k_s (in mm). Thus

$$n = 0.032 k_s^{1/6} \qquad (2.37)$$

Typical values of Manning's n are given in Table 2.3. In winding creeks with weeds the loss of flow energy can be as much as an order of magnitude greater than in straight (and clean) channels.

Data from a 100 m long flume lined with beds of kaolinite clay and mud from the Atchafalaya Bay in Louisiana were analyzed for Manning's n [Dixit *et al.*, 1982]. These values along with the corresponding friction factors are given in Table 2.4. The three sets of tests for beds of kaolinite clay differed in mean discharge, but yielded similar values of n. The natural mud had the highest value, suggesting the presence of surface asperities not found on kaolinite beds. Nevertheless, the overall low values of n indicate that all the beds could be characterized as hydraulically smooth.

Table 2.3. Typical values of Manning's n in channels.

Channel configuration and bottom	n
Straight, clean	0.025–0.030
Straight, clean, alluvial	$0.08 d_{75}^{1/6}$ (d_{75} in mm)[a]
Winding with pools and shoals	0.033–0.040
Winding, very weedy and overgrown	0.075–0.150

[a] d_{75} is the 75th percentile diameter of the cumulative size distribution; see Chapter 3.
Source: Henderson [1966].

Table 2.4. Bed friction coefficients for fine sediment beds[a].

Bed sediment	Manning's n	Friction factor f_c
Kaolinite	0.010	0.026
Kaolinite	0.010	0.030
Kaolinite	0.011	0.032
Atchafalaya Bay mud	0.012	0.035

[a] For mud in general, Soulsby [1983] states $z_0 = 0.2$ mm. Assuming the applicability of Eq. 2.37 (with $k_s = 29.7z_0$) would yield $n = 0.0092$, which is close to 0.010.

The relationship between n and f_c is obtained from Eqs. 2.30 and 2.35 as

$$f_c = \frac{8gn^2}{R_h^{1/3}} \qquad (2.38)$$

For open channels it has been argued that n is the true measure of bed roughness. If so, f_c would become a derived quantity varying inversely with the cubic root of R_h. However, the dependence of f_c on R_h is weak, and in many cases permits one to ignore the effect of R_h without introducing a significant error in the estimation of energy loss.

2.6 Interfacial Shear Stress

In a flow stratified by a dense layer of suspended fine sediment, the shear stress between the surface of this layer and water layer above can induce a relative motion between the two layers, and mixing when this motion is significant. Much of the investigative work on interfacial shear stress has been with saltwater as the lower layer. The relevance of densimetric Froude number is found in the calculation of the interfacial friction factor and intrusion of a fluid density wedge into a fluid of lighter density. An early problem of interest was the stationary (steady-state) salt wedge in estuaries [Schijf and Schoenfeld, 1953]. Seawater in the wedge is denser than fresh river water flowing downstream. The intrusive wedge becomes stationary at a certain distance of penetration L_i when the excess upstream-directed hydrostatic force, due to the difference in densities of seawater and freshwater, is balanced by the drag force acting downstream along the water–wedge interface. Similar wedges containing non-depositable colloidal particles or very slowly settling suspended matter can occur in tidal docks and as lock exchange flows [McDowell, 1971; Gole *et al.*, 1973; Lin and Mehta, 1997]. As we shall see in Chapter 4, stratified flows have also been created by producing auto-suspensions from deflocculated, non-settling fine particles.

Referring to Fig. 2.9, the shear stress at the wedge interface is

$$\tau_i = \frac{f_i}{4} \rho \frac{u_1^2}{2} \qquad (2.39)$$

where f_i is the interfacial friction factor, ρ is freshwater density, $u_1(x)$ is the local freshwater velocity above the wedge and x is the distance coordinate. The overbar denoting a turbulence–mean quantity has been conveniently dropped. The quantity $\overline{f_i}$ representing locally varying f_i averaged over the wedge length L_i is determined under four assumptions: (1) one-dimensional water flow over the wedge, (2) no mixing between the two layers, (3) infinite channel width, and (4) zero net velocity within the wedge. The resulting expression for $\overline{f_i}$ is

$$\frac{\overline{f_i}}{8} \tilde{x} = \tilde{h}_1 \left(\frac{1}{5F_0'^2} \tilde{h}_1^4 + \frac{1}{4F_0'^2} \tilde{h}_1^3 - \frac{1}{2} \tilde{h}_1 + 1 \right)$$

$$+ 3F_0'^{2/3} \left(\frac{1}{10} F_0'^{2/3} - \frac{1}{4} \right) \qquad (2.40)$$

and the wedge length

$$L_i = \frac{2h_0}{\overline{f_i}} \left(\frac{1}{5F_0'^2} + 3F_0'^{2/3} - \frac{6}{5} F_0'^{4/3} - 2 \right) \qquad (2.41)$$

In these expressions, $\tilde{x} = x / L_i$, h_0 is the total water depth, $\tilde{h} = h_1 / h_0$ and $h_1(x)$ is the water depth above the wedge. From Eq. 2.16 the densimetric Froude number is defined as

$$F_0' = \frac{u_0}{\sqrt{g' h_0}} \qquad (2.42)$$

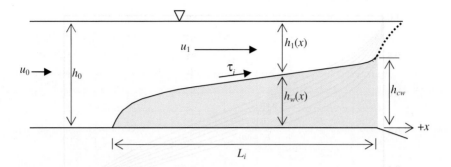

Figure 2.9. Steady, non-uniform two-layer flow and turbid wedge.

where u_0 is the river velocity upstream of the wedge and g' is reduced gravity. The following relationship is obtained between $\tilde{h}$, the local wedge height $h_w(x)$ and the wedge height h_{cw} at $x = 0$

$$\tilde{h} = 1 - \tilde{h}_w \left(1 - F_0'^{2/3}\right) \tag{2.43}$$

where $\tilde{h}_w = h_w/h_{cw}$. This expression highlights the role of the densimetric Froude number in configuring the wedge.

Example 2.2: Show that the shape of the density wedge in terms of the variation of the dimensionless wedge thickness $\tilde{h}_w$ with dimensionless distance x/L_i is uniquely determined by the densimetric Froude number F_0'.
    By eliminating $\tilde{f}_i$ between Eqs. 2.40 and 2.41 and introducing Eq. 2.43 yields

$$\tilde{x} = \frac{\tilde{h}_w^3 \left\{ \left[4 - 2\tilde{h}_w \left(1 - F_0'^{2/3}\right) \right]^2 + \left[4 + \tilde{h}_w \left(1 - F_0'^{2/3}\right) \right] \right\} - 10\tilde{h}_w^2 \left(F_0'^{4/3} + F_0'^{2/3} + 1 \right)}{6F_0'^{4/3} + 3F_0'^{2/3} + 1} + 1 \tag{2.44}$$

Dimensionless wedge profiles based on this relationship are plotted in Fig. 2.10. There is an evident dependence of the profiles on F_0'. However, over a wide range of F_0', from about 0.20 to 0.80, its effect on shape is not significant. In flume experiments of Keulegan [1966] on stationary salt wedges over the range of F_0' from 0.18 to 0.40, the effect of F_0' could not be detected as the data points fell on a single curve at about its mean position in Fig. 2.10.

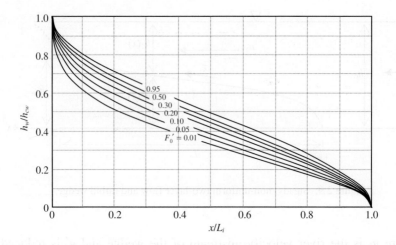

Figure 2.10. Dimensionless shape of the stationary wedge (from Partheniades *et al.,* [1975]).

As we shall see in Chapter 12, in contrast to the salt wedge, the behavior of a suspended sediment wedge depends on whether the particles remain in suspension over the distance of wedge penetration. If they do not have the potential to deposit, *i.e.,* the material is a wash-load (Chapter 6) of colloid-sized or very light-weight particles, wedge characteristics will be analogous to those of a salt front. When the particles are able to settle the character of the wedge becomes complex because at the upstream terminus of the wedge the difference between the suspension density and water density practically vanishes and the wedge becomes ill-defined.

Gosselin [1993] conducted stratified flow tests in a Plexiglas tank shown schematically in Fig. 2.11. The experiments were modeled on field observations at the mouth of the Rappahannock River in Chesapeake Bay, an area characterized by wind-driven backflows of saline undercurrent over a natural sill [Kuo and Park, 1992]. In the experiments salt water was replaced by tap water and freshwater by denatured ethyl alcohol. Airflow from a blower caused the free-surface to tilt upward in the down-wind direction, while the interface tilted up-wind, causing water to flow over the sill and mix with alcohol. Sample velocity profiles measured at a fixed air speed (1.02 m s^{-1}) are shown in Fig. 2.12.

The following two equations were used to estimate the interfacial friction factor $\bar{f_i}$:

$$\frac{dH}{dx} = \frac{\left(\dfrac{dh_1}{dx} + \dfrac{f_c h_1}{4B_c} + \dfrac{f_u}{8}\right)\dfrac{U^2}{gh_1} + \left(\dfrac{h_0 - h_1}{h_1}\right)\dfrac{\Delta\rho}{\rho}\dfrac{dh_1}{dx}}{\left[1 + \left(1 - \dfrac{\Delta\rho}{\rho}\right)\left(\dfrac{h_0 - h_1}{h_1}\right)\right]} \tag{2.45}$$

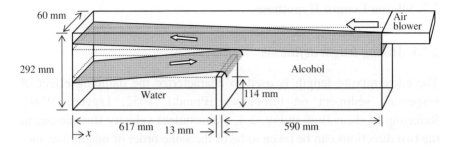

Figure 2.11. Schematic drawing of experimental tank to generate up-wind interfacial flow.

$$\overline{f}_i = \left[\frac{\Delta\rho}{\rho} \frac{dh_1}{dx} - \left(1 - \frac{\Delta\rho}{\rho}\right) \frac{dH}{dx} \right] \frac{8g(h_0 - h_1)}{U^2} \qquad (2.46)$$

in which H denotes the elevation of the energy grade line, h_1 is the mean depth of the upper fluid, h_0 is the mean total depth, B_c is the channel width, U is the mean velocity of the upper fluid (assuming the lower fluid to be stationary), ρ is the density of the upper fluid, $\Delta\rho$ is the density difference between the two fluids, f_c is the wall friction factor and f_u is the wind-induced friction factor at the water surface [Partheniades and Dermissis, 1978]. If the distance of penetration of the wedge is known, $\overline{f}_i$ can be more easily obtained from Eq. 2.41; see Exercise 2.5.

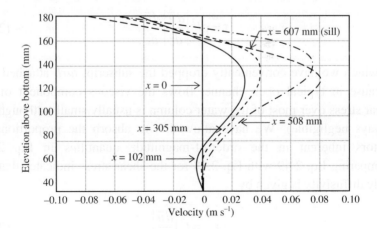

Figure 2.12. Wind-driven velocity profiles in a stratified flow tank.

2.7 Mixing Length Hypothesis

2.7.1 *Mixing length definition*

The eddy mixing length is used to characterize the damping effect of suspended sediment on turbulence [Prandtl, 1952; Liggett, 1994]. Referring to shear flow in Fig. 2.4, the turbulent velocity fluctuations in the two directions can be taken to be of the same order of magnitude, *i.e.*,

$$O\left(\overline{|u'|}\right) = O\left(\overline{|w'|}\right) \tag{2.47}$$

which implies local isotropy. We can express this equality in terms of the readily measurable time–mean velocity gradient

$$O\left(\overline{|u'|}\right) = O\left(\overline{|w'|}\right) = l_t \left|\frac{d\bar{u}}{dz}\right| \tag{2.48}$$

where l_t is the mixing length, a measure of the size of turbulent eddies. Taking the absolute value of the velocity gradient permits the inclusion of its positive and negative values. Substituting the equalities of Eq. 2.48 into Eq. 2.19 yields

$$\tau = \rho l_t^2 \left|\frac{d\bar{u}}{dz}\right| \frac{d\bar{u}}{dz} \tag{2.49}$$

in which we have conveniently dropped the subscript *turb* attached to τ because in typical open channel flows the viscous component of the shear stress over most of the water column is usually small (although not always negligible). We have allowed l_t to absorb the proportionality factors inherent in the order-of-magnitude quantities in Eq. 2.48. Comparing Eq. 2.49 with Eq. 2.18 for the shear stress indicates that the eddy diffusivity is given by

$$\varepsilon_m = l_t^2 \left|\frac{d\bar{u}}{dz}\right| \tag{2.50}$$

Using a somewhat different set of assumptions concerning turbulent motion, von Kármán [Liggett, 1994] arrived at the following expression for the shear stress

$$\tau = \kappa \frac{(d\bar{u}/dz)^4}{d^2\bar{u}/dz^2} \qquad (2.51)$$

where κ is the Karman constant. Comparing this expression with Eq. 2.49 of Prandtl leads to

$$l_t = \sqrt{\frac{\kappa}{\rho}} \frac{d\bar{u}/dz}{\sqrt{d^2\bar{u}/dz^2}} \qquad (2.52)$$

which permits determination of the turbulent mixing length provided the velocity profile $\bar{u}(z)$ is known for calculation of the first and second derivatives of $\bar{u}$ with respect to z.

2.7.2 *Boundary layer velocity profile*

The generic velocity profile in the boundary layer shown in Fig. 2.13a is based on data obtained over a hydraulically smooth bed [Dyer, 1986]. It consists of a lowermost viscous profile ($\tilde{z} = zu_*/\nu \leq 5$), a log-velocity profile in the inner boundary layer, a transition or buffer zone between the two profiles (for which no equation is prescribed) and an outer boundary layer profile ($\tilde{z} > 1,000$). The viscous profile, which is characteristically linear because molecular viscosity is constant, appears as a curve on a log-linear scale plot. The $\tilde{z} = 1,000$ value at which the outer profile begins is not precise. It is expressed as an empirical one-seventh power law. As an approximation of the velocity in the buffer zone, the viscous layer and the log–layer may be extrapolated to their point of intersection at $\tilde{z} = 11.6$.

The derivation of the log-velocity profile, which is most commonly employed in open channel flow descriptions, is recapped here. We will assume that in the turbulent boundary layer the shear stress τ is equal to its value τ_b at the bed (or wall). Furthermore, the mixing length is assumed to increase linearly with elevation above the bed, *i.e.*,

$$l_t = \kappa z \qquad (2.53)$$

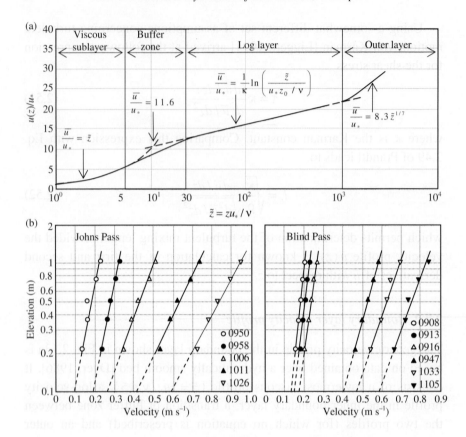

Figure 2.13.   Boundary layer velocity profiles: (a) Generic description based on measurements over a hydraulically smooth bed; (b) bottom log-layer velocity profiles on August 7, 1974 at Johns Pass and August 14, 1974 at Blind Pass (from Mehta [1978] and Mehta *et al.* [1976]).

Thus, from Eq. 2.49

$$\tau_b = \rho \kappa^2 z^2 \left| \frac{d\overline{u}}{dz} \right| \frac{d\overline{u}}{dz} \qquad (2.54)$$

Since by definition $\tau_b = \rho u_*^2$, the above equation can be written as

$$\frac{\kappa z}{u_*} \frac{d\overline{u}}{dz} = 1 \qquad (2.55)$$

As we will note later, when the flow is stratified, the right-hand side of Eq. 2.55 is no longer unity and must be replaced by a function dependent on the type and degree of stratification.

Integration of Eq. 2.55 yields

$$\frac{\bar{u}}{u_*} = \frac{1}{\kappa} \ln z + C' \tag{2.56}$$

in which the constant C' is evaluated by setting $z = z_0$ at $\bar{u} = 0$. Using this condition we can rewrite Eq. 2.56 as

$$\frac{\bar{u}}{u_*} = \frac{1}{\kappa} \ln\left(\frac{z}{z_0}\right) \tag{2.57}$$

Since the generally accepted value of κ for boundary layer in clear water is 0.4, Eq. 2.57 becomes

$$\frac{\bar{u}}{u_*} = 2.5 \ln\left(\frac{z}{z_0}\right) = 5.75 \log\left(\frac{z}{z_0}\right) \tag{2.58}$$

From integration of this equation over depth h the depth–mean velocity $\bar{u}_m$ is found to be

$$\bar{u}_m = \frac{u_*}{\kappa}\left[\ln\left(\frac{h}{z_0}\right) - 1\right] \tag{2.59}$$

The height z_0 is the virtual origin of the velocity profile. In this derivation we have assumed that $z_0/h \ll 1$. Examples of such profiles in the bottom 1 m of flow are shown in Fig. 2.13b. The tidal channels of Johns Pass and Blind Pass on the Gulf of Mexico coast of Florida are lined with flat beds of shells interspersed with well-sorted sand of 0.22 mm median diameter (d_{50}). Typically there is little freshwater outflow from these channels, and salinity-wise the water column is usually well-mixed. These conditions are conducive to the occurrence of log-velocity profiles. Based on data collected over two tidal cycles, mean values of the bed resistance defining coefficients f_c, n, C_z and k_s are given in

Table 2.5. At both channels we find that $k_s \gg d_{50}$, because the bed roughness reflects high flow-resistance due to shells.

Equation 2.59 is commonly stated as

$$\frac{\bar{u}}{u_*} = \frac{1}{\kappa}\ln\left(\frac{z}{k_s}\right) + B_s = \frac{1}{\kappa}\ln\left(\frac{\tilde{z}}{Re_*}\right) + B_s \qquad (2.60)$$

where $\tilde{z} = zu_* / \nu$. The coefficient B_s depends on the roughness Reynolds number $Re_* = k_s u_* / \nu$, in which k_s is empirically related to z_0 in the fully rough range of flow ($Re_* > \sim70$) by

$$k_s = 29.7 z_0 \qquad (2.61)$$

When the bed is flat, Re_* and $\tilde{z}$ are synonymous. In that case, when $Re_* > \sim 70$, $B_s = 8.5$. In the transition range ($5 \lesssim Re_* \lesssim \sim70$) k_s is not prescribed by Eq. 2.61 (see *e.g.*, Christoffersen and Jonsson [1985], for an extended empirical expression defining k_s) and B_s varies non-monotonically with Re_*. When $Re_* < 5$ the flow is non-turbulent, so the velocity u varies linearly with elevation z (Fig. 2.13a and Chapter 6).

From Eq. 2.50 we find that

$$\varepsilon_m = \kappa u_* z \qquad (2.62)$$

which indicates that the eddy diffusivity increases linearly with elevation, and is maximum at the top of the log boundary layer. However, data from laboratory flumes and shallow channels in which the boundary layer height equals the water depth do not wholly support this inference. These data indicate that the diffusivity reaches its maximum value at some elevation below the surface, and decreases above that elevation. The cause of the deviation between Eq. 2.60 and observations

Table 2.5. Composite grain size and bed resistance coefficients at two tidal channels in Florida.

Channel	Grain size (mm)	f_c	n	C_z (m$^{1/3}$ s^{-1})	k_s (mm)
Johns Pass	2.7–3.3	0.027	0.026	54	95
Blind Pass	0.35–5.90	0.021	0.020	61	21

Source: Mehta [1978].

is the Prandtl assumption of a constant shear stress in the boundary layer. In contrast, the von Kármán development leads to linearly increasing shear stress (Fig. 2.5) given by

$$\tau = \tau_b\left(1 - \frac{z}{h}\right) \tag{2.63}$$

in which τ is taken as τ_{turb}, and τ_{visc} (Eq. 2.18) is ignored. This stress variation yields

$$\varepsilon_m = \kappa u_* z\left(1 - \frac{z}{h}\right) \tag{2.64}$$

in which $z_0^2 \ll 1$ is assumed. This expression indicates a parabolic dependence of ε_m on z, which more faithfully reproduces data on ε_m obtained in flumes than Eq. 2.62. However Eq. 2.64 is applicable only to uniformly dense flows, and is sensitive to stratification, which modifies the structure of turbulence as we shall see in Chapter 6.

The depth–mean value of ε_m from Eq. 2.62 is

$$\bar{\varepsilon}_m = \frac{1}{2}\kappa u_* h \tag{2.65}$$

and from Eq. 2.64

$$\bar{\varepsilon}_m = \frac{1}{6}\kappa u_* h \tag{2.66}$$

2.7.3 *Viscous contribution to velocity profile*

A deficiency in Eq. 2.58 is that at $z = 0$, *i.e.*, the theoretical bed plane, $\bar{u} \to -\infty$, which does not satisfy the required no-slip ($\bar{u} = 0$) boundary condition for real fluids. At distances of up to a few grain diameters next to the bed the error introduced in the velocity due to this limitation can be significant, as the profile becomes unrealistic with a negative velocity when $z < z_0$. To correct for this discrepancy, the log-velocity profile of

Eq. 2.49 can be revised by including the viscous contribution to the total shear stress as an extension of the Prandtl derivation. We may write

$$\tau_b = \rho l_t^2 \left(\frac{d\bar{u}}{dz} \right)^2 + \rho l_v^2 \left(\frac{d\bar{u}}{dz} \right)^2 \qquad (2.67)$$

where l_v is a notional laminar (viscous) mixing length. Thus we obtain the dimensionless expression

$$1 = \left(\frac{l_t^2}{u_*^2} + \frac{l_v^2}{u_*^2} \right) \left(\frac{d\bar{u}}{dz} \right)^2 = \frac{l_{tv}^2}{u_*^2} \left(\frac{d\bar{u}}{dz} \right)^2 \qquad (2.68)$$

where $l_{tv} = \sqrt{l_t^2 + l_v^2}$ is the composite (turbulent and laminar) mixing length. We may represent it conveniently in the linear form

$$l_{tv} = 1 + \kappa Z + \beta_f Re_* \qquad (2.69)$$

in which the first term on the right-hand side (unity) is the dimensionless composite mixing length, the second term is the dimensionless turbulent mixing length and the third term is a dimensionless "forced" mixing length [Christensen, 1972]. The latter length accounts for the dependence of flow velocity on the roughness Reynolds number Re_* through the coefficient β_f. Equation 2.69 is valid only in the fully rough range of flow with $Re_* \geq 70$. Introducing Eq. 2.69 into 2.68 and integrating yields

$$\frac{\bar{u}}{u_*} = \frac{1}{\kappa} \ln \left(\frac{\nu}{u_*} + \kappa \beta_f + \kappa z \right) + C'' \qquad (2.70)$$

The coefficient β_f and the constant of integration C'' are evaluated by using the following two boundary conditions:

$$\bar{u}(z = 0) = 0 \qquad (2.71)$$

and

$$\frac{\bar{u}(z \to \infty)}{u_*} = B_s + \frac{1}{\kappa} \ln \left(\frac{z}{k_s} \right) \qquad (2.72)$$

Taking $B_s = 8.5$ and $\kappa = 0.4$ changes the above condition to

$$\frac{\bar{u}(z \to \infty)}{u_*} = 8.5 + 2.5\ln\left(\frac{z}{k_s}\right) \tag{2.73}$$

Applying the conditions in Eqs. 2.71 and 2.73 to Eq. 2.70 yields

$$C'' = 10.8 - 2.5\ln k_s \tag{2.74}$$

and

$$\beta_f = \frac{1}{74} - \frac{1}{Re_*} \tag{2.75}$$

which indicates that β_f is a representation of the roughness (or wall) Reynolds number. When $Re_* = 74$ (which is close to 70) $\beta_f = 0$, which marks the transition to fully rough flow. Substitution of Eqs. 2.74 and 2.75 into Eq. 2.70 followed by integration leads to

$$\frac{\bar{u}}{u_*} = \frac{1}{\kappa}\ln\left(\frac{z}{z_0} + 1\right) = \frac{1}{\kappa}\ln\left(\frac{29.7z}{k_s} + 1\right) \tag{2.76}$$

A relationship of the same form was introduced by Rossby [1932] invoking kinematic similarity for boundary layer wind flow over the sea surface. Extensions of Eq. 2.76 have been developed including

$$\frac{\bar{u}}{u_*} = B_s + \frac{1}{\kappa}\ln\left(\frac{z + z_0}{k_s}\right) \tag{2.77}$$

Values of B_s have been reported to range from 1.46 to 7.56 for a creek lined with clayey sand and taking $k_s = 4z_0$ [Kironto and Graf, 1994; Papanicolaou *et al.*, 2002].

The deviation in velocity values close to the bed ($z < 0.001$ m) using the uncorrected Eq. 2.57 is apparent from Table 2.6, in which $\bar{u}$ values from Eqs. 2.57 and 2.76 have been calculated for $u_* = 0.06$ m s^{-1} and $z_0 = 1$ mm. As expected, without the viscous correction the velocities below 1 mm are negative. In the immediate neighborhood above this

Table 2.6. Velocities using Eqs. 2.57 and 2.76 for $u_* = 0.06$ m s^{-1} and $z_0 = 0.001$ m.

z (m)	z/z_0	$\bar{u}$ (Eq. 2.57) (m s^{-1})	$(z/z_0)+1$	$\bar{u}$ (Eq. 2.76) (m s^{-1})	Error in $\bar{u}$ (%)
1×10^{-4}	1×10^{-1}	0.345	1.100	0.014	Very large
5×10^{-4}	5×10^{-1}	0.104	1.500	0.608	Very large
1×10^{-3}	1×10^{0}	0	2	0.104	100
5×10^{-3}	5×10^{0}	0.241	6	0.268	10
1×10^{-2}	1×10^{1}	0.345	11	0.359	4
5×10^{-2}	5×10^{1}	0.586	51	0.589	0.5
1×10^{-1}	1×10^{2}	0.690	101	0.691	0.2
5×10^{-1}	5×10^{2}	0.931	501	0.931	0.03

elevation the error continues to be high. As mentioned, a practical significance of this discrepancy is that the hydrodynamic lift on sandy or silty particles at the bed cannot be calculated from the uncorrected equation.

For fitting Eq. 2.76 to elevation-versus-velocity data, as a starting point it is necessary to assume the applicability of Eq. 2.57 using the method of least-squares. The value of z_0 thus obtained is a first estimate, which is improved by iteration until Eq. 2.76 matches the data. Examples of velocity profiles obtained in a flume and fitted to Eq. 2.76 are shown in Fig. 2.14. These data were taken over beds of bivalve shells of

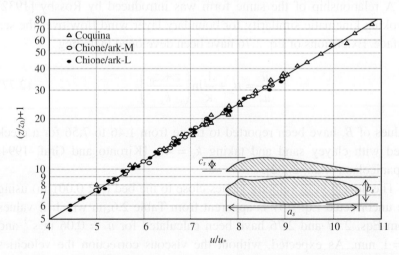

Figure 2.14. Rossby–Christensen representation of log-velocity profile using data of Lee [1978] on beds of bivalve shells.

coquina, chione and ark from beaches in southwestern Florida. Two of the beds were 1:1 (by volume) mixtures of chione and ark of different sizes (L for large and M for medium). The third bed consisted of coquina. In Table 2.7 the mean dimensions (a_s, b_s and c_s) and the material density of the shells are given. Table 2.8 gives flume-derived values of the friction factor f_c and the bed roughness k_s. Each value is the mean of four experimental runs in which the roughness Reynolds number Re_* was varied. The f_c and k_s values reflect energy losses due to form drag associated with strong wakes characteristic of flows past bivalve-type shells.

Example 2.3: Calculate the difference in the velocity $\bar{u}$ obtained from Eqs. 2.57 and 2.76 1 mm above the theoretical bed. Take u_* to be 0.02 m s^{-1}, k_s = 10 mm and (assume) κ = 0.4.

From Eqs. 2.57 (Prandtl) and 2.76 (Rossby–Christensen) the velocity plots of Fig. 2.15 are deduced. With $v = 10^{-6}$ m^2 s^{-1} we obtain $\tilde{z} = 20$, $Re_* = 200$ and $z_0 = 3.4 \times 10^{-4}$ m. From Eq. 2.57, $\bar{u} = 0.055$ m s^{-1} and from Eq. 2.76, $\bar{u} = 0.070$ s^{-1}, which is 28% higher. Taking the bed shear stress to be proportional to $\bar{u}^2$ would mean that neglecting the viscous contribution would lead to a 39% smaller bed shear stress.

Table 2.7. Dimensions and density of bivalve shells.

Bivalve	a_s [a] (mm)	b_s (mm)	c_s (mm)	Density (kg m^{-3})
Coquina	15	8	3	2780
Chione	20	18	6	2830
Ark	21	14	6	2800

[a] a_s, b_s and c_s are mean values of the three mutually perpendicular dimensions of a bivalve shell.

Table 2.8. Roughness Reynolds number and resistance coefficients.

Bed	Re_*	f_c	k_s (m)
Coquina	690–2140	0.097	0.054
Chione/ark-M	1220–1920	0.098	0.061
Chione/ark-L	1820–3200	0.121	0.061

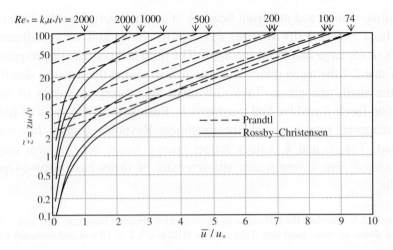

Figure 2.15. Time–mean velocity profiles near bed in the rough range of flow (adapted from Christensen [1972]).

2.7.4 *Effect of buoyancy on velocity profile*

As mentioned, Eq. 2.57 (or Eq. 2.76) characterizes the velocity profile only in uniformly dense boundary-layer flows. In stratified flows the mixing length is modified by the buoyancy effect. The dependence of the shape of velocity profile on the mode of stratification is as shown schematically in Fig. 2.16. The log-velocity profile under the neutral (*i.e.,* uniform density) condition becomes concave-upward as the density increases with elevation. This is the case of progressively heavier fluid layers from bottom up, which is an unstable state due to the raised center of gravity of the fluid relative to the neutral condition. The other case is one of stable stratification when the center of gravity is lower than neutral. In this case the velocity profile is convex-upward.

The effect of stratification on the velocity gradient is introduced into Eq. 2.55 by replacing the value 1 on the right-hand side by a function of the normalized elevation z/L_m according to

$$\frac{\kappa z}{u_*}\frac{d\bar{u}}{dz} = f^n\left(\frac{z}{L_m}\right) \qquad (2.78)$$

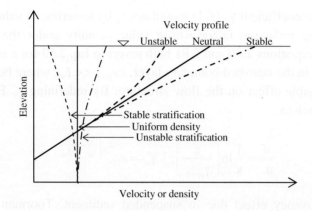

Figure 2.16. Dependence of velocity profile on fluid density gradient.

where L_m is the Monin–Obukhov turbulence length scale. This scale depends on the ratio of inertia to buoyancy, and therefore on the densimetric Froude number (or the Richardson number). In Chapter 10 we will refer to this dependence in connection with the transport of fluid mud.

We may conveniently express $f^n(z/L_m)$ as a series and truncate it to retain just the first two (linear) terms as

$$\frac{\kappa z}{u_*}\frac{d\bar{u}}{dz} = 1 + \alpha_m \frac{z}{L_m} \tag{2.79}$$

in which a typical value of the coefficient α_m is 5. Integration of this equation gives

$$\frac{\bar{u}}{u_*} = \frac{1}{\kappa}\left(\ln\frac{z}{z_0} + \alpha_m \frac{z - z_0}{L_m}\right) \tag{2.80}$$

This relationship is similar to one obtained by Barenblatt [1953]

$$\frac{\bar{u}}{u_*} = \frac{1}{\kappa}\left(\ln\frac{z}{\gamma_m z_0} + 4.7\frac{z}{L_m}\right) \tag{2.81}$$

in which the coefficient γ_m (≤ 1) modulates z_0 by lowering its value. Thus γ_m is a drag reduction factor whose value is unity under the neutral condition. Equations 2.80 and 2.81 both revert to Eq. 2.57 for a neutrally stable flow in the near-bed boundary layer, *i.e.*, $z \ll L_m$, where buoyancy has no tangible effect on the flow structure. By redefining γ_m Eq. 2.80 can be written as

$$\frac{\overline{u}}{u_*} = \frac{1}{\kappa} \ln\left(\frac{z}{\gamma_m z_0}\right); \quad \gamma_m = e^{-\alpha_m\left(\frac{z}{z_0}-1\right)\frac{z_0}{L_m}} \tag{2.82}$$

For buoyancy effect due to suspended sediment, Toorman [2002] derived Eq. 2.80 by arguing that the Karman constant κ depends on stratification. Experimental studies [Toorman 2000a, 2000b] led to

$$\gamma_m = e^{-\left(1+a_m \frac{w_s}{u_*}\right)\left(1-e^{-b_m Ri_g^{n_m}}\right)} \tag{2.83}$$

in which w_s is the settling velocity of the sediment, a_m, b_m and n_m are empirical coefficients and Ri_g is the gradient Richardson number (Eq. 10.74). The coefficient values were found to be $a_m = 7.7$, $b_m = 1.7$ and $n_m = 0.85$. This expression explicitly accounts for the effect of the ratio w_s/u_*, which is proportional to the Rouse number (Chapter 6) governing the degree of non-uniformity of the suspended sediment concentration profile.

Equation 2.83 along with the three coefficient values is applicable to dense suspensions with concentrations close to the limit above which turbulence collapse can be expected, resulting in viscous-dominated slurry flow. It has been argued that since the actual concentration is typically far below this limiting or saturation value, Eq. 2.83 predicts excessive reduction (70–80%) in the bed shear stress, in contrast to expected reduction of no more than about 10–15% [Winterwerp and van Kesteren, 2004].

For a linear density profile, *i.e.*, a uniform density gradient, the following velocity profile has been obtained for salt-induced stratification:

$$\frac{\bar{u}}{u_*} = \frac{1}{\kappa}\ln\sqrt{\frac{\left(\dfrac{z}{z_0}\right)^2 + \dfrac{\alpha_m\beta_m g}{\rho_m u_*^2}\kappa^2 z^2}{1 + \dfrac{\alpha_m\beta_m g}{\rho_m u_*^2}\kappa^2 z^2}} \qquad (2.84)$$

where $\beta_m = d\rho/dz$ is the density gradient and ρ_m is the mean density. In Fig. 2.17, velocity data of McCutcheon [1981] are compared with Eqs. 2.57 and 2.84. Given the density profile in Fig. 2.17a in a flume using salt, Eq. 2.84 is observed in Fig. 2.17b to be a better predictor of the measured velocity profile than Eq. 2.57. It is noteworthy that this equation is useful only as long as $z / L_m \leq 0.025$.

The majority of past studies have been on stratification induced by salt or heat. A flume study by Vanoni [1941] used suspended sediment for establishing a stable density gradient. Winterwerp [2001] compared the neutral (clear-water) velocity profile with one based on the data of Coleman [1981] showing buoyancy effect due to sand particles in suspension. Studies of the type reported by Toorman [2000a, 2000b, 2002] are relatively recent contributions to fine sediment transport.

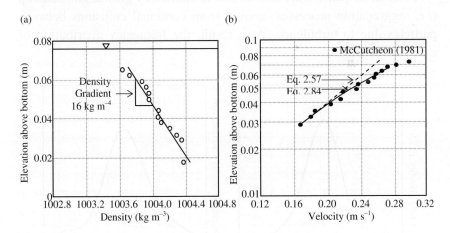

Figure 2.17. (a) Uniform density gradient; (b) stable-stratified and neutrally stable velocity profiles and flume data (from McCutcheon [1981]).

2.7.5 *Spectral description of turbulence*

The time-series of a surface water wave or the turbulent length-scale, $\eta(t)$, can be expressed as an infinite sum of monochromatic harmonic components of amplitude a_n, angular frequency σ_n and phase δ_n

$$\eta(t) = \sum_{n=0}^{n \to \infty} a_n \cos(\sigma_n t - \delta_n) \qquad (2.85)$$

This sum is conveniently represented as a density spectrum derived from energy contained at discrete values of σ_n (Fig. 2.18a). Since a continuous range of frequencies occurs in nature, the above expression changes to

$$\eta(t) = \text{Re}\left\{ \int_0^\infty a(\sigma) e^{i[\sigma t - \delta(\sigma)]} d\sigma \right\} \qquad (2.86)$$

where Re indicates "real part of," $a(\sigma)$ is the amplitude density function, σ is the angular frequency and $a(\sigma)d\sigma$ is the amplitude of each wave (Fig. 2.18b). An advantage of this representation is that the area under the curve is a measure of total wave energy [Dean and Dalrymple, 1991].

Flocs in suspension are formed and re-formed by growth and breakup (*i.e.*, aggregation processes) arising from continual collisions between particles due to turbulence. As a result, the frequency distribution of

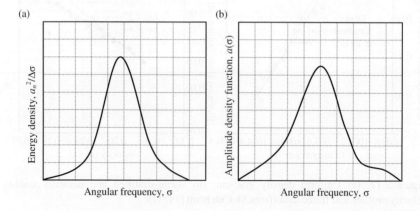

Figure 2.18. Spectral representations: (a) discrete; (b) continuous. Scales are arbitrary.

turbulent eddy length-scale is of considerable importance for the estimation of floc size. Since the wave number is inversely proportional to eddy length, this distribution is characteristically described by the continuous wave number spectrum of the TKE.

Turbulence frequency is sometimes represented as cycles per unit time $f = \sigma/2\pi$, rather than as the angular frequency σ, because wave representation is not the objective. Let $E(f)$ be the density distribution of TKE (Eq. 2.4) defined in such a way that $E(f)df$ is the fraction of the total kinetic energy in the frequency band from f to $f+df$. Thus, $E(f)$ must satisfy the condition

$$\int_0^\infty E(f)df = 1 \qquad (2.87)$$

The spectrum in Fig. 2.19 is based on representative data on turbulence in flumes. The ordinate is a measure of the spectral energy density $E_k(k) = \overline{u}E(f)/2\pi$, and the abscissa is the wave number $k = \sigma/\overline{u} = 2\pi f/\overline{u}$. Increasing wave number indicates decreasing eddy size, and the transfer of energy from large to smaller eddies can be thought of as a cascading process. Large eddies are scaled by water depth and therefore tend to be anisotropic, with the two longer axes in the horizontal plane and the shorter axis in the vertical direction. These eddies survive on energy supply from the mean flow such as a stream, tide or the wave boundary layer, and because they are not directly affected by viscous energy loss. The smallest eddies transfer their energy to molecular motion where viscous energy is converted into heat. Between these two eddy scales is the size range in which eddies are intermediate and change quite rapidly with the local conditions. This is the Kolmogorov equilibrium range in which the law $E_k \sim k^{-5/3}$ is found to be in agreement with theory. The law $E_k \sim k^{-7}$ applies to smaller eddies, also in agreement with theory [Tennekes and Lumley, 1972].

Floc growth and breakup occur throughout the boundary layer, in which the turbulence intensity and isotropy tend to vary widely. In order-of-magnitude analyses of floc dynamics it is convenient to assume that turbulence, as it affects a floc, is either locally isotropic or homogeneous.

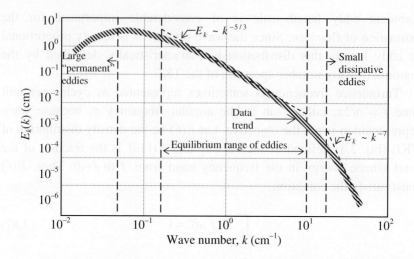

Figure 2.19. Experimental wave number spectrum of turbulent energy in the boundary layer over a rigid flat bed (based on Tennekes and Lumley [1972]).

This means that turbulent intensities at a given point in the three directions are equal. The rate of TKE dissipation ε_t is found from [Taylor, 1935]

$$\varepsilon_t = 15\nu \left(\frac{du'}{dz} \right)^2 \qquad (2.88)$$

Based on the scaling law of Kolmogorov the eddy length in the equilibrium range can be shown to be

$$\lambda_0 = \left(\frac{\nu^3}{\varepsilon_t} \right)^{1/4} \qquad (2.89)$$

where ν is the kinematic viscosity of the fluid. This length is useful as a measure of the diameter of flocs that have sustained presence in the equilibrium range.

2.8 Energy Balance Equation

Estimation of the rate at which fine sediment is suspended by turbulent flow in channels and other geometries requires the use of TKE balance at steady-state. Since this balance involves all three spatial dimensions, for convenience we will replace the coordinates x, y and z by x_i, and velocity components u, v and w by u_i, with $i = 1$, 2 and 3. Accordingly, the equation of motion becomes

$$\frac{Du_i}{Dt} = \frac{\partial u_i}{\partial t} + u_j \frac{\partial u_i}{\partial x_j} = -\frac{1}{\rho}\frac{\partial p}{\partial x_i} + \nu \frac{\partial^2 u_i}{\partial x_j^2} + F_i \tag{2.90}$$

where the fluid density ρ and the kinematic viscosity ν are assumed constant. The term F_i represents the summation of all external forces per unit volume of fluid. For a given subscript i, j takes on the values 1, 2 and 3. With $i = 1$ and F_3 representing gravity, the x-component of the equation of motion becomes

$$\frac{Du}{Dt} = \frac{\partial u}{\partial t} + u\frac{\partial u}{\partial x} + v\frac{\partial u}{\partial y} + w\frac{\partial u}{\partial z} =$$

$$-g\frac{\partial h}{\partial x} - \frac{1}{\rho}\frac{\partial p}{\partial x} + \nu\left(\frac{\partial^2 u}{\partial x^2} + \frac{\partial^2 u}{\partial y^2} + \frac{\partial^2 u}{\partial z^2}\right) \tag{2.91}$$

If as before we write $u = \bar{u} + u'$, *etc.*, and carry out time-averaging, Eq. 2.5 is obtained for mean flow in the x-direction.

Disregarding the external forces, we will multiply Eq. 2.90 by u_i and carry out the summations with respect to index i to yield

$$\underset{I}{\frac{1}{2}\frac{\partial}{\partial t}u_i u_i} = \underset{II}{-\frac{\partial}{\partial x_i}u_i\left(\frac{p}{\rho} + \frac{u_j u_j}{2}\right)} + \underset{III}{\nu\frac{\partial}{\partial x_i}\left(\frac{\partial u_i}{\partial x_j} + \frac{\partial u_j}{\partial x_i}\right)} - \underset{IV}{\nu\left(\frac{\partial u_i}{\partial x_j} + \frac{\partial u_j}{\partial x_i}\right)\frac{\partial u_j}{\partial x_i}} \tag{2.92}$$

where, for every i (1, 2 and 3), j takes on the values 1, 2 and 3. Hinze [1959] used a different approach to obtain Eq. 2.92 from Eq. 2.91 in order to highlight the physical meaning of the four terms: $I \equiv$ local change of kinetic energy per unit mass and time, $II \equiv$ change in convective transport of total energy per unit mass and time, or work done per unit mass and time by the total dynamic pressure, $III \equiv$ work done per unit mass and time by the viscous stresses, and $IV \equiv$ energy loss per unit mass.

We will now separate Eq. 2.92 into its mean and fluctuating parts based on:

$$u_i = \bar{u}_i + u_i'$$
$$p = \bar{p} + p' \tag{2.93}$$
$$u_i u_i = \bar{u}_i \bar{u}_i + 2\bar{u}_i u' + u_i' u_i' = \bar{u}_i \bar{u}_i + 2\bar{u}_i u_i' + q^2$$

where $q = \sqrt{u_i' u_i'}$. Substituting these quantities in Eq. 2.92 and time-averaging yields, after a fair amount of manipulation, the energy balance equation

$$\frac{D}{Dt}\left(\frac{q^2}{2}\right) = -\frac{\partial}{\partial x_i}\overline{u_i'\left(\frac{p}{\rho} + \frac{q^2}{2}\right)} - \overline{u_i' u_j'}\frac{\partial \bar{u}_i}{\partial x_j} + \frac{1}{2}\nu\frac{\partial}{\partial x_i \partial x_i}\left(\overline{q^2}\right) - \nu\overline{\frac{\partial u_j'}{\partial x_i}\frac{\partial u_j'}{\partial x_i}} \tag{2.94}$$
$$I \qquad\qquad II \qquad\qquad III \qquad IV \qquad\qquad V$$

in which $I \equiv$ change in TKE per unit mass and time, $II \equiv$ convective diffusion of the total turbulent energy, $III \equiv$ energy transferred from mean motion through the turbulent shear stresses, $IV \equiv$ work done by the viscous stresses of turbulent motion, and $V \equiv$ energy loss by turbulent motion. The last term does not represent energy loss if turbulence is not isotropic.

2.9 Flow Through Porous Media

Seepage of water through bed sediment containing silt or sand-sized particles in appreciable fractions relative to clay-sized particles plays an important role in the dynamic response of the bed to oscillatory forcing. Consider the simplest case of creeping steady flow of an incompressible fluid, *i.e.*, viscous flow at very low Reynolds numbers, in a porous medium. In this case flow acceleration as well as energy loss may be ignored, thus yielding from the momentum balance the Laplace equation for total pressure

$$\nabla^2(p + \rho g H) = 0 \qquad (2.95)$$

where p is the flow-induced pressure, ρ is the fluid density and H (denoted as h in Eq. 2.23 for open channel flow) is the hydrostatic pressure head. With total pressure $p_t = p + \rho g H$, the above equation can be restated as

$$\frac{\partial^2 p_t}{\partial x^2} + \frac{\partial^2 p_t}{\partial y^2} + \frac{\partial^2 p_t}{\partial z^2} = 0 \qquad (2.96)$$

For flow in a narrow tube in the z-direction, *e.g.*, through a drainage channel in silty mud, the mean water velocity w_w due to the vertical gradient of p_t is obtained from the law due to Darcy [1856]

$$w_w = -\frac{K_p}{\eta} \frac{\partial p_t}{\partial z} \qquad (2.97)$$

where η is the dynamic viscosity of the fluid and K_p is the specific permeability of the porous medium. Note that w_w is obtained by dividing discharge by the total cross-section. If only the cross-section occupied by the voids is used the resulting (higher) velocity becomes the seepage velocity. Equation 2.97 may be alternatively written as

$$w_w = -\frac{K_p \rho g}{\eta} \frac{\partial \left(\dfrac{p}{\rho g} + H \right)}{\partial z} = -k_p \frac{\partial H_t}{\partial z} \qquad (2.98)$$

where $k_p = K_p \rho g / \eta$ is the coefficient of permeability, or hydraulic conductivity, and H_t is the piezometric head (Fig. 2.20a). Ranges of values of K_p and k_p are found in Table 2.9. Additional data on permeability are given in Chapter 11.

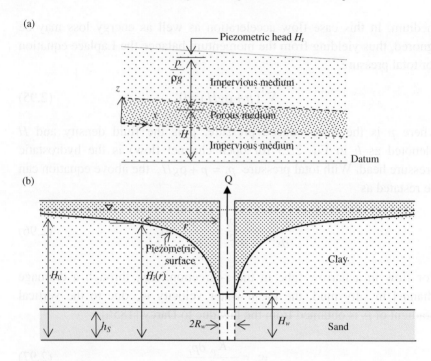

Figure 2.20. Flow through a porous medium: (a) Definition of total pressure or piezometric head (the porous medium is an aquifer confined between two impervious media), (b) radial flow into a well (adapted from Daily and Harleman [1966]).

Example 2.4: Consider the well-known problem of radial flow in a well (Fig. 2.20b). The clay layer above sand layer of thickness h_S is impervious. The sand layer rests on bedrock, which is also impervious. Thus the sand layer is a confined aquifer. Before drawdown of water the phreatic surface relative to bedrock and water level in the well are both H_0, and the piezometric head H_t is constant throughout the medium.[d] At steady-state the water level falls to H_w. The well radius R_w is small compared to H_0. Determine the steady-state piezometric surface profile $H_t(r)$ as water is drawn at a constant rate Q.

From continuity the discharge Q is

$$Q = 2\pi r h_S u_r \qquad (2.99)$$

---

[d] If we dig a hole into an aquifer to the depth of the water table or the phreatic surface, we can actually "touch" the surface. The phreatic surface is analogous to the free surface of a surface-water body. The piezometric surface cannot be "touched" [*e.g.*, Bear 1988].

Table 2.9. Ranges of permeability values.

Material	Specific permeability, K_p (m^2)	Coefficient of permeability, k_p (m s^{-1})
Clay	$10^{-16} <$	$10^{-9} <$
Silt	10^{-12} to 10^{-16}	10^{-5} to 10^{-9}
Sand	10^{-9} to 10^{-12}	10^{-2} to 10^{-5}
Gravel	10^{-7} to 10^{-9}	10^{0} to 10^{-2}

in which the radial discharge velocity u_r outward from the well centerline is obtained from Eq. 2.98 as

$$u_r = -k_p \frac{dH_t}{dr} \tag{2.100}$$

Therefore

$$Q = 2\pi r h_s k_p \frac{dH_t}{dr} \tag{2.101}$$

which flows towards the well. To obtain the radial variation of $H_t = (p/\rho g) + H$ we carry out the following integration

$$\int_{H_{trw}}^{H_{tr}} dH_t = \frac{Q}{2\pi h_s k_p} \int_{R_w}^{r} \frac{dr}{r} \tag{2.102}$$

which yields

$$H_{tr} - H_{tR_w} = \frac{Q}{2\pi h_s k_p} \ln\left(\frac{r}{R_w}\right) \tag{2.103}$$

where the product $h_s k_p = T_{tm}$ is the transmissivity. The profile of H_t is shown schematically in Fig. 2.20b. Consider the following values: $Q = 2{,}166$ m^3 d^{-1}, $T_{tm} = 360$ m^2 d^{-1}, $R_w = 0.6$ m. Thus for instance at a distance of $r = 40$ m from the well Eq. 2.103 yields 4.02 m.

2.10 Water Waves

2.10.1 *Wave celerity and other properties*

For the basic properties of surface water waves over a rigid bottom, we will refer to the two-dimensional sketch of Fig. 2.21. The role of a non-rigid or compliant bottom is considered in Chapter 11.

The forces that restore an initially perturbed surface are gravity and surface tension. Although water is typically assumed to be a Newtonian fluid with viscosity independent of flow shear rate (see Chapter 5), in order to obtain expressions for the wave speed (or celerity) and related properties we will consider water to be inviscid. This can be done without losing too much accuracy of the wave properties except very close to the bottom, where boundary layer effects cannot be ignored. These effects are reviewed in Chapter 11 in conjunction with wave damping due to energy loss in the boundary layer. We will also assume the fluid to be incompressible, *i.e.,* one in which the total rate of change of density with respect to time and space is nil. The only case in which compressibility of a medium is accounted for in this text is in the treatment of consolidation of bottom sediment (Chapter 8). In that process freshly deposited sediment is compressed (by overburden or self-weight) while water remains incompressible.

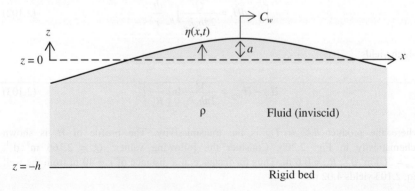

Figure 2.21. Wave in an inviscid fluid over a rigid bed: $\eta(x,t)$ is the surface elevation, a is the amplitude, h is the water depth, ρ is the water density and C_w is the wave speed (celerity).

The wave is assumed to have a small amplitude a (or one-half the wave height) such that $a/L \ll 1$, where $L = C_w T$ is the wave length, C_w is the wave celerity and T is the wave period. In other words, terms of order $(a/L)^2$ and higher are ignored. In shallow water, terms of order $(a/h)^2$ and higher are ignored as well. The small-amplitude assumption permits linearization of the surface boundary conditions resulting in a significant simplification of the problem setup. However, at the same time additional approximations are introduced especially at the free surface, which for purposes of application of the surface boundary conditions is assumed to be at $z = 0$, rather than $z = \eta$, because $\eta \ll L$. The reader may refer to a basic text on wave mechanics for the Airy [1842] small-amplitude linear wave theory [*e.g.,* Holthuijsen, 2007] which is only briefly reviewed here. An effect of relaxing the linear assumption is mentioned.

For the simple case of a wave traveling in the *x*-direction (only), the flow continuity equation is treated in the vertical *x-z* plane, with an unbounded extent of fluid in the lateral (*y*) direction (Fig. 2.1b). From Eq. 2.11 we obtain

$$\frac{\partial u}{\partial x} + \frac{\partial w}{\partial z} = 0 \qquad (2.104)$$

where u, w are the x, z coordinate components, respectively, of the velocity vector $\vec{U} = u\vec{i} + v\vec{j} + w\vec{k}$, and $\vec{i}$, $\vec{j}$, $\vec{k}$ are the associated unit vectors. Since $v = 0$, the problem is limited to $\vec{U} = u\vec{i} + w\vec{k}$. Equation 2.104 can also be stated as

$$\nabla \cdot \vec{U} = 0 \qquad (2.105)$$

Considering $\vec{U}$ as the gradient of the velocity potential ϕ, we define

$$u = -\frac{\partial \phi}{\partial x}; \quad w = -\frac{\partial \phi}{\partial z} \qquad (2.106)$$

Combining Eqs. 2.104 and 2.106 yields

$$\nabla^2 \phi = \frac{\partial^2 \phi}{\partial x^2} + \frac{\partial^2 \phi}{\partial z^2} = 0 \qquad (2.107)$$

which is the Laplace equation for ϕ.[e] Depending on the boundary conditions, numerous solutions of this equation are possible [*e.g.,* Lamb, 1945]. For the present purpose, specification of only three boundary conditions is required, as the fluid is assumed to be infinite in extent in the $\pm x$ directions. Two of these, the kinematic and the dynamic conditions, are applied at the free surface, and the third at the bottom.

At the rigid bed ($z = -h$) the velocity normal to the bed is zero, *i.e.,*

$$\frac{\partial\phi}{\partial z} = 0 \qquad (2.108)$$

Note that $\partial\phi/\partial x$ cannot be zero (except in the absence of flow), because the fluid is inviscid. As a result, the horizontal velocity at the bottom ($u = -\partial\phi/\partial x\big|_{z=-h}$) is not required to be zero as in real fluids, *i.e.,* flow slippage can occur (which is unrealistic).

For a linear wave, at the free surface ($z = 0$) the normal velocity is given by the time-rate of change of surface elevation η, *i.e.,*

$$\frac{\partial\eta}{\partial t} = -\frac{\partial\phi}{\partial z} \qquad (2.109)$$

which is the required kinematic free-surface boundary condition.

---

[e] The existence of the velocity potential ϕ implies that the flow is irrotational, *i.e.,* it has no vorticity. This is explained as follows. In the *x-z* plane the vorticity is

$$\omega_y = \frac{1}{2}\left(\frac{\partial w}{\partial x} - \frac{\partial u}{\partial z}\right) \qquad (F2.3)$$

which represents the mean rate of rotation of a fluid element of dimensions Δx and Δz about the *y*-axis. When *u* and *v* from Eq. 2.106 are substituted into Eq. F2.3 the vorticity vanishes. It follows that Eq. 2.107 applies to irrotational flows only. In general, vorticity gives rise to circulation (Γ), which is the line integral of the tangential velocity component about any closed contour. Accordingly, it is defined by the area integral

$$\Gamma = \iint_A 2\omega_y \, dA \qquad (F2.4)$$

For the second condition at the surface we note that the linearized Bernoulli equation relating pressure p to velocity (or ϕ and surface elevation η) is

$$\frac{p}{\rho} + \frac{\partial \phi}{\partial t} + g\eta = \frac{p_{atm}}{\rho} \qquad (2.110)$$

where p_{atm} is the atmospheric pressure. Since ϕ and η are harmonic functions of time, Eq. 2.110 averaged over a wave period yields zero time–mean values of $\partial \phi / \partial t$ and $g\eta$. This in turn implies that $p/\rho = p_{atm}/\rho$. As a result Eq. 2.110 reduces to

$$\eta = -\frac{1}{g} \frac{\partial \phi}{\partial t} \qquad (2.111)$$

which is the required dynamic boundary condition at the free (water) surface.

When waves are merely small ripples at the water surface, pressure is modified by the effect of surface tension. This is given by

$$p = p_{atm} - \gamma_s \frac{\partial^2 \eta / \partial x^2}{[1 + (\partial \eta / \partial x)^2]^{3/2}} \qquad (2.112)$$

By combining the conditions represented by Eqs. 2.110, 2.111 and 2.112 we obtain

$$\rho g \frac{\partial \phi}{\partial z} + \rho \frac{\partial^2 \phi}{\partial t^2} - \gamma_s \frac{\partial^3 \phi}{\partial z \partial x^2} = 0 \qquad (2.113)$$

Now to determine ϕ we will assume it to be of the following harmonic form

$$\phi = F_z \sin(kx - \sigma t) \qquad (2.114)$$

where $k = 2\pi/L$ is the wave number, $\sigma = 2\pi/T$ is the wave angular frequency and T is the wave period. Substituting Eq. 2.114 into Eq. 2.107 and integrating twice leads to

$$F_z = A' \cosh[k(h+z)] + A'' \sinh[k(h+z)] \qquad (2.115)$$

By applying the bottom boundary condition of Eq. 2.108 we find that the constant $A'' = 0$. Then, making use of Eq. 2.113, the constant A' is evaluated and the equation for the wave speed or celerity $C_w = L/T$ is obtained as

$$C_w^2 = \left(\frac{g}{k} + \frac{\gamma_s k}{\rho} \right) \tanh kh \qquad (2.116)$$

This expression is known as the wave dispersion equation because of its property that causes waves of different celerities to travel independently of each other, leading to their spreading or dispersal from any common point of origin.

With A' known and $A'' = 0$, from Eq. 2.114 it can be shown that

$$\phi = -\frac{ag}{\sigma} \frac{\cosh k(h+z)}{\cosh kh} \sin(kx - \sigma t) \qquad (2.117)$$

Substituting this expression into Eq. 2.109 and integration of the resulting equation yields the surface wave form

$$\eta(x,t) = a\cos(kx - \sigma t) \qquad (2.118)$$

Example 2.5: Determine the celerity of a 0.05 s wave and that of a 2 s wave in 0.75 m of water ($\rho = 1{,}000$ kg m^{-3}), both with and without the effect of surface tension ($\gamma_S = 0.0728$ N m^{-1}).

Equation 2.116 is solved iteratively for the wave length L ($= 2\pi/k$). The starting estimate of L is the gravity-controlled deep water (defined later) wave length $L_0 = gT^2 / 2\pi$. Values of L_0 for 0.05 s and 2 s periods are 0.004 m and 6.25 m, respectively. The corresponding celerity values are as follows:

T (s)	γ_S (N m^{-1})	C_w (m s^{-1})
0.05	0.0728	0.24
0.05	0	0.08
2	0.0728	2.37
2	0	2.37

For the 0.05 s wave surface tension changes the wave celerity from 0.24 to 0.08 m s^{-1}, *i.e.*, by 67%. The 2 s wave is too long to be affected.

As a rule, surface tension dependent ripples or capillary waves are important only when the wave length is less than about 0.1 m, and will not be considered further. After ignoring surface tension Eq. 2.116 becomes

$$C_w = \sqrt{\frac{g}{k} \tanh kh} \qquad (2.119)$$

which, along with $L = C_w T$ indicates that a relationship exists between any three of the four parameters C_w, L, T and h. For given water depth the celerity increases with the wave period and length. Figure 2.22 is a plot of C_w as a function of T and h.

The trends in Fig. 2.22 can be explained by shallow water and deep water approximations of Eq. 2.119. These approximations result from the limiting behavior of the hyperbolic functions of any variable x. They are

$$\sinh x = \frac{1}{2}(e^x - e^{-x}); \quad \cosh x = \frac{1}{2}(e^x + e^{-x});$$

$$\tanh x = \frac{\sinh x}{\cosh x} \qquad (2.120)$$

The respective limiting values of tanh are

$$\tanh x \rightarrow 1 \quad \text{for large values of } x$$
$$\tanh x \rightarrow x \quad \text{for small values of } x \qquad (2.121)$$

Given $x = 2\pi h/L = kh,$ we obtain the following expressions for the celerity:

Shallow water (assumed to be specified by the condition $kh \leq \pi/10$, *i.e.*, $h/L \leq 0.05$):

$$C_w = \sqrt{gh} \qquad (2.122)$$

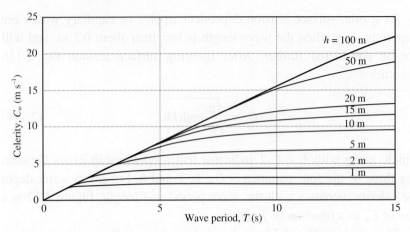

Figure 2.22. Variation of wave celerity with period and water depth based on the Airy theory.

Deep water ($kh > \pi$, *i.e.*, $h/L > 0.5$):

$$C_w = \sqrt{\frac{g}{k}} = \sqrt{\frac{gL}{2\pi}} \qquad (2.123)$$

According to Eqs. 2.122 and 2.123, all shallow-water waves travel at the same speed independent of wave length, and deep-water waves are independent of water depth. Waves in-between ($\pi < kh < \pi/10$) are in "intermediate water depth," for which Eq. 2.121 must be solved. Shallow and deep water trends are apparent in Fig. 2.22. Each constant-depth curve has two features. For large wave periods the curve approaches a horizontal line in shallow water. As the period becomes small the curve merges into a line of constant slope $C_w/T = g/2\pi$ corresponding to deep water.

Water particle orbital velocities u (horizontal) and w (vertical) are obtained from:

$$u = -\frac{\partial \phi}{\partial x} = aC_w k \,\frac{\cosh k(z+h)}{\sinh kh}\cos(kx - \sigma t) \qquad (2.124)^{f}$$

---

[f] Alternatively,

$$u = \frac{agk}{\sigma}\frac{\cosh k(h+z)}{\cosh kh}\cos(kx - \sigma t) \qquad (F2.5)$$

$$p = \rho gz + a\rho g \frac{\cosh k(h+z)}{\cosh kh} \cos(kx - \sigma t) \qquad (2.130)$$

The first term on the right-hand side is the hydrostatic pressure and the second term is the dynamic pressure due to water particle acceleration. A submerged pressure gauge located at a depth less than one-half wave length would detect the dynamic pressure and permit calculation of the wave amplitude from Eq. 2.130. Wells and Kemp [1986] carried out such measurements off the muddy coast of Suriname, which we will refer to in Chapter 11.

The total wave energy per unit water surface area E is the sum of the kinetic and the potential energies $E_{kinetic}$ and $E_{potential}$, respectively, *i.e.*,

$$E = E_{kinetic} + E_{potential} = \frac{1}{4}\rho ga^2 + \frac{1}{4}\rho ga^2 = \frac{1}{2}\rho ga^2 \qquad (2.131)$$

Note that the total wave energy per unit *crest width* is *EL*. The wave power per unit water surface area is EC_g and wave power per unit crest width is $EC_g L$, where C_g is the wave group velocity, which is the speed at which energy is transmitted. It is given by

$$C_g = \frac{d\sigma}{dk} = \frac{C_w}{2}\left(1 + \frac{2kh}{\sinh 2kh}\right) \qquad (2.132)$$

In deep water $2kh/\sinh 2kh = 0$ and therefore $C_g = C_w/2$, whereas in shallow water $2kh/\sinh 2kh = 1$ and $C_g = C_w$. At intermediate water depths Eq. 2.132 must be used. As the wave approaches the shoreline from deep water, the rate of wave energy transmission increases with decreasing depth.

Example 2.7: A 1.2 m high and 7 s deep wave travels towards the shore of a large freshwater lake. The wave crest is parallel to the shoreline. Determine the wave power per unit crest width. If the same wave were in 5 m of water, what would be the maximum dynamic pressure at a depth of 3 m?

The deep water wave power per unit crest width is equal to $E_0 C_{w0} L_0/2$, where subscript "0" refers to deep water. This yields 7.39×10^5 J s^{-1} m^{-1}. The pressure amplitude is obtained from Eq. 2.130 as

$$p_{max} = a\rho g \frac{\cosh k(h+z)}{\cosh kh} \qquad (2.133)$$

Its value at 3 m depth is $p_{max} = (0.6\text{m}) \times (1000 \text{ kg m}^{-2}) \times (9.81 \text{ m s}^{-2}) \times \cosh [0.1376$ $(5–3)]/\cosh (0.1376 \times 5) = 4.9 \times 10^{3}$ Pa.

2.10.2 *Shoaling and refraction*

Once waves leave deep water their heights change. This phenomenon of shoaling can be simply examined by assuming that the wave energy between two adjacent, arbitrarily spaced wave orthogonals, or rays (Figs. 2.24a, b) is conserved, *i.e.*, no energy leaks outside the water column between the two orthogonals and there is no energy loss in the direction of travel. If in deep water the spacing between the rays is B_{w0} and in shallower water B_w, conservation of the rate of energy transport implies the balance

$$E_0 C_{g0} B_{w0} = E C_g B_w \qquad (2.134)$$

By substitution of Eqs. 2.131 and 2.132 into the above equality we obtain

$$\frac{a}{a_{H0}} = \frac{H_s}{H_{s0}} = \sqrt{\frac{\sinh 2kh + 2kh}{\sinh 2kh}} \sqrt{\frac{B_{w0}}{B_w}} \qquad (2.135)$$

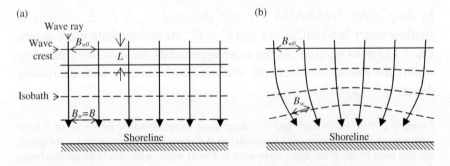

Figure 2.24. Wave shoaling: (a) Without refraction; (b) with refraction due to non-parallel isobaths.

where $H_s = 2a$ is the wave height and $H_{s0} = 2a_{H0}$ is its value in deep water. When the rays approach the shoreline in the shore-normal direction and the isobaths are parallel to the shoreline, the refraction coefficient $\sqrt{B_{w0}/B_w} = 1$. In other words there is no bending of the rays (Fig. 2.24a), which occurs only when the celerity changes along the wave crest (Fig. 2.24b). When the wave is traveling in the shore-normal direction, H_s varies solely due to changing water depth, *i.e.*, due to shoaling.

When wave approach is not normal to the isobaths the rays diverge or converge depending on the curvature of the isobaths, because changing depth changes the wave celerity by different amounts along the crest. As the local celerity is determined by local depth and wave length, it follows that Snell's law of wave refraction approximately applies between any two points A and B along the ray. The celerities C_{WA} and C_{WB} at these two points are obtained from Eq. 2.119, given the respective water depths and the wave period. Then, knowing one of the two angles between the wave crest and the isobath, say α_A, the angle at B is obtained from

$$\alpha_B = \sin^{-1}\left(\frac{C_{wB}}{C_{wA}} \sin \alpha_A \right) \qquad (2.136)$$

Next, Eq. 2.135 is used to obtain the variation of wave height H_s along the ray. The computation usually begins in deep water, where $\alpha_A = \alpha_0$ is specified, and proceeds along each ray.

Example 2.8: A wave train approaches land, with isobaths that are straight and parallel to the shoreline. At 4 m depth the wave length and the height are 100 m and 1.5 m, respectively, and the wave crest forms an angle of 10° with the shoreline. Determine the wave height and direction in deep water.

From Eq. 2.136 (with subscript 0 for deep water replacing subscript A, and subscript B deleted)

$$\alpha_0 = \sin^{-1}\left(\frac{C_{w0}}{C_w} \sin \alpha \right) \qquad (2.137)$$

which, with the use of Eq. 2.119 and $\alpha = 10°$ yields $\alpha_0 = 44.3°$. Then from the geometric relationship

$$\frac{B_{w0}}{B_w} = \frac{\cos\alpha_0}{\cos\alpha} \tag{2.138}$$

we obtain $\sqrt{B_{w0}/B_w} = 0.853$. Finally, from Eq. 2.135, $H_{s0} = 1.23$ m.

2.10.3 *Finite-amplitude effect*

The Airy wave theory does not hold when the wave amplitude is non-negligible compared to the wavelength or depth. A relevant finite-amplitude theory is the deep-water analysis of Stokes [1847], which involves series expansion in powers of a/L. Another theory is the shallow water solitary wave analysis of Russell [1844], which invokes series expansion in powers of a/h. The solitary-wave form is entirely above the still water level. The celerities of the two types of waves are

Stokes wave:

$$C_w = \sqrt{\frac{g}{k}\tanh kh} \left\{ 1 + \left(\frac{\pi a}{2L}\right)^2 \left(\frac{8 + \cosh 4kh}{8\sinh^4 kh}\right) + \dots + O\left(\frac{a}{L}\right)^4 \right\} \tag{2.139}$$

Solitary wave:

$$C_w = \sqrt{gh}\left(1 + \frac{a}{h}\right) \tag{2.140}$$

for which the wave number is given by $k = \mathrm{sech}^2\sqrt{3a/4h^3}$.

Due to the peaked shape of the finite-amplitude wave crest, the celerity is higher than for a linear wave of the same period. In Stokes wave the square-root of the term within the brackets { } amounts to a correction to the linear celerity given by $\sqrt{g\tanh kh/k}$. However, terms involving fourth and higher order of a/L are usually ignored because their contributions to water surface variation are typically minor. For the solitary wave the correction to the linear shallow water celerity $\sqrt{gh}$ is $[1+(a/h)]$. Solitary-type waves have been recorded in the nearshore muddy coast environment [Wells and Kemp, 1986].

In contrast to the Airy wave, the water parcel orbit during the wave period from the Stokes theory is not truly elliptical. Due to the co-variance of the oscillating water surface and orbital velocity, the crest position of the water parcel at the end of the cycle shifts in the direction of wave travel compared to the beginning of the cycle. Thus there is a net transport of water mass with an associated purely horizontal velocity given by

$$u_s = \frac{1}{2} C_w k^2 a^2 \frac{\cosh 2k(h+z)}{\sinh^2 kh} \qquad (2.141)$$

The corresponding velocity averaged over depth h is

$$u_{sm} = \frac{1}{2} \frac{g a^2}{h} \qquad (2.142)$$

In deep water Eq. 2.141 reduces to

$$u_{s0} = C_w k^2 a^2 e^{2kz} \qquad (2.143)$$

Example 2.9: Calculate the mass transport velocity for a 3.1 m high wave of 12 s period in deep water, and also in 9 m deep water.

From Eq. 2.119 we obtain $k_0 = 224.83$ m^{-1} in deep water and $k = 108.01$ m^{-1} in 9 m depth. Values of u_s obtained from Eqs. 2.143 and 2.141 are as follows:

(m)	u_s (m s^{-1})	
	Deep water	9 m depth
0	0.035	0.195
−1	0.033	0.179
−2	0.031	0.165
−3	0.030	0.153
−4	0.028	0.143
−5	0.027	0.135
−6	0.025	0.129
−7	0.024	0.125
−8	0.022	0.123
−9	0.021	0.122
−10	0.020	
−11	0.019	
−12	0.018	
−13	0.017	
−14	0.016	

(m)	u_s (m s^{-1}) Deep water	9 m depth
−15	0.015	
−16	0.014	
−17	0.014	
−18	0.013	
−19	0.012	
−20	0.011	

The vertical profile of u_s derived from Eq. 2.141 is shown in Fig. 2.25.

2.10.4 *Breaking waves*

In deep water the water parcel velocity u_{pc} at the crest of a wave of given period can be taken to be proportional to the local wave height. As the wave grows either due to energy input from wind or due to shoaling, at some water depth the crest velocity u_{pc} will increase to a critical value u_{pcmax} that is equal to the wave celerity, *i.e.*, $u_{pcmax}/C_w = 1$. At that location the forward momentum of the crest will cause the wave to become unstable and break soon thereafter.

In Eq. 2.124 setting $z = 0$ (and $kx - \sigma t = 0$) yields

$$\frac{u_{pcmax}}{C_w} = \pi \left(\frac{H_s}{L} \right)_{max} \cotanh \left(\frac{2\pi h}{L} \right) = 1 \qquad (2.144)$$

where H_s/L is the wave steepness, and $(H_s/L)_{max}$ is its value at breaking.

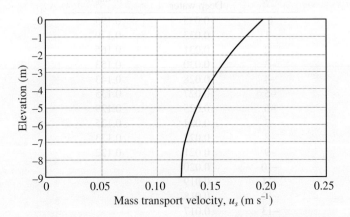

Figure 2.25. Mass transport velocity profile in 9-m water depth.

Therefore,

$$\left(\frac{H_s}{L}\right)_{max} = \frac{1}{\pi}\tanh\left(\frac{2\pi h}{L}\right) \tag{2.145}$$

For a qualitative description of wave breaking the Stokes wave form (Eq. 2.139) is more realistic than the Airy linear wave because the former consists of peaked crests and flat troughs just before breaking. Accordingly, Michell [1893] determined the condition

$$\left(\frac{H_s}{L}\right)_{max} = \frac{1}{7}\tanh\left(\frac{2\pi h}{L}\right) \tag{2.146}$$

which indicates that the steepness at breaking of a Stokes wave is about one-half ($\pi/7 = 0.45$) the value predicted by the Airy theory. Experiments carried out in a wave tank with a horizontal bottom have indicated Eq. 2.146 to be sufficiently accurate for many engineering purposes [Danel, 1952]. In deep water this equation reduces to

$$\left(\frac{H_{s0}}{L_0}\right)_{max} = \frac{1}{7} \tag{2.147}$$

In other words, the wave breaks when the deep water wave height increases to one-seventh of wave length. Deep water breaking modifies the wave spectrum during a storm, as the growth of short-period waves is limited by breaking.

In shallow water, from Eq. 2.146

$$\left(\frac{H_s}{L}\right)_{max} = \frac{H_b}{L_b} = \frac{1}{7}\frac{2\pi h_b}{L_b} \tag{2.148}$$

where subscript b refers to depth-limited breaking. Thus,

$$\frac{H_b}{h_b} = \kappa_b = 0.90 \tag{2.149}$$

where κ_b is the breaking wave index. From the solitary wave theory McCowan [1891] obtained $\kappa_b = 0.78$. This value is commonly used in engineering calculations even though the actual number varies somewhat with bottom slope and wave period.

From their experiments on a mildly sloping beach, Thornton and Guza [1982] observed that in the surf zone (between breaking water depth h_b and the shoreline) the local root–mean–square wave height, H_{rms}, was proportional to the local depth h, *i.e.*,

$$H_{rms} = \alpha_h h \qquad (2.150)$$

The proportionality constant α_h was determined to be 0.42. Although this value is site-specific, in general Eq. 2.150 enables one to model wave height in the surf zone simply by knowing the bottom depth profile.

The height H_{rms} is calculated from

$$H_{rms} = \sqrt{\frac{1}{N} \sum_{i=1}^{N} H_{si}^2} \qquad (2.151)$$

where N is the total number of waves H_{si} in the time-series. When the distribution of wave height is approximated by the common Raleigh probability density function, the significant wave height is obtained from

$$H_s = 1.416 H_{rms} \qquad (2.152)$$

This relationship is often used for a rough estimation of H_s from H_{rms} even when the wave frequency distribution is unknown.

2.10.5 *Wind-generated waves*

In general, the complex nature of energy transfer from wind to water producing surface waves, wave–wave interactions and energy loss require the use of numerical models for prediction of wave properties. All non-numerical approaches for prediction of wind fetch-limited and duration limited waves are essentially empirical.

When the wind duration is sufficiently long, waves are fetch-limited. These waves are considered here. The significant wave height H_s and the modal period (corresponding to the peak of the wave energy spectrum)

T_p depend on the wind speed U_{10} at the standard meteorological reference elevation of 10 m, water depth h and the fetch distance x.[g] These five variables are represented as dimensionless numbers

$$\hat{H} = \frac{gH_s}{U_{10}^2}, \quad \hat{T} = \frac{gT_p}{U_{10}}, \quad \hat{h} = \frac{gh}{U_{10}^2}, \quad \hat{F} = \frac{gx}{U_{10}^2} \qquad (2.153)$$

Empirical relationships involving these numbers were developed by Young and Verhagen [1996] based on extensive wave data collected at Lake George in Australia. Revised forms of these relationships reported in Holthuijsen [2007] are

$$\hat{H} = \hat{H}_\infty \left\{ \tanh\left(k_3 \hat{h}^{m_3}\right) \tanh\left[\frac{k_1 \hat{F}^{m_1}}{\tanh\left(k_3 \hat{h}^{m_3}\right)}\right] \right\}^p \qquad (2.154)$$

with $\hat{H}_\infty = 0.24$, $k_1 = 4.14 \times 10^{-4}$, $k_3 = 0.343$, $m_1 = 0.79$, $m_3 = 1.14$ and $p = 0.572$, and

$$\hat{T} = \hat{T}_\infty \left\{ \tanh\left(k_4 \hat{h}^{m_4}\right) \tanh\left[\frac{k_2 \hat{F}^{m_2}}{\tanh\left(k_4 \hat{h}^{m_4}\right)}\right] \right\}^q \qquad (2.155)$$

with $\hat{T}_\infty = 7.69$, $k_2 = 2.77 \times 10^{-7}$, $k_4 = 0.10$, $m_2 = 1.45$, $m_4 = 2.01$ and $q = 0.187$.

---

[g] For fetch-limited waves the wind must blow for a minimum duration t_d, which depends on the wind speed, water depth and fetch. The following is an approximate relationship in which the coefficient c_d varies implicitly with water depth

$$t_d = c_d \left(\frac{x^2}{g U_{10}}\right)^{1/3} \qquad (\text{F2.6})$$

For a mine tailings pond, Lawrence *et al.* [1991] estimated c_d to be 68.8.

2.11 Exercises

2.1   Show that the rate of flow-energy loss per unit volume of a fluid is equal to $\tau du/dz$, where τ is the local shear stress and du/dz is the local shear rate. Hint: Consider the energy expended in shearing an elemental fluid volume (of unit thickness) shown in Fig. E2.1. Let A be the area of the top and bottom faces of the element. The shear stress acts over A for an incremental time dt.

2.2   Determine the distance by which the plane at the top of spherical particles of diameter d_p must be lowered such that the plane becomes the plane of zero velocity (Fig. E2.2). This plane is set at an elevation at which the volume (not projected area!) of each particle protruding above the plane equals the volume of the void below the plane. Hint: Volume-based integration is required (Prof. Bent Christensen personal communication).

2.3   Consider a steady stream of 3.5 m depth. The flow is stratified due to suspended sediment with a surface density of 1,005 kg m^{-3} and a bottom density of 1,125 kg m^{-3}. Assume friction velocity $u_* = 0.055$ m s^{-1}, bed roughness $k_s = 5$ cm, Karman constant $\kappa = 0.4$ and $\alpha_m = 5$. (1) Plot the velocity profile using Eq. 2.84. (2) Plot the log-velocity profile $u(z)$ over the bottom 0.1 m, assuming water density to be uniform and equal to the mean of the two given values. Compare the results of (1) with (2) by plotting them together.

2.4   The complete set of equations for calculating seawater salinity S (psu) are as follows [Lewis, 1980; Fofonoff and Millard, Jr., 1983]. The relevant conductivity ratio is defined as

$$R_s = \frac{Con(S, \theta, p)}{Con(35, 15, 0)} \qquad (E2.1)$$

where Con denotes electrical conductivity, which is a function of S, temperature θ (°C) and pressure p (decibars). $Con(35,15,0)$ is the reference conductivity of the standard potassium chloride (KCl) solution. A parameter R_θ is next introduced

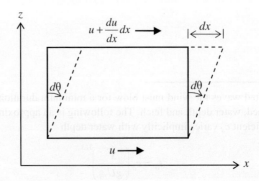

Figure E2.1. Shearing of an elemental fluid volume of unit thickness.

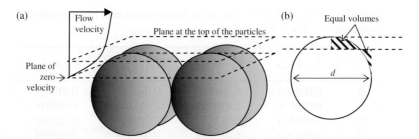

Figure E2.2. Plane of zero velocity: (a) Definition sketch; (b) two-dimensional sketch showing *volumes* to be equated in three-dimensional space.

$$R_\theta = \frac{R_s}{R_p \, r_\theta} \qquad (E2.2)$$

where

$$R_p = 1 + \frac{p - (A_1 + A_2 \, p + A_3 \, p^2)}{1 + B_1 \, \theta + B_2 \, \theta^2 + B_3 \, R + B_4 R \theta} \qquad (E2.3)$$

$$r_\theta = c_0 + c_1 \theta + c_2 \theta^2 + c_3 \theta^3 + c_4 \theta^4 \qquad (E2.4)$$

$A_1 = 2.070 \times 10^{-5}$	$B_1 = 3.426 \times 10^{-2}$	$c_0 = 6.766097 \times 10^{-1}$	$a_0 = 0.0080$	$b_0 = 0.0005$
$A_2 = -6.370 \times 10^{-10}$	$B_2 = 4.464 \times 10^{-4}$	$c_1 = 2.00564 \times 10^{-2}$	$a_1 = -0.1692$	$b_1 = -0.0056$
$A_3 = 3.989 \times 10^{-15}$	$B_3 = 4.215 \times 10^{-1}$	$c_2 = 1.104259 \times 10^{-4}$	$a_2 = 25.3851$	$b_2 = -0.0066$
	$B_4 = -3.107 \times 10^{-3}$	$c_3 = -6.9698 \times 10^{-7}$	$a_3 = 14.0941$	$b_3 = -0.0375$
		$c_4 = 1.0031 \times 10^{-9}$	$a_4 = -7.0261$	$b_4 = 0.0636$
$k_S = 0.0162$			$a_5 = 2.7081$	$b_5 = -0.0144$

Finally, the salinity is obtained from

$$S = \sum_{j=0}^{5} a_j \, R_\theta^{j/2} + \frac{(\theta - 15)}{1 + k_s(\theta - 15)} \sum_{j=0}^{5} b_j \, R_\theta^{j/2} \qquad (E2.5)$$

Calculate S at $\theta = 22°C$. Select $R_s = 0.8$ and $p = 0.2$ atm. For general information the electrical conductivity of seawater at different temperatures and salinities are given in Table E2.1.

2.5 A turbid wedge at steady-state penetrates upstream into a 4.2 m deep canal carrying freshwater with an outflow velocity of 0.25 m s^{-1}. If the density of turbid water is 1,050 kg m^{-3} and the interfacial friction factor is 0.007, determine the distance of penetration. If now the canal is dredged to 5.5 m, what will be the impact on penetration?

Table E2.1. Conductivity of seawater in Siemens per cm ($S\ cm^{-1}$) at different temperatures and salinities.

θ (°C)	$S = 5$ psu	$S = 10$ psu	$S = 15$ psu	$S = 20$ psu
0	0.004808	0.009171	0.013357	0.017421
5	0.005570	0.010616	0.015441	0.020118
10	0.006370	0.012131	0.017627	0.022947
15	0.007204	0.013709	0.019905	0.025894
20	0.008068	0.015346	0.022267	0.028948
25	0.008960	0.017035	0.024703	0.032097
30	0.009877	0.018771	0.027204	0.035330

θ (°C)	$S = 25$ psu	$S = 30$ psu	$S = 35$ psu	$S = 40$ psu
0	0.021385	0.025257	0.029048	0.032775
5	0.024674	0.029120	0.033468	0.037734
10	0.028123	0.033171	0.038103	0.042935
15	0.031716	0.037391	0.042933	0.048355
20	0.035438	0.041762	0.047934	0.053968
25	0.039276	0.046267	0.053088	0.059751
30	0.043123	0.050888	0.058373	0.065683

2.6   Darcy's law can be obtained directly from the force balance for creeping motion of a fluid

$$\nabla(p + \rho gH) = \eta\nabla^2\vec{U} \qquad (E2.6)$$

where $\vec{U} = u\vec{i} + v\vec{j} + w\vec{k}$ is the flow velocity vector. Derive the form of Darcy's law by showing that $\nabla^2(p + \rho gH) = 0$, making use of the identity

$$\nabla^2\vec{U} = \nabla(\nabla\cdot\vec{U}) - \nabla\times(\nabla\times\vec{U}) \qquad (E2.7)$$

Hints: (1) Obtain $\nabla^2(p + \rho gH)$ from Eq. E2.6; (2) show that $\nabla\cdot[\nabla\times(\nabla\times\vec{U})] = 0$.

2.7   Consider a tube of radius r_t in which water is filled to a certain level. Show that the excess pressure p at the meniscus due to surface tension is given by $p = 2\gamma_S/r_t$.

2.8   Evaluate the coefficient A' in Eq. 2.115.

2.9   A 1.1 m high and 6.7 s deep-water wave travels toward the shore without refraction. (1) Determine the wave length, celerity and power in deep water. (2) Determine the wave height, length, celerity and power in a water depth of 2.5 m.

2.10 A wave train from deep water approaches a shore. The isobaths are straight and parallel to the shoreline. At the water depth of 3 m the wave length and amplitude are 100 m and 0.75 m, respectively, and the wave crest forms a 5° angle with the shoreline. Determine the wave height and direction in deep water.

2.11 A wave train of 6.5 s period and 0.7 m amplitude is traveling in deep water. Calculate wave celerity assuming the wave to be: (1) linear and (2) Stokes (up to second order). Use linear theory to calculate the wave number.

2.12 In a laboratory flume in which the still water depth was 20 cm, 1 s waves with 3.5 cm amplitude (*i.e.*, 7 cm height) were produced. At a certain time (*t* = 0), Rhodamine red dye was injected from a syringe to form an instantaneous vertical streak at a location (distance 0) along the flume centerline. The shapes of the streak at *t* = 1 s (*i.e.*, after passage of one wave) and at 5 s (after five waves) are shown in Fig. E2.3. Could one attribute this behavior to wave-induced mass transport? On what basis?

2.13 At the center of Lake Apopka in central Florida, So [2009] reported 3-min average wind speeds at 10 m elevation ranging from 0 to 16 m s^{-1} during a one-year period of observation. Given a mean water depth of 1.3 m over a distance of 5.6 km from the shoreline to the center in the direction of the predominant wind, plot the significant wave height and the modal period as functions of wind speed. Also plot the same for water levels of 1.9 m and 0.7 m corresponding to the seasonal range of stage variation in the lake.

2.14 For non-breaking waves in water of depth *h* the representative value of the bed shear stress is

$$\tau_b = \frac{f_w}{4} \rho u_b |u_b| \tag{E2.8}$$

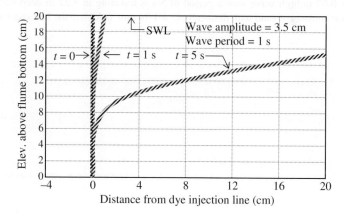

Figure E2.3. Displacement of vertical line-source of dye with time in a flume.

where f_w is the wave friction factor, ρ is the water density and u_b is the amplitude of the wave orbital velocity at the bottom. The quadratic representation of u_b preserves the sign of the flow direction, although direction is immaterial in the present exercise. The friction factor depends on the wave Reynolds number $Re_w = a_b^2\sigma/\nu$ and the relative bed roughness k_s/a_b, where k_s is the bed roughness and a_b is the amplitude of wave-induced water parcel excursion corresponding to u_b. Expressions for f_w in the fully rough range of flow $(Re_w > 5\times10^4)$ are as follows [Kamphuis, 1975]:

$$\frac{1}{2\sqrt{f_w}} + \ln\frac{1}{2\sqrt{f_w}} = -0.35 - \frac{4}{3}\ln\frac{k_s}{a_b}; \qquad \frac{k_s}{a_b} < 0.02$$

$$f_w \simeq 0.1\left(\frac{k_s}{a_b}\right)^{3/4} \; ; \qquad \frac{k_s}{a_b} > 0.02 \tag{E2.9}$$

For these expressions the roughness is defined as $k_s = 2d_{90}$, where d_{90} = sand size in the cumulative size distribution such that 90% is finer by weight (as defined in Chapter 3). The amplitude a_b is obtained by integrating the wave velocity in Eq. 2.124. Setting $z = -h$ in the resulting expression yields

$$\zeta_b = -a_b\sin(kx - \sigma t) \tag{E2.10}$$

where

$$a_b = \frac{agk}{\sigma^2} \tag{E2.11}$$

A 0.67 m high wave with a period of 5 s is traveling in 3.22 m deep water. If the bed consists of sand with diameter d_{90} = 1.24 mm, calculate the bed shear stress.

Chapter 3

Methods of Sediment Classification

3.1 Chapter Overview

The main method of classification of sediment is in terms of size. Classification of fine-grained sediment, a subset of a wider range of natural particles, requires parameterization due to the tendency of certain types of fine particles to agglomerate as flocs. In addition to size, classifications based on mineral composition and on plasticity are briefly described. Particle surface area is noted because of its association with cohesion arising from electrostatic charge on the particle surface. Information on the surface area as well as volume of particles is also essential because the basis for measurement of particle dimensions varies with the instrument. For instance, optical devices are more sensitive to smaller particles, with backscatter response proportional to their surface area. Acoustic devices are more sensitive to larger particles, with response proportional to volume [Grey and Gartner, 2010].

The method for determination of mineral composition relies to a great extent on x-ray powder diffractometry. This topic is briefly reviewed along with crystallographic description of minerals in general, and basic structures of clays. Reference has been made to differential thermal analysis, which is another technique for identification of clay minerals.

Organic matter is mentioned in the context of its role in intrinsically altering the chemical structure of clay, and also as an extrinsic agent influencing floc formation. It is difficult to characterize extrinsic organic matter in terms of size except when it occurs in the aggregated form in water. The final topic on classification by plasticity is relevant to both surface and mass erosion of cohesive beds and design of stable (non-eroding) channels lined with generally stiff clayey beds.

3.2 Classification by Size

3.2.1 *Size and sediment type*

The term "size" is commonly used to represent particle diameter. For coarse sediment such as sand, size may refer to the sieve diameter. For fine sediment, since sieving is not feasible, size often refers to the so-called Stokes diameter (Chapter 7) or some other diameter depending on the principle underlying measurement.

Several methods are available for measuring particle size. Electron micrographs are used to image and automatically count particles of different sizes. In the method involving a Coulter counter the change in the conductivity of a suspension, which occurs when (non-conducting) particles pass through an orifice, is measured. The number density of particles is obtained by counting the pulses, and particle size from the strength of each pulse. A sedigraph uses x-ray absorption by particles for their detection.

In sedimentation methods, the terminal settling velocity is determined as a proxy for size in a gravity-settling column. This is done conventionally by measuring the suspended sediment concentration based on sample volume withdrawal at different times and elevations. Concentration is often obtained by the accurate but tedious method of gravimetric analysis. However, the use of calibrated optical or acoustic sensors is common as they permit sampling at high frequencies and extended periods. These sensors do not require withdrawal of samples, which changes the initial volume of suspension and may complicate the interpretation of data, particularly in laboratory columns in which the total available volume of suspension is not sufficiently large.

The laser diffraction method is popular for fine sediment size determination. It relies on the detection of a halo of diffracted light produced when a focused laser beam passes through an ensonified dispersion of suspended particles. Since the angle of diffraction increases as particle size decreases, this method is particularly useful for measuring sizes less than about 1 μm [Law *et al.*, 1997; Cramp *et al.*, 1997; Law and Bale, 1998]. Calibration of instruments based on this principle appears to hold almost independently of the effects of changing particle color, composition or size [Agrawal and Pottsmith, 2000].

Classification by size often refers to individual or dispersed particles rather than flocs because floc properties are dependent of the flow field in which they occur.[a] As particle size and cohesion are interdependent to a degree, dispersed particle size can be a useful measure in qualitative interpretations of fine sediment transport data. In that regard, an important use of size is for the division of sediment as coarse- or fine-grained. This distinction is commonly based on the classification given in Table 3.1. Note that the term "clay" is also used to describe fine-grained soil having plasticity. Any likely confusion can be avoided by using "clay size" rather than "clay" to denote particles smaller than 2 microns (1 μm = 10^{-3} mm = 10^{-6} m), because clays are specific (crystalline) inorganic minerals that dominantly occur in that size range.

Table 3.1 separates particles at the 0.063 mm (or, more precisely, 0.0625 mm) boundary such that particles less than this size are classified as fine, and above as coarse.[b] The finest sediment is characteristically taken to be 0.1 μm. As mentioned in Chapter 1, a suspension of particles smaller than 1 μm is called a sol or colloidal suspension. These particles typically do not settle easily due to momentum imparted by the movement of water molecules possessing thermal energy (Chapter 4). Therefore sol particles, particularly those smaller than 0.1 μm, are not classified as constituents of sediment. However, they are commonly present in flocs of larger particles. In fact, particles down to the size of

---

[a] As a result, the size distribution of flocs is reported in the same way as that for dispersed particles, but with additional variables characterizing the flow.

[b] Since most natural sediment size distributions are skewed with a preponderance of finer sizes, classification has been made convenient by using the logarithm with base 2. For particle diameter d_p in mm, the ϕ (phi) unit of size is defined as

$$\phi = -\log_2 d_p = -\frac{\log d_p}{\log 2} = -3.3219 \log d_p \qquad \text{(F3.1)}$$

where the minus sign is used so that the more common sediment sizes ($d_p < 1$ mm) have a positive ϕ value [Krumbein, 1936]. From Eq. F3.1 $d_p = 2^{-\phi}$. The value 62.5 μm (= 0.0625 mm) arises from the choice of $\phi = 4$ as the boundary between sand and silt.

Table 3.1. Classification of particles as coarse or fine.

Name	Size (mm)	Designation
Boulder	> 305	
Cobble	52 to 305	
Gravel	2 to 52	Coarse
Sand	0.063 to 2	
Silt	0.002 (2 μm) to 0.063 (63 μm)	Fine
Clay	0.0001 (0.1 μm) to 0.002 (2 μm)	
Sol particles	< 0.001 (1 μm)	Fine (but not considered to be sediment, especially particles < 0.1 μm)

molecules and atoms can influence floc behavior, with the effects of bacteria and viruses complicated by their variety and concentration.

For historical reasons geologists, soil scientists and engineers have followed different classifications of size, although the differences are minor. The system indicated in Table 3.2 and accepted by the American Geophysical Union (AGU) is similar to the one in Table 3.1. Since most natural sediments occur as mixtures, *i.e.*, with heterogeneous (or polydisperse) particle populations, different approaches have been adopted to represent the particle size distribution and parameters such as the expected values of diameter, area and volume. These representations employ weight or volume as a basis and include the frequency distribution or the cumulative frequency distribution. Both are usually plotted with respect to the ith-class diameter d_{pi} using log-scale axis. The cumulative distribution is popular because it permits easy identification of sediment in terms of sand, silt and clay in hydraulic and geotechnical applications. In Fig. 3.1, cumulative percent finer by weight, $F(d_p)$, is plotted against d_{pi} for three fine sediments and a typical beach sand. All sediments, two of which are natural muds (from the Maracaibo estuary in Venezuela and the San Francisco Bay), characteristically span more than single categories defined by clay, silt or sand. The difference between the distributions for dispersed bay mud

Table 3.2. AGU's soil technology classification of particles.

Name	Size (mm)	Designation
Gravel	64–32	Very coarse
	32–16	Coarse
	16–8	Medium
	8–4	Fine
	4–2	Very fine
Sand	2–1	Very coarse
	1–0.5	Coarse
	0.5–0.25	Medium
	0.25–0.125	Fine
	0.125–0.062	Very fine
Silt	0.062–0.031	Coarse
	0.031–0.016	Medium
	0.016–0.008	Fine
	0.008–0.004	Very fine
Clay	0.004–0.002	Coarse
	0.002–0.001	Medium
	0.001–0.0005	Fine
	0.0005–0.0002	Very fine

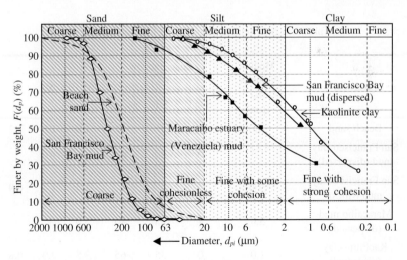

Figure 3.1. Cumulative particle size distributions of five sediments. This framework is often referred to as the MIT classification. Degree of cohesion as defined applies only to dispersed particle distribution, *i.e.*, not to floc size distribution of San Francisco Bay mud (from Mehta [1973] and Letter [2009]).

versus naturally flocculated San Francisco Bay mud is remarkable. The flocs, except the largest ones, are even larger than beach sand. However, the settling velocity of flocs is considerably lower than sand because they are lighter (Chapters 6 and 7).

For sediment load calculations, measures of the central tendency (mean or median value) and the spread (standard deviation) based on the cumulative size distribution are commonly used.

Example 3.1: Determine the median size d_{50} and the fractions of sand, silt and clay in the sediments of Fig. 3.1. Also calculate: (1) the uniformity coefficient $S_u = \sqrt{d_{90}/d_{40}}$; (2) the sorting coefficient $S_o = \sqrt{d_{75}/d_{25}}$; and (3) the skewness coefficient $S_k = d_{75}\, d_{25}/\, d_{50}^2$. The quantities S_u and S_o characterize the spread and S_k the asymmetry of the distribution. For uniform sediment $S_u = S_o = 1$, and $S_k = 1$ for a symmetric distribution.

The results are given in Table 3.3. Due to the logarithmic scale for diameter (Fig. 3.1), the degree of uniformity of sediment is visually not evident. Beach sand and kaolinite distributions are much more uniform (low S_u) than Maracaibo mud, which is highly graded (large S_u), *i.e.*, non-uniform. The dispersed size distributions of San Francisco Bay mud and kaolinite are comparable, while the Bay flocs are considerably larger. The transport behavior of these flocs and Maracaibo flocs are known to differ [Mehta and Partheniades, 1975], underscoring the need to characterize natural cohesive sediment transport in terms of floc properties in addition to individual particles.

Table 3.3. Size-related properties of four sediments.

Sediment	Median diameter, d_{50} (μm)	Size fraction (%)			S_u	S_o	S_k
		Sand	Silt	Clay			
Beach sand	180	94	6	0	1.8	1.5	1.2
San Francisco Bay mud (dispersed)	1.3	0	40	60	-	-	
San Francisco Bay mud (flocs)	300	-	-	-	1.4	1.4	0.9
Maracaibo mud (dispersed)	3.0	7	53	40	5.0	-	-
Kaolinite clay (dispersed)	1.0	0	35	65	3.8	2.0	0.8

3.2.2 *Volume-based size distribution*

A common representation of size distribution is in terms of particle volume concentration C_v (volume of dry particles divided by volume of suspension) as a function of particle diameter d_p [*e.g., Kranck,* 1986]. An example is shown in Fig. 3.2 for a carbonate mud consisting of only non-clay minerals from a Pacific atoll. Floc and dispersed particle concentrations were determined from micrographs. The dispersed particle concentration was also obtained by an automated size analyzer. The fine-end tail of the micrograph data is truncated as it was difficult to analyze very small (less than about 2 µm) particles. The calculation of C_v is developed as follows.

Consider N spherical particles per unit volume of suspension. The particles are subdivided into i size-classes, and the volume of each particle is V_{pi}. The differential volume of particles of the i-class is

$$dV_i = V_{pi}dN_i \qquad (3.1)$$

We may relate the number range dN_i to the corresponding volume range dV_{pi} by

$$dN_i = n_v dV_{pi} \qquad (3.2)$$

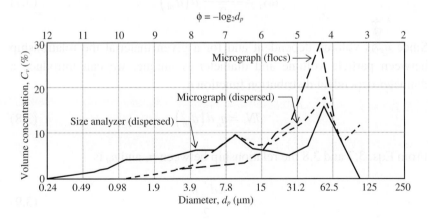

Figure 3.2. Size distribution (by volume) of carbonate mud from the Mataiva Atoll in French Polynesia. For flocs, $d_p = d_f$, the floc diameter (adapted from Wolanski *et al.* [1994]).

where n_v is a function dependent on particle volume. Combining Eqs. 3.1 and 3.2 yields

$$dV_i = V_{pi}dN_i = n_v V_{pi} dV_{pi} \qquad (3.3)$$

Thus the total volume V of all particles is

$$V = \int_0^V n_v V_{pi} dV_{pi} \qquad (3.4)$$

Next, the spherical volume of the i-class particle is

$$V_{pi} = \frac{\pi}{6} d_{pi}^3 \qquad (3.5)$$

Therefore

$$dV_{pi} = \frac{\pi}{2} d_{pi}^2 \, d\left(d_{pi}\right) \qquad (3.6)$$

and from Eq. 3.2

$$dN_i = \frac{\pi d_{pi}^2 n_v}{2} \, d\left(d_{pi}\right) \qquad (3.7)$$

Since n_v is volume dependent and for a given material the relationship between particle volume and diameter is unique, we can introduce a diameter dependent distribution function n_d

$$dN_i = n_d d\left(d_{pi}\right) \qquad (3.8)$$

From Eqs. 3.7 and 3.8 the relationship between n_v and n_d is

$$n_d = \frac{\pi d_{pi}^2 n_v}{2} \qquad (3.9)$$

We now insert Eqs. 3.5, 3.6 and 3.8 into Eq. 3.4 and obtain

$$V = \int_0^{d_{pi}} \frac{\pi}{6} n_d d_{pi}^3 \, d\left(d_{pi}\right) \tag{3.10}$$

If we further assume that n_d is independent of diameter, integration of the above equation yields

$$\frac{dV_i}{d(\log d_{pi})} = \frac{V_{pi} dN_i}{d(\log d_{pi})} = \frac{2.303\pi^2}{12} d_{pi}^6 \tag{3.11}$$

where $C_{vi} = V_{pi} dN_i / d(\log d_{pi}) \approx V_{pi} \Delta N_i / \Delta(\log d_{pi})$ and 2.303 is the conversion factor between logarithm with respect to base e and base 10. Using Eq. 3.11, C_v for each differential interval $\Delta(\log d_p)$ can be obtained from measurements of V_p and ΔN. The diameter d_p is obtained from Eq. 3.5 [Friedlander, 2000].

3.2.3 *Moments of size distribution*

The moments of particle size distribution are used to characterize a population of suspended particles. The general moment is defined as

$$M_n(\vec{x},t) = \int_0^\infty n_d d_p^n \, d\left(d_p\right) \tag{3.12}$$

where n is the order of moment. The 0th moment ($n = 0$) is the total number concentration of particles in suspension

$$N = M_0 = \int_0^\infty n_d \, d\left(d_p\right) \tag{3.13}$$

The first moment is

$$M_1 = \int_0^\infty n_d d_p \, d\left(d_p\right) \tag{3.14}$$

The ratio M_1 to M_0 is an estimate of the average particle diameter

$$\bar{d}_p = \frac{M_1}{M_0} \tag{3.15}$$

The second moment is a measure of the total surface area of the particles A_{pt} in a given volume of suspension

$$A_{pt} = \pi M_2 = \pi \int_0^\infty n_d d_p^2 \, d\left(d_p\right)$$
(3.16)

The average surface area of a particle is then obtained as

$$\bar{A}_p = \pi \frac{M_2}{M_0}$$
(3.17)

The third moment is proportional to the total volume of particles

$$V = \frac{\pi M_3}{6} = \frac{\pi}{6} \int_0^\infty n_d d_p^3 \, d\left(d_p\right)$$
(3.18)

The average particle volume is

$$\bar{V}_p = \frac{\pi M_3}{6 M_0}$$
(3.19)

The fourth moment is proportional to the total surface area of sediment settling per unit time. Inserting the dynamic viscosity η and particle density ρ_s from the Stokes settling velocity equation (Eq. 7.15) into the moment integral yields settling coverage, a characteristic settling time-constant (in unit of inverse time)

$$A_w = \int_0^\infty \left(\frac{\pi d_p^2}{4}\right)\left(\frac{\rho_s d_p^2 g}{18\eta}\right) n_d \, d\left(d_p\right)$$

$$= \left(\frac{\pi g}{72\eta}\right) \int_0^\infty \rho_s d_p^4 n_d \, d\left(d_p\right) \approx \left(\frac{\pi g \rho_s}{72\eta}\right) M_4$$
(3.20)

The fifth moment is proportional to the settling flux of the suspended matter

$$F_s = \int_0^\infty \left(\frac{\rho_s \pi d_p^3}{6} \right) \left(\frac{\rho_s d_p^2 g}{18\eta} \right) n_d \, d(d_p)$$

$$= \left(\frac{\pi g}{108\eta} \right) \int_0^\infty \rho_s^2 d_p^5 n_d \, d(d_p) \approx \left(\frac{\pi g \rho_s^2}{108\eta} \right) M_5$$

(3.21)

For evaluating the moments analytically, the integral in Eq. 3.12 is approximated as a summation over particle size classes

$$M_n(\vec{x}, t) = \int_0^\infty n_d d_p^n \, d(d_p) \approx \sum_{i=1}^{n_c} n_d d_p^n \Delta(d_p)$$

(3.22)

The quantities V_{pi} and dN_i within constant intervals of $\Delta(\log d_p)$ for suspended matter in South San Francisco Bay [Letter, 2009] were derived from *in situ* video-images and mass concentration data. Calculated values N, $\overline{d}_p$, $\overline{A}_p$, $\overline{V}_p$ and F_s for five sets of data are given in Table 3.4. The corresponding distributions of volume concentration are shown in Fig. 3.3. The viscosity η of the suspension was taken as 0.00115 Pa.s, and particle density 2,650 kg m^{-3}.

Table 3.4. Particle parameters from San Francisco Bay.

Time (June 30 2007)	Particle concentration, N (# m^{-3})	Average diameter, $\overline{d}_p$ (μm)	Average surface area, $\overline{A}_p$ (μm^2)	Average volume, $\overline{V}_p$ (μm^3)	Settling flux, F_s (kg m^{-2} s^{-1})
08:59:19	1.08×10^8	81	2.37×10^4	4.33×10^5	0.48×10^{-3}
09:06:03	4.22×10^8	75	1.98×10^4	3.25×10^5	1.22×10^{-3}
09:08:25	4.39×10^8	92	3.06×10^4	6.37×10^5	4.04×10^{-3}
09:59:58	2.36×10^8	103	3.97×10^4	9.90×10^5	3.93×10^{-3}
10:02:20	1.84×10^8	123	5.55×10^4	1.51×10^5	5.43×10^{-3}

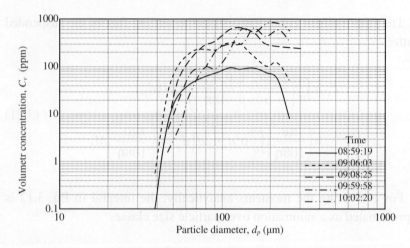

Figure 3.3. Five distributions of volume concentration of suspended sediment from San Francisco Bay using measurements on June 30, 2007 (based on Letter [2009]).

3.2.4 *Size, specific surface area and cohesion*

In Table 3.5 the degree of electrochemical cohesion has been qualitatively related to particle size. As fine (especially clay mineral) particles become characteristically flatter with decreasing nominal diameter, it serves as an approximate measure of the ratio of particle's surface area to mass, *i.e.,* the specific surface area. As the particles become smaller their specific surface area increases and therefore

Table 3.5. Sediment size and its relationship with cohesion.

Size range (μm)	Classification	Qualitative degree of cohesion
> 62.5	Coarse	Cohesionless
62.5 to 40	Coarse silt	Practically cohesionless
40 to 20	Coarse silt	Cohesion increasingly important with decreasing size
20 to 2	Medium and fine silts	Cohesion important
2 to 0.1	Coarse, medium and fine clays	Cohesion very important

Source: Mehta and Lee [1994].

cohesion also increases. Silt-size particles in the range of 20 μm to 40 μm are practically cohesionless, although they are often embedded in biofilms forming effectively "cohesive" aggregates in which electrochemical cohesion may not play a primary binding role. In highly organic flocs the mineral particles may be so far apart that electrochemical cohesion may have no tangible role in floc growth or breakup (Fig. 3.4a, b).

Clay minerals are crystalline with particles generally less than 2 μm, and most are plate-like (Figs. 3.5 and 3.6). As mentioned, they have a high specific surface area (Table 3.6). This in turn means that electrochemical surface forces dominate over particle weight, and cohesion plays an

(a)   (b)

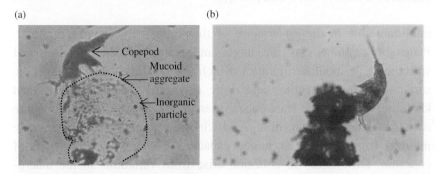

Figure 3.4. Non-electrochemical binding: (a) The aggregate includes a few solid inorganic particles bound by exopolymeric biofilm (mucopolysaccharide or mucus). A copepod (length ~ 1 mm) is ingesting the mucus; (b) a copepod is devouring an organic aggregate (from Wolanski [2007] and personal communication).

(a)   (b)

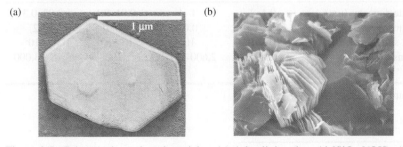

Figure 3.5. Primary clay–mineral particles: (a) A kaolinite plate $Al_4[SiO_{10}](OH)_8$ (from Lambe and Whitman [1979]); (b) well-crystallized kaolinite stacks. Magnification 2000x (from Welton [1984]).

(a)   (b)

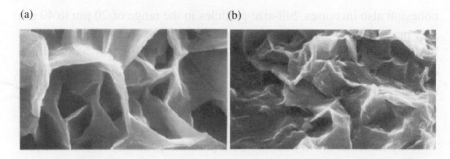

Figure 3.6. (a) Well-developed, highly crenulated smectite, $(0.5Ca, Na)_{0.7}(Al, Mg, Fe)_4[Si, Al]_8O_{20}].nH_2O$. Magnification 10,000×; (b) bentonite clay consisting of crenulated, sodium-rich smectite, $(0.5Ca, Na)_{0.7}(Al, Mg, Fe)_4[Si, Al]_8O_{20}](OH)_4.nH_2O$. Magnification 8,000× (from Welton [1984]).

important role in the governance of their transport behavior in water. Plates of kaolinite clay are larger and better defined than those of smectite (*e.g.*, montmorillonite[c]), which are much thinner. The specific surface areas of smectites and illitic clays are generally higher than kaolinite, which is the least cohesive among these clays, while smectite is more cohesive than illite (and chlorite). These observations are inferred from the ion exchange capacity (Chapter 4), a measure of cohesion. Differences in cohesion arise from variations in the macromolecular structure of clay minerals.

Table 3.6. Properties of clays common in the marine environment.

Property	Kaolinite	Chlorite	Smectite	Illite
Plate diameter (μm)	0.1–4	1–4	0.01–0.1	0.03–0.3
Plate thickness (μm)	0.05–2	0.03	≤ 0.01	0.02
Sp. surf. area ($m^2 kg^{-1}$)	$10^3–10^4$	10^4	$10^4–10^5$	$10^4–10^5$
Density (kg m^{-3})	2,600–2,680	2,600–2,960	2,200–2,700	2,760–3,000

Source: Dade and Nowell [1991].

---

[c] *Montmorillonite* is the most abundant mineral in the *smectite* group of minerals. Both terms have been used interchangeably in the engineering literature. However, smectite is the correct group designation.

In some clays such as attapulgite or palygorskite[d] and halloysite, the plates are rolled into needle-like tubes or laths (Figs. 3.7 and 3.8). Figure 3.9 shows a variety of plate-like and tube-like particles from a dispersed sample of Chesapeake Bay sediment. Specific gravities of common non-clay and clay minerals are given in Table 3.7. When such data are unavailable, a value ranging between 2.60 and 2.70 is commonly used for clays (and quartz sand) in sediment transport calculations.

Figure 3.7. Palygorskite (attapulgite) clay tubes (from van Olphen [1977]).

Figure 3.8. Palygorskite (attapulgite), $(OH_2)_4(OH)_2Mg_5Si_8O_{20}4H_2O$. Magnification 10,000× (from Welton [1984]).

---

[d] Attapulgite has been renamed palygorskite. However, in older literature both terms may refer to the same material. In this text, in some instances the term attapulgite has been retained.

Figure 3.9. Dispersed particles from Chesapeake Bay (from Zabawa [1978a]).

Example 3.2: Calculate the specific surface areas of: (1) 1 mm (cohesionless) quartz sand assumed to be spherical, and (2) a unit cell of (cohesive) montmorillonite clay in the sodium form (*i.e.,* Na-montmorillonite). Since clay molecules are crystalline structures of connected molecules, the unit cell is equivalent to a single (macro) molecule.

For a sphere of diameter d_p and density ρ_s the specific surface area A_p is

$$A_p = \frac{\pi d_p^2}{\rho_s \frac{1}{6}\pi d_p^3} = \frac{6}{\rho_s d_p} \tag{3.23}$$

Table 3.7. Specific gravities of some clay minerals and non-clay minerals.

Mineral		Specific gravity [a]
Clay	Kaolinite	2.60–2.68
	Halloysite (2H$_2$O)	2.55–2.56
	Illite	2.64–3.10
	Smectite	2.20–2.70
	Vermiculite	~2.80–3.20
	Chlorite	2.60–2.96
	Attapulgite (palygorskite)	2.29–2.36
Non-clay	Quartz	2.65
	K-Feldspars	2.54–2.57
	Na-Ca-Feldspars	2.62–2.76
	Calcite	2.72
	Dolomite	2.85
	Muscovite (Mica)	2.70–3.10

[a] Unit weight of mineral divided by unit weight of water.
Sources: Grim [1968]; Lambe and Whitman [1979].

which indicates that A_p is inversely proportional to particle diameter. For quartz sand taking $\rho_s = 2,650$ kg m^{-3} and $d_p = 10^{-3}$ m yields $A_p = 2.264$ m^2 kg^{-1}.

The following calculation is provided for sodium montmorillonite [van Olphen, 1977]. Given the molecular formula $(Si_8)(Al_{3.33}Mg_{0.67})O_{20}(OH)_4Na_{0.67}$, the unit-cell weight is 734 (Table A.1, Appendix A). Thus, 734 g of the clay contain 6.022×10^{23} (Avogadro's number, N_A)[e] unit cells. Each cell has a surface area of 45.8Å^2 (1Å $= 10^{-10}$m) on each side. Therefore the total surface area (excluding the thin edge) per gram is

$$A_p = \frac{6.02 \times 10^{23} \times 2 \times 45.8}{734 \times 10^{20}} = 752 \text{ m}^2 \text{ g}^{-1} = 7.52 \times 10^5 \text{ m}^2 \text{ kg}^{-1}$$

Taking the surface density of the electrochemical-electrostatic force of attraction to be constant, cohesion will be $7.52 \times 10^5 / 0.231 = 3.3 \times 10^6$ times more important for the clay than sand.

Due to their low specific surface areas, fine non-clay mineral particles do not display cohesion to any significant degree. In laboratory tests on incipient particle movement under steady flows, sediment consisting of crushed silica particles in the silt size range tends to show little cohesion and behaves like sand. The responses of four uniform-slope beach profiles to wave action were tested in a wide basin — two beaches made of clay mixtures (kaolinite + attapulgite and kaolinite + montmorillonite), one of loess (wind-deposited loam) from Vicksburg, Mississippi, and the fourth 0.1 mm sand. The profile of loess, which was practically cohesionless, showed patterns of erosion and accretion that were qualitatively similar to those of the sandy profile. The two clayey profiles showed behaviors that were noticeably different from those of the other two profiles [Mantz, 1977; Lee, 1995, Chapters 6 and 12].

An indirect measure of the relationship between particle size and cohesion can be derived from tests on fine sediments in a settling column. In Fig. 3.10 the so-called flocculation factor F_f is plotted against the median dispersed (*i.e.*, individual) particle size d_{50}. This factor is the ratio of floc settling velocity w_s to the settling velocity w_d of the individual particles comprising the floc. Several sediments were used

---

[e] Avogadro's number N_A (Table A.2, Appendix A) is the number of molecules in a mole, *i.e.*, molecular weight expressed in grams. For example, the molecular weight of calcium is 40, so 1 mole of calcium equals 40 grams.

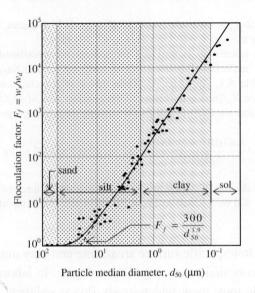

Figure 3.10. Flocculation factor as a function of dispersed particle median diameter (from Migniot [1968]).

including those of marine, estuarine, fluvial and lacustrine origin, sludge and powders. Their fine fractions were obtained by size separation of the raw (heterogeneous) materials. The factor F_f increased from a little over 1 for 40 μm particles to 30,000–40,000 for 0.1 μm particles because with decreasing particle size and increasing cohesion the floc size and the settling velocity increased (Fig. 3.10). In other words, a 40-fold decrease in the dispersed particle size resulted in a remarkable increase in the settling velocity of the floc relative to the dispersed particle [Migniot, 1968].

When extrapolated, the linear segment in Fig. 3.10 intersects the size axis at about 20 μm and is represented by the relation

$$F_f = \frac{300}{d_{50}^{1.9}} \tag{3.24}$$

in which d_{50} is in microns. It is remarkable that so many sediments show an affine behavior represented by Eq. 3.24 over nearly five decade-fold variation in F_f. A possible explanation for this trend is that clay minerals are chemically similar, and natural muds often contain similar non-clay minerals as well. The fractal character of flocs (Chapter 4) may underpin the physical basis of Eq. 3.24.

Example 3.3: Calculate the diameter of flocs composed of 2 μm individual particles having a mineral density of 2,650 kg m^{-3} and water density 1,000 kg m^{-3}. Assume a floc density of 1,100 kg m^{-3}. Further assume that the viscosities of the flocculated and dispersed particle suspensions are both equal to that of water.

From Eq. 3.24, with $d_{50} = 2$ μm we obtain $F_f = 80.4$. Stokes law (Eq. 7.15) for the settling velocity w_s is

$$w_s = \frac{g}{18v}\left(\frac{\rho_s}{\rho}-1\right)d_p^2 \tag{3.25}$$

where v is the kinematic viscosity of the fluid of density ρ. Thus

$$F_f = \left(\frac{d_f}{d_{50}}\right)^2 \frac{\rho_f - \rho}{\rho_s - \rho} \tag{3.26}$$

where d_f is the floc diameter and ρ_f is the floc density. Therefore, with $\rho_s = 2,650$ kg m^{-3} and $\rho = 1,000$ kg m^{-3}, we obtain $d_f = 73$ μm.

3.3 Classification by Composition

3.3.1 *Minerals*

Clay minerals are formed by disintegration of rock, the primary material, due to weathering accompanied by chemical alteration of the resulting mineral, often mica.[f] This process is known as diagenesis. When minerals precipitate directly from solution the process is called genesis. The formation of crystals implies that the rate of precipitation is slow enough to permit the precipitating molecules to form self-organized structures consistent with thermodynamic requirements.

Weathering is not merely a physical process of rock disintegration; it involves the release, or liberation, of alkalies (K^+, Na^+) and alkaline

---

[f] In Table C.1 of Appendix C, common minerals and rock types encountered in x-ray diffraction (XRD) analysis are listed.

earths (Ca^{++}, Mg^{++}) from the rock via cation exchange. In this process, a reaction occurs by which certain cations present in the clay in an exchangeable state are preferentially replaced by other ions. These ions become lodged in the cleavage surfaces present in the mineral due to its layered structure. The type of material formed from weathering depends on whether the released cations are retained in the secondary material that is formed. Age is also therefore an important factor. Thus, smectite is abundant in many Mesozoic (225M to 65M years BP) and Cenozoic (65M years BP to the present) sediments. However, smectite tends to disappear in sediments of increasing age, and is not found in sediments older than Mesozoic. The same is true of kaolinite. Thus very old sediments largely contain illites and chlorites. Attapulgite-sepiolite also appears to be limited to relatively young sediments. Halloysite is generally absent in rocks, and is believed to be formed from kaolinite. In Florida the clay minerals, mainly smectite, kaolinite and chlorite (and non-clay calcareous, silicate and phosphatic sediments), are thought to have deposited during the Miocene Epoch (24M to 2M BP) by alluvial transport from southern Georgia to the then-submerged shallow Florida platform [Weaver and Beck, 1977].

The total amount of exchangeable cations in a clay sample can be determined analytically. This amount, expressed in milliequivalents[g] (mEq) per 100 grams of the dry clay sample, is the cation exchange capacity (CEC), or the base exchange capacity of the clay. It is an operational (rather than theoretically meaningful) quantity, and is measured as the amount of a cation that can be removed by a specific substance once the clay and the solution have reached chemical equilibrium. Exchangeable cations compensate the unbalanced electric charge in the interior of the crystal lattice by what is called isomorphous substitution.[h] As a result, the CEC is a measure of the degree of this substitution. Its use is not limited to clay minerals. Table 3.8 provides

[g] Equivalent weight is the weight of an element that combines with 1 g of hydrogen, 8 g of oxygen or 35.5 g of chlorine. Milliequivalent is the weight in grams multiplied by 1,000.

[h] Such substitutions occur when the replacing cation has the same molecular volume as the one being replaced.

typical ranges of the CEC of clay minerals and three non-clay materials — iron hydroxide, soil humic acids and manganese oxides. Anion exchange also occurs but is not significant or relevant to the marine environment in the present context.

The CEC of a clay mineral reflects its reactive nature; increasing CEC implies increasing reactivity, which also relates to increasing cohesion. However, this connection is not rigorously proportional, because the CEC can vary due to a variety of factors of inorganic as well as organic origin. Among inorganic factors the availability of ions for adsorption is important. When several competing species are present the CEC is dependent on two effects that come into play. Firstly, if all the ions are of the same valence, the affinity of adsorption decreases with the ionic diameter, *e.g.*, within the alkali group of metallic elements the sequence is $Li^+ < Na^+ < K^+ < Rb^+ < Cs^+$. Secondly, adsorption increases with the valence. Thus, with regard to the propensity for adsorption, we have $Al^{+++} > Ca^{++} > K^+ > Na^+$. Since higher valence ions are commonly present in water, an outcome of hierarchical adsorption is that at concentrations of salts (*e.g.*, those of Ca^{++} and also Mg^{++}) far lower than that of NaCl, clayey sediments are found in the flocculated state even in non-brackish or fresh water. In fact, colloidal suspensions of clay

Table 3.8. Cation exchange capacity of clay and non-clay minerals.

Mineral	Clay/Non-clay	Cation exchange capacity (mEq 100g^{-1})
Kaolinite		3–15
Halloysite, $2H_2O$		5–10
Halloysite, $4H_2O$		40–50
Montmorillonite		80–120
Smectites	Clay	80–150
Illite		10–40
Vermiculites		120–200
Chlorite		20–50
Attapulgite–palygorskite–sepiolite		3–15
Iron hydroxide		10–25
Soil humic acids	Non-clay	170–590
Manganese oxides		200–300

Source: Horowitz [1991].

minerals are rare in nature due to this effect, and also because even small concentrations of organic products such as mucus can limit the freedom of movement of individual clay particles.

Since the specific surface area depends on particle size, even for clayey sediment from a single source, the CEC tends to vary with size as illustrated for a kaolinite in Table 3.9.

Chemical pretreatment can substantially alter the CEC and nullify any physicochemical connection between CEC and cohesion. Clays meant for industrial uses such as the production of ceramics are often chemically pre-treated. CEC values of such clays can be quite different from those of their mineral constituents. Organic matter, when present in high fractions (*e.g.,* > ~10%), tends to take up exchangeable cations and may increase the CEC to hundreds of mEq $100g^{-1}$, which does not imply enhanced cohesion.

As expected, clays reflect the environment of diagenesis (or genesis). For example, an acidic environment favors the formation of kaolinite. Another important factor is the presence or absence of certain ions. The presence of Mg^{++} and Al^{+++} is conducive to the formation of smectite, but if these ions are leached out in the long run, *e.g.,* by rainfall and percolation, kaolinite may result. The presence of K^+ favors the formation of illite. On the surface of the earth kaolinite, smectite, illite and chlorite are abundant followed by halloysite and vermiculite, while attapulgite-palygorskite and sepiolite are comparatively rare.

Clay minerals contain a limited number of chemicals as illustrated in Table 3.10 by the constituents of three clays from specific sources. The trace elements, *i.e.,* elements present in very small but detectable amounts, vary from site to site. Seemingly minor quantities of elements can measurably influence the flocculation behavior and the color of clays.

Table 3.9. Variation of CEC with particle size for kaolinite.

Size (μm)	10–20	5–10	2–4	1–0.5	0.5–0.25	0.25–0.1	0.0–0.05
CEC (mEq $100g^{-1}$)	2.4	2.6	3.6	3.8	3.9	5.4	9.5

Source: Grim [1968].

Table 3.10. Composition in percent by weight of three clays of specific origin.

Chemical	Kaolinite (%)	Attapulgite (%)	Bentonite[a] (%)
SiO_2	46.50	55.20	63.02
Al_2O_3	37.62	9.67	21.08
Fe_2O_3	0.51	2.32	3.25
MgO	0.16	8.92	2.67
Na_2O	0.02	0.10	2.57
K_2O	0.40	0.10	
CaO	0.25	0.65	1.65
FeO	-	0.19	0.35
H_2O	-	-	5.64
NH_2O	-	9.48	-
SO_3	0.21	-	-
V_2O_5	< 0.001	-	-
Trace elements	-	-	0.72

[a] A smectite.
Source: Lee [1995].

The inorganic constituents of mud often include non-clay minerals that dominate the silt size range while, as noted, clay minerals occur mainly in the clay range. Two commonly found non-clay minerals in coastal and estuarine sediments are quartz and calcium carbonate; the latter being the dominant material in biologically active waters. Some marine muds contain mainly carbonates, principally calcium carbonate (Table 3.11), and have the color of white chalk. Numerous non-clay constituents can influence sediment appearance and transport behavior. For example, San Francisco Bay mud contains some structural iron due to the replacement of aluminum by iron in illite present in this sediment. Suspended or recently deposited Bay mud has a light brown color, while a few millimeters to a centimeter below the mud surface the sediment tends to be dark grey due to ferrous sulfide. If a sample of this dark material is mixed with water its color changes to light-brown, as ferrous sulfide is rapidly oxidized by aeration to ferric hydroxide (Fig. 3.11). If allowed to stand for a few days, uptake of bound oxygen by bacteria in the sediment first reduces ferric iron to ferrous iron which is greenish,

Table 3.11. Percent by weight composition of two carbonate sediments.

Element	Mataiva (French Polynesia) (%)	Myrmidon Reef (Australia) (%)
Calcium	37.4	36.7
Magnesium	0.18	1.47
Strontium	0.72	0.42
Sodium	0.23	0.31
Sulfur	0.18	0.18
Potassium	0.01	0.14
Barium	8×10^{-4}	7×10^{-4}
Total carbon	11.94	9.88
Organic carbon	0.34	0.012
Total nitrogen	0.044	-
Total hydrogen	0.26	0.20
Total phosphorus	0.019	0.027

Source: Wolanski *et al.* [1994].

then back to grey ferrous sulfide [Krone, 1962]. This mud also contains biochemical constituents which influence the deposition and erosion fluxes.

Since clay minerals are commonly transported by water or wind to areas outside their sources, a mud may contain a mixture of clays from a

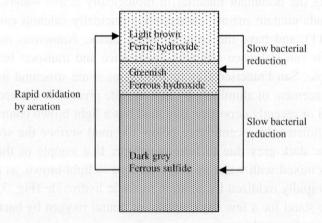

Figure 3.11. Effects of oxidation by aeration and reduction by bacteria in mud from San Francisco Bay.

variety of sources. A river receiving sediments from multiple tributaries is in this category. Examples are mixtures of kaolinite, montmorillonite and illite deposited by the Mississippi River and its tributaries (Table 3.12). Similar data for marine muds compiled in Table 3.13 also include chlorite, whose presence is notable on the West Coast.

Just as there is a mismatch between common particle size classifications and cohesion (Table 3.5), partitioning of clay size from silt at 2 μm may misrepresent the distribution of clay mineral. Plate-like clays can occur up to sizes as high as ~10 μm or larger [Gibbs, 1971; McCave and Hall, 2006]. This is illustrated by data from the Mississippi River Basin in Fig. 3.12. An advantage of the 2 μm limit is that the fraction of non-clay minerals is negligible below this size. Accordingly, this limit, although quite approximate, remains convenient for separating clay minerals from non-clay minerals.

Table 3.12. Clay mineral distribution in sediment deposited by the Mississippi River and its tributaries.

Rivers	Kaolinite (%)	Montmorillonite (%)	Illite (%)
Mississippi River	5–15	25–45	40–60
Eastern tributaries of Mississippi River: Ohio, Cumberland, Tennessee, Duck and Clinch	10–20	10–15	65–75
Western tributaries of Mississippi River: Milk, Yellowstone, Missouri, Platte and Arkansas	10–20	20–45	40–60

Source: Holmes and Hearn [1942].

Table 3.13. Percent by weight clay minerals in marine muds.

Location	Kaolinite (%)	Chlorite (%)	Smectite (%)	Illite (%)
All ocean basins (mean)	14	14	35	38
Georgia Inner Shelf	36	3	45	17
Mississippi Delta	20	8	48	24
San Francisco Bay	23	23	20	35
Astoria Canyon (WA)	-	25	52	23

Sources: Weaver [1989] cited by Dade and Nowell [1991].

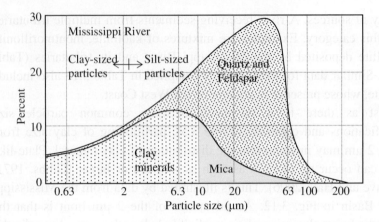

Figure 3.12. Distribution of particle size of sediment from the Mississippi River Basin (adapted from Potter *et al.* [1975]).

3.3.2 X-ray diffraction pattern

Most clay minerals are white when pure, and their (Debye–Scherrer) x-ray diffraction (XRD) signatures are used to identify minerals. The clay mineral lattice can be viewed as made of stacked unit-cells in the shape of cubes[i], the most basic of which is the primitive or simple cube with an atom, ion or molecule at the eight (corner) lattice points (Fig. 3.13). The other arrangements shown are a face-centered cube with

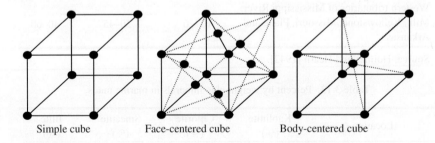

Figure 3.13. Basic cubic lattices in crystalline structures.

---

[i] A cube is just one of several forms of crystal symmetry. Others include prism, pyramid, dipyramid, *etc.*, that are characterized by Miller Indices.

six additional lattice points at the centers of the six faces, and a body-centered cube in which there is an additional lattice point at the center of the cube [Moore and Reynolds, 1997].

The response of the crystalline structure to monochromatic x-radiation can be illustrated with the simple cube. In Fig. 3.14 three two-unit cells of a simple cubic lattice are shown. Three sets of parallel or basal planes are designated by three numbers or Miller Indices for each set of parallel planes. The indices of a set are obtained by counting the number of each set *crossed* between a lattice point and the next point going in the directions of the *a*, *b* and *c* axes, respectively. Thus, in the left-hand cell we cross 1 set of planes along the *a*-axis. Along the *b*-axis, since we travel along a plane, we cross 0 set, and the same along the *c*-axis. Therefore the Miller Indices are 1, 0, 0, or simply 100. Likewise the indices for the middle and the right-hand cells are 110 and 111, respectively.

For general Miller Indices *h*, *k* and *l* the distance d_M between the parallel planes of a set is

$$d_M = \frac{a_c}{\sqrt{h^2 + k^2 + l^2}} \qquad (3.27)$$

where a_c is the cube length. Thus d_M values between planes in a simple cubic lattice are: a_c, $a_c/\sqrt{2}$, $a_c/\sqrt{3}$, $a_c/\sqrt{4}$, $a_c/\sqrt{5}$, $a_c/\sqrt{6}$, $a_c/\sqrt{8}$, *etc.* The ratio $a_c/\sqrt{7}$ does not exist because 7 cannot be obtained from $\sqrt{h^2 + k^2 + l^2}$, as *h*, *k* and *l* take on integer values only. The lattice points in a crystal plane are shown schematically in Fig. 3.15.

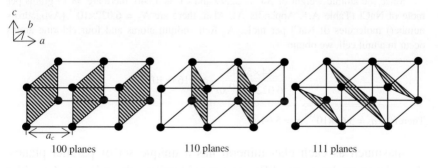

100 planes   110 planes   111 planes

Figure 3.14. Two-unit cells of simple cubic lattice.

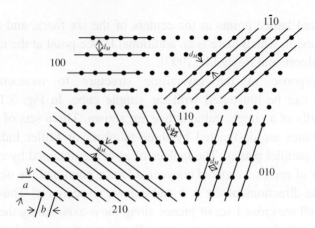

Figure 3.15. Planes through lattice points as seen along the *c*-axis of a crystal (from Daniels and Alberty [1955]).

This plane may be thought of as an end-view of the crystal along the *c*-axis perpendicular to the plane of *a* and *b*-axes. The parallel sets of planes are in their edgewise configurations. Although each type of plane is a possible crystal face, the actual faces of simple crystals correspond to planes that have a high density of lattice points _and wide inter-planar spacing. One set of planes has been identified as $1\bar{1}0$. The overbar, which represents a negative sign, indicates that, if a plane were intercepted by going in the positive direction along *a*, it would be necessary to go in the negative direction along *b* in order to intercept the same plane.

Example 3.4: Calculate the length a_c of the unit cell of sodium chloride. The density of salt is 2,163 kg m^{-3}. Each unit cell contains four sodium atoms and four chlorine atoms.

Since the atomic weight of Na$^+$ is 22.99 and Cl$^-$ is 35.46, there are 58.45 grams per mole of NaCl (Table A.1, Appendix A). Also, there are N_A = 6.022×10^{23} (Avogadro's number) molecules of NaCl per mole. As four sodium atoms and four chlorine atoms occur in a unit cell, we obtain

$$\frac{4 \times 58.45}{6.022 \times 10^{23} \times a_c^3} = 2.163 \times 10^6$$

Therefore a_c = 5.64×10^{-10} m = 5.64 Å.

Inasmuch as each clay mineral has a unique set of parallel planes, their detection by x-ray diffractometry identifies the mineral. Consider the impingement and reflection of a monochromatic x-radiation on a set of basal planes with spacing d_M. Taking two consecutive planes, ray *BSM* will constructively interfere with ray *ARL* (Fig. 3.16), provided the extra distance *FSG* that *BSM* is required to travel is an integral multiple of the x-ray wave length L_λ, *i.e.*, nL_λ, where *n* is an integer. We note that

$$FS = SG = d_M \sin\theta \qquad (3.28)$$

where θ is $\angle FRS$, which is equal to $\angle SRG$. Accordingly, for the wave to be detected upon reflection the following law of XRD must hold

$$nL_\lambda = 2d_M \sin\theta_M \qquad (3.29)$$

where θ_M is the value of θ that satisfies Eq. 3.29. This equation is named after William Bragg, who pointed out that although x-rays are diffracted by the crystal lattice just as visible light is diffracted by a grating, it is convenient to consider that x-rays are "reflected" from the planes. The reflection corresponding to $n = 1$ is of the first-order, $n = 2$ of the second-order and so on. Each successive order of reflection exhibits a wider angle θ_M. Setting $n = 1$ and considering that the second-order reflection is from planes separated by half the lattice distance we may restate Eq. 3.29 as

$$L_\lambda = 2\left(\frac{d_M}{n}\right)\sin\theta_M = 2d_{hkl}\sin\theta_M \qquad (3.30)$$

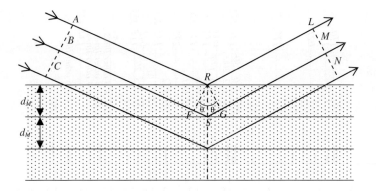

Figure 3.16. Diagram illustrating Bragg's law of XRD.

where d_{hkl} is the distance between planes having Miller Indices *hkl*. Thus, the second-order reflection from 100 planes is labeled 200, from 110 planes 220, and from 111 planes 222.

Example 3.5: The wave length of the Copper-Kα x-ray used in XRD measurement is 1.5405Å. The 001 set of planes of kaolinite show a reflected wave at angle $\theta_M = 6.17°$. Determine the distance of separation d_{001}.
  From Eq. 3.30 we obtain

$$d_{001} = \frac{1.5405}{2\sin(6.17°)} = 7.17 \text{ Å}$$

There are two methods by which a clay sample can be subjected to XRD analysis. In one the clay is sedimented from suspension onto a rectangular porous porcelain slide. The water is drained out by suction, leaving a residue of clay particles flat on the surface of the slide. The slide is then dried and set sideways, resting along its longer edge on a turn-table. The x-ray impinges on the slide which is turned slowly to change the angle θ. At values of θ (= θ_M) for which sets of planes satisfy Eq. 3.30, distinct peaks occur in the reflected radiation. Since the angles for the various planes are known by calibration using known clay structures (*e.g.*, Fig. 3.17), the unknown clay is identified (Table C.2, Appendix C). In the second method a paste of clay powder in water is smeared on the slide. This slide, containing randomly oriented particles, is subjected to XRD without turning the sample. Peaks in reflected

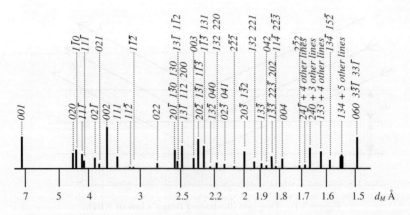

Figure 3.17. Inter-planar spacing for a well crystallized kaolinite (from Grim [1968]).

radiation are identified in the same way as in the first method [Moore and Reynolds, 1997].

Figure 3.18 shows the output in terms of the relative intensity of diffracted CuKα x-radiation versus diffraction angle 2θ for a mud sample from the Maracaibo estuary in Venezuela. A smear slide was used for the < 63 μm size fraction containing silt and clay. An abundance of quartz is present, and a calcite (also a non-clay mineral) peak is found. Clay mineral peaks are weak because the sample contained low amounts of clays compared to silt, and also because a smear slide does not enhance clay mineral peak intensities compared to those of the non-clay minerals as much as a sedimented slide.

The XRD signature for the < 2 μm fraction of Maracaibo mud (Fig. 3.19) is obtained by separating this fraction from the original sample by differential sedimentation. In this method the initially suspended material is permitted to settle up to an estimated time by which (silt sized) particles greater than 2 μm settle out. The remaining suspension is then allowed to sediment on a slide. Intense peaks for the (001 and 002) basal reflections are observed for kaolinite and illite. This slide was specially treated by soaking in ethylene glycol followed by drying. Ethylene glycol molecules inter-leave the clay layers and cause the resulting

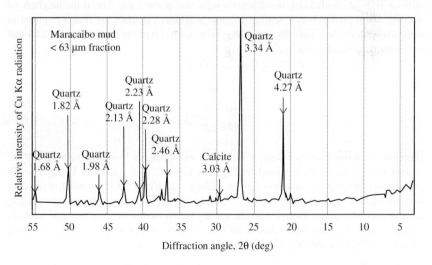

Figure 3.18. XRD pattern for Maracaibo mud; < 63 μm size fraction.

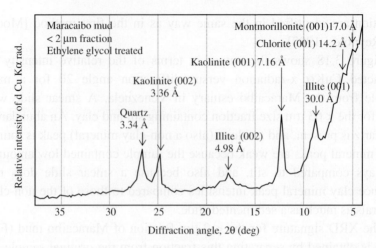

Figure 3.19.   XRD pattern for Maracaibo mud; ≤ 2 μm size fraction (treated with ethylene glycol).

Montmorillonite-glycol complex to exhibit a peak at 17 Å; a characteristic signature of this mineral. A small chlorite peak is visible at 14.2 Å, and quartz is also present.

Example 3.6: For XRD analysis, a natural inorganic mud containing silt- and clay-sized particles is to be divided into two fractions separated at the 2 μm size. If the height h_s of the suspension poured into a settling column is 20 cm, calculate the time t required to settle the silt particles. Take the specific gravity $s = \rho_s / \rho$ of the material to be 2.65 and the viscosity of water as 10^{-6} m^2 s^{-1}.

From the Stokes law of settling (Eq. 7.15) the time is

$$t = \frac{h_s}{\dfrac{gd^2}{18v}\left(\dfrac{\rho_s}{\rho}-1\right)}\qquad(3.31)$$

Therefore, with diameter $d_p = 2$ μm $= 2 \times 10^{-6}$ m, we obtain $t = 15.4$ h. If the height of the column were 1 m, the time would be 77 h. In Table C.3 of Appendix C the time (in minutes) is indicated for particles of given settling velocity (at 20°C) to fall 1 m in water of temperature ranging from 5 to 30°C. Particle diameters corresponding to the settling velocities are also given. Particle specific gravity is assumed to be 2.65. For $d_p = 1.95$ μm (which is close to 2 μm) and water at 20°C, $t = 4,895$ min, or 81.24 h. Therefore, for a fall height of 0.20 m, the time would be 16.25 h.

3.3.3 *Clay structural units*

3.3.3.1 *Basic units*

Clay minerals have different arrangements of two basic crystalline structural units; a tetrahedral unit and an octahedral unit. The tetrahedral unit is made up of four oxygen atoms at the four corners and a silicon at the center. The tetrahedra are arranged to form a hexagonal network that is repeated indefinitely to form a sheet of composition $Si_4O_6(OH)_4$ (Fig. 3.20 top). The hydroxyl (OH^-) groups attached to the structure balance its charge. The tetrahedra are so arranged that their apices point in the same direction, and the bases are in the same plane. The unit is 4.65 Å in thickness.

The octahedral unit consists of two planar sheets of closely packed hydroxyls in which an aluminum, iron or magnesium atom is embedded in octahedral coordination such that the metal is equidistant from the six hydroxyls (Fig. 3.20). When aluminum is present, only two-thirds of the possible positions are filled to electrochemically balance the structure, which has the formula $Al_2(OH)_6$ and is known as gibbsite. When magnesium is present all positions are filled to balance the structure, which is $Mg_3(OH)_6$, known as brucite. Each unit is 5.05 Å thick.

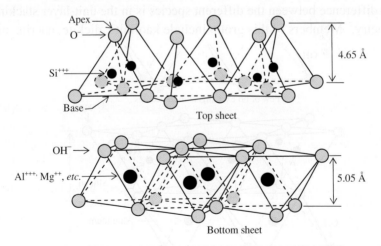

Figure 3.20. Sheet structures of clay minerals. Top: Sheet made of silica tetrahedra. Bottom: Sheet made of octahedra (based on Grim [1968]).

The tetrahedral and octahedral structural units combine to yield a variety of clay minerals. The more important ones are as follows.

3.3.3.2 *Kaolinites*

The structural unit of kaolinites is made up of an aluminum octahedral layer with a superposed inverted tetrahedral layer such that the tips of the tetrahedra and one of the hydroxyl layers of the octahedral sheet, a gibbsite or a brucite layer, form a common plane (Fig. 3.21). The unit thus formed is 7 Å in thickness. Successive structural units are held together by electrochemical attraction. Specifically, the hydroxyl group of the octahedron is linked to the nearest oxygen of the tetrahedron to form a comparatively strong but secondary bond because no high-strength primary valence bonds are available. As the individual units are electrically nearly neutral, it is difficult for water molecules or cations to penetrate between these units. The result is that the clay is chemically stable and not expandable as smectite. As a result the cation exchange capacity is in the low range of 3 to 15 mEq $100g^{-1}$ (Table 3.8). Kaolinite particles are plate-like with well defined hexagonal boundaries (Figs. 3.4 and 3.5).

Nearly perfect two-layer clay lattices are found in kaolinites. The main difference between the different species is in the unit-layer stacking geometry. Members of this group include kaolinite, dickite, nacrite and

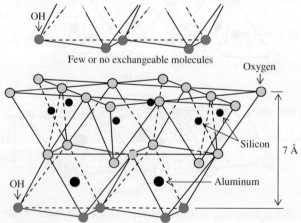

Figure 3.21. Structure of a kaolinite layer (based on Grim [1968]).

halloysite. Halloysite additionally contains interlayer water in two common forms (Table 3.8), and unlike kaolinite swells in water. Halloysite is apparently formed from kaolinite by weathering which involves rolling up of the kaolinite hexagonal plate (Fig. 3.22). When heated halloysite is irreversibly dehydrated to form metahalloysite.

The low activity of kaolinite means that it is far less cohesive than montmorillonite and many natural muds. Pure kaolinite has a peculiar flocculation behavior (Chapter 4). At the same time, because of its wide industrial use, it is abundantly available, which is a reason why it has been commonly used in laboratory experiments meant to investigate cohesive sediment transport. As the larger size fractions in kaolinite are barely cohesive, significant segregation of flocculated and non-flocculated particles tends to occur when kaolinite particles are subjected to turbulent shear flow. This behavior is not entirely absent even in other clays. As a result, attempts to explain the transport behavior of particles in suspension by assuming that the sediment can be effectively described by a single representative (floc) size has led to erroneous inferences about fine-sediment erosion and deposition processes (Chapters 7 and 9).

3.3.3.3 *Smectites*

As mentioned, montmorillonite is the most common clay mineral within the smectite group. For example, it is the main mineral in a bentonite

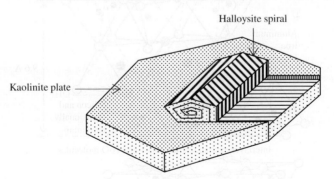

Figure 3.22. Schematic rendering of the formation of halloysite from kaolinite (adapted from Moore and Reynolds [1997], based on Robertson and Eggleton [1991]).

clay used as drilling mud. Hectorite and vermiculite are two other members of the group.

The basic unit of smectites consists of two tetrahedral layers with their tips pointing towards each other, and an octahedral sheet in between. The oxygens of the tetrahedra are shared with the oxygens of the octahedral layer in such a way that the three layers form a 9.6 Å thick macromolecular unit (Fig. 3.23). Part of the silicon (Si^{++++}) of the tetrahedral layer is typically replaced by aluminum (Al^{+++}). There can be replacement of Al^{+++} by magnesium (Mg^{++}), without complete filling of the third vacant octahedral position. Al^{+++} may also be replaced by iron, chromium, zinc, lithium or other atoms. The small size of these atoms allows them to take the place of the small silicon and aluminum atoms by isomorphous substitution. Since Si^{++++} is also subject to isomorphous substitution by, for example, iron (Fe^{+++}), the unit acquires a net negative charge, which in turn causes the electrically unbalanced unit to attract exchangeable cations between units. The resulting dipole bond only weakly holds the negative units together. Specifically, the nearest oxygens of the tetrahedral bases are linked by weak secondary bonds. Due to weak bonding, water molecules can get lodged between the lamellas in addition to the exchangeable cations, which cause this active clay to swell when wet and shrink when dry. The clay forms undulating

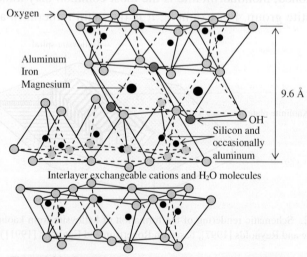

Figure 3.23. Structure of smectite (based on Grim [1968]).

mosaic sheets, which upon any disturbance break into flake-like fragments (Fig. 3.6). The high activity of this clay is manifested as a large cation exchange capacity in the range of 80 to 150 mEq $100g^{-1}$ (Table 3.8).

3.3.3.4 *Illites*

Illites are formed from mica minerals muscovite and phlogopite (magnesium mica). The structural unit of illite is similar to that of smectite, but the negative charge deficiency is greater and is almost always compensated by a potassium ion. However, when water is added this flake-shaped clay does not swell. The cation exchange capacity is in the range of 10 to 40 mEq $100g^{-1}$ (Table 3.8).

3.3.3.5 *Chlorites*

Chlorites are structurally related to the three-layer minerals. The charge-compensating cations between smectite-type unit layers are replaced by a layer of octahedral magnesium hydroxide (brucite). The cation exchange capacity is in the range of 10 to 40 mEq $100g^{-1}$ (Table 3.8).

3.3.3.6 *Palygorskite–sepiolite*

Palygorskite (attapulgite) and sepiolite are related clays formed by a different mode of superposition of the tetrahedral and the octahedral units than in smectites, illites or chlorites. The unit cell has very narrow channels in which water molecules and exchangeable cations are located. These channels are formed by the inversion of alternate pairs of silica tetrahedra. The cation exchange capacity is in the range of 3 to 15 mEq $100g^{-1}$ (Table 3.8). Both clays occur as long lath-shaped units and bundles (Figs. 3.7 and 3.8).

3.3.4 *Differential thermal analysis*

In addition to XRD, a useful tool for the identification of clay minerals is differential thermal analysis (DTA). This method is based on water loss and phase changes that take place in the clay as it is heated. The temperatures at which such changes occur are characteristic of the clay structure. Figure 3.24 shows the DTA plot of a calcium montmorillonite. The temperature difference ΔT between the clay sample and an inert powder (which does not change its properties) is plotted against the temperature of the inert sample. In the region of 100°C and 200°C a negative (endothermic) peak occurs as heat is required for desorption of the interlayer water (Fig. 3.24). The loss of the hydroxyls manifests as another endothermic peak at about 600°C. Finally, at about 900°C an exothermic peak occurs due to a phase change in the crystal lattice. The small endothermic peak just before that marks the loss of the last traces of hydroxyl water.

3.3.5 *Electric charge on clay particles*

Sol particles, *i.e.,* non-settling colloidal particles in a fluid, are hydrophobic, *i.e.,* they do not dissolve in water, and unlike hydrophilic substances, do not form true solutions. However, like an ionic (hydrophilic) solution, hydrophobic sol is electrically neutral. When an electric field is applied to a sol, the charged particles move towards one of the electrodes. This transport is called electrophoresis. Particles in clay sols as well as those containing silica and quartz are negatively charged, so they move towards the positive electrode (anode). Sols of metal

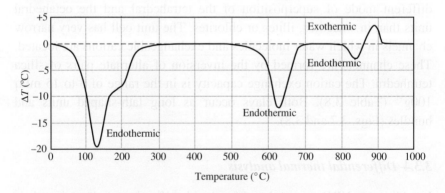

Figure 3.24. DTA plot for a Ca-montmorillonite (from van Olphen [1977]).

hydroxide such as ferric hydroxide have positively charged particles which move towards the negative electrode (cathode).

The negative charge on clay surfaces is compensated by positive ions in the fluid, as these ions are drawn toward the particle surface (Fig. 3.25). The envelope of ions is called an electric double layer, and the electrically neutral combination of particle and double layer is called a clay micelle. As the number of positive and negative ions are equal in the solution the sol remains neutral overall, even as the positive ions redistribute themselves to cancel out the negatively charged clay particles.

There are two causes of the negative charge on particle surfaces: (1) imperfections within the interior of the clay crystal lattice, which cause a net negative charge, and (2) preferential adsorption of certain specific ions, called peptizing ions, on the particle surface. The peptizing ions make up the inner coating of the double layer. Clay surfaces have active sites for adsorption, because the valences of the lattice atoms are not entirely compensated at the surface, even though they are compensated in the interior (Fig. 3.26). The surface charge deficit manifests as the universal van der Waals electrostatic force of attraction. As we will see in Chapter 4, this force is the main cause of electrochemical cohesion.

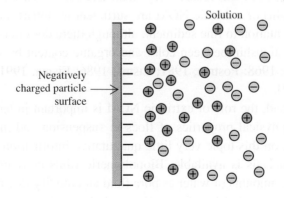

Figure 3.25. Ions in solution adjacent to a negatively charged clay surface.

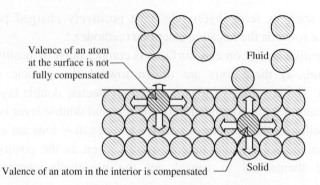

Figure 3.26. Schematic diagram highlighting uncompensated atomic valence at the particle surface.

3.3.6 *Organic matter*

Organic matter can influence the charge distribution on clay particles and may even overwhelm electrochemical attraction. As mentioned, natural organic matter (NOM) is either extrinsic or intrinsic to the clay mineral. Its extrinsic presence (Figs. 3.27a, b) is in the form of discrete particles of wood, leafy matter, spores, biopolymers, diatoms, diatomic frustules (hard shells), worm tubes, fecal pellets, *etc.* [Håkanson and Jansson, 1983]. Intrinsically the mineral-associated matter is either in the form of adsorbed molecules and/or fossil remains within the matrix of particles weathered from sediment rocks. Very small fractions (*e.g.*, on the order of 1% by weight) of extrinsic NOM are sufficient to impart a dark-grey or black pigmentation to fine sediment, although there does not appear to be a direct relationship between color and organic content by weight or volume [Grim, 1968; Postma, 1981; Eisma, 1986; Eisma, 1991; Hedges and Keil, 1999].

As mentioned, the role of extrinsic NOM is important in terms of its effect on the physical properties of flocs in suspension and in the bed. Unfortunately, on this topic very little quantitative information useful in engineering analysis is available. Biopolymeric films or filaments can sequester large amounts of water as pore fluid and modify floc properties through adhesive bridge formations. Filaments of mucus, or mucopolysaccharide formed by bacteria, can coat flocs and reinforce the

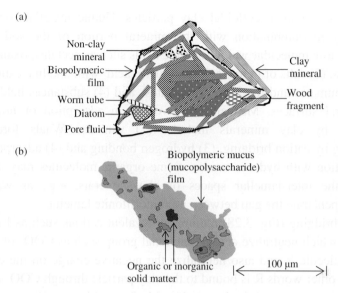

Figure 3.27. (a) Schematic rendering of a floc containing extrinsic organic matter; (b) structure of a typical organic floc from Lake Apopka in central Florida (from Mehta *et al.* [2009]).

electrochemical bonds holding particles together. In many of Florida's lakes suspended flocs contain minor amounts of inorganic substances. The solids are held together by embedment within a spacious mucus film (Fig. 3.27b). Such flocs do not conform to the fractal description commonly used to treat inorganic flocs because there is no agglomerate formation that can be identified as the primary building block (Chapter 4).

Flocculation by zooplankton filtering can be significant compared to inorganic processes alone. Kranck and Milligan [1980] reported that a mixture of 50% organic and 50% inorganic sediments settled an order of magnitude faster than an equivalent concentration of 100% inorganic sediment. In areas of high bio-deposition, large suspended aggregates that are mainly organic have been recorded *in situ* using focused-beam laser devices [McCave, 1984; Kranck, 1986; Wells and Goldberg, 1993; Law and Bale, 1998].

Humic substances, the end product of decayed matter, usually contain large quantities of trace elements. In soil, they are not found in a free state but are bound to colloidal clay particles. Humic materials occur intrinsically in combination with the mineral portion of the soil as: (1) salts of low molecular weight organic acids such as acetates, oxalates and lactates, (2) salts of humic acid[j] and fulvic acid[k] with alkaline cations including humates and fulvates, (3) chelates[l], and (4) substances held on clay mineral surfaces. Mechanisms involved in adsorption of humic substances by clay minerals include: (1) van der Waals forces, (2) bonding by cation bridging, (3) hydrogen bonding and (4) adsorption by association with hydrous oxides. Some organic molecules may also penetrate the inter-lamellar spaces of clay minerals, *e.g.*, as water molecules penetrate the gap between montmorillonite lamella.

Cation bridging (Fig. 3.28) requires polyvalent cations such as Fe^{+++} and Al^{+++}, which neutralize acidic functional group such as COO^- of the organic molecule R and also neutralize the negative charge on the clay surface. In other words R is bound to the clay particle through COO^- and the polyvalent cation. A hydrogen bond is formed by the H^+ ion and

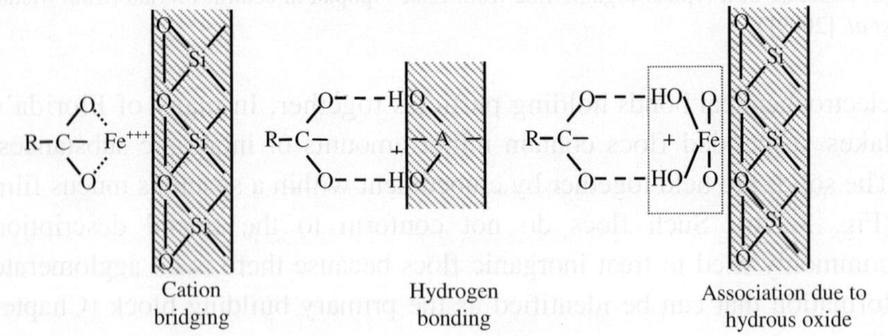

Figure 3.28. Three modes by which organic molecules of humic substances are attached to clay particles (adapted from Fotyma and Mercik [1992]).

[j] An organic substance in soil, peat and coal formed from the decomposition of vegetable matter (humus). It is responsible for imparting much of the color of surface water.

[k] A low molecular-weight organic substance of the same family as humic acid derived from humus, and often found in surface water.

[l] These are substances in which organic and metallic atoms are combined. Chelates possess an aromatic ring structure in which a metal ion is attached to two non-metal ions by covalent bonds, *i.e.*, bonds formed by sharing electrons.

two oxygen atoms; one from the clay silicon tetrahedron and the other from the COO^- group. Thus for bonding between an organic molecule and a clay particle two hydrogen bonds are required per COO^- group. In another instance shown, a hydrous oxide coats the clay particle. As a result the surface properties of the particle are determined by the hydrous oxide. The COO^- group remains adsorbed on the particle surface by van der Waals attraction.

Some measures of the influence of NOM on the erodibility of fine-grained sediment have been investigated. Examples include the effects of Chlorophyll-*a*, biopolymeric films and colloidal carbohydrate on the erosion shear strength of cohesive beds. Brief overviews of the interaction between physical and biological parameters that affect erodibility are found elsewhere [*e.g.,* Dade and Nowell, 1991; Black, 1991; Black *et al.,* 2002; Montague *et al.,* 1993; Paterson, 1997; Sutherland *et al.,* 1998; Amos *et al.,* 1998]. Unfortunately, far less information is available on the effect of NOM on other transport processes such as deposition and consolidation. In any event, these overviews reveal two significant features of the role of NOM as follows.

Firstly, as mentioned, due to the nearly ubiquitous presence and abundance of NOM, marine flocs are almost always coated partly or wholly with organic matter, and their response to flow is only partially governed by electrochemical cohesion. It is conceivable that at least in some cases results from studies of natural mud have been explained in terms of the effect of cohesion, when in fact the role of biopolymeric films may have been more important or even paramount. For instance, studies on the San Francisco Bay mud indicate it to be more cohesive than several other marine sediments [Krone, 1963]. As the mineral content of Bay mud does not reveal uncommonly high fractions of cohesive clays compared to similar sediments, one is led to infer that the role of one or more biochemical constituents as binding agents is likely to be important. Following the same argument we may note that some marine fine sediments appear to have the ability to form unusually large floc structures, sometimes called marine snow. Inter-floc binding in marine snow has been shown to be dependent on NOM, as the abiotic constituents by themselves do not replicate the flake-like open structures [Wolanski, 2007].

Secondly, due to variability in NOM properties, site-specific correlative relationships such as between sediment erodibility and NOM may not be suitable for wider applications. Therefore it becomes essential to test several samples from a site to characterize the role of NOM. Interestingly enough, similarities in the dependence of the bed density on organic content, and the dependence of bed shear strength on density, have been reported for sediments from several locations in peninsular Florida [Gowland, *et al.,* 2007 and Chapter 9]. These data suggest that, depending on the sources of bed material, it may be feasible to develop at least some workable regional-scale relationships for fine sediment transport with regard to the effect of NOM.

3.4 Classification by Plasticity

Mixtures of fine sediment particles and water can exist in any of four states depending on the water content *W* defined as

$$W = \frac{M_w}{M_s} \tag{3.32}$$

where M_w is the mass of water and M_s is the mass of solids in a sample. Upon the addition of water to a dry clayey powder the mixture proceeds through the solid, semi-solid, plastic, and liquid states (Fig. 3.29). The water content at the boundaries (or the Atterberg limits) between adjacent states are referred to as shrinkage limit, W_{sh} (solid/semi-solid

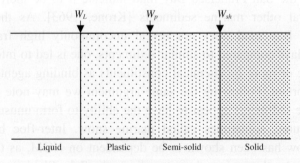

Figure 3.29. Diagrammatic representation of Atterberg limits.

boundary), plastic limit, W_P (semi-solid/plastic boundary) and liquid limit, W_L (plastic/liquid boundary).

The liquid limit is determined by two methods. One uses the Casagrande device, which consists of a dish filled with the sediment sample of known water content. A groove of a certain width is made in the sample, and by turning a handle (at 2 rps) the dish is made to bump cyclically on a hard base. The liquid limit is the water content corresponding to the groove collapsing at 25 bumps. The second device is the cone penetrometer, in which the sample is placed in a metal cup. A steel cone is then dropped onto the sample from a certain height. The liquid limit is defined when the cone penetrates a certain distance into the sample.

The plastic limit is determined by measuring the water content when 3.2 mm diameter threads of the sediment begin to crumble as they are rolled. The shrinkage limit is the water content at which further loss of moisture will not result in additional volume reduction. This limit is determined as the water content after just enough water is added to fill all the voids of a dry pat of sediment [Bowles, 1991]. Shrinkage limit is not as commonly used as the liquid limit and the plastic limit, and has not found noteworthy use in fine sediment transport characterization.

As mentioned further in Chapter 9, the Plasticity Index *PI* is at times used to characterize the stability of beds against mass erosion

$$PI = W_L - W_P \qquad (3.33)$$

The Atterberg limits and *PI* values of four clay minerals and natural clayey sediments are given in Tables 3.14 and 3.15, respectively. Their utility as indicators of erosion has also been advocated for the design of stable channels lined with cohesive soils [*e.g.,* Smerdon and Beasley, 1959; Task Committee on Erosion of Cohesive Materials, 1968; Winterwerp *et al.*, 2012]. The approach, which is briefly reviewed in Chapter 9, is analogous to one for coarse beds with the difference that the critical shear stress (or tractive stress) for erosion is correlated not with fine particle size but with plasticity related indices. As noted the most commonly used index is *PI*.

Table 3.14. Atterberg limits of clay minerals.

Mineral	Exchangeable Ion	W_L (%)	W_P (%)	PI (%)	W_{sh} (%)
Montmorillonite	Na	710	54	656	9.9
	K	660	98	562	9.3
	Ca	510	81	429	10.5
	Mg	410	60	350	14.7
	Fe	290	75	215	10.3
	Fe	140	73	67	-
Illite	Na	120	53	67	15.4
	K	120	60	60	17.5
	Ca	100	45	55	16.8
	Mg	95	46	49	14.7
	Fe	110	49	61	15.3
	Fe	79	46	33	-
Kaolinite	Na	53	32	21	26.8
	K	49	29	20	-
	Ca	38	27	11	24.5
	Mg	54	31	23	28.7
	Fe	59	37	22	29.2
	Fe	56	35	21	-
Attapulgite	H	270	150	120	7.6

Source: Lambe and Whitman [1979].

Table 3.15. Liquid limit, plastic limit and Plasticity Index of natural clayey sediments.

Sediment	Particle size (μm)	Liquid limit (%)	Plastic limit (%)	Plasticity Index (%)
Lake Charles K319	1.9	56.4	22.0	34.4
Lufkin K116	8.4	49.4	15.9	33.5
Houston K177	3.3	43.7	20.5	23.2
Houston K177A	3.3	44.7	17.7	27.0
Houston K177B	3.3	48.7	18.0	30.7
San Saba	2.0	47.7	22.0	25.7
Taylor Marl	4.8	47	21	26

Sources: Espey [1963], Rektorik and Smerdon [1964], Task Committee [1968].

3.5 Exercises

3.1 Grab sampling of bottom sediment in the Loxahatchee River in Florida yielded data from 54 sites given in Table E3.1. For each site, percent clay is equal to 100 minus percent sand plus percent silt. Plot these data on a ternary diagram (Fig. E3.1) and comment on the observed trend.

3.2 Derive Eq. 3.11 from Eq. 3.10.

Table E3.1. Sand and silt in grab samples from the Loxahatchee River, Florida.

Sample No.	Sand (%)	Silt (%)	Sample No.	Sand (%)	Silt (%)
1	32.2	49.3	28	56.2	39.7
2	25.6	61.6	29	----	----
3	27.1	63.9	30	22.3	71.0
4	15.4	72.9	31	71.5	26.5
5	78.5	19.8	32	29.9	64.8
6	69.8	28.0	33	----	----
7	69.6	27.7	34	62.9	33.8
8	60.0	35.1	35	75.5	22.8
9	46.9	46.4	36	74.7	23.5
10	21.4	71.3	37	70.7	27.3
11	17.3	74.3	38	38.1	55.1
12	14.8	76.8	39	68.8	29.1
13	18.8	73.5	40	40.9	54.0
14	15.7	76.2	41	57.6	38.2
15	16.3	75.6	42	73.9	23.8
16	----	----	43	67.0	30.6
17	19.9	73.7	44	54.2	42.2
18	28.7	64.9	45	----	----
19	23.6	69.4	46	27.2	30.6
20	23.9	69.2	47	72.3	25.2
21	58.9	37.8	48	71.0	26.7
22	39.5	55.2	49	69.6	28.4
23	18.4	73.6	50	67.2	30.0
24	20.5	73.1	51	47.6	44.6
25	36.5	57.4	52	40.0	55.9
26	18.1	74.5	53	----	----
27	27.4	65.7	54	52.0	42.0

Source: Jaeger and Hart [2001].

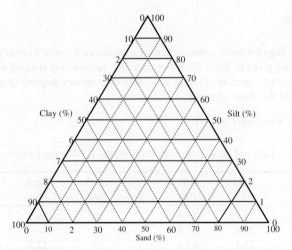

Figure E3.1. Ternary diagram for percentages of sand, silt and clay (from Flemming [2000]).

3.3   The data in Table E3.2 were obtained from an analysis of a high-resolution video imaging of falling particles in South San Francisco Bay. Plot the volume concentration $C_{vi} = V_{pi} \Delta N_i / \Delta(\log d_{pi})$ in ppm against d_{pi}. In the table $\Delta(\log d_{pi})$ $= \log (d_{pi+1} / d_{pi}) = 0.075$ Note that floc density is not required to obtain the plot. What type of distribution is obtained? What are the characteristic particle sizes?

3.4   The molecular formula of a trioctahedral smectite known as saponite is

$$(Si_{6.67}Al_{1.33})(Mg_{5.33}Al_{0.67})O_{20}(OH)_4Na_{0.67}$$

Similarly, hectorite belonging to the same subgroup is

$$Si_8(Mg_{5.33}Li_{0.67})O_{20}(OH)_4Na_{0.67}$$

Calculate the molar weights of these clays. Atomic weights are given in Table A.1 of Appendix A.

3.5   The XRD signature of the clay fraction of a sediment sample is shown in Fig. E3.2. Identify its mineral composition. Make use of d_M spacing values in Table C.2 of Appendix C. Note: You need not consider the relative peak intensity values.

3.6   The XRD signatures in Fig. E3.3 are from locations in the Loxahatchee River estuary in southern Florida. All six signatures are similar and imply that the sediment is homogeneous in the sampled region. Three clays are present and quartz is prominent as well. Identify these minerals. Can you detect two additional non-clay minerals? Note: You need not consider the relative peak intensities.

Table E3.2. Floc size data from San Francisco Bay.

Floc diameter, d_{pi} (μm)	No. of flocs per unit vol., ΔN_i (# m^{-3})	Floc density, $\rho_{fi}{}^{b}$ (kg m^{-3})
32.8	0^{a}	-
39.0	0	-
46.4	7754551	1092.4
55.1	27815237	1179.3
65.5	32703976	1184.5
77.9	32872553	1135.2
92.6	27478083	1098.0
110.1	18037760	1076.2
130.8	13823330	1082.9
155.5	10114632	1060.4
184.8	5563047	1082.2
219.7	1854349	1052.9
261.1	1348618	1050.5
310.4	505732	1062.9
369.0	168577	1038.9
438.6	0	-
521.3	0	-

^{a} 0 ≡ No particles of representative size of 32.8 μm.
^{b} Water density was 1,025 kg m^{-3}.
Source: Letter [2009].

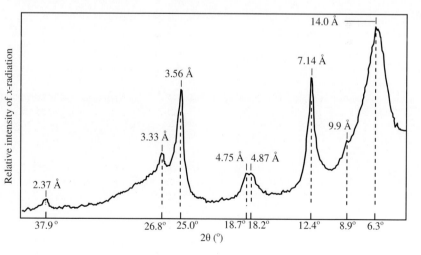

Figure E3.2. XRD signature of a sediment sample (from Griffin [1971]).

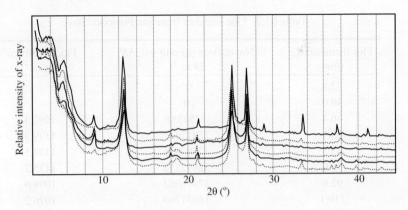

Figure E3.3. XRD signatures of six sediment samples from the Loxahatchee River estuary in Florida (from Jaeger *et al.* [2009]).

3.7 Using data derived from submerged soil samples in Table E3.3, determine if a correlation exists between the finer than 2 μm fraction (percent), the liquid limit, the plastic limit or the Plasticity Index. Briefly comment on your results.

Table E3.3. Physical properties of soils from waters within U.S. Army Corps of Engineers Districts.

U.S. Army Corps of Engineers District	Liquid limit	Plastic limit	Finer than 2 µm (%)	Specific gravity of solids	Organic content (%)
Albuquerque	31.2	23.1	13	2.60	5.1
Albuquerque	30.0	21.0	22	2.66	7.0
Albuquerque	22.1	22.1	13	2.67	7.8
Fort Worth	41.1	14.5	40	2.67	11.0
Fort Worth	29.8	15.5	25	2.61	10.4
Galveston	75.0	25.3	53	2.63	7.1
Kansas City	31.5	20.4	33	2.62	4.7
Kansas City	36.8	20.3	30	2.67	5.0
Little Rock	24.0	22.1	17	2.69	4.0
Little Rock	20.3	17.1	17	2.66	3.3
Los Angeles	42.8	19.1	30	2.69	5.2
Los Angeles	24.0	13.0	25	2.67	4.2
Memphis	56.0	19.9	42	2.70	7.5
Memphis	33.8	17.7	36	2.66	5.6
Mobile	42.0	23.2	37	2.65	5.8
New Orleans	53.1	21.8	46	2.64	5.5
New Orleans	55.5	21.4	25	2.63	4.0
Omaha	54.0	28.7	42	2.68	8.6
Omaha	54.0	25.0	40	2.62	11.6
Philadelphia	25.6	25.6	12	2.72	5.1
Sacramento	29.2	20.4	14	2.73	5.5
Sacramento	29.2	20.6	12	2.71	5.7
St. Louis	27.8	21.0	18	2.71	3.0
St. Louis	44.9	23.6	34	2.65	9.1
San Francisco	24.5	24.5	5	2.71	4.9
San Francisco	27.0	24.0	6	2.70	5.0
Savannah	38.3	27.8	29	2.67	5.1
Savannah	38.3	23.0	28	2.65	6.4
Tulsa	32.4	18.1	29	2.66	5.7
Tulsa	33.0	17.5	30	2.68	6.2

Source: Arulanandan *et al.* [1980].

Chapter 4

Flocculation and Floc Properties

4.1 Chapter Overview

This chapter begins with a brief survey of the electrochemical theory of flocculation, which includes the equilibrium response of clay particles to ionic charge in the fluid. Modes of flocculation are reviewed and the concept of critical salinity is introduced. Floc properties include size, density and shear strength, and mechanisms for inter-particle collisions responsible for floc formation, growth and breakup. Floc geometry is conveniently described in terms of fractal self-similarity which, while an approximation, is exceptionally useful in harmonizing the transport equations for clay mineral sediments. As a continuation of particle size distribution analysis introduced in Chapter 3, descriptions of floc size dynamics, equilibrium floc size and size spectra are summarized.

4.2 Nomenclature

We will consider the term "flocculation" to be synonymous with the coagulation of primary clay particles by attractive electrochemical forces. The resulting sedimentary units are flocs or aggregates. The term "floc" will be generally used, and "aggregate" will be used mainly in relation to flocs grouped into fractal "orders" depending on their strength, void ratio and density, and agglomerates containing large fractions of natural organic matter (NOM). Floc formation by electrochemical effects decreases in significance as the NOM fraction increases and the role of biochemical bonding becomes important. A dilemma for the modeler is that while the role of NOM in affecting flocculation cannot be ignored, no robust theories exist to enable the quantification of this role. As a result, biochemical effects are accounted for, to the extent they can be, by empirical approaches.

Transport processes dependent on floc properties include erosion and resuspension, turbulent diffusion and mixing, settling and deposition, deformation (such as the flattening of a floc upon impact with the bottom followed by rebound) and consolidation (Fig. 4.1). The properties of flocs in transport depend on the mechanisms of particle aggregation and disaggregation. The term aggregation has often been used in the literature to imply the processes of growth *and* disaggregation, *i.e.*, breakup, of flocs. We will refer to floc growth and breakup collectively as *aggregation processes*. The term aggregation will generally mean growth only and disaggregation only breakup (Table 4.1). Growth and breakup require particles to collide; growth occurs when the particles adhere, and breakup is due to shearing.

4.3 Cohesion and Flocculation

4.3.1 *Electric double layer*

Clays are special because they are the main relevant minerals whose particle surfaces exhibit cohesion — contingent upon the mineral and the

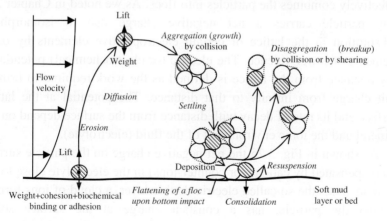

Figure 4.1. Cohesive sediment transport components dependent on aggregation processes.

Table 4.1. Definitions relevant to aggregation processes.

Term	Meaning
Particle	Unless qualified, this term in general can refer to a sedimentary unit, a sol particle or even a floc — depending on the frame of reference.
Grain	Coarse particle such as sand.
Floc	Agglomeration of a large number of primary clay mineral particles cohering by attractive electrochemical forces or biochemical bonding or binding.
Aggregate	Agglomeration of a large number of particles of organic origin along with some inorganic (*e.g.*, clay) mineral by biochemical (and electrochemical) bonding or adhesion. At times also a floc made up of smaller flocs, including a floc in reference to the fractal order of aggregation.
Flocculation	Coagulation of primary clay mineral particles by attractive electrochemical forces.
Aggregation	Growth of flocs by inter-particle collisions.
Aggregation processes	Processes causing growth or breakup of flocs.
Disaggregation	Processes causing flocs to break up.

surrounding fluid. It is essential to know how the particles and the surrounding fluid interact, since this interaction, in terms of the net effect of the forces of attraction and repulsion, determines whether cohesion effectively combines the particles into flocs. As we noted in Chapter 3, a clay particle carries a net negative charge due to isomorphous substitution in the lattice of certain electropositive elements by other elements of lower valence. The electric (or electrochemical) potential at any distance from the lattice is defined as the work required to bring a unit charge from infinity to that distance. The potential at the lattice surface and its rate of decay with distance from the surface depend on the mineral and the ionic environment of the fluid (electrolyte).

As shown in Fig. 3.25 the net negative charge on the particle surface is compensated by cations (or counter-ions) in the electrolyte close to the particle. Thus the so-called electric double layer, a cloud of ions formed around the particle, has a constant charge at the surface solely determined by the type and extent of isomorphous substitution. In other words, the negative charge density at the surface is independent of the counter-ions.

As we noted in reference to Fig. 3.25, the double layer cloud also contains negative (anionic) charges such that, overall, the clay micelle is electrically neutral. Within the double layer the positive (cationic) and negative charges are not distributed evenly as the positive charge density is higher next to the particle surface. These charge-compensating counter-ions also have a tendency to diffuse away from the particle as their concentration decreases with distance from the surface, even as they are attracted towards it.

The condition of equilibrium between the two opposing ionic fluxes determines the thickness of the double layer. In Fig. 4.2a the counter-ion concentration in the electrolyte is low, and as a result the concentration gradient outward from the surface is weak, with the double layer extending well into the solution. In Fig. 4.2b the double layer is compressed because counter-ion concentration is high and the concentration gradient is steep.

The decay rate of the electric potential within the double layer has a major role in governing the cohesive behavior of the particle. Before we state the decay rate, it is useful to note that the double layer is separated from the particle surface by a thin layer of sorbed water which has

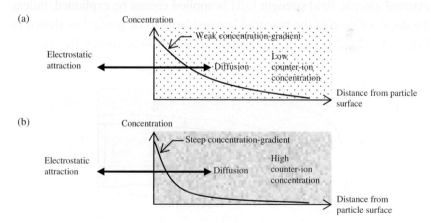

Figure 4.2. Opposing ionic fluxes (due to electrostatic attraction to the left and ionic diffusion to the right) determine the equilibrium thickness of the ionic double layer: (a) low counter-ion concentration; (b) high counter-ion concentration.

molecules organized as a pseudo-solid matrix. As an illustration, a 5 Å thick layer surrounding a kaolinite particle is sketched in Fig. 4.3. The water content in this layer is equal to (specific surface area) × (matrix layer thickness) × (density of water). For a typical kaolinite the pseudo-solid water content is $(10^4 \, \text{m}^2 \, \text{kg}^{-1}) \times (5 \times 10^{-10} \, \text{m}) \times (10^3 \, \text{kg} \, \text{m}^{-3})$ = 5×10^{-5} or 0.5%. For an illite the value increases to 5% and for a smectite to 50%. The pseudo-solid layer influences the electric potential and therefore cohesion. It also contributes to the lubricating effect of clay particles and thus the viscosity of the suspension [Lambe and Whitman, 1979].

The simplest description of decay of the electric potential Φ_x with distance x from particle surface is the Helmholtz–Perrin (H–P) model (Fig. 4.4a). According to it the decay is linear from $\Phi_x = \Phi_0$, at $x = 0$ to $\Phi_x = 0$, at $x = \delta_d$, the effective thickness of the double layer. When the particle is in motion, the fluid within thickness δ_d adheres to the particle. At distance δ_d a sheared surface called the slipping plane occurs. The double layer is a parallel-plate condenser in this model, with one plate at the particle surface and the other at the slipping plane.

Gouy [1910] and Chapman [1913] independently showed that the H–P model is an over-simplification of the character of the double layer. Due to this, the transport rate of the particle in a solution to which external electric field strength (*efs*) is applied cannot be explained, unless the decay or drop in the potential is considered to be gradual as shown in Fig. 4.4b. In this case, the potential which determines the rate of

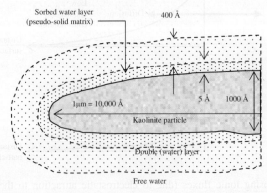

Figure 4.3. Schematic drawing of the double layer of a kaolinite particle (modified from Migniot [1968]).

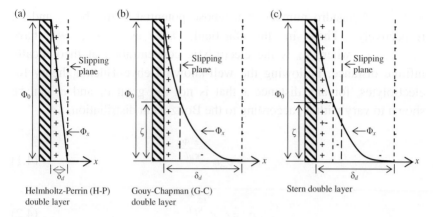

Figure 4.4. Three models of the double layer: (a) Helmholtz–Perrin; (b) Gouy–Chapman; (c) Gouy–Chapman–Stern.

movement of the particles under an *efs* is not Φ_0 but the ζ-potential (see Sec. 4.3.5) at the slipping plane. In the H–P model the ζ-potential is always zero, which is contrary to observations.

An assumption in the combined Gouy–Chapman (G–C) model is that the counter-ions are points theoretically occupying no volume, which would not explain why the properties of the double layer depend on counter-ion size. In fact, due to their finite volumes the counter-ions form a thin sorbed layer at the particle surface (which is also coated with pseudo-solid water as noted). To include the effect of sorption, Stern [1924] added a sorbed layer to the G–C double layer. This (Stern) layer in a sense amounts to adding an H–P double layer to the G–C double layer (Fig. 4.4c). The resulting model is sometimes known as the Gouy–Chapman–Stern model.

The rate of decay of the electrical potential in the double layer is the matter of main interest, as explained next.

4.3.2 Double layer thickness

To determine the variation of the electrical potential Φ_x as a function of distance x, and from it the thickness of the double layer, we will adopt the Gouy–Chapman model. Let the concentrations of positive and negative counter-ions at a point (in units of ions per cm^{-3} = normality

$\times 10^{-3} \times$ Avogadro number N_A), whose potential is Φ_x, be c_+ and c_-, respectively. Then in the far-field, *i.e.*, as $x \rightarrow \infty$, we have $c_+ = c_- = c_\infty$, where c_∞ is the electrolytic concentration at theoretically infinite distance. Following the well-known Debye–Hückel theory for electrolytes, for any distance x that is not too great c_+ and c_- can be shown to vary with Φ_x according to the Boltzmann distribution, *i.e.*,

$$c_+ = c_\infty e^{-\frac{e_c z^+ \Phi_x}{\kappa_B T}} \tag{4.1}$$

$$c_- = c_\infty e^{\frac{e_c z^+ \Phi_x}{\kappa_B T}} \tag{4.2}$$

where e_c is the electronic charge (1 e_c = 1.602×10^{-19} Coulomb; see Appendix A, Table A.2), z^+ is the ionic valence, κ_B is the Boltzmann constant and T is the absolute temperature.[a] The product $\kappa_B T$ represents the thermal energy of the counter-ion [Voyutsky, 1978; Lyklema, 1995].

The volume density ρ_c of charge in the electrolytic solution is

$$\rho_c = e_c z^+ (c_+ - c_-) \tag{4.3}$$

Substituting for c_+ and c_- from Eqs. 4.1 and 4.2 into the above expression yields

$$\rho_c = e_c z^+ c_\infty \left(e^{-\frac{e_c z^+ \Phi_x}{\kappa_B T}} - e^{\frac{e_c z^+ \Phi_x}{\kappa_B T}} \right) \tag{4.4}$$

---

[a] The Boltzmann constant in the present context is associated with the outcome of the Maxwell–Boltzmann analysis which describes the distribution of material particles over various energy microstates in thermal equilibrium. Let i designate a microstate, n_i the number of particles with energy ε_i and n_T the total number of particles. According to the Boltzmann distribution

$$\frac{n_i}{n_T} = \frac{e^{-\frac{\varepsilon_i}{\kappa_B T}}}{Z_c} \tag{F4.1}$$

where Z_c is a constant when the absolute temperature T is constant. The Boltzmann constant κ_B and T are associated with the thermodynamic behavior of ions.

The potential Φ_x is related to ρ_c by the Poisson equation in electrostatics

$$\nabla^2 \Phi_x = -\frac{4\pi\rho_c}{\varepsilon_c} \tag{4.5}$$

where ε_c is the dielectric constant of the electrolyte.[b] Since we are interested in the variation of the potential with x only, the above equation reduces to

$$\frac{d^2\Phi_x}{dx^2} = -\frac{4\pi\rho_c}{\varepsilon_c} \tag{4.6}$$

Then, combining this equation with Eq. 4.4 we obtain

$$\frac{d^2\Phi_x}{dx^2} = -\frac{4\pi}{\varepsilon_c} e_c z^+ c_\infty \left(e^{-\frac{e_c z^+ \Phi_x}{\kappa_B T}} - e^{\frac{e_c z^+ \Phi_x}{\kappa_B T}} \right) \tag{4.7}$$

For the present problem this equation must satisfy the boundary conditions

$$x=0: \quad \Phi_x = \Phi_0; \quad \frac{d\Phi_x}{dx} = -\frac{4\pi\sigma_c}{\varepsilon_c} \tag{4.8}$$

$$x \to \infty: \quad \Phi_x = 0; \quad \frac{d\Phi_x}{dx} = 0; \quad \rho_c = 0 \tag{4.9}$$

where σ_c is the surface charge density. Since the double layer as a whole is electrically neutral, σ_c must be equal to the total volume charge in the solution, *i.e.*,

$$\sigma_c = -\int_0^\infty \rho_c dx \tag{4.10}$$

---

[b] The dielectric constant of a material ε_c is equal to $\varepsilon_s/\varepsilon_0$, where ε_s is the static permittivity and ε_0 is the permittivity in vacuum. Permittivity characterizes a material's ability to transmit an electric field. Therefore, the dielectric constant is also called relative permittivity. The dielectric constant in a vacuum is unity, and its value for air is nearly unity.

Integration of Eq. 4.7 by taking into account Eq. 4.9, *i.e.*, the condition that the volume charge density ρ_c is zero in the far-field, one obtains

$$\frac{d\Phi_x}{dx} = -\sqrt{\frac{8\pi\kappa_B T c_\infty}{\varepsilon_c}} \left(e^{\frac{e_c z^+ \Phi_x}{2\kappa_B T}} - e^{-\frac{e_c z^+ \Phi_x}{2\kappa_B T}} \right) \tag{4.11}$$

For small values of Φ_x we may introduce the following approximation

$$e^{\frac{e_c z^+ \Phi_x}{2\kappa_B T}} - e^{-\frac{e_c z^+ \Phi_x}{2\kappa_B T}} \approx \frac{e_c z^+ \Phi_x}{\kappa_B T} \tag{4.12}$$

Thus

$$\frac{d\Phi_x}{dx} = -\frac{e_c z^+}{\kappa_B T} \sqrt{\frac{8\pi\kappa_B T c_\infty}{\varepsilon_c}} \Phi_x \tag{4.13}$$

Integration of this equation using the condition (Eq. 4.8) that at $x = 0$, $\Phi_x = \Phi_0$ gives

$$\Phi_x = \Phi_0 e^{-\kappa_D x} \tag{4.14}$$

where

$$\kappa_D = \frac{e_c z^+}{\kappa_B T} \sqrt{\frac{8\pi\kappa_B T c_\infty}{\varepsilon_c}} = \sqrt{\frac{8\pi e_c^2 z^{+2} c_\infty}{\kappa_B T \varepsilon_c}} \tag{4.15}$$

which has the unit of inverse distance. Thus the electrical potential decays exponentially with distance from the particle. The quantity κ_D^{-1}, known as the Debye–Hückel parameter, is equal to the distance x at which $\Phi_x = \Phi_0/e = 0.368\Phi_0$.

To obtain the surface charge, substitution of Eq. 4.6 into Eq. 4.10 gives

$$\sigma_c = \int_0^\infty \frac{\varepsilon_c}{4\pi} \frac{d^2\Phi_x}{dx^2} dx \tag{4.16}$$

Therefore,

$$\sigma_c = -\frac{\varepsilon_c}{4\pi}\frac{d\Phi_x}{dx}\bigg|_{x=0} \tag{4.17}$$

Substituting for $d\Phi_x/dx$ from Eq. 4.13 yields

$$\sigma_c = \frac{e_c z^+ \Phi_0}{\kappa_B T}\sqrt{\frac{\varepsilon_c \kappa_B T c_\infty}{2\pi}} = \frac{\varepsilon_c \kappa_D}{4\pi}\Phi_0 \tag{4.18}$$

Thus the surface charge σ_c is proportional to the potential Φ_0, and $\kappa_D^{-1} \approx \delta_d$ is the distance of separation of the Gouy–Chapman double layer parallel-plate condenser. In other words κ_D^{-1} is a measure of double layer thickness. Table 4.2 provides values of δ_d, taken to be equal to κ_D^{-1}, and Φ_0 for monovalent (*e.g.*, NaCl) and divalent (*e.g.*, MgSO$_4$) electrolytes. As the ionic concentration in the electrolyte increases, the thickness of the double layer decreases.

The Gouy–Chapman estimate of the double layer thickness is often adequate for dilute suspensions but not for concentrated ones, for which a Stern correction to the adsorption of the ions at the particle surface is required. In recent decades significant advances have been made in methods to calculate the double layer thickness [Lyklema, 1995; Raveendran and Amreethrajah, 1995].

Table 4.2. Gouy–Chapman double layer thickness and surface potential of electrolytes.

Concentration (N)[a]	Monovalent		Divalent	
	δ_d (cm)	Φ_0 (mv)	δ_d (cm)	Φ_0 (mv)
10^{-5}	10^{-5}	355	0.5×10^{-5}	178
10^{-3}	10^{-6}	240	0.5×10^{-6}	120
10^{-1}	10^{-7}	130	0.5×10^{-7}	65

[a] Concentration is expressed in normality unit. 1 N (Normal) = equivalent weight of the compound in grams. For example, the equivalent weight of NaCl is 23 (Na$^+$) + 35.5 (Cl$^-$) = 58.5. Therefore, in a 0.1 N NaCl (solution in water) there will be $0.1\times58.5 = 5.85$ grams of NaCl in 1 liter.
Source: van Olphen [1977].

4.3.3 *Double layer repulsion*

When two clay particles in suspension approach each other, their double layers begin to interact as the electrical potentials of the counter-ion clouds overlap. Their deformation changes the distribution of the counter-ions in such a way as to increase the free energy of the system.[c] Work must therefore be performed to bring about these changes, *i.e.*, effective repulsion must occur between the two clay micelles. The energy of repulsion is obtained as follows.

The distribution of the electric potential Φ_x between two double layers representing two clay micelles close to each other is shown in Fig. 4.5 according to the Gouy–Chapman model. The distance of separation between the two particles is 2Δ. The surface potential Φ_0 decreases midway between the particles to Φ_c.

The surface charge density σ_c given by Eq. 4.17 is expressed as

$$\sigma_c = -\frac{\varepsilon_c}{4\pi}\frac{d\Phi_x}{dx}\bigg|_{x=0} = -\frac{\varepsilon_c}{4\pi}\left(\frac{d\widehat{\Phi}}{d\hat{x}}\right)_0 \frac{\kappa_D \kappa_B T}{e_c z^+} \tag{4.19}$$

in which we have introduced the dimensionless electrical potential $\widehat{\Phi}$ and the dimensionless distance $\hat{x}$ defined as

---

[c] A meaningful definition of the so-called Gibbs free energy is in terms of a change in this energy. When the temperature and pressure are constant, the *decrease* in free energy is equal to the maximum work that can be done, *i.e.*, energy expended, by a process other than pressure-volume work. Thus an *increase* in free energy represents energy gained by the system. A simple example of change in free energy is during dilution of a solution, which increases the distance between solute molecules.

To calculate free energy change, *molality* must be defined. In general, three measures of the strength of a solution are noteworthy. Molality is the concentration of a solution expressed as the number of moles of a dissolved substance in a kilogram of solvent. *Molarity* is the number of moles of a solute dissolved *in a liter of solution*. Finally, *normality* = molarity $\times n$, where n is the number of protons exchanged in a reaction. See also footnote of Table 4.2.

Consider the free energy change when a 0.01 ($=c_{m1}$) molal aqueous solution of a solute at 25°C is transferred to a $c_{m2} = 0.001$ molal solution. The free energy change $\Delta F = RT \ln(c_{m2}/c_{m1}) = 8.314$ (J° K^{-1} mole^{-1}) $\times$ 298.15 (°K) $\times$ ln(0.001/0.01) $= -5707.7$ J mole^{-1} [Daniels and Alberty, 1955].

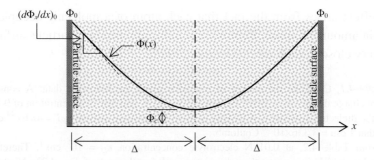

Figure 4.5. Distribution of electric potential between two interacting particles.

$$\widehat{\Phi} = \frac{e_c z^+}{\kappa_B T} \Phi_x; \qquad \widehat{x} = \kappa_D x \qquad (4.20)$$

In Eq. 4.19, $d\Phi_x/dx\big|_{x=0}$ is the potential gradient at the particle surface and $\left(d\widehat{\Phi}/d\widehat{x}\right)_0$ is its dimensionless value. Equation 4.19 can also be expressed as

$$\sigma_c = \sqrt{\frac{\varepsilon_c \kappa_B T c_\infty}{2\pi}} \sqrt{2\cosh\Omega_0 - 2\cosh\Omega_c} \qquad (4.21)$$

where

$$\Omega_0 = \frac{e_c z^+ \Phi_0}{\kappa_B T}; \qquad \Omega_c = \frac{e_c z^+ \Phi_c}{\kappa_B T} \qquad (4.22)$$

The energy of repulsion V_R is the work required to bring two particles from an infinite separation to a distance 2Δ. This energy is approximately given by

$$V_R = \frac{64\kappa_B T c_\infty}{\kappa_D} \left(\frac{e^{\Omega_0/2} - 1}{e^{\Omega_0/2} + 1}\right)^2 e^{-2\kappa_D \Delta} \qquad (4.23)$$

and is applicable when interaction between the particles is weak, which is the case when $\kappa_D \Delta > 1$. At closer distances ($\kappa_D \Delta < 1$) it is necessary to account for Born repulsion. More generally known as steric hindrance, this effect arises from the fact that each atom of a molecule occupies a certain amount of space. Repulsion is induced when two particle surfaces are very close to each other [Gregory, 1975].

Example 4.1: Calculate the energy of repulsion given the following data: A constant surface charge density $\sigma_c = 3 \times 10^4$ esu cm^{-2}, $2\Delta = 80$ Å, and NaCl concentration of 0.1 N. Take the dielectric constant $\varepsilon_c = 80$, and at room temperature assume $\kappa_B T = 4 \times 10^{-14}$ ergs. Note that 1 esu = 3.336×10^{-10} Coulomb.

From Table 4.2, at 0.01 N electrolyte concentration $\kappa_D = 10^7$ cm^{-1}. Therefore, $\kappa_D \Delta = 4.0$, which is greater than 1, as required for the use of Eq. 4.23. Next, for NaCl, $z^+ = 1$. Thus, from Eq. 4.19

$$\left(\frac{d\hat{\Phi}}{d\hat{x}} \right)_0 = \frac{4\pi e_c z^+}{\varepsilon_c \kappa_B T} \frac{\sigma_c}{\kappa_D} = \frac{4\pi \times 1.602 \times 10^{-19} \times 3 \times 10^4}{80 \times 4 \times 10^{-14} \times 10^7 \times 3.336 \times 10^{-10}} = 5.65$$

For the constant charge density case, van Olphen [1977] provides tables of values of the potential Ω_0 as a function of $\left(d\hat{\Phi}/d\hat{x} \right)_0$ and $\kappa_D \Delta$. We will select a portion of the table applicable to $\left(d\hat{\Phi}/d\hat{x} \right)_0 = 5.6$.

$\kappa_D \Delta$	Ω_0
15.545	3.51
13.242	3.51
10.940	3.51
8.637	3.51
6.335	3.51
4.030	3.51
2.386	3.51
1.629	3.54
1.162	3.58

Given $\kappa_D \Delta = 4.0$, we obtain $\Omega_0 = 3.51$. Then from Eq. 4.23

$$V_R = \frac{64 \times 4 \times 10^{-14} \times 0.1 \times 10^{-3} \times 6.02 \times 10^{23}}{10^7} \left(\frac{e^{1.76} - 1}{e^{1.76} + 1} \right)^2 e^{-8}$$

$$= 0.00363 \text{ erg cm}^{-2} = 3.65 \times 10^{-6} \text{ J m}^{-2}$$

4.3.4 *van der Waals attraction*

As we noted in Chapter 3, the van der Waals (or London–van der Waals) electrostatic force of attraction exists at all particle surfaces. Several factors are responsible for this force, chief among them being an interaction between the dipoles of neighboring atoms and molecules in a particle. A dipole is generated by the oscillation of the electron cloud surrounding the nucleus of each atom or a molecule, which as a whole is electrically neutral. The van der Waals force acts over very short distances as it decays rapidly at a rate which according to the equation for the well-known Lennard–Jones potential is proportional to x^{-6}, where x is the distance from the atom or molecule.

As a reasonable assumption the forces exerted by individual atoms or molecules can be considered to be additive, and therefore the force due to a particle is obtained by integration over all atoms and molecules in the particle. The energy of interaction V_A between two particles is equal to the spatial gradient of this force. For two spherical particles of diameter d_p at a short distance 2Δ from each other (Fig. 4.6), V_A is approximated by

$$V_A = -\frac{A_H}{6}\left[\frac{d_p^2}{2\Delta_p^2} + \frac{d_p^2}{2(\Delta_p^2 - d_p^2)} + 2\ln\left(1 - \frac{d_p^2}{\Delta_p^2}\right)\right] \qquad (4.24)$$

where $\Delta_p = 2\Delta + d_p$ is the distance between the particle centers and A_H is the Hamaker constant characterizing resonant interactions between the electronic orbits of the two particles and the intervening medium. This constant typically ranges between 10^{-19} and 10^{-21} J. Its exact value depends on the particle and on the suspending medium, *e.g.*,

Mineral	A_H in water (J)	A_H in vacuum (J)
Crystalline quartz	1.70×10^{-20}	8.8×10^{-20}
Calcite	2.23×10^{-20}	10.1×10^{-20}

When the diameters of the two particles are much greater than the distance of separation between them, V_A decays at a rate proportional to the square of the distance between particle edges at the closest point of approach. For three-layer clays such as smectites, van Olphen [1977, 1987] gives

$$V_A = -\frac{A_H}{48\pi}\left[\frac{1}{\Delta^2} + \frac{1}{(\Delta+\Delta_c)^2} - \frac{2}{(\Delta+\Delta_c/2)^2}\right] \qquad (4.25)$$

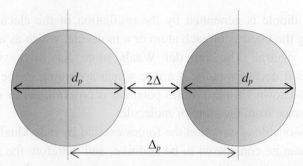

Figure 4.6. Two identical particles of diameter d_p separated by distance 2Δ.

where Δ is one-half the distance separating the plate-like particles measured between the planes of the centers of the oxygen atoms of the tetrahedral sheet, and Δ_c is the thickness of the three layered unit measured between the same planes (Δ_c = 6.60 Å). In this case the Hamaker constant $A_H \sim 10^{-12}$ J. Figure 4.7 illustrates the rapid fall in the absolute value of V_A with Δ in accordance with Eq. 4.25.

For a sense of the magnitudes of forces involved, we may conveniently use Eq. 4.24 under the condition that as the particles approach each other, $\Delta \to 0$. Ignoring second-order terms, this equation reduces to

$$V_A = -\frac{A_H d_p}{48\Delta} \qquad (4.26)$$

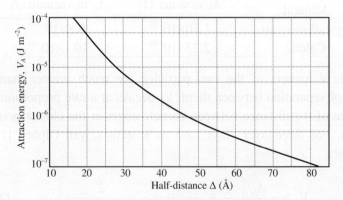

Figure 4.7. Variation of the absolute value of the attraction energy with half-distance of separation between particles (from van Olphen [1977]).

An estimate of the distance over which V_A is effective can be obtained from the consideration that at greater distances the effect of thermal energy $\kappa_B T$ of the molecule becomes dominant. At a distance $\Delta = x_c$ the two energies balance, and therefore

$$\kappa_B T = \frac{A_H d_p}{24 x_c} \tag{4.27}$$

or

$$x_c = \frac{A_H}{24 \kappa_B T} d_p \tag{4.28}$$

Example 4.2: Calculate the distance x_c for a 1 μm diameter particle. You are given $A_H = 10^{-20}$ J, $\kappa_B = 1.381 \times 10^{-16}$ erg °K^{-1} mol^{-1} and $T = 293$°K.

From Eq. 4.28 we obtain $x_c = 1.03 \times 10^{-7}$ m $\simeq$ 0.1 μm. In freshwater (defined with respect to the ionic strength[d] of NaCl, $I < 10^{-4}$ M) this distance ranges between 0.1 and 0.01 μm, in brackish water ($I = 1$ M NaCl) 10^{-3} μm, and much less in seawater ($I > 0.5$ M NaCl) [Dade and Nowell, 1991].

From Eq. 4.26 the van der Waals force F_A is obtained as

$$F_A = \frac{dV_A}{dx} = \frac{A_H}{24 d_p} \left(\frac{d_p}{x} \right)^2 \tag{4.29}$$

Example 4.3: Calculate the van der Waals force due to a 1 μm particle at a distance of 100 Å. Compare this force with the drag force on the particle falling in freshwater. Let $A_H = 10^{-20}$ J (or N m^{-1}) and assume a particle density of 2,830 kg m^{-3}.

We have $d_p = 10^{-6}$ m and $x = 1 \times 10^{-8}$ m. From Eq. 4.29 $F_A = 4.17 \times 10^{-12}$ N. From Eq. 3.25 the settling velocity of the particle is 1×10^{-6} m s^{-1}. The drag force $F_d = 3\pi \times 10^{-3} \times 10^{-6} \times 10^{-6} = 9.4 \times 10^{-15}$ N. Thus $F_A/F_d = 443$, which highlights the significance of cohesion relative to drag on the 1 μm particle.

---

[d] The ionic strength is a measure of the intensity of the electric field due to ions in solution. It is defined as one-half the sum of the terms obtained by multiplying the concentration (molality) of each ionic species in the solution by the square of its valence. Thus it is equal to $0.5 \sum z^{+2} c_\infty$.

4.3.5 *Modes of particle association*

As shown in Fig. 4.8a, with increasing electrolytic concentration the double layer is compressed as the V_R curve is pushed toward the particle surface while the V_A curve remains practically unchanged. At some "intermediate" concentration the effects of V_R and V_A reach the same distance from the particle, which is a threshold condition. With the slightest increase in concentration the sol changes from the dispersed state to incipient flocculation as V_A is able to attract and bind the particles. Thus the sol is destabilized and flocs are formed. When the counter-ion valence also changes the *net* energy varies between repulsive and attractive depending on the combined effects of concentration and valence. This is observed in the calculated curves in Fig. 4.8b, in which the cationic valence is changed from one to three and concentration from 0.01 N and 0.08 N [Kandiah, 1974].

Flocculation of plate-like clay particles causes three modes of particle association: face-to-face (F-F), edge-to-face (E-F) and edge-to-edge (E-E). In the double layer, the net potential curve of interaction and the rate of diffusion are different for these three orientations and, therefore, the associations do not necessarily occur simultaneously or to the same extent when a dispersed clay suspension is flocculated. The F-F

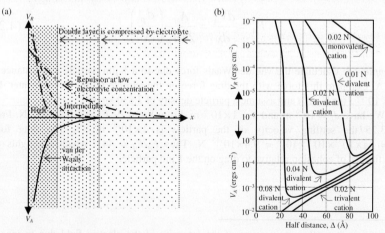

Figure 4.8. (a) Compression of the double layer by addition of electrolytic ions; (b) net effect of cationic concentration and valence (after Kandiah [1974]).

association leads to thicker, and at times larger, particles or flakes like a deck of cards, and are often not called flocs (*e.g.*, Fig. 3.5b). Only E-F and E-E yield three-dimensional, voluminous card-house structures that can be appropriately referred to as flocs.

Compared to most other clays, kaolinite has the unusual property that even in deionized or distilled water it forms flocs of the E-F type. In this non-salt flocculation the agglomerates have an open structure and are surprisingly robust. As the absence of salt prevents corrosion of hydraulic equipment, the use of kaolinite is popular in laboratory experiments. The presence of a very small quantity of salt can actually lead to dispersion, so a bed formed in the absence of salt has a much denser matrix with F-F orientation. At higher concentration salt flocculation occurs, in which the flocs are weaker than those formed without salt. The bed volume is intermediate between non-salt and low-salt states. It is important to point out that these differences in bed volume result from *independently formed* suspensions at different salt concentrations. Bed volumes obtained by incremental addition of salt without changing the suspension follow a different sequence. In either case kaolinite flocs formed at seawater salinity are smaller than those formed without salt. Up to two orders of magnitude reduction in the rate of deposition of suspended kaolinite in flowing freshwater was reported after salt was added [Mehta and Partheniades, 1975].

The ζ-potential (Fig. 4.4) is a measure of the state of suspended particles as it represents inter-particle repulsion. As the double layer is compressed and flocculation occurs the ζ-potential falls (Table 4.3). Its measurement in the laboratory is based on electrophoresis, *i.e.*, the transport of sol particles when they are subjected to an electric field (Chapter 3). A relationship exists between the ζ-potential and the (electrophoretic) velocity v_e of the moving particles. It can be derived by considering the Gouy–Chapman parallel-plate condenser analogy of the double layer. The ionized particle in the condenser attains the velocity v_e under a balance between viscous drag and the electrostatic force. This balance is embodied in the Helmholtz–Smoluchowski equation

$$\zeta = \frac{4\pi\eta v_e}{\varepsilon_s E_f} \qquad (4.30)$$

Table 4.3. Fine particle associations and floc characteristics.

Particle association	Edge-to-face (E-F)	Edge-to-edge (E-E)	Face-to-face (F-F)
Flocculated particles (low ζ-potential)			
Dispersed particles (high ζ-potential)			
Forces in particle interaction	Electrostatic + Edge-to-face	Electrostatic + van der Waals	van der Waals between faces
Area of particle contact	Low	Low	High
Bed matrix	Voluminous, porous, high void ratio	Voluminous, high void ratio	Dense, tightly packed
Water content	High	Intermediate to high	Low

Sources: van Olphen [1977]; Dennett *et al.* [1998].

where η is the dynamic viscosity of the liquid, ε_s is the static permittivity of the liquid and E_f is the *efs* applied between a positive and a negative electrode. The ratio $v_e/E_f = M_B$ is called electrophoretic mobility. Therefore

$$\zeta = \frac{4\pi\eta}{\varepsilon_s} M_B \qquad (4.31)$$

The electrophoretic mobility of colloidal particles is typically in the range of 1.1×10^{-4} to 4.0×10^{-4} cm^2 s^{-1} V^{-1}. Equation 4.31 is a means to determine the ζ-potential by measuring the mobility of particles in a liquid of known viscosity and dielectric constant. The mobilities are usually negative, which means that the particles are negatively charged and travel towards the positive electrode.

Example 4.4: For a suspension of 1 μm clay particles in water you are given $M_B = -2 \times 10^{-4}$ cm^2 s^{-1} V^{-1} and $\eta/\varepsilon_s = 20$ V^2 s. cm^{-2}. Calculate the ζ-potential.
From Eq. 4.31 $\zeta = -4\pi \times 20 \times 2 \times 10^{-4} = -0.0503$ V $= 50.2$ mV. The association between ζ-potential and the stability of the (colloidal) suspension is as follows.

ζ-potential range (mV)	Colloidal stability
0 to ± 5	Rapid flocculation (coagulation)
± 10 to ± 30	Incipient stability
± 30 to ± 40	Moderate stability
± 40 to ± 60	High stability
> 60	Very high stability

This indicates that the suspension with $\zeta = 50.2$ mV is highly stable, *i.e.,* the particles are fully dispersed.

4.3.6 *Solution effects on flocculation*

The role of salts forming electrolytic solutions is evident from Table 4.3. The relative effect of different dissolved salts on flocculation is described by the Schulze–Hardy rule according to which the coagulative power of a salt is determined by the valence of one of its ions. This ion is either negative or positive, depending on whether the sol particles move down or up the electric potential gradient. The coagulating ion is always of the opposite electric sign to the particle [Hardy, 1900]. This rule is valid for electrolytes that do not chemically interact with the sol particles. Such electrolytes therefore must not contain ions that are adsorbed on the particles that would react with ions in the double layer.

The flocculating power of cations for negatively charged sol particles decreases slightly in the order Cs > Rb > NH$_4$ > K > Na > Li. Similarly, the order for anions and positively charged sol particles is F > Cl > Br > NO$_3$ > I. These sequences are called lyotropic series. From the hydraulic engineering point of view, sodium chloride is the most important salt although ions of higher valence can enhance the effect of NaCl. Even water-miscible organic solvents such as alcohols and acetone increase flocculation by further compressing the double layer.

An effective method to obtain dispersed particles from a suspension of flocs, also called an unstable suspension because these flocs tend to eventually settle out, is by reversing the positive-edge charge and creating a negative-edge clay micelle. The positive-edge-to-negative-face attraction is eliminated, and a strong edge-to-edge as well as edge-to-face repulsion is created, resulting in a breakdown of the floc. This process producing a stable suspension of dispersed particles is called peptization. A chemical that is most commonly used as a peptizer is sodium hexametaphosphate, whose commercial name is Calgon. Other sodium containing peptizers are also used.

Figure 4.9a shows the effect of sodium hexametaphosphate on the critical concentration of NaCl for flocculation of Na-montmorillonite clay. At zero peptizer concentration the critical concentration is 20 milliequivalents per liter. With increasing peptizer concentration, the critical concentration increases due to the effect of the peptizer in preventing flocculation. Finally, at a peptizer concentration of about 120 $mEq \ L^{-1}$ the critical concentration reaches a peak of about 340 $mEq \ L^{-1}$ and flocculation is minimized. Beyond that point the sodium in the peptizer assists flocculation so the critical concentration decreases with increasing peptizer concentration. This is called over-treatment. Beyond a peptizer concentration of 600 $mEq \ L^{-1}$ the peptizer alone is sufficient for flocculation without the addition of NaCl.

In laboratory studies on turbidity currents in which settling and erosion play no role in advection, *i.e.,* auto-suspensions (Chapter 6), saltwater has been used as a substitute for the turbid suspension (*e.g.,* Garcia and Parker [1993]). For similar experimental purposes, Abraham and Diephuis [1959] stabilized naturally occurring Black Swamp clay by adding sodium hydroxide (NaOH) to the clay in water. Sodium (Na^+) displaces calcium (Ca^{++}) absorbed on the clay particles and causes calcium to precipitate out. The outcome is that the particles are dispersed. The stability diagram of Fig. 4.9b was developed from experiments in a cylindrical vessel in which NaOH and calcium chloride ($CaCl_2$) were added in different proportions to a mixture of the Black Swamp clay in water. Stability of the resulting suspension was assessed

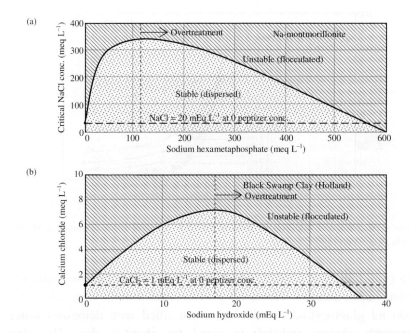

Figure 4.9. (a) Effect of peptizer on NaCl concentration for flocculation of a sodium-montmorillonite clay (adapted from van Olphen [1977]); (b) effects of NaOH and $CaCl_2$ on the flocculation of Black Swamp clay (adapted from Abraham and Diephuis [1959]).

in terms of the settling ability of the particles — a stable suspension being one which did not show significant settling of particles after 24 h, and an unstable suspension as one from which the particles settled out within four hours and the supernatant liquid was clarified. The initial concentration of the Black Swamp clay was chosen to form a stable suspension with a density of 1,030 kg m^{-3}, which was then used in hydraulic model tests in lieu of seawater to obviate the corrosive effects of salt on the equipment.

The effect of salt (NaCl) flocculation on a peptized sample of natural mud from Fernandina Municipal Marina in Florida is shown in Fig. 4.10. The wet sample was peptized by Calgon after the sample was "washed" with deionized water. Washing is necessary when the native water has a

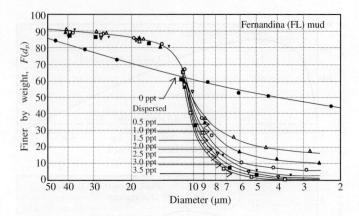

Figure 4.10. Effect of addition of sodium chloride on mud from Fernandina Marina, Florida (from Yeh [1979]).

high (~30 ppt[e]) salt content, which interferes with peptization, and was carried out as follows. About 50 g of the wet sample were taken in a 1,000 ml glass cylinder which was then filled with deionized water, thoroughly shaken and left to stand for about a day. The clear supernatant liquid was decanted and the process of adding water, shaking *etc.,* was repeated two more times. Thus washing took three days, following which Calgon was added. The suspension was thoroughly shaken and its size distribution was measured. The resulting 0 ppt dispersed sample size distribution shows highly graded fine-grained material with a median size of 4 μm.

The addition of the first 0.5 ppt increment of NaCl led to re-flocculation accompanied by a drastic change in the size distribution which became far more uniform. The median grain size increased to about 10 μm, indicating the presence of flocs. With further additions of

_____

[e] Salinity S of water (Chapter 2) used to be expressed as parts per thousand (ppt or ‰), *i.e.,* grams of salt per liter of solution. Later it was also expressed as ‰ based on the ratio of electrical conductivity of the sample to "Copenhagen water", which is reference sea water from the North Sea. In 1978, S was redefined in terms of Practical Salinity Units (psu) based on the ratio R_s of conductivity of the sample to a standard potassium chloride (KCl) solution [Lewis, 1980; Culkin and Ridout, 1989]. This definition is used to calculate the salinity in Exercise 2.4. We have retained the use of ppt in those cases in which NaCl concentration was gravimetrically measured. Elsewhere the term salinity has been used freely to mean NaCl concentration and also the conductivity-based value.

salt the median size did not increase very much, even as the fines at the tail of the distribution became increasingly incorporated into the flocs. The uniformity coefficient $U_c = d_{60}/d_{10}$ plotted in Fig. 4.11 as a function of salt concentration indicates that at first this coefficient decreased rapidly with increasing salinity, then began to level off once the salinity reached 1.5 ppt. Thus, salinity of about 1 ppt appeared to be sufficient to cause the particles to flocculate.

The term "dispersed" in reference to clay particles must be understood with the recognition that peptization of clay in water does not ensure complete dispersion of flocs into primary particles. The size distribution which actually results depends on the chemical or physical pre-treatment due to natural causes or intervention in the laboratory. Thus in principle the dispersed particle size distribution in Fig. 4.10 must be used only for comparison with the other distributions in the same plot. Dispersion can be accentuated by the use of acoustic waves in the laboratory. Sonification causes disruption of flocs through the process of cavitation. Small bubbles are formed during the rarefaction phase of the sound wave. These bubbles collapse during the compression phase, releasing shock waves through the liquid and resulting in floc disruption [Doulah, 1977; Biggs and Lant, 2000].

Different methods to determine particle size in general result in different size distributions, and none may produce even nearly complete

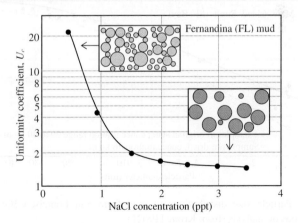

Figure 4.11. Variation of uniformity coefficient with NaCl concentration for Fernandina mud (from Yeh [1979]).

dispersion. The size distributions of a sample of mud from the San Francisco Bay given in Fig. 4.12 were obtained after pre-drying the sample in an oven and also without pre-drying, *i.e.*, using the original wet sample taken from the bay. In both methods a peptizing agent was added. However, dispersion was not effective in the dried sample even though it was mechanically pulverized. The outcome was that the sample median diameter was 2.5 μm compared to 1.2 μm for the wet sample. No verification such as by electron micrography was carried out to ascertain if the wet sample was fully dispersed. In general, when organic matter is present even in small quantities as in the Bay mud, it is practically impossible to regenerate the original matrix of the combined organic and inorganic solids once the material is dried.

Among other solution effects a noteworthy factor is the role of fluid pH. Unstable-stable suspension or flocculation-dispersion transition boundary curves for a smectite based on the Sodium Adsorption Ratio (*SAR*) are shown in Fig. 4.13. In these experiments ions were present in the pore water only; the ambient fluid being deionized water. *SAR*, a non-dimensional measure of soil stability, is the ratio of concentration of sodium ions (c_{Na+}) divided by (a measure of) the sum of concentrations of calcium (c_{Ca++}) and magnesium ions (c_{Mg++}) in the pore fluid. It is defined as

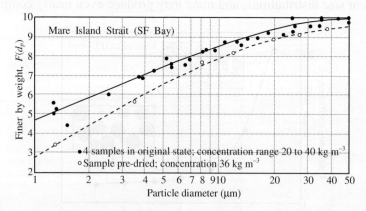

Figure 4.12.   Particle size distributions of mud from San Francisco Bay using two different methods of analysis (from Krone [1962]).

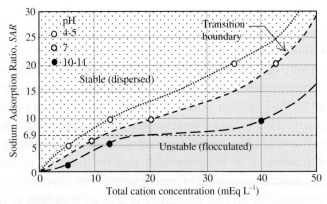

Figure 4.13. Flocculation-dispersion transition boundary curves for a smectite at three pH ranges. The ambient fluid was free of ions (adapted from Kandiah [1974]).

$$SAR = \frac{c_{Na^+}}{\sqrt{\frac{1}{2}\left(c_{Ca^{++}} + c_{Mg^{++}}\right)}} \qquad (4.32)$$

where the concentrations are in milliequivalents per liter.

Example 4.5: A 3-liter water solution contains 430 mg of sodium ions, 70.2 mg of calcium ions and 18.2 mg of magnesium ions. Calculate the Sodium Adsorption Ratio.

The molecular weight of sodium in milligrams is 23,000, calcium is 40,100 and magnesium is 24,300. Since the valence of Na is +1, Ca +2 and Mg +2, their respective equivalent weights are 23,000 mg, 20,050 mg and 12,150 mg. Therefore, there are 23,000/1,000 = 23 mg of sodium in 1 mEq of this element. For calcium the conversion is 20.5 mg = 1 mEq, and for magnesium 12.15 mg = 1 mEq. One liter of water contains 143.3 mg of sodium, 23.4 mg of calcium and 6.07 mg of magnesium. Therefore, $c_{Na^+} = 6.23$ mEq L^{-1}, $c_{Ca^{++}} = 1.14$ mEq L^{-1} and $c_{Mg^{++}} = 0.5$ mEq L^{-1}. Finally, $SAR = 6.23/\sqrt{0.5(1.14+0.5)} = 6.9$.

Since the flocculation-dispersion boundary varies with the sediment, Fig. 4.13 must not be used for purposes other than as an illustration of the sensitivity of this boundary to pH. In general, it can be inferred that sensitivity may be potentially significant with regard to the stability of aquatic cohesive sediment beds relative to the eroding force. With reference to Fig. 4.13, if for some reason at $SAR = 6.9$ the pH value of the pore fluid were to decrease from, say, neutral 7 to acidic 5, the floc structure of the bed could crumble and the resulting material swept away by, say, water current.

In general, low (<~5) *SAR* promotes inter-particle attraction and, therefore, instability of the suspension due to flocculation. At high (>~30) *SAR* repulsion prevails and initially dispersed particles remain as such, *i.e.*, the suspension is stable. The smectite in Fig. 4.13 could be altered between the stable and unstable states merely by changing the pH of the pore fluid either by holding *SAR* constant or the total concentration of cations (Na^+, Ca^{++} and Mg^{++}) in the pore fluid. This indicates that sediment composition alone is not a unique or even dominant determinant of bed erosion potential, which also depends on the pore fluid chemistry. This behavior underscores the need to use native water in laboratory tests to characterize fine sediment erosion or settling. When this is not feasible, it is recommended that the fluid be reconstituted based on the ionic composition of native water. The importance of fluid chemical composition on floc properties is also an argument in favor of *in situ* testing of bed erodibility and settling of suspended sediment. This is especially the case in natural environments where salt composition and concentration show significant spatial and temporal variability [Kandiah, 1974; Ariathurai and Arulanandan, 1978].

4.3.7 *Critical salinity*

As we have noted the behavior of cohesive particles in water is complicated by a variety of solution effects. In hydraulic engineering practice the critical salinity for incipient flocculation (or coagulation) is a commonly used approximate threshold of NaCl concentration (in ppt or mEq L^{-1}). The critical salinity of three common clays is given in Table 4.4. These values are consistent with the effect of salinity of the San Francisco Bay sediment in Fig. 4.14, in which the median settling velocity of flocs is plotted against salinity measured in a 1-liter glass cylinder at four suspended sediment concentrations. Bay sediment contains all three clays of Table 4.4 and also chlorite. The effect of

Table 4.4. Critical NaCl concentration for flocculation.

Clay mineral	Critical NaCl concentration	
	(ppt)	mEq L^{-1}
Kaolinite	0.6	10
Illite	1.1	19
Smectite (or montmorillonite)	2.4	36

Source: Revised from Ariathurai *et al.* [1977].

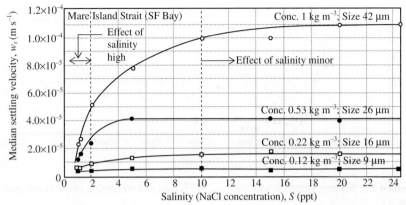

Figure 4.14. Effect of NaCl on the settling velocity of San Francisco Bay mud (adapted from Krone [1962]).

increasing salinity on the settling velocity is cumulative, with only minor coagulation below about 1 ppt. Increasing salinity from about 1 to 2 ppt rapidly increased the settling velocity indicating that the flocs grew (by edge-to-face bonding). The growth rate then decreased and above about 10 ppt the settling velocity increase was comparatively small. At 15 ppt the effect of salinity was virtually complete. The maximum floc size given at each concentration has been calculated using Stokes law (Eq. 3.25) with an assumed floc density of 1,140 kg m^{-3} and saltwater density of 1,025 kg m^{-3}. A characteristic feature of the trends is that at a constant sediment concentration the settling velocity changes monotonically with salinity. However, as we will see in Chapter 7, the effect of salinity can be more complex than implied by the behavior in Fig. 4.14.

An inference one may draw from the trends in Fig. 4.14 is that since the salinity in San Francisco Bay is usually well above 1 ppt, it does not limit flocculation, except perhaps when salt is flushed out during very high river runoffs. An example of a dramatic reduction in salinity during a high river flood event is shown in Fig. 4.15. Over the four-day period of the initial flood and its aftermath the salinity dropped rapidly to near-zero values in the first two days, then began a slow recovery, but at the end of the period had reached only about one-half the pre-storm value (20 ppt). The link-node numerical model of Orlob *et al.* [1967], simple by present modeling capabilities, was able to reproduce the salinity variation reasonably well.

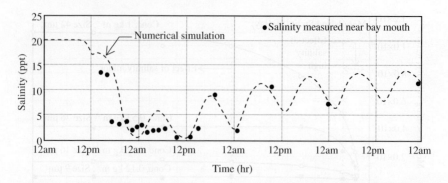

Figure 4.15. Salinity variation with time ca. 1955; data and simulation (from Orlob *et al.* [1967]).

Unsteady physicochemical process dynamics which influences floc properties, especially in the transient sense during a river flood (*e.g.,* Fig. 4.15), is not well understood. Two effects are particularly noteworthy. Firstly, much of the available information on the dependence of flocs on salinity is based on steady-state laboratory experiments. Synoptic field data do include unsteady effects but, possibly with some exceptions, these effects are assessed mainly in a qualitative sense with regard to the influence of hydrodynamic factors on flocs. Secondly, information on the effect of a salinity shock, *i.e.,* a sudden and significant change in the salinity of the ambient fluid, on bed stability is not well investigated. This is so because typical laboratory experiments related to the effect of fluid chemistry on bed erosion have been carried out under conditions of constant fluid chemical composition.

A noteworthy phenomenon studied in the laboratory is the role of osmotic pressure.[f] An osmotic pressure gradient exists between the ambient fluid and the pore fluid when their salt concentrations differ.

---

[f] Osmosis is the process by which diffusion of a solvent occurs through a semi-permeable membrane from a less concentrated solution into a more concentrated solution. The osmotic pressure that must be applied on the side of the concentrated solution to prevent the flow of the pure solvent into the solution is $p_\pi = gRT/MV_t$, where M is the molecular weight, R is the gas constant, T is the absolute temperature and V_t is the volume of the solution. If 4 g of a substance having a molecular weight of 342 were dissolved in 100 ml of solution at 25°C, p_π would be 4(g)×8.314 (L kPa °K^{-1} mole^{-1})×298 (°K)/[342 (g mole^{-1})×0.1 (L)] = 290 kPa.

This gradient causes a flow of water between the fluids. In estuaries a salt concentration anomaly can arise due to tidal variation, change in the river outflow, or rainfall. Benthic flux can be another significant cause especially if the groundwater is rich in, say, dissolved nutrients [King *et al.*, 2010]. The change in solute concentration is comparatively prominent in the ambient fluid because the pore fluid is partially encapsulated within the low-permeability bed pores. To an extent the floc matrix close to the bed surface acts as a semi-permeable membrane constraining the diffusion of the solute but allowing the exchange of water (Figs. 4.16a, b, c).

Osmosis alters the surface fabric of the cohesive bed and its stability. For example, when the pore fluid initially contains high levels of dissolved salts and ambient water seeps into the bed, the rate at which the affected bed erodes due to a current varies with the ionic composition and concentration in the pore fluid [Kandiah, 1974]. Since osmosis is a molecular transport process it is slow in the same sense as molecular diffusion of a solute in a solvent. An outcome is that the rate of change of solute concentration in the ambient fluid influences bed stability. Thus, depending on the properties of the bed the effect of low-frequency changes in estuarine salinity, *e.g.*, due to tidal variation, can be more significant than higher frequency changes associated with wind-wave oscillations.

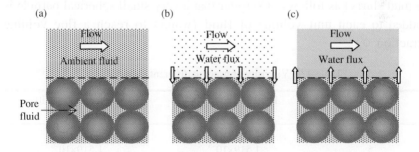

Figure 4.16. Schematic drawings showing the effect of changing salt concentration in the ambient fluid on the pore fluid due to osmosis: (a) Pore fluid and ambient fluid have the same salt concentration; (b) ambient salt concentration decreases and brings about water flux into the bed; (c) ambient salt concentration increases and brings about water flux out of the bed.

4.4 Floc Concentration Measures and Shear Strength

4.4.1 *Floc density*

Floc density (*i.e.*, bulk density including the constitutive particles and interstitial water) is commonly determined by assuming the validity of Stokes law of settling (Eq. 3.25) or its extension (Eq. 7.22). In this approach the settling velocity and the diameter are measured by, say, high-resolution video imaging, leaving the floc density as the unknown to be calculated [Manning, 2004]. A different approach was developed by Krone [1962] to obtain approximate values of the floc density and strength from tests in a viscometer.

For density determination let η denote the dynamic viscosity of a suspension of spheres at infinite dilution, *i.e.*, when the effect of spheres on the shearing fluid nearby does not affect shearing in the neighborhood of other spheres. Einstein [1911] derived the equation

$$\eta = \eta_w (1 + 2.5\phi_{vf}) \qquad (4.33)$$

where η_w is the viscosity of water (Table 4.5) treated as a suspending Newtonian liquid and ϕ_{vf} is the floc volume fraction, *i.e.*, volume of flocs divided by volume of suspension.

Equation 4.33 can be extended to suspensions having higher ϕ_{vf} (as in a mud slurry) as follows. Consider that a very small spherical particle is added to each unit volume of fluid (water) to reach a floc volume fraction ϕ_{vf}'.

Table 4.5. Viscosity of water at different temperatures.

Temperature ($^\circ$C)	Dynamic viscosity (Pa.s)	Kinematic viscosity (m^2 s^{-1})
0	1.787×10^{-3}	1.787×10^{-6}
5	1.519×10^{-3}	1.519×10^{-6}
10	1.307×10^{-3}	1.307×10^{-6}
20	1.002×10^{-3}	1.004×10^{-6}
30	0.798×10^{-3}	0.801×10^{-6}
40	0.653×10^{-3}	0.658×10^{-6}

The viscosity of suspension then becomes

$$\eta_1 = \eta_w (1 + 2.5 \phi'_{vf}) \tag{4.34}$$

Following this relationship, given $n = \phi_{vf} / \phi'_{vf}$ we obtain the nth equation

$$\eta_n = \eta_{n-1} (1 + 2.5 \phi'_{vf}) \tag{4.35}$$

Substituting the viscosity after the first addition (η_1) into the equation for the second (η_2) and so on yields (after dropping the subscript of η on the l.h.s.)

$$\eta = \eta_w (1 + 2.5 \phi'_{vf})^{\phi_{vf} / \phi'_{vf}} \tag{4.36}$$

In the limit, as ϕ'_{vf} becomes very small this relation reduces to

$$\eta = \eta_w e^{2.5 \phi_{vf}} \tag{4.37}$$

A similar but more general expression is

$$\eta = \eta_w e^{2.5 \phi_{vf} / (1 - \chi_f \phi_{vf})} \tag{4.38}$$

where χ_f is an empirical coefficient [Mooney, 1951]. This expression reduces to Eq. 4.37 when $\phi_{vf} \ll \chi_f^{-1}$.

Equation 4.37 can be expanded into a series

$$\eta = \eta_w \left[1 + 2.5 \phi_{vf} + \frac{(2.5 \phi_{vf})^2}{2!} + \frac{(2.5 \phi_{vf})^3}{3!} + \dots \right] \tag{4.39}$$

which can be visualized as a weighted sum of the interactions of 1, 2, 3,... particles at a time. If this equation is restricted to the second order and $(2.5)^2 / 2!$ is replaced by 6.2, we obtain an equation derived analytically by Batchelor [1977].

In Fig. 4.17, data obtained in a capillary viscometer using estuarine sediment from North Carolina follows the relationship

$$\eta = \eta_w e^{k_f C} \tag{4.40}$$

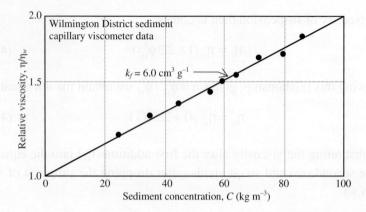

Figure 4.17. Viscosity of suspension relative to water versus concentration of Wilmington District (NC) sediment measured in a capillary viscometer (adapted from Krone [1963]).

where k_f (= 6 cm^3 g^{-1}) depends on the makeup of the sediment and the fluid and C is the sediment mass concentration, *i.e.*, dry sediment mass per unit volume of suspension. Thus, by comparison with Eq. 4.37

$$k_f C = 2.5 \phi_{vf} \qquad (4.41)$$

Therefore the floc volume fraction is obtained from

$$\phi_{vf} = \frac{k_f C}{2.5} \qquad (4.42)$$

provided k_f is measured at a given sediment concentration. The value 2.5 is for spheres and for natural sediment the value depends on floc shape.

The floc density is readily obtained from mass balance

$$\rho_f = \frac{C}{\phi_{vf} \rho_s} (\rho_s - \rho_w) + \rho_w \qquad (4.43)$$

where ρ_s is the particle material (*e.g.*, mineral) density. Densities of several sediments obtained in this way are given in Table 4.6, in which $\phi_v = C / \rho_s$ is the solids volume fraction, which would be equivalent to the floc volume fraction ϕ_{vf} if the particles were

Table 4.6. Floc densities from capillary viscometer data.

Sediment	$k_f{}^a$ $(cm^3 g^{-1})$	C/ϕ_{vf} $(g\ cm^{-3})$	$\phi_v/\phi_{vf}{}^b$	$\rho_f{}^c$ $(kg\ m^{-3})$
Wilmington District estuary	6.00	0.417	0.138	1,281
Brunswick Harbor	11.1	0.225	0.085	1,164
Gulfport Channel	8.56	0.292	0.110	1,205
San Francisco Bay	6.31	0.396	0.150	1,269
Delaware Bay	5.49	0.455	0.172	1,305
Potomac River	3.73	0.670	0.253	1,437
White River	8.24	0.304	0.115	1,212

[a] For tests with a water salinity of 33.8 ppt.
[b] $\phi_v/\phi_{vf} = (\rho_f - \rho_w)/(\rho_s - \rho_w)$.
[c] With $\rho_w = 1,025$ kg m^{-3} and $\rho_s = 2,650$ kg m^{-3}.
Source: Krone [1963].

(fully) dispersed. The densities range between 1,164 and 1,437 kg m^{-3} with a mean of 1,263 kg m^{-3}. These values are comparable with the densities of beds formed from deposition in estuaries (Chapters 5, 8).

4.4.2 *Volume-based and mass-based concentrations*

In Chapter 3 we introduced the volume concentration C_v as volume of dry particles divided by volume of suspension. In general, sediment concentration C is commonly expressed as a volume-based quantity, *e.g.*, g L^{-1} or mg L^{-1}, and sometimes as a mass concentration C_{mass}, *e.g.*, mg kg^{-1} or parts per million. In this text C has also been referred to as mass concentration. The terms volume-based concentration and mass concentration are considered synonymous. The units of C vary according to the method and needs of measurement, and require the user to be able to readily convert units. A volume-based concentration of 1 g L^{-1} can also be expressed as 1,000 mg L^{-1} = 1 kg m^{-3} = 1 mg ml^{-1} = 10^{-3}g cm^{-3}.

At low concentrations, when the suspension density ρ is close to that of water, the difference between C and C_{mass}, *e.g.*, mg L^{-1} versus mg kg^{-1}, is small. At higher concentrations, when the suspension density differs from water, C can be converted to C_{mass}, as sometimes required, for instance in dredging calculations. Thus

$$C_{mass} = \frac{C}{\rho} \qquad (4.44)$$

Based on the mass balance relationship

$$C = \left(\frac{\rho - \rho_w}{\rho_s - \rho_w}\right)\rho_s \qquad (4.45)$$

we obtain the suspension density (in Eq. 4.44)

$$\rho = \frac{C(\rho_s - \rho_w)}{\rho_s} + \rho_w \qquad (4.46)$$

Example 4.6: What are the suspension density and the mass-based concentration for volume-based concentrations of 100 and 10,000 mg L^{-1}? Assume $\rho_w = 1,000$ kg m^{-3} and $\rho_s = 2,650$ kg m^{-3}.

Equation 4.46 yields

$$\rho = \frac{100 \, (\text{mg L}^{-1}) \times 1000 \left(\dfrac{\text{kg m}^{-3}}{\text{mg L}^{-1}}\right)(2650 - 1000)}{2650} + 1000 = 1000.062 \text{ kg m}^{-3}$$

From Eq. 4.44

$$C_{mass} = \frac{100 \, (\text{mg L}^{-1}) \times 1000 \, (\text{L m}^{-3})}{1000.062 \, (\text{kg m}^{-3})} = 99.994 \text{ mg kg}^{-1}$$

which is within 0.01 percent of the volume-based concentration. Extra significant figures are given to illustrate the small difference. Applying the same equations to a volume-based concentration of 10,000 mg L^{-1} yields a suspension density of 1,006 kg m^{-3} and a mass-based concentration of 9,938 mg kg^{-1}.

Example 4.7: Calculate the volume fraction of flocs ϕ_{vf} having a density of 1,100 kg m^{-3} in freshwater. Assume a particle density of 2,650 kg m^{-3}.

The water volume fraction ϕ_w is determined from the mass balance relationship

$$\phi_w = \frac{\rho_s - \rho_f}{\rho_s - \rho_w} \qquad (4.47)$$

We obtain $\phi_w = 0.94$, or 94%, which indicates that the floc is mostly water (which is typical) with only 6% solids, *i.e.*, the floc volume fraction ϕ_{vf} is 0.06.

4.4.3 *Floc shear strength*

A variety of methods has been used to measure the shear strength (and viscosity) of flocs. As a result, and also because the definition of strength depends on the technique, measures of strength are not always consistent. In Fig. 4.17 the dependence of viscosity on the floc volume fraction is characterized by a single value of k_f (Eq. 4.40). This would imply that sediment concentration is the sole determinant of floc density, with no tangible effect of flow-induced shear. As the capillary viscometer used is limited by the low velocity gradients it produces (in the viscous range), Krone [1963] subsequently used a rotating concentric-cylinder viscometer to generate higher velocity gradients comparable to those in estuaries and a more uniform velocity gradient within the shearing volume. For each sediment suspension, the outer cylinder was rotated at high *rpm*, following which the speed was decreased to a lower *rpm*. The *rpm* value is proportional to the velocity gradient, *i.e.*, the shear rate $\dot{\gamma}$, and from the resulting torque on the inner cylinder the applied shear stress τ can be calculated. The plot showed definable ranges of the shear rate within which the stress-shear rate (flow curve) relation was approximately linear, but the slope changed from range to range, *e.g.*, Fig. 4.18 in which three ranges are shown. Each range defines a viscosity η (and therefore floc volume fraction ϕ_{vf}) and a shear strength τ_f

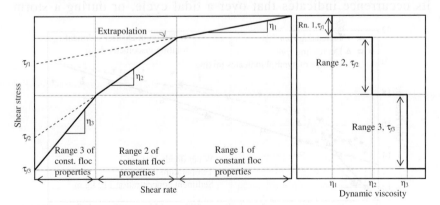

Figure 4.18. Schematic plot of shear stress versus flow shear rate (or velocity gradient) and viscosity of cohesive sediment derived from measurements in a concentric-cylinder viscometer.

analogous to the upper-Bingham yield strength obtained by extrapolation of the flow curve (Chapter 5) equation:

$$\tau = \eta\dot{\gamma} + \tau_f \qquad (4.48)$$

Each constant-slope range corresponds to a floc structure which survives over that range. At high shear rates ($\dot{\gamma}$) the structure breaks down into units that are tightly packed with high shear strength. At the same time their viscosity is low. At lower $\dot{\gamma}$ the structure is loosely packed with low shear strength and high viscosity.

In Fig. 4.19 data are shown from the concentric-cylinder viscometer using Wilmington District sediment (Table 4.6). The shear stress is proportional to the dial reading, and the shear rate to cylinder *rpm*. The data show a more complex flow behavior than the idealized description in Fig. 4.18. A noteworthy feature is hysteresis, which occurs because when the *rpm* value is increased and then decreased, the rising and falling pathways of the flow curve do not coincide due to different sequential formations of floc structures. The area enclosed between the rising and falling curves represents the extent of hysteresis effect. Under an oscillatory stress due to waves the rate of energy loss is dependent on hysteresis [Isobe *et al.*, 1992; Robillard, 2009, Chapter 12].

While in fine sediment transport modeling hysteresis is often ignored, its occurrence indicates that over a tidal cycle, or during a storm

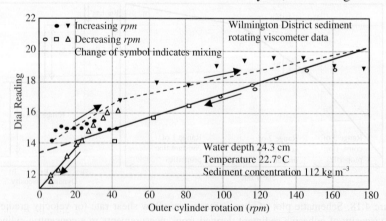

Figure 4.19. Raw concentric-cylinder viscometer data for Wilmington District sediment (from Krone [1963]).

characterized by rising then falling wave action, the strength, density and size of flocs formed during accelerating flows differ from those during deceleration. In a high-concentration environment this difference can be significant because high inter-particle collision frequency accentuates the effect of floc dynamics on sediment load.

In Fig. 4.20, the relative viscosity η / η_w obtained from the shear rate ranges during the falling shear rate phase (in Fig. 4.19) is plotted against concentration C of the Wilmington District sediment. Each line corresponds to a specific floc structure characterized by the k_f -value (Eq. 4.40).

The shear strength τ_f (Pa) for the same sediment in Fig. 4.21 conforms to the following relationship with C (kg m^{-3})

$$\tau_f = k_{ff} C^{5/2} \tag{4.49}$$

with k_{ff} = 2.83×10^{-6} (m$^{17/2}$ kg$^{-3/2}$ s^{-2}). Although k_{ff} varies with the sediment, all sediments tested (Table 4.6) had the exponent 5/2. This constancy suggests that self-similarity plays a significant role in the development of structural hierarchies during flocculation of clay mineral particles.

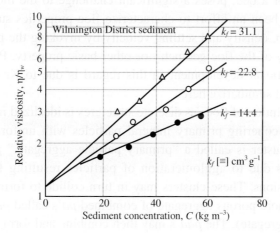

Figure 4.20. Relative viscosity versus concentration for Wilmington District sediment (from Krone [1963]).

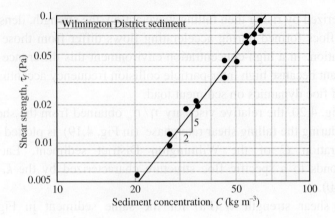

Figure 4.21.   Shear strength of Wilmington District sediment flocs as a function of concentration (adapted from Krone [1963]).

4.5 Flocs as Discrete Structures

4.5.1 *Order of aggregation concept*

The wide size range of cohesive flocs even in a small sample of a natural suspension, or a bed, poses a significant challenge to the modeler. The outcome has been an effort to characterize floc properties such as size, shear strength, density and settling velocity by invoking the principle of self-similarity of the floc geometry or other basic property. Possibly the earliest significant development in this regard is due to Krone [1963], which we will summarize here.

The description of flocs as discrete structures is idealized in Fig. 4.22. A cluster of cohering primary mineral particles with uniform porosity within the cluster is called a "primary particle aggregate," *pa*. Such a cluster forms due to agglomeration of particles resulting from inter-particle collisions. These clusters may in turn collide to form a weaker, lighter and more porous aggregate of combined *pa*'s called *paa* (particle aggregate-aggregate). The *paa*'s may then combine and form yet weaker and even more loosely packed *paaa* (particle aggregate-aggregate-aggregate). Subsequently, *paaaa* may result from *paaa*'s and so on. These structures have been called "orders" of aggregation. The

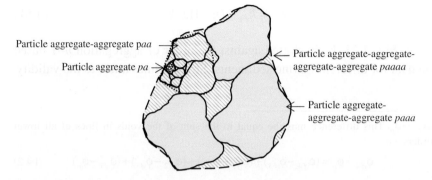

Particle aggregate-aggregate p*aa*

Particle aggregate *pa*

Particle aggregate-aggregate-aggregate-aggregate *paaaa*

Particle aggregate-aggregate-aggregate *paaa*

Figure 4.22. Particle aggregate-aggregate-aggregate-aggregate, *paaaa*, is made up of one-order lower aggregates *paaa*, each consisting of aggregates *paa*, that are an order lower than *paaa*, and so on (adapted from Krone [1963]).

number of orders that occur in a suspension depends on its concentration, physicochemical properties, influence of biology, and the flow field. In Fig. 4.18, if we select subscript 1 to represent *pa*, *i.e.*, consider the flocs in Range 1 to be primary flocs, then 2 would be *paa*, 3 would be *paaa* and so on. As an aside, we note that measurements of floc properties in the Adriatic Sea reveal that the architecture of flocs may be complicated by non-uniformities as flocs include not only aggregated networks but also large individual particles [Mikkelsen *et al.*, 2006].

Relative to the Fig. 4.22 model, let n be the index defining the order of aggregation, $n-1$. Assuming that the primary aggregates (*pa*) retain their density in successive orders of aggregation, the volume fraction ϕ_{vfn} of aggregate order $n-1$ will be related to the volume concentration ϕ_{vf1} of *pa* by the volume balance

$$\phi_{vfn} = \phi_{vf1}[1 + (n-1)e_{v1}] \qquad (4.50)$$

where e_{v1} is the void ratio of *pa*.[g] Using Eq. 4.41 the above expression becomes

---

[g] Let us derive Eq. 4.50 for a floc of order $n = 4$, *i.e.*, a floc with volume fraction ϕ_{vf4}. The floc is made up of the next lower order flocs of volume fraction ϕ_{vf3} per floc. This floc consists of flocs with volume fraction ϕ_{vf2} per floc, which in turn consists of flocs of volume fraction ϕ_{vf1} per floc. Finally, the latter floc is made up of primary particles that are solid, *i.e.*, they do not have voids. Let the volume fraction of each primary particle be ϕ_v. The total volume fraction of voids in the floc of volume fraction ϕ_{vf4} is

$$k_{fn} = k_{f1} + (n-1)e_{v1}k_{f1} \qquad (4.51)$$

In Fig. 4.23, k_{fn} is plotted against the order of aggregation $n-1$ based on data for different marine sediments. Straight lines attest to the validity

---

$\phi_{vf4} - \phi_p$. This difference must be equal to the sum of the voids in flocs of all lower orders, *i.e.*,

$$\phi_{vf4} - \phi_p = (\phi_{vf4} - \phi_{vf3}) + (\phi_{vf3} - \phi_{vf2}) + (\phi_{vf2} - \phi_{vf1}) + (\phi_{vf1} - \phi_p) \qquad (F4.2)$$

which reduces to

$$\phi_{vf4} = (\phi_{vf4} - \phi_{vf3}) + (\phi_{vf3} - \phi_{vf2}) + (\phi_{vf2} - \phi_{vf1}) + \phi_{vf1} \qquad (F4.3)$$

The void ratio e_v is the volume of voids divided by the volume of solids (Eq. 5.10). The e_v values of the first, the second and the third order flocs, e_{v1}, e_{v2} and e_{v3}, respectively, are

$$e_{v1} = \frac{\phi_{vf2} - \phi_{vf1}}{\phi_{vf1}}; \quad e_{v2} = \frac{\phi_{vf3} - \phi_{vf2}}{\phi_{vf2}}; \quad e_{v3} = \frac{\phi_{vf4} - \phi_{vf3}}{\phi_{vf3}} \qquad (F4.4)$$

Substitution into Eq. F4.3 results in

$$\phi_{vf4} = \phi_{vf1} + e_{v1}\phi_{vf1} + e_{v2}\phi_{vf2} + e_{v3}\phi_{vf3} \qquad (F4.5)$$

in which each $e_v \phi_{vf}$ represents the volume fraction contribution of the flocs of a given order. Equation F4.5 can be generalized as

$$\phi_{vfn} = \phi_{vf1} + \sum_{i=1}^{i=n-1} e_{vi}\phi_{vfi} \qquad (F4.6)$$

In order to proceed the following self-similarity assumption relating the ratio of volume fractions of two consecutive orders to the corresponding ratio of the void-ratios is introduced [Krone 1963]:

$$\frac{\phi_{vf\,i+1}}{\phi_{vf\,i}} = \frac{e_{v\,i}}{e_{v\,i+1}} \qquad (F4.7)$$

i.e., the volume fraction of flocs of a given order divided by the same quantity representing the next lower order is inversely proportional to the ratio of the respective void ratios. This assumption, which is independent of the shape or the size of the floc, leads to $e_{vi}\phi_{vfi} = e_{vi+1}\phi_{vfi+1}$, or, in other words, in the present case ($n = 4$) $e_{v1}\phi_{vf1} = e_{v2}\phi_{vf2} = e_{v3}\phi_{vf3}$. Equation F4.5 then reduces to

$$\phi_{vf4} = \phi_{vf1}(1 + 3e_{v1}) \qquad (F4.8)$$

which in the generalized form yields Eq. 4.50. Since this analysis is based on self-similarity associated with the aggregate volume fraction and the void ratio based on Eq. 4.7, it does not invoke geometric similarity. Therefore, Eq. 4.50 as such does not address the issue of fractal relationship among diameters of different orders.

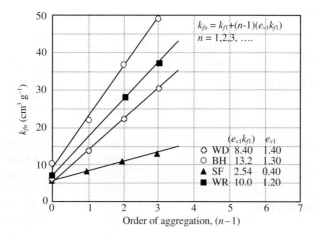

$$k_{fn} = k_{f1}+(n-1)(e_{v1}k_{f1})$$
$$n = 1,2,3, \ldots$$

	$(e_{v1}k_{f1})$	e_{v1}
✪ WD	8.40	1.40
○ BH	13.2	1.30
▲ SF	2.54	0.40
■ WR	10.0	1.20

Order of aggregation, $(n-1)$

Figure 4.23. Dependence of k_{fn} (cm^3 g^{-1}) on order of aggregation (adapted from Krone [1963]).

of Eq. 4.51. The following example illustrates the way in which the plot is obtained and the void ratio e_{v1} is calculated.

Example 4.8: Calculate the void ratio e_{v1} of the primary aggregates (*pa*) of the Wilmington District (WD) sediment.

Starting with Eq. 4.51 we will take k_{f1} to be 6.0 cm^3 g^{-1} (Fig. 4.17) obtained in the capillary viscometer, considering the flocs to be primary (zero-order) aggregates, *pa*. Thus $n-1 = 0$, *i.e.*, $n = 1$. In Fig. 4.20 this sediment has three orders of aggregates higher than 1, which are, therefore, $n = 2$, 3 and 4. The respective k_{fn} values (cm^3 g^{-1}) are $k_{f2} = 14.4$, $k_{f3} = 22.8$ and $k_{f4} = 31.1$. In the plot of k_{fn} versus $n-1$ given in Fig. 4.23, the slope of the line $e_{v1}k_{f1}$ is equal to 8.4. Therefore, $e_{v1} = 8.4/6.0 = 1.4$.

In Table 4.7, floc densities and shear strengths of the four sediments from Fig. 4.23 are given along with the orders. For each sediment, the density and the shear strength decrease with increasing order. With the exception of the sample (SF) from the San Francisco Bay, the flocs conform to three orders of aggregation ($n-1 = 3$). The six orders of SF aggregates suggest highly cohesive sediment, as a result of which very loosely packed high-order flocs were formed and sustained (at low rates of shear). At the other end, zero-order flocs of SF were more tightly packed than zero-order flocs of other sediments, as reflected by the void ratio e_{v1}; the value for SF was only 0.40, as opposed to 1.2 to 1.4 for the

Table 4.7. Floc densities and shear strengths of four sediments.

Sediment	Order of aggregation	Floc density[a] (kg m^{-3})	Floc shear strength[c] (Pa)
Wilmington District (WD)	0	1,250	2.1×10^{0}
	1	1,132	9.4×10^{-1}
	2	1,093	2.6×10^{-1}
	3	1,074	1.2×10^{-1}
Brunswick Harbor (BH)	0	1,164	3.4×10^{0}
	1	1,090	4.1×10^{-1}
	2	1,067	1.2×10^{-1}
	3	1,056	6.2×10^{-2}
San Francisco Bay (SF)	0	1,269	2.2×10^{0}
	1	1,179	3.9×10^{-1}
	2	1,137	1.4×10^{-1}
	3	1,113	1.4×10^{-1}
	4	1,098	8.2×10^{-2}
	5	1,087	3.6×10^{-2}
	6	1,079	2.0×10^{-2}
White River (WR)[b]	0	1,212	4.9×10^{0}
	1	1,109	6.8×10^{-1}
	2	1,079	4.7×10^{-1}
	3	1,065	1.9×10^{-1}

[a] Density of seawater = 1,025 kg m^{-3}.
[b] In seawater.
[c] Assuming that this is equal to the upper Bingham yield stress.
Source: Krone [1963].

others. Tight packing at high shear rates is also the outcome of a high degree of cohesion. There is the likelihood that in San Francisco Bay cohesion is augmented by mucus coating surrounding and binding the mineral particles. If so, cohesion would not be entirely electrochemical; it would include the effect of biochemical adhesion [Kranck and Milligan, 1992; Wolanski, 2007].

Broadly speaking floc orders can be generated in two ways. Perikinetic flocculation refers to weak flocs resulting from diffusion associated with the random motion of colloidal (sol) particles bombarded rapidly by the fluid molecules. This is also called diffusion-limited aggregation (DLA). In orthokinetic flocculation collisions result from bulk fluid motion, such as by flow shearing. Strong flocs are formed when collisions overcome strong inter-particle repulsion. This is reaction-limited aggregation (RLA). Since successful outcomes tend to be infrequent, floc growth is slow. This is the mode by which flocs are formed in turbulent flows [Wells and Goldberg, 1993; Lin *et al.*, 1990a, 1990b; Winterwerp and van Kesteren, 2004].

Figure 4.24 shows colloidal flocs in waters off the coast of southern California. The large floc in Fig. 4.24a is from 80 m depth in the Santa Monica Basin of the coastal ocean. It displays an open structure associated with high-order aggregates. Notwithstanding limitations in the interpretation of the two-dimensional micrograph, if we assign this floc the order $n-1$, within the rectangular frames there is the suggested presence of at least three lower-order building blocks. It is postulated that these weak structures are formed by DLA. The much more compact and lower-order flocs in Fig. 4.24b are from a depth of 125 m in the San Clemente Basin. These stronger flocs are believed to be formed by RLA.

4.5.2 *Fractal representation*

The order of aggregation concept is consistent with the postulate that properties of the flocs of clay minerals can be approximately described in terms of self-similarity. For floc size, representation by geometric similarity, *i.e.*, fractal description, to the extent it is obeyed, would

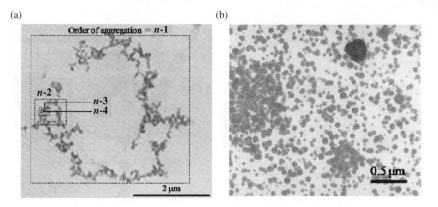

Figure 4.24. Electron micrographs of colloidal flocs from waters off southern California. (a) At a depth of 80 m in the Santa Monica Basin; (b) at a depth of 125 m in the San Clemente Basin (adapted from Wells and Goldberg [1993]).

obviously be most helpful. Fractal shapes display consistent patterns at multiple scales, *i.e.*, they are independent of scale. For example, at a high altitude the shoreline of many tidal creeks commonly appears irregular with crenulations. At a lower altitude a segment of the same shoreline will show similar crenulations at a smaller scale. At an even higher resolution the same mode of irregularity will be repeated at the scale of individual rocks and eroded notches [Kranenburg, 1994].

Consider a simple cube of size a_c as a representation of a zero-order aggregate (Fig. 4.25). Combining 8 such units gives rise to a cubic aggregate of size $a_f = 2a_c$. Further combining 8 aggregates of size $2a_c$ results in a cubic "aggregate-aggregate" of size $a_f = 4a_c$ with 64 primary units, and so on. In general, the number N_f of particles in a fractal aggregate of size a_f is given by

$$N_f = \left(\frac{a_f}{a_c}\right)^{D_f} \tag{4.52}$$

where D_f is the fractal dimension, which has no units. Thus,

$$D_f = \frac{\log N_f}{\log\left(\dfrac{a_f}{a_c}\right)} \tag{4.53}$$

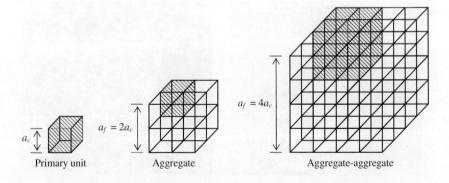

Figure 4.25. Cubic fractal structures formed from a primary cube.

In the present example, for an aggregate made of 8 primary units, we have $N_f = 8$ and $a_f/a_c = 2$. Therefore $D_f = \log 8/\log 2 = 3$, and $N_f = (2a_c/a_c)^3 = 8$. Taking the aggregate-aggregate, $N_f = (4a_c/a_c)^3 = 64$ and $D_f = \log 64/\log 4 = 3$. Thus the fractal dimension D_f remains constant. The density of the aggregate is that of the primary unit and remains independent of particle size. However, if some of the primary units are conveniently removed from the aggregate without changing its overall size, the fractal dimension would decrease (see Exercise 4.8). As mentioned the order of aggregation model of Krone [1963] does not depend on geometric self-similarity. Instead, void ratio similarity is assumed. For consistency between this model and the fractal description, the fractal dimension must be permitted to decrease as the order of aggregation increases. As described later, this allowance improves model prediction of floc growth in turbulent flows.

The range of D_f is from 1 (linear self-similarity) to 3 (volumetric self-similarity), with 2 corresponding to areal self-similarity. The case of $D_f = 3$ is one of pure coalescence, and corresponds to the continuum viewpoint, *i.e.*, the material remains the same no matter how finely it is divided. This viewpoint works well in most fluid mechanics problems, breaking down only when the length scale of interest approaches the molecular size. In marine flocs D_f is found to be less than 3, with its exact value dependent on the mechanism by which aggregation occurs, and on the type of aggregating material. Visually also there is a difference; a structure with $D_f > 2$ appears tightly packed, whereas structures with $D_f < 2$ are more open. If the flocs are subject to consolidation, restructuring may result in a D_f value of up to about 2.7. In contrast, offshore of southern California in 80 m depth, Wells and Goldberg [1993] reported very open colloidal flocs in suspension with $D_f = 1.3$ formed by perikinetic aggregation (Fig. 4.24a).

In the following two sections we will review fractal relationships between floc shear strength, density and size.

4.5.3 *Floc shear strength and density*

Although one cannot expect flocs to be ideally fractal given the heterogeneity in the particle population and the variety of complex mechanisms for flocculation, bulk properties that depend on the structure of flocs mainly of inorganic matter do support the fractal hypothesis to a reasonable degree. For example, as described next, floc density and shear strength can be empirically related to floc diameter raised to a power dependent on the fractal dimension.

Based on the data in Table 4.7 we may restate Eq. 4.49 in the form

$$\tau_f = K_F \Delta \rho_f^{5/2} \tag{4.54}$$

in which the excess density of flocs $\Delta \rho_f = \rho_f - \rho_w$, where ρ_f is the floc density, ρ_w is the water density and K_F is a sediment dependent constant [Partheniades, 1993]. Recall that this relationship was obtained for laminar flows. To assess its fractal basis let η_v be the number of particles or flocs per unit volume of suspension. The associated solids volume fraction is

$$\phi_v = \eta_v N_f V_p \tag{4.55}$$

where V_p is the particle volume. Similarly, if V_f is the floc volume, the volume fraction of flocs will be

$$\phi_{vf} = \eta_v V_f \tag{4.56}$$

Since from geometry,

$$\frac{V_f}{V_p} = \left(\frac{a_f}{a_c}\right)^3 \tag{4.57}$$

by combining Eqs. 4.52, 4.55, 4.56 and 4.57 we obtain

$$\frac{\phi_{vf}}{\phi_v} = \left(\frac{a_f}{a_c}\right)^{3-D_f} \tag{4.58}$$

Now, as $\phi_v = C / \rho_s$, we can restate Eq. 4.43 as

$$\rho_f = \frac{\phi_v}{\phi_{vf}}(\rho_s - \rho_w) + \rho_w \tag{4.59}$$

which may be combined with Eq. 4.58 to yield

$$\Delta\rho_f = \Delta\rho_s \left(\frac{a_f}{a_c}\right)^{D_f-3} = \Delta\rho_s \left(\frac{d_f}{d_p}\right)^{D_f-3} \tag{4.60}$$

where $\Delta\rho_s = \rho_s - \rho_w$ and a_f/a_c is replaced by the ratio of diameters d_f/d_c for (nominally) spherical particles. This equation provides a method for calculation of the fractal dimension from $D_f = [\log(\Delta\rho_f / \Delta\rho_s)/\log(d_f / d_p)] + 3$, when all the parameters on the right-hand side are known. Once $\Delta\rho_f$ is obtained from Eq. 4.60, Eq. 4.54 yields τ_f. Note that $\Delta\rho_s$ refers to the excess density of the mineral particle of diameter d_p. This particle for clays is typically plate-like or tubular and bears no self-similar relationship with the primary aggregate, either with respect to the diameter or the void ratio. In other words, the use of the diameter of the mineral particle, and its density for the primary fractal unit, results in a fractal dimension that is constrained by the inherent assumption of self-similarity between the clay particle and the resulting floc.

It can be argued that Eq. 4.54 may be used even when the flow is turbulent because at the scale of floc-floc interactions the flow closely surrounding the floc can well be effectively laminar. A caveat is that this condition holds only as long as the particles are smaller than the Kolmogorov turbulent eddy length scale λ_0 (Eq. 2.89).

Example 4.9: Estimate the Kolmogorov eddy length scale for turbulence-mean flow shear rate $\bar{G} = 1/15$ s^{-1}.

From Eq. 2.88, with $\bar{G} = \sqrt{(du'/dz)^2}$ the energy dissipation rate $\varepsilon_t = v$. Therefore from Eq. 2.89 $\lambda_0 = \sqrt{v}$. Taking $v = 10^{-6}$ m^2 s^{-1} we obtain $\lambda_0 = 1,000$ µm, which is the scale of a macrofloc. In the marine environment the typical range of λ_0 is 100 to 1,000 µm. Thus it would appear that, at least in principle, Eq. 4.54 would be more appropriate for calculation of the shear strength of microflocs than macroflocs.

In turbulent flows the shear force F_{dmax} at the yield point or break point of the floc may be stated as

$$F_{dmax} = \tau_{max} a_f^p = \tau_f a_f^p \tag{4.61}$$

where p is an unknown exponent. It can be postulated that the floc shear strength is determined mainly by the relatively small number of particles that form the weakest inter-floc links. As an approximation we may consider F_{dmax} to be independent of floc size.[h] Then, since the floc cross-sectional area is proportional to a_f^2, p will be 2. Thus we find that [Bremer *et al.*, 1989; Kranenburg, 1994]

$$\tau_f a_f^2 = \text{constant} \tag{4.62}$$

Eliminating a_f between Eqs. 4.60 and 4.62 yields

$$\tau_f = \text{constant}\left(\frac{\Delta\rho_s^{-2/(3-D_f)}}{a_c^2}\right)\Delta\rho_f^{2/(3-D_f)} = K_F \Delta\rho_f^{2/(3-D_f)} \tag{4.63}$$

For given sediment the term within parentheses is a constant, hence the logarithm of τ_f should vary linearly with the logarithm of $\Delta\rho_f$. A comparison between Eqs. 4.54 and 4.63 indicates that $2/(3-D_f) = 5/2$, *i.e.*, $D_f = 2.2$, which is consistent with the structure of orthokinetic macroflocs *e.g.*, as noted by Alldredge and Gotschalk [1988].

If each sediment in Table 4.7 were treated separately, the mean D_f value would deviate from 1.95 (Fig. 4.26), but the correlation between

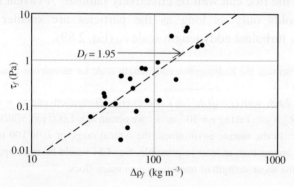

Figure 4.26. Floc shear strength versus excess floc density. Data are based on the analysis of U.S. Army Corps of Engineers (USACE) District sediment samples by Krone [1963].

---

[h] As we will note later this assumption is overly restrictive.

data and Eq. 4.63 would improve considerably (Table 4.8). Hawley [1982] reported a much lower value of 1.4 from lake flocs, suggesting the effect of organic matter in the production of flat, at times chain-like, flocs with highly open structures.

4.5.4 *Floc size and density*

A characteristic increase in the floc void ratio or decrease in the density with increasing floc size is unique to cohesive particles, and arguably plays a critical role in governing the dynamics of floc exchange at the bed surface. For mono-size particles this relationship is given by Eq. 4.60. The following description is an extension applicable to multi-size particles [Khelifa and Hill, 2006].

From Eq. 4.52 the diameter d_f of flocs of primary particles of size d_p is given by

$$d_f = d_p \, N_f^{1/D_f} \tag{4.64}$$

This definition can be extended to include multi-size primary particles as

$$d_f = \left(\sum_{i=1}^{i=N_f} d_{pi}^{D_f} \right)^{1/D_f} \tag{4.65}$$

in which the fractal dimension D_f is generally defined as

Table 4.8. Fractal dimension of USACE District sediment samples.

Data set	Regression exponent, $2/(3-D_f)$ of Eq. 4.70	Correlation coefficient, r^2	Fractal dimension, D_f
All data	1.91	0.559	1.95
Wilmington District	1.89	0.952	1.94
Brunswick Harbor	2.70	0.998	2.26
Gulfport Bay	2.05	0.976	2.02
San Francisco Bay	2.86	0.967	2.30
White River	2.02	0.974	2.01

Source: Letter [2009].

$$D_f = \alpha_D \left(\frac{d_f}{d_p} \right)^{\beta_D} \qquad (4.66)$$

This implies that D_f can vary with the ratio of floc diameter d_f to the diameter d_p of the constituent particles. The coefficients α_D and β_D are specified from two limiting cases. Firstly, D_f should approach a maximum value of 3 (volumetric self-similarity) when the floc size approaches the particle size. Secondly, D_f should reach a lower value D_{fc} when the floc size reaches some characteristic value d_{fc}. Applying these conditions to Eq. 4.66 yields

$$D_f = 3 \left(\frac{d_f}{d_p} \right)^{\frac{\log(D_{fc}/3)}{\log(d_{fc}/d_p)}} \qquad (4.67)$$

Example 4.10: For use with Eq. 4.67, Khelifa and Hill [2006] suggest illustrative values of D_{fc} and d_{fc} to be 2 and 2,000 µm, respectively. Select $d_p = 0.1$ µm and determine the fractal dimension for $d_f/d_p = 1$, 10, 100 and 1000. Comment on the results.

The general equation is $D_f = 3(d_f / d_p)^{-0.041}$, in which the small value of β_D (–0.041) is commensurate with 0.1 recommended by Maggi *et al.* [2007]. The answers are $D_f = 3$, 2.73, 2.48 and 2.26. For $d_f/d_p = 1$, full cubic self-similarity ($D_f = 3$) exists as expected. With increasing d_f/d_p the fractal dimension approaches 2, consistent with laboratory and field observations. The fractal dimension of 1.95 in Fig. 4.26 implies that, since $d_p = 0.1$ µm is a typical value of the primary particle diameter, flocs in the six estuaries listed in Table 4.8 are, on average, about three orders of magnitude larger than the respective primary particles, which is common.

The multi-size form of Eq. 4.60 can be stated by maintaining the assumption of a constant fractal dimension for a floc containing N_f primary particles of diameter d_{pi}. Thus

$$\Delta \rho_f = \Delta \rho_s \frac{\sum_{i=1}^{i=N_f} d_{pi}^3}{\left(\sum_{i=1}^{i=N_f} d_{pi}^{D_f} \right)^{\frac{3}{D_f}}} \qquad (4.68)$$

The following mean variables are now introduced

$$m_v = \frac{\sum\limits_{i=1}^{i=N_f} d_{pi}^3}{N_f} \quad \text{and} \quad m_f = \frac{\sum\limits_{i=1}^{i=N_f} d_{pi}^{D_f}}{N_f} \tag{4.69}$$

Combining Eqs. 4.68 and 4.69 yields

$$\Delta\rho_f = \Delta\rho_s N_f^{\frac{D_f - 3}{D_f}} \psi \tag{4.70}$$

where $\psi = m_v / \left(m_f^{3/D_f} \right)$. For mono-size particles $\psi = 1$ and Eq. 4.70 reduces to Eq. 4.60.[i] Equation 4.70 is intractable because the number of particles within a floc is generally unknown. To obtain a useful relationship the following modified form of this equation is introduced

$$\Delta\rho_f = \Delta\rho_s \left(\frac{d_f}{d_{50}} \right)^{D_f - 3} \psi \tag{4.71}$$

When using the median diameter d_{50} to represent the primary particle size distribution the value of ψ is taken as unity. Curves of excess density as a function of floc size from Eq. 4.71 are plotted in Fig. 4.27 for three values of D_f. These curves roughly encompass the domain of data compiled from different experimental studies by Khelifa and Hill [2006].

4.5.5 *Floc size and shear strength*

As mentioned, the order of aggregation concept does not lead to quantitative conclusions concerning floc size, although in general the nominal diameter of the floc can be expected to increase with increasing order, along with a tendency towards increasing flatness. These trends imply that as the floc size d_f increases its strength τ_f can be expected to decrease, which is supported by the laboratory data in Table 4.9. The

---

[i] We note that $N_f^{(D_f-3)/D_f} = \left(N_f^{1/D_f} \right)^{D_f-3} = \left(d_p N_f^{1/D_f} / d_p \right)^{D_f-3} = \left(d_f / d_p \right)^{D_f-3}$.

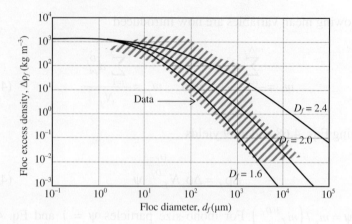

Figure 4.27. Effect of fractal dimension on the variation of floc excess density with diameter (adapted from Khelifa and Hill [2006]).

strengths were determined by using materials and methods different from the ones described in the previous section. However, the trend is consistent with that for cohesive sediment. Field observations of flocs in the Adriatic Sea do not seem to corroborate the trend of decreasing strength with increasing size [Mikkelsen *et al.*, 2006]. It is postulated that a significant cause may be the disparate stress-history of flocs, all sampled at a given time and location. There also are some unavoidable differences between data collected in laterally confined laboratory settling columns and unconfined waters of the sea (*e.g.*, Curran *et al.*

Table 4.9. Floc size and strength.

Investigators	Type of floc	Floc diameter, d_f (µm)	Shear Strength, τ_f (Pa)
Bache *et al.* [1999]	Alumino-humic	238	0.08
		182	0.16
		143	0.29
		120	0.42
Yeung and Pelton [1996]	Polymer A + calcium carbonate	25	100
	Polymer B + calcium carbonate	10	1000

Source: Data compiled by Jarvis *et al.* [2005].

[2003]). Accordingly, the following analysis should be considered to be approximate.

Representing the relationship between τ_f (Pa) and d_f (μm) generally as

$$\tau_f = K_f d_f^{-m_f} \tag{4.72}$$

yields the values $K_f = 10^6$ and $m_f = 3$. A relationship between τ_f and d_f can also be obtained by recognizing that at higher rates of flow shear a size-limiting effect evolves, once the floc breakup rate begins to exceed the growth rate. This development can be elaborated as follows [Overbeek, 1952; Krone, 1962].

Consider a two-dimensional representation of a floc of diameter $d_f = 2r_f$ in laminar flow (Fig. 4.28a, b). The linear velocity gradient, *i.e.*, shear rate, is du/dz and the torque available to spin the sphere about a central axis is

$$\Gamma_f = 2\int_0^{r_f} \frac{3\eta}{2r_f} z \frac{du}{dz} 2\pi r_f dz . z \tag{4.73}$$

which upon integration gives

$$\Gamma_f = 2\pi \eta r_f^3 \frac{du}{dz} \tag{4.74}$$

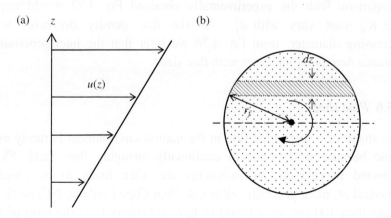

Figure 4.28. Torque on a nominally spherical floc due to laminar (linear) velocity gradient.

or

$$\Gamma_f = 4\pi \int_0^{r_f} \tau z \sqrt{(r_f^2 - z^2)} dz \qquad (4.75)$$

where $\tau = \eta du/dz$ is the shear stress. When two non-rigid spherical particles or flocs collide their common area of contact can be estimated by taking Δr as the inter-penetration distance. Then the contact area is

$$\pi[r_f^2 - (r_f - \Delta r)^2] \approx 2\pi r_f \Delta r \qquad (4.76)$$

Since the shear force is Γ_f / r_f, we obtain the size-limiting maximum collisional shear stress τ_{max} over the contact area

$$\tau_{max} = \frac{2\pi \eta r_f^3 \dfrac{du}{dz}}{2\pi r_f \Delta r} \qquad (4.77)$$

From this the floc diameter is

$$d_f = \left[\frac{2\Delta r}{\eta (du/dz)}\right] \tau_{max} = K_f' \tau_{max} \qquad (4.78)$$

If in this model τ_{max} is interpreted as the floc shear strength τ_f, comparison with the experimentally obtained Eq. 4.72 would imply that K_f' must vary with d_f^4. As the floc density decreases with increasing diameter, from Eq. 4.78 we infer that the inter-penetration distance Δr rapidly increases with floc size.

4.5.6 *Floc shape*

The shape of a suspended floc in the natural environment is rarely ever static because it responds to continually changing flow field. Flocs collected in energetic environments are often likely to be roughly spherical or, more generally, elliptical. Near Cape Lookout, NC, particles larger than 100 μm were found to have sphericity (*i.e.,* the ratio of the surface area of a sphere to the surface area of particle having the same volume) of 0.6 to 0.7. About 80% of measured flocs at a salinity of 2 ppt from upper Chesapeake Bay were found to have cylindrical shapes with the long axis (parallel to the direction of settling), that was on average 1.6 times as long as the narrow axis. The drag coefficient for cylinders best fit the observed settling velocities [Gibbs, 1985; Luettich *et al.*, 1993].

In low shear rate conditions non-spherical shapes are commonly produced. Flocs formed by differential settling in the laboratory appear crescent-shaped, while "marine snow" flocs in deep estuaries and shallow seas tend to be long and chain-like. Marine snow consists of very large, exceptionally fragile and typically planar macroflocs often seen by divers or recorded instrumentally [Hill, 1998]. These are high-order flocs with very open structure, weak inter-particle bonding and density close to water. As mentioned before, there is a strong indication that organic matter coating mineral particles play a key role in enabling the formation of extended structures of the type shown in Fig. 4.24a [*e.g.*, Jackson, 1990].

Kranck and Milligan [1980] produced large floc structures using glacial mud from the Southwest Miramichi River near Sillikers in New Brunswick, Canada, mixed with organic matter from samples collected in the Bedford Basin. The mixture was placed inside a 100-cm tall and 80-cm diameter glass cylinder subjected to an upward flow of recirculating seawater at a speed ranging between 0.01 to 0.25 cm s^{-1}. Initially, this flow prevented the suspended particles from settling, and they recirculated with water. However, due to shear-induced aggregation, after about a day to several days depending on mixture composition and concentration, the particles became large enough to have settling velocities equaling or exceeding the counterflow. The outcome was that the aggregates began to accumulate at the bottom.

Figure 4.29 shows large flakes formed using a mixture containing 50% organic matter by weight. They are orthokinetic by virtue of the procedure used to create them. The initial suspension concentration was 237 ppm and counterflow speed 0.1 cm s^{-1}. Details related to this type of flocculation in the laboratory are found in a range of publications [Krone, 1986; Kranck *et al.*, 1993; Lick and Huang, 1993; Heffler *et al.*, 1991; Wells and Goldberg, 1993; Zabawa 1978a, 1978b]. Michaels and Bolger

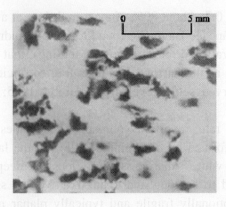

Figure 4.29. Orthokinetically produced marine snow in a laboratory apparatus (from Kranck and Milligan [1980]).

[1962] point out that, as we have noted, selection of the diameter of the column becomes an important consideration when floc structures are likely to grow to sizes comparable to the dimensions of the vessel, as the walls interfere and contaminate the growth data [see also Curran *et al.*, 2003].

4.6 Collision Mechanisms

4.6.1 *Collision frequency functions*

As mentioned, floc growth (aggregation) occurs when colliding particles or their parts cohere and increase in mass, whereas breakup (disaggregation) is due to shearing of the floc by flow and collisions.

As to collisions, the number of flocs formed per unit volume N_c by (two-body) collisions between an i-class particle and a j-class particle (Fig. 4.30) per unit volume of suspension and time must be calculated. In general

$$N_C = \alpha_c \beta_c n_{vi} n_{vj} \qquad (4.79)$$

where α_c [0, 1] represents collision efficiency, β_c is a collision frequency function which depends on particle size and the mechanism of collision,

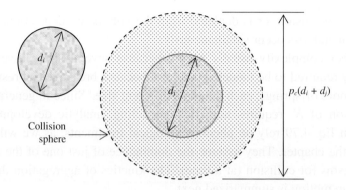

Figure 4.30. Definition diagram for two-body encounters.

n_{vi} is the number concentration of i-particles and n_{vj} that of j-particles. Two particles are considered to experience a collision if their relative motion causes the i-particle to intrude within the collision sphere of the j-particle. This sphere has a diameter $p_c(d_i + d_j)$, where the collision diameter function p_c ranges between 0 and 1. A common mean value derived from laboratory data is 0.75, which would indicate that on average interpenetration of two particles during a collision effectively reduces the particle diameter by 25% [McAnally, 1999]. We will refer to the significance of p_c in connection with mechanisms for aggregation.

In the simplest calculation of N_c, particles can have two size classes. Assuming $d_2 > d_1$, the diameter d_1 would be the basic particle and d_2 typically a floc, with the volume (and mass) of the second particle an integral multiple of the first particle. There can be three types of two-body collisions, 1-1, 2-2 and 1-2. Three-body collisions can also occur but are rarer and are sometimes ignored. They have also been treated as two successive two-body encounters.

Holding α_c in Eq. 4.79 constant, for β_c there are three main mechanisms, all of which depend on d_1 and d_2. If an external supply of sediment to the suspension volume or loss from the volume is absent, then the initial number concentrations corresponding to n_{v1} and n_{v2} will change until an equilibrium condition occurs between the two concentrations, while the total mass concentration of sediment in the volume remains constant. Thus in due course N_c will become constant. Other cases can arise if new sediment is permitted into the volume, *e.g.,* by bed erosion, or is depleted from the volume by deposition, or if both gain and loss occur simultaneously.

Further complexity usually occurs because several size classes are typically required to be modeled, and also because breakage can result in two or more offsprings, each having a different size.[j] Since in general the calculation of N_c requires numerical modeling, analytic developments based on Eq. 4.79 rely on phenomenological treatments, as we will see later in the chapter. They assume the dominance of just one of the three mechanisms for collision (at a time). The kinetics of aggregation due to Brownian motion is summarized next.

4.6.2 *Brownian motion*

In 1827 Robert Brown noted that the movements of pollen and other microscopic particles suspended in a liquid "arose neither from currents in the fluid nor from its gradual evaporation, but belonged to the particle itself." This type of motion occurs because at any given instant the sum of forces due to thermally induced scattering of the fluid molecules is typically not zero, *i.e.,* the forces do not balance. Perikinetic aggregation by Brownian mechanism is slow and the resulting flocs tend to be fragile. It is the weakest of the three mechanisms of inter-particle collisions and is active in, *e.g.,* shallow estuaries when the current is close to slack and the settling particles accumulate in the lower water column. Although this mechanism is of minor importance in many coastal sediment transport models, its kinetics, described here in a brief

---

[j] Sediment classes can be defined by a progression of particle/floc diameter according to

$$d_i = \beta_d d_{i-1} \tag{F4.9}$$

Since the coefficient of progression β_d is taken to be a constant, the difference between the logarithms of successive size classes is also constant. Given the total number of size classes M_c, β_d is given by

$$\beta_d = \left(\frac{d_{max}}{d_{min}} \right)^{1/(M_c-1)} \quad ; \quad M_c > 1 \tag{F4.10}$$

where d_{min} and d_{max} refer to the (selected) minimum and the maximum diameters, respectively [Letter, 2009].

way, leads to an equation for the rate of decrease of suspended particles that has been used in simple explanations of aggregation processes when modeling deposition of cohesive sediment (Chapter 7).

For a dispersed particle of diameter d_p, its dimensionless representation can be defined as $\tilde{d}_p = d_p(\overline{G}/K_T)^{1/3}$, where $\overline{G}$ is the turbulence-mean flow shear rate. The quantity $K_T = \kappa_B T/\eta$ is the ratio of thermal to viscous energy dissipation. When $\tilde{d}_p \ll 2^{1/3}$ Brownian aggregation dominates, and when $\tilde{d}_p \gg 2^{1/3}$ aggregation is controlled by flow shear [Hunt, 1982]. Given $d_n = 4$ μm, $T = 300°$ K, $\eta = 0.5$ Pa.s, $\kappa_B = 1.381 \times 10^{-23}$ J°K^{-1} and $\tilde{d}_p = 2^{1/3}$ yields $\overline{G} = 0.26$ s^{-1} as the threshold value of the shear rate. According to this criterion Brownian motion may be ignored at values of $\overline{G}$ greater than 0.26 s^{-1}.

The Brownian collision frequency function β_B is treated as arising from random-walk of particles according to the law

$$\beta_B = 4\pi D_c p_c (d_i + d_j) \qquad (4.80)$$

where D_c, the diffusion coefficient associated with the two particles d_i and d_j, is given by

$$D_c = D_B d_p \left(\frac{1}{d_i} + \frac{1}{d_j} \right) \qquad (4.81)$$

in which the so-called Brownian diffusion coefficient D_B is

$$D_B = \frac{K_T}{3\pi d_p} \qquad (4.82)$$

Substitution of Eqs. 4.81 and 4.82 into Eq. 4.80 results in

$$\beta_B = \left(\frac{2}{3} p_c K_T \right) \frac{(d_i + d_j)^2}{d_i d_j} \qquad (4.83)$$

The model for (perikinetic) aggregation is obtained as follows [Smoluchowski, 1917; Overbeek, 1952]. Let the coordinate origin of a randomly selected particle be its center. The coordinate of a displacement Δ' of any other i-particle with respect to the randomly selected particle will be equal to the actual displacement Δ_i of the i-particle minus the displacement Δ_1 of the randomly selected particle, *i.e.,*

$$\Delta' = \Delta_i - \Delta_1 \tag{4.84}$$

Squaring both sides and time-averaging yields

$$(\overline{\Delta'})^2 = \overline{(\Delta_i - \Delta_1)^2} = \overline{\Delta_i^2} - 2\overline{\Delta_i \Delta_1} + \overline{\Delta_1^2} \tag{4.85}$$

The displacements of both particles are random and therefore independent of each other. As a result

$$\overline{\Delta_i \Delta_1} = \overline{\Delta_i} \cdot \overline{\Delta_1} = 0 \tag{4.86}$$

As Δ_1 can also be considered to be the displacement of an i-particle, we have

$$(\overline{\Delta'})^2 = \overline{\Delta_i^2} + \overline{\Delta_1^2} = 2\overline{\Delta_i^2} \tag{4.87}$$

Thus the mean-squares of the displacements of all the particles relative to the first particle are doubled. Since particle displacements are due to random thermal scattering, Brownian motion of the other particles relative to the first particle is represented by the double diffusivity D_B', which governs the probability of i-particle successfully coalescing with the first particle. Accordingly, the rate of diffusion is given by

$$N_B = D_B' 4\pi R_v^2 \frac{d\,n_v(R_v)}{dR_v} \tag{4.88}$$

where N_B is the number of particles that pass per unit time towards the central particle through the surface area $4\pi R_v^2$ of a spherical volume of radius R_v, and $n_v(R_v)$ is the number concentration of particles. The first particle is at the center of this sphere, and the approaching particles coalesce with this particle at a distance of separation r_v ($< R_v$) (Fig. 4.31).

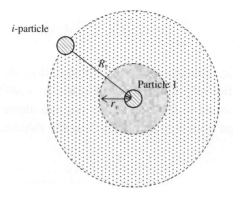

Figure 4.31. Distances relevant to i-particle approaching particle 1.

We may now assume that $n_v(r_v) = 0$, because once the particles are at a distance r_v from the central particle they are instantly absorbed by this particle. With this condition, integration of Eq. 4.88 yields

$$n_v(R_v) = \frac{N_B}{4\pi D_B'}\left(\frac{1}{r_v} - \frac{1}{R_v}\right) \qquad (4.89)$$

As $n_v(\infty) = 0$, we obtain

$$N_B = 8\pi D_B'' n_{v0} r_v \qquad (4.90)$$

in which n_{v0} is the initial value of n_v and D_B' has been conveniently replaced by $2D_B''$. The total number of approaches of all particles per unit volume must be equal to $N_B n_v/2$, where division by 2 is required because every pair of particles would otherwise be counted twice. Replacing n_{v0} in the above equation by $n_v(R_v)$ to represent N_B at any time, the total number of approaches per unit volume becomes $4\pi D_B'' r_v n_v^2$.

When the number of particles at the start of their growth by collisions is large, the rate at which their number decreases is rapid. Thus for instance when collisions are forced, as in a stirred tank reactor, the rate equation has a high order (*e.g.*, Mehta [1969]). For our "natural" case we will assume that the rate of decrease of n_v due to aggregation is a second-order process, and is accordingly represented as

$$\frac{dn_v}{dt} = -4\pi D_B'' r_v n_v^2 = -k_v n_v^2 \tag{4.91}$$

where $k_v = 4\pi D_B'' r_v$ characterizes the probability of two particles approaching sufficiently close to each other to coalesce into a new particle of mass equal to the sum of the original masses. Integration of the above equation from $n_v = n_{v0}$ at $t = 0$ to any value n_v at time t yields

$$\frac{1}{n_v} - \frac{1}{n_{v0}} = k_v t \tag{4.92}$$

or

$$n_v = \frac{n_{v0}}{1 + k_v n_{v0} t} \tag{4.93}$$

Introducing a characteristic time of aggregation $t_c = 1/k_v n_{v0}$ we obtain

$$n_v = \frac{n_{v0}}{1 + \dfrac{t}{t_c}} \tag{4.94}$$

This expression may be restated as

$$n_v^{-1} = k_v t + n_{v0}^{-1} \tag{4.95}$$

which indicates that n_v^{-1} is linearly proportional to t. Experimental evidence confirms this aggregation model (Fig. 4.32). Aggregation ends when collisions cannot occur any further.

4.6.3 *Flow shear*

In the natural environment turbulent shear is the most important mechanism for inter-particle collision because it produces tightly packed, lasting flocs of low orders of aggregation. Referring to Fig. 4.33a, at time $t = 0$ an i-particle moving in the x-direction with a velocity u_i relative to

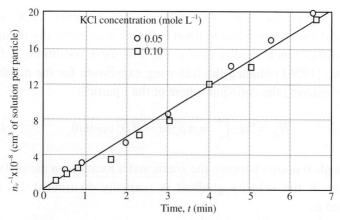

Figure 4.32. Dependence of the inverse of number concentration of particles of a blue hydrosol of gold on time of aggregation (fromVoyutsky [1978]).

the j-particle makes initial contact with the collision sphere of the j-particle. The mean velocity u_i in shear flow is taken as

$$u_i = \frac{(d_i + d_j)}{2}\overline{G} \qquad (4.96)$$

where $\overline{G}$ is the turbulence-mean shear rate. It is estimated by averaging the velocity gradient du/dx over the short time interval $t = t_*$ it takes to complete the collision. At this point the distance of penetration of the i-particle into the collision sphere of the j-particle is Δr (Fig. 4.33b). The two masses m_i and m_j of the i- and j-particles, respectively, coalesce to yield a new mass $m_k = m_i + m_j$. It is assumed that the two particles are

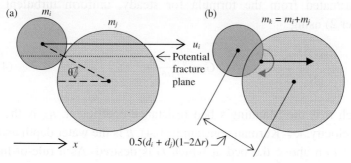

Figure 4.33. Collision between two particles: (a) Initial contact at time $t = 0$; (b) collision complete at time $t = t_*$.

approximately of the same size, they do not influence each other's motion, and the energy released upon collision is isotropically dissipated through eddies much smaller than $d_i + d_j$. For these conditions Saffman and Turner [1956] obtained the following expression for the number of i-particles entering the collision sphere of the j-particle

$$N_{sh} = 2n_{vi} \int_0^{\pi/2} u_i \pi p_c^2 (d_i + d_j)^2 \cos\theta d\theta \qquad (4.97)$$

The angle θ occurs between the x-axis and a location on the surface of the sphere. If the velocity gradient is normally distributed, $\overline{G}$ can be determined as

$$\overline{G} \approx \sqrt{\left(\frac{du}{dz}\right)^2} = \sqrt{\frac{2\varepsilon_t}{15\pi v}} \approx 0.8 \sqrt{\frac{\varepsilon_t}{15v}} \qquad (4.98)$$

where ε_t is the turbulent energy dissipation rate [McAnally, 1999]. This expression differs from Eq. 2.88 of Taylor [1935] by the factor 0.8.

Combining Eqs. 4.96, 4.97 and 4.98 and integrating the resulting equation yields the collision frequency function

$$\beta_{sh} = \left(\frac{\pi p_c^2}{4} \sqrt{\frac{2}{15\pi}}\right) \sqrt{\frac{\varepsilon_t}{v}} (d_i + d_j)^3 \qquad (4.99)$$

Table 4.10 gives estimates of ε_t based on Eq. 4.98. The shear rate $\overline{G}$ was estimated from the formula for steady, uniform turbulent flow (Chapter 2) using

$$\overline{G} \approx \frac{\sqrt{g}\,n u_m}{\kappa h^{1/6} z_b} \qquad (4.100)$$

in which n is the Manning's bed resistance coefficient, u_m is the local mean velocity, κ = Karman constant (= 0.4), h is the water depth and z_b is the elevation above the bed at which $\overline{G}$ is desired. As a rule-of-thumb, flocs grow when $\overline{G}$ is of order 1 Hz or less. With increasing $\overline{G}$ breakup kinetics plays an increasing role.

Table 4.10. Estimates of energy dissipation rate for typical estuarine conditions.

Location	u_m (m s^{-1})	$\bar{G}$ (Hz)	ε_t (m^2 s^{-3})
Mean over	0.05	0.1	10^{-7}
water column	0.5	4	10^{-4}
	1.0	10	10^{-3}
Near-bed	0.05	3	10^{-4}
	0.5	200	10^{0}
	1.0	1000	10^{1}

Source: McAnally [1999].

Example 4.11: You are given the following parameters applicable to an estuarine channel: current speed 0.8 m s^{-1}, water depth 6.7 m, Manning's coefficient 0.028 and kinematic viscosity of water 10^{-6} m^2 s^{-1}. Estimate the energy dissipation rate 0.2 m above the bed.

With h = 6.7 m, n = 0.028, u_m = 0.8 m s^{-1}, v = 10^{-6} m^2 s^{-1} and z_b = 0.2 m, from Eq. 4.100 $\bar{G}$ = 0.639 s^{-1}. From Eq. 4.98 ε_t = 9.56×10^{-6} m^2 s^{-3}, which, due to turbulence, is an order of magnitude larger than v.

4.6.4 Differential settling

Let the i and j-particles (Fig. 4.33) have different settling velocities, which cause the particles to collide as they fall. The number of i-particles entering the collision sphere of the j-particle solely by differential settling is described by Eq. 4.97 with the velocity u_i replaced by the absolute value of the difference in settling velocities $|w_{si} - w_{sj}|$ between the two particles, *i.e.*,

$$N_{ds} = 2n_{vi}\int_0^{\pi/2} |w_{si} - w_{sj}| \pi p_c^2 (d_i + d_j)^2 \cos\theta \, d\theta \qquad (4.101)$$

from which the collision frequency function is obtained as

$$\beta_{ds} = \left(\frac{\pi p_c^2}{4}\right)(d_i + d_j)^2 |w_{si} - w_{sj}| \qquad (4.102)$$

The settling velocities can be calculated from Eq. 3.25 under the assumption that local processes influencing settling are viscous dominated. Other formulas for the settling velocity are given in Chapter 7.

4.6.5 *Net collision frequency*

The net collision frequency function may now be assumed to be the sum of collisions due to Brownian motion, shear and differential settling in turbulent flow [Han, 1989; Lick *et al.*, 1992], *i.e.*,

$$\beta_c = \beta_B + \beta_{sh} + \beta_{ds} \qquad (4.103)$$

Equation 4.79 therefore becomes

$$N_C = \alpha_c (\beta_B + \beta_{sh} + \beta_{ds}) n_{vi} n_{vj} \qquad (4.104)$$

A process which limits collisional encounters is flow divergence. Particle 1 approaching particle 2 is deflected due to the presence of 2, and if as a result 1 misses the encounter it will contribute to a decrease in α_c [Hill and Nowell, 1990; Hill, 1992].

Example 4.12: You are given the following parameters applicable to a saltwater channel: water temperature $20°C$, $p_c = 0.75$ and $\varepsilon_t = 5\times10^{-3}$ m^2 s^{-3}, $\rho_s = 2,600$ kg m^{-3}, $\rho = 1,025$ kg m^{-3}, $d_{so} = 3$ μm, $v = 1.15\times10^{-6}$ m^2 s^{-1} and $D_f = 2.1$. Take the diameter of the first particle to be 10 μm, and vary the second particle diameter from 0.1 to 1,000 μm. Use Eq. 3.25 to calculate the settling velocity. Determine the collision frequency functions β_B, β_{sh} and β_{ds} and plot them against the diameter of the second particle.
The results are shown in Fig. 4.34. As expected the differential settling function is nil when the two particles are of the same size. The well, which is an artifact of the deterministic approach, vanishes if the settling velocities are treated probabilistically, as the likelihood that any instant two particles have exactly the same settling velocity becomes very low [Letter, 2009]. For particles larger than about 2–4 μm Brownian motion has a negligible effect compared to the other two mechanisms.

4.6.6 *Collision efficiency and probability*

The collision efficiency α_c in Eq. 4.79 and the probability p_c in Eq. 4.80 are evaluated as follows.

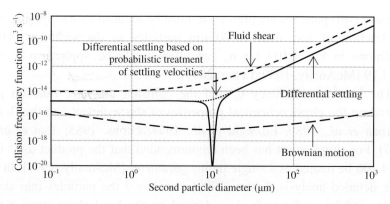

Figure 4.34. Variation of collision frequency functions with second particle diameter (for first particle diameter of 10 μm) (revised from McAnally [1999]).

We may restate Eq. 4.79 as

$$N_c = \alpha' \left(\frac{\beta_{sh}}{p_c^2} \right) n_{vi} n_{vj} \qquad (4.105)$$

where

$$\alpha' = \alpha_a \alpha_d \, p_c^2 \qquad (4.106)$$

By selecting $\beta_c = \beta_{sh}$ we have assumed the dominant role of shear-induced collisions in governing the collision efficiency. We will further consider α_a to be due to floc growth only, and separate from the floc breakup efficiency α_d.

Based on an inspection of collision efficiency data from laboratory experiments, the following empirical formulas for α_a have been proposed (for flocs containing particles less than 20 μm in size)

$$\alpha_a = 1; \qquad S \geq S_o$$
$$\alpha_a = \frac{S}{S_o}; \quad S < S_o \qquad (4.107)$$

where S is the salinity and S_o is a reference salinity, taken as 2 psu. As a practical matter, it is reasonable to assume $S/S_o \approx 1$ and $\alpha_a = 1$. If non-cohesive (silt) particles are present, α_a must be reduced by multiplying by the fraction of less than 10-20 μm particles in the sediment. The conditions in Eq. 4.107 for α_a can alternatively be applied to α_c in Eq. 4.79 [McAnally, 1999; Letter, 2009].

The breakup efficiency α_d and the collision diameter function p_c depend on the physicochemical properties of the sediment and the fluid [Burban *et al.*, 1989; Edzwald *et al.*, 1974; Gibbs, 1985; Tsai *et al.*, 1987]. For simplicity it has been recommended that the product $\alpha_d p_c^2$ in Eq. 4.106 be treated as a single tuning parameter [McAnally, 1999]. In a more detailed analysis, Letter [2009] distributed the particles into size classes and let α_d for each class depend on the local shear stress τ in excess of the floc shear strength τ_f. By treating the shear stress, the floc shear strength and the floc settling velocity as probabilistic variables, the effect of p_c could also be incorporated, implicitly, in the calculations of the three collision frequency functions in Eq. 4.79.

4.6.7 *Flow shearing and collisional stresses on flocs*

Floc breakup by shearing will occur when flow stress τ exceeds the floc shear strength τ_f. To obtain τ we may conveniently consider flow surrounding the particle to be locally viscous. Restating Eq. 4.74 as

$$\Gamma_f = 8\pi\eta r_f^3 \omega_r \qquad (4.108)$$

where the angular velocity of rotation is

$$\omega_r = \frac{1}{4}\frac{du}{dz} \qquad (4.109)$$

it can be shown (*e.g.*, Krone [1963]) that the shear stress τ around a thin disc at the equator is

$$\tau = \eta\frac{\omega_r}{2} \qquad (4.110)$$

Substituting for ω_r from Eq. 4.109 yields

$$\tau = \frac{1}{8}\eta\frac{du}{dz} \tag{4.111}$$

This expression permits a rough estimation of τ from the measured velocity gradient du/dz, or from the log-velocity profile (Chapter 2), or from model simulation [McAnally, 1999; Letter, 2009].

Floc growth or breakup due to particle-particle collisions are dependent on the collisional shear stress τ_k. We note (Fig. 4.33) that at time t_*, the duration of collision, the new mass m_k acquires a velocity u_k by the conservation of linear momentum

$$u_k = \frac{u_i m_i}{m_i + m_j} \tag{4.112}$$

and the force exerted on and by the particles during a collision is

$$F_k = \frac{u_k m_j}{t_*} \tag{4.113}$$

where t_* is approximately

$$t_* = \frac{p_c'(d_i + d_j)}{2u_i} \tag{4.114}$$

The quantity p_c' is the relevant fraction of interpenetration of the two flocs with a recommended value of 0.1 [McAnally, 1999; Krone, 1963]. The definition of p_c' differs from the collision diameter function p_c. Substitution of Eqs. 4.112 and 4.114 into 4.113 yields

$$F_k = \frac{2u_i^2 m_i m_j}{p_c'(d_i + d_j)(m_i + m_j)} \tag{4.115}$$

In the event a particle fractures, the fracture plane will be the weakest surface between the point where the force is applied and the center of mass of the particle. However, for a given force the location of the highest shear stress will be where the plane has the least area, which is between the applied force and the edge of the particle. In a particle of homogeneous composition the probable fracture plane will pass through the contact point between the particles. Dividing F_k by the area of the fracture surface (Fig. 4.33a) the collisional shear stress experienced by the j-particle is

$$\tau_k = \frac{8u_i^2 m_i m_j}{\pi d_j^2 p_c'(d_i + d_j)(m_i + m_j)} \tag{4.116}$$

When τ_k exceeds the shear strength τ_{fj} of the j-particle, it will break into two smaller particles. The shear strength may be approximated by the particle (floc) yield strength τ_y estimated from Eq. 4.135.

At typical turbulent energy dissipation rates ε_t in estuaries (Table 4.10) both, the collisional stress τ_k and the flow stress τ, can be important for aggregation processes operating within the full cohesive particle size range. For macroflocs and high energy dissipation rates τ_k tends to dominate [McAnally, 1999].

The outcome of the aggregation process is simply-defined by mass balance

$$m_j = m_{j1} + m_{j2}$$
$$m_{j1} = m_j - \Delta m_j \tag{4.117}$$
$$m_{j2} = \Delta m_j$$

The respective mean rate of change of mass per unit volume, C, will be

$$\frac{dC_1}{dt} = \left(m_j - \frac{1}{2} m_j \right) n_j$$
$$\frac{dC_2}{dt} = \frac{1}{2} m_j n_j \tag{4.118}$$

which implies that the combined mass of the two newly formed particles will be equal to that of the original, and *on average* each will have half the mass of the original. It also indicates that if the shear strength of the j-particle is exceeded by τ_k within any small fluid volume, *all* j-particles in that volume will be broken.

Expressions for τ_k due to Brownian motion, shear and differential settling can be obtained by substituting the respective velocity u_i into Eq. 4.116. For a conservative (maximum) estimate of shear stress, u_i can be represented as the sum of those three velocities. Depending on whether τ_k exceeds the shear strength of one or both particles, one or both colliding particles will break up, forming multiple particles. Outcomes of two-particle (and three-particle) collisions and mass change rates for determining the time-history of floc properties are obtained from numerical solutions involving Eqs. 4.117 and 4.118 [McAnally, 1999; Letter, 2009] or somewhat different parameterization [Lick *et al.*, 1992].

4.7 Floc Size Evolution

In place of numerical modeling of aggregation processes, floc growth and the final equilibrium size can be roughly estimated by using a simple approach for mono-size particles. The time-rate of change of a representative floc diameter d_f is described by the balance

$$\frac{dd_f}{dt} = R_{sh} + R_{ds} + R_{br} + R_{er} + R_{dp} \qquad (4.119)$$

where the (algebraically added) rate terms are: R_{sh} for shear-induced growth, R_{ds} for differential settling-induced growth, R_{br} for breakup, R_{er} for the effect of addition of particles by erosion and R_{dp} for depletion due to deposition. From laboratory tests Davis [1993] showed that where marine infaunal activity is high, R_{er} may be taken as the sum of three contributions: R_{erp} due to physical resuspension, R_{erb} due to bioturbation and R_{erph} due to physical-biological interaction. Other terms may be added to Eq. 4.119 such as those representing wave breaking, water exchange due to tide, river flow, and so on. However, in general, Eq. 4.119 can be simplified without a significant loss of accuracy by including only two terms, a composite growth function R_{gr} and a breakup function R_{br}

$$\frac{dd_f}{dt} = R_{gr} - R_{br} \qquad (4.120)$$

in which the negative sign identifies the opposite sense of breakup relative to growth. The growth and breakup functions are obtained as follows.

Growth Function: For R_{gr} the expression (equivalent to Eq. 4.91 for Brownian motion) giving the rate of change of the number concentration of flocs n_v due to shear represented by the turbulence-mean shear rate $\bar{G}$ is

$$\frac{dn_v}{dt} = -\frac{3}{2}\alpha_c\alpha_b\pi\bar{G}d_f^3 n_v^2 \qquad (4.121)$$

where α_c and α_b represent the collision efficiency and a diffusion-related efficiency, respectively, both limiting the rate of merger of colliding particles [Levich, 1962]. The above second-order rate expression is equivalent to Eq. 4.91 with $k_v = 3\pi\alpha_c\alpha_b\bar{G}d_f^3/2$. Based on Eq. 4.43 we obtain the floc volume fraction

$$\phi_{vf} = \left(\frac{\rho_s - \rho_w}{\rho_f - \rho_w}\right)\frac{C}{\rho_s} = \frac{\Delta\rho_s}{\Delta\rho_f}\frac{C}{\rho_s} = \alpha_s n_v d_f^3 \qquad (4.122)$$

where α_s is a volume shape factor and d_f is the nominal diameter of flocs. Combining Eq. 4.122 and Eq. 4.60 yields

$$n_v = \frac{C}{\alpha_s\rho_s}\frac{d_p^{D_f-3}}{d_f^{D_f}} \qquad (4.123)$$

Differentiation of Eq. 4.123 by applying the chain rule and use of Eq. 4.121 results in

$$R_{gr} = \frac{dd_f}{dt} = k_A'\frac{C}{\rho_s}\bar{G}d_p^{D_f-3+\beta_D}d_f^{-D_f+4-\beta_D}\frac{1}{\beta_D\ln\left(\dfrac{d_f}{d_p}\right)+1} \qquad (4.124)$$

where $k_A' = \alpha_c\alpha_b\pi/2\alpha_s$, and use is made of Eq. 4.66 for the representation of variable fractal dimension D_f.

Breakup Function: The function R_{br} is based on dimensional considerations following Winterwerp [1998]

$$\frac{dn_v}{dt} = \alpha_d n_v \overline{G} a_b \left(\frac{d_f - d_p}{d_p} \right)^{p_b} \left(\frac{\eta \overline{G}}{F_y / d_f^2} \right)^{q_b} \quad (4.125)$$

where α_d represents the breakup efficiency, η is the dynamic viscosity of fluid and F_y is the yield force (such that F_y / d_f^2 is the yield stress τ_y). Of the three empirical coefficients a_b, p_b and q_b, the latter two (exponents) measurably influence the breakup rate. By combining Eqs. 4.125 and 4.123 we obtain

$$R_{br} = \frac{dd_f}{dt} = k_B' \frac{\overline{G}}{3} \left(\frac{\eta \overline{G}}{F_y} \right)^{q_b} d_p^{\beta_D - p_b} d_f^{-\beta_D + 1 + 2q_b} (d_f - d_p)^{p_b} \frac{1}{\beta_D \ln \left(\dfrac{d_f}{d_p} \right) + 1}$$

$$(4.126)$$

where $k_B' = \alpha_d a_b$. From sensitivity analysis Son [2009] concluded that the values $p_b = 1$ and $q_b = 0.5$ selected by Winterwerp [1998] were indeed appropriate. With these values Eq. 4.126 can be stated as

$$\frac{dd_f}{dt} = \frac{\overline{G} d_p^{\beta_D}}{\beta_D \ln \left(\dfrac{d_f}{d_p} \right) + 1} \left[k_A' \frac{C}{\rho_s} d_p^{D_f - 3} d_f^{-D_f + 4 - \beta_D} - \frac{k_B'}{3} \sqrt{\frac{\eta \overline{G}}{F_y}} d_p^{-1} d_f^{-\beta_D + 2} (d_f - d_p) \right]$$

$$(4.127)$$

When $\beta_D = 0$, by virtue of Eq. 4.66 the fractal dimension becomes independent of the floc order. Under this assumption Eq. 4.127 changes to

$$\frac{dd_f}{dt} = \overline{G} \left[k_A' \frac{C}{\rho_s} d_p^{D_f - 3} d_f^{-D_f + 4} - \frac{k_B'}{3} \sqrt{\frac{\eta \overline{G}}{F_y}} d_p^{-1} d_f^2 (d_f - d_p) \right] \quad (4.128)$$

Further selecting $D_f = 2$, Eq. 4.128 reduces to the form derived by Winterwerp [1998]

$$\frac{dd_f}{dt} = k_A C\overline{G}d_f^2 - k_B \overline{G}^{3/2} d_f^2 (d_f - d_p) \qquad (4.129)$$

where $k_A = k_A'/\rho_s d_p$ and $k_B = k_B'\sqrt{(\eta/F_y)}/d_p$. At equilibrium $dd_f/dt = 0$, and one solution of Eq. 4.129 for the equilibrium diameter of the floc d_{fe} is trivial. The other is

$$d_{fe} = d_p + \frac{k_A}{k_B}\frac{C}{\sqrt{\overline{G}}} \qquad (4.130)$$

Taking C to be a constant, Eq. 4.129 is solved with the initial condition $d_f = d_{f0}$ at $t = 0$ yielding

$$\ln\left(\frac{d_{fe} - d_{f0}}{d_{fe} - d_f}\frac{d_f}{d_{f0}}\right) + \frac{d_{fe}}{d_{f0}} - \frac{d_{fe}}{d_f} = \frac{t}{t_f} \qquad (4.131)$$

which is implicit in d_f with the characteristic time given by

$$t_f = \left(k_B \overline{G}^{3/2} d_{fe}^2\right)^{-1} \qquad (4.132)$$

Two end-member time-scales emerge. One is for floc growth of highly cohesive sediment such as mud from San Francisco Bay, in which flocs with open structures are formed from small particles at low shear rates ($\overline{G}$). Thus $d_{fe} \gg d_{f0}$, and from Eq. 4.132

$$t_{f1} = t_f \frac{d_{fe}}{d_{f0}} \qquad (4.133)$$

In the second case the same flocs are broken at high shear rates, in which case $d_{fe} \ll d_{f0}$, and

$$t_{f2} = 2t_f \qquad (4.134)$$

Example 4.13: For a marine sediment you are given $k_B = 1.4 \times 10^4$ s$^{1/2}$ m^{-2}, $d_{f0} = 1.5$ μm and $d_{fe} = 200$ μm. Floc growth occurs when $\overline{G} = 10$ Hz is applied to the suspension. Then $\overline{G}$ is increased to 1,000 Hz due to which the grown flocs break up and revert to 1.5 μm particles. Calculate the characteristic time-scales of the two processes.

From Eq. 4.132 we obtain $t_f = 56.5$ s and 1,004 s for the first and the second cases, respectively. From Eq. 4.133 $t_{f1} = 7,529$ s, and from Eq. 4.134 $t_{f2} = 2,008$ s. Such a high variability in the characteristic time can occur from the surface to the bottom in open channel flow. Floc growth would be promoted in the low-shear upper water column and breakup at high shear close to the bed. Large flocs settle while the broken units tend to rise due to turbulent diffusion. Thus the presence of the boundary layer and variation in the shear rate may set up a convective cell of flocs moving down and up the column. To track changes in a particular cell one must follow the cell, which would mean that the description of floc properties in the cell would require a Lagrangian reference frame. It also means that the size, density and strength of a captured floc depend on its stress history.

Example 4.14: Tests using suspended sediment (concentration 1.2 kg m^{-3}) from the Ems–Dollard estuary in the Netherlands were carried out in a 4.25 m tall laboratory settling column in which grid-generated turbulence was produced. The following values were derived: $k_A = 14.6$ m^2 kg^{-1}, $k_B = 1.4 \times 10^4$ s$^{1/2}$ m^{-2}, $d_{f0} = 4$ μm, $\overline{G} = 81.7$ s^{-1}. Plot d_f as a function of time. Note that $d_p = d_{f0}$.

The plot of Fig. 4.35a is obtained from Eq. 4.131. Macroflocs (size > ~100 μm) are generated in this illustration. Figure 4.35b shows the measured growth of an initially deflocculated sludge activated by bacteria in a laboratory tank. The shear rate was 19.4 s^{-1}, and macroflocs were generated within the first 20 min. A limitation of the model is that the trend of floc evolution shows concavity (Fig. 4.35a) in contrast to the measured trend (Fig. 4.35b). The latter is consistent with data based on other sediments [van Leussen, 1994; Winterwerp, 1998; Biggs and Lant, 2000; Son, 2009].

Son [2009] argued that a limitation of Eq. 4.126 is that the yield force F_y is assumed to be independent of floc properties (and therefore floc order of aggregation). It is logical to consider that two flocs having the same diameter but different cohesive bond densities would have different magnitudes of F_y for fracture. It was therefore suggested that the yield stress τ_y based on the corrected representation of F_y would be

$$\tau_y = b_y \left(\frac{d_f}{d_p} \right)^{\frac{2D_f}{3}} d_f^{-2} \qquad (4.135)$$

in which the coefficient b_y varies with the sedimentary material. Equation 4.126 is accordingly revised as

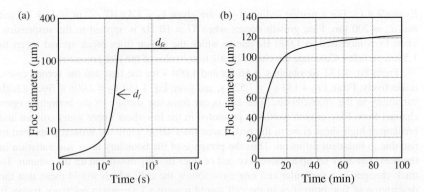

Figure 4.35. (a) Ems–Dollard equilibrium floc size data and predicted time-dependent growth curve (adapted from Winterwerp [1998]); (b) growth of bacteria-activated sludge flocs (adapted from Biggs and Lant [2000]).

$$R_{br} = \frac{dd_f}{dt} = k_B' \frac{\overline{G}}{3} \left(\frac{\eta \overline{G}}{b_y} \right)^{q_b} d_p^{\beta_D - p_b + \frac{2}{3} q_b D_f} d_f^{-\beta_D + 1 + 2 q_b - \frac{2}{3} q_b D_f} (d_f - d_p)^{p_b} \frac{1}{\beta_D \ln \left(\dfrac{d_f}{d_p} \right) + 1}$$

(4.136)

Thus the complete rate equation becomes

$$\frac{dd_f}{dt} =$$

$$\frac{\overline{G} d_p^{\beta_D}}{\beta_D \ln \left(\dfrac{d_f}{d_p} \right) + 1} \left[k_A \frac{C}{\rho_s} d_p^{D_f - 3} d_f^{-D_f + 4 - \beta_D} - \frac{k_B'}{3} \sqrt{\frac{\eta \overline{G}}{b_y}} d_p^{-1 + \frac{D_f}{3}} d_f^{-\beta_D + 2 - \frac{D_f}{3}} (d_f - d_p) \right]$$

(4.137)

A noteworthy advantage of Eq. 4.137 over Eq. 4.127 is that while both account for variability in the fractal dimension with floc order, Eq. 4.137 may include a more realistic representation of the breakup mechanism. In Fig. 4.36 data are shown from an experimental run of Spicer *et al.* [1998], who measured the aggregation of polystyrene particles in a mixing tank. The initial particle diameter was 1 μm, material density 1,050 kg m^{-3}, mass concentration 0.015 kg m^{-3} and the mean dissipation parameter in the tank 50 s^{-1}. We will assume $\beta_D = -0.041$ (from

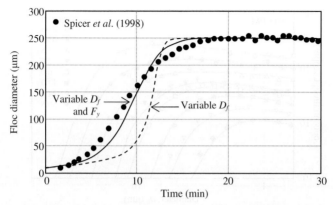

Figure 4.36. Floc growth data of Spicer *et al.* [1998] compared with two flocculation models (from Son [2009]).

Example 4.10). Equations 4.127 and 4.137 were solved with an explicit Runge–Kutta scheme. For Eq. 4.127 (variable D_f) the selected coefficients are $k'_A = 2.5$, $k'_B = 1.72 \times 10^{-6}$ and $F_y = 10^{-10}$ N. For Eq. 4.137 (variable D_f and F_y) $k'_A = 6.74$, $k'_B = 4.59 \times 10^{-6}$ and $b_y = 2.63 \times 10^{-14}$. The value 1 μm was chosen for d_p; however the initial diameter d_{f0} for simulation purposes was taken as 10 μm. Improvement in the description of the breakup function is the main factor contributing to time-evolution of floc size. Hill *et al.* [2000] discuss concentration effects on flocs in the field environment.

4.8 Size Spectra

For measurement of size spectra of suspended particles, settling experiments were conducted in a 26 cm tall 4-liter column using dispersed suspensions of a glacial marine mud from a river near Sillikers in Canada mentioned earlier [Kranck, 1986]. The column was placed on a shaker table and after vigorous stirring the sediment was allowed to settle while being shaken at a constant rate to produce turbulence. Suspension samples were withdrawn from the bottle at different times and analyzed for particle size and concentration. Tests conducted at different rates of shaking produced similar spectral plots, one of which is shown in Fig. 4.37. The shaking frequency was 80 excursions per minute (EPM).

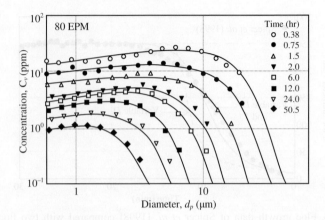

Figure 4.37. Volume-based concentration versus dispersed particle diameter from a settling experiment of Kranck [1986]).

The curves at different times show a reasonable degree of self-similarity and conform to the relationship

$$\frac{C_v}{C_{vl}} = d_p^{m_p} e^{-n_p d_p^2} \qquad (4.138)$$

where C_v is the particle volume concentration (Chapter 3), C_{vl} is the value of C_v at a diameter d_p of 1 μm, and m_p and n_p are sediment-specific coefficients. For tests at 80 EPM, values of C_v, m_p and n_p are given in Table 4.11. Tests at different EPMs showed that C_{vl}, m_p and n_p varied with EPM in response to changing level of turbulence.

Kranck and Milligan [1992] noted that despite the dramatic shift in particle size spectra due to coagulation of dispersed particles into flocs, with suitable adjustments of the coefficients Eq. 4.138 may be made to fit both individual particle spectra and floc spectra. This empirical finding simplifies the method to estimate floc size from individual particles. Unfortunately, relationships between the sets of coefficients for flocs and dispersed particles vary with the sediment and turbulence.

Table 4.11. Parameters for concentration-size spectra.[a]

Time (hr)	C_v (ppm)	m_p	n_p
0.38	16.240	0.293	0.0037
0.75	10.737	0.200	0.0053
1.5	6.896	0.231	0.0105
2.0	4.202	0.365	0.0225
6.0	3.649	0.416	0.0347
12	2.829	0.505	0.0766
24	1.873	0.404	0.1130
50.5	1.348	0.602	0.2465

[a] m_p and n_p values are applicable when C_{vl} is measured in ppm (parts by volume of sediment per million parts by volume of suspension) and d_p in µm.
Source: Kranck [1986].

Example 4.15: The following values of the coefficients in Eq. 4.138 are obtained from Kranck and Milligan [1992] and Kranck [1986]:

Parameter	Particle spectrum	Floc spectrum
C_{vl} (ppm)	3.123	0.001
m_p	0.608	2.72
n_p	0.0047	0.0007

Plot the particle and floc size-distributions.

The results are shown in Fig. 4.38. An increase in the modal diameter (value at peak of the spectrum) from $O(10^1)$ µm to $O(10^2)$ µm occurs and macroflocs (> ~100 µm) are formed. A wide distribution of dispersed particles plays an important role in floc formation inasmuch as small particles tend to capture larger ones, resulting in a comparatively narrow distribution of floc size. A significant contribution in this development can be from biochemical binding, because as mentioned mucus films are

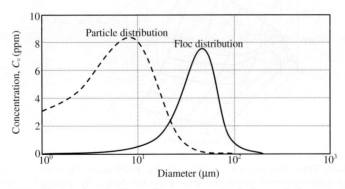

Figure 4.38. Size spectra of a fine sediment suspension before and after coagulation.

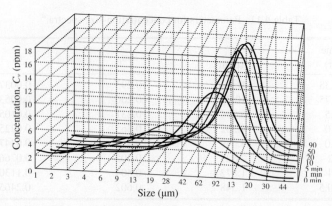

Figure 4.39. Evolution of floc size distribution of an initially dispersed activated sludge (adapted from Biggs and Lant [2000]).

efficient in embedding large particles that may otherwise remain detached due to weak electrochemical cohesion relative to gravity. An example is shown in Fig. 4.39, which plots size spectra every ten minutes of a bacteria-activated sludge, which after dispersion was allowed to re-flocculate in a tank at a mean flow-shear rate of 19.4 s^{-1} (see also Fig. 4.35b). Flocculation was reported to be entirely due to a biopolymer.

Compilation of data from different sources resulted in the diagram of Fig. 4.40 relating the equilibrium modal size of flocs to the suspended sediment concentration and shear stress as a surrogate for the shear rate $\bar{G}$ (see also Eq. 4.130). Floc size increases with shear stress, reaches a maximum at ~0.1 Pa and then decreases with a further increase in stress

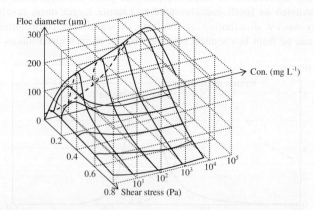

Figure 4.40. Three-dimensional rendition of the relationship between floc diameter, suspended sediment concentration and shear stress for shear-only aggregation (from Dyer [1989]).

as breakup overtakes growth. Floc size also increases with concentration at low shear stresses. It then decreases as settling is hindered (Chapter 7). At high shear stresses the role of concentration decreases as the effects of breakup forces become dominant. Thus a domain is defined by middle ranges of shear stress and concentration within which macroflocs occur [Dyer, 1989].

Attempts to measure equilibrium floc size in the laboratory can suffer from uncontrolled effects of settling, especially at low levels of turbulence. In fact, on account of this effect a floc may not get enough time while in suspension to achieve equilibrium [Winterwerp, 1998]. In flocculation experiments using a kaolinite at different values of pH, salt concentration and rate of agitation, Mietta *et al.* [2009] found that regardless of fluid chemistry, when the shear rate $\bar{G}$ was less than 35 s^{-1}, deposition prevented flocs from achieving equilibrium properties and they became smaller as $\bar{G}$ decreased. If so, contrary to the description in the previous paragraph, it may not be permissible to interpret data such as those in Fig. 4.40 in regard to the role of shear stress on floc size because the effect of settling on size is not identified in the plot.

When the effects of Brownian motion and differential settling are included in flocculation models, likely trends at very low shear stresses in nearly quiescent flow are shown in Fig. 4.41. At low rates of energy dissipation ε_t (Eq. 2.88) the equilibrium floc diameter increases rapidly as shear-induced collisions, the main destructive mechanism, cease to be a factor limiting the diameter. For the same reason the diameter becomes sensitive to changes in suspended sediment concentration.

A feature of natural floc size spectra not described thus far is their occasional multimodal character. Such spectra can result from a simple mixing of different sediment populations, such as when river flow brings material into the estuary with its own residential suspended and bed matter, or when turbid fronts in the seaward zone of the estuary or the coastal sea collide during flood or ebb flows. Recently mixed particle populations can change with cycles of erosion, aggregation and deposition. The effect of aggregation processes on the spectrum depends on the constituents. Lee *et al.* [2011] numerically solved the population balance equations representing bimodal distribution of particles and

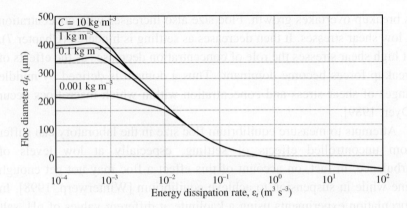

Figure 4.41. Simulated variation of floc diameter with energy dissipation rate and concentration for San Francisco Bay sediment (from McAnally [1999]).

tracked spectral evolution starting with initial populations parameterized by experimental data [van Leussen, 1994]. It was argued that aggregation processes act selectively depending on particle composition. For example, biopolymeric substances tend to produce large but light-weight and "floppy" aggregates (*e.g.,* marine snow, Chapter 3) in contrast to more tightly packed three-dimensional mineral flocs. These differences can lead to distinctly bimodal size spectra.

4.9 Exercises

4.1 Let M denote mass and V the volume along with subscript t meaning total, s for solids and w for water. For a saturated sediment bed we define the wet bulk density $\rho = M_t/V_t$, the dry density $\rho_D = M_s/V_t$ and the particle material (mineral) density $\rho_s = M_s/V_s$. Note that $M_t = M_s + M_w$ and $V_t = V_s + V_w$. Show by mass balance that the following identity holds for the solids volume fraction

$$\phi_v = \frac{\rho_D}{\rho_s} = \frac{\rho-\rho_w}{\rho_s-\rho_w} \tag{E4.1}$$

where ρ_w is the density of water. You may carry out the exercise by adding the intermediate steps, then prove the following development

$$\rho_D = \frac{M_s}{V_s+V_w} = \frac{M_s}{V_s+V_w}\left(\frac{M_s V_w - M_w V_s}{M_s V_w - M_w V_s}\right)$$

$$= \frac{M_s}{V_s+V_w}\left[\frac{(M_s + M_w)V_w - M_w(V_s+V_w)}{M_s V_w - M_w V_s}\right] \tag{E4.2}$$

4.2 Show how Eq. 4.26 is obtained from Eq. 4.25.

4.3 Derive Eq. 4.37 from Eq. 4.36.

4.4 Calculate the energy of repulsion V_R for all values of $\kappa_D\Delta$ tabulated in Example 4.1. Plot V_R against $\kappa_D\Delta$. Briefly explain the physical significance of the trend.

4.5 You are required to prepare a bed of smectite in water containing 15 mg L^{-1} of calcium and 5 mg L^{-1} of magnesium (both are cations). Using the stability diagram of Fig. 4.13, determine the concentration (mg L^{-1}) of sodium you would have to add to ensure that the bed remains flocculated.

4.6 Given Fig. 4.20 showing viscometer data on relative viscosity versus concentration for the Wilmington District sediment, find the floc density for $k_f = 31.1$ cm^3 g^{-1}. Take the density of water as 1,025 kg m^{-3} and mineral density 2,650 kg m^{-3}.

4.7 A floc traveling with a velocity of 0.9 m s^{-1} in the x-direction has a diameter of 6 μm and density 1,750 kg m^{-3}. This floc collides with a 10 μm floc with a density of 1,650 kg m^{-3} traveling –0.1 m s^{-1} in the x-direction. The interpenetration fraction is 0.3. Assume that both particles can be represented as spheres.

(1) Determine the approximate time required to complete the collision.

(2) Determine the shear stress necessary to break the floc formed by collision.

(3) Comment on the likelihood of floc breakup based on your answer for (2).

4.8 From a cubic aggregate (dimension d_f) containing 64 primary cubes of dimension d_p (Fig. E4.1a), every other primary unit has been removed such that the aggregate now contains only 32 primary cubes (Fig. E4.1b). Note that the dimension d_f of the new aggregate has remained unchanged. Determine the fractal dimension of the new aggregate. What is its porosity (see Chapter 5 for definition)?

4.9 The size and the settling velocity of kaolinite flocs in water (of density 998.7 kg m^{-3} and kinematic viscosity 1.095×10^{-6} m^2 s^{-1}) were measured by optical imaging in a settling column (Table E4.1). Determine the density and the fractal dimension of each floc size class, assuming the primary particle size to be 1.2 μm and mineral density 2,357 kg m^{-3}. Plot the settling velocity, density and fractal dimension against size. Briefly interpret the results. Hint: Assume the applicability of Stokes law (Eq. 3.25).

4.10 Calculate the collision frequency function due to shear for a 6 μm and a 10 μm particle. You are given: water depth of 2.3 m, Manning's $n = 0.015$, temperature 30°C and current velocity 1.2 m s^{-1} 0.3 m above the bed. Assume $p_c = 0.65$, $v = 10^{-6}$ m^2 s^{-1} and $\kappa = 0.4$.

(a)   (b)

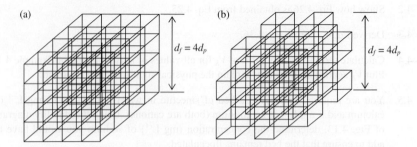

Figure E4.1. Two cubic aggregates with the same dimension: (a) Consisting of 64 primary cubes; (b) consisting of 32 primary cubes (from Son [2009]).

Table E4.1. Settling of kaolinite.

Floc size (μm)	Settling velocity (mm s^{-1})
126	0.71
86	0.38
102	0.38
85	0.54
96	0.53
68	0.57
90	0.48
89	0.74
153	0.31

4.11   Show that the well in the graph for differential settling in Fig. 4.34 is "filled up" if the first and the second particle diameters are assumed to vary relative to their mean values. Consider the mean size of the first particle to be 10 μm and the range of the second particle from 2 μm to 50 μm. Other variables remain the same as those in Example 4.12. Assume that the velocities of both particles are normally distributed and that the coefficient of variation is 0.15. Use Monte Carlo simulation. Consider the following steps:

(1) Generate a lookup table of particle diameters and corresponding settling velocities (mean values) for sizes from 2 μm to 50 μm.

(2) Generate samples of second particle diameter assuming a discrete uniform distribution with 2 μm – 50 μm as range (at least 1000 samples).

(3) Using values in the lookup table calculate the normally distributed settling velocity for each particle sample from the previous step. Similarly generate random values of settling velocities for the first particle. Standard deviation of settling velocities required to generate the random numbers can be obtained

from the coefficient of variation, which is the ratio of standard deviation to mean.

(4) Using random values of settling velocities calculate the collision frequency for differential settling.

4.12 Plot the growth of floc diameter with time using Eq. 4.127 for shear rates $\bar{G}$ = 1 and 5 s^{-1}, and concentrations C = 0.005 and 0.5 kg m^{-3} (*i.e.*, four plots in all). Select k_A' = 4, k_B' = 3×10^{-6}, β_D = 0.05, D_f = 2.2, ρ_s = 2,300 kg m^{-3}, η = 10^{-3} Pa.s, F_y = 10^{-10} N, d_p = 1 μm and d_{f0} = 5 μm. Provide a brief explanation for the observed trends. Then repeat the calculations using Eq. 4.137 with b_y = 2×10^{-14} N.

4.13 Derive a relationship between flow shear rate $\bar{G}$ and the Kolmogorov eddy length scale λ_0 (Eq. 2.89). Assume v = 1.16×10^{-6} m^2 s^{-1}. Hint: Use Eq. 4.98.

4.14 From their field investigation, Fugate and Friedrichs [2003] obtained a relationship between the median floc size and the Kolmogorov eddy length scale λ_0 as shown in Fig. E4.2. Based on the answer in Exercise 4.13 and data in Table E4.2, plot floc size against shear rate. Assume v = 1.16×10^{-6} m^2 s^{-1}. Briefly comment on the observed trend with respect to aggregation processes.

4.15 For mud from the Atchafalaya Bay, LA, you are given three sets of flow-curve data from a concentric-cylinder rheometer (Figs. E4.3a, b, c and Table E4.2), each at a different mud wet bulk density.

(1) Determine the viscosity and the floc shear strength following the graphical method in Fig. 4.18 (left) for up to three ranges. Select seawater density 1,025 kg m^{-3}, density of solids 2,580 kg m^{-3} and kinematic viscosity 1.165×10^{-6} m^2 s^{-1}. Assume that this method is applicable at the high densities tested.

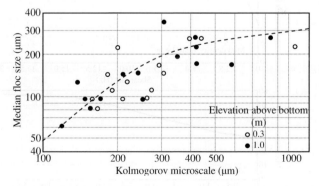

Figure E4.2. Floc size plotted against Kolmogorov eddy length scale derived from data of Fugate and Friedrichs [2003].

Table E4.2. Atchafalaya mud flow curve data.

$\rho = 1{,}190$ (kg m^{-3})		$\rho = 1{,}240$ (kg m^{-3})		$\rho = 1{,}350$ (kg m^{-3})	
$\dot{\gamma}$ (s^{-1})	τ (Pa)	$\dot{\gamma}$ (s^{-1})	τ (Pa)	$\dot{\gamma}$ (s^{-1})	τ (Pa)
0	0.6	0	4.5	0	22
0.5	1.3	1	7.4	0.2	28
1	2.5	1.5	8.2	0.5	30
1.3	2.8	2.5	8.8	1	33
2	2.83	5	9.1	1.5	38
3	2.87	7.5	9.2	2	42
4	2.88	20	9.4	2.5	47
4.5	2.9			3	52
5	2.91			3.4	54
				4	56
				4.5	57
				5	58

Source: Robillard [2009].

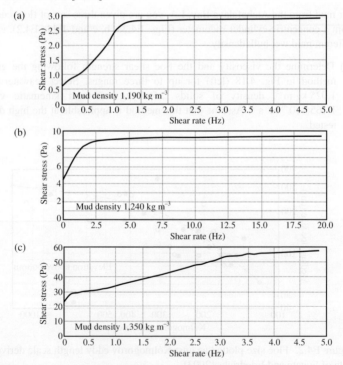

Figure E4.3. Flow curves for Atchafalaya mud: (a) Density 1,190 kg m^{-3}; (b) density 1,240 kg m^{-3}; (c) density 1,350 kg m^{-3} (from Robillard [2009]).

(2) For each range plot the relative viscosity against concentration as in Fig. 4.20. In each case include an additional point at $C = 0$, where the relative viscosity is 1. Calculate k_f for each range using Eq. 4.40.

(3) Assume that Range 1 corresponds to 0th order of aggregation, Range 2 to 1st order and Range 3 to 2nd order. Using Eqs. 4.59 calculate the floc density for each order.

(4) Plot floc shear strength τ_f against concentration C as in Fig. 4.21. Does it follow the 5/2 power law?

Chapter 5

Characterization of Mud Properties

5.1 Chapter Overview

Common bulk parameters characterizing mud transport are reviewed first. They include quantities that describe the physical and physicochemical properties of the sediment and the fluid. When water waves travel over a bottom of soft or fluid mud they are damped, and the dynamic character of the bottom changes. Modeling these changes requires an understanding of the deformation behavior of muds under static and dynamic loading. Accordingly, the remainder of the chapter reviews basic shear rheology and rheometry applicable to cohesive muds. For beds of silt (and sand), oscillatory motion over poroelastic media is reviewed in Chapter 11.

5.2 Sediment and Fluid Characterization

5.2.1 *Bulk descriptors*

Inter-particle forces operate over distances of the scale of a nanometer, whereas the length scale associated with hydraulics-related parameters characterizing cohesive sediment transport is on the order of a micrometer, *i.e.*, a thousand-fold larger. This disparity results in a considerable gap in our ability to model the movement of fine particles, because there are few methods in hydraulic engineering that accurately measure fine sediment properties at scales smaller than about a micron. As a result, for inter-comparisons of transport-related data from different sites, it is common to determine bulk property descriptors that influence mud movement and sequestration. Early efforts to identify bulk parameters related to cohesive soil erosion were summarized by ASCE [Task Committee on Erosion of Cohesive Materials, 1968]. This was followed by studies at the University of California, Davis in which emphasis was on the role of clay minerals and pore fluid chemistry in governing erosion [Arulanandan *et al.*, 1973; Kandiah, 1974; Arulanandan *et al.*, 1975]. Based in part on this effort, a summary of erosion related parameters followed [Mehta, 1981]. Table 5.1 includes parameters used to characterize mud in studies carried out under the European MAST G6M investigations mainly during the 1990's [Berlamont *et al.*, 1993]. These parameters reflect the wide ranges of physical and chemical factors that influence the state and transport of mud, in particular with respect to erosion.

From a review of literature, Parchure [1984] compiled a list of over a hundred factors and parameters known to be influential in cohesive

Table 5.1. Characteristic descriptors of fine sediment transport in MAST G6M studies.

Fluid properties	*Mud properties*
Chlorinity[a]	Chlorinity[a]
Temperature	Temperature
Dissolved oxygen	Dissolved oxygen
Redox potential[b]	Redox potential[b]
pH	pH
Na^+, K^+, Mg^{++}, Ca^{++}, Fe^{+++}, Al^{+++} ions	Na^+, K^+, Mg^{++}, Ca^{++}, Fe^{+++}, Al^{+++} ions
Sodium Adsorption Ratio	Cation exchange capacity
Suspended sediment concentration	Gas content
	Organic content
Bed structure	Particle size distribution
Density profile	Particle specific surface area
Void ratio	Sediment mineral composition
Permeability	
Pore pressure	*Water-bed exchange processes*
Effective stress	Settling
Liquid limit	Bed erosion shear strength
Plastic limit	Erosion rate constant

[a] Chlorinity (*Cl*) in parts per thousand (ppt or $^o/_{oo}$) and salinity (*S* in ppt or $^o/_{oo}$) do not refer to the same quantity, although the former is often confused for the latter. The relationship between the two is $S (^o/_{oo}) = 1.80655 \ Cl \ (^o/_{oo})$.
[b] A measure of the state of oxidation of a system. Its symbol is *Eh*.
Source of characteristic descriptors: Berlamont *et al.* [1993].

sediment erosion. In general, there are practical reasons for limiting the number of parameters. Firstly, their number is so large that it is unrealistic to expect useful predictive correlations between such parameters and transport-related quantities. Secondly, the time and cost of determination of a large number of parameters is prohibitive in most studies. Given these caveats, as a practical approach we will consider only seven descriptors: (1) particle size, (2) settling velocity, (3) sediment mineral composition, (4) organic content, (5) bottom density, (6) fluid salinity and (7) fluid temperature. The significance of these parameters is briefly reviewed next.

5.2.2 *Particle size and settling velocity*

Coarse-grain sediment transport is conventionally described in terms of particle diameter, a static measure, because it is convenient to do so even though the basic particle property governing transport is the settling velocity, a kinematic quantity. In addition to particle and fluid densities, the relationship between settling velocity and diameter includes the drag coefficient, which is usually unknown and must be obtained by direct measurement of the settling velocity. In the absence of the correct value of the drag coefficient, the settling velocity of clayey flocs calculated from measured diameter can be inaccurate. On the other hand, as large silt particles in the size range of about 40 to 63 μm remain unaggregated when no significant quantities of clays (> ~20%) or organic matter (> ~10%) are present, it is reasonable to characterize their transport in terms of diameter.[a]

In hydraulic engineering the median value of the dispersed[b] particle size distribution is commonly reported, whereas in many scientific

---

[a] The No. 200 Tyler sieve with an opening of 74 μm, which is close to 62 μm (No. 230 sieve), is typically used to separate coarse from fine particles. This is done by wet–sieving, in which the original sample is taken on the sieve and the fine fraction washed with water [ASTM, 1993a]. The filtrate contains the fine fraction. Native water, *i.e.,* water from the site where the sediment is collected, must be used in the washing process to avoid unforeseen chemical effects of changing the water chemistry on flocs.

[b] As mentioned in Chapter 4, there is no certainty that dispersion will be complete, unless electrolytic ions in the fluid are almost completely removed by washing.

studies the entire size distribution is given (Chapters 3, 4). Dispersed particle size distribution provides a qualitative basis for comparing the transport behavior of different cohesive sediments. In regard to transport load calculations, primary particle size occurs in some empirical formulas (*e.g.*, Eq. 3.2). As mentioned in Chapter 4, fractal representation of mineral flocs is useful in the development of scaling laws for floc size as a function of primary particle size. The primary particle is the basic building block of flocs and specification of its size distribution is essential for modeling aggregation dynamics.

For calculation of cohesive sediment flux it is preferable to know the settling velocity of flocs. For very small or light-weight flocs at fall Reynolds numbers on the order of unity or less, the validity of Stokes law (Eq. 3.25) can be assumed provided floc density is known. However, this is permissible only when settling is measured by recording the fall of a floc over short (~seconds) time intervals (*e.g.*, Manning [2006]). When the particles are larger, a general force balance (Eq. 7.16) is available for settling velocity calculation provided the density and the drag coefficient are known in addition to diameter. Given this limitation, direct measurement of settling velocity becomes essential for its value at the least level of uncertainty.[c]

Settling velocity of flocs measured in the laboratory under standard conditions does not address the time and space varying behavior of flocs in nature. Some semi-predictive formulas are available (Chapter 7); however, as mentioned the best approach is to carry out *in situ* measurements at selected locations and times in the study area. With the use of high-resolution optical devices it is feasible to measure the settling velocity and the concentration associated with each floc visualized in the photo-frame. This technique, when repeated for a large number of flocs, permits representation of sediment as a discrete spectrum of floc size [Syvitski *et al.*, 1995].

---

[c] Uncertainty can result from inadequate knowledge of a process or inability to measure it accurately (epistemic uncertainty), or from the random nature of the process (natural uncertainty). Epistemic uncertainty can be reduced through improvements in data collection techniques. Natural uncertainty cannot be reduced, just better sensed [Merz and Thieken, 2005; Letter, 2009].

At very low concentrations ($\ll 0.1$ kg m^{-3}), uncertainty in measurement is likely to decrease with increasing concentration. When the concentration exceeds about 0.1 kg m^{-3}, the effects of inter-particle collisions and fluid shear rate on the settling velocity becomes important. Given this threshold, a reasonable value of a standard concentration at which measurements from different sediments can be conveniently compared is 0.1 kg m^{-3}. Likewise, 15–25°C can be the standard temperature range, although it is recognized that over a wider range the effect of temperature on settling is not negligible (Chapter 7).

5.2.3 *Mineral composition and organic content*

In Chapter 3 we referred to the cation exchange capacity (CEC) of clay minerals as an approximate measure of cohesion. Standardized methods are available for the determination of minerals and CEC [Soil Conservation Service, 1992; ASTM, 1993b; Moore and Reynolds, 1997]. The number of minerals in sediment can be large, while their cumulative effect on floc properties is generally unknown. As a practical matter in engineering investigations it is sufficient to report only the main clay and non-clay mineral constituents.

Biochemicals play a role in the transport of flocs through aggregation processes, even when they constitute a small fraction of total mass. Unfortunately, quantification of these influences remains an unachieved research goal due to the diversity of biological factors. Where biological effects dominate transport, physicochemical descriptors unrelated to these effects may yield little benefit with regard to floc transport. A case in point is the CEC, which is altered by organic molecules lodged within the clay structure. The outcome is that the CEC of natural sediment, from which organic matter is removed by oxidation, say, with hydrogen peroxide, does not bear a fixed relationship with the floc forming potential of the original sediment (Chapter 3). On the other hand, if the CEC is measured without removing the organic matter, the resulting (high) CEC values are not meaningful in terms of cohesion.

Natural organic matter (NOM) is expressed in two ways. One is as milligrams of total organic carbon per gram of sediment. The second is as loss of organic mass on ignition of a sediment sample at 450°C. Both quantities are determined by standard methods [ASTM, 1993c]. In hydraulic laboratories it is cumbersome but feasible to reproduce muds with pre-set organic content. For flume-based erosion experiments, Dennett [1990] reported a procedure for the addition of NOM to a kaolinite to create beds of "designer" aquatic sediments with variable organic content. A suspension of pure kaolinite was prepared by mixing 800 g of dry clay with 3.2 L of tap water (pH 6.6 and ionic strength 0.001 M) and up to 2.8 L of concentrated dissolved NOM. The suspension was mixed for 1 h and allowed to settle for 24 h. The amount of NOM adsorbed on kaolinite, expressed as percent organic carbon, was determined from mass balance by measurement of the organic carbon concentration in the supernatant liquid using a carbon analyzer. Eleven beds were prepared in this way with organic carbon ranging from 0 to 0.91 mg C g^{-1} of sediment.

5.2.4 *Density and volumetric measures*

We defined water content $W = M_w/M_s$ (Eq. 3.32), where M_w is the mass of water and M_s is the mass of solids. In what follows the description of other density related measures is limited to saturated beds, *i.e.*, beds in which the degree of saturation $S_s = V_w/(V_w+V_g) = 1$ or 100%, where V_g is the volume of gas (*i.e.*, volume of void not occupied by water). For saturated beds ($V_g = 0$) mass balance relationships are required only for the solids and the liquid.

The densities of water ρ_w and solid (particle) ρ_s are

$$\rho_w = \frac{M_w}{V_w} \qquad (5.1)$$

$$\rho_s = \frac{M_s}{V_s} \qquad (5.2)$$

where V_w is the volume of water and V_s is the volume of solids. For mud the wet bulk density ρ and the dry bulk density ρ_D, respectively, are

$$\rho = \frac{M_s + M_w}{V_s + V_w} \qquad (5.3)$$

or

$$\rho = \frac{M_s + M_w}{V_t} \qquad (5.4)$$

where V_t is the total volume, and

$$\rho_D = \frac{M_s}{V_t} \qquad (5.5)$$

For bed sediment we will use the symbol ρ_D denoting dry bulk density, *i.e.*, dry sediment mass per unit bed area multiplied by the height of deposit. Deposition and erosion fluxes discussed in Chapters 7, 9 are defined in terms of change in dry sediment mass per unit bed area and unit time. Therefore, knowing the porosity of the deposit permits calculation of the change in bed thickness due to shoaling (Chapter 1) or scour.

For sediment in suspension, we will retain the symbol C (in place of ρ_D) to indicate concentration defined as dry sediment mass M_s in suspension divided by volume V_t of suspension. The solids volume fraction is ϕ_v and the floc volume fraction ϕ_{vf} (Chapter 4). The symbol for solids volume fraction is retained for the suspension and the bed. Thus, in the suspension $\phi_v = C/\rho_s$, and in the bed $\phi_v = \rho_D/\rho_s$. The definitions of specific weight (or specific gravity) of water s_w and solid s, respectively, are

$$s_w = \frac{\rho_w}{\rho_{w0}} \qquad (5.6)$$

$$s = \frac{\rho_s}{\rho_w} \qquad (5.7)$$

where ρ_{w0} is density of water at 4°C and ρ_w is density at the actual temperature of water, *e.g.*, room temperature. The unit weights of water and solid, respectively, are

$$\gamma_w = \rho_w g; \quad \gamma_s = \rho_s g \qquad (5.8)$$

The unit weight of water at 4°C is $\rho_{w0} g$.

Example 5.1: Calculate the solids volume fraction and the floc volume fraction of a cohesive sediment suspension, given $\rho_w = 1,007$ kg m^{-3}, $\rho_s = 2,685$ kg m^{-3}, $C = 10$ kg m^{-3} and $\rho_f = 1,115$ kg m^{-3}.

The solids volume fraction $\phi_v = C/\rho_s = 0.0037$. From Eq. 4.43 $\phi_{vf} = 0.058$, which is an order of magnitude greater than ϕ_v because flocs take up considerably greater space than solid particles. However, as we will see in Chapter 7, at this value of ϕ_{vf} the packing of flocs is sufficiently dense to result in hindered settling characteristic of high-density sediment suspensions.

Two related volumetric measures of saturated bed structure are the porosity n_r and the void ratio e_v defined as

$$n_r = 1 - \phi_v = 1 - \frac{V_s}{V_t} = \frac{V_w}{V_t} \tag{5.9}$$

$$e_v = \frac{V_w}{V_s} \tag{5.10}$$

Both are expressed either as decimal values or percent. The void ratio expressed as a decimal value can be greater than unity, *i.e.*, > 100%. The quantities n_r and e_v are related by

$$n_r = \frac{e_v}{1 + e_v} = \frac{\rho_s - C}{\rho_s} \tag{5.11}$$

$$e_v = \frac{n_r}{1 - n_r} = \frac{1 - \phi_v}{\phi_v} = \frac{\rho_s - C}{C} \tag{5.12}$$

Values of n_r and e_v depend on the sediment, *i.e.*, on sand, silt or clay, mineral composition, organic content and the packing arrangement of particles. For a given soil several states of packing can occur. In Table 5.2 two end-states are considered. Dense packing represents the closest possible packing state, *i.e.*, the lowest porosity n_{min} and the lowest void ratio e_{min}. Simple cubic packing (Fig. 3.13) is the loosest of packing arrangements with the highest porosity n_{max} and the highest void ratio e_{max}. These end-state values have been calculated from the packing

Table 5.2. Void ratio, porosity and density ranges for granular soils.

Description	Void ratio[a]		Porosity		Dry density	
	e_{max}	e_{min}	n_{max} (%)	n_{min} (%)	ρ_{Dmin} (kg m^{-3})	ρ_{Dmax} (kg m^{-3})
Uniform spheres	0.92	0.35	47.6	26.0	–	–
Uniform sand	1.0	0.40	50	29	1,328	1,888
Uniform (inorganic) silt	1.1	0.40	52	29	1,280	1,888
Silty sand	0.90	0.30	47	23	1,392	2,032

[a] These end-values of e_v define the relative density of soil $D_r = (e_{max} - e_v)/(e_{max} - e_{min})$.
Source: Lambe and Whitman [1979].

geometries. The table also gives the densities of typical granular (*i.e.,* coarse) beds in both the densest (ρ_{Dmax}) and the loosest (ρ_{Dmin}) states. It should be noted that: (1) the smaller the range of particle sizes (*i.e.,* the more uniform the soil), (2) the smaller the particles, and (3) the more angular the particles, the smaller the minimum density. Thus there is a greater opportunity for building a loose arrangement of particles under these three conditions. On the other hand, for instance, the greater the range of particle size the greater is the maximum density, because the voids among large particles are filled with smaller particles.

The data in Table 5.2 do not apply to cohesive muds because it is not practicable to calculate the limiting values of n_r and e_v purely from geometry. Particle packing depends on the state of flocculation and the ability of flocs to withstand self-weight or overburden. The degree of consolidation of the bed plays a significant role because dewatering of a fresh fine sediment deposit may continue for months. Therefore n_r and e_v may vary significantly with time and depth (Chapter 8).

5.2.5 *Fluid salinity and temperature*

As noted in Chapter 3, the effect of salinity on floc structure can be minor once NaCl concentration exceeds about 10 ppt. At lower salinities settling, consolidation and erosion are measurably influenced by salt concentration. In Chapters 7 and 9 we will refer to the effect of temperature on fine sediment settling and erosion, respectively. This effect has not been examined extensively, even though, as mentioned earlier, settling and erosion flux calculations for water at "room" temperature are not suitable for transport at high or low temperatures. A reason for this deficiency is that most studies are conducted for applications in the mid-latitudes. Furthermore, maintenance of temperature control in laboratory experiments requires costly equipment and time-consuming tests.

5.3 Mud Deformation Behavior

5.3.1 *Cohesive shear strength*

The yield strength of (cohesive) mud indicates its resistance to failure. In Chapter 4 the floc shear strength was introduced. Here we will consider the plastic-yield shear strength and the vane shear strength.

5.3.2 *Yield under normal and shear stresses*

Consider the solid soil element in Fig. 5.1. The normal stress σ_n and the normal strain ε_n are

$$\sigma_n = \frac{F_n}{A}; \qquad \varepsilon_n = \frac{dz}{h_z} \qquad (5.13)$$

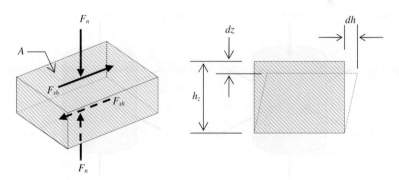

Figure 5.1. Normal and shear stresses and strains in a soil element.

where F_n is the normal force over area A, h_z is the height of the element and dz is the resultant (small) vertical displacement. The shear stress and strain, respectively, are

$$\tau = \frac{F_{sh}}{A} ; \qquad \gamma = \frac{dh}{h_z} \qquad (5.14)$$

where F_{sh} is the shear force and dh the horizontal displacement. Compressive stresses and strains are considered positive.

Let the soil element be cylindrical in shape such as the segment of a sedimentary core extracted from the bed (Fig. 5.2a). At equilibrium the normal stresses and strains in the two radial directions are equal, *i.e.*, $\sigma_{nx} = \sigma_{ny}$ and $\varepsilon_{nx} = \varepsilon_{ny}$. The three sets of σ_n, ε_n pairs for this triaxial state of the element are the principal stresses and strains, as they are in the directions of the principal axes.

In Fig. 5.2b the x and y-components of σ_n and ε_n are conveniently replaced by their radial representations σ_r and ε_r, and the z-components by σ_a and ε_a. In order to determine the principal normal and shear stresses along any arbitrary plane passing through the cylindrical sample, the well-known Mohr circle representation of stresses is used. In the x-z plane in Fig. 5.3, both normal stresses (σ_{nx} and σ_{nz}) and shear stresses (τ_{xz} and τ_{zx}), occur. The first subscript of τ represents the direction along which the stress acts and the second the plane that is sheared.

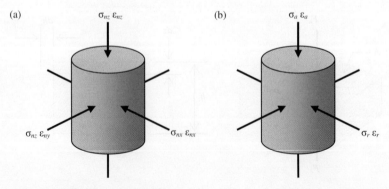

Figure 5.2. Principal normal stresses on a cylindrical soil sample: (a) Cartesian (x,y,z) coordinate representation; (b) equivalent axial (r,a) coordinate representation.

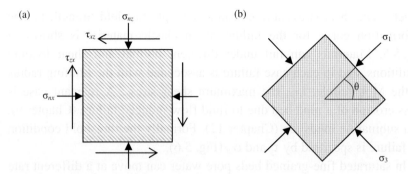

Figure 5.3. (a) Principal normal and shear stresses in the x-z plane; (b) normal stresses on a plane rotated by angle θ.

The Mohr circle representing the elements in Figs. 5.3a and b is drawn in Fig. 5.4. There are specific planes on which the shear stress is zero because the applied stress is normal to those planes. As a result, the stresses acting on the planes can be described by the peak axial normal stress σ_1 and the confining pressure σ_3. We will refer to $\sigma_{pm} = (\sigma_1 + \sigma_3)/2$ as the peak value of the mean normal stress and $\sigma_{qm} = (\sigma_1 - \sigma_3)/2$ the peak value of the deviator normal stress.

In the elastic range of the deformation of a soil element, when the shear stress τ is increased, the shear strain γ initially increases linearly in accordance with Hooke's law. However, there is a limiting condition at which the strain becomes very large and independent of the stress. This condition is one of plastic yield which causes the material to fail or

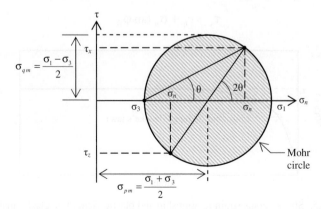

Figure 5.4. Mohr circle representation of stresses in Fig. 5.3.

liquefy. The shear stress at that instant is the plastic yield strength τ_y. The deformation curve for the failure of an elastic material is shown in Fig. 5.5. Materials can fail under different normal and shear loading conditions, and in each case failure is associated with the limiting radius of the Mohr circle, *i.e.,* the maximum shear stress. A relevant case is mass erosion of a mud bed due to fluid flow-induced failure (Chapter 9), or a submarine mudslide (Chapter 12). Formally the threshold condition for failure is specified by τ_y and σ_{nf} (Fig. 5.6).

In saturated fine-grained beds pore water can move at a different rate than the solid particles, and as a result stress loading is considered in the "short term" and "long term". Initially, excess pore water pressure is generated in the fluid (water) phase (Chapter 8) and the bed volume remains unchanged, because the low bed permeability prevents rapid drainage of water. The undrained shear strength s_u in this state is independent of the normal stress as the response to loading merely increases the pore pressure. In this short-term response called the Tresca criterion, changing the normal stress does not change the radius (s_u) of the Mohr circle (Fig. 5.7). Characteristic ranges of s_u are given in Table 5.3.

Over a longer term, once water drains and the pore pressure dissipates the drained shear (yield) strength τ_y, which is greater than zero when the normal stress σ_n is zero, increases with increasing σ_n. The τ_y envelope is nearly linear (Fig. 5.8) and is approximated by the Mohr–Coulomb equation

$$\tau_y = c_h + \sigma_n \tan \varphi_{fr} \tag{5.15}$$

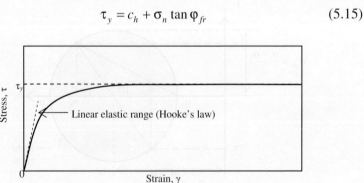

Figure 5.5. Stress versus strain relationship and plastic yield of an elastic material.

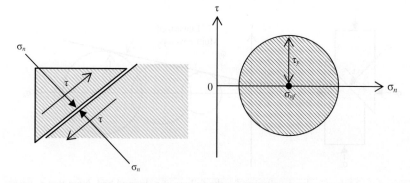

Figure 5.6. Bed failure by shearing.

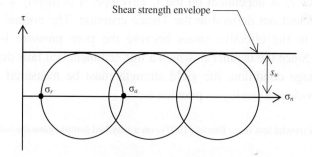

Figure 5.7. Tresca criterion for the failure of an undrained bed.

Table 5.3. Undrained shear strength of soils.

Soil	Undrained shear strength, s_u (kPa)
Hard	> 150
Stiff	75–150
Firm	40–75
Soft	20–40
Very soft	< 20

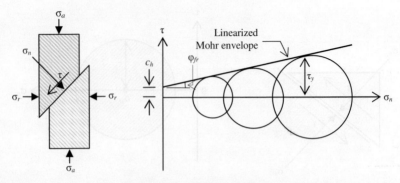

Figure 5.8. Mohr–Coulomb criterion for the failure of a drained bed. Note that τ_y passes through the point of intersection of the circle and the envelope.

where the intercept c_h is called cohesion and φ_{fr} is the angle of internal friction. Since c_h is dependent on the τ_y envelope, it is merely a curve-fitting coefficient not defined in the Tresca criterion. The normal stress σ_n is equal to the effective stress because the pore pressure is zero (Chapter 8). Since the manner in which a fine sediment bed fails depends on the drainage condition, the yield strength must be measured under conditions as close to reality as practicable.

Example 5.2: A triaxial test (*e.g.*, Bardet [1997]) on a silty bed sample gave the following data:

Confining pressure, σ_3 (kPa)	Peak axial stress, σ_1 (kPa)
100	520
200	1010
400	2100
800	4170

Determine the internal angle of friction φ_{fr} and cohesion c_h.

The stresses σ_{qm} and σ_{pm} are as follows. The Mohr diagram is constructed in Fig. 5.9. The answer is $\varphi_{fr} = 37°$ and $c_h = 80$ kPa.

σ_{qm} (kPa)	σ_{pm} (kPa)
210	310
405	605
850	1250
1685	2485

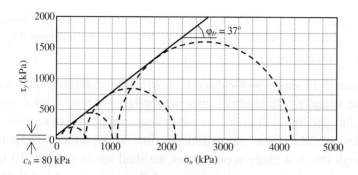

Figure 5.9. Mohr diagram.

In the natural environment the state of drainage of marine mud with heterogeneous properties and subject to variable stresses is usually unknown, even though this state can be identified in the laboratory. Thus the state of drainage is commonly ignored in spite of its importance in erosion.

5.3.3 *Vane shear strength*

The vane shear strength τ_v measured by turning a metallic vane in the fine-grained soil sample is used to estimate bed yield strength under shear loading. Vane shear testing is usually done *in situ* in the field, or on mud cores from field. A vane of diameter D_v and length H_v (Fig. 5.10) is pushed into the sample and the torque required to rotate the vane is recorded. The sample is sheared along the vertical and horizontal edges

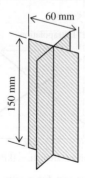

Figure 5.10. Vane geometry for shear measurement. The same (but typically smaller) artifact is one of the geometries used in rheometers.

of the vane [Bardet, 1997]. The shear strength τ_v, typically in kilopascals (kPa), is calculated from the maximum shear-induced moment M_{ms}, *i.e.*,

$$\tau_v = M_{ms} / \left\{ (\pi D_v^2 / 2)[H_v + (D_v / 3)] \right\}.$$

Figure 5.11 shows an idealized plot of depth-variations of mud water-content W and τ_v. Quite often, since the top layer of mud is weak with W exceeding the liquid limit W_L, its vane shear strength is not registered by the device. The shear strength-depth line is therefore extrapolated. Although this is a crude approach, as we shall see in Chapter 10 it has been used to obtain a rough estimate of the thickness of the fluid-like mud layer.

5.4 Deformation by Static and Dynamic Shear Loading

5.4.1 *Mud rheology*

Mud rheology concerns the deformation and flow behavior of mud. In shear rheology, the typically soft sample is shear loaded. Extensional rheology deals with deformation induced by normal stress. An example of extensional deformation is the biaxial straining produced when a balloon is inflated. Any small rectangular patch of the balloon (Fig. 5.12) stretches in the two principal planar directions and its thickness decreases. In Chapter 8 reference is made to extentional-compressive

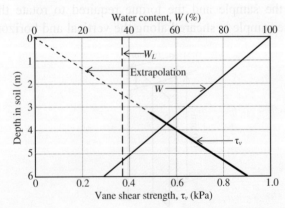

Figure 5.11. Result from a vane shear test for a silty clay (adapted from Lambe and Whitman [1979]).

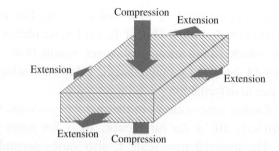

Figure 5.12. Normal stresses causing biaxial straining of an elastic material (adapted from Barnes *et al.* [1989]).

deformation strength of flocs, which is typically about two orders of magnitude greater than the shear yield strength, and plays a significant role in the development of floc orders of aggregation (Chapter 4) during consolidation. In Chapter 10, extensional rheology has been mentioned in connection with oscillatory straining of the cohesive bed under water waves.

The dynamic behavior of mud of given composition and water content can range from that of a stiff solid to a soft solid, or even viscous fluid. This behavior is qualitatively characterized by the Deborah number

$$D_e = \frac{\tau_e}{T_e} \tag{5.16}$$

where τ_e is a characteristic time intrinsic to the material and T_e is a characteristic time of the deformation process, an extrinsic quantity. The name was coined by Professor Marcus Reiner [1969]. In the Old Testament (*Judges 5:5*), at one point the prophetess and judge Deborah says, "The mountains melted from before the Lord, even that Sinai...." The laic meaning of the first clause, provided we replace "mountains melted" by "mountains flowed," is that all materials including a mountain can behave as a solid or as a liquid depending on the Deborah number.[d] For a Newtonian liquid $\tau_e = 0$, and therefore $D_e = 0$. For a

---

[d] Reiner [1969] noted that in Deborah's famous song after victory over the Philistines she sang, "The mountains flowed before the Lord." The original word in Hebrew for "flowed" also means "melted" or "shook". The Septuagint translation from Hebrew and

purely elastic (Hookean) solid, $\tau_e \rightarrow \infty$ and $D_e \rightarrow \infty$. For liquid water $\tau_e = 10^{-12}$ s [Barnes *et al.*, 1989]. Thus if $T_e = 0.1$ s, we obtain $D_e = 10^{-11}$, a small value, which would mean that water would flow. If T_e were 10^{-14} s, D_e would be 100, a much larger value indicating solid-like response of water to deformation.

The cyclic forcing time-scale T_e in the marine environment can vary widely between say, 10^{-3} s for turbulence, 10^{1} s for water waves and 10^{4} s for tide. The material time-scale τ_e also varies depending on the consistency of mud. Taking $\tau_e = 1$ s as a notional value the Deborah number range would be 10^{-4} to 10^{3}. Thus, in general, both viscous fluid as well as elastic solid behaviors of (viscoplastic) mud are naturally important.

For viscoelastic response of mud under applied oscillatory shear forcing, the time-scale of forcing relative to strain response dictates the methods of measurement and their interpretation. For example, given sufficient time, say on the order of days, many estuarine muds creep when sheared at very low rates by tidal residual flows. On the other hand, mud erosion can occur over periods on the order of hours during which creep may be so small that for erosion it could be reasonable to describe mud rheology by models without creep. At the other end, creep can be the principal mechanism by which dredged navigation channels fill up with fluid mud over periods of months to years. In this instance the rheology of creeping mud would have to be modeled.

Relative to the viscoelastic behavior we will introduce the shear modulus of elasticity, and the dynamic viscosity of liquids.

5.4.2 *Shear modulus of elasticity and viscosity*

If a uniaxial normal (compressive) stress σ_{nz} is applied to an elastic cylinder (Fig. 5.2a), there will be vertical compression and lateral expansion such that the compressive strain is

---

Chaldee into Greek in ~300 BCE, and translated from Greek to English by Charles Thomson (secretary of the Continental Congress of the USA, 1774–1789) uses "shook". Reiner's use of "flowed" is for the simple reason that humans in their short lifetime usually do not experience flowing mountains.

$$\varepsilon_{nz} = \frac{\sigma_{nz}}{E_n} \tag{5.17}$$

where E_n is the Young's modulus, a measure of the stiffness of the material[e], and

$$\varepsilon_{nx} = \varepsilon_{ny} = -\mu_p \varepsilon_{nz} \tag{5.18}$$

where μ_p is Poisson's ratio. For a perfectly incompressible material μ_p is 0.5, whereas typical engineering materials have values between 0 and 0.5. If a shear stress τ_{zx} is applied to a cube (Fig. 5.3), the strain produced is given by Hooke's law

$$\gamma_{zx} = \frac{\tau_{zx}}{G} \tag{5.19}$$

where G is the shear or rigidity or elastic modulus. For materials with isotropic properties

$$G = \frac{E_n}{2(1+\mu_p)} \tag{5.20}$$

In the absence of data any parameter in this relationship can be estimated from the other two. However, given the non-isotropic state of the actual material, *e.g.*, a sediment bed stratified by consolidation, its response to stress tends to be directional to a significant extent. Thus, Eq. 5.20 is only useful for calculating G if E_n and μ_p are known.

For stress and strain in one-dimensional representation, Eq. 5.19 can be stated as

$$\tau = G\gamma \tag{5.21}$$

The corresponding relationship for a fluid (liquid) is

$$\tau = \eta\dot{\gamma} \tag{5.22}$$

---

[e] For clays, E_n ranges are: 0.05–0.5 kPa for very soft clay, 0.5–2 kPa for soft clay, 2–5 kPa for medium-strength clay and 5–10 kPa for stiff clay. Young's modulus is also expressed as the product of the undrained shear strength s_u and a correlation factor K_{cf}, *i.e.*, $E_n = K_{cf} s_u$. The correlation factor depends on the overconsolidation ratio and the Plasticity Index (PI).

where η is the dynamic viscosity and $\dot{\gamma}$ is the time-rate of change of strain γ. For a Newtonian fluid η is independent of $\dot{\gamma}$.

5.4.3 *Viscoplastic behavior*

In the flow curves of Fig. 5.13a from tests applying non-oscillatory forcing, *e.g.*, by rotational flow in a cylindrical viscometer, Newton's law is represented as a straight line of slope η passing through the origin. The flow behavior of soft mud beds and fluid muds is viscoplastic. A simple viscoplastic model is the Herschel–Bulkley equation

$$\tau = \tau_y + K_h \dot{\gamma}^n \tag{5.23}$$

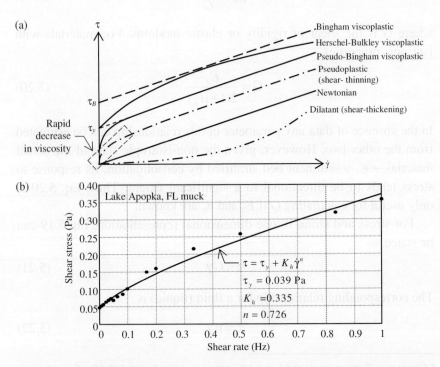

Figure 5.13. Stress versus rate of strain flow curves in non-oscillatory forcing: (a) Definition sketch; (b) data for Lake Apopka (Florida) muck and Herschel–Bulkley equation fit (from Mehta *et al.* [2009]).

where the plastic yield stress or shear strength τ_y and coefficients K_h and n are constants for a sediment and $n \neq 1$ (*e.g.*, curve in Fig. 5.13a). Rheometric tests on muds have shown that n typically ranges between 0.33 and 0.90. In other words, τ varies non-linearly with $\dot{\gamma}$. Both n and K_h are sensitive to fluid temperature, pH, electrolyte concentration, solids concentration and clay type [Cousot and Piau, 1994].

An example of data and fit of Eq. 5.23 is given in Fig. 5.13b for bottom muck from Lake Apopka in central Florida containing about 80% organic matter (based on loss on ignition). These data were obtained in a Brookfield (applied shear rate) rotational viscometer by increasing the shear rate in steps. Each step was maintained for 10 min, as preliminary tests showed that the viscosity achieved a nearly constant value within this duration. Shear stress was recorded at the end of each step. When the highest shear rate (1 s^{-1}) was attained it was stepped down in reverse order. The rising arm of the flow curve was non-monotonic as the viscosity was seemingly influenced by initial state of sediment in the viscometer. The falling arm was more consistent with the Herschel–Bulkley flow behavior. The equation fit in Fig. 5.13b is for the falling arm.

Coefficients derived from equation fit tend to vary with the selected range of shear rate. Therefore, it is essential to agree upon a reference range for comparison of flow curves for different muds. Since the flow curve is typically most sensitive to low shear rates, the range chosen in Fig. 5.13b is 0 to 1 Hz. A physical basis is that within this range floc growth usually dominates breakup. If a different range were chosen, *e.g.*, 0 to 10 Hz, the values of K_h and n would be different than 0.335 and 0.726, respectively.

Another useful viscoplastic model is the Casson [1959] equation

$$\sqrt{\tau} = \sqrt{\tau_y} + \sqrt{\eta_\infty \dot{\gamma}} \tag{5.24}$$

where η_∞ is the viscosity at theoretically infinite shear rate. This equation, originally developed for inks and later adopted as a standard for blood rheology, has been applied to fluid muds, and at least in some instances shown to be in better agreement with data than other relationships such as Eq. 5.23.

For denser muds granuloviscous models may be more suitable, but are rarely used [James *et al.,* 1987; Kirby, 1988; Dade and Nowell, 1991].

Equation 5.23 can be written as

$$\tau = \tau_y + (K_h \dot{\gamma}^{n-1}) \dot{\gamma} = \tau_y + \eta \dot{\gamma} \qquad (5.25)$$

where

$$\eta = K_h \dot{\gamma}^{n-1} \qquad (5.26)$$

denotes non-Newtonian viscosity. When $n = 1$, this equation describes the Bingham viscoplastic model characterized by viscosity η

$$\tau = \tau_B + \eta \dot{\gamma} \qquad (5.27)$$

in which τ_B is the Bingham yield stress. When τ is equal to or less than τ_B the material is an elastic solid, and when τ is greater than τ_B the material is a Newtonian fluid (Fig. 5.13a).[f]

In Fig. 5.14a a qualitative perspective is given in terms of the domains of material behaviors identified by the combined effects of floc volume fraction ϕ_{vf} and inter-particle interaction. Newtonian behavior is defined by low values of ϕ_{vf} and very low, almost negligible, inter-particle collisions. With increasing collisions the probability of forming strong flocs from weak ones and dispersed particles increases. Thus the horizontal axis can also be interpreted as representing a change from deflocculated or weakly flocculated particles on the left to increasingly flocculated particles towards the right. Irrespective of the level of inter-particle interaction, the viscosity rises slowly at first, but rapidly at higher values of ϕ_{vf} when particle packing approaches its space-filling value of 1. In that condition inter-particle bonds are formed throughout the space containing water and particles. For randomly packed *rigid*

---

[f] The shear strength of a material, *i.e.,* shear resistance per unit area, is determined by applying a shear stress, *i.e.,* shear force per unit area. Here the terms yield stress and yield strength are used interchangeably because for example, Bingham mud yield stress is also equal to the associated mud shear strength.

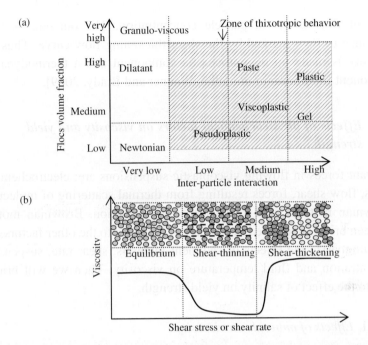

Figure 5.14. Sediment behavior: (a) Combined effects of inter-particle interaction and solids volume fraction (adapted from Cheng [1980]) — the hatched area is the zone in which thixotropic effect can be significant; (b) Change in the microstructure of a colloidal suspension associated with equilibrium state and transitions to shear-thinning and shear-thickening behaviors (adapted from Wagner and Brady [2009]).

spheres the volume fraction at which viscosity begins to increase rapidly is about 0.62 [Heywood, 1991].

In the hatched zone material behavior is thixotropic, *i.e.,* the material rapidly loses its strength when disturbed and slowly regains strength when left undisturbed. Shear-thinning and thickening effects in the zones of pseudoplastic and dilatant behaviors, respectively, are brought about by changes in the microstructure of fine particles. As shown in the flow curve of Fig. 5.14b, at equilibrium, when the (Newtonian) viscosity is independent of the shear rate, stochastically random collisions between particles cause them to resist flow. However, as the shear stress (or shear rate) increases the particles become organized as sheets and the viscosity decreases (shear-thinning). At higher shear rates hydrodynamic interactions dominate over stochastic ones, and lead to clustering as transient fluctuations in particle concentration. The outcome is high resistance to flow marked by the shear-thickening flow curve. Thus the viscosity is made up of a stochastic component and a thermodynamic component [Schreuder *et al.*, 1986; Wagner and Brady, 2009].

5.4.4 *Effects of physicochemical factors on viscosity and yield strength*

Relevant forces in flowing viscoplastic suspensions are: electrochemical forces, flow shear, forces resulting from thermal scattering of molecules (Brownian motion), and viscous energy dissipation. Brownian motion has been briefly addressed in Chapter 4. In regard to the other factors, we will summarize the effects of sediment minerals, shear rate, suspension concentration and fluid temperature on viscosity. Then we will briefly refer to the effect of salinity on yield strength.

5.4.4.1 *Effects of minerals on viscosity*

By setting the yield strength $\tau_y = 0$ in Eq. 5.23 we obtain the equation for a pure power-law fluid

$$\tau = K_h \dot{\gamma}^n \qquad (5.28)$$

which is a representation of Eq. 5.26. For power $n < 1$ the material has a shear-thinning behavior, whereas $n > 1$ represents shear-thickening. A common variant of Eq. 5.26 is the Sisko [1958] model

$$\eta = c_c \dot{\gamma}^{n-1} + \eta_\infty \qquad (5.29)$$

where c_c is called the consistency of the power-law fluid. As mentioned, when $n = 1$ the suspension becomes Newtonian with a constant viscosity η_∞, as in this case c_c is zero and the suspension has a non-deformable structure. At high, theoretically infinite, shear rate a pseudoplastic also approaches Newtonian behavior with viscosity η_∞.

The viscosity of "designer" (flocculated) clay mineral slurries was measured by pumping them through a specially made horizontal pipe viscometer. The slurries consisted of different proportions of a kaolinite (K), an attapulgite (A) and a bentonite (B) all in tap water. Power-law coefficients for these mixtures are given in Table 5.4, and an example of the Sisko representation is shown in Fig. 5.15a. The shear rate range is limited because data could not be obtained in the lower range $(< ~0.008 \text{ s}^{-1})$ of the shear rate over which the viscosity decreases rapidly with increasing shear rate. In Fig. 5.13a this zone is enclosed within the rectangle. Notwithstanding the narrow range of shear rates, Fig. 5.15a confirms the potential utility of Eq. 5.29. The coefficients characterizing this equation were found to approximately follow relations dependent on the cation exchange capacity:

$$\eta_\infty = 0.05\,CEC_s + 0.001$$
$$n = 0.033\,CEC_s + 0.28 \qquad (5.30)$$
$$\log c_c = 0.13\,CEC_s + 0.22$$

Table 5.4. Sisko coefficients for designer clays.

Sediment	Density (kg m^{-3})	η_∞ (Pa.s)	c_c	n
Kaolinite	1,300	2.10	7.08	0.106
Kaolinite + 5°/$_{oo}$ NaCl	1,300	2.06	3.31	0.117
Bentonite	1,050	0.41	48.7	0.207
Bentonite + 5°/$_{oo}$ NaCl	1,030	2.46	28.3	−0.009
Attapulgite	1,100	6.34	6.86	−1.0
Attapulgite + 5°/$_{oo}$ NaCl	1,080	5.00	11.5	0.038
Kaolinite + Bentonite (1:1)[a]	1,160	0.61	12.3	−0.057
Kaolinite + Bentonite (1:1) + 5°/$_{oo}$ NaCl	1,160	4.69	20.6	−0.114
Attapulgite + Bentonite (1:1)	1,050	4.28	45.2	0.002
Attapulgite + Bentonite (1:1) + 5°/$_{oo}$ NaCl	1,050	7.06	45.1	−0.039
Attapulgite + Kaolinite (1:1)	1,190	4.44	0.76	−1.083
Attapulgite + Kaolinite (1:1) + 5°/$_{oo}$ NaCl	1,190	3.35	8.02	0.059

[a] Equal proportion by weight.
Source: Jinchai [1998].

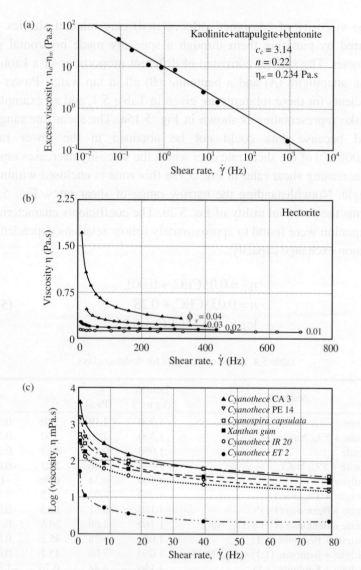

Figure 5.15. Pseudoplastic behavior of materials in water: (a) Sisko representation of a mixture of kaolinite, attapulgite and bentonite (from Jinchai [1998]); (b) dilute suspensions of hectorite (based on Wood *et al.* [1955]); (c) polysaccharides (from De Philipps and Vicenzini [1998]).

in which CEC_s is defined as

$$CEC_s = f_K CEC_K + f_A CEC_A + f_B CEC_B \qquad (5.31)$$

where f is the weight fraction of each clay and CEC with a subscript denotes (pure) clay cation exchange capacity. Note that

$$f_K + f_A + f_B = 1 \qquad (5.32)$$

The CEC values (in mEq 100 g^{-1}) were taken as 6 for K, 28 for A and 105 for B. The water content in the tests ranged from 86% to 423%, and CEC_s from 1.9 to 10.4 mEq 100 g^{-1}. Such a representation of CEC must not be extended to muds containing significant fractions of NOM ($\sim> 0.1$), which has an effect on the CEC unrelated to cohesion (Chapters 3 and 4).

Viscosity data for a hectorite (a smectite) in water shown in Fig. 5.15b reveal that even dilute suspensions of this clay show power-law type behavior, which expectedly increases with the solids volume fraction [Wood *et al.*, 1955]. Equally noteworthy is the pseudoplasticity of natural polysaccharides, which plays an important role in the locomotion of small aquatic animals. In shallow-water environments containing high fractions of NOM, the power-law behavior of the suspended mucoid matter is shown to govern the resuspension of sediment during storms (*e.g.*, Mehta *et al.* [2009]).

5.4.4.2 *Effects of concentration and shear on viscosity*

Based on data compiled from several sources the following approximate relationship for η_∞ with cohesive slurry concentration C has been proposed

$$\eta_\infty = \eta_w \left(1 + \alpha_r C^{\beta_r} \right) \qquad (5.33)$$

where η_w is the viscosity of water (see also Eq. 4.36). Viscometric data for kaolinite in fresh water have been found to be commensurate with coefficient values $\alpha_r = 1.68$ and $\beta_r = 0.346$, given C in kg m^{-3}. The range of C over which Eq. 5.33 may be used varies with slurry composition.

For kaolinite the range has been found to be 5 to 400 kg m^{-3}, which includes fluid mud (Chapter 10) and the concentration range over which mud can exist as a very soft bed as long as it is left undisturbed [Engelund and Zhaohui, 1984; Ross, 1988].

Based on a viscosity equation of Odd *et al.* [1993], the following empirical relations were developed for fluid mud viscosity in seawater using sediment from the Amazon River mouth

$$\eta = \rho_w C e^{-(0.78\,\dot{\gamma} + 10.24)} + \eta_w \; ; \qquad \dot{\gamma} < 3.55 \text{ s}^{-1} \qquad (5.34)$$

$$\eta = \rho_w C e^{-(0.017\,\dot{\gamma} + 12.95)} + \eta_w \; ; \qquad \dot{\gamma} > 3.55 \text{ s}^{-1} \qquad (5.35)$$

The maximum concentration over which these equations were applied was on the order 100 kg m^{-3}. They include the effect of shear rate, which can be important in the energetic tidal environment [Faas, 1986; Faas, 1995; Vinzon, 1998].

Example 5.3: You are given a mud suspension in seawater ($\rho_w = 1{,}027$ kg m^{-3}, $\eta_w = 10^{-3}$ Pa.s) at a concentration of 2 kg m^{-3}. (1) Calculate the viscosity at shear rates of 0, 1, 4 and 6 s^{-1}. (2) Repeat the calculations for a concentration of 10 kg m^{-3}.

Based on Eqs. 5.34 and 5.35 the answers are given below. The ratio of viscosity at 10 kg m^{-3} to 2 kg m^{-3} is a high value of about 4, which explains why dense slurries settle slowly. In a crude experiment on the settling of a kaolinite slurry, retardation in the settling rate due to the effect of viscosity was estimated to be about one-third, with the remaining two-thirds from the effect of (upward) return flow of water and the buoyancy of settling flocs [Sampath, 2009, Chapter 7].

Concentration 2 kg m^{-3}		Concentration 10 kg m^{-3}		
Shear rate (s^{-1})	Viscosity (η_2) (Pa.s)	Shear rate (s^{-1})	Viscosity (η_{10}) (Pa.s)	η_{10}/η_2
0	0.0744	0	0.3678	4.95
1	0.0346	1	0.1691	4.88
4	0.0056	4	0.0238	4.28
6	0.0054	6	0.0230	4.26

Figure 5.16 illustrates the effect of high concentration on the viscosity of different muds. They range from cohesive sediments such as Provins clay to the non-cohesive Provins limestone [Migniot, 1968]. The rate of rise of viscosity remains low until the concentration reaches its space-filling value. This transition is marked by the onset of a measurable effective normal stress due to tenable inter-particle contact. As described

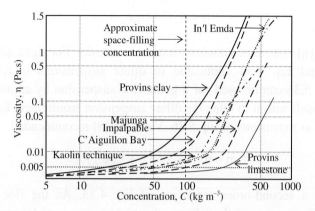

Figure 5.16. Dependence of viscosity on suspension concentration (adapted from Migniot [1968]).

in Chapter 8, above that concentration the total hydrostatic load at any point is shared between the fluid (water) and the particle matrix. While the space-filling concentration varies with sediment composition as well as the dynamic state of mud, in Fig. 5.16 the lowest value is on the order of 100 kg m^{-3}, *i.e.,* a solids volume fraction of about 0.04, which is consistent with the onset of a rapid rise in viscosity. Note however that individually the space-filling concentration is seen to vary over a wide range with the highest value of about 300 kg m^{-3}.[g]

Equation 4.33 relates suspension viscosity to the floc volume fraction. This equation can be more generally stated as

---

[g] As mentioned earlier, the space-filling concentration must be defined in terms of the floc volume fraction ϕ_{vf} rather than dry sediment concentration C_s (or the solids volume fraction ϕ_v). Given a particle mineral density of 2,650 kg m^{-3}, C_s = 100 kg m^{-3}, floc density 1,100 kg m^{-3} and water density 1,000 kg m^{-3}, from Eq. 4.43 we obtain the floc volume fraction as 1,650/2650 = 0.62, which is commensurate with randomly packed rigid spheres. Space-filling by flocs occurs when the entire space is a uniform matrix, *i.e.,* when ϕ_{vf} = 1, or C_s = 160 kg m^{-3}. If the floc density were 1,050 kg m^{-3}, C_s would be 80 kg m^{-3} and for 1,200 kg m^{-3} it would be 320 kg m^{-3}. In other words C_s can have a wide range of values depending on the floc structure. It has also been called gelling concentration [Winterwerp and van Kesteren, 2004]. This is the threshold concentration at which gelation occurs, *i.e.,* a sol changes to a gel. A gel is *jelly-like* state of matter which is mostly liquid, but tends to behave like a solid due to its particle-supported matrix, which gives gel the strength it possesses. Mud gels are usually quite soft, at times with yogurt-like consistency.

$$\eta = \eta_w (1 + [\eta]\phi_{vf})$$ (5.36)

in which $[\eta]$ is a (dimensionless) intrinsic viscosity. For rigid spheres $[\eta]$ = 2.5, and Eq. 4.33 applicable to dilute suspensions is recovered. Equation 5.36 can be used for a given floc suspension by adjusting $[\eta]$. Other models meant to relax the dilute suspension requirement have been proposed as extensions of Eq. 5.36. Vand [1948] recommended

$$\eta = \eta_w (1 + 2.5\phi_{vf} + 7.3\phi_{vf}^2)$$ (5.37)

which is a second-order extension of Eq. 4.33. As the floc volume fraction ϕ_{vf} increases the effect of neighboring particles cannot be ignored. Most treatments dealing with this effect appear to be for rigid particles, and their original application is not associated with flocculated sediments, which tend to have high resistance to flow at high concentrations. For elastic spheres Goddard and Miller [1967] give

$$\eta = \eta_w \left\{ 1 + \phi_{vf} \left[\frac{[\eta]\left(1 - \frac{3}{2}Se^2\right)}{1 + \left(\frac{3}{2}Se\right)^2} \right] \right\}$$ (5.38)

in which the shearoelastic number Se is given by

$$Se = \frac{\eta_w \dot{\gamma}}{G}$$ (5.39)

with G the shear modulus of the elastic particles. Krieger and Dougherty [1959] obtained a simpler expression

$$\eta = \eta_w \left(1 - \frac{\phi_{vf}}{\phi_m}\right)^{-[\eta]\phi_m}$$ (5.40)

When ϕ_{vf} attains a value ϕ_m, the volume fraction at maximum packing, η is infinitely large and the suspension is immobile.

Pal [2003] extended the application of Eq. 5.36 to deformable elastic spheres. Consider a suspension of such particles at a volume fraction ϕ_{vf} and viscosity $\eta(\phi_{vf})$. Into this suspension an infinitesimally small amount of new particles is added. The resulting incremental increase in viscosity, $d\eta$, can be calculated from Eq. 5.38 by treating the suspension into which the new particles are added as an equivalent homogeneous medium of viscosity $\eta(\phi_{vf})$. This development can be shown to yield

$$d\eta = \eta \left[\frac{[\eta] \left\{ 1 - \frac{3}{2}\left(\frac{\eta\dot{\gamma}}{G}\right)^2 \right\}}{1 + \left(\frac{3}{2}\left(\frac{\eta\dot{\gamma}}{G}\right)\right)^2} \right] d\phi_{vf} \tag{5.41}$$

After rearrangement and integration with the limit $\eta \to \eta_w$ as $\phi_{vf} \to 0$ one obtains

$$\eta = \eta_w \left[\frac{1 - \frac{3}{2}Se^2}{1 - \frac{3}{2}Se^2\left(\frac{\eta}{\eta_w}\right)^2} \right]^{5/4} e^{[\eta]\phi_{vf}} \tag{5.42}$$

When $Se \to 0$ this expression reduces to the form of Eq. 4.37

$$\eta = \eta_w e^{[\eta]\phi_{vf}} \tag{5.43}$$

Eq. 5.42 is applicable when $\phi_{vf} < 0.20$. Pal [2003] developed two additional models for higher concentrations.[h]

---

[h] Creeping hyper-concentrated mud flow occurs in the Yellow River in China [Wan, 1982]. This regime is characterized by concentrations in the range of about 400 to 800 kg m^{-3} [Bradley and McCutcheon, 1987]. Thomas [1963] proposed the following formula for the viscosity of slurries having particles in the size range of 0.1 to 20 μm and applicable to concentrations up to about 800 kg m^{-3}:

$$\eta = \eta_w \left(1 + 2.5\phi_{vf} + 10.05\phi_{vf}^2 + 0.062e^{\frac{1.875\phi_{vf}}{1-1.595\phi_{vf}}} \right) \tag{F5.1}$$

As mentioned, formulas for viscosity variation with concentration not derived for mud flocs tend to under-predict the viscosity. Consider the following illustration.

Example 5.4: You are given the following data for a clay suspension: floc density $\rho_f = 1,100$ kg m^{-3}, particle density $\rho_s = 2,650$ kg m^{-3}, water density $= 1,000$ kg m^{-3}, water viscosity $\eta_w = 0.001$ Pa.s and suspension concentration $C = 100$ kg m^{-3}. Calculate the viscosity using Eq. 5.43, compare it with the following equation of Engelund and Zhaohui [1984], and obtain the value of the intrinsic viscosity [η]:

$$\eta = \eta_w \left(1 + \frac{0.206}{\eta_w} \phi_{vf}^{1.68} \right) \qquad (5.44)$$

We obtain $\phi_{vf} = C(\rho_s - \rho_w)/[\rho_s (\rho_f - \rho_w)] = 0.623$. Therefore from Eq. 5.44 $\eta = 0.094$ Pa.s. Then from Eq. 5.43

$$[\eta] = \frac{1}{\phi_{vf}} \ln\left(\frac{\eta}{\eta_w} \right) = 7.30$$

Comparison of the [η] value of 7.3 with 2.5 in Eq. 4.33 indicates the highly viscous and therefore energy dissipative nature of mud. The reason for high viscosities of natural mud is the heterogeneity of matter with respect to size, shape and mineral density as well as binding effects of organic matter.

5.4.4.3 *Effect of temperature on viscosity*

There appears to be sparse information on the effect of fluid temperature on the viscosity of fine sediment suspensions, even though the viscosity of liquids is known to vary significantly with temperature. To review this effect for dilute suspensions let us consider Newtonian liquids. The well-known Arrhenius relationship, also known as the Andrade correlation between dynamic viscosity η and absolute temperature T, is

$$\eta = A_T e^{\frac{B_T}{T}} \qquad (5.45)$$

where A_T and B_T are constants of the liquid. The basis of this relationship has been explored by many, with original contributions by Eyring and co-workers and considerable subsequent developments [Viswanath *et al.*, 2007]. Here we will review a phenomenological interpretation by Krone

[1983] for organic liquids. To accept the explanation for fine sediment suspensions, they must be thought of as fluid continua. Although flocculated suspensions are not quite like single-phase fluids, one can anticipate qualitative agreement between Eq. 5.45 and the dependence of the viscosity of suspensions on temperature.

Consider thermally agitated layers of liquid molecules (Fig. 5.17). These laminas with inter-layer spacing δ are in relative motion. The lowest layer is moving with velocity u_1, the layer above it with velocity u_2 and so on. Thus a velocity gradient exists between the layers. Each molecule occasionally vibrates towards the adjacent layer, temporarily interacts with molecules in that layer, and reverts to vibration about its mean position, amounting to a "cage effect" in the force field of its neighbors. This interaction produces a stress between the layers that is relieved by transitory tearing of the layer initiated where a molecule has sufficient instantaneous thermal energy to disrupt its attraction to a neighbor. The total molecular mass between such "rips" or defects, causes a transfer of momentum between adjacent layers.

Using the linear impulse–momentum principle, *i.e.*, equality between the integral of force over a short time interval and the change of momentum over that interval, the viscosity resulting from the transfer of momentum between adjacent layers in relative motion is obtained from

$$\eta = \frac{\tau}{du/dz} \approx \frac{n_A f^r m_m \overrightarrow{(u_2 - u_1)}}{(u_2 - u_1)/\delta} \tag{5.46}$$

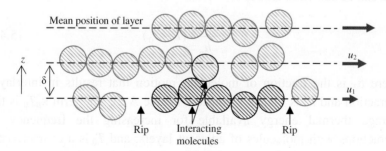

Figure 5.17. Schematic description of momentum transfer between inter-molecular layers.

where τ is the apparent shear stress, du/dz is the velocity gradient, n_A is the number of molecules per unit area of a molecular sheet, f^r is the frequency of interaction of each molecule with the adjacent layer and m_m is the mass of the molecules in the sheet acting as a unit. The vectors $\overrightarrow{u_2 - u_1}$ cancel, leaving only the scalar terms, i.e.,

$$\eta = n_A \delta f^r m_m \tag{5.47}$$

As we noted in Chapter 4, for molecules in statistically significant numbers, energy is assumed to follow the Boltzmann distribution. Thus the fraction of molecules in a layer having sufficient threshold thermal energy E_e to break with a neighbor and initiate a rip will be $\exp(-E_e/\kappa_B T)$, where κ_B is the Boltzmann constant (Appendix A, Table A.2). This distribution provides the required link between the properties of individual molecules and the thermodynamic bulk properties of the fluid medium. Accordingly, the average number of molecules per molecular unit of mass m_m that initiates this rip is $1/\exp(-E_e/\kappa_B T) = \exp(E_e/\kappa_B T)$. This means that

$$m_m = \frac{M}{N_A} e^{\frac{E_e}{\kappa_B T}} \tag{5.48}$$

where M is the molecular weight and N_A is Avogadro's number. The frequency of interaction f^r is taken to be proportional to the average energy of thermal motion of the molecular unit above a threshold level. This can be shown to lead to

$$f^r = \frac{\alpha_T \kappa_B}{h_p} (T - T_0) \tag{5.49}$$

where α_T is the fraction of molecular motion that results in interlayer interaction and h_p is Planck's constant (Table A.2). The term $\kappa_B T_0$ is the average thermal energy available for increasing the frequency of interactions with molecules of adjacent layers, and T_0 is a characteristic (Debye) temperature. By combining Eqs. 5.47, 5.48 and 5.49 we obtain

$$\eta = \frac{\alpha_T n_A \delta \kappa_B M}{h_p N_A}(T-T_0)e^{\frac{E_e}{\kappa_B T}}$$ (5.50)

Comparison with Eq. 5.45 yields

$$A_T = \frac{\alpha_T n_A \delta \kappa_B M}{h_p N_A}(T-T_0); \quad B_T = \frac{E_e}{\kappa_B}$$ (5.51)

which indicates that A_T varies linearly with T. Equation 5.50 can now be written as

$$\eta = A_T'(T-T_0)e^{\frac{B_T}{T}}$$ (5.52)

where $A_T' = \alpha_T n_A \delta \kappa_B M / h_p N_A$ is another constant. Equation 5.52 provides a physicochemical basis for the behavior of the viscosity of benzene and other organic liquids, which according to Krone [1983] differs from the Andrade correlation because A_T is a temperature dependent quantity. The effect of temperature on the viscosity of a dilute suspension ($\phi_v = 0.05$) of attapulgite in water at different shear rates is given in Table E5.4 (see also Exercise 5.6).

5.4.5 *Effects of salinity on yield strength*

A simple apparatus for determination of approximate values of the Bingham yield stress is the Ostwald capillary-tube viscometer. Tests using this apparatus at constant applied rates of strain resulted in the data in Fig. 5.18. They were meant to shed light on the transport of dense turbid currents in Lake Mead upstream of Hoover Dam in Nevada [Einstein, 1941]. The addition of very small amounts of salt (NaCl) in a suspension of the lake clay initially dispersed in distilled water coagulated the particles. As a result the flow behavior changed from Newtonian to Bingham plastic with measurable yield strength. The product ηu_m, where u_m is the mean velocity of the suspension in the tube, is plotted against the hydrostatic pressure loss Δh driving the flow. There is proportionality between ηu_m and Δh because u_m is a measure of the

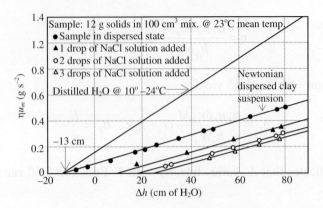

Figure 5.18. Effect of NaCl on yield strength (from Einstein [1941]).

mean shear rate and Δh is proportional to the wall shear stress τ. Since in its unmodified form the viscometer had to be operated at a constant pressure head, it was provided with a simple means to vary the head. After subtracting from all readings a head of 13 cm of water in the tube leading to the viscometer, the remaining pressure loss was proportional to the Bingham yield stress τ_B. The pressure loss increased with increasing addition of NaCl, starting with no yield strength in the dispersed suspension.

The sensitivity of the yield strength to fluid composition reflects changes in floc structure and strength. As we noted in Chapter 4, pore fluid composition strongly influences flocculation. Fig. 5.19 shows as much as an order of magnitude change in the Bingham yield stress for three clay suspensions with increasing NaCl concentration. From this it is evident that, when modeling fluid mud flow in nature, laboratory data on the yield strength of mud must not be used without calibration.

Careful examinations of viscometric data show that reported yield strengths of some marine muds may in fact be apparent quantities. This is illustrated in Fig. 5.20, in which flow curves for two samples of Rotterdam Harbor mud at different solids volume fractions have been plotted at different scales. While the upper plot suggests nearly Bingham behavior, *i.e.,* there is a reasonably well defined yield stress, the middle and the lower figures indicate that when the same data are plotted at

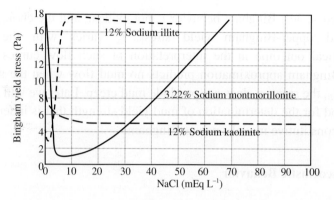

Figure 5.19. Variation of yield strength with salt concentration (from van Olphen [1977]).

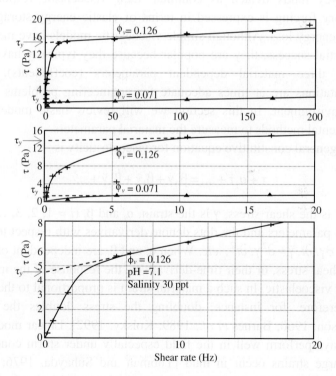

Figure 5.20. Flow curves for samples of Rotterdam harbor mud plotted at different scales. Data reported by Williams and James [1978] (from Parker and Kirby [1982]).

expanded scales, Bingham behavior is not evident, and the choice of the so-called upper-Bingham yield stress τ_y varies with the scale. A practical outcome in the interpretation of such flow curves is that while Bingham approximation predicts no mud flow when the stress is below τ_B, the stress τ_y does not preclude mud creep. Later we will refer to a method for the determination of mud yield strength that is independent of any constitutive model, such as Bingham.

5.5 Viscoelastic Behavior

5.5.1 *General behavior*

For clayey muds treated as continua, their viscoelastic response to oscillatory forcing is expressed in terms of elastic energy storage and viscous energy loss. Although simple viscoelastic models have met with only partial success, they are popular because they typically have only two or three material dependent parameters (coefficients). Such representations are at times adequate for engineering problems in the wave environment. In this section we will review basic models and experimental methods to measure the material parameters.

The general constitutive equation for viscoelasticity is

$$\tau + \alpha_1 \dot{\tau} + ... = \beta_0 \gamma + \beta_1 \dot{\gamma} + \beta_2 \ddot{\gamma} + ... \tag{5.53}$$

where τ is the shear stress, γ is the strain, α_i and β_i ($i = 1, 2, 3, ...$) are material parameters and the dots denote derivatives with respect to time, *i.e.*, $\dot{\gamma} = \partial\gamma/\partial t, \ddot{\gamma} = \partial^2\gamma/\partial t^2$, *etc.* When α_i and β_i are independent of shear strain, shear stress, or their time-derivatives, the response of the material is linear viscoelastic. In such a material strain is proportional to the stress and therefore, for instance, doubling the stress doubles the strain [Wilkinson, 1960; Barnes *et al.*, 1989; Kolsky, 1992]. Linear models do not always perform well in the field especially under storm conditions when large strains occur in mud [Tubman and Suhayda, 1976; Jiang, 1993]. However, their simplicity and given that empiricism is usually unavoidable in field representation of mud rheology, the use of linear viscoelastic models with calibrated coefficients can be reasonable.

In Eq. 5.53, if β_0 were the only non-zero parameter, one would obtain Hooke's law with $\beta_0 = G$, the elastic or rigidity modulus (Eq. 5.21). If β_1 were the only non-zero parameter, Eq. 5.53 would indicate (Newtonian) viscous fluid response with $\beta_1 = \eta$, the dynamic viscosity (Eq. 5.22). Hooke's law is mechanically represented as a spring with elastic constant G and Newton's law as a dashpot containing fluid of viscosity η. Thus, Eq. 5.53 can be described by a model made up of combinations of springs and dashpots. The force on a spring is proportional to strain γ, and on a dashpot to strain rate $\dot{\gamma}$.

Simple mechanical analogs are described next. Each analog is represented by a set of the coefficients α_i and β_i, but because not all models are reduced to a common form, coefficients of two models need not have the same meaning. In other words, α_i and β_i from one analog may not be applicable to another.

5.5.2 Basic mechanical analogs

In Eq. 5.53, when β_0 and β_1 are both non-zero while all others are zero, the simplest material analog, the viscoelastic Kelvin–Voigt (KV) model, is obtained. This model is a combination of a spring and a dashpot in parallel configuration (Eq. 5.54 and Fig. 5.21). Both elements have the same deformation (strain), and the total stress is equal to the sum of the stresses in each element. The constitutive equation for this behavior is

$$\tau = G\gamma + \eta\dot{\gamma} \qquad (5.54)$$

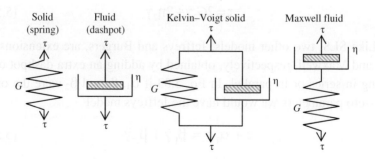

Figure 5.21. Mechanical analogs of a solid, a fluid and combinations.

If one were to stretch a KV element at rest from both ends, the spring would prevent further stretching after it is fully loaded. Therefore this model represents a solid which withstands an applied stress. In the Maxwell model the spring occurs in series with dashpot (Fig. 5.21). Both elements experience the same stress and the total strain rate is the sum of strain rates of the two elements. From Eq. 5.53 with $\alpha_1 = \eta/G$ and $\beta_1 = \eta$ as the only non-zero coefficients, the constitutive equation reduces to

$$\tau + \left(\frac{\eta}{G}\right)\dot{\tau} = \eta\dot{\gamma} \qquad (5.55)$$

Since this element can be stretched indefinitely it represents a fluid. Mud liquefaction by waves has been treated as a process in which a KV element representing the solid bed changes to a Maxwell element representing liquefied mud (Chapter 11).

Mud can also be modeled as a three-parameter "standard linear solid" (SLS). These parameters include two rigidity moduli, G_1 and G_2, and viscosity η. The constitutive equation is

$$\tau + \alpha_1\dot{\tau} = \beta_0\gamma + \beta_1\dot{\gamma} \qquad (5.56)$$

This model is one of the simplest choices after the basic KV/Maxwell representations, and involves the 0th and the 1st order derivatives of stress and strain. It has been used to model the consolidation behavior of soils [Keedwell, 1984]. When G_1 approaches infinity, the model reduces to KV representation

$$\tau = 2G_2\gamma + 2\eta_2\dot{\gamma} \qquad (5.57)$$

Like SLS, two other models, Jeffreys and Burgers, are extensions of KV and Maxwell, respectively, obtained by adding an extra dashpot or a spring in series or in parallel. In Eq. 5.53 if α_1, β_1 and β_2 were the only non-zero parameters we would have the Jeffreys model

$$\tau + \alpha_1\dot{\tau} = \beta_1\dot{\gamma} + \beta_2\ddot{\gamma} \qquad (5.58)$$

Inclusion of the term $\alpha_2 \ddot{\tau}$ in the above equation yields the Burgers model

$$\tau + \alpha_1 \dot{\tau} + \alpha_2 \ddot{\tau} = \beta_1 \dot{\gamma} + \beta_2 \ddot{\gamma} \qquad (5.59)$$

Compared with a two- or a three-parameter model, the Burgers model is more complex with four parameters, and requires extensive rheometric tests that may not yield expected benefits, especially if the response of the material does not meet the assumption of linearity.

5.5.3 *Viscoelasticity characterization*

Two rheometric methods, static testing and the dynamic testing, are used to determine the material parameters of viscoelastic behavior (Fig. 5.22). The types of tests for KV, Maxwell and other models is given in Fig. 5.22. As we will see later, if the test data do not conform to any of the canonical models (*e.g.,* Fig. 5.21), an entirely empirical viscoelastic model, or one based on Eq. 5.53, can be constructed for the desired stress or strain ranges.

5.5.3.1 *Static testing*

Static testing in general includes creep test at constant stress, and stress relaxation at constant strain. The input stress or strain is applied instantaneously. In reality this means that the time required for the input

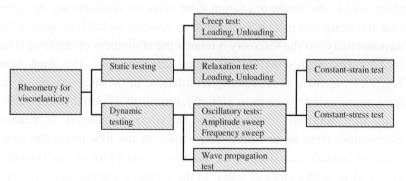

Figure 5.22. Rheometric test protocol for determination of viscoelastic parameters.

signal to reach its steady value must be very short compared to the time over which the output is to be recorded.

5.5.3.1.1 Creep test

In the creep test there are two phases of the input stress, the loading phase and the unloading phase. The output of the loading phase ($0 \leq t \leq t_{ch}$, where t_{ch} is the time interval during which the sample is loaded) is also called the creep curve. The unloading phase ($t > t_{ch}$) amounts to rebound or creep recovery. The input stress is represented as

$$\tau = \begin{cases} \tau_0, & \text{for } 0 \leq t \leq t_{ch} \\ 0, & \text{for } t_{ch} < t \end{cases} \qquad (5.60)$$

For the KV element the creep curve is described by

$$\gamma = \frac{\tau_0}{G}\left(1 - e^{-\frac{G}{\eta}t}\right) \qquad 0 \leq t \leq t_{ch} \qquad (5.61)$$

When t_{ch} is large enough, γ approaches a steady value. Creep recovery is described by

$$\gamma = \frac{\tau_0}{G}e^{-\frac{G}{\eta}(t-t_{ch})} \qquad t_{ch} < t \qquad (5.62)$$

where η/G is the strain relaxation time constant representing the time-scale for strain creep. The final value of γ, equal to τ_0/G, is approached "exponentially" as the viscosity η retards the attainment of the final state. Hooke's law represents instantaneous attainment of the final state (Fig. 5.23), as there is no viscous retardation. In the recovery phase the strain, represented by the ratio $\gamma/(\tau_0/G)$, returns to its initial (zero) value.

In the Maxwell element, due to viscosity-induced continuous deformation, there is no equilibrium strain. In the first phase the strain ramps up linearly at a slope τ_0/η. In other words a plot of $\gamma\eta/\tau_0$ against time is a line with a slope of unity. In the second phase the strain remains constant (Fig. 5.24). These responses are

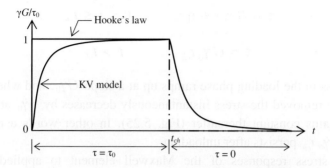

Figure 5.23. Strain response of a KV element in creep test.

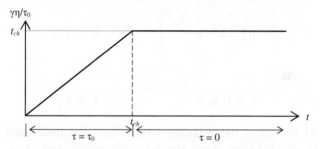

Figure 5.24. Strain response of a Maxwell element in creep test.

$$\gamma = \frac{\tau_0}{\eta} t \qquad 0 \leq t \leq t_{ch} \qquad (5.63)$$

$$\gamma = \frac{\tau_0}{\eta} t_{ch} \qquad t_{ch} < t \qquad (5.64)$$

5.5.3.1.2 Relaxation test

The input strain rate $\dot{\gamma}$ in the relaxation test is

$$\dot{\gamma} = \begin{cases} \dot{\gamma}_0 \,, & \text{for } 0 \leq t \leq t_{ch} \\ 0 \,, & \text{for } t_{ch} < t \end{cases} \qquad (5.65)$$

where $\dot{\gamma}_0$ is the applied strain rate. For the KV element the stress responses to the strain loading and the unloading phases are, respectively,

$$\tau = G\,\dot{\gamma}_0\, t + \eta\,\dot{\gamma}_0 \qquad 0 \le t \le t_{ch} \qquad (5.66)$$

$$\tau = G\,\dot{\gamma}_0\, t_{ch} \qquad t > t_{ch} \qquad (5.67)$$

The stress in the loading phase ramps up at a slope $G\dot{\gamma}_0$, and when $\dot{\gamma}$ is suddenly removed the stress instantaneously decreases by $\eta\dot{\gamma}_0$ at $t = t_{ch}$, and remains constant thereafter (Fig. 5.25). In other words a residual stress $G\dot{\gamma}_0\, t_{ch}$ persists after unloading.

The stress responses of the Maxwell element to applied strain rate $\dot{\gamma}_0$ are

$$\tau = \eta\,\dot{\gamma}_0 \left(1 - e^{-\frac{G}{\eta}t} \right) \qquad 0 \le t \le t_{ch} \qquad (5.68)$$

$$\tau = \eta\,\dot{\gamma}_0\, e^{-\frac{G}{\eta}(t - t_{ch})} \qquad t > t_{ch} \qquad (5.69)$$

where η/G ($= \tau_M$) is the relaxation time, *i.e.*, the time-scale of stress release. In the loading phase the stress increases and an equilibrium stress is attained after a long time. In the limit $\eta/G \to \infty$ the response is instantaneous and in accordance with Hooke's law (Eq. 5.21). After the removal of strain rate, *i.e.*, setting $\dot{\gamma}_0 = 0$, the stress reduces to the initial (zero) value (Fig. 5.26). This response (of a fluid element) is different from what occurs in the KV model (for a solid), in which a residual stress remains.

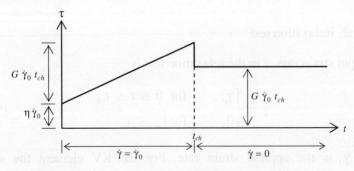

Figure 5.25. Stress response of a KV element in relaxation test.

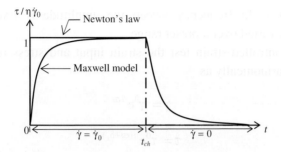

Figure 5.26. Stress response of a Maxwell element in relaxation test.

5.5.3.2 *Dynamic testing*

5.5.3.2.1 Types of tests

Dynamic testing includes two types of sub-tests — the oscillatory test and the shear wave propagation test. A rheometer and a shearometer, respectively, are used for these tests.[i]

5.5.3.2.2 Oscillatory test

Method: Two types of oscillatory tests can be performed depending on the input. One is the controlled-strain test in which the input is oscillating strain. The other is controlled-stress test in which the input is oscillating stress. In either representation the stress wave is ahead of the strain wave by a phase δ_s ranging between 0 and $\pi/2$ radians. When $\delta_s = 0$ the response is purely elastic, whereas $\delta_s = \pi/2$ indicates a purely viscous response. In the intermediate range of $0 < \delta_s < \pi/2$, the response is viscoelastic.

There are two ways to apply the dynamic load. In amplitude sweep at a fixed frequency the stress or strain is ramped up linearly or in another

---

[i] A rheometer is any instrument measuring rheological properties. A special kind of rheometer is known as rheogoniometer, which measures normal as well as shear components of the stress tensor. A viscometer is a rheometer limited to measuring the viscosity. A shearometer only measures the shear (elastic) modulus of a viscoelastic material.

preset manner. In frequency sweep the amplitude is fixed and the frequency is varied over a preset range.

In the controlled-strain test the strain input and stress response are expressed harmonically as

$$\gamma = \gamma^0 e^{-i\sigma t} \tag{5.70}$$

$$\tau = \tau^0 e^{-i(\sigma t + \delta_s)} \tag{5.71}$$

where τ^0 is the oscillatory stress amplitude, γ^0 is the corresponding strain amplitude and σ is the angular frequency. Stress input and strain response in the controlled-stress test are represented as

$$\tau = \tau^0 e^{-i\sigma t} \tag{5.72}$$

$$\gamma = \gamma^0 e^{-i(\sigma t - \delta_s)} \tag{5.73}$$

For convenience of further analysis we will introduce Hooke's law in the form

$$\tau = G^* \gamma \tag{5.74}$$

Phase difference between stress and strain requires that G^* be represented as a complex number expressed as

$$G^* = G' - iG'' \tag{5.75}$$

where G' and G'' are the storage modulus and the loss modulus, respectively. The absolute value $|G^*| = \sqrt{G'^2 + G''^2}$ is the effective shear modulus of the material based on Eq. 5.74.

We may also introduce a complex viscosity η^* defined by

$$\tau = \eta^* \dot{\gamma} \tag{5.76}$$

where

$$\eta^* = \eta' + i\eta'' \tag{5.77}$$

The real part η' represents viscous energy loss and η'' corresponds to elastic energy storage. The absolute value $\left|\eta^*\right| = \sqrt{\eta'^2 + \eta''^2}$ represents the effective viscosity of the material based on Eq. 5.76. Note that $\left|\eta^*\right|$ does not provide any indication of the significance of viscous energy loss relative to elastic energy storage.

With regard to G' and G'', from Eqs. 5.72 to 5.74 we obtain

$$G' = \frac{\tau^0}{\gamma^0}\cos\delta_s \qquad (5.78)$$

$$G'' = \frac{\tau^0}{\gamma^0}\sin\delta_s \qquad (5.79)$$

which is consistent with $\delta_s = 0$ representing elastic response and $\delta_s = \pi/2$ as viscous response. The relationship between the two moduli and the phase is

$$\frac{G''}{G'} = \tan\delta_s \qquad (5.80)$$

Equations 5.78, 5.79 and 5.80 permit determination of the viscosity η and the shear modulus G of any viscoelastic model. Based on Eq. 5.54 representing the KV model, for the controlled-stress as well as controlled-strain tests

$$G' = G; \qquad G'' = \eta\sigma; \qquad \eta' = \eta; \qquad \eta'' = \frac{G}{\sigma} \qquad (5.81)$$

and

$$\eta = \frac{\tau^0}{\gamma^0}\frac{\sin\delta_s}{\sigma} \qquad (5.82)$$

$$G = \frac{\tau^0}{\gamma^0}\cos\delta_s \qquad (5.83)$$

With respect to the Maxwell model (Eq. 5.55), for both types of tests

$$G' = \frac{\eta^2 \sigma^2 G}{\eta^2 \sigma^2 + G^2}; \quad G'' = \frac{\eta \sigma G^2}{\eta^2 \sigma^2 + G^2}; \quad \eta' = \frac{\eta G^2}{\eta^2 \sigma^2 + G^2}; \quad \eta'' = \frac{\eta^2 \sigma G}{\eta^2 \sigma^2 + G^2} \quad (5.84)$$

and

$$\eta = \frac{\tau^0}{\gamma^0} \frac{1}{\sigma \sin \delta_s} \quad (5.85)$$

$$G = \frac{\tau^0}{\gamma^0} \frac{1}{\cos \delta_s} \quad (5.86)$$

For the KV model or the Maxwell model, η and G can be determined by the oscillatory test alone from Eqs. 5.82 and 5.83, or Eqs. 5.85 and 5.86, respectively. For the three-parameter SLS and Jeffreys models as well as the four-parameter Burgers model, the static test is also required for an unequivocal evaluation of η and G.

5.5.3.2.3 Shear-wave propagation test

Method: Ultrasound shear wave is used to measure the rheological parameters of fluids and soft solids [Dixon and Lanyon, 2005]. The test for finding the shear modulus can be carried out in a pulse shearometer. A small-amplitude, high-frequency shear wave is made to pass through a gap containing the sample between two parallel metal plates that are a small distance ΔD apart. Each plate is connected to a piezoelectric crystal (Fig. 5.27). At one plate a pulse generator is used to initiate a shear wave which travels through the sample and is detected at the other plate after a short interval Δt. Thus the shear wave celerity C_{sh} is

$$C_{sh} = \frac{\Delta D}{\Delta t} \quad (5.87)$$

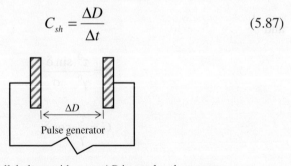

Figure 5.27. Parallel plates with a gap ΔD in a pulse shearometer.

Determination of G' and G'' : The relevant momentum equation for wave motion in the *x-z* plane is

$$\frac{\partial^2 x}{\partial t^2} = \frac{1}{\rho}\frac{\partial \tau}{\partial z} \tag{5.87}$$

which represents a balance between the inertia force and viscous energy loss. Using Eq. 5.74 yields.

$$\frac{\partial^2 x}{\partial t^2} = \frac{G^*}{\rho}\frac{\partial \gamma}{\partial z} \tag{5.88}$$

The strain is given by

$$\gamma = \frac{\partial x}{\partial z} \tag{5.89}$$

Combining Eqs. 5.88 and 5.89 we obtain the wave equation

$$\frac{\partial^2 x}{\partial t^2} = \frac{G^*}{\rho}\frac{\partial^2 x}{\partial z^2} \tag{5.90}$$

The solution for amplitude *x* is

$$a_{sz} = a_{s0}e^{-k^* z}e^{-i\sigma t} \tag{5.91}$$

where a_{s0} is the value of *x* at $z = 0$. The complex wave number k^* is expressed as the sum

$$k^* = k_r + ik_i \tag{5.92}$$

where $k_r = 2\pi/L$ is the wave number, *L* is the wave length and k_i is the wave damping coefficient. The latter is easily shown by substitution of Eq. 5.92 into Eq. 5.91 to yield

$$a_{sz} = a_{s0}e^{-k_i z}e^{-i(\sigma t + k_r z)} \tag{5.93}$$

which represents a wave traveling in the $-z$ direction with angular frequency σ and wave number k_r. The amplitude $a_{s0}e^{-k_i z}$ decays exponentially with *z;* the rate being dependent on k_i such that, at a distance z_0, the initial amplitude a_{s0} decreases to a_{s0}/e (Fig. 5.28).

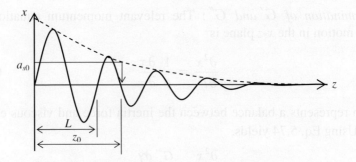

Figure 5.28. Exponentially-damped shear wave.

From Eqs. 5.90, 5.91 and 5.92 we find that

$$k^* = i \frac{\sigma}{\sqrt{G^*/\rho}} \qquad (5.94)$$

Substitution of Eqs. 5.74 and 5.75 into the above expression yields

$$G' = \rho \sigma^2 \frac{k_r^2 - k_i^2}{\left(k_r^2 + k_i^2\right)^2} \qquad (5.95)$$

$$G'' = \rho \sigma^2 \frac{2 k_r k_i}{\left(k_r^2 + k_i^2\right)^2} \qquad (5.96)$$

For a given material, wave attenuation depends mainly on the wave frequency. Since the pulse shear-wave frequency is high, k_i can be expected to be quite small.[j] When k_i is on the order of 10^{-1}, the order of k_r is 10^3. Therefore it is reasonable to assume that $k_i/k_r \ll 1$. Thus Eqs. 5.95 and 5.96 are simplified as

$$G' = \rho C_{sh}^2 \qquad (5.97)$$

[j] Even so, the shear wave signal damps out rapidly in mud and may be undetectable after a few centimeters of passage.

and

$$G'' = 2\rho C_{sh}^2 \frac{k_i}{k_r} \qquad (5.98)$$

in which the shear wave celerity is

$$C_{sh} = \frac{\sigma}{k_r} \qquad (5.99)$$

and

$$k_i = \frac{1}{2} k_r \tan \delta_s \qquad (5.100)$$

Model Parameters: For the KV model, starting with Eq. 5.54 and using Eqs. 5.70, 5.74, 5.75, 5.80 and 5.99 we obtain

$$G = G' \qquad (5.101)$$

Therefore

$$G = \rho C_{sh}^2 \qquad (5.102)$$

or $C_{sh} = \sqrt{G/\rho}$ and

$$\eta = \frac{\tan \delta_s}{\sigma} \rho C_{sh}^2 \qquad (5.103)$$

For the Maxwell model, using Eq. 5.55 and following the same approach it can be shown that

$$G = (1 + \tan^2 \delta_s) \rho C_{sh}^2 \qquad (5.104)$$

$$\eta = \frac{1 + \tan^2 \delta_s}{\sigma \tan \delta_s} \rho C_{sh}^2 \qquad (5.105)$$

Accordingly, once C_{sh} and δ_s are obtained from experiments, G and η can be calculated.

5.6 Mud Rheometry

5.6.1 *Viscometer flow geometries*

We will consider the basic principles of some of the viscometer geometries, which are also used in rheometers. Even though differences in the physicochemical composition of sediment among mud samples taken from a small bed area, perhaps no more than a square meter, may be seemingly minor, their shear strengths and viscosities may differ widely. In addition, interpretation of viscometric data requires an understanding of the hydromechanics peculiar to each device.

Commercial rheometers include automated testing protocols and software for analysis which display a range of test outputs. For example, oscillatory (stress or strain) frequency sweep enables identification of the response of mud to different wave periods and wave durations. The sample is exposed to small-strain oscillations over the selected range of frequencies and response recorded over different durations. Amplitude sweep is similarly applied.

5.6.1.1 *Coaxial cylinders viscometer*

A rotational device of this type is also known as a Couette–Hatschek viscometer. The fluid is contained in the narrow annular space and the outer cylinder of radius R_c rotates at angular speed ω_o (Fig. 5.29a). The inner cylinder of radius $m_c R_c$, where m_c is the ratio of the radius of the inner cylinder to the outer cylinder, is stationary. The viscosity η, which varies with the torque Γ_c required to turn the shaft driving the outer cylinder, is calculated by measuring Γ_c.

At steady-state there is no pressure gradient in the (tangential) θ-direction, and the relevant equation of motion (in cylindrical coordinates, Appendix B, Eq. B.13) reduces to

$$0 = \frac{d}{dr}\left(\frac{1}{r}\frac{d}{dr}ru_\theta\right) \qquad (5.106)$$

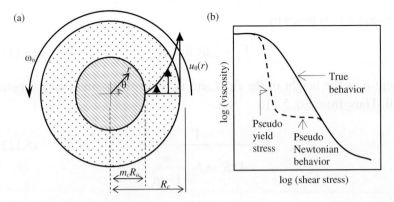

Figure 5.29. (a) Laminar flow in the annular space between two coaxial cylinders; (b) Wall depletion effect (from Barnes [1995]).

where $u_\theta(r)$ is the tangential velocity. The appropriate boundary conditions for flow are

$$\text{At } r = m_c R_c, \quad u_\theta = 0$$
$$\text{At } r = R_c, \quad u_\theta = \omega_o$$
(5.107)

Upon integration of Eq. 5.106 using these conditions we obtain

$$u_\theta = \omega_o R_c \frac{\left(\dfrac{m_c R_c}{r} - \dfrac{r}{m_c R_c} \right)}{\left(m_c - \dfrac{1}{m_c} \right)}$$
(5.108)

For a Newtonian fluid the tangential stress is

$$\tau = -\eta r \frac{\partial}{\partial r} \left(\frac{u_\theta}{r} \right)$$
(5.109)

which after substitution of u_θ from Eq. 5.108 yields

$$\tau = -2\eta \omega_o \left(\frac{R_c}{r} \right)^2 \left(\frac{m_c^2}{1 - m_c^2} \right)$$
(5.110)

The torque Γ_c is given by

$$\Gamma_c = 2\pi R_c h_s \tau_R R_c \qquad (5.111)$$

where h_s is the height of the fluid and τ_R is the shear stress on the outer wall. Thus, from Eq. 5.110

$$\eta = \frac{\Gamma_c}{4\pi R_c^2 \omega_0 h_s \left(\dfrac{m_c^2}{1 - m_c^2} \right)} \qquad (5.112)$$

In measurements of the flow curves of fluids extraneous effects can arise, because the fluid height h_s is finite and the resulting end-effects may be significant. This is minimized in practice by ensuring that h_s is at least 100 times greater than the gap width between the two cylinders [Barnes *et al.*, 1989]. In suspensions, problems may also occur due to phase separation resulting from sedimentation, which may enhance the viscosity. This is common when dilute mud slurries are being tested.

Migration of particles away from the walls tends to deplete the concentration at the solid boundaries and reduces the viscosity. Wall depletion is especially important at high solids volume fractions and low shear rates. Evidence of depletion is sketched in Fig. 5.29b. A false yield stress effect is observed in the flow curve and a (pseudo) Newtonian plateau appears. It is possible that the material actually undergoes these changes due to its structural response to forcing. However, in regard to marine muds, it is likely to be an artifact of the measurement technique. Use of the vane geometry (*e.g.*, Fig. 5.10) attempts to minimize wall depletion. This is because the rotating inner boundary is almost entirely replaced with the material itself so the possibility of depletion is significantly reduced. Under most test conditions, this geometry may make wall depletion insignificant because the response of this boundary is what is measured during the test. The outer cylinder wall, though a static boundary, may still contribute to significant wall depletion under certain conditions [Barnes, 1995].

Among other potential problems in rheometry a noteworthy effect, when testing complex materials, is due to secondary flow. This is most likely to occur with low solids volume fractions and high shear rates (the opposite of conditions conducive to wall depletion). Inertia-driven secondary flow appears in the form of Taylor vortices, which distort the data and their onset marks the upper limit of reliable data for that sample. It appears that there is no direct way to compensate for this effect defined by the Taylor number (for concentric cylinder geometries) representing the importance of the rotational force relative to the viscous drag [Barnes, 2000].

5.6.1.2 *Capillary tube viscometer*

Let us consider capillary flow of a Bingham fluid mud. In a narrow tube of radius R_c the velocity distribution is characterized by plug flow (Fig. 5.30) occupying the core of the tube in which the shear stress $\tau(r)$ is less than the Bingham yield stress τ_B. In other words, within plug flow the velocity u_{plug} is constant because the applied shear is insufficient to liquefy the non-sheared core.

The equation of motion for Bingham flow (Eq. 5.27) is[k]

$$\tau_B + \eta \frac{du}{dr} = \left(\frac{p_0 - p_L}{2L_p} \right) r \qquad (5.113)$$

---

[k] The steady-state stress distribution in the capillary tube is obtained by integrating the equation of motion (in cylindrical coordinates). Equation B.12 in Appendix B applicable to this case is reduced to

$$\frac{d}{dr}(r\tau) = \frac{\Delta p}{L_p} r \qquad (F5.2)$$

Upon integration we obtain

$$\tau = \frac{\Delta p}{2L_p} r + \frac{C'}{r} \qquad (F5.3)$$

The constant of integration $C' = 0$ because the momentum flux cannot be infinite at $r = R_c$. Therefore

$$\tau = \frac{\Delta p}{2L_p} r \qquad (F5.4)$$

This relation indicates that at any distance r from the tube centerline, fluid stress is proportional to the pressure gradient $\Delta p/L_p$ across the tube.

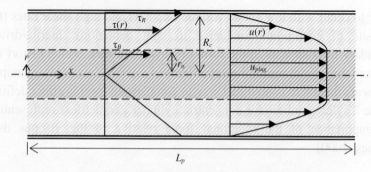

Figure 5.30. Bingham fluid mud flow in a capillary tube.

or

$$\tau_B + \eta \frac{du}{dr} = \frac{\Delta p}{2L_p} r \qquad (5.114)$$

where the fluid pressure drop is

$$\Delta p = p_0 - p_L \qquad (5.115)$$

with pressures p_0 and p_L at the beginning and the end, respectively, of the tube of length L_p. Making use of the boundary condition $u = 0$ at $r = R_c$, and integration of Eq. 5.115 yields

$$u = \frac{R_c^2 \Delta p}{4\eta L_p}\left[1-\left(\frac{r}{R_c}\right)^2\right]-\frac{\tau_B R_c}{\eta}\left[1-\left(\frac{r}{R_c}\right)\right]; \qquad r \geq r_0 \qquad (5.116)$$

and given the core radius $r_0 = 2\tau_B L_p / \Delta p$, we obtain

$$u_{plug} = \frac{R_c^2 \Delta p}{4\eta L_p}\left(1-\frac{r_0}{R_c}\right)^2; \qquad r < r_0 \qquad (5.117)$$

The discharge is then obtained from

$$Q = 2\pi \int_0^{r_0} u_{plug}\, r dr + 2\pi \int_{r_0}^{R_c} u r dr \qquad (5.118)$$

After substituting for u and u_{plug} from Eqs. 5.116 and 5.117, respectively, the viscosity is found to be

$$\eta = \frac{\pi R_c^4 \Delta p}{8QL_p}\left[1 - \frac{4}{3}\left(\frac{\tau_B}{\tau_R}\right) + \frac{1}{3}\left(\frac{\tau_B}{\tau_R}\right)^4\right] \qquad (5.119)$$

This is the Buckingham–Reiner equation [Reiner, 1956], in which the wall shear stress $\tau_R = R_c \Delta p / 2L_p$. By setting $\tau_B = 0$ we recover the well-known Hagen–Poiseuille equation for viscous flow of a Newtonian fluid

$$\eta = \frac{\pi R_c^4 \Delta p}{8QL_p} \qquad (5.120)$$

which is the basis of an Ostwald-type viscometer.

Example 5.5: A sediment slurry of 1,260 kg m^{-3} density flows through a horizontal capillary tube of 0.3 mm radius and 0.3 m length. For a pressure drop $\Delta p = 275$ kPa the flow rate is 2×10^{-7} m^3 s^{-1}. Find the slurry viscosity assuming Newtonian flow.

From Eq. 5.120 we obtain $\eta = 0.015$ Pa.s. The Reynolds number $Re = 2Q\rho / \pi R_c \eta$ is equal to 36.7, which confirms laminar flow in the capillary.

5.6.1.3 *Cone-and-plate viscometer*

In a cone-and-plate viscometer geometry the sample is held in the space between a stationary flat plate of radius R_c and a rotating conical geometry (Fig. 5.31a). The cone subtends an aperture angle θ_o with respect to the plate. Bird *et al.* [1960] give the complete solution of the boundary value problem and obtain the following relationship between the viscosity and the torque Γ_c applied to the conical geometry rotating at angular speed ω_o

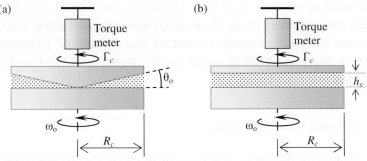

Figure 5.31. (a) Cone-and-plate geometry; (b) parallel-plate geometry (adapted from Barnes *et al.* [1989]).

$$\eta = \frac{3\Gamma_c}{4\pi R_c^2 \omega_o \sin\theta'}\left[\cot\theta' + \frac{1}{2}\left(\ln\frac{1+\cos\theta'}{1-\cos\theta'}\right)\sin\theta'\right] \quad (5.121)$$

where $\theta' = (\pi/2) - \theta_o$.

5.6.1.4 Parallel-plate viscometer

In a parallel-plate viscometer a flat plate replaces the conical geometry of the cone-and-plate viscometer (Fig. 5.31b). The gap width between the two plates is h_s. The relationship between the viscosity η and torque Γ_c applied to the upper plate rotating at angular speed ω_o is

$$\eta = \frac{3\Gamma_c h_s}{2\pi R_c^4 \omega_o \left(1+\dfrac{n}{3}\right)} \quad (5.122)$$

This expression is applicable to power-law fluids which obey Eq. 5.28 with n as the exponent [Barnes *et al.*, 1989]. Since most mud slurries are generally pseudoplastic they fall within this fluid category.

5.6.1.5 Virtual gap rheometer

A virtual gap rheometer (VGR), which is a miniature shearometer, has been used to measure the shear modulus G of soft clayey sediment beds subjected to monochromatic water waves. In a flume study the VGR was embedded in the sediment bed, thus allowing the time-variation of G to be measured as waves worked the bed [Mehta *et al.*, 1995]. In a follow-up study, the capability of the shearometer was improved and it was used to measure the storage modulus G' and the loss modulus G'' [Babatope *et al.*, 1999]. Since the VGR is useful for the detection of fluid mud, it is further described in Chapter 10.

5.6.2 Rheometry for yield stress

Controlled-stress rheometry can be used in creep tests to determine the yield stress τ_y free from ambiguities associated with estimation of the upper-Bingham yield stress from rotational flow tests (Fig. 5.20). The stress τ_y has the same physical meaning as the Bingham yield stress τ_B in the sense that no flow can occur at lower applied stresses. Characteristic test curves are shown schematically in Fig. 5.32, where creep compliance $J(t)$ is defined as

$$J(t) = \frac{\gamma(t)}{\tau_0} \tag{5.123}$$

The quantity J_0 is the instantaneous value of J resulting from a purely elastic initial response to applied stress τ_0, and J_0^{-1} is the instantaneous rigidity modulus (= G_0). As J increases above J_0 its rise is retarded by viscous drag. When $\tau_0 \geq \tau_y$ the particle matrix collapses and elastic response vanishes.

For mud samples, a miniature vane-geometry similar to one in Fig. 5.10 is commonly used with the rheometer to obtain the response curves in Fig. 5.32. An advantage of using a vane is that it avoids wall-slip which occurs when for example a bob (a solid right-circular metallic cylinder) is used in place of the vane. The instantaneous compliance J_0 thus obtained is plotted against the applied stress. At small stresses the response tends to be predominantly elastic. In this event, J_0 is almost independent of stress, which permits identification of the range of linear response to stress [James *et al.*, 1987; 1988].

Buscall *et al.* [1987] measured J_0 of dense flocculated suspensions by creep and wave propagation tests. Comparison of the results at small

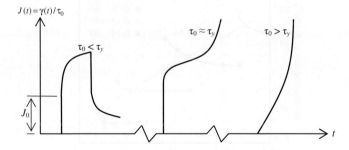

Figure 5.32. Schematic diagram showing changing creep-compliance with increasing applied stress (from James *et al.* [1987]).

strains showed good agreement between the two approaches, implying that J_0 was independent of the forcing frequency.

Creep tests were carried out by Jiang [1993] on a 1:1 mixture of an attapulgite and a kaolinite in tap water (AK sediment, Table 5.5) in a controlled-stress rheometer. In Fig. 5.33, the instantaneous compliance J_0

Table 5.5. Composition of selected sediments.

Sediment	Median size (µm)	Particle density (kg m^{-3})	Main constituents
Okeechobee mud (OK)	9	2,140^a	Kaolinite, sepiolite, montmorillonite, organic matter (40%)^b
Attapulgite + kaolinite (AK)	1	2,510	Attapulgite (50%) + kaolinite (50%)
Kerala India mud (KI)	2	2,650^c	Montmorillonite, kaolinite, illite, gibbsite, organic matter (5%)
Mobile Bay mud (MB)	15	2,650^c	Clayey silt of undetermined composition, quartz (sand)
Atchafalaya Delta mud (AD)	1	2,580	Smectite, illite, kaolinite, quartz (sand), organic matter (7%)

^{a} The low density reflects the presence of 40% organic matter.
^{b} By weight.
^{c} Assumed.

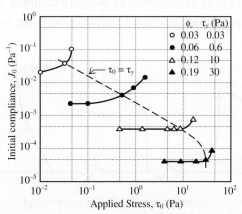

Figure 5.33. Initial compliance versus applied stress for AK sediment (from Jiang [1993]).

is plotted against applied stress amplitude τ_0 for four samples at different solids volume fractions ϕ_v. The curve for each ϕ_v is approximately divided by the dashed line representing the yield threshold ($\tau_0 = \tau_y$). At stresses below τ_y, J_0 increases only gradually with τ_0, with the rate of increase decreasing with increasing ϕ_v. Except for the $\phi_v = 0.03$ curve, J_0 is nearly independent of τ_0, indicating a practically linear response. Once yield commences J_0 rapidly increases with τ_0. As shown in Fig. 5.34, which is based on the $\tau_0 = \tau_y$ trajectory in Fig. 5.33, when ϕ_v exceeds ~0.05 the rise in τ_y is dramatic due to the development of a space-filling matrix. Thus the space-filling value of ϕ_v appears to be about 0.05. This limit possibly explains why the curve for $\phi_v = 0.03$ in Fig. 5.33 does not show a distinct value of τ_y, as space-filling may have been incomplete at that volume fraction.

5.6.3 *Rheometry for standard linear solid*

Tests on soft mixtures of a kaolinite and a montmorillonite in controlled-strain tests showed that under constant strain loading ($0 < t < t_{ch}$), the resulting stress reached a constant value [Chou, 1989]. This is contrary to the response of a KV solid in which the stress continues to increase with time (Eq. 5.66). When loading was removed a residual stress was recorded, indicating that the response was not that of a Maxwell fluid

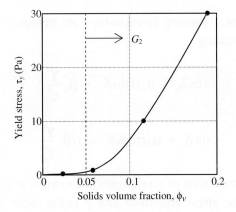

Figure 5.34. Yield stress versus solids volume fraction for AK mud. The quantity G_2 refers to one of the two shear moduli in the SLS model (from Jiang [1993]).

in which the stress vanishes. Thus neither the KV model nor the Maxwell model was suitable for the tested clay mixtures.

To circumvent some of the limitations of two-coefficient models, the three-coefficient SLS model (Eq. 5.56) has been used. Tests on several mud beds showed that SLS can be a better descriptor of mud response to applied stress than KV [Jiang, 1993]. An advantage of SLS is its greater flexibility in terms of data fit, although this is not surprising because an additional material coefficient must be evaluated from measurements. The coefficients in Eq. 5.56 are

$$\alpha_1 = \frac{\eta}{G_1 + G_2} \tag{5.124}$$

$$\beta_0 = \frac{G_1 G_2}{G_1 + G_2} \tag{5.125}$$

$$\beta_1 = \frac{\eta G_1}{G_1 + G_2} \tag{5.126}$$

Therefore

$$\tau + \frac{\eta}{G_1 + G_2} \dot{\tau} = \frac{G_1 G_2}{G_1 + G_2} \gamma + \frac{\eta G_1}{G_1 + G_2} \dot{\gamma} \tag{5.127}$$

From Eq. 5.56 the following relationships are obtained for controlled-stress oscillatory testing

$$\cos \delta_s - \alpha_1 \sigma \sin \delta_s = \beta_0 \frac{\gamma^0}{\tau^0} \tag{5.128}$$

$$\sin \delta_s + \alpha_1 \sigma \cos \delta_s = \sigma \beta_1 \frac{\gamma^0}{\tau^0} \tag{5.129}$$

where δ_s is the phase shift of strain relative to stress, σ is the oscillation frequency, τ^0 is the stress amplitude and γ^0 is the strain amplitude. The viscoelastic parameters G_1, G_2 and η can be calculated from

$$G_1 = \frac{\tau^0}{\gamma^0} \frac{\sigma}{\sigma \cos \delta_s - \beta_s \sin \delta_s} \tag{5.130}$$

$$G_2 = \frac{\tau^0}{\gamma^0} \frac{\sigma \beta_s}{\sin \delta_s \left(\sigma^2 + \beta_s^2\right)} \tag{5.131}$$

$$\eta = \frac{\tau^0}{\gamma^0} \frac{\sigma}{\sin \delta_s \left(\sigma^2 + \beta_s^2\right)} \tag{5.132}$$

with

$$\beta_s = \frac{\beta_0}{\beta_1} \tag{5.133}$$

The complex equivalent viscosity $\eta^* = \tau / \dot{\gamma}$ (Eq. 5.76) is

$$\eta^* = \frac{(\beta_1 - \beta_0 \alpha_1)\sigma + i(\beta_0 + \alpha_1 \beta_1 \sigma^2)}{\sigma(1 + \alpha_1^2 \sigma^2)} \tag{5.134}$$

A comprehensive set of rheometric tests is required to fully assess the degree to which SLS can predict the behavior of a given mud. A partial confirmation of the model assumption that the second order derivative of shear strain γ is expectedly negligible is illustrated in Fig. 5.35 based on creep tests. The rate of strain-rate $\ddot{\gamma}$ is plotted against the strain rate $\dot{\gamma}$ for mud from Lake Okeechobee in Florida (OK mud, Table 5.5). As observed $\ddot{\gamma}$ is practically independent of $\dot{\gamma}$ (and is nearly zero).

Figure 5.36 shows creep responses of OK and AD muds (Table 5.5) in terms of the time-variation of strain during the loading and relaxation phases. In all three tests, in the loading phase the strain shows an "exponential-type" increase with time under a constant applied stress. At the end of the phase the stress was removed and relaxation took place. In the case of OK mud it is conceivable that the continuation of each test phase would have shown that at the end of the loading phase the strain attained a constant value and at the end of relaxation it vanished. However, practically speaking the sediment was neither KV nor Maxwell.

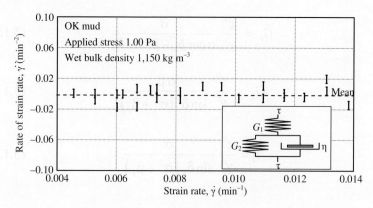

Figure 5.35. Second derivative of strain (rate of strain rate) against strain rate. Inset shows the mechanical analog of SLS (from Jiang [1993]).

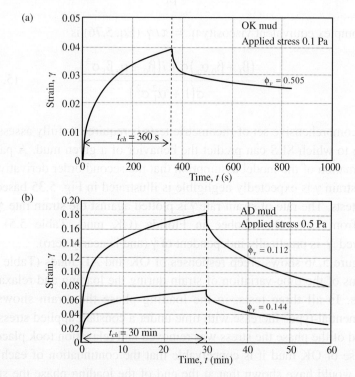

Figure 5.36. Strain response in creep test: (a) OK mud (from Jiang [1993]); (b) AD mud (from Robillard [2009]).

The AD mud (at both volume fractions) shows a drop in strain immediately upon stress removal. The drop is higher at the lower volume fraction because of greater fluidity (*i.e.,* lower viscosity) of the mud sample. In both cases the subsequent decline in strain is gradual and suggests that given a longer duration of relaxation the strain could have vanished.

For OK mud the following relationship was found between β_s and the (wet bulk) density ρ (between 1,050 and 1,150 kg m^{-3})

$$\beta_s = 1.62(\rho - 1)^{0.65} \qquad (5.135)$$

which indicates that as density increased the elastic response represented by β_0 increasingly dominated the viscous effect embodied in β_1 [Jiang, 1993].

In oscillatory tests, for input stress amplitude τ^0 the strain response amplitude γ^0 and the phase shift δ_s were obtained. Then the SLS material parameters G_1, G_2, and η were calculated from Eqs. 5.133, 5.134 and 5.135, respectively. In Fig. 5.37 the mean trends for the parameters have been plotted for OK mud at a density of 1,120 kg m^{-3}. With increasing frequency f, G_2 and η decrease while G_1 increases slightly. At high frequencies the response of SLS approaches that of KV (as G_1 becomes large relative to G_2).

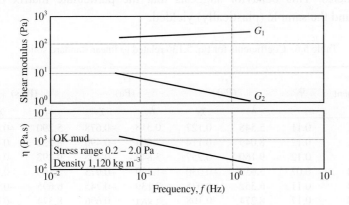

Figure 5.37. Material parameters for OK mud against oscillation frequency. The lines represent stress-averaged values (from Jiang and Mehta [1995]).

The instantaneous rigidity modulus $G_0 = J_0^{-1}$ for OK mud was found to be $1/0.003 = 333$, which is equal to G_1 at the high frequency of 10 Hz. This equality occurs because G_0 represents high-frequency response as mud behavior is dominated by its elastic property.

In Fig. 5.36 the material parameters for OK mud have the functional form

$$G_1, G_2, \eta = e^{\lambda_v} f^{\chi_v} \qquad (5.136)$$

Values of, λ_v and χ_v, are given for G_1, G_2 and η in Table 5.6. A likely explanation for the dependence of these parameters on the oscillation frequency is that they have a stochastic component, a component due to effects of clustering and also a component dependent on the thermodynamic state. At sufficiently low frequencies energy loss tends to be isothermal because adequate time is available to attain thermodynamic equilibrium. At high frequencies the process becomes adiabatic as there is little time to dissipate heat generated by Coulomb friction [Krizek, 1971; Wagner and Brady, 2009; Chapter 11.]

The parameters G_1, G_2 and η depend on the stress amplitude τ^0 once it exceeds a threshold value, as illustrated in Fig. 5.38 for viscosity η at three selected frequencies. The viscosity is reasonably independent of τ^0 until it reaches about 2 Pa. Beyond this linear response η falls at all three frequencies. This behavior suggests that the particulate matrix broke down and the sample (plastically) yielded.

Table 5.6. Coefficients for Eq. 5.136 related to linear standard solid.

Sediment	ϕ_v	G_1 (Pa)		G_2 (Pa)		η (Pa.s)	
		λ_v	χ_v	λ_v	χ_v	λ_v	χ_v
OK	0.11	5.548	0.127	0.318	−0.678	5.290	−0.687
AK	0.12	8.049	0.114	2.604	−0.490	8.222	−0.490
KI	0.12	9.160	0.257	3.843	−0.405	9.292	−0.405
MB	0.07	3.659	−0.030	−1.439	−0.975	3.165	−0.975
MB	0.11	6.352	0.075	2.139	−0.745	6.695	−0.745
MB	0.17	8.274	0.108	−3.864	−0.696	8.374	−0.696

Source: Jiang and Mehta [1995].

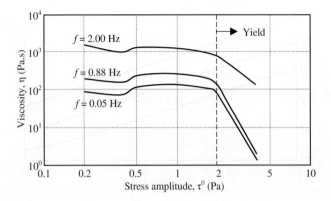

Figure 5.38. Variation of viscosity with stress amplitude at three frequencies for OK mud (from Jiang and Mehta [1995]).

In Table 5.6, coefficients λ_v and χ_v characterizing G_1, G_2 and η are given for the AK sediment, mud from Kerala in India (KI) and from Mobile Bay in Alabama (MB). Comparison of AK with OK indicates that the latter is more fluid-like than AK, with a lower value of G_2 possibly due to a high fraction of organic matter (Table 5.5). The role of organic matter can be analogous to that of long-chain polymers added to increase the viscosity and reduce the rigidity of, for example, slurries flowing in pipes.

For MB mud Eq. 5.134 is used to plot the complex equivalent viscosity $|\eta^*|$ against frequency f for three samples at three values of the solids volume fraction ϕ_v (Fig. 5.39). The viscosity decreases with f and increases with ϕ_v. For a 10 s wave ($f = 10^{-1}$ s^{-1} or Hz) at $\phi_v = 0.11$, $|\eta^*| = 100$ Pa.s, which for a 1 s ($f = 1$ Hz) wave reduces to 9 Pa.s. The change in viscosity with ϕ_v is equally significant. Reducing ϕ_v from 0.17 to 0.07 at $f = 10^{-1}$ Hz decreases $|\eta^*|$ from 900 Pa.s to 15 Pa.s. This change in $|\eta^*|$ illustrates the high sensitivity of mud properties, and therefore their role in absorbing wave energy, to changing mud density. In the range of ϕ_v from 0.07 to 0.17, space-filling is complete and increasing particle packing leads to a rapid increase in rigidity.

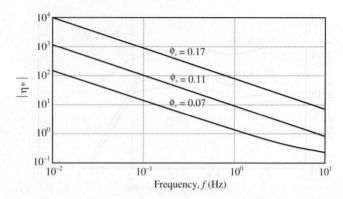

Figure 5.39. Dependence of complex viscosity on solids volume fraction and frequency for MB mud.

5.6.4 *Rheometry for empirical modeling*

A limitation of mechanical analogs such as KV, Maxwell and SLS is that they do not account for the continuous distribution of relaxation time constants characteristic of real materials. For instance, the only time constant in the Maxwell model is $\tau_M = \eta/G$. By integrating Eq. 5.55 we can obtain

$$\tau(t) = \frac{\eta}{\tau_M} \int_{-\infty}^{t} e^{-\left(\frac{t-t'}{\tau_M}\right)} \dot{\gamma}(t')dt' \tag{5.137}$$

which is an equivalent representation of the Maxwell model, and is a special form of the integral

$$\tau(t) = \int_{-\infty}^{t} R(t-t')\dot{\gamma}(t')dt' \tag{5.138}$$

where $R(t-t')$ is a general relaxation function defined as

$$R(t-t') = \int_{-\infty}^{t} \frac{f^N(\tau_M)}{\tau_M} e^{-\left(\frac{t-t'}{\tau_M}\right)} dt' \tag{5.139}$$

In Eq. 5.139, given the function $f^N(\tau_M)$, the differential area $f^N(\tau_M)\,d\tau_M$ represents the contribution to the total viscosity of all Maxwell elements having relaxation times in the range of τ_M and $\tau_M+d\tau_M$. The function $R(t-t')$, which must be obtained from rheometry, permits the use of empirical constitutive models [Barnes *et al.*, 1989].

An empirical model of a marine mud tested in a (coaxial cylinders) viscometer is due to Isobe *et al.* [1992]. The measured stress-shear strain rate flow curve in Fig. 5.40 shows a characteristic hysteresis loop indicating a phase difference between applied stress τ and induced shear rate $\dot\gamma$. As we noted in reference to the flow curve in Fig. 4.19, for cohesive sediments the phase difference can be interpreted in terms of differences in the floc structures formed during the rising and falling phases of the flow curve.

Based on an idealized representation of data in Fig. 5.41, the set of (rotational test) flow curve equations is as follows:

(1) Backbone curve (representing the basic flow curve in the absence of hysteresis):

$$\tau = \tau_p\left(\frac{|\dot\gamma|}{\dot\gamma_c}\right)^n \cdot \mathrm{sign}(\dot\gamma); \quad |\dot\gamma| \le \dot\gamma_c$$

$$\tau = \tau_y \cdot \mathrm{sign}(\dot\gamma) + \eta_1\dot\gamma; \quad |\dot\gamma| > \dot\gamma_c \tag{5.140}$$

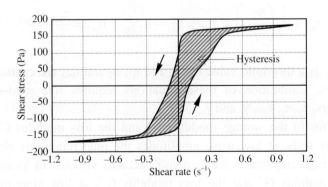

Figure 5.40. Applied stress versus shear rate flow curve showing hysteresis.

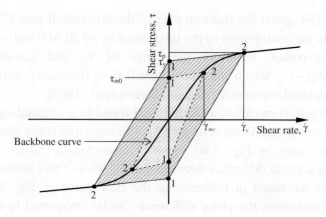

Figure 5.41. Stress versus shear rate empirical model of Isobe *et al.* [1992].

(2) Curve when $\dot{\gamma} > \dot{\gamma}_c$:

$$\tau = \tau_y \cdot \text{sign}(\dot{\gamma}) + \eta_1 \dot{\gamma}; \ |\dot{\gamma}| \leq \dot{\gamma}_c; \ \dot{\gamma}\ddot{\gamma} \leq 0; \ \text{or} \ |\dot{\gamma}| > \dot{\gamma}_c$$
$$\tau = -\tau_y \cdot \text{sign}(\dot{\gamma}) + \eta_2 \dot{\gamma}; \ |\dot{\gamma}| \leq \dot{\gamma}_c; \ \dot{\gamma}\ddot{\gamma} > 0 \quad (5.141)$$

(3) Curve when $\dot{\gamma} \leq \dot{\gamma}_c$:

$$\tau = \tau_{m0} \cdot \text{sign}(\dot{\gamma}) + \eta_1 \dot{\gamma}; \ \dot{\gamma}\ddot{\gamma} \leq 0$$
$$\tau = -\tau_y \cdot \text{sign}(\dot{\gamma}) + \eta_{m2} \dot{\gamma}; \ \dot{\gamma}\ddot{\gamma} > 0 \quad (5.142)$$

$$\tau_{m1} = \tau_p \left(\frac{\dot{\gamma}_{mc}}{\dot{\gamma}_c} \right)^n ; \ \tau_{m0} = \tau_{m1} - \eta_2 \dot{\gamma}_{mc} ; \ \eta_{m2} = \frac{\tau_{m0} + \tau_{m1}}{\dot{\gamma}_{mc}} \quad (5.143)$$

where $\dot{\gamma}_c$ is a threshold value of $\dot{\gamma}$, and η_1, η_2 and η_{m2} are characteristic viscosities. Calibration of these equations requires five mud-specific parameters, τ_p, τ_y, τ_{m0}, $\dot{\gamma}_c$ and $\dot{\gamma}_{mc}$.

In terms of the storage modulus G' and the loss modulus G'' from oscillatory tests, at low wave amplitudes the mud retains its elasticity, while at high amplitudes the response becomes viscous. In terms of the storage modulus G' and the loss modulus G'', at low amplitudes G' dominates and at high amplitudes G'' takes over. An example of a graphical way in which the resulting elastic, viscous and viscoelastic regimes can be delineated is given in Fig. 5.42. This is a plot of strain amplitude γ^0 against mud wet bulk density ρ (in terms of relative density $\tilde{\rho} = \rho/\rho_w$) for a kaolinite in water of 35 ppt salinity tested in the oscillatory applied-strain mode. Purely viscous response occurs when γ^0 exceeds γ_v at which $G' = 0$. The strain γ_v increases exponentially with $\tilde{\rho}$. The response is purely elastic once γ^0 falls below the threshold γ_e (which also increases exponentially with density) at which $G' = 0$ [Chou, 1989; Foda *et al.*, 1993].

A shortcoming of the plots in Fig. 5.42 is that complex nomograms are not easily described by simple equations. Robillard [2009] obtained an empirical model in which the experimental coefficients could be assigned physical meaning. Tests on mud from the Atchafalaya Delta (AD in Table 5.5) subjected to oscillatory strain yielded the schematic description given in Fig. 5.43. Three zones of mud behavior are evident. At low rates of strain $\dot{\gamma}$, mud behaved as a linear elastico-viscous liquid with elastic component η'' of the complex viscosity η^* (Eq. 5.77) prevailing over the viscous component η'. Both components are independent of $\dot{\gamma}$ in this zone, which implies that mud structure remained intact, even as some viscous energy loss occurred. The linear elastico-viscous behavior continued down to $\dot{\gamma}$ approaching (although not exactly) zero.

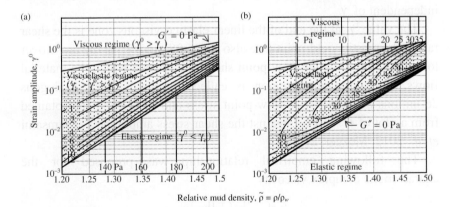

Relative mud density, $\tilde{\rho} = \rho/\rho_w$

Figure 5.42. Viscous, viscoelastic and elastic responses of a kaolinite in saltwater. Regimes identified by contours of: (a) storage modulus G', and (b) loss modulus G'' (from Foda *et al.* [1993]).

An Introduction to Hydraulics of Fine Sediment Transport

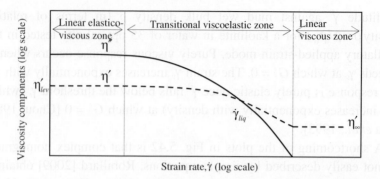

Figure 5.43. Schematic drawing of the dependence of the components η' and η'' of complex viscosity η^* on strain rate $\dot{\gamma}$ (adapted from Robillard [2009]). Note the log-log scale.

With increasing $\dot{\gamma}$, mud structure began to breakdown in the transitional viscoelastic liquid region, with both viscosity components having the same order of magnitude. In this zone, with increasing $\dot{\gamma}$ mud behavior was increasingly that of a fluid concurrent with a rapid decrease in η''. The two components crossed over and η' became dominant by the end of the zone indicating a liquefied state. At higher rates of strain, mud structure further broke down and the fluid became purely viscous. Since there were no subsequent changes to the structure, the viscosity became independent of $\dot{\gamma}$.

The value $\dot{\gamma}_y$ at the end of the linear elastico-viscous zone is the shear rate at the plastic yield-point (see also Fig. 5.38) and $\dot{\gamma}_{liq}$ at the cross-over location is the so-called flow-point shear rate beyond which the material flows. The intervening region is called the yield zone. Analogous definitions of yield stress, flow-point stress and yield zone are obtained from a plot of G' and G'' against the shear stress amplitude τ^0 [Samsami *et al.*, 2012; Mezger, 2006].

The following empirical relationship was proposed for the characteristic complex viscosity

$$\left|\eta^*\right| = \frac{\eta'_\infty \dot{\gamma} + \eta'_{lev} \dot{\gamma}_{liq}}{\dot{\gamma} + \dot{\gamma}_{liq} \sin \delta_{lev}} \qquad (5.144)$$

in which η'_∞ is the asymptotic viscosity at a high (theoretically infinite) rate of strain in the linear viscous zone, η'_{lev} is the asymptotic real viscosity in the low rate of strain linear elastico-viscous zone, $\dot{\gamma}_{liq}$ is the rate of strain at the cross-over location and $\delta_{lev} = \tan^{-1}(\eta' / \eta'')\big|_{lev}$ is the asymptotic value of the phase angle in the linear elastico-viscous zone. Note that the viscosity $|\eta^*|$ varies with $\dot{\gamma}$.

In Table 5.7, values of the coefficients in Eq. 5.144 are given for AD mud samples at four solids volume fractions. The strain sweep frequency was maintained at 0.628 rad s^{-1} corresponding to a period of 10 s relevant to the natural environment. Two sets of tests were carried out, the first with a forward sweep (increasing $\dot{\gamma}$ amplitude) followed by the second with a reverse sweep (decreasing $\dot{\gamma}$ amplitude). The first sequence, which began with an effectively undisturbed "bed" placed in the rheometer, can be thought of as representing the effect of the onset and growth of storm-wave induced oscillatory forcing in a "pre-storm" bed. In that context the second sequence would mimic the decaying phase of storm starting at the end of the growth phase.

The sweep sequence had a noteworthy effect on η'_{lev}, which in general was greater at the beginning of the forward sweep than at the end of the reverse sweep. While there was no measurable effect at the lowest volume fraction (ϕ_v) of 0.05, the space-filling threshold value, at the highest volume fraction of 0.2 there was a substantial decrease in η'_{lev}, from 285 Pa.s to 40 Pa.s. This is consistent with a breakdown of the

Table 5.7. Coefficients in Eq. 5.144 for AD mud, sweep period 10 s.

ϕ_v	η'_∞ (Pa.s)	η'_{lev} (Pa.s)	$\sin \delta_{lev}$	$\dot{\gamma}_{liq}$ (s^{-1})
		"Pre-storm bed"		
0.05	0.01	0.56	0.26	0.40
0.11	0.03	8.6	0.20	0.30
0.14	0.04	25	0.20	0.30
0.20	0.07	285	0.20	0.35
		"Post-storm bed"		
0.05	0.01	0.56	0.26	0.45
0.11	0.03	4.0	0.40	1.10
0.14	0.04	8.5	0.35	1.40
0.20	0.07	40	0.25	2.60

Source: Robillard [2009].

initial "pre-storm" floc order by the end of the storm due to oscillatory agitation of the bed.

Krotov and Mei [2007] proposed an analytic approach to obtain a representative viscosity by stating Eq. 5.56 more succinctly as

$$\left(1 + \sum_{k=1}^{n} \alpha_k \frac{\partial^k}{\partial t^k}\right) \tau = \left(\beta_0 + \sum_{k=1}^{m} \beta_k \frac{\partial^k}{\partial t^k}\right) \gamma \qquad (5.145)$$

where k is the time-derivative order and $n = m$ or $m-1$ [Oldroyd, 1964]. For oscillatory motion the strain rate and stress are

$$\dot{\gamma} = \dot{\gamma}^0 e^{-i\sigma t}; \quad \tau = \tau^0 e^{-i\sigma t} \qquad (5.146)$$

where the expression for the strain rate is analogous to Eq. 5.70 and that for stress is the same as Eq. 5.71. Thus from Eq. 5.145 we obtain

$$\tau^0 = \eta \dot{\gamma}^0 \qquad (5.147)$$

with

$$\eta = \frac{\beta_0 + \displaystyle\sum_{k=1}^{2N-1} \beta_k (-i\sigma)^k}{1 + \displaystyle\sum_{k=1}^{2N-1} \alpha_k (-i\sigma)^k} \qquad (5.148)$$

These $\eta(\sigma_n)$ values are now equated to the respective $|\eta^*(\sigma_n)|$ values derived from measurements at $2N$ sweep frequencies ($n = 1, 2, \ldots 2N$). The resulting $2N$ complex, linear algebraic equations are solved for the $2N$ coefficients α_k and β_k. Defining a new coefficient $\delta_k = \beta_k/\beta_0$ ($k = 1, 2 \ldots 2N-1$), for η to be a characteristic constant of the material tested, α_k and δ_k must be zero such that $\eta = \beta_0$. Using the rheometric data on muds KI, OK and AK (Table 5.5) reported in Jiang and Mehta [1995], values of α_k and δ_k are plotted against k in Fig. 5.44. Barring the values at $k = 3$ and 4, the coefficients are generally small to nil. Accordingly, assuming that $\alpha_k = 0$ and $\delta_k = 0$ are reasonable assumptions, the viscosities are:

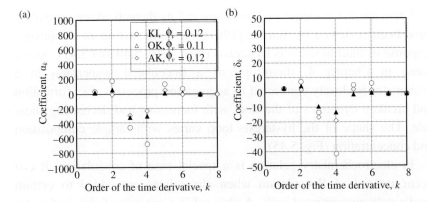

Figure 5.44. Coefficients α_k and δ_k plotted against k (= 1, 2....8) for mud samples KI, OK and AK (from Krotov and Mei [2007]).

Mud	Volume fraction (ϕ_v)	Viscosity (Pa.s)
KI	0.12	2.18×10^5
OK	0.11	1.00×10^4
AK	0.12	7.11×10^4

This analysis represents an improvement over Eq. 5.135, in which η is dependent on the sweep frequency and is not a characteristic property of the material. On the other hand, the representative viscosity is an artifact of the analysis and may not have a distinct physical meaning.

5.7 Thixotropy and Gelation

According to definition, if the viscosity of a material sheared at a constant rate progressively decreases the material is said to be thixotropic. Recovery follows stress release. Formally, thixotropy is an isothermal, reversible and time-dependent process occurring under conditions of constant material composition and volume, whereby a material stiffens or gels while at rest and softens or liquefies upon remolding [Mitchell, 1993]. The process of thickening and gelation is also called rheopexy, as demonstrated by rheopectic fluids (which are rare).

Experimental evidence on thixotropy and its theoretical representation have been reviewed by Barnes [1997]. A manifestation of thixotropic response of mud to stress is hysteresis (Figs. 4.19, 5.40 and 5.41). As we noted in Chapter 4, thixotropy of a flocculated suspension in a viscometer is closely connected to aggregation processes, *i.e.*, formation and breakup dynamics of the flocs with increasing and decreasing shear rate. The shape of the hysteresis loop varies with sample composition and concentration (Fig. 5.45).

In electrochemistry gelation is a special case of coagulation. It can occur instead of flocculation when electrolytes are added to certain moderately concentrated soils. A clay gel is a mixture of clay and water that is homogeneous in appearance and has some rigidity. In mud dynamics gelation has a somewhat different meaning and need not indicate the absence of flocs, although floc structure is altered from its non-gelled state. When gelation of mud occurs its effect is observed in the flow curve. Starting at a very low shear rate the stress decreases with increasing shear rate as the gel breaks down. Thereafter the stress rises as the shear rate continues to increase. Attapulgite and bentonite slurries show measurable gelation effect. Figure 5.46 shows the flow curve for an attapulgite obtained in a Brookfield viscometer. The slurry density was 1,100 kg m^{-3} indicating that it was gelled fluid mud. The maximum effect of breakdown is at a shear rate of about 1.8 s^{-1}.

In mud gelation, pore-water molecules close to the clay surface form a "solid" layer through hydrogen bonding. If the sample is disturbed the structure collapses and the molecules are randomized. Progressive gelation therefore can be detected through changes in pore water pressure.

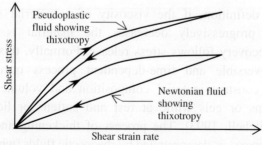

Figure 5.45. Newtonian and pseudoplastic (shear-thinning) fluids with thixotropic flow curves (from Wilkinson [1960]).

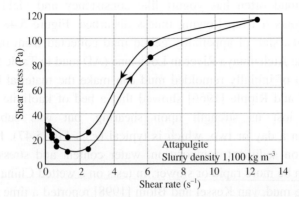

Figure 5.46. Stress versus shear rate flow curve for an attapulgite (from Feng and Mehta [1992]).

Starting with a disturbed sample, as the material gels pore water pressure drops with increasing ordering of water molecules. These thixotropic changes can be recorded by a pore-pressure gage, or by a tensiometer, a transducer for *in situ* measurement of the negative pore pressure, or tension, of water [Kirkham and Powers, 1972]. Experimental data on the gelation of a montmorillonite paste in water are given in Fig. 5.47. The letter S indicates shearing produced by inserting a rotating blade in the paste. Day and Ripple [1966] observed that even minor shearing, no more than 2° or 3°, was sufficient to initiate a significant drop in tension (increase in pore pressure), which rose rapidly after shearing was stopped.

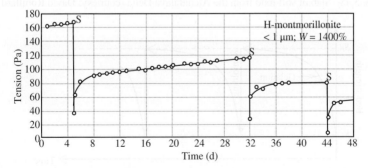

Figure 5.47. Thixotropic effect of pore water tension on H-montmorillonite (adapted from Day and Ripple [1966]).

Gelled mud often has yogurt like consistency and yield strength which prevents it from flowing unless disturbed. Figure 5.48 shows a chocolate-like slab of apparently gelled mud collected from the deltaic region of the Atchafalaya River in Louisiana (AD mud in Table 5.5).

Gelation of initially remolded mud can make the material harder to erode. Day and Ripple [1966] showed that a bed of kaolinite in water practically lost its strength upon shearing, but was substantially recovered in a day or two, which is typical (*e.g.*, Fig. 5.47). However, depending on sediment composition, water content and stress history, gelation can be more rapid or slower. In tests on a wetted China clay and an estuarine mud, van Kessel and Blom [1998] reported a time range on the order of 10^4 to 10^5 s for recovery. This process is schematized in Fig. 5.49, in which S_P is the peak undisturbed strength of mud and S_R is the remolded strength. These values can be determined from an unconfined

Figure 5.48. Slab of soft mud from the Atchafalaya Delta (courtesy: David Robillard).

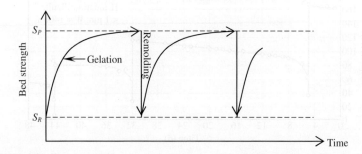

Figure 5.49. Schematic drawing of changes in bed strength due to repeated cycles of gelation and remolding (based on Mitchell [1993]).

compression test, or from the vane shear test for low strength muds [Lambe and Whitman, 1979].

A measure of gelation referred to as sensitivity S_t is defined as

$$S_t = \frac{S_P}{S_R} \qquad (5.149)$$

Nearly all normally consolidated and lightly overconsolidated clays show some sensitivity. However, with a few exceptions, some of the most sensitive large deposits of quick clay, *i.e.,* a clay that turns into a viscous fluid upon remolding (Table 5.8), are found in previously glaciated areas of North America and Scandinavia [Mitchell, 1993].

5.8 Exercises

5.1 Calculate the void ratio, the porosity and the (fresh) water content of a saturated marine mud sample. The ceramic container weight including wet sample is 227.1 g before drying and 105.7 g after drying, the weight of the container is 45.7 g and the specific gravity of the mud is 2.65.

5.2 A triaxial test on a bed sediment sample yields the following data:

Peak axial stress, σ_1 (kPa)	Confining pressure, σ_3 (kPa)
650	100
1300	200
2625	400
5312	800

Determine the angle of internal friction as well as cohesion.

Table 5.8. Sensitivity of clays.

Clay type	S_t
Insensitive	~1.0
Slightly sensitive	1–2
Medium sensitive	2–4
Very sensitive	4–8
Slightly quick	8–16
Medium quick	16–32
Very quick	32–64
Extra quick	> 64

Source: Rosenqvist [1953].

5.3   In a vertical tube a Bingham fluid mud of given thickness and uniform density is retained by a plate held over the lower end (Fig. E5.1). When the plate is removed the fluid may or may not flow out. Establish the criterion for the inception of flow.

5.4   Viscosity data for cohesive marine sediment Slurry A and a practically non-cohesive fine-grained Slurry B are given in Table E5.1. In a more general form, Eq. 5.44 can be written as

$$\eta = \eta_w \left(1 + \frac{a}{\eta_w} \phi_{vf}^b \right)$$ (E5.1)

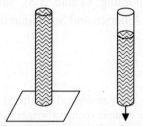

Figure E5.1.   Vertical tube with Bingham mud (from Bird *et al.* [1960]).

Table E5.1.   Viscosity data for two fine sediment slurries.

Slurry A		Slurry B	
Concentration (kg m^{-3})	Viscosity (Pa.s)	Concentration (kg m^{-3})	Viscosity (Pa.s)
5.0	3.54×10^{-3}	5.0	3.34×10^{-3}
8.4	3.97×10^{-3}	7.6	3.40×10^{-3}
13.2	4.53×10^{-3}	12.9	3.57×10^{-3}
24.4	5.59×10^{-3}	21.1	3.63×10^{-3}
36.7	7.11×10^{-3}	32.8	3.77×10^{-3}
50.6	9.75×10^{-3}	50.9	3.84×10^{-3}
65.8	1.40×10^{-2}	78.2	3.99×10^{-3}
82.8	2.18×10^{-2}	129.9	4.22×10^{-3}
101.4	3.66×10^{-2}	193.0	4.64×10^{-3}
124.9	6.57×10^{-2}	265.6	5.79×10^{-3}
157.2	1.32×10^{-1}	338.8	8.10×10^{-3}
187.1	2.40×10^{-1}	394.9	1.40×10^{-2}
232.2	5.07×10^{-1}	470.1	2.79×10^{-2}
259.6	8.27×10^{-1}	548.0	5.11×10^{-2}
375.7	1.50×10^{0}	680.2	1.21×10^{-1}

Calibrate Eq. E5.1 for these two slurries by obtaining best-fit coefficients a and b for each slurry. The following values are applicable to both slurries: particle density $\rho_s = 2,650$ kg m^{-3}, seawater density $\rho_w = 1,027$ kg m^{-3} and viscosity $\eta_w = 0.001$ Pa.s. For Slurry A, the floc density is 1,200 kg m^{-3}. For Slurry B the wet bulk density is 1,150 kg m^{-3}.

5.5 Experimental data are given in Tables E5.2 and E5.3 on the variation of viscosity η of dilute suspensions of a bentonite and an attapulgite with solids volume fraction ϕ_v and shear rate $\dot{\gamma}$. For each clay, plot η against $\dot{\gamma}$ for different values of ϕ_v. Interpret the results using relevant relationships among these three parameters given in the chapter.

5.6 Experimental data are given in Table E5.4 on the variation of viscosity η of a dilute suspension ($\phi_v = 0.05$) of an attapulgite with shear rate $\dot{\gamma}$ and temperature. Plot η against $\dot{\gamma}$ for different temperatures. Then interpret the results using relevant relationships from the chapter.

5.7 Mud slurry is to be pumped though a 3 m long pipe of 2.5 cm radius. The slurry is a Bingham plastic with a yield strength of 3.5 Pa and a viscosity of 550 cp. If the desired flow rate in the pipe is 0.01 m^3 s^{-1}, determine the required pressure loss across the pipe.

Table E5.2. Viscosity (at 30° C) of Wyoming bentonite at different volume fractions and shear rates.

$\phi_v = 0.005$		$\phi_v = 0.01$		$\phi_v = 0.02$		$\phi_v = 0.03$		$\phi_v = 0.04$	
η (Pa.s)	$\dot{\gamma}$ (s^{-1})	η (Pa.s)	$\dot{\gamma}$ (s^{-1})	η (Pa.s)	$\dot{\gamma}$ (s^{-1})	η (Pa.s)	$\dot{\gamma}$ (s^{-1})	η (Pa.s)	$\dot{\gamma}$ (s^{-1})
0.1052	82.67	0.1568	55.45	0.6635	18.35	2.156	8.60	5.750	6.00
0.1039	178.5	0.1414	133.2	0.4859	35.79	1.438	16.94	3.450	11.76
0.1027	271.0	0.1337	212.4	0.4107	52.93	1.150	26.21	2.875	16.33
0.1033	353.6	0.1317	281.8	0.3750	71.88	0.9583	36.29	2.464	21.41
0.1039	435.1	0.1287	360.3	0.3450	90.74	0.8625	47.05	1.917	31.15
0.1039	507.7	0.1268	429.6	0.3255	111.3	0.7841	58.04	1.816	36.40
0.1039	585.8	0.1259	497.2	0.3053	132.9	0.7188	74.20	1.643	43.76
0.1033	670.7	0.1250	570.4	0.2828	170.1	0.6161	103.51	1.438	54.44
0.1027	745.3	0.1232	639.9	0.2614	219.6	0.5391	139.80	1.380	61.12
				0.2464	272.9	0.4929	173.49	1.232	73.40
				0.2363	320.2	0.4539	212.0	1.150	83.93
				0.2270	375.5			1.113	92.72
								1.078	101.1

Source: Wood *et al.* [1955].

Table E5.3. Viscosity (at 30°C) of attapulgite at different volume fractions and shear rates.

$\phi_v = 0.01$		$\phi_v = 0.02$		$\phi_v = 0.03$		$\phi_v = 0.04$		$\phi_v = 0.05$	
η (Pa.s)	$\dot{\gamma}$ (s^{-1})	η (Pa.s)	$\dot{\gamma}$ (s^{-1})	η (Pa.s)	$\dot{\gamma}$ (s^{-1})	η (Pa.s)	$\dot{\gamma}$ (s^{-1})	η (Pa.s)	$\dot{\gamma}$ (s^{-1})
0.2917	12.47	1.021	4.16	0.8376	13.02	1.719	10.75	5.44	4.01
0.2438	17.40	0.8167	6.68	0.5833	28.57	1.167	20.78	3.27	8.53
0.2042	22.26	0.5939	10.20	0.4414	49.42	0.8167	35.62	2.04	16.62
0.1815	26.72	0.5269	13.80	0.3670	72.65	0.6533	53.34	1.56	25.72
0.1667	36.36	0.4804	13.88	0.3141	100.3	0.5444	73.47	1.17	40.00
0.1571	40.52	0.4083	18.55	0.2816	129.1	0.4601	97.48	0.93	58.17
0.1471	49.42	0.3712	21.22	0.2552	161.5	0.4083	122.4	0.76	81.37
0.1256	96.47	0.3439	24.68	0.2317	205.4	0.3630	151.9	0.58	126.2
0.1184	143.38	0.2816	43.04	0.2055	275.8	0.3267	192.9	0.48	179.1
0.1134	192.36	0.2178	77.92	0.1899	351.8	0.2816	259.3	0.40	243.4
0.1089	256.03	0.1856	117.6	0.1766	418.7	0.2552	324.2	0.36	298.1
0.1061	345.7	0.1667	160.0	0.1667	494.5	0.2402	384.8		
0.1030	435.2	0.1556	202.6						
0.1008	529.0	0.1439	261.1						
0.0984	616.0	0.1361	340.6						
0.0967	702.3	0.1307	422.1						
0.0950	788.2	0.1271	493.5						
		0.1237	575.5						
		0.1219	646.4						

Source: Wood et al. [1955].

Table E5.4. Viscosity of attapulgite at different shear rates and temperatures.

20°C		35°C		50°C	
η (Pa.s)	γ (s^{-1})	η (Pa.s)	γ (s^{-1})	η (Pa.s)	γ (s^{-1})
2.50	9.85	1.81	13.63	1.27	19.31
1.10	33.51	0.81	45.44	0.58	63.62
0.87	57.51	0.59	85.12	0.43	116.6
0.63	99.95	0.47	130.7	0.35	173.4
0.54	136.3	0.40	182.5	0.30	243.5
0.48	181.5	0.36	239.1	0.27	314.2
0.45	219.8	0.32	303.7	0.24	388.4
0.40	276.4	0.29	369.6	0.22	460.1

Source: Wood et al. [1955].

5.8 A parallel-plate viscometer with a plate radius of 10 cm and gap height of 0.1 cm is used to test a Sisko power fluid with exponent $n = 0.3$. If the angular speed of rotation is 0.2 rad s^{-1}, calculate the viscosity for an applied torque of 0.0336 Nm. Could this fluid represent fluid mud? Why?

5.9 For the KV model derive Eqs. 5.102 and 5.103 from Eq. 5.54. For the Maxwell model (Eq. 5.55) obtain Eqs. 5.104 and 5.105.

5.10 Try to fit the KV model (Eqs. 5.61 and 5.62) and the Maxwell model (Eqs. 5.63 and 5.64) to the strain-versus-time data in Fig. 5.36b for $\phi_v = 0.112$ and 0.144 given numerically in Table E5.5. Note that $\tau_0 = 0.5$ Pa and $t_{ch} = 30$ min. Assume $G = 2.75$ Pa (a very low value) and find the best value of η. What would be the outcome if G were increased to 6.9 Pa? Discuss model comparisons with data.

5.11 For the AD mud solids volume fraction ϕ_v values of 0.11 and 0.20 given in Table 5.7, plot the stress as a function of the rate of strain varying between 0.001 and 1,000 s^{-1}. Do this for the "pre-storm" and "post-storm" beds and explain the trends.

5.12 For the Maxwell model, the variation of the normalized loss component of viscosity η'/η and also $\log \sigma \tau_M$ are given in Table E5.6. Note that $\tau_M = \eta/G$. Plot the corresponding normalized quantities G'/G and G''/G, and the phase angle δ_s.

Table E5.5. Variation of strain with time (Fig. 5.36b, $\phi_v = 0.112, 0.144$).

Time (min)	Strain $\phi_v = 0.112$	$\phi_v = 0.144$	Time (min)	Strain $\phi_v = 0.112$	$\phi_v = 0.144$
0	0	0	30.01	0.168	0.0665
2.5	0.11	0.044	32.5	0.146	0.057
5	0.128	0.054	35	0.135	0.05
7.5	0.14	0.0585	37.5	0.125	0.046
10	0.148	0.0605	40	0.118	0.043
12.5	0.155	0.063	42.5	0.112	0.04
15	0.16	0.066	45	0.105	0.038
17.5	0.166	0.069	47.5	0.1	0.035
20	0.17	0.07	50	0.097	0.033
22.5	0.175	0.0705	52.5	0.092	0.031
25	0.1775	0.072	55	0.088	0.03
27.5	0.18	0.0745	57.5	0.083	0.027
29.99	0.183	0.0755	60	0.08	0.0245

316   *An Introduction to Hydraulics of Fine Sediment Transport*

Table E5.6. Variation of η'/η with $\sigma\tau_M$ for Maxwell model.

$\sigma\tau_M$	η'/η	$\sigma\tau_M$	η'/η	$\sigma\tau_M$	η'/η	$\sigma\tau_M$	η'/η
0.1	0.990099	2.2	0.171233	5.2	0.035663	8.2	0.014654
0.2	0.961538	2.4	0.147929	5.4	0.033156	8.4	0.013974
0.3	0.917431	2.6	0.128866	5.6	0.030902	8.6	0.01334
0.4	0.862069	2.8	0.113122	5.8	0.028868	8.8	0.012749
0.5	0.800000	3.0	0.100000	6.0	0.027027	9.0	0.012195
0.6	0.735294	3.2	0.088968	6.2	0.025355	9.2	0.011677
0.7	0.671141	3.4	0.079618	6.4	0.023832	9.4	0.011191
0.8	0.609756	3.6	0.071633	6.6	0.022442	9.6	0.010734
0.9	0.552486	3.8	0.064767	6.8	0.021169	9.8	0.010305
1.0	0.500000	4.0	0.058824	7.0	0.020000	10.0	0.009901
1.2	0.409836	4.2	0.053648	7.2	0.018925		
1.4	0.337838	4.4	0.049116	7.4	0.017934		
1.6	0.280899	4.6	0.045126	7.6	0.017018		
1.8	0.235849	4.8	0.041597	7.8	0.016171		
2.0	0.200000	5.0	0.038462	8.0	0.015385		

Chapter 6

Transport Load Definitions

6.1 Chapter Overview

There are three reasons for including an introduction to coarse-grain sediment transport at this stage. Firstly, the hydraulics of fine sediment transport is better understood within the framework of transport mechanics that includes fine as well as coarse particles. Secondly, in the majority of engineering projects dealing with mud, one is required to address issues on coarse-grained materials, and as we have defined mud in Chapter 1, it may include fine sand. The presence of (cohesionless) coarse silt is even more common. Thirdly, the basic principles of sediment transport are germane to particles of all sizes either directly or indirectly. Just as texts on sand transport usually include a chapter on fine sediment transport, it is essential to cover the elements of coarse-grain transport in a general description of transport loads.

Wherever convenient, we will examine coarse- and fine-grained transport in tandem, starting with the condition of incipient particle movement. As for coarse particles, although there is agreement on the physical basis of the condition for incipient movement, approaches for calculation of the critical shear stress have differed. Deterministic as well as stochastic approaches based on force balance for the particle about to move have been proposed. Similarly, for the calculation of sediment load, the same two modes of analyses have been used. The aim in this chapter is to review selected formulas for the critical shear stress and the sediment load. In addition to the citations given, an extensive coverage of the subject is found in Garcia [2008], a large work, and concise descriptions in compendia such as Dyer [1986], Julien [1995], Soulsby [1997] and van Rijn [2007a, b]. The last section is in a sense the most

317

important one as it provides the descriptive setting within which fine sediment transport is viewed in the remaining chapters.

6.2 Bed Material Load and Wash Load

Sediment load is the mass, weight or volume of dry sediment in transport per unit time. Parsing total sediment load as bed load, suspended load and wash load (Fig. 6.1) is partly notional, but helps in the conceptualization of transport. In general total load includes the entire size range of particles moving in water.

Bed material consists of particles whose statistical properties are fully represented in the bed. Bed material moving close to the bed due to rolling, sliding or jumping (or saltation) is known as bed (material) load transport. There is continual exchange between particles of the load and those on the bed. Since the rate of exchange depends on the current velocity, bed load varies with water discharge.

When each particle in motion is entirely surrounded by the fluid (*e.g.*, water or air) it is said to move in suspension, *i.e.*, such particles make up suspended load. The tendency of particles to settle during transport is countered by upward diffusion due to the turbulent movement of fluid parcels. Thus whether a given size fraction is sustained as a suspension

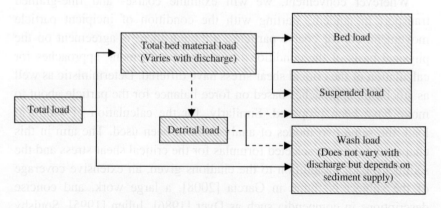

Figure 6.1. Sediment load definitions.

depends on the current velocity.[a] Particles present in suspension at a given instant may at another time become part of the bed and vice versa.

---

[a] The general empirical relationship between sediment concentration C (mean value over the flow cross-section and expressed in mg L^{-1}) and water discharge Q ($m^3 s^{-1}$), also known as the sediment rating curve, is of the form

$$C = a_q Q^{b_q} \qquad \text{(F6.1)}$$

or, alternatively,

$$G_t = a_q Q^{b_q+1} \qquad \text{(F6.2)}$$

where $G_t = CQ$ is the sediment mass discharge (kg s^{-1}). The regression coefficient a_q ranges from zero (clear water) to as high as 80,000, and the exponent b_q between 0 (defined as wash load) and about 2.5 [Müller and Förstener, 1968]. For given water discharge, increasing a_q implies increasing turbidity (in terms of concentration C). When a_q is held constant, the trend of the rating curve depends on whether b_q is greater or less than unity. As b_q decreases below unity, the rating curve becomes flatter with increasing discharge and implies that the load is supply-limited at high discharges.

Values of a_q and b_q from some rivers in Florida and California are given in Table F6.1. Small values of a_q indicate low turbidity levels. Turbidity in the Cedar and the Loxahatchee Rivers is supply limited. The Ortega becomes turbid only at high discharges during heavy rainfall.

In some rivers such as those fed in the spring season by water containing suspended matter from snowmelt, the rating curve during rising discharge differs from that during falling discharge. This difference is manifested as a time-lag between discharge and concentration, *i.e.*, hysteresis [Dyer, 1986].

Table F6.1. Sediment rating curve coefficients for some Florida and California rivers.

River	a_q	b_q	Source
Cedar near Ortega River, FL	1.7×10^{-2}	0.5	Paramygin [2002], Stoddard [2002]
Loxahatchee into Central Embayment, FL	1.2×10^{-2}	0.5	Ganju [2001]
Ortega near St. Johns River, FL	2.0×10^{-5}	2.2	Marván [2001]
Yolo Bypass near Woodland, CA	1.2×10^{-1}	0.09	Wright and Shoellhamer [2005]
Consumnes near Michigan Bar, CA ($Q \le 20$ $m^3 s^{-1}$)	2.0×10^{-3}	0.18	Wright and
($20 < Q \le 50$ $m^3 s^{-1}$)	9.7×10^{-7}	2.73	Shoellhamer [2005]
($Q > 50$ $m^3 s^{-1}$)	8.0×10^{-5}	1.6	
Mokelumne at Woodbridge, CA ($Q \le 25$ $m^3 s^{-1}$)	6.2×10^{-3}	0.05	Wright and
($Q > 25$ $m^3 s^{-1}$)	3.8×10^{-5}	1.6	Shoellhamer [2005]

As to whether one may refer to moving material as bed load or suspended load close to the bed depends on the operational definition of bed load because it also consists of particles that are essentially in suspension.

Wash load typically includes colloidal particles (< 0.1 μm) of clayey origin and light-weight organic particles that do not deposit at a given discharge. Oftentimes these particles are supplied to the stream by overland flows, or are derived from indigenous aquatic vegetation or biota. Wash load particles are not present in statistically significant quantities in the bed, and therefore wash load differs from bed material load, and its concentration does not depend on the current velocity. A sufficient decrease in discharge (and current) can change wash load into bed material load. For example, ebb flow through many tidal entrance channels in Florida with sandy beds carries suspended fine sediment from river discharge [Mehta *et al.*, 2005]. This material usually does not deposit in the channel due to strong currents and is not contained in the bed. However, it deposits offshore where the velocity is low. In general an increase in discharge may entrain bed material without permitting its redeposition. This increase would produce wash load provided no non-depositable particles remain in the bed.

Detrital load is transferred to the stream from the floodplain of a river or the intertidal flat in an estuary. As defined here it is mainly organic[b], and its contribution to total load depends on the hydroperiod[c] and supply rate from the floodplain (or tidal flat) vegetation and biota. Once in stream detrital matter is transported either as suspended load or as wash load. A change in the rate of detrital transfer from the floodplain may occur when there is a change in the hydroperiod due to natural or human causes. This change can be a matter of interest due to its potential effect on sediment load. A decrease in detrital transfer may occur when the

---

[b] According to Watt [1982], detrital, which has the same meaning as clastic or allogenic, refers to sediment fragments produced by breaking-up of igneous, sedimentary or metamorphic rocks. Thus the term as used in the present context does not conform to the Watt definition.

[c] Hydroperiod refers to the duration and frequency of flooding and soil saturation at a specified floodplain elevation.

maximum water level during the annual hydrologic cycle is reduced due to removal of water for say, urban or agricultural uses.

A reduction in the supply of detrital material due to human factors may be an environmental concern. For each bank of a stream the simplest model considers a reduction in the detrital transfer rate due to lowering of the water level to be proportional to a power m of the loss of wetted floodplain area $(A_o - A_N)$, where A_o is the old wetted area, A_N is the new area and K_d is the sediment load-scaling proportionality constant (Fig. 6.2). The value of K_d is site dependent. The simplest choice of m is unity.

Total load is the sum of bed load and suspended load. Depending on its definition, total load can be total bed material load. However, if wash load is included, total load will be larger than bed material load. Since wash load depends on its supply rate and not on the prevailing discharge, we will not consider it further except to acknowledge its presence where necessary. We will mainly pay attention to particles that actively interchange between the water column and the bed.

According to some investigators, bed load denotes sediment discharge determined from the bed load equation, such as that of Einstein [1950; see also Partheniades, 1977]. As we will see in Chapters 7 and 9, according to this definition there would be no bed load of fine sediment, except when it is transported as densely packed units of mud such as fecal pellets rather than as flocs, and provided these units are present in sufficiently large numbers to be considered as sediment load.

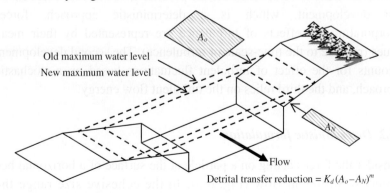

Figure 6.2. Reduction in detrital transfer due to lowering water level and decrease in the hydroperiod. For the other bank the K_d and m values can be different.

In treatments of sediment transport it is commonly assumed that the continuum behavior applies to the mixture of water and particles when they are smaller than about 0.25 mm. When the particles are large (> ~2 mm), inertial effects become important and particle transport acquires a two-phased (solid and liquid) character [Bagnold, 1966]. These thresholds are somewhat arbitrary and do not define the transition range between 0.25 mm and 2 mm, which includes numerous coastal and estuarine regions. An advantage of the use of two-phased transport equations (momentum and mass balance equations for the solid and the liquid phases) is that they bypass the need to separate bed load from suspended load. In general these equations are solved numerically. Reference is made to them later in the chapter.

6.3 Threshold of Movement

6.3.1 *Critical shear stress*

The concept of critical bed shear stress as the threshold of bed particle movement was originally associated with the design of stable channels free of sediment load. The principal variables in design are flow discharge, channel cross-section dimensions, sediment size and bed slope. Inter-relationships between these variables depend on the flow condition at which the bed shear stress equals its critical value. We will summarize three developments for the prediction of this condition. In the first development, which is a deterministic approach, forces incorporating the effects of turbulence are represented by their mean values (relative to the time-scale of turbulence). The second development accounts for the effect of turbulent fluctuations based on a stochastic approach, and the third relies on the turbulent flow energy.

6.3.2 *Deterministic formulation*

Consider the forces acting on a particle at the surface of a horizontal bed subject to turbulent flow (Fig. 6.3). In the cohesive size range the

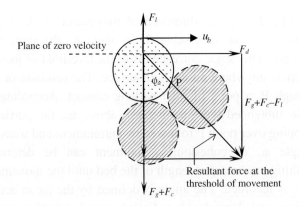

Figure 6.3. Forces on a particle or floc at a horizontal bed surface subject to turbulent flow.

material is assumed to be flocculated. Exposed flocs at the bed surface tend to deform elastically by flow shear, which can measurably change bed resistance [Gust, 1976]. For the present purpose, the particles will be considered to be non-deformable, each with an identity when at rest or in motion. These assumptions permit us to include (idealized) cohesive flocs along with cohesionless particles in the same general treatment.

The analysis involves turbulence–mean flow variables without explicit reference to the fluctuating quantities. In a fine sediment bed the electrochemical (cohesive) and biochemical (adhesive) bridges that bond each particle to its neighbors can be broken by normal or shear forces. We may consider the combined cohesive and adhesive force to be analogous to electromagnetic attraction in the sense that to rupture the inter-particle connection a normal tensile force, in this case due to hydro-dynamic lift, is required. The corresponding fluid drag is opposed by resistance due to cohesion, adhesion, friction and interlocking. The resistive effect of friction and interlocking can be incorporated within cohesion including adhesion. The weight of the particle is additive to cohesion, while the buoyancy force is additive to the lift force.

The lift force is F_l and the drag force F_d, with the latter acting at the level of the (spatial-mean) zero-velocity plane representing the bed surface (Chapter 2). The cohesive-adhesive force F_c, combined with particle buoyant weight F_g resists the lift force. Thus the net normal force is $F_g + F_c - F_l$. At the threshold of movement of the top particle (Fig. 6.3), the resultant force subtends an angle φ_a, the angle of repose, with the normal. This angle corresponds to the threshold of movement of the first particle anywhere on the bed surface. The resultant of all forces passes through P, a point of inter-particle contact. Accordingly, if the actual angle subtended were to increase above φ_a, the particle would move by tipping over point P followed by entrainment and transport.

The angle φ_a of cohesionless sediment can be determined by gradually tilting a submerged length of the bed until the movement of the first particle is recorded. The angle is defined by the mean bed slope at this condition, and depends on sediment size distribution and particle angularity (Table 6.1).[d] When the sediment is cohesive, φ_a loses its original physical meaning and becomes a coefficient embodying resistance to shear. The bed slope depends on cohesion and has been related to the bed's yield strength (Chapter 9).

Referring to Fig. 6.3, force balance at the threshold of movement yields

$$\tan \varphi_a = \frac{F_d}{F_g + F_c - F_l} \tag{6.1}$$

in which the submerged or buoyant weight is given by

Table 6.1. Angle of repose of selected non-cohesive sediments.

Sediment	Angle of repose φ_a (deg)	
	Very rounded grains	Very angular grains
Silt (non-plastic)	26	30
Uniform fine to medium sand	26	30
Well-graded sand	30	34

Source: Hough [1957].

---

[d] If the bed were tilted further, sliding motion of particles would occur *en masse*. The bed slope defined by this condition is also commonly called the angle of repose, and is somewhat larger than the angle corresponding to the movement of the first particle [Lambe and Whitman, 1979; Mehta and Rao, 1985]. Data on the angle of repose (*e.g.*, Table 6.1) usually refer to this angle, which has been conventionally (albeit erroneously) used also to characterize first particle movement.

$$F_g = \alpha_1 g (\rho_{sf} - \rho) d_p^3 \qquad (6.2)$$

where α_1 is a volume shape factor $\{ = (4\pi d_p^3 / 8) / d_p^3 = \pi/6$ for a spherical particle$\}$, ρ_{sf} is the coarse particle, fine particle or floc density and d_p is the diameter. The forces F_d and F_l are specified in terms of the reference velocity u_b at the top of the particle for which the condition at incipient movement is sought. Accordingly, the drag force is obtained from

$$F_d = \alpha_2 d_p^2 \tau_c \qquad (6.3)$$

in which the critical shear stress τ_c is related to u_b by the quadratic expression (Chapter 2)

$$\tau_c = C_D \rho \frac{u_b |u_b|}{2} \qquad (6.4)$$

where C_D is the particle-based drag coefficient and α_2 is an area shape factor $\{ = (\pi d_p^2 / 4) / d_p^2 = \pi/4$ for a spherical particle$\}$. The lift force is given by

$$F_l = \alpha_2 d_p^2 L_l \qquad (6.5)$$

where the lift L_l, *i.e.*, F_l per unit area, is related to u_b by

$$L_l = C_L \rho \frac{u_b^2}{2} \qquad (6.6)$$

in which C_L is the particle-based lift coefficient. Setting $\alpha_3 = \alpha_2 C_L / C_D$ yields

$$F_l = \alpha_3 d_p^2 \tau_c \qquad (6.7)$$

Then, substitution of Eqs. 6.2, 6.6 and 6.7 into Eq. 6.1 gives

$$\frac{\tau_c}{g(\rho_{sf} - \rho) d_p} = \theta_{ec} = \frac{\alpha_1 \tan \varphi_a}{(\alpha_2 + \alpha_3 \tan \varphi_a)} + \frac{F_c \tan \varphi_a / (\alpha_2 + \alpha_3 \tan \varphi_a)}{g(\rho_{sf} - \rho) d_p^3} \qquad (6.8)$$

where the Shields' entrainment parameter, $\theta_{ec} = \tau_c / [g(\rho_{sf} - \rho)d_p]$, embodies the ratio of drag force to buoyant weight [Shields, 1936]. The first term on the right hand side is a sediment-dependent parameter, and the second term represents the ratio of cohesion (*i.e.,* cohesive force) to buoyant weight [Mehta and Lee, 1994].

For cohesive sediment Eq. 6.8 has similarities with the Mohr–Coulomb equation for bed failure. In fact, Eq. 6.8 can be thought of as a form of Eq. 5.15 applicable to the bed surface, where the effective normal stress (Chapter 8) is nil. When cohesion is negligible (*i.e.,* $F_c = 0$), the second term on the right hand side vanishes. Equation 6.8 then reduces to Shields' entrainment relationship for coarse particles

$$\theta_{ec} = f^n(Re_*) \qquad (6.9)$$

in which we have replaced the first term on the right hand side of Eq. 6.8 by a function of the roughness Reynolds number $Re_* = u_{*c}d_p / v$, where $u_{*c} = \sqrt{\tau_c / \rho}$ is the critical value of the friction velocity. The density ρ_{sf} has been substituted by the coarse particle mineral density ρ_s.

The condition for incipient particle movement can also be developed using moments rather than forces [Mirtskhoulava, 1991; Righetti and Lucarelli, 2007]. Equation 6.9 is recovered when all moment arms are assumed to be proportional to the particle diameter.

Flume data along with the curve representing Eq. 6.9 are shown in Fig. 6.4. Along the curve the condition $u_* = u_{*c}$ applies. There can be fairly wide variability in the critical shear stress at a given value of Re_*. In uniform or well-sorted sediment the shape and texture of particles influence the angle of repose. In graded (*i.e.,* non-uniform) sediments larger particles may hide smaller ones from turbulent eddies and reduce their chances of entrainment [Einstein, 1950]. Following Mantz [1977] the diagram has been extended to include data for beds of cohesionless fine-grained particles in the range of Re_* from 0.02 to 0.4. The smallest particles included were made of crushed silica with a median diameter of about 10 μm.

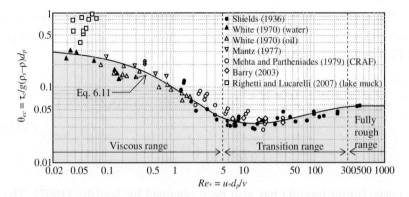

Figure 6.4. Extended Shields' diagram for cohesionless particles (adapted from Mantz [1977]).

The roughness Reynolds number can be written in the form of a dimensionless particle diameter $Re_* = d_p / (v / u_*)$. Given thickness δ_l of the viscous boundary layer, we have $v / u_* = \delta_l / 11.6$ (Fig. 2.13a). Therefore,

$$Re_* = 11.6 \frac{d_p}{\delta_l} \qquad (6.10)$$

In the range $Re_* < 5$ the condition $\delta_l \gg d_p$ applies and viscous stresses and energy loss due to skin friction prevail. In the transition range (Re_* from 5 to 300)^{e} the ratio d_p / δ_l increases with increasing Re_* until the viscous boundary layer practically vanishes. When $Re_* > 300$, $\delta_l \ll d_p$ and viscous effects are negligible. Energy loss is due to form drag associated with particles at the bed surface. In this fully rough turbulent flow range θ_{ec} is nearly independent of Re_*. It is evident that a causal similarity exists between the Shields' diagram and the well-known Moody or Stanton diagram for friction factor in pipe flows [Liggett, 1994]. The entrainment parameter and the friction factor qualitatively vary with the Reynolds number the same way.

---

^{e} In the literature the upper limit of Re_* (300) has been assigned values ranging from 70 to 400, which makes the choice of the threshold between the transition range and the fully rough range somewhat variable.

The following empirical data-fit equation from Fig. 6.4, modified from Brownlie [1981], permits the estimation of θ_{ec} in the range of Re_* from 0.02 to 1000:

$$\theta_{ec} = \left(\frac{0.1050}{Re_*^{0.6}} - \frac{0.0142}{Re_*^{1.03}}\right) + \left(0.770e^{-102.7Re_*} - 0.135e^{-15.95Re_*}\right)$$

$$+ 0.068 \cdot 10^{-\frac{4}{Re_*^{0.6}}}$$

(6.11)

A curve-fitting formula has also been obtained by Soulsby [1997]. The data of Righetti and Lucarelli [2007] in Fig. 6.4 are derived from erosion tests on seven (fine-grained) lake muds with NOM (loss of organic mass on ignition) ranging between 8 and 25%. The basic particle size range was 4 to 6 μm, and a mineral density value of 2,650 kg m^{-3} has been assumed in plotting the data. The data seemingly demonstrate the effects of cohesion and adhesion on the critical shear stress, which is three- to four-fold larger than for cohesionless particles. The investigators attempted to define and isolate the contribution of adhesion (from cohesion) to the critical shear stress.

Based on Eq. 6.1 modified by permitting the bed to be sloping and excluding cohesion[f], Wiberg and Smith [1987] developed the following expression for the critical shear stress

$$\tau_c = \frac{2}{\alpha_4 C_D \left[f^n\left(\dfrac{z}{z_0}\right)\right]^2} \cdot \frac{(\tan\varphi_a \cos\theta_b - \sin\theta_b)}{1 + \left(\dfrac{F_l}{F_d}\right)_c \tan\varphi_a}$$

(6.12)

where α_4 is equal to particle cross-sectional area multiplied by diameter divided by volume ($\alpha_4 = 1.5$ for a sphere). The force ratio $(F_l/F_d)_c$ is defined for the condition of incipient movement. The function $f^n(z/z_0)$ is

---

[f] This makes the result applicable to coarse particles or fine cohesionless particles as small as 10 μm.

determined from the log-velocity profile with a virtual origin at elevation z_0, $\theta_b = \sin^{-1} S_o$ which is the bed angle relative to the horizontal plane and S_o is the bed slope.

For the log-velocity profile of Eq. 2.57, the function $f^n(z/z_0) = \kappa^{-1} \ln(z/z_0)$ is obtained. To model the transition between the viscous boundary layer and the turbulent boundary layer, Wiberg and Smith selected the function of Reichardt [1951]

$$f^n\left(\frac{z}{z_0}\right) = \frac{1}{\kappa}\ln\left(1 + \kappa\tilde{z}\right) - c_z\left(1 - e^{-\tilde{z}-11.6} - \frac{\tilde{z}}{11.6}e^{-0.33\tilde{z}}\right) \qquad (6.13)$$

where $\tilde{z} = u_* z/\nu = Re_k z/k_s$, Re_k is the bed roughness (k_s) Reynolds number, $c_z = \kappa^{-1}\left(\ln z_0^+ + \ln \kappa\right)$, $z_0^+ = u_* z_0/\nu = Re_k z_0/k_s$ and $z_0 = k_s/29.7$ in the fully rough flow range (Chapter 2). In the transition range $c_z = -7.78$, and z_0, which varies with Re_*, is determined from velocity profile. The angle of repose was obtained from

$$\varphi_a = \cos^{-1}\left(\frac{\dfrac{d_p}{k_s} + z_*}{\dfrac{d_p}{k_s} + 1}\right) \qquad (6.14)$$

where z_* is the average level of the bottom of the almost moving particle. Its value was taken as -0.02. Importantly, Eq. 6.14 introduces the effect of particle size-to-roughness ratio on the critical shear stress.[g]

---

[g] For $d_p/k_s = 1$ corresponding to a flat bed, we obtain $\varphi_a = 60.66°$. This value is unrealistically high compared to those given in Table 6.1, but must be used with Eq. 6.12. The reason for the discrepancy may be that Miller and Byrne [1966], on whose experimental work Eq. 6.14 is based, reported the angle of repose of a single particle on a fixed bed with glued roughness elements. As the bed was tilted, grains were able to remain lodged in the fixed rough bed for much steeper angles than would have been possible over a loose bed.

6.3.3 Stochastic formulation

Consider the instantaneous velocity u_b at the top of a cohesionless particle to be the sum of its time–mean value $\bar{u}_b$ and a turbulent fluctuation u' (Fig. 6.5). The condition for the incipient movement of particle at the surface analogous to Eq. 6.1 is

$$\tan\varphi_a = \frac{|F_d|}{F_g - F_l} \tag{6.15}$$

where the absolute value of F_d accounts for two possible threshold conditions. Particle movement will occur to the right when u_b is positive, and to the left, against the mean velocity $\bar{u}_b$, when u_b is negative. Introducing Eqs. 6.2, 6.6 and 6.7 into Eq. 6.15 followed by rearrangement results in

$$|\tau_c| = \frac{\alpha_1 g(\rho_s - \rho)d_p}{\alpha_2 \dfrac{L_l}{|\tau_c|} + \cot\varphi_a} \tag{6.16}$$

We may conveniently replace the ratio $L_l/|\tau_c|$ of the instantaneous values of lift to the critical stress by the ratio $\bar{L}_l/\bar{\tau}_c$ of the corresponding time–mean values. This substitution is reasonable because both L_l and τ_c are proportional to the square of the same velocity u_b. The stress $\bar{\tau}_c$ is always positive because the mean flow is taken to be unidirectional. Therefore, the absolute sign is removed and

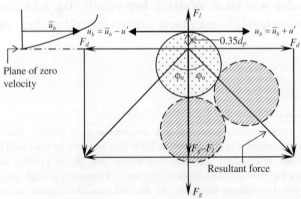

Figure 6.5. Forces on a cohesionless particle at the bed surface subject to turbulent flow.

$$|\tau_c| = \frac{\alpha_1 g(\rho_s - \rho)d_p}{\alpha_2 \dfrac{\overline{L_l}}{\tau_c} + \cot \varphi_a} \tag{6.17}$$

The absolute value of the instantaneous critical bed shear stress can also be expressed by the quadratic law

$$|\tau_c| = C_D \rho \frac{u_b^2}{2} \tag{6.18}$$

or

$$|\tau_c| = C_D \rho \frac{\overline{u}_b^2}{2}\left(1 + \frac{u'}{\overline{u}_b}\right)^2 = C_D \rho \frac{\overline{u}_b^2}{2}\left(1 + \frac{\sigma_u}{\overline{u}_b}\frac{u'}{\sigma_u}\right)^2 \tag{6.19}$$

where σ_u is the standard deviation of the velocity fluctuation u'. Thus the corresponding time–mean critical bed shear stress is

$$\overline{\tau}_c = \overline{|\tau_c|} = C_D \rho \frac{\overline{u}_b^2}{2}\overline{\left(1 + \frac{\sigma_u}{\overline{u}_b}\frac{u'}{\sigma_u}\right)^2} = C_D \rho \frac{\overline{u}_b^2}{2}\left[1 + \left(\frac{\sigma_u}{\overline{u}_b}\right)^2\right] \tag{6.20}$$

Dividing Eq. 6.19 by 6.20 yields

$$|\tau_c| = \frac{\overline{\tau}_c}{1 + \sigma_{su}^2}\left(1 + \sigma_{su}n_u\right)^2 \tag{6.21}$$

where $n_u = u'/\sigma_u$ is the normalized value of u' and $\sigma_{su} = \sigma_u/\overline{u}_b$ is the dimensionless standard deviation. We will consider u', and therefore n_u, to be a random variable, i.e., with normal frequency distribution [Christensen, 1975].[h]

---

[h] This choice is based on Christensen [1965], who reanalyzed the experimental data of Einstein and El-Samni [1949] on the frequency distribution of lift fluctuation L_l'. The standard deviation of σ_u was obtained by taking the velocity at the top of the particle to be proportional to the square-root of the lift. It was found that the data yielded constant values of σ_u, whereas the corresponding standard deviation $\sigma_{L_l'}$ of lift showed greater

Combining Eqs.6.17 and 6.21, followed by rearrangement we obtain

$$\bar{\tau}_c = \left[\frac{\alpha_1}{\alpha_2 \dfrac{\bar{L}_l}{\bar{\tau}_c} + \cot\varphi_a} \frac{1+\sigma_{su}^2}{\left(1+\sigma_{su}n_u\right)^2} \right] g(\rho_s - \rho)d_p \qquad (6.22)$$

The frequency distribution $\Phi(n_u)$, along with the velocity u_b and the critical bed shear stress τ_c, are shown in Fig. 6.6. The probability of erosion p_e is the sum of the areas A_r and A_l when $\bar{\tau}_c$ is given by Eq. 6.22. Since σ_{su} typically has small values, *e.g.*, 0.18 obtained by Laufer [1949], the area A_l can be ignored. If so, p_e can be approximated as

$$p_e = A_r + A_l \approx A_r = \int_{n_u}^{\infty} \Phi(\omega)d\omega = \frac{1}{\sqrt{2\pi}} \int_{n_u}^{\infty} e^{-\frac{\omega^2}{2}} d\omega \qquad (6.23)$$

where ω is a dummy variable.

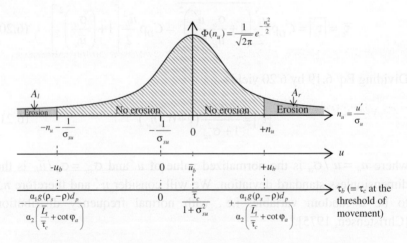

Figure 6.6. Diagrammatic representation of the probability of erosion.

variability. Based on this finding it was argued that u' rather than L_l' must be treated as a random variable. This conclusion was supported by Letter [2009]; see Chapter 7.

For evaluating the term within brackets in Eq. 6.22 the ratio $\overline{L}_l / \overline{\tau}_c$ must be known. We will assume that $\overline{L}_l$ is adequately represented by the time-mean quadratic relationship

$$\overline{L}_l = C_L \rho \frac{\overline{u}_b^2}{2} \qquad (6.24)$$

Dividing by $\overline{\tau}_c = \rho u_{*c}^2$ yields

$$\frac{\overline{L}_l}{\overline{\tau}_c} = \frac{C_L}{2} \left(\frac{\overline{u}_b}{u_{*c}} \right)^2 \qquad (6.25)$$

Following Einstein [1950], we will select $C_L = 0.178$ at elevation $z = 0.35 d_p$ specifying the top of the particle above the plane of zero velocity (see Exercise 2.2). The velocity profile is assumed to be logarithmic according to Eq. 2.76 due to its accuracy at small heights on the order of a grain diameter above the bed. Thus

$$\frac{\overline{u}_b}{u_{*c}} = \frac{1}{\kappa} \ln \left(\frac{0.35 d_p}{z_0} + 1 \right) \qquad (6.26)$$

Taking $\kappa = 0.4$ and $z_0 = k_s/29.7$ and substituting the resulting expression into Eq. 6.25 yields

$$\frac{\overline{L}_l}{\overline{\tau}_c} = 0.556 \left[\ln \left(1 + \frac{10.4}{k_s/d_p} \right) \right]^2 \qquad (6.27)$$

Introducing this expression in Eq. 6.22 gives

$$\frac{\overline{\tau}_c}{g(\rho_s - \rho)d_p} = \overline{\theta}_{ec} = \frac{\alpha_1}{\alpha_2} \frac{1}{0.556 \left[\ln \left(1 + \dfrac{10.4}{k_s/d_p} \right) \right]^2 + \cot \varphi_a} \frac{1 + \sigma_{su}^2}{\left(1 + \sigma_{su} n_u \right)^2} \qquad (6.28)$$

where $\overline{\theta}_{ec}$ is the time-mean value of the entrainment parameter θ_{ec}. Equation 6.28, which is applicable in the range of $Re_k > 70$, highlights the dependence of critical shear stress on the probability of erosion (or pickup probability) and therefore on n_u, and also on the ratio of bed roughness to particle size, k_s/d_p. From experimental data on $\overline{\tau}_b$ for given values of k_s/d_p, it has been observed that a reasonable value of n_u is 3.09 corresponding to the probability of erosion (Appendix D) equal to 1×10^{-3} [Christensen, 1975].

The pickup probability has attracted considerable attention due to its dominant role in determining the exact value of the time-mean critical bed shear stress in stochastic development. Bayesian statistical formulations such as those based on the random walk model, a Markov chain process, have been used to improve the prediction of the critical shear stress. For fine-grained sediment the critical shear stress depends on the joint probability representation of the turbulent shear stress and the bed shear strength [Letter, 2009].

In flume experiments under steady flows p_e can be estimated by measuring the short time interval Δt during which a single particle moves anywhere on the bed. Given $\Sigma \Delta t$ to be the cumulative duration of movement as part of the total duration T_D of the experiment, we have $p_e = \Sigma \Delta t / T_D$. Inasmuch as it is easier to observe the movement of larger particles, Dixit [1982] estimated p_e using a bed of bivalve shells in a flume. Their nominal diameter was 11 mm and density 2,770 kg m^{-3}. Four tests were run with conditions summarized in Table 6.2, over a total duration T_D of 3,600 s in each test. The probability of erosion p_e was found to vary between 6.0×10^{-4} to 1.1×10^{-3}. Values reported in the literature range more widely from 0.8×10^{-5} to 6×10^{-3} and indicate the effect of particle packing [Papanicolaou, 1999; Papanicolaou *et al.*, 2002].

Table 6.2. Tests to estimate the pickup probability and critical stress.

Test no.	Mean velocity (m s^{-1})	Re_k	$\Sigma \Delta t$ (s)	$p_e = \Sigma \Delta t / T_D$	$\overline{\tau}_c$ (Pa)	$\overline{\theta}_{ec}$
1	0.89	2220	9.0	0.0025	5.94	0.031
2	0.84	2140	9.2	0.0026	5.29	0.028
3	1.06	2420	40.6	0.0113	8.48	0.044
4	0.95	2160	2.3	0.0006	6.77	0.035

When a particle is far from spherical, its nominal diameter may be an inappropriate measure for the specification of lift and drag forces. In the shell bed experiments of Dixit the nominal diameter was obtained by measuring the volume of water displaced by the shell and equating that volume to the volume of a sphere. In Table 6.2 the entrainment parameter $\bar{\theta}_{ec}$ values (with a mean of 0.035) based on the nominal diameter are on average lower by a factor of 1.6 compared to those of Shields (with a mean of 0.057 in Fig. 6.4 at high values of Re_k).[i] An equivalent diameter value for shells would be $11 \times 0.035/0.057 \approx 7$ mm. Bivalve shells on the bed tend to have a "convex-upward" attitude, and their stream-wise cross-section resembles an airfoil. Hydrodynamic lift therefore occurs easily and may explain why the equivalent diameter, as defined, is smaller than the nominal diameter [Mehta and Christensen, 1977].

Example 6.1: Calculate the critical shear stress $\bar{\tau}_c$ for a bed of spherical particles of 2 mm diameter, particle density 2,650 kg m^{-3}, water density 1,020 kg m^{-3}, angle of repose 30°, σ_{su} = 0.18 and k_s/d_p = 10. Assume that the probability of erosion is 0.001 (n_u = 3.09). Note that Eq. 6.28 is applicable only in the fully rough range of flow. Is this condition satisfied?

For spherical particles the shape factors are $\alpha_1 = \pi/6$ and $\alpha_2 = \pi/4$. With k_s/d_p = 10, from Eq. 6.28, $\bar{\theta}_{ec}$ = 0.141 and $\bar{\tau}_c$ = 4.5 Pa. Therefore, u_* = 0.066 m s^{-1}, and Re_k = 133, which is fully turbulent or nearly so, depending on the choice of the threshold value, *e.g.*, 70 or 400.

As mentioned, for k_s/d_p = 1 the entrainment parameter $\bar{\theta}_{ec} = \bar{\tau}_c / g(\rho_s - \rho)d_p$ = 0.057 in fully rough flow (Fig. 6.4). For k_s/d_p = 10, $\bar{\theta}_{ec}$ is 2.5 times larger and can be construed to indicate sheltering by bed features which increase the roughness k_s ten-fold relative to the grain size. A somewhat similar effect has been reported in tidal channels in which the roughness of patchy sandy beds is influenced by large shells that are immobile at bed shear stresses sufficient to entrain sand. At John's Pass and Blind Pass, two entrance channels on the Gulf of Mexico coast of Florida, the incipient movement of 0.2 mm sand (in patchy beds) residing in the interstitial spaces formed by the shells was recorded by divers. The corresponding bed shear stress $\bar{\tau}_c$ was determined from concurrently measured velocity profiles. The respective values of k_s were 95 mm and 21 mm in the

---

[i] The roughness Reynolds numbers Re_* and Re_k are equal only when $d_p = k_s$, *i.e.*, for a flat bed of uniform particles. In other cases, since Re_k is larger, the threshold of transition to fully rough range of flow is not the same as that for Re_*. In general it is appropriate to define this transition by Re_* unless bed forms or other features modulate the incipient movement of the particle. The use of Re_k for a sandy bed whose stability is compromised by the presence of shells in the neighborhood is an example of modulation.

two channels. The particle density was 2,650 kg m^{-3} and water density 1,030 kg m^{-3}. The value of $\overline{\theta}_{ec}$ was found to range between 0.14 and 0.22 [Mehta and Christensen, 1983].

6.3.4 *Threshold of suspension*

Early observations on the threshold condition for suspended load are due to Knapp [1938], to which Bagnold [1966] made a substantive contribution. This condition can be defined as

$$\frac{\sqrt{\overline{w_u'^2}}}{w_s} \geq 1 \tag{6.29}$$

where $\sqrt{\overline{w_u'^2}}$ is the *rms* value of the upward velocity fluctuation w_u' due to turbulence and w_s is the particle settling velocity. The *rms* velocity is evaluated as follows.

Consider a unit volume of fluid of density ρ. Let a small fraction of the fluid mass $\rho(0.5 - \delta_m)$ contained in this volume move upward at a velocity $\sqrt{\overline{w_u'^2}}$ and the remaining (larger) mass, $\rho(0.5 + \delta_m)$, move downward at a velocity $\sqrt{\overline{w_d'^2}}$. The term δ_m is introduced to permit asymmetry between the two moving masses. Since the overall system is considered to be in force equilibrium, the net vertical momentum per unit volume must be zero, *i.e.*,

$$\rho(0.5 - \delta_m)\sqrt{\overline{w_u'^2}} - \rho(0.5 + \delta_m)\sqrt{\overline{w_d'^2}} = 0 \tag{6.30}$$

or

$$(0.5 + \delta_m)\sqrt{\overline{w_d'^2}} = (0.5 - \delta_m)\sqrt{\overline{w_u'^2}} \tag{6.31}$$

i.e.,

$$\sqrt{\overline{w_d'^2}} = \sqrt{\overline{w_u'^2}} \, \frac{0.5 - \delta_m}{0.5 + \delta_m} \tag{6.32}$$

The vertical momentum flux, *i.e.*, the rate of change of vertical momentum per unit area of a horizontal shear plane, in the two directions will be unequal, with the net momentum flux M_f given by

$$M_f = \rho \overline{w_u'^2}(0.5 - \delta_m) - \rho \overline{w_d'^2}(0.5 + \delta_m) \tag{6.33}$$

Using Eq. 6.32 we obtain

$$M_f = 2\delta_m \rho \overline{w_u'^2} \frac{0.5 - \delta_m}{0.5 + \delta_m} \tag{6.34}$$

According to this equation, M_f varies non-monotonically with δ_m and reaches a maximum at $\delta_m = 0.5(\sqrt{2} - 1) = 0.2071$. This fraction can be taken as the characteristic value of δ_m.

The *rms* velocity fluctuation $\sqrt{\overline{w'^2}}$ is derived from the summation

$$\overline{w'^2} = \overline{w_u'^2}(0.5 - \delta_m) + \overline{w_d'^2}(0.5 + \delta_m) \tag{6.35}$$

which, after using Eq. 6.32 yields

$$\sqrt{\overline{w_u'^2}} = \sqrt{\frac{0.5 + \delta_m}{0.5 - \delta_m}} \sqrt{\overline{w'^2}} = 1.56 \sqrt{\overline{w'^2}} \tag{6.36}$$

Experimental evidence indicates that $\sqrt{\overline{w'^2}}$ is approximately proportional to the friction velocity $u_* = \sqrt{\tau_b / \rho}$. For instance, Laufer [1954] found the proportionality constant to be 0.8. Based on this value we have

$$\sqrt{\overline{w'^2}} = 0.8 u_* \tag{6.37}$$

Accordingly, from Eq. 6.36

$$\sqrt{\overline{w_u'^2}} = 1.25 \sqrt{\frac{\tau_b}{\rho}} \tag{6.38}$$

Taking the condition of equality in Eq. 6.29 we obtain

$$\tau_b = 0.64 \rho w_s^2 \tag{6.39}$$

or

$$\frac{\tau_b}{\left(\rho_s - \rho\right)gd_p} = \theta_e = \frac{0.64w_s^2}{\left(\dfrac{\rho_s}{\rho} - 1\right)gd_p} \qquad (6.40)$$

where θ_e is a dimensionless bed shear stress or entrainment parameter. For quartz sand in seawater, selecting ρ_s = 2,650 kg m^{-3} and ρ = 1,027 kg m^{-3} yields

$$\theta_e = 0.4\frac{w_s^2}{gd_p} \qquad (6.41)$$

This relationship is plotted in Fig. 6.7 along with the Shields' curve for the incipient movement of bed particles. For particles exceeding about 2 mm in diameter, w_s is proportional to $\sqrt{d_p}$ (Eq. 7.16) with the result that θ_e becomes independent of particle size. Taking w_s = 0.2 m s^{-1} for the 2 mm particles yields θ_e = 0.82 as the entrainment parameter. Thus, as a rule of thumb, for sand transport $\tau_b/(\rho_s - \rho)gd_p$ is on the order of 1.

Roberts *et al.* [2003] provided experimental evidence in support of Fig. 6.7 by showing that the bed load decreases from nearly full (95%) to a minor (< ~1%) fraction of the total load as particle size decreases from about 0.2 mm (200 μm) to about 0.05 mm (50 μm). We may notionally

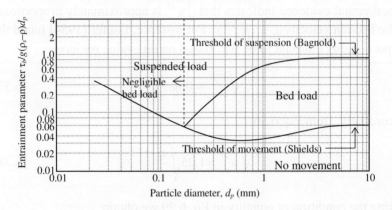

Figure 6.7. Threshold curves for bed particle incipient movement and suspended sediment transport (adapted from Bagnold [1966]).

add a "minor bed load" curve (Fig. 6.8). Its exact shape is unknown but it can be expected to intersect Shields' curve at about 0.015 mm (15 μm) marking the onset of cohesion when particle size decreases below about 10–20 μm (Chapter 3).

Figure 6.7 suggests a heuristic basis linking the formulas for cohesionless and cohesive ($<$ ~10–20 μm) sediment erosion fluxes as follows. The potential energy of a particle of diameter d_p at a height z_a of the bed load layer relative to an arbitrary datum within the bed (Fig. 6.9) is

$$\alpha_1 d_p^3 g (\rho_s - \rho)(z_a - k_s) \tag{6.42}$$

The shear work done to raise the particle to height z_a is

$$\alpha_2 d_p^2 (\tau_b - \tau_c) d_p \tag{6.43}$$

At equilibrium we may equate the above two quantities to obtain

$$z_a = \frac{\alpha_2}{\alpha_1} \frac{(\tau_b - \tau_c)}{g(\rho_s - \rho)} + k_s \tag{6.44}$$

or

$$\frac{z_a}{d_p} = \frac{\alpha_2}{\alpha_1}(\theta_e - \theta_{ec}) + \frac{k_s}{d_p} \tag{6.45}$$

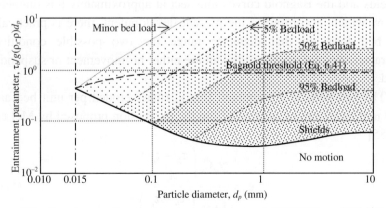

Figure 6.8. Entrainment diagram based on Roberts *et al.* [2003] (from Alkhalidi and Mehta [2005]).

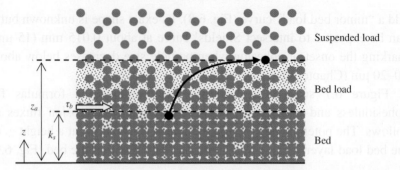

Figure 6.9. Loose boundary layer differentiated by bed load and suspended load.

where

$$\theta_e = \frac{\tau_b}{g(\rho_s-\rho)d_p}; \qquad \theta_{ec} = \frac{\tau_c}{g(\rho_s-\rho)d_p} \qquad (6.46)$$

Simply taking $k_s/d_p = 1$, Eq. 6.45 becomes

$$z_a - d_p = \frac{\alpha_2}{\alpha_1}(\theta_e - \theta_{ec})d_p \qquad (6.47)$$

where θ_{ec} identifies Shields' curve, and θ_e is associated with any of the bed load percent curves in Fig. 6.8. Taking the minor bed load curve, the thickness of the bed load layer decreases with decreasing d_p and becomes zero at $z_a = d_p = \sim15$ μm (or more generally 10–20 μm). In Fig. 6.8 the Shields and the Bagnold curves intersect at approximately this diameter, below which, according to the present development, there is practically no bed load. In that diameter range the two possible conditions characteristic of cohesive sediment are either no-movement or suspended load.

The bed particle entrainment flux ε_e (mass eroded per unit bed area and unit time) can be obtained from the expression proposed by McLean [1985]

$$\varepsilon_e = \frac{\gamma_a C_a \left(\dfrac{\tau_b - \tau_c}{\tau_c}\right)}{1 + \gamma_a \left(\dfrac{\tau_b - \tau_c}{\tau_c}\right)} \qquad (6.48)$$

where C_a is the reference concentration at elevation $z = z_a$ and γ_a is a constant with units of velocity. A heuristic interpretation of Eq. 6.48 for cohesive sediment is as follows. For $d_p < 10\text{--}20$ μm z_a is equal to d_p, and C_a is the concentration of cohesive flocs recently detached from the bed. Therefore, for a given bed density, C_a and the product $\gamma_a C_a = M_e$ can be taken as constants. The quantity γ_a is of the scale of the settling velocity, on the order of 10^{-4} to 10^{-5} m s^{-1} (Chapter 7). Moreover, the ratio $(\tau_b - \tau_c)/\tau_c$ is of order 10^0 to 10^2 (Chapter 9). Thus we may infer that $\gamma_a(\tau_b - \tau_c)/\tau_c \ll 1$, and Eq. 6.48 reduces to

$$\varepsilon_e = M_e \left(\frac{\tau_b - \tau_c}{\tau_c} \right) \tag{6.49}$$

This expression for the cohesive sediment erosion flux was proposed by Kandiah [1974] based on extensive tests on clayey beds in a laboratory apparatus. Its use is examined in Chapter 9.

6.4 Bed Load

6.4.1 *Deterministic formulation*

The unit bed load is either expressed on weight basis (sediment dry weight per unit time and unit bed width, symbolized by g_s) or volume basis (sediment volume per unit time and unit bed width, q_s). The relationship between these quantities is $g_s = \gamma_s q_s = \rho_s g q_s$ where γ_s is the unit weight of sediment.

In the early analytic development of the bed load equation of Du Boys [1879] it is assumed that granular transport of near-bed sediment driven by the bed shear stress τ_b is a composite of n number of thin sublayers, each of constant thickness Δd (Fig. 6.10). The total layer thickness is $n\Delta d$ and at the bottom-most ($n = 1$) sublayer τ_b balances the total resistance force between sublayers. Thus

$$\tau_b = \rho g h S_f = C_F n g \Delta d(\rho_s - \rho) \tag{6.50}$$

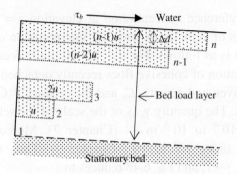

Figure 6.10. Definition sketch of Du Boys granular transport model.

where C_F is a friction coefficient, h is the water depth and S_f is the slope of the energy grade line (Chapter 2). The fastest moving sublayer is closest to the water layer and has a velocity $(n-1)u$. With the velocity assumed to decrease linearly with depth, the mean velocity of the entire moving bed layer is $(n-1)u/2$. Accordingly, the unit volumetric bed load q_s is given as

$$q_s = \frac{(n-1)u}{2} n\Delta d \tag{6.51}$$

Setting $n = 1$ in Eq. 6.50 yields the critical shear stress for the incipient movement of grains at the top of the bed load layer

$$\tau_c = C_F \Delta d g (\rho_s - \rho) \tag{6.52}$$

Dividing Eq. 6.50 by 6.52 results in the definition of n

$$n = \frac{\tau_b}{\tau_c} \tag{6.53}$$

Eliminating n between Eqs. 6.51 and 6.53 yields

$$q_s = \left(\frac{u\Delta d}{2\tau_c^2} \right) \tau_b (\tau_b - \tau_c) \tag{6.54}$$

or

$$q_s = \chi_s \tau_b (\tau_b - \tau_c) \tag{6.55}$$

where $\chi_s = u \Delta d / 2 \tau_c^2$ is a characteristic coefficient to be determined by calibration against data. Finally, the total volumetric rate of sediment transport Q_s in a channel of width B_c is

$$Q_s = \int_0^{B_c} q_s dy = \chi_s \int_0^{B_c} \tau_b (\tau_b - \tau_c) dy \tag{6.56}$$

where y is the lateral coordinate.

Equation 6.55 can be derived without referring to the layered model of Fig. 6.10 as follows [Graf, 1971]. In general, the (unit) bed load transport is a function of the bed shear stress, *i.e.*,

$$q_s = f^n(\tau_b) \tag{6.57}$$

which we may express as the series

$$q_s = a_0 + a_1 \tau_b + a_2 \tau_b^2 + \ldots \ldots \tag{6.58}$$

where a_0, a_1, a_2, *etc.*, are coefficients to be evaluated from the boundary conditions. Neglecting terms of order higher than two we will apply the following two conditions to Eq. 6.58:

(1) $q_s = 0$ when $\tau_b = 0$. This means that $a_0 = 0$ and

$$q_s = a_1 \tau_b + a_2 \tau_b^2 \tag{6.59}$$

(2) $q_s \approx 0$ when $\tau_b = \tau_c$. Therefore

$$a_1 = -a_2 \tau_c \tag{6.60}$$

Eliminating a_1 between Eqs. 6.59 and 6.60 results in

$$q_s = a_2 \tau_b (\tau_b - \tau_c) \tag{6.61}$$

in which by replacing a_2 by χ_s Eq. 6.55 is recovered. For given particle diameter χ_s and τ_c may be estimated from Fig. 6.11 [Zeller, 1963].

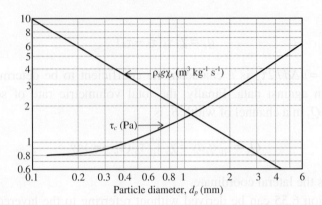

Figure 6.11. Parameters in Du Boys bed load equation (revised from Zeller [1963]).

6.4.2 *Stochastic formulation*

In the formulation of Einstein [1950], bed load occurs when the number flux of depositing particles N_d (*i.e.*, number of particles in transport depositing per unit bed area and unit time) equals the number flux due to erosion N_e (number of particles eroding from the bed per unit bed area and unit time), *i.e.*,

$$N_d = N_e \tag{6.62}$$

This assumption of equilibrium is reasonable because in steady flows the equality of number fluxes is rapidly established starting with any arbitrarily selected initial particle concentration, *e.g.*, zero (clear water).

To find N_d, each particle of diameter d_p is considered to move in steps of length proportional to d_p, and to deposit over a bed area $A_L d_p$ (of unit width), where A_L is a proportionality constant (Fig. 6.12). Since the particle weight is $\gamma_s \alpha_1 d_p^3$, N_d is obtained as

$$N_d = \frac{g_s}{(A_L d_p)(\gamma_s \alpha_1 d_p^3)} = \frac{g_s}{A_L \gamma_s \alpha_1 d_p^4} \tag{6.63}$$

where the unit (bed) load g_s is based on sediment weight rather than mass. To evaluate N_e, let t_e be the characteristic time of exchange between a particle on the bed and one in suspension. Thus,

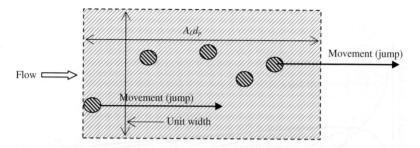

Figure 6.12. Definition of deposition distance $A_L d_p$ (based on Graf [1971]).

given $1/\alpha_2 d_p^2$ as the total number of particles available for erosion per unit bed area, we have

$$N_e = \frac{1}{\alpha_2 d_p^2}\frac{1}{t_e}p_e = \frac{p_e}{\alpha_2 d_p^2 t_e} \tag{6.64}$$

where p_e is the probability of erosion. The time t_e may be taken to be proportional to d_p/w_s, where w_s is the particle settling velocity. In turbulent flow, from Eq. 7.16 for w_s we obtain

$$t_e \propto \frac{d_p}{w_s} = \alpha_5 \sqrt{\frac{d_p}{g\left(\dfrac{\rho_s}{\rho} - 1\right)}} \tag{6.65}$$

where α_5 is a proportionality constant. Therefore Eq. 6.64 becomes

$$N_e = \frac{p_e}{\alpha_2 \alpha_5 d_p^2 \sqrt{\dfrac{d_p}{g\left(\dfrac{\rho_s}{\rho} - 1\right)}}} \tag{6.66}$$

The probability p_e is the fraction of time during which the instantaneous lift force exceeds the buoyant or submerged weight of the particle. Therefore p_e can be related to $A_L d_p$ (Fig. 6.13). When p_e is small ($\ll 1$), the distance of travel is virtually constant, and $A_L d_p = \lambda_b d_p$, where λ_b is a single step of bed load considered to have a value of about 100. When

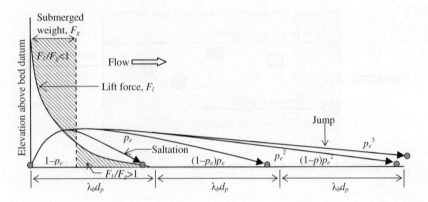

Figure 6.13. Relationship between submerged weight of the particle, instantaneous lift force and probabilistic particle movement.

p_e is large, only $1-p_e$ particles will be able to deposit after having traveled a distance $\lambda_b\, d_p$, while p_e particles will remain in transport. Of these, $p_e(1-p_e)$ particles will deposit after traveling a distance $2\lambda_b\, d_p$ while p_e^2 particles will continue to move, and so on. The total travel distance can be expressed as a series

$$A_L d_p = \sum_{n=0}^{n=\infty} (1 - p_e) p_e^2 (n+1)\lambda_b d_p = \frac{\lambda_b d_p}{1 - p_e} \qquad (6.67)$$

The probability p_e is determined by noting that erosion occurs when the instantaneous lift force F_l exceeds the submerged weight F_g (Fig. 6.13), *i.e.*, the probability that[j]

$$\frac{F_l}{F_g} \ge 1 \qquad (6.68)$$

---

[j] In this derivation the drag force has been ignored by assuming its effect to be minor compared to the lift force. Since drag and lift both vary with the same flow velocity at the top of the particle, the inclusion of drag enhances the effect of lift and causes particle entrainment at a lower velocity than when drag is ignored. Since the sediment-specific coefficients in the final equation are obtained from experiments, their magnitudes indirectly, albeit approximately, account for the effect of the ignored drag. However, the deficiency in the Einstein model remains [Christensen 1975].

The lift force F_l is now expressed as the sum of its time–mean value $\overline{F}_l$ and the turbulence induced fluctuation F_l', *i.e.*,

$$F_l = \overline{F}_l + F_l' = \overline{F}_l\left(1 + \frac{F_l'}{\overline{F}_l}\right) = \overline{F}_l\left(1 + \eta_l\right) \qquad (6.69)$$

in which $\eta_l = F_l' / \overline{F}_l$ is the lift force fluctuation made dimensionless by the respective mean. The fluctuation F_l' is positive when the velocity fluctuation which induces it is in the mean flow direction, and negative when the velocity fluctuation is opposed to mean flow. In the latter case if $F_l' > \overline{F}_l$, F_l will be negative (Fig. 6.14). For erosion, as interest is in the positive value of η_l, we will introduce Eq. 6.69 into Eq. 6.68 by taking the absolute value of F_l / F_g, *i.e.*,

$$\left|\frac{F_l}{F_g}\right| = \frac{\overline{F}_l}{F_g}\left|1 + \eta_l\right| \geq 1 \qquad (6.70)$$

Therefore

$$\left|1 + \eta_l\right| \geq \frac{F_g}{\overline{F}_l} \qquad (6.71)$$

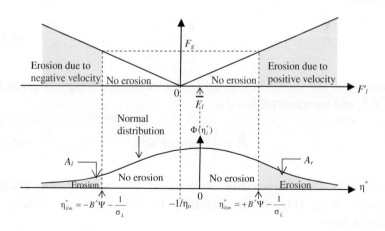

Figure 6.14. Areas of erosion and deposition specified by particle submerged weight and hydrodynamic lift force.

We will now define $\eta_l^* = \eta_l / \sigma_L$, where σ_L is the standard deviation of η_l. Measurements of lift were made on a large, immobile hemispherical "particle" placed at the bottom of a flume in turbulent flow. It was found that the distribution of the lift force could be approximated as Gaussian (normal) with a best-fit value of $\sigma_L = 0.5$ (*e.g.*, Graf [1971]). We accordingly obtain

$$\left| \frac{1}{\sigma_L} + \eta_l^* \right| \geq \frac{F_g}{\overline{F}_l \sigma_L} \qquad (6.72)$$

Squaring both sides

$$\left(\frac{1}{\sigma_L} + \eta_l^* \right)^2 \geq \left(\frac{F_g}{\overline{F}_l \sigma_L} \right)^2 \qquad (6.73)$$

The exact limit defined at $\eta_l^* = \eta_{lim}^*$, when incipient erosion occurs, is given by the equality

$$\left(\frac{1}{\sigma_L} + \eta_{lim}^* \right)^2 = \left(\frac{F_g}{\overline{F}_l \sigma_L} \right)^2 \qquad (6.74)$$

Therefore

$$\eta_{lim}^* = \pm \frac{F_g}{\overline{F}_l \sigma_L} - \frac{1}{\sigma_L} \qquad (6.75)$$

These limits are shown in Fig. 6.14. The submerged weight is given by Eq. 6.2, and the mean lift force is

$$\overline{F}_L = \frac{1}{2} C_L \rho \alpha_2 d^2 \overline{u}_\delta^2 \qquad (6.76)$$

where $\overline{u}_\delta$ is the characteristic velocity at the edge of the viscous boundary layer (Fig. 6.15). Referring to Fig. 2.13 this velocity is obtained from

$$\overline{u}_\delta = 11.6 u_* \qquad (6.77)$$

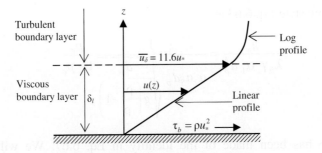

Figure 6.15. Definition sketch of viscous boundary layer.

where u_* is the friction velocity based on bed particle-induced, rather than the usual bedform-induced, resistance to flow.

Therefore Eq. 6.75 becomes

$$\eta^*_{lim} = \pm B^* \Psi - \frac{1}{\sigma_L} \qquad (6.78)$$

where $B^* = \alpha_1 / 67.3 C_L \alpha_2 \sigma_L$ and the dimensionless quantity

$$\Psi = \frac{(\rho_s - \rho)g d_p}{\tau_b} \qquad (6.79)$$

is known as the flow intensity. Its inverse, $1/\Psi$, equals the entrainment parameter θ_e when τ_b represents bedform-induced resistance. For further treatment and following a common approximation we will consider the bed shear stress to be based on bedform-induced resistance.

The probability of erosion p_e is now obtained as the sum of the two shaded end areas A_l and A_r defining the zones of erosion in Fig. 6.13 as

$$p_e = A_1 + A_2 \qquad (6.80)$$

or

$$p_e = 1 - \frac{1}{\sqrt{2\pi}} \int_{-B^*\Psi - \frac{1}{\sigma_L}}^{+B^*\Psi - \frac{1}{\sigma_L}} e^{-\frac{\omega^2}{2}} d\omega \qquad (6.81)$$

We can now state Eq. 6.63 as

$$\frac{g_s}{\lambda_b \gamma_s \alpha_1 d_p^4} = \frac{1}{\alpha_2 \alpha_5 d_p^2 \sqrt{\dfrac{d_p}{g\left(\dfrac{\rho_s}{\rho}-1\right)}}} \frac{p_e}{1-p_e} \qquad (6.82)$$

where use has been made of the identity in Eq. 6.67. We will now introduce the dimensionless bed load, or the bed load function

$$\varphi_s = \frac{g_s}{\gamma_s} \sqrt{\frac{\rho}{\rho_s - \rho} \frac{1}{g d_p^3}} \qquad (6.83)$$

As we have noted, in this expression g_s (weight basis) must be divided by g to yield g_s on mass basis.

With the introduction of $A^* = \alpha_2 \alpha_5 / \alpha_1 A_L$, Eq. 6.82 becomes

$$\frac{A^* \varphi_s}{1 + A^* \varphi_s} = 1 - \frac{1}{\sqrt{2\pi}} \int_{-B^* \psi - \frac{1}{\sigma_L}}^{+B^* \psi - \frac{1}{\sigma_L}} e^{-\frac{\omega^2}{2}} d\omega \qquad (6.84)$$

Experimental evidence based on the transport of 0.80 mm and 28 mm sands in flumes yielded $A^* = 43.5$ and $B^* = 0.143$ [Gilbert, 1914; Meyer–Peter et al., 1934; Einstein, 1950].

For values of the bed load function φ_s greater than about 0.05, commonly the range of interest in nature, Eq. 6.84 can be approximated by

$$\varphi_s = 40\left(\frac{1}{\psi}\right)^3 = 40\theta_e^3 \qquad (6.85)$$

which is the Einstein–Brown equation [Brown, 1950; Henderson, 1966]. This bed load function is observed to vary with power 3 of the inverse of flow intensity. This dependence means that, for instance, in the tidal environment bed load transport can be expected to rise steeply with increasing current speed soon after slack. A rapid decline in water clarity can occur due to increased turbidity within the first few minutes after flow reversal.

Equation 6.85 and the Kalinske [1947] equation in Table 6.3 permit bed load calculation at all flow intensities. The other three equations include a notional critical flow intensity $\Psi_c=(\rho_s-\rho)gd_p / \tau_c$ below which there can be no bed load. In the Einstein, the Einstein–Brown and the Kalinske equations Ψ_c is zero because the lift fluctuation is assumed to be normally distributed, which means that theoretically there is no finite bound on its value. Since the tails of distribution of the lift fluctuation cover the entire range of lift values from $-\infty$ to ∞, a large fluctuation can dislodge the particle from the bed even when the mean lift is low.

Table 6.3. Selected bed load equations.

Investigator(s)	Equation
Kalinske [1947]	$\varphi_s = 10\left(\dfrac{1}{\Psi}\right)^2$
Einstein–Brown [Brown, 1950]	$\varphi_s = 40\left(\dfrac{1}{\Psi}\right)^3$
Meyer–Peter and Müller [1948]	$\varphi_s = 8\left(\dfrac{1}{\Psi}-\dfrac{1}{\Psi_c}\right)^{3/2}$
Engelund and Fredsøe [1976]	$\varphi_s = 11.6\left(\dfrac{1}{\Psi}-\dfrac{1}{\Psi_c}\right)\left(\dfrac{1}{\sqrt{\Psi}}-0.7\dfrac{1}{\sqrt{\Psi_c}}\right)$
van Rijn [1984a]	$\varphi_s = \dfrac{0.053\sqrt{(s-1)gd_{50}}\,d_*^{-0.3}}{w_s}\left(\dfrac{\Psi_c}{\Psi}-1\right)^{2.1}$ for $\left(\dfrac{\Psi_c}{\Psi}-1\right)<3$ $\varphi_s = \dfrac{0.1\sqrt{(s-1)gd_{50}}\,d_*^{-0.3}}{w_s}\left(\dfrac{\Psi_c}{\Psi}-1\right)^{3/2}$ for $\left(\dfrac{\Psi_c}{\Psi}-1\right)\geq 3$

Symbols: h = water depth, k_s = bed roughness, w_s = particle settling velocity, u_* = friction velocity, d_{50} = particle median diameter, s = specific weight of particle (= γ_s/γ = ρ_s/ρ), $d_* = d_{50}\{[(s-1)g]/v^2\}^{1/3}$ = the dimensionless particle size and v = fluid (water) kinematic viscosity.

It has been argued that since an unbounded distribution of lift is physically unrealistic, upper and lower limits of the lift fluctuation must be introduced for consistency with the notion of critical flow intensity (*e.g.*, Partheniades [1977]). On the other hand we observe that, for instance, from the data of Shields (Fig. 6.4) in the fully rough range of flow $\Psi_c^{-1} = 0.056$, which corresponds to $\varphi_s = 0.007$ from the Einstein–Brown equation, a negligible value in natural flows. In other words, even without including the critical flow intensity, bed load equations predict negligible sediment transport at low flows.

Einstein [1950] considered the thickness of the bed load layer to be merely two particle diameters using the d_{35} size, *i.e.*, $2d_{35}$. It has been suggested that as long as the particles are saltating as bed material load, the height of flow boundary layer should not matter, *i.e.*, bed load transport can permeate the entire boundary layer. If so the Einstein bed load equation can justifiably be used to represent total bed material load [Einstein, 1950; Chiu, 1972; Christensen and Chiu, 1973].

Equations 6.84 and 6.85 are plotted in Fig. 6.16 along with selected data sets. The total load data analyzed by Bishop *et al.* [1965] were for quartz sand with representative sizes of 0.19 mm and 0.27 mm and specific weight 2.65. Also used were granular plastic beads (light-weight aggregates) having a diameter of 0.37 mm and specific weight 1.87.

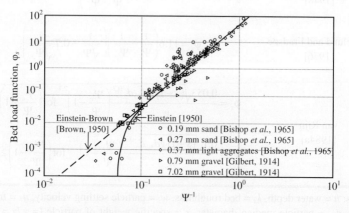

Figure 6.16. Einstein [1950] bed load equation and Einstein–Brown [Brown, 1950] approximation (revised from Alkhalidi [2004]).

Among a variety of sediments tested by Gilbert [1914], gravel-size particles of 0.79 mm and 7.02 mm and specific weight of 2.65 were used. The total load in gravel tests would be nearly the same as the bed load, because at the velocities set in the experiments the contribution from gravel to suspended load was minor. Notwithstanding the spread of data points, the applicability of Eq. 6.85 over the range of $\varphi_s = 10^{-2}$ to 10^2 can be considered to be reasonable. Below $\varphi_s \approx 10^{-2}$ the two equations diverge significantly; however, this is the zone of negligible transport.[k]

6.4.3 *Energy-based formulation*

A sediment load equation was derived by Bagnold [1966] based on the stream-energy flux needed to transport particles in the bed load layer relative to energy flux supplied by the turbulent flow. This physical basis differs from that of Einstein [1950], who considered forces, as opposed to energy, as being directly responsible for particle movement.

As a starting point the unit weight-based bed load g_s is conveniently represented as bed load transport g_s' defined as

$$g_s' = \frac{\rho_s - \rho}{\rho_s} g_s \qquad (6.86)$$

in units of power (energy per unit time).

At the base of the bed load layer the shear stress τ_{bt} is the sum of the granular stress τ_{bs} due to inter-particle collisions and momentum transfer, and the inter-granular fluid stress τ_{bf}, *i.e.*,

$$\tau_b = \tau_{bs} + \tau_{bf} \qquad (6.87)$$

When measurable bed load occurs the granular stress is on the order of 80% of the total stress τ_b. Thus, ignoring the fluid stress, the stream-energy flux (in units of power per unit area) consumed by the bed load is the product $\tau_{bs}u_B$, where u_B is the turbulence–mean velocity of the particle. The stream-energy flux available for bed load is $\tau_b u_m$, where u_m

---

[k] Madsen [1991] found that Eq. 6.85 is applicable to bed load transport under waves when the bed shear stress for waves (Chapter 2) is used.

is the depth–mean flow velocity. The ratio $\tau_{bs}u_B/\tau_b u_m$ defines transport efficiency e_b, *i.e.*,

$$\tau_{bs}u_B = e_b \tau_b u_m \qquad (6.88)$$

The unit volumetric bed load q_s ($=g_s/\rho_s g$) can be represented as $V_B u_B$, *i.e.*,

$$q_s = V_B u_B \qquad (6.89)$$

where V_B is the volume of bed load per unit area. As $\tau_{bs}u_B$ acts horizontally and $(\rho_s-\rho)gV_B u_B$ is normal to it, shear stress equilibrium at the base of the bed load layer requires

$$\tan\varphi_{fr} = \frac{\tau_{bs}u_B}{(\rho_s-\rho)gV_B u_B} \qquad (6.90)$$

where φ_{fr} is the angle of internal friction (for cohesionless particles). Thus

$$\tan\varphi_{fr} = \frac{\tau_{bs}u_B}{(\rho_s-\rho)gq_s} \qquad (6.91)$$

Using Eq. 6.88

$$\tan\varphi_{fr} = \frac{e_b \tau_b u_m}{(\rho_s-\rho)gq_s} \qquad (6.92)$$

or, with Eqs. 6.88 and 6.89

$$\tan\varphi_{fr} = \frac{\rho_s}{(\rho_s-\rho)g}\,e_b \tau_b u_m \qquad (6.93)$$

Further, using Eq. 6.86

$$\tan\varphi_{fr} = \frac{e_b \tau_b u_m}{g_s'} \qquad (6.94)$$

Therefore,

$$g_s = \frac{\rho_s}{\rho_s - \rho} \frac{e_b \tau_b u_m}{\tan \varphi_{fr}} \qquad (6.95)$$

Plots for e_b and $\tan\varphi_{fr}$ are given in Figs. 6.17a, b for different particle sizes. The efficiency is observed to vary over a narrow range (0.11 to 0.16).

Bagnold identified θ_{ex} as the "critical stage" value of θ_e at which bedforms diminish and do not contribute much to form drag. In the strict sense Eq. 6.95 is applicable when $\theta_e > \theta_{ex}$ and the bed is flat. The transition over which bedforms disappear in reality occurs over a range of θ_{ex} identified by the dark band (Fig. 6.17b). At lower values of θ_e the uncertainty in the $\tan\varphi_{fr}$ values is greater, as suggested by the dashed line extrapolations.

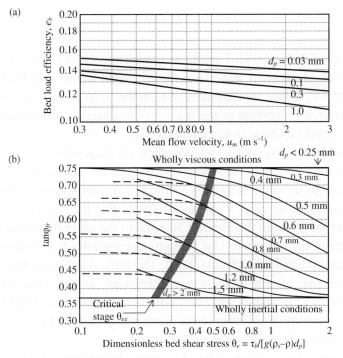

Figure 6.17. For quartz sand, dependence of: (a) Bed load efficiency e_b on flow velocity and particle size; (b) $\tan\varphi_{fr}$ on the dimensionless bed shear stress and particle size (from Bagnold [1966]).

6.5 Suspended Load

6.5.1 *Sediment mass balance*

The settling behavior of suspended sediment was formalized by McLaughlin [1959] in terms of a comprehensive mass balance approach. Following it the present description assumes a single-phase continuum of particles in water. Although bed load is restricted to cohesionless particles, suspended load includes cohesionless as well as cohesive sediment. Wash load, *e.g.*, load consisting of sol particles less than about 0.1 μm is excluded. However, sol particles may be present as part of a floc.

The unit suspended load g_{ss} (dry sediment mass per unit time and unit flow width) in the *x*-direction is

$$g_{ss} = \int_{z_a}^{h} C(z,t)u(z,t)dz \qquad (6.96)$$

where C is the suspended sediment concentration (dry mass per unit volume of suspension) and $z = z_a$ is the height of the bed load layer where $C = C_a$. Since measurement of C_a at this small elevation is usually impractical, it is common to select $z_a = 0.05h$, where h is the water depth, as a reference elevation (*e.g.*, Garcia [2008]).

Referring to Fig. 6.18, determination of suspension concentration requires the solution of the sediment mass balance (continuity) equation applicable to turbulent flows for a differential fluid volume element of dimensions dx_1, dx_2, dx_3. Continuity requires that the rate of change of sediment mass within this element be equal to the net sediment mass into the element per unit time. Due to multi-size particles it is necessary to apply mass balance to each size class *i*. The vertical and horizontal fluxes into and out of the element are referenced to the fluid velocity components u_1, u_2 and u_3, and gravity-settling flux to the settling velocity w_{si}. The quantity S_i includes terms representing gain or loss of particle mass in units of concentration per unit time.

For each class, adding all the flux terms algebraically, equating them to the rate of change of sediment mass within the volume, and dividing

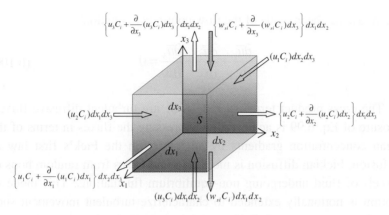

Figure 6.18. Mass fluxes through an elemental fluid volume.

the resulting equation by the element volume $dx_1\, dx_2\, dx_3$ yields (excluding for brevity the molecular diffusion terms)

$$\frac{\partial C_i}{\partial t} = -\frac{\partial}{\partial x_1}\left(C_i u_1\right) - \frac{\partial}{\partial x_2}\left(C_i u_2\right) - \frac{\partial}{\partial x_3}\left(C_i u_3\right) + \frac{\partial}{\partial x_3}\left(w_{si} C_i\right) + S_i \quad (6.97)$$

Each velocity is now decomposed into a time-average term and a fluctuating term (Chapter 2). The i-class concentration and the settling velocity are decomposed in the same way, *i.e.*,

$$\begin{aligned} C_i &= \overline{C}_i + C_i' \\ w_{si} &= \overline{w}_{si} + w_{si}' \end{aligned} \quad (6.98)$$

Substituting all the decompositions into Eq. 6.97 and averaging each resulting term with respect to a time interval much longer than the time-scale of turbulent fluctuations leads to

$$\frac{\partial \overline{C}_i}{\partial t} + \overline{u}_1 \frac{\partial \overline{C}_i}{\partial x_1} + \overline{u}_2 \frac{\partial \overline{C}_i}{\partial x_2} + \overline{u}_3 \frac{\partial \overline{C}_i}{\partial x_3} - \frac{\partial}{\partial x_3}\left(\overline{w}_{si}\overline{C}_i\right) - \frac{\partial}{\partial x_3}\left(\overline{w_{si}' C'_i}\right)$$

$$= \overline{S}_i - \frac{\partial}{\partial x_1}\left(\overline{u_1' C_i'}\right) - \frac{\partial}{\partial x_2}\left(\overline{u_2' C_i'}\right) - \frac{\partial}{\partial x_3}\left(\overline{u_3' C_i'}\right) \quad (6.99)$$

where use has been made of the flow continuity equation[1]

$$\frac{\partial \overline{u}_1}{\partial x_1} + \frac{\partial \overline{u}_2}{\partial x_2} + \frac{\partial \overline{u}_3}{\partial x_3} = 0 \tag{6.100}$$

The cross-product terms such as $\overline{u_1' C_i'}$ are turbulent diffusive fluxes. Closure of Eq. 6.99 is achieved by expressing the fluxes in terms of the mean concentration gradients by analogy with the Fick's first law of diffusion. Fickian diffusion is molecular and arises from random hops of parcels of fluid undergoing non-equilibrium fluctuations. This mode of motion is notionally extended to characterize turbulent movement such that the dimensions of fluid parcels span all the length scales from molecular up to macroscopic that are orders of magnitude larger [Brogioli and Vailati, 2001].

The turbulent mass fluxes are now expressed as

$$\overline{u_1' C_i'} = -D_{si1}\frac{\partial \overline{C}_i}{\partial x_1}, \quad \overline{u_2' C_i'} = -D_{si2}\frac{\partial \overline{C}_i}{\partial x_2}, \quad \overline{u_3' C_i'} = -D_{si3}\frac{\partial \overline{C}_i}{\partial x_3} \tag{6.101}$$

where D_{si1}, D_{si2} and D_{si3} are the i-class turbulent mass diffusivities in the x_1, x_2 and x_3 directions, respectively. These fluxes are summed over the contributions made by the particles or flocs of each size. Thus Eq. 6.99 becomes, after summing over all M_c size classes

$$\frac{\partial\left(\sum_{i=1}^{M_c}\overline{C}_i\right)}{\partial t} + \frac{\partial\left(\overline{u}_j\sum_{i=1}^{M_c}\overline{C}_i\right)}{\partial x_j} - \frac{\partial\left(\delta_{j3}\sum_{i=1}^{M_c}\overline{w}_{si}\overline{C}_i\right)}{\partial x_j} - \frac{\partial\left(\delta_{j3}\sum_{i=1}^{M_c}\overline{w_{si}'C_i'}\right)}{\partial x_j}$$

$$= \sum_{i=1}^{M_c}\overline{S}_i + \frac{\partial}{\partial x_j}\left[D_{sj}\frac{\partial\left(\sum_{i=1}^{M_c}\overline{C}_i\right)}{\partial x_j}\right] \tag{6.102}$$

---

[1] In the decomposed equation of continuity, the quantities $\overline{\partial u_1'/\partial x_1}$, $\overline{\partial u_2'/\partial x_2}$ and $\overline{\partial u_3'/\partial x_3}$ are identically zero. This leaves only Eq. 6.100.

where the coordinate designation is x_j (j = 1, 2, 3) and δ_{j3} is the Kronecker delta (0 for j = 1, 2, and 1 for j = 3).

The fluctuation in S_i is considered to be randomly generated and therefore its time-average value has been set to zero. When S_i represents a source of particles merely of size i derived from the growth or breakage of particles of other sizes present in the volume element, such as when the particles are cohesive flocs, the sum

$$\sum_{i=1}^{M_c} \overline{S_i}$$

must be zero for overall mass balance. Next, a composite settling velocity is defined based on concentration-weighted class settling velocities as

$$\overline{w}_s \equiv \frac{\sum_{i=1}^{M_c} \overline{w}_{si}\overline{C}_i}{\sum_{i=1}^{M_c} \overline{C}_i} = \frac{\sum_{i=1}^{M_c} \overline{w}_{si}\overline{C}_i}{C} \qquad (6.103)$$

With these two changes Eq. 6.102 becomes

$$\frac{\partial \overline{C}}{\partial t} + \frac{\partial \overline{u}_j \overline{C}}{\partial x_j} - \frac{\partial \left(\delta_{j3}\overline{w}_s\overline{C}\right)}{\partial x_j} - \frac{\partial \left(\delta_{j3}\sum_{i=1}^{M_c} \overline{w'_{si}C'_i}\right)}{\partial x_j} = \frac{\partial}{\partial x_j}\left(D_{sj}\frac{\partial \overline{C}}{\partial x_j}\right) \qquad (6.104)$$

The remaining summation term in this equation is non-trivial, except when $w'_{si} = 0$, which is not a reasonable assumption. We note however that this term can be conveniently incorporated into the eddy diffusion term without loss of generality according to

$$D_{sj}\frac{\partial \overline{C}}{\partial x_j} = \sum_{n=1}^{M_c}\left(\delta_{j3}\overline{w'_{sn}C'_n} - \overline{u'_jC'_n}\right) \qquad (6.105)$$

Then, without changing the symbol D_{sj} for the diffusivity, Eq. 6.104 reduces to

$$\frac{\partial \overline{C}}{\partial t} + \frac{\partial \overline{u}_j \overline{C}}{\partial x_j} - \frac{\partial \left(\delta_{j3} \overline{w}_s \overline{C} \right)}{\partial x_j} = \frac{\partial}{\partial x_j} \left(\nu + D_{sj} \right) \frac{\partial \overline{C}}{\partial x_j} \qquad (6.106)$$

The general solution of this mass balance equation for $\overline{C}$ requires numerical integration, with values of $\overline{u}_j$ obtained from the corresponding equations of flow continuity and momentum conservation (Chapter 2).

For use with Eq. 6.96, Eq. 6.106 is assumed to be laterally (x_2) averaged, which simplifies it to

$$\frac{\partial C}{\partial t} + u_1 \frac{\partial C}{\partial x_1} + u_3 \frac{\partial C}{\partial x_3} - \frac{\partial}{\partial x_3} \left(w_s C \right) = \frac{\partial}{\partial x_1} \left(D_{s1} \frac{\partial C}{\partial x_1} \right) + \frac{\partial}{\partial x_3} \left(D_{s3} \frac{\partial C}{\partial x_3} \right) \qquad (6.107)$$

in which the overbars have been dropped for convenience. We will further assume that $w \ll w_s$.[m] To examine the contribution of each term in this equation we will consider D_{s1} to be independent of the x_1-coordinate, and w_s and D_{s3} independent of the x_3-coordinate. Equation 6.107 then reduces to

$$\frac{\partial C}{\partial t} + u_1 \frac{\partial C}{\partial x_1} - w_s \frac{\partial C}{\partial x_3} = D_{s1} \frac{\partial^2 C}{\partial x_1^2} + D_{s3} \frac{\partial^2 C}{\partial x_3^2} \qquad (6.108)$$

We will now make this equation scale-independent by introducing non-dimensional quantities

$$\tilde{t} = \sigma t, \quad \tilde{C} = \frac{C}{C_0}, \quad \tilde{u} = \frac{u_1}{U}, \quad \tilde{x}_1 = \frac{x_1}{L}, \quad \tilde{x}_3 = \frac{x_3}{h_0} \qquad (6.109)$$

in which σ is the flow angular frequency, C_0 is a reference suspended sediment concentration, U is a reference flow velocity, L is a reference length and h_0 is a reference depth. Thus, Eq. 6.105 becomes

$$\frac{\partial \tilde{C}}{\partial \tilde{t}} + \left(\frac{U}{\sigma L} \right) \tilde{u} \frac{\partial \tilde{C}}{\partial \tilde{x}_1} - \left(\frac{w_s}{\sigma h_0} \right) \frac{\partial \tilde{C}}{\partial \tilde{x}_3} = \left(\frac{D_{s1}}{\sigma L^2} \right) \frac{\partial^2 \tilde{C}}{\partial \tilde{x}_1^2} + \left(\frac{D_{s3}}{\sigma h_0^2} \right) \frac{\partial^2 \tilde{C}}{\partial \tilde{x}_3^2} \qquad (6.110)$$

---

[m] At times this may not be reasonable. In estuaries and over shelves with strong tides, the vertical velocity can at times be of the same order of magnitude as the settling velocity of the smaller particles (*e.g.*, Vinzon and Mehta [2001]).

By conveniently allowing L to represent the shallow water wave length and recognizing that $h_0 \ll L$, it is readily seen that the advective terms involving the horizontal gradients are much smaller than the convective terms associated with the vertical gradients. Discarding the former terms leads to

$$\frac{\partial C}{\partial t} = \frac{\partial}{\partial z}(w_s C) + \frac{\partial}{\partial z}\left(D_{sz} \frac{\partial C}{\partial z} \right) = \frac{\partial}{\partial z}\left(w_s C + D_{sz} \frac{\partial C}{\partial z} \right) \qquad (6.111)$$

where x_3 is now the z-coordinate and the symbol D_{sz} replaces D_{s3}. This mass balance indicates that the rate of change of the concentration $C(z,t)$ is determined by the net difference (algebraic sum) in the divergence of the settling flux and the eddy diffusive flux. The simplicity of Eq. 6.111 makes it convenient for application to both cohesionless and cohesive sediment transport in which the divergence of the advective flux is small and particle movement is mainly in the vertical direction [Luettich *et al.*, 1990; Hawley and Lesht, 1992; Winterwerp, 1999; Teeter, 2001b]. This assumption, which has been invoked in what follows, permits us to focus on the physical state of suspended fine sediment moved by a current or waves.

6.5.2 Concentration profile

Analytic solutions of Eq. 6.111 depend on the choices of w_s and D_{sz}. For a given mud, representations of the settling velocity of the form $w_s = w_{so} f^n(C,\bar{G})$ are assumed (Chapter 7), where w_{so} is a characteristic value of w_s and $f^n(C,\bar{G})$ denotes a function of suspended sediment concentration C and the depth-averaged flow shear rate $\bar{G}$. In some situations, *e.g.*, lakes and reservoirs with low levels of turbulence, the effect of $\bar{G}$ may be minor and therefore ignored [So, 2009].

A common representation of the mass diffusivity is $D_{sz} = D_{so}\Phi_s$, in which D_{so} is the diffusivity under neutral (*i.e.*, uniform fluid density) flow condition, and the diffusion damping coefficient Φ_s depends on the length-scale of turbulence, which changes with stratification (Chapter 10). Substituting these relations in Eq. 6.111 yields

$$\frac{\partial C}{\partial t} = \frac{\partial}{\partial z}\left(w_{s0}Cf^{n'}(C,\bar{G}) + D_{so}\Phi_s \frac{\partial C}{\partial z} \right) \qquad (6.112)$$

where $f^{n'}(C,\bar{G})$ is another function of concentration and shear rate.

When the upward and downward mass fluxes are in balance throughout the water column, or are assumed as such, the resulting equilibrium profile is obtained by setting $\partial C/\partial t$ to zero in Eq. 6.112. This condition can exist only under a steady flow or wave motion, and its invocation in the natural environment is rarely justified in the true sense. Therefore, whenever a steady flow is assumed the error introduced, *e.g.*, in the estimation of the settling velocity, must be borne in mind. On the other hand, an advantage in using the equilibrium approach is that integration of Eq. 6.112, typically numerical, is obviated. Integration of the steady form of Eq. 6.112 gives

$$w_{s0}Cf^{n'}(C,\bar{G}) + D_{so}\Phi_s \frac{\partial C}{\partial z} = C' \qquad (6.113)$$

For the common case of no-net input of sediment into or removal from the water column, the constant of integration $C' = 0$. Therefore,

$$w_{s0}Cf^{n'}(C,\bar{G}) + D_{so}\Phi_s \frac{\partial C}{\partial z} = 0 \qquad (6.114)$$

which can be integrated from reference values $C = C_a$ at $z = z_a$ to give

$$\int_{C_a}^{C} \frac{dC}{Cf^{n'}(C,\bar{G})} = -w_{so}\int_{z_a}^{z} \frac{dz}{D_{so}\Phi_s} \qquad (6.115)$$

When stratification is absent, *i.e.*, the flow is uniform, $\Phi_s = 1$ and Eq. 6.115 reduces to

$$\int_{C_a}^{C} \frac{dC}{Cf^{n'}(C,\bar{G})} = -w_{so}\int_{z_a}^{z} \frac{dz}{D_{so}} \qquad (6.116)$$

Let us now assume a simple case in which the settling velocity is independent of concentration and the shear rate, *i.e.*, $f^{n'}(C,\bar{G}) = 1$. Then from Eq. 6.116,

$$\ln\left(\frac{C}{C_a}\right) = -w_{so} \int_{z_a}^{z} \frac{dz}{D_{so}} \qquad (6.117)^n$$

We will further consider that $D_{so} = \bar{D}_{so}$ is independent of height. Accordingly the solution of Eq. 6.117 is

$$\frac{C}{C_a} = e^{-\frac{w_s}{\bar{D}_{so}}(z-z_a)} \qquad (6.118)$$

where w_{so} is denoted as w_s for simplicity. The choice of $\bar{D}_{so}$ depends on the flow structure. In general, particles do not follow the turbulent movement of water parcels, *i.e.*, particle motion is not iso-kinetic. This discrepancy in movements increases with increasing frequency of turbulence. Lack of iso-kineticity has been attributed to the effect of the centrifugal force causing the ejection of particles from turbulent eddies. The ability of particles trapped within eddies to settle out also contributes to the difference in the movements of the solid and fluid phases. As a result mass and eddy or momentum diffusivities are usually unequal. As we shall see in Chapter 10, the neutral mass diffusivity is defined as $\bar{D}_{so} = Sc^{-1}\bar{\varepsilon}_m$, where $\bar{\varepsilon}_m$ is the neutral depth–mean momentum diffusivity (Chapter 2), and Sc is the turbulent Schmidt–Prandtl (or Schmidt) number. In the absence of data on $\bar{\varepsilon}_m$, its approximate value may be estimated from Eq. 2.65 or 2.66 [Sumer and Diegaard, 1981; van Rijn, 1984b; Rose and Thorne, 2001].

Example 6.2: The following data for an equilibrium suspension of a kaolinite were obtained in a counter-rotating annular flume (CRAF) [Mehta, 1973]. In this apparatus the vertical diffusivity may be assumed constant within the central zone of the water column corresponding to the height over which the data points were obtained. The total water depth was 33 cm and bed shear stress 0.185 Pa. Assume $z_a = 4$ cm and estimate the settling velocity.

Elevation (cm)	Concentration (kg m^{-3})
4.0	1.69
12.7	1.54
22.2	1.41
30.8	1.31

[n] Leliavsky [1955] attributes this equation to H.E. Hurst, W. Schmidt, M.P. O'Brien and Th. von Kármán.

We will use Eq. 2.66 for momentum diffusivity, $\rho = 1,000$ kg m^{-3}, $\kappa = 0.4$ and assume $Sc = 1$. Equation 6.118 becomes

$$\frac{C}{C_a} = e^{-\frac{15w_s(z-z_a)}{u_*h}} \qquad (6.119)$$

Calculations using the above equation are given below. The resulting line in Fig. 6.19 shows acceptable agreement with the data points when $w_s = 2.9 \times 10^{-4}$ m s^{-1}.

$(z-z_a)/h$	C/C_a (Data)	C/C_a (Equation)
0	1	1
0.26	0.91	0.92
0.55	0.83	0.84
0.81	0.78	0.77

In Eq. 6.116 we will now introduce the parabolic form of momentum diffusivity (Eq. 2.62, the depth-average value of which is used in Example 6.2) along with a constant settling velocity, and equality between mass and momentum diffusivities as before. Thus we obtain

$$\frac{C}{C_a} = \left[\frac{z_a(h-z)}{z(h-z_a)} \right]^{w_s/\kappa u_*} \qquad (6.120)$$

The dimensionless ratio $w_s/\kappa u_*$ is the Rouse number R_n (named after Professor Hunter Rouse [1951]). It is a measure of the effect of

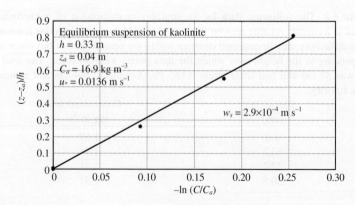

Figure 6.19. Estimation of settling velocity from measured suspended sediment concentration profile.

gravitational settling to eddy diffusion represented by the friction velocity (or equivalently the bed shear stress). The smaller the value of the R_n, the greater will be the uniformity of the concentration profile.

The number R_n or equivalently the ratio w_s / u_* is a significant quantity influencing coarse or fine sediment load in turbulent flows. In the present context, for example, $R_n = 0.8$ has been proposed as the threshold separating wash load ($R_n < 0.8$) from bed material load, in place of the more empirical particle size based threshold ($< {\sim}0.1$ μm). As we will note later, stratification of the suspension is also defined in terms of R_n.

Example 6.3: The following data were obtained from the Cumbarjua Canal in Goa, India, where the total water depth was 6.7 m and the friction velocity 0.027 m s^{-1} [Mehta *et al.*, 1983]. Select $z_a = 0.3$ m and estimate the settling velocity.

Elevation (m)	Concentration (kg m^{-3})
0.3	0.043
1.7	0.031
3.4	0.026
5.0	0.022
6.0	0.018

Calculations for the settling velocity using Eq. 6.120 are given below. The curve in Fig. 6.20 agrees with the data when $w_s = 1.78 \times 10^{-3}$ m s^{-1}.

z (m)	C (kg m^{-3})	$z_a (h-z)/z(h-z_a)$	C/C_a (Data)	C/C_a (Equation)
0.3	0.043	1	1	1
1.7	0.031	0.138	0.72	0.72
3.4	0.026	0.046	0.60	0.60
5.0	0.022	0.016	0.51	0.51
6.0	0.018	0.006	0.42	0.42

Now let us consider a case in which high suspended sediment concentration results in hindered settling. In this event, and ignoring the effect of turbulence let

$$f^{n'}(C,G) \approx f^{n'}(C) = (1 - K'C)^m \qquad (6.121)$$

where K' and m are constants (Chapter 7). Selecting Eq. 2.64 to represent the diffusivity, from Eq. 6.116 we obtain

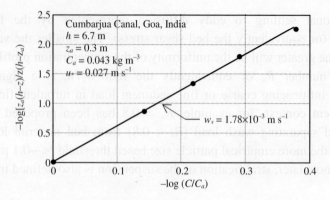

Figure 6.20. Estimation of settling velocity from measured suspended sediment concentrations in Cumbarjua Canal.

$$\int_{C_a}^{C} \frac{dC}{C(1-K'C)^m} = -15 \frac{w_s}{u_* h}(z - z_a) \qquad (6.122)$$

Taking $m = 5$ (Chapter 7), the integral can be evaluated in the form of a series as

$$\left[\ln \left| \frac{(1-K'C)}{C} \right| + \frac{4K'C}{(1-K'C)} - \frac{6K'^2 C^2}{2(1-K'C)^2} + \frac{4K'^3 C^3}{3(1-K'C)^3} - \frac{K'^4 C^4}{4(1-K'C)^4} + \dots \right]_{C_a}^{C}$$

$$= 15 \frac{w_s}{u_* h}(z - z_a) \qquad (6.123)$$

For small values of $K'C$ the above equation reduces to Eq. 6.119 [O'Connor and Tuxford, 1980].

6.5.3 *Reference concentration*

The reference concentration C_a can be roughly estimated from

$$C_a = 0.015 \frac{d_{50}}{z_a} \frac{\tilde{\tau}^{3/2}}{d_*^{0.3}} \qquad (6.124)$$

Other formulas are summarized by Soulsby [1997]. For cohesive sediment see Eq. F.7.7. The dimensionless shear stress $\tilde{\tau}$ is defined as

$$\tilde{\tau} = \frac{e_\tau \tau_b - \tau_c}{\tau_c} \qquad (6.125)$$

where e_τ is a stress reduction parameter (to be taken equal to unity unless determined by calibration in accordance with the method of van Rijn [1984b]). The dimensionless particle diameter is obtained from

$$d_* = \left[\frac{(s-1)g}{v^2} \right]^{1/3} d_{50} \qquad (6.126)$$

where $s = \rho_s / \rho$. The elevation z_a must be taken as one-half the height of the bed form (ripple or dune). When the bedform height is unknown one may select $z_a = k_s$, the bed roughness. The minimum value of z_a must not be less than $0.01h$, where h is the water depth.

Example 6.4: Calculate the reference concentration, given the following experimental values: $d_{50} = 0.05$ mm, $z_a = 0.1$ m, $e_\tau = 1$, $\tau_b = 2.13$ Pa, $\tau_c = 0.22$ Pa, $s = 2.65$ and $v = 10^{-6}$ m^2 s^{-1}.
From Eqs. 6.125 and 6.126 we obtain $\tilde{\tau} = 8.68$ and $d_* = 1.265$. Then from Eq. 6.124, $C_a = 1.8 \times 10^{-4}$ kg m^{-3}.

6.6 Two-Phase Flow Equations

When the particles are large, two-phase flow analysis can be carried out to improve the accuracy of calculated transport loads. Even though cohesive primary particles are small, their flocs are large enough to benefit, in terms of their transport behavior, from the two-phase treatment [Hsu et al., 2003, 2004, 2007; Krishna, 2009].

Numerical approaches for multiphase transport are broadly classified as Euler-Lagrange and Euler-Euler. In the Euler-Lagrange approach the fluid is treated as a continuum and the time-averaged Navier-Stokes equations are solved. Particles are treated by tracking a large number in the flow. For dilute suspensions the assumption made is that since the second phase has a low volume-concentration, it has no direct impact either on the production or dissipation of turbulence.

In the Euler–Euler approach both phases are treated as inter-penetrating continua. The mass continuity and momentum conservation equations for the phases are coupled through pressure or the inter-phase exchange coefficient. This approach is computationally intensive as it evaluates more equations than the Euler–Lagrange approach, but it is also more accurate.

For the sediment load, ensemble averaging of the mass and momentum equations is carried out over the scale of particle diameter. However, since sediment concentration varies over a scale much larger than the diameter, another level of averaging is required [Hsu, 2002]. This process, Favre averaging, is carried out on the already averaged two-phase flow equations. Here we will state the final equations in terms of the solids volume fraction ϕ_v (z,t). For flocs the volume fraction ϕ_{vf} along with floc diameter and density are used. In this simplified approach transport occurs in the vertical z-direction only, but fluid pressure gradient in the x-direction is permitted to influence the kinematics.

The Favre averaged fluid-phase mass continuity equation is

$$\frac{\partial}{\partial t}\rho^f(1-\phi_v)+\frac{\partial}{\partial z}\rho^f(1-\phi_v)\overline{w}^f = 0 \qquad (6.127)$$

where superscript f refers to the fluid phase and the bar denotes an averaged quantity. The sediment-phase mass continuity equation is

$$\frac{\partial}{\partial t}\rho^s\phi_v+\frac{\partial}{\partial z}\rho^s\phi_v\overline{w}^s = 0 \qquad (6.128)$$

where superscript s refers to the solid phase. Since the horizontal (x-direction) pressure gradient drives the fluid, momentum conservation is considered in both the x and the z directions. Since the volume fraction can change with height and time, changes in the fluid and solid phase densities account for the compressibility of floc volume.

The fluid-phase momentum equation in the x-direction is

$$\frac{\partial}{\partial t}\rho^f(1-\phi_v)\overline{u}^f = -\frac{\partial}{\partial z}\rho^f(1-\phi_v)\overline{u}^f\overline{w}^f - (1-\phi_v)\frac{\partial \overline{p}^f}{\partial x}$$

$$+\frac{\partial \tau^f_{xz}}{\partial z} - \rho^f(1-\phi_v)S_f g - \beta_D\phi_v\left(\overline{u}^f - \overline{u}^s\right)$$

(6.129)

and in the z-direction

$$\frac{\partial}{\partial t}\rho^f(1-\phi_v)\overline{w}^f = -\frac{\partial}{\partial z}\rho^f(1-\phi_v)\overline{w}^f\overline{w}^f - (1-\phi_v)\frac{\partial \overline{p}^f}{\partial z}$$

$$+\frac{\partial \tau^f_{zz}}{\partial z} + \rho^f(1-\phi_v)g - \beta_D\phi_v\left(\overline{w}^f - \overline{w}^s\right) + \beta_D\varepsilon_m\frac{\partial \phi_v}{\partial z}$$

(6.130)

where S_f is the slope of the energy grade line, $\overline{p}^f$ is the ensemble averaged fluid pressure, ε_m is the eddy diffusivity and τ^f_{xz}, τ^f_{zz} are the average fluid-phase stresses including, in both cases, the viscous stress and the fluid-phase Reynolds stress. In Eqs. 6.129 and 6.130, apart from the first (unsteady) term on the left hand side and the first (convective) term on the right hand side, the remaining terms represent Favre averaged drag forces due to fluid–sediment interaction and the relative mean velocity between the two phases. In the development of Hsu [2002] the drag coefficient β_D is given by

$$\beta_D = \frac{\rho^f\Delta u}{d_p}\left(\frac{18}{Re_p} + 0.3\right)\frac{1}{(1-\phi_v)^{4.45/Re_p}}$$

(6.131)

in which the term $(1-\phi_v)^{4.45/Re_p}$ accounts for hindered settling effect characteristic of dense slurries (Chapter 7). The velocity between the fluid and sediment phase velocities is

$$\Delta u = \sqrt{\left(\overline{u}^f - \overline{u}^s\right)^2 + \left(\overline{w}^f - \overline{w}^s\right)^2}$$

(6.132)

and the particle-based Reynolds number Re_p is

$$Re_p = \frac{d_p\Delta u}{\eta}$$

(6.133)

The sediment-phase momentum equation in the x-direction is

$$\frac{\partial}{\partial t}\rho^s\phi_v\overline{u}^s = -\frac{\partial}{\partial z}\rho^s\phi_v\overline{u}^s\overline{w}^s - \phi_v\frac{\partial\overline{p}^f}{\partial x}$$

$$+\frac{\partial\tau^s_{xz}}{\partial z} - \rho^s\phi_v S_f g + \beta_D\phi_v\left(\overline{u}^f - \overline{u}^s\right) \tag{6.134}$$

and in the z-direction

$$\frac{\partial}{\partial t}\rho^s\phi_v\overline{w}^s = -\frac{\partial}{\partial z}\rho^s\phi_v\overline{w}^s\overline{w}^s - \phi_v\frac{\partial\overline{p}^f}{\partial x} + \frac{\partial\tau^s_{zz}}{\partial z}$$

$$+\rho^s\phi_v g + \beta_D\phi_v\left(\overline{u}^f - \overline{u}^s\right) - \beta_D\varepsilon_m\frac{\partial\phi_v}{\partial z} \tag{6.135}$$

The parameters τ^s_{xz} and τ^s_{zz} are the average shear stresses in the sediment phase. Each stress includes contributions from inter-particle collisions, sediment–fluid interaction and the Reynolds stress due to particle movement induced by velocity fluctuations. Numerically based evaluations of the fluid stresses and inter-particle stresses are given in Hsu [2002].

Solution of the equations along with initial and boundary conditions requires numerical schematization. An application to simulate gravity-driven turbid flows is briefly mentioned in Chapter 12.

6.7 Concentration Determining Processes

6.7.1 *Layered concentration structure*

The vertical profile of suspended sediment concentration can have a complex structure when the particles are fine-grained, particularly when they are cohesive and the suspension is not dilute. Recognition of the complex behavior of stratified fine-sediment suspensions is evident in Fig. 6.21 from the work of Owen [1977] on English estuaries (Thames, Great Ouse and Avon). Although the Thames and the Great Ouse differ from the Avon in velocity and concentration structures, in all cases they

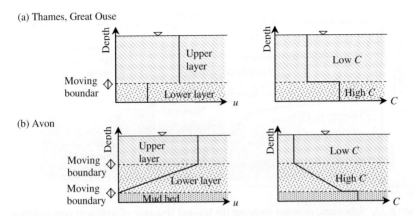

Figure 6.21. Schematic drawings of the vertical structures of tidal velocity and suspended sediment concentration in English estuaries: (a) Thames and Great Ouse; (b) Avon (adapted from Owen [1977]).

are conditioned by a strong interaction between flow and fine sediment. The significance of Fig. 6.21 lies in its originality at a time when attempts at such descriptions were often based on sandy estuaries, in which interaction between the flow and sand is different and often less dramatic.

Cohesive sediment transport processes in the movements of sediment-stratified layers can be highlighted by a generalization of Fig. 6.21, as sketched in Fig. 6.22. The term "mobile" implies that the layer can move horizontally or heave under wave motion. "Stationary" indicates a layer which neither moves horizontally nor heaves. Starting from the water surface, in the top layer the sediment is well-mixed and mobile. Sediment concentration (or wet bulk density) is low and the fluid rheology is practically Newtonian, *i.e.,* fluid viscosity is independent of flow shear, and the suspension does not have a yield strength. The frequency of inter-particle collisions is so low that outcomes leading to the growth or breakup of flocs are rare. As a result they settle more-or-less independently in the so-called free settling mode. The settling velocity does not depend on the concentration or the shear rate. The concentration profile varies smoothly, and turbulent mass diffusion is practically neutral, *i.e.,* diffusive transport of sediment is not significantly influenced by stratification.

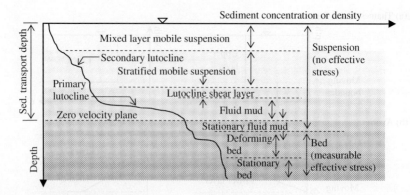

Figure 6.22. Schematic drawing showing the layered structure of sediment concentration (or wet bulk density) in suspension and in the bed.

Below the mixed layer the suspension is nearly Newtonian at the top and has been called Concentrated Benthic Suspension (CBS) [Toorman, 2001]. Alternatively, it is termed Benthic Suspended Sediment Layer (BSSL) [Jain *et al.*, 2007] because it is not always concentrated enough to have significantly non-Newtonian properties. With increasing depth (and concentration) the suspension becomes increasingly non-Newtonian. The frequency of inter-particle collisions increases and the settling velocity usually increases with concentration as the flocs become larger in the so-called flocculation settling mode. Mass diffusion is retarded by the negative buoyancy associated with gradients in concentration.

The bed, with characteristically low permeability, can be differentiated from the suspension with respect to effective normal stress (Chapter 8), which is practically nil above the bed and increases with concentration within the bed. Thus the bed surface is also the boundary at which the concentration has its space-filling value. The upper part of the bed may undergo deformations due to oscillatory motion by waves, whereas the bed below remains stationary. These deformations can dissipate the effective stress if plastic yield occurs. If the effective stress vanishes the oscillating bed layer would be considered liquefied (Chapter 10).

Layering of suspension is the outcome of coupling between the concentration-dependent settling velocity and the concentration-gradient-dependent diffusion. A step-like concentration gradient between any two layers is called a lutocline, which is a sediment-induced pycnocline [Kirby, 1986; Parker, 1987]. The main or primary lutocline occurs near the base of the stratified mobile suspension. This gradient encompasses a layer which resembles the boundary layer above a rigid bed, with high shear production and energy loss. However, unlike a rigid bed, the "bottom" below the primary lutocline is fluid-like, and is dragged along with the flow above. Thus the flow velocity becomes zero at some depth below the lutocline. Any lutocline above this primary lutocline is called a secondary lutocline.

A noteworthy feature of the primary lutocline is that due to the high negative buoyancy of the suspension, the current may not be strong enough to destabilize the concentration gradient (Chapter 10). As sketched in Fig. 6.23, a fluid parcel (containing particles at high concentration) from the lower layer and ejected upward by a turbulent eddy falls rapidly and merges with the lower layer. Thus the lower layer displays high stability. Furthermore, as the concentration increases with depth the permeability of water decreases and slows down the rate of dewatering of fluid mud below the lutocline. The outcome is that in a depositional flow-environment densification of fluid mud occurs slowly compared to the less dense suspension above the lutocline. Similar processes occur at a secondary lutocline. However, due to weaker stratification coupled with a stronger effect of eddy diffusion, a secondary lutocline may weaken or even vanish within minutes compared to possibly hours of stability of the primary lutocline (and fluid mud beneath).

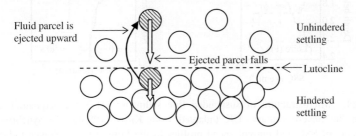

Figure 6.23. Stability of a lutocline.

Figure 6.24a shows three profiles of suspended sediment concentration in the Severn estuary, UK. They were obtained by a free-falling optical sensor aboard a vessel, and as a result each profile measurement was practically instantaneous. Changes in the microstructure of the stratified mobile suspension are evident over a short period of 16 min. Within that duration the lutocline was relatively stable. Posmentier [1977] pointed out a similar staircase microstructure of haloclines in the Hudson River, NY, which was stable over periods on the order of 90 min. It was observed that the current velocity was smoother than the halocline, suggesting that the density difference between the saltier and fresher water masses was responsible for the microstructure. Carpenter and Timmermans [2012] argue that the

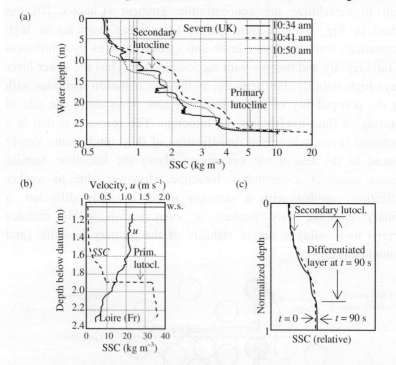

Figure 6.24. Microstructure of suspended fine sediment: (a) Three sequential SSC profiles in the Severn estuary, UK (adapted from Kirby [1986]); (b) synchronous measurements of SSC and current speed profiles in the Loire estuary, France (adapted from Le Hir [1997]); (c) simulated generation of a differentiated layer of SSC (from Scarlatos and Mehta [1993]).

generation of staircase also requires the right combination of salinity and temperature changes with depth. In the high turbidity estuarine environment, SSC profiles show stratified structures even when there is practically no salinity variation with depth [Kirby, 1986].

In Fig. 6.24b, observe that in the Loire estuary (France) the current (u) profile is smooth and does not reflect the presence of the primary lutocline. A one-dimensional analysis by Scarlatos and Mehta [1990] including the damping effect of high SSC under conditions similar to the Loire appears to reproduce the SSC microstructure (Fig. 6.24c). The staircase-like SSC profile after just 90 s of simulation is a more stable state of the stratified system than the initial smooth profile. The initial SSC profile responds rapidly to changes in concentration when it is high, and turbulence may even collapse leading to flow laminarization and disappearance of the microstructure. This is expected to occur when the concentration exceeds a threshold value as mentioned in Chapter 10 [Winterwerp, 1999].

Returning to Fig. 6.22, the base of the (mobile) fluid mud layer is the plane of zero velocity. Under eroding conditions fluid mud can rise to a height at which the turbulent kinetic energy from mean flow supplies the potential energy of the risen fluid mud. This mud is diluted by downward entrainment of water. If conditions conducive to entrainment of interfacial sediment occur, fluid mud particles may gradually "leak" from the risen layer into water above the interface.

Figure 6.25a shows concentration profiles from a location along the Gulf of Mexico coast of Louisiana where fluid mud occurs in shallow waters close to the shoreline. Turbidity rose due to waves during the passage of a winter cold front. A noteworthy feature is the upward leakage of sediment. In terms of sediment mass flux, leakage is described by an entrainment function (Chapter 11). Despite entrainment, the concentration in the upper water column remained very low compared to concentration below the lutocline. As mentioned, while waves can liquefy mud and sustain it as such within the wave boundary layer, they do not provide an efficient mechanism to entrain sediment compared to a steady current in which eddy diffusion tends to be much stronger. In Fig. 6.25b a one-dimensional numerical model has been used to

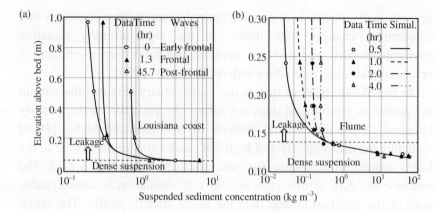

Figure 6.25. Illustration of fluid mud leakage: (a) measurements offshore of Louisiana [Wells and Kemp, 1986]; (b) measurements in a laboratory flume and numerical simulation [Maa, 1986; Ross 1988].

simulate measured entrainment in a laboratory flume. After 4 h the overwhelming mass of sediment remained under the (primary) lutocline.

Below the zero-velocity plane (Fig. 6.22), low-permeability mud in a fluid-like state is stationary, *i.e.*, there is practically no horizontal movement. This layer may show vertical movement due to settling or entrainment. It may also heave under wave motion.° The total thickness of (mobile plus stationary) fluid mud depends on the current or waves, availability of sediment and its composition. In low concentration (*e.g.*, a few tens of milligrams per liter) and low energy (*e.g.*, microtidal and low wind) environments fluid mud may be absent or form thin, ephemeral, layers. They may achieve an order of magnitude greater thickness under episodic conditions. In contrast, in the highly energetic tidal environment such as in the Amazon estuary, persistent fluid mud is found to be as much as several meters in thickness [Kineke, 1993].

To summarize, fluid mud is an ephemeral state of mud whose thickness and height at the bottom of the water column depend on the rate of supply of fluid energy and the rate of loss of energy. Fluid mud

---

° In a field investigation in the Atchafalaya River (LA) delta area by Sahin *et al.* (2012), it was found that during storm waves a zero wave-velocity plane occurred below the zero current-velocity plane by as much as 5 cm. It appeared that this 5 cm thick mud layer underwent elastic deformation due to wave oscillation.

itself may be the main energy sink, and therefore as its thickness and density change the rate of energy loss also changes. If the source of energy (waves or current) is removed fluid mud will revert to a bed.

In nature fluid mud is rarely, if ever, found in a state of true equilibrium. Accordingly, to characterize fluid mud behavior a time-series of relevant measurements is required. Instantaneous data can be misinterpreted and lead to erroneous conclusions concerning the viability of fluid mud in a given environment.

6.7.2 *Transport processes*

The physical description in Fig. 6.22 is an instantaneous one, since the concentration profile along with the elevations and thicknesses of the various layers tend to change in response to a current or waves. For model simulation of these changes it is necessary to know the vertical and horizontal sediment transport fluxes. Since the horizontal transport load depends on vertical sediment exchange mechanisms, we will identify the vertical unit transport processes (in qualitative analogy with the terminology once coined at MIT for chemical engineering processes) and fluxes which must be modeled to calculate the sediment load.[p] The following description of the processes partly overlaps Fig. 6.22.

Consider the schematic plot of the concentration (or wet bulk density) profile in Fig. 6.26. Also shown are profiles of the horizontal velocity due to a current and waves. At the boundary between mobile fluid mud and mobile suspension also indicated in Fig. 6.27a, sediment entrainment, mixing and settling can occur. As mentioned, upward entrainment of fluid mud (Fig. 6.27b) depends on the kinetic energy of the turbulent eddies. The lutocline interface is destabilized, interfacial waves and wave breaking occur and the lower fluid, with its higher sediment content, entrains, or leaks, into the fluid above the lutocline,

---

[p] An exception is the case in which exchange of sediment at the bed is nil or minor, and advection dominates. In that event, vertical sediment transport descriptors such as the bed shear stress or the Richardson number require different interpretations (*e.g.*, Lamb *et al.* [2004]).

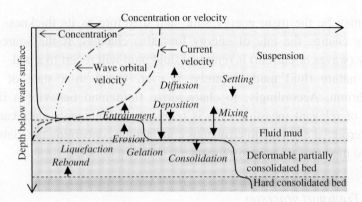

Figure 6.26. Unit transport processes associated with concentration profile dynamics.

where sediment concentration gradually increases. Due to sediment entrainment the lutocline interface is lowered or eroded (scoured).

In the formal sense, mixing differs from upward entrainment defined above because mixing involves a second process by which the upper fluid (sediment and water) is entrained into the lower fluid (Fig. 6.27c). When the rate of downward entrainment is lower than the rate of upward entrainment and observations are limited to measurements of the interfacial elevation, the process may appear as erosion (or scouring) of the interface. The density of suspension below the interface is not affected.

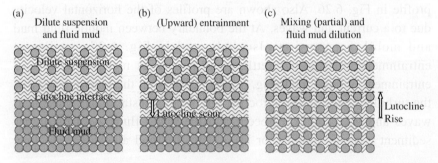

Figure 6.27. Schematic plots highlighting processes at or close to the lutocline interface: (a) Position of the interface between dilute suspension and fluid mud; (b) upward entrainment of sediment and scouring of the interface; (c) mixing of fluid mud and water layers.

In the opposite case (downward entrainment greater than upward) the interface rises, but at the same time fluid below the interface is diluted by water from above. Therefore, in order to identify (upward) entrainment as a separate process it is necessary to measure the change in sediment concentration both above and below the interface (lutocline) along with the movement of the interface itself. Mass balances are carried out for the upper and lower layers to obtain the fluxes of sediment and water in both directions. If entrainment occurs without mixing the eventual outcome could be complete mixing of the two layers, as may occur as well due to mixing alone. In the natural environment it is often difficult to differentiate between entrainment and mixing, and the term entrainment is often used to cover both in the informal sense. This sense is retained in some cases in the subsequent chapters.

The term erosion is conventionally considered to denote bed scour (as opposed to fluid mud scour) and the transport of detached particles. At the lower level of the fluid mud layer the rate of erosion depends on the bed shear stress in excess of the bed shear strength. Bed erosion occurs either by a gradual dislodgement and entrainment of the flocs at the bed surface, or by a more traumatic mass failure of a sizeable thickness of the bed and entrainment of the failed material.

Accumulation of a deposit on the bed can also be called sedimentation.[q] The deposition flux at the lutocline or the bed is usually the highest when there is no flow, and decreases as the bed shear stress increases until none of the suspended matter is able to deposit. Deposition involves sorting of sediment as heavier and more sticky particles/flocs that arrive close to the bed become attached to the bed, while the remaining material stays in suspension, or elastically rebounds from the bed surface. When the rate of deposition is high, fluid mud can form because settling is hindered. At low rates of deposition, in case no fluid mud initially exists, the settling sediment may deposit to form a bed without forming fluid mud if the rate of hindered settling combined with the rate of bed consolidation is greater than the rate of deposition.

---

[q] In common usage sedimentation has a broader meaning without a scientifically unique definition. Here it means *net* accumulation of deposit.

Consolidation of the deposit, *i.e.,* reduction in its volume by removal of pore water, occurs due to the weight of the deposit itself. However, a deposit that is fully consolidated can further reduce in volume if a surcharge (or overburden) in the form of a new deposit occurs. If overburden is removed, the volume of the consolidated bed will expand somewhat due to the elastic properties of the bed particle matrix. This is known as bed swelling or rebound. Consolidation is accompanied by gelation due to a rearrangement of water molecules within the pores and also electrochemical bonds between particles. Organic polymers may also contribute to the gelled state.

The mechanics of formation of fluid mud by waves is dependent on the material properties of the bed, its initial state, and wave-induced stresses. Bed failure by shearing is often called liquefaction, whereas failure by breakup due to excess pore pressure buildup has been called fluidization [Toorman, 2001]. Since it is difficult to distinguish between these processes without careful experimentation, liquefaction will be considered to occur by normal and shear stresses, which weaken and eventually disrupt the bed. The effective normal stress vanishes, and we may conveniently assume that the bed undergoes plastic yield more or less concurrently and changes to fluid mud. Once waves cease, fluid mud may revert to the bed state almost immediately, or start to dewater before reverting to bed. Dewatering of fluid mud or partially consolidated bed is by hindered settling and consolidation. [r]

In the above description the identified unit transport processes are: settling, deposition, consolidation (and gelation), erosion, entrainment, diffusion and mixing of eroded/entrained sediment. In a general sense the difference between turbulent diffusion and turbulent mixing is a matter of scale. Turbulent mixing can be thought of as a larger scale phenomenon involving the overturning of water masses larger than in diffusion.

---

[r] Fluidization also refers to the process by which a flow introduced from bottom-up counteracts particles that are settling or have settled in otherwise still water. When the current velocity exceeds the particle settling velocity fluidization is said to take place as the particles acquire an upward velocity. In the present context it is also appropriate to use the term "fluidization" to mean "fluid mud generation".

A limitation of the descriptions of fine-sediment layering (stratification) in plots such as Figs. 6.22 and 6.26 is that since the thickness and position of each layer is time-dependent, variables related to layer dynamics must be represented along the abscissa as well as the ordinate. A preliminary effort was made by Ozdemir [2010] to characterize the role of turbulence on fine particle transport in an oscillatory channel along with the feedback effect of particles on turbulence. The hydrodynamic equations were solved in the three-dimensional space for direct numerical simulation of turbulent flow, inclusive of the effects of particle-induced stratification. Figure 6.28 is a schematic representation, showing layering to be dependent on the Richardson number and the particle settling velocity. A third parameter, the wave boundary-layer Reynolds number (defined in terms of the Stokes boundary layer thickness, Chapter 11) has been held constant. This description assigns the layers hydrodynamic and sediment related variables that can be easily estimated.

6.8 Exercises

6.1 Derive the force balance at incipient (or threshold of) movement of a particle when the bed is tilted at an angle θ_b relative to the horizontal plane.

(1) Show that this balance is reduced to Eq. 6.1 for a flat bed.

(2) Derive Eq. 6.12.

(3) Show how the cohesionless form of Eq. 6.8 may be obtained from Eq. 6.12.

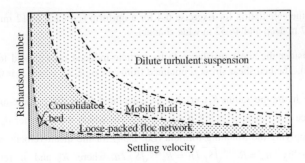

Figure 6.28. Schematic characterization of sediment-induced stratification on relevant dynamic variables (adapted from Ozdemir [2010]).

6.2   (1) Calculate the critical shear stress based on the Shields parameter from Eq. 6.28, and the particle Reynolds number $Re^* = u_{*c}d_p/v$ (where $u_{*_c} = \sqrt{\tau_c / \rho}$) for: (a) k_s/d_p = 1 and 10, and (b) angle of repose $\varphi_a = 35°$ and 26°. Assume spherical quartz particles in freshwater and particle diameter in the range of 1.5 mm to 10 mm. Hint: For a normally distributed probability density function, $n_u = 3.09$ when $p_e = 10^{-3}$ (Appendix D).

(2) For the same diameter range, determine θ_{ec} corresponding to the Bagnold curve for the threshold of suspension.

6.3   Prove the identity

$$\sum_{n=0}^{n=\infty} (1 - p_e)p_e^n(n+1) = \frac{1}{1 - p_e}$$ (E6.1)

Hint: Note the following relationships:

$$\sum_{n=0}^{n=\infty} (1 - p_e)p_e^n(n+1) = \frac{\partial}{\partial p_e}\sum_{n=0}^{n=\infty} p_e^{n+1} - \left(\frac{\partial}{\partial p_e}\sum_{n=0}^{n=\infty} p_e^{n+2} - \sum_{n=0}^{n=\infty} p_e^{n+1}\right)$$ (E6.2)

and

$$\sum_{n=0}^{n=\infty} p_e^{n+1} = \frac{p_e}{1 - p_e}; \qquad \sum_{n=0}^{n=\infty} p_e^{n+2} = \frac{p_e^2}{1 - p_e}$$ (E6.3)

6.4   A freshwater stream with a sandy bed is characterized by $d_{35} = 3.5$ mm and $d_{65} = 6$ mm. Sand density is 2,650 kg m^{-3}. The stream is 52 m wide and approximately rectangular in cross-section with a mean depth of 4.7 m. The bed slope is 1 m in 3.6 km. Assume that the bed is composed of mono-size particles whose equivalent diameter is obtained by taking the logarithmic mean of d_{35} and d_{65}, i.e., $A\log[(\log d_{35} + \log d_{65})/2]$. You may approximate R'_h by R_h. Determine the bed load using the Einstein equation (Eq. 6.84). Express bed load in metric tons per hour.

6.5   Repeat the calculation for Exercise 6.4 for a uniform sediment of 12 mm diameter in a 5.87 m deep and 46.52 m wide channel.

(1) Consider the shear resistance to be particle-induced (as opposed to bedform-induced). The bed shear stress is obtained from $\tau_b = \rho g R'_h S_f$, where S_f is 0.00065.

(2) The hydraulic radius R'_h can also be calculated from the particle-induced bed resistance coefficient n' using $n' = 0.0131d_p^{1/6}$, in which the diameter d_p is in mm. Then, applying the Manning formula (Eq. 2.34) for the mean velocity u_m, assume the equality $u_m = R'^{2/3}_h \sqrt{S_f}/n' = R^{2/3}_h \sqrt{S_f}/n$, where R_h and n refer to the bedform-induced resistance. Therefore, $R'_h = (n'/n)^{3/2} R_h$. Select $n = 0.030$.

6.6   For Exercise 6.5, compare the Einstein result for bed load (but based on bedform resistance) with values obtained from the bed load formulas of Du Boys, Bagnold and those given in Table 6.3. For Du Boys assume $\tau_c = 11$ Pa and $\rho_s g \chi_s = 0.3 \text{ m}^2 \text{ N}^{-1} \text{ s}^{-1}$.

6.7   In text Example 6.2 indicate how the settling velocity would change if:

(1) The fluid density is increased from 1,000 kg m^{-3} to 1,025 kg m^{-3}.

(2) Total depth is increased from 33 cm to 150 cm. (Note that this change redefines the original concentration profile.)

(3) Schmidt number is increased from 1 to 1.5.

(4) Bed shear stress is: (a) reduced (from 0.185 Pa) to 0.1 Pa, and (b) increased to 0.3 Pa. Briefly give physical reasons for the trends obtained.

6.8   You are given $h = 6.7$ m, $u_* = 0.3$ m s^{-1}, $z_a = 0.3$ m, $C_a = 4.1$ kg m^{-3} and $w_s = 1.8 \times 10^{-5}$ m s^{-1}. Plot C on the horizontal axis against $(z-z_a)/(h-z_a)$ on the vertical axis, similar to Fig. 6.20. Use Eq. 6.123 along with $K' = 0.001$, 0.01 and 0.1.

6.9   A harbor navigation channel is to be protected from excessive deposition of cohesive sediment by construction of two caisson dikes (Fig. E6.1). The ambient seawater depth h is 10 m and sediment settling velocity 2.9×10^{-3} m s^{-1}. Due to the requirement to allow vessels to navigate the channel safely, the crest of the dike must not exceed 6 m above ambient bottom. Determine the maximum wave-induced bed shear stress at which 75% of the total suspended sediment mass in the water column remains below the dike crest. Assume that Eq. 6.119 applies.

6.10 If the wave period in the harbor in Exercise 6.9 is 4.5 s, determine the allowable (maximum) wave height corresponding to the calculated bed shear stress. Assume that the bed shear stress is obtained from $\tau_b = 0.01 \rho u_{1m}^2$, where ρ is seawater density and u_{1m} is the amplitude of the horizontal component of wave-induced velocity (Eq. 2.115) at 1 mab (meter above bottom).

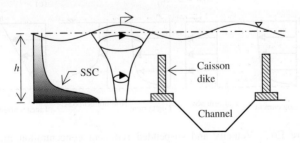

Figure E6.1. Protective caisson dikes and wave-induced suspended sediment in a harbor. The channel is perpendicular to the view, and waves are from left to right.

6.11 In uniform flow the depth–mean velocity u_{msed} at which sediment particles are collectively transported is obtained approximately from

$$u_{msed} = \left(h^2 \frac{\int_{z_a}^{h} C(z)u(z)dz}{\int_{z_a}^{h} C(z)dz \cdot \int_{z_a}^{h} u(z)dz} \right) u_m = \Lambda u_m \qquad \text{(E6.4)}$$

where u_m is the depth–mean current velocity. The proportionality coefficient Λ is equal to 1 when SSC is uniform (Fig. E6.2a). Consider two cases:

(1) For a non-uniform distribution of concentration (Fig. E6.2b), determine the value of Λ when $u(z)$ is given by Eq. 2.55 and $C(z)$ by Eq. 6.119. Assume $h = 0.5$ m, $u_* = 0.03$ m s^{-1}, $w_s = 1 \times 10^{-5}$ m s^{-1}, $C_a = 3.3$ kg m^{-3} $\kappa = 0.4$ and $z_a = 0.001$ m.

(2) Assume linear velocity and SSC profiles (Fig. E6.2c) with the same values of h, u_*, w_s, C_a and z_a along with $u_m = 1.0$ m s^{-1}, $u_b = 0.01$ m s^{-1} and $C_m = 0.1$ kg m^{-3}. What can you conclude from the different values of Λ obtained in the two cases?

6.12 Wave-induced velocity amplitude of 0.25 m s^{-1} is required to initiate sand grain movement at the seafloor. Determine the farthest distance offshore from the shoreline where this movement would begin at the bottom with a uniform slope of 1/25, given a deep-water wave of 1.5 m height and period of 9 s. Assume that the direction of wave approach is normal to the shoreline. Hint: Use Eq. 2.214.

6.13 A spherical particle in a stream acquires momentum which causes it to spin (Magnus effect). The equilibrium elevation of the i-particle depends on the diameter d_{pi} and therefore its buoyant weight W_i (Fig. E6.3). Given lift force on the particle F_{li}, the equilibrium elevation z of the particle is determined by the condition

$$\frac{F_{li}}{W_i} = 1 \qquad \text{(E6.5)}$$

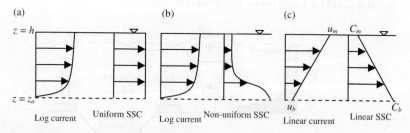

(a)   (b)   (c)

Figure E6.2. Velocity and suspended sediment concentration in uniform flow: (a) velocity profile is logarithmic and concentration profile is uniform; (b) velocity profile is logarithmic and concentration profile is non-uniform (bottom-heavy); (c) velocity profile and concentration profile are both linear.

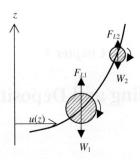

Figure E6.3. Schematic drawing showing particles in *boundary layer* flow.

In *inviscid flow* the stream function is

$$\psi_f = -U\left(r - \frac{d_{pi}^2}{4r}\right)\sin\theta - \frac{\Gamma}{2\pi}\ln r \qquad (E6.6)$$

where U is the mean stream velocity and Γ is the circulation (Chapter 2). Based on the Bernoulli equation, the pressure distribution at the particle surface is

$$p = \rho\frac{U^2}{2}\left[1 - \left(2\sin\theta + \frac{\Gamma}{\pi dU}\right)^2\right] \qquad (E6.7)$$

Is it possible to argue that at equilibrium the particles will be vertically sorted by size? What assumptions are necessary for this explanation? How reasonable are they?

Chapter 7

Settling and Deposition

7.1 Chapter Overview

This chapter introduces particle settling and deposition processes. Whereas settling refers to the falling rate of particles in water as the carrier fluid, deposition implies removal of particles from flow as they become part of the bottom material. This means that in the absence of flow, settling and deposition are synonymous. The chapter begins with basic aspects of settling in the absence of imposed flow, which however does not preclude circulation induced by the settling particles. A section is added on the settling of coarse particles. The inclusion of sand is a continuation of the review of the transport properties of coarse sediment in Chapter 6. Shells are mentioned because they occur in marine mud and also because their transport is an extension of the behavior of coarse particles larger than sand. Modes of settling of fine sediment including free settling, flocculation settling and hindered settling are reviewed. Whereas free settling is common to all particles, flocculation settling is unique to cohesive sediments and is governed by floc growth and breakup, *i.e.*, aggregation processes. Hindered settling becomes increasingly pronounced with decreasing size and is significant when the sediment is flocculated. Hindered settling is examined in some detail because of its closeness to the mechanics of consolidation. The effects of shear rate, salinity, temperature and organic matter on the settling velocity are mentioned.

The latter part of the chapter deals with sediment deposition flux in turbulent flows. This flux is commonly expressed as the time-rate of decrease of suspended dry sediment mass per unit bed area. Uniform particles in suspension are treated first, followed by multi-size particles.

The last section introduces the physical basis for considering cohesive sediment deposition and erosion either as mutually exclusive or as simultaneous processes. This development, which is treated further in Chapter 9, is relevant to modeling bed sediment exchange in the tidal environment.

7.2 Settling Velocity

7.2.1 *Free settling*

7.2.1.1 *Settling at low particle Reynolds number*

The rate at which a particle falls in a still fluid depends on the properties of the particle, whether other particles or objects such as a solid boundary or vegetation canopy are present, and on the carrier fluid. Any intervention retarding the fall is a resistance force which, in principle, can be estimated from hydromechanics. Submicron-size particles are especially sensitive to a variety of effects such as due to Brownian motion and pycnoclines. Gravitational settling of small particles is influenced by resistance so much so that on average the particles may not fall at all or fall at exceptionally slow speeds, as for instance at oceanic depths [Sheng, 1986]. Our interest is in particles that show measurable settling over time-scales as short as the wave period to as large as the astronomical tide, river flood or storm duration.

As noted in Example 4.7, a typical floc contains mainly water. To ascertain if the floc's high porosity permits flow through the solid card-house matrix while falling, Moudgil and Vasudevan [1989] examined macroflocs isolated from a suspension of kaolinite in water containing non-ionic polyacrylamide as a particle binder. Each floc was dropped in a water-filled column both in frozen and unfrozen states. Table 7.1 indicates that there was little effect of freezing on the settling velocity. It was concluded that the floc matrix and enclosed water settled as a single unit, *i.e.*, there was no measurable flow through the matrix. To the extent that this limited experiment is generally valid, it supports the common assumption that particles and water in the floc move at the same rate, at

Table 7.1. Effect of floc porosity on settling velocity.

Floc diameter (μm)		Settling velocity (m s^{-1})	
Unfrozen	Frozen	Unfrozen	Frozen
1,128	1,128	0.056	0.056
1,673	1,635	0.056	0.058
2,257	2,257	0.057	0.062

Source: Moudgil and Vasudevan [1989].

least at low Reynolds numbers. Indeed, if this was not so, the inter-particle bonds would easily rupture and the floc would disintegrate during its fall. We will assume that there is no water motion inside flocs.

The complete force balance for a particle falling at a velocity $w_s(t)$ in flow with a vertical velocity $w(t)$ has been presented by Tchen [1947]. Since there is no interference from neighboring particles, we may refer to this as free settling. We will assume that w_s and w are independent of time (see however Exercise 7.1). When all particles fall at a constant velocity, their concentration remains unchanged and the settling flux (product of velocity and concentration) remains constant.

Consider a non-deformable spherical particle of radius r_p in a fluid of density ρ and viscosity η (Fig. 7.1). The fluid is flowing at a low and constant upward-velocity W_v. At this velocity creeping flow occurs, as defined by the condition that the particle-based Reynolds number $Re_p = 2r_p W_v \rho / \eta$ is less than about 0.1. In this (viscous) flow range, the shear stress and pressure distributions in spherical coordinates (r,θ,φ) are given by

$$w_r = W_v \left(1 - \frac{3}{2}\tilde{R} + \frac{1}{2}\tilde{R}^3 \right)\cos\theta \qquad (7.1)$$

$$w_\theta = -W_v \left(1 - \frac{3}{4}\tilde{R} - \frac{1}{4}\tilde{R}^3 \right)\sin\theta \qquad (7.2)$$

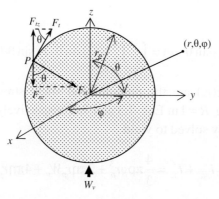

Figure 7.1. Stationary spherical particle in upward flow.

$$\tau = \frac{3}{2}\frac{\eta W_v}{r_p}\tilde{R}^4 \sin\theta \tag{7.3}$$

$$p = -\rho g z - \frac{3}{2}\frac{\eta W_v}{r_p^2}\tilde{R}^2 \cos\theta \tag{7.4}$$

where w_r and w_θ are the vertical flow velocity components in the r- and θ-directions, respectively, $\tilde{R} = r_p / r$, and p is the (static plus dynamic) relative pressure, *i.e.*, the difference between the actual pressure at a point on the particle surface and the ambient pressure far away from the particle. At the surface the pressure force is normal and the shear force is tangential. We are interested in calculating the total force exerted by the fluid on the particle in the z-direction. This force is the sum of F_{nz} and F_{tz} due to pressure and shear, respectively.

Consider an elemental surface area dA at point P on which F_{nz} and F_{tz} act. From geometry

$$dA = r_p^2 \sin\theta d\theta d\varphi \tag{7.5}$$

Therefore, we obtain the following volume integrals

$$F_{nz} = \int_V -p(r_p)\cos\theta \cdot dA = \int_0^{2\pi}\int_0^{\pi} -p(r_p)\cos\theta \cdot r_p^2 \sin\theta d\theta d\varphi \tag{7.6}$$

and

$$F_{tz} = \int_V \tau(r_p)\sin\theta \cdot dA = \int_0^{2\pi}\int_0^{\pi}\tau(r_p)\sin\theta \cdot r_p^2\sin\theta d\theta d\varphi \qquad (7.7)$$

where $p(r_p)$ and $\tau(r_p)$ are the pressure and shear stress at the surface obtained by setting $\tilde{R}=1$ in Eqs. 7.3 and 7.4, respectively. The above two equations are easily solved to yield

$$F_T = F_{nz} + F_{tz} = \frac{4}{3}\pi\rho g r_p^3 + 2\pi\eta r_p W_v + 4\pi\eta r_p W_v \qquad (7.8)$$

or

$$F_T = \frac{4}{3}\pi\rho g r_p^3 + 6\pi\eta r_p W_v \qquad (7.9)$$

The first term on the right hand side is the buoyancy force, and the second term is the linear or Stokes drag F_D. It is the sum of a contribution $2\pi\eta r_p W_v$ from the pressure distribution around the sphere and $4\pi\eta r_p W_v$ from shear.

Resetting the reference framework by considering a particle of diameter d_p ($=2r_p$) falling with a velocity w_s in a quiescent fluid, the drag force becomes

$$F_D = 3\pi\eta d_p(w_s - w) \qquad (7.10)$$

in which W_v has been replaced by $w_s - w$, where w is the fluid velocity component in the z-direction. The force due to gravity is represented by the buoyant weight

$$F_g = \frac{1}{6}\pi d_p^3 g\Delta\rho \qquad (7.11)$$

where $\Delta\rho = \rho_s - \rho$ and ρ_s is the particle density. The terminal, *i.e.*, final steady-state, settling velocity is then obtained from the force balance $F_g = F_D$, *i.e.*,

$$3\pi v d_p w_s \left(1 - \frac{w}{w_s}\right) = \frac{1}{6}\pi d_p^3 g \frac{\Delta\rho}{\rho} \qquad (7.12)$$

where $v = \eta/\rho$ is the kinematic viscosity of the fluid. Strictly speaking, the balance between linear drag and buoyant weight is applicable only at values of the fall Reynolds number $Re_p = w_s d_p / v < 0.01$ in purely laminar flow (a more stringent condition than the threshold of 0.1 for creeping flow). For a more general force balance we must replace linear drag by quadratic drag (Chapter 2)

$$\frac{1}{2}C_D \frac{\pi d_p^2}{4}\rho w_s^2 \left(1 - \frac{w}{w_s}\right)\left|1 - \frac{w}{w_s}\right| = \frac{1}{6}\pi d_p^3 g \Delta\rho \qquad (7.13)$$

In most cases it is reasonable to assume $w_s \gg w$ and ignore w. An exception (as we noted in Chapter 6) is energetic tidal environments in which the two velocities are of comparable magnitudes.

Example 7.1: In an estuary assume linearly decreasing vertical velocity with depth below the water surface

$$w = \frac{z}{h}\frac{\partial h}{\partial t} \qquad (7.14)$$

where h is the water depth and z is the coordinate for the vertical elevation. Based on Eq. 7.14, how significant can be the fluid velocity?

The characteristic value of $\partial h / \partial t$ is the ratio of tidal range to tidal period, which can be taken as 1×10^{-4} m s^{-1}. Let us consider mid-depth, *i.e.*, $z/h = 0.5$, at which $w = 0.5 \times 10^{-4}$ m s^{-1}. This value is comparable to the settling velocity of small, *e.g.*, 5–10 μm, flocs.

Neglecting w in Eq. 7.12 yields

$$w_s = \frac{g\Delta\rho}{18\eta}d_p^2 = \frac{g(\rho_s - \rho_w)}{18\eta}d_p^2 = \frac{g(\rho_s - \rho_w)}{18\rho_w v}d_p^2$$

$$\qquad (7.15)$$

$$= \frac{g(s-1)}{18v}d_p^2 = \frac{g}{18v}\left(\frac{\gamma_s - \gamma_w}{\gamma_w}\right)d_p^2$$

which is Stokes law in its commonly used form for very slowly settling particles in an unbounded fluid (water).[a,b,c] Note that $s = \rho_s/\rho_w$, $\gamma_s = g\rho_s$ and $\gamma_w = g\rho_w$. From Eq. 7.13 we obtain a more general form of Eq. 7.15

$$w_s = \sqrt{\frac{4g}{3C_D}\frac{\Delta\rho}{\rho}}\sqrt{d_p} \qquad (7.16)$$

When Eq. 7.15 or Eq. 7.16 is applied to individual particles, the excess density relative to water density ρ_w is obtained from $\Delta\rho = \rho_s - \rho_w = (\rho_s/C)(\rho - \rho_w) = \phi_v^{-1}(\rho - \rho_w)$, in which C is the suspended sediment concentration (dry mass per unit volume), ρ is the wet bulk density of the suspension (Eq. 4.46) and ϕ_v is the solids volume fraction. When applied to flocs, ρ_s must be replaced by ρ_f from Eq. 4.43.[d]

When the sediment is flocculated, Eqs. 7.15 and 7.16 may be used in the free settling range in which the suspended sediment concentration

---

[a] When Eq. 7.15 is used with measured values of w_s and $\Delta\rho$, the resulting value of d_p is called Stokes diameter.

[b] Equation 7.15 can be restated as $18w_s\nu/g'd_p^2 = 1$. The ratio $w_s\nu/g'd_p^2$ is known as Stokes number [Liggett, 1994].

[c] When the particles are non-spherical, Eq. 7.15 must be modified. For instance for prolate spheroids (shaped like an airship or dirigible) the equation is

$$w_s = \frac{\left(1 - \dfrac{1}{a_R}\right)ga_s^2\Delta\rho}{\left(1 - \dfrac{A_r}{a_R}\right)18\eta} \qquad (F7.1)$$

in which the aspect ratio $a_R = a_s/\sqrt{b_s c_s}$, a_s is the longest dimension of the particle, b_s is the intermediate dimension and c_s is the smallest dimension along three mutually perpendicular axes. The constant A_r depends on settling orientation; $A_r = 0.2$ for parallel settling and 0.4 for perpendicular settling [Lerman *et al.*, 1974].

[d] The symbol ρ is used to denote the density of any fluid including water, and also the wet bulk density of a sediment suspension or a bed. When it is necessary to differentiate between water density and suspension or bed density, water density has been designated as ρ_w.

$C < C_f$, limit below which the inter-particle collision frequency is considered too low to affect floc properties. Krone [1962] estimated C_f to be 0.3 kg m^{-3} from measurements of the settling velocity of cohesive sediment from San Francisco Bay over a wide range of concentrations. When C was above C_f the settling velocity had a recognizable dependence on concentration. As mentioned earlier this is flocculation settling. Careful measurements by Ozturgut and Lavelle [1986] using North Pacific clay in saltwater indicated that the transition between practically free and flocculation settling can be gradual, and may occur at a concentration as low as 0.05 kg m^{-3} (Fig. 7.2). In other words, $C = C_f$ is essentially a notional (or operational) boundary.

7.2.1.2 *Effect of higher Reynolds number*

Comparing Eq. 7.15 with Eq. 7.16 we find that

$$C_D = \frac{24}{Re_p} \qquad (7.17)$$

The dependence of the drag coefficient C_D on the Reynolds number Re_p not only embodies the effect of flow state (laminar or turbulent; the latter including inertia effects), but also the effect of fluid temperature. The effect of a change in the kinematic viscosity v on the settling velocity due

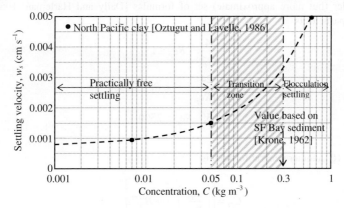

Figure 7.2. Variation of median settling velocity with suspended sediment concentration below 1 kg m^{-3}.

to a change in temperature can be significant depending on the temperature change. For example, for water at 30°C, $v = 0.8\times10^{-6}$ m^2 s^{-1} and at the lower temperature of 3.5°C the value is 1.6×10^{-6} m^2 s^{-1}. Thus the drop in temperature doubles the drag coefficient and therefore flow resistance during particle fall.

The applicability of Eq. 7.15 becomes increasingly tenuous as Re_p increases above 0.01. Beyond $Re_p \approx 1$ this equation cannot be used because the condition for creeping motion is greatly exceeded once the effect of drag associated with inertia becomes non-negligible. Therefore, for higher values of Re_p, Eq. 7.16 is used by introducing empirical corrections for the drag coefficient in Eq. 7.17. This results in much higher values of C_D. The revised expression for C_D typically has the form

$$C_D = \frac{24}{Re_p}(1+X) \qquad (7.18)$$

where X is the correction factor. Early corrections introduced by Oseen [1927] and Goldstein [1929] are mentioned in Exercise 7.1. Clift *et al.* [1978] established the following ranges of C_D for spherical particles[e]

---

[e] A simpler (but more approximate) set of formulas [Daily and Harleman, 1966] for spherical particles is

$$C_D = \begin{cases} \dfrac{24}{Re_p}; & Re_p \leq 1 \\[2ex] \dfrac{24}{\sqrt{Re_p}}; & 1 < Re_p \leq 2300 \\[2ex] 0.5; & Re_p > 2300 \end{cases} \qquad (F7.2)$$

For $Re_p > 10^5$ another general formula is

$$C_D = \left[\left(\frac{A_D}{Re_p}\right)^{1/m} + B_D^{1/m}\right]^m \qquad (F7.3)$$

Experimental values of coefficients A_D and B_D and exponent m are given in Table F7.1.

$$X = \begin{cases} \dfrac{3}{16} Re_p \; ; & Re_p \leq 0.01 \\[2mm] 0.1315 Re_p^{(0.82-0.05\Upsilon)} \; ; & 0.01 < Re_p \leq 20 \\[2mm] 0.1935 Re_p^{0.6305} \; ; & 20 < Re_p \leq 260 \end{cases} \tag{7.19}$$

$$C_D = 10^{(1.6435-1.1242\Upsilon+0.1558\Upsilon^2)} \; ; \quad 260 < Re_p \leq 1500$$

where $\Upsilon = \log Re_p$. In the highest range of Re_p (260 to 1,500) the formula is wholly empirical. Note also that equations for C_D developed by different investigators are not always compatible; see Exercise 7.2.

A limitation of using drag coefficient formulas is that flocs tend to be non-spherical, and deviation from sphericity, as might occur with increasing floc size, increases the drag coefficient. Thus, for reliable estimates of C_D, *in situ* measurements of the settling velocity of flocs in the natural environment are preferable over laboratory tests.

The excess floc density $\Delta \rho_f = \rho_f - \rho$ varies with the floc diameter according to Eq. 4.71. Introduction of this variation (with d_{50} more generally characterized by the individual or dispersed particle diameter d_s, and $\psi = 1$) and Eq. 7.18 (in terms of floc diameter d_f) into Eq. 7.16 yields

$$w_s = \frac{g \Delta \rho_s d_s^{3-D_f}}{18\eta(1+X)} d_f^{D_f-1} \tag{7.20}$$

Table F7.1. Experimental values of coefficients in Eq. F7.3.

Investigator	A_D	B_D	m
Dallavalle [1948][a]	24.0	0.40	2.0
Julien [1995]	24.0	1.50	1.0
Soulsby [1997]	26.4	1.27	1.0
Cheng [1997a]	32.0	1.00	1.5

[a] For spherical particles.
Source: Camenen [2007].

where the subscript s denotes dispersed particles and D_f is the fractal dimension. Winterwerp [1998] chose the following relationship for X

$$X = 0.15 Re_p^{0.687} \qquad (7.21)$$

which is applicable for $Re_p < 800$ [Schiller and Naumann, 1933] and is similar to the equation for X in the Re_p range of 20 to 260 in Eq. 7.19. Thus,

$$w_s = \frac{g \Delta \rho_s d_s^{3-D_f}}{18 \eta (1 + 0.15 Re_p^{0.687})} d_f^{D_f - 1} \qquad (7.22)$$

Example 7.2: You are given: $d_s = 4$ μm, $\rho_s = 2,650$ kg m^{-3}, $\rho_w = 1,020$ kg m^{-3} and $\eta = 10^{-3}$ Pa.s. Plot Eq. 7.22 for $D_f = 1.7$, 2.0 and 2.3.

The three curves are plotted in Fig. 7.3, after iteratively solving for w_s (also contained in Re_p). Data points are from the Chesapeake Bay, the Tamar estuary in UK and the Ems estuary in the Netherlands. Some of the Ems data were obtained in a laboratory column, while others were derived from field tests using the Video In Situ (VIS) camera [Gibbs, 1985; van Leussen, 1994; Fennessy et al., 1994].

As floc size increases, the settling velocity increases at a decreasing rate because floc density decreases. The Stokes solution does not display this effect.

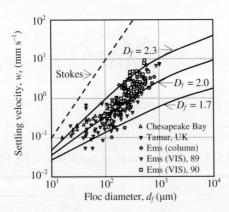

Figure 7.3. Variation of settling velocity with floc size (data points from Winterwerp [1998]).

7.2.1.3 *Settling velocity distribution*

Settling velocities are commonly represented in terms of the frequency distribution function $f(w_s)$, or by the cumulative distribution function

$$F(w_s) = \int_0^{w_s} f(w_s)\, dw_s \qquad (7.23)$$

Figure 7.4 shows cumulative distributions of the settling velocity of several marine and some non-marine sediment samples. The corresponding distributions for the dispersed sediments are shown in Fig. 7.5. As expected the settling velocities of flocs are considerably larger than those of dispersed particles. We noted in Chapters 2 and 4 that during their growth flocs tend to capture smaller particles, thus narrowing the size and settling velocity distributions, especially their lower tails.

An empirical equation for data on $F(w_s)$ is based on the frequency distribution

$$f(w_s) = \frac{\lambda_\Gamma^{r_\Gamma} w_s^{r_\Gamma - 1} e^{-\lambda_\Gamma w_s}}{\Gamma(r_\Gamma)}; \quad 0 \leq w_s < \infty \qquad (7.24)$$

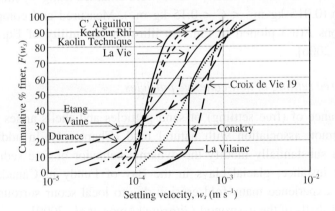

Figure 7.4. Cumulative distributions of the settling velocity of non-dispersed (*i.e.*, without addition of a peptizer) sediments (from Migniot [1968]).

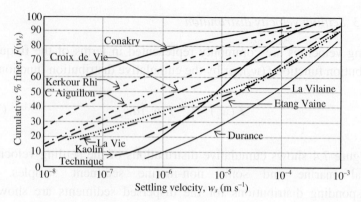

Figure 7.5. Cumulative distributions of the settling velocity of dispersed sediments (from Migniot [1968]).

where $\Gamma(r_\Gamma)$ is the gamma function. For positive integers, $\Gamma(r_\Gamma) = (r_\Gamma - 1)!$. The coefficients r_Γ and λ_Γ are

$$r_\Gamma = \frac{\bar{w}_s^2}{\sigma_{w_s}^2}, \qquad \lambda_\Gamma = \frac{r_\Gamma}{\bar{w}_s} \qquad (7.25)$$

where $\bar{w}_s$ is the mean and σ_{w_s} the standard deviation of w_s values. From settling velocity measurements by Burt [1986] in the Thames River, England, $\bar{w}_s = 2.7$ mm s^{-1} and $r_\Gamma = 0.65$ were obtained using $\bar{w}_s$ (m s^{-1}) = $0.025 C^{0.92}$ (0.015 kg m^{-3} < C < 0.15 kg m^{-3}). Measured and computed distributions $F(w_s)$ plotted in Fig. 7.6 illustrate the utility of Eq. 7.24 [Sanchez, 2006].

7.2.1.4 *Free settling of non-cohesive particles*

The relevance of (free settling) of sand particles and shells arises from their common association with natural mud (Chapter 1). In addition, shells can substantially modify the dynamic behavior of fine sediment beds. For instance, glacial clays in the Bay of Fundy in Canada are known to experience many-fold erosion due to local scour surrounding embedded shells of the gastropod *Littorina* [Amos *et al.*, 2000].

Free settling velocities of quartz sand spheres in water at 20°C as a function of diameter are given in Fig. 7.7. For diameters exceeding about 0.1 mm corresponding to fall Reynolds number $Re_p \approx 1$, the deviation

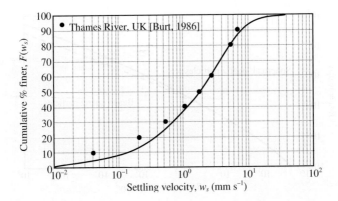

Figure 7.6. Measurement-based [Burt, 1986] and analytic cumulative frequency distributions of settling velocity of mud from the Thames River, England (from Sanchez [2006]).

between the curve based on measurements and Stokes law is significant, and increases with the diameter [Graf, 1971]. For quartz beach-sand an empirical (dashed) curve is plotted in Fig. 7.8 in terms of Re_p against Buoyancy Index BI defined as

$$BI = \frac{g\left(\dfrac{\rho_s - \rho}{\rho}\right)d_p^3}{\nu^2} = \frac{g'd_p^3}{\nu^2} \qquad (7.26)$$

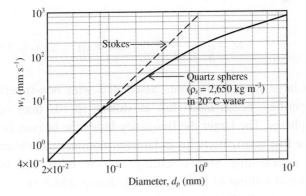

Figure 7.7. Settling velocity of quartz spheres in 20°C water.

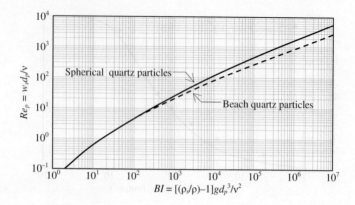

Figure 7.8. Variation of settling velocity with Buoyancy Index for spheres and natural particles [Hallermeier, 1981].

where $g' = g(\rho_s - \rho)/\rho$ is reduced gravity. The solid curve represents spherical (quartz) particles [Hallermeier, 1981].

For common beach sands three segments of the dashed curve are approximated by the expressions

$$w_s = \frac{g'd_p^2}{18\,v}\;; \qquad BI < 39 \qquad\qquad (7.27a)$$

$$w_s = \frac{g'^{0.7}d_p^{1.1}}{6v^{0.4}}\;; \qquad 39 < BI < 10^4 \qquad (7.27b)$$

$$w_s = \sqrt{\frac{g'd_p}{0.91}}\;; \qquad BI > 10^4 \qquad\qquad (7.27c)$$

Due to large drag coefficients associated with non-sphericity and surface asperities, beach-sands fall at slower speeds than equivalent (smooth) spheres. The larger the Buoyancy Index the heavier is the sand particle. For fine quartz particles in laminar motion, w_s is obtained from Eq. 7.27a representing Stokes settling. At the other end Eq. 7.27c represents rapid settling of heavy particles during which fluid motion is turbulent. Equation 7.27b is intermediate.

Cheng [1997a] proposed the formula

$$w_s = \frac{v}{d_p} \left(\sqrt{25 + 1.2 d_{p*}^2} - 5 \right)^{1.5} \tag{7.28}$$

where

$$d_{p*} = \left(\frac{g'}{v^2} \right)^{1/3} d_p \tag{7.29}$$

A limitation common to Eqs. 7.26 through 7.29, all of them empirical, is that they do not specify the degree of sphericity of particles, which cannot be ignored if for instance the particles are flat. For increasingly non-spherical particles the Corey shape factor SF is defined as

$$SF = \frac{c_s}{\sqrt{a_s b_s}} \tag{7.30}$$

in which a_s is the longest, b_s the intermediate and c_s the smallest dimension of the particle along three mutually perpendicular axes [Garcia, 2008; Wu and Wang, 2006; Graf, 1971]. Thus SF is a measure of the flatness of the particle, which allows the drag coefficient C_D to be related to particle diameter as well as flatness (Fig. 7.9).[f]

For oddly shaped natural particles such as shells the Corey shape factor is not fully descriptive of the flatness effect because SF does not account for the distributions of surface area and volume of the particle [Dietrich, 1982; Jimenez and Madsen, 2003]. A simple illustration of the ambiguity inherent in SF is a cubic particle. This particle and a sphere of diameter equal to the length of the cube have the same SF, namely 1.00, but very different settling velocities. Thus a correction that takes into account the shape of the surface and maintains a shape parameter equal

---

[f] For non-spherical natural, as opposed to crushed, particles Swamee and Ojha [1991] recommend the drag coefficient formula

$$C_D = \frac{128}{(1 + 4.5 \cdot SF^{0.35}) Re_p^{0.8}} \tag{F7.4}$$

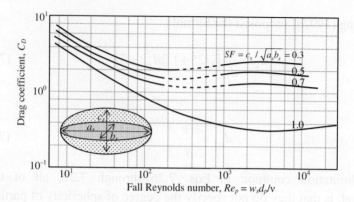

Figure 7.9. Variation of drag coefficient with fall Reynolds number and Corey shape factor (based on Albertson [1953]).

to 1.00 for spheres but not cubes is required. This correction is a multiplier of SF specified by the ratio d_n/d_A in which d_n is the nominal diameter, *i.e.*, diameter of a sphere having the same volume as the particle, and d_A is the diameter of a sphere having the same surface area as the particle. Given particle volume V_p and surface area A_A

$$\frac{d_n}{d_A} = \frac{\left(\dfrac{6V_p}{\pi}\right)^{1/3}}{\sqrt{\dfrac{A_A}{\pi}}} \qquad (7.31)$$

Thus the modified shape factor is

$$SF_A = SF\frac{d_n}{d_A} = \frac{c_s}{\sqrt{a_s b_s}}\frac{\left(\dfrac{6V_p}{\pi}\right)^{1/3}}{\sqrt{\dfrac{A_A}{\pi}}} \qquad (7.32)$$

As with the Corey shape factor, SF_A approaches unity with increasing sphericity. The corresponding fall Reynolds number is $Re_A = w_s d_A / v$ [Alger and Simons, 1968; Mehta *et al.*, 1980].

Figure 7.10 defines a_s, b_s and c_s dimensions of a typical bivalve shell of mollusks. As an example, patches of the sandy beach at Sanibel Island on the east coast of Florida are often covered by such shells [Lee, 1978]. The shortest length c_s is the height of shell measured from the base plane to the top of the convex surface. The base is approximated by an ellipse with major axis a' and minor axis b'. For an ellipse with the same area A as the base,

$$A = \frac{1}{4} \alpha_{s1} \pi a' b' \tag{7.33}$$

in which α_{s1} is an area shape factor. The longest axis a_s of the base may be taken to be equal to the axis a' of the ellipse. The middle axis b_s usually passes through the umbo region, where the two shell halves are hinged when they are part of the mollusk. Therefore, selecting $b' = \alpha_{s2} b_s$, in which α_{s2} is another shape factor we obtain

$$A = \frac{1}{4} \alpha_{s1} \alpha_{s2} \pi a_s b_s \tag{7.34}$$

Bivalve shells of the same species tend to have shapes that are approximately self-similar [Mehta and Christensen, 1977]. Accordingly, as their surface is typically flat ($c_s << a_s$ and b_s), one may relate the base area A to the surface area A_A by

$$A = \frac{1}{2} \alpha_{s3} A_A \tag{7.35}$$

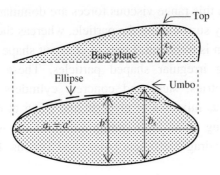

Figure 7.10. Three mutually perpendicular axes of a half shell.

in which α_{s3} is a proportionality constant. The factor 1/2 is introduced because A_A incorporates both inner and outer surface areas of the shell, whereas A is the area of only one side of the base (if it were a solid surface). In other words, if the shell were to become flatter, *i.e.*, as $c_s \to 0$, one would have $A_A \to 2A$, which is a reasonable approximation because the thickness of shell is typically small, amounting to an area of only about 5% of A_A in many cases.

The drag coefficient C_D, the shape parameter SF_A and the diameter d_A of bivalve shells are respectively expressed as

$$C_D = \frac{8g\left(\dfrac{\rho_s}{\rho} - 1\right)}{\alpha_{s1}\alpha_{s2}\pi} \frac{V_p}{a_s b_s w_s^2} \qquad (7.36)$$

$$SF_A = 0.39\pi^{1/3}\alpha_{s4}^{1/3}\frac{c_s}{V_p^{1/3}} \qquad (7.37)$$

$$d_A = 0.71\sqrt{\alpha_{s4}}\sqrt{a_s b_s} \qquad (7.38)$$

in which $\alpha_{s4} = \alpha_{s1}\alpha_{s2}/\alpha_{s3}$.

Shells of three bivalve species including Coquina clam (*donax variabilis*), Cross-barred chione (*chione cancellata*) and Ponderous ark (*noetia ponderosa*) were tested in a settling column containing fresh water. In Fig. 7.11 the zone of the drag coefficient values for these shells occurs in the Reynolds number Re_A range of 700 to 6,000. For a freely falling particle in this range viscous forces are dominant in the boundary layer immediately surrounding the particle, whereas the wake region due to flow separation is turbulent. Lines of constant shape factor are derived from tests using irregular shaped particles. These particles include machined geometries, steel and concrete cylinders and gravel-sized particles in various mixtures of glycerin and water at room temperature. Flat particles bring about high energy losses in the wake embodied in high values of the drag coefficient [Alger and Simons, 1968].

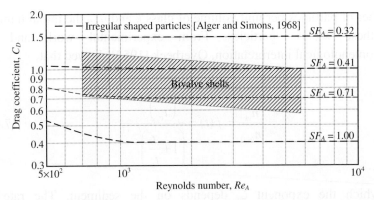

Figure 7.11. Dependence of drag coefficient on fall Reynolds number for tested bivalve shells (grey box) and other particles (from Mehta *et al.* [1980]).

7.2.2 *Effect of aggregation processes*

For the settling velocity of suspended flocs at concentrations in excess of C_f, aggregation processes (floc growth and breakup) must be accounted for. Let there be n_{v0} flocs and (perhaps) primary particles in a unit volume of suspension at time $t = 0$. As aggregation proceeds at low shear rates conducive to floc growth, the number of particles $n_v(t)$ at a later time will be reduced, *i.e.*,

$$n_v(t) = n_{v0} \cdot f(t) \qquad (7.39)$$

where the decay function $f(t) \leq 1$. From Eq. 4.94, $f(t)$ may be taken as

$$f(t) = \frac{1}{1 + \dfrac{t}{t_c}} \qquad (7.40)$$

where t_c is the characteristic time of aggregation. If we start with primary particles, floc growth would involve collisions only between these particles at the very beginning. Once flocs are formed in sufficiently large numbers, their collisions with primary particles must be considered along with those between flocs. It follows that, although Eq. 7.39 holds for the total number of particles at any time in the unit volume, the function $f(t)$ varies with particle class. Let n_{vk} be the instantaneous number of particles per unit volume of class k and n_{vk0} the initial number of these k-class particles. From experimental work and its phenomenological interpretation, Overbeek [1952] showed that

$$\frac{n_{vk}}{n_{vk0}} = f(t) = \frac{\left(\dfrac{t}{t_c}\right)^{\zeta_a - 1}}{\left(1 + \dfrac{t}{t_c}\right)^{\zeta_a + 1}} \tag{7.41}$$

in which the exponent ζ_a depends on the sediment. The rate of aggregation characterized by t_c is such that when $t_c \ll t$, i.e., $t/t_c \gg 1$, aggregation is rapid, and when $t/t_c \ll 1$, it is slow. Measurements included in Tables 7.2 and 7.3 indicate rapid aggregation of a kaolinite suspension and slow aggregation of a sol of selenium particles coagulated by potassium chloride (KCl).

Krone [1962] assumed that $f(t)$ in Eq. 7.41 can be substituted into Eq. 7.39, which includes all, not just k-class, particles. For rapid aggregation of present interest,

$$1 + \frac{t}{t_c} \approx \frac{t}{t_c} \tag{7.42}$$

Table 7.2. Aggregation of kaolinite suspension.

Time (s)	No. of particles per $cm^3 \times 10^{-8}$
0	5.00
105	3.90
180	3.18
255	2.92
335	2.52
420	2.00
510	1.92
600	1.75
1020	1.54
2340	1.15

Source: Overbeek [1952].

Table 7.3. Aggregation of selenium sol with KCl.

Time (h)	No. of particles per $cm^3 \times 10^{-8}$
0	29.7
0.66	20.90
4.25	19.10
19	14.40
43	10.70
73	7.70
167	6.45

Source: Overbeek [1952].

Therefore, after conveniently dropping subscript k in Eq. 7.41, we obtain

$$\frac{n_v}{n_{v0}} = \frac{1}{f(t)} = \left(\frac{t_c}{t}\right)^2 \qquad (7.43)$$

Next we note that the ratio n_v/n_{v0} is proportional to floc volume $\pi d_f^3 / 6$. Therefore,

$$d_f^2 \propto \left(\frac{n_{v0}}{n_v}\right)^{2/3} \propto \left(\frac{t}{t_c}\right)^{4/3} \qquad (7.44)$$

Since from Stokes law $w_s \propto d_f^2$ (Eq. 7.15 with $d_p = d_f$), it follows that

$$w_s \propto \left(\frac{t}{t_c}\right)^{4/3} \propto \left(\frac{1}{t_c}\right)^{4/3} t^{4/3} \qquad (7.45)$$

The quantity $1/t_c$ is the collision frequency which can be expected to vary with concentration C. Assigning a fixed value to time t at which the degree of aggregation is to be assessed, we may interpret Eq. 7.45 as

$$w_s \propto C^{4/3} \qquad (7.46)$$

or, in general,

$$w_s = a'_w C^{n_w} = a''_w \phi_v^{n_w} \tag{7.47}$$

where a'_w and $a''_w = a'_w \rho_s^{n_w}$ are velocity scaling coefficients dependent on the sediment and $n_w = 4/3 = 1.33.$[g] As we will see, a'_w and n_w vary with flow shear and the fractal dimension D_f of flocs. Equation 7.47 is applicable in the flocculation settling range. Confirmatory tests were conducted by Krone [1962] using sediment from the San Francisco Bay in graduated cylinders and in a flume under turbulent flow (Fig. 7.12). In both environments n_w was shown to be 1.33 for that sediment. Subsequently, numerous *in situ* measurements in the Thames estuary in England yielded $n_w = 1.37$ as the most reliable value, which is close to 1.33 [Burt, 1986]. More field examples of the dependence of w_s on C are found in Dyer [1986] and Whitehouse *et al.* [2000a].

7.2.3 *Hindered settling*

When the suspended sediment concentration exceeds a value C_h defining the upper limit of flocculation settling (Eq. 7.47), which may range between ~1 and ~10 kg m^{-3} depending on the composition of

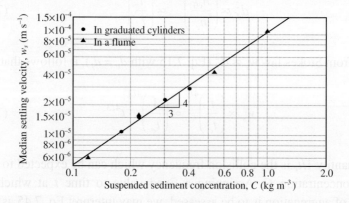

Figure 7.12. Flocculation settling of San Francisco Bay sediment (from Krone [1962]).

---

[g] Strictly speaking, $1/t_c$ depends on the flocs volume fraction ϕ_{vf}, rather than C or the solids volume fraction $\phi_v = [\phi_{vf}(\rho_f - \rho_w)/(\rho_s - \rho_w)]$. Equation 7.47 was used by Krone [1962] because C and therefore ϕ_v can be measured readily compared to ϕ_{vf}.

sediment[h], w_s reduces rapidly with increasing C. This reduction is due to the decreasing rate at which the settling slurry dewaters as its permeability decreases with time. We will refer to the settling velocity from Eq. 7.47 at C_h as w_{sm} (Fig. 7.13). Assuming that w_{sm} can be calculated from Stokes law (Eq. 7.15), the hindered settling velocity w_s at a given concentration can be represented by the product $K_w w_{sm}$, where K_w is a retardation factor applicable to Stokes settling. From settling column experiments, Richardson and Zaki [1954] found $K_w = (1 - \phi_{vf})^{4.65}$ (see also Eq. 7.101). Thus,

$$w_s = w_{sm} K_w; \quad K_w = \tilde{w}_s = (1 - \phi_{vf})^{4.65} \quad (7.48)$$

The factor K_w can be attributed to the effects of increased viscosity, particle buoyancy and return-flow of fluid due to continuity as particles settle [Winterwerp, 2002]. In order to account for these effects, K_w has been revised as

$$w_s = w_{sm} K_w; \quad K_w = \tilde{w}_s = \frac{(1 - \phi_{vs})(1 - \tilde{\phi}_v)^m}{(1 + 2.5\tilde{\phi}_v)} \quad (7.49)$$

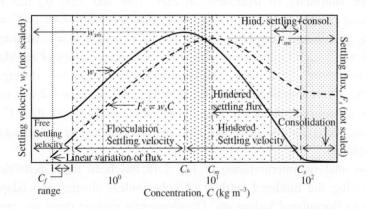

Figure 7.13. Variations of settling velocity and settling flux with concentration. Concentration values are included for illustrative purposes.

---

[h] A trend associated with organic matter appears to be that the higher its content the lower is the value of C_h, *i.e.*, sooner commencement of hindered settling with increasing concentration.

in which $\phi_v = C/\rho_s$, $\phi_{vs} = C_s/\rho_s$, C_s is the space-filling concentration (Chapter 5) and

$$\tilde{\phi}_v = \frac{C}{C_s} \qquad (7.50)$$

The fraction $(1 - \phi_{vs})$ accounts for a decrease in the settling velocity due to return-flow through the slurry pores and micro-channels. The fraction $(1 - \tilde{\phi}_v)^m$ represents the buoyancy effect which acts against the weight of falling particles, and $(1 + 2.5\tilde{\phi}_v)$ accounts for an increase in drag force on the particle (in opposition to the weight of particles) due to the viscosity of suspension being higher than water.

From the settling of a kaolinite suspension, Sampath [2009] crudely estimated K_w by measuring the return-flow velocity using Rhodamine dye as a tracer. Let the suspension concentration C be equal to 95 kg m^{-3}, $\rho_s = 2,650$ kg m^{-3} and $C_s = 156$ kg m^{-3}. We then have $\phi_{vs} = 0.059$ and $\tilde{\phi}_v = 0.61$. Substitution of these values into Eq. 7.49 (with $m = 1$ assumed) yields $K_w = 0.145$, which indicates a substantial reduction in the Stokes free-settling velocity.

For simplicity of treatment, in Eqs. 7.49 and 7.50 ϕ_{vf} has been replaced by ϕ_v. The upper-limit concentration for Eq. 7.49 is $C = C_s$, i.e., $\tilde{\phi}_v = 1$. Winterwerp [2002] selected the exponent $m = 1$. Settling of slurries in which hindrance is high can yield higher values of m [Maude and Whitmore, 1958]. For simulating the density profiles during the settling of a fine-grained slurry tested by Kynch [1952], $m = 2$ was found to be appropriate [Letter, 2009]. At concentrations larger than C_s the settling behavior is governed by the mechanics of consolidation (Chapter 8).

An analytic interpretation of Eq. 7.48 has been made possible by examining the hindered settling of cohesionless slurries and adapting them to flocculated sediments.[i] Developments include those by, among

---

[i] Although the exponent 4.65 is found to be reasonable for fine sediments, experimental data on the settling of cohesionless particles indicates that the exponent decreases with increasing particle diameter. For 1 mm particles a range of values between about 2.5 and 3 has been reported [Baldock *et al.*, 2004].

others, Richardson and Zaki [1954] and Cheng [1997b]. The former is summarized here.

As we have noted, the fall of densely spaced particles in a suspension is complicated by a variety of effects. According to the description in Fig. 7.14 they include: (1) upward entrainment of water through preferential micro-channels governed by continuity, (2) downward entrainment of water dragged by falling flocs, (3) interfering effects of mucus forming macro-flocs, and (4) the so-called sinking ship effect in which a large ship (macrofloc) draws so much surface water volume with it as it sinks that life-boats (microflocs) and sometimes even larger boats caught up in this volume sink as well. This effect can be treated by accounting for the settling fluxes of particles of sizes both smaller and larger than the size-class being modeled [Letter, 2009]. A point to note also is that the return flow mentioned in connection with Eq. 7.49 is essentially the net effect of opposing movements of particles and water indicated in Fig. 7.14.

The above description of falling particles must be simplified in any analytic treatment of the settling of a dense suspension. Since sediment concentration depends on particle arrangement in the suspending medium, we will consider two simple configurations. In Configuration *I* (Fig. 7.15a) particles occur in parallel planes. In each plane a particle of radius r_p (= $0.5d_p$) is at the center of each hexagon of the carrier fluid. The distance between any two neighboring particles in the horizontal

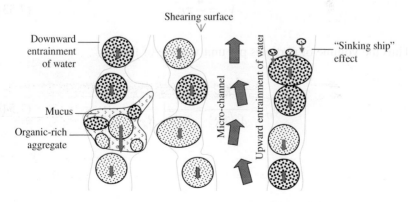

Figure 7.14. Fluid movements associated with falling flocs (adapted from Wolanski [2007]).

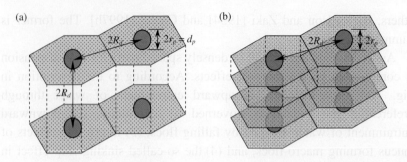

Figure 7.15. Idealized distributions of falling particles in a dense suspension: (a) Configuration *I* (two prisms are shown); (b) Configuration *II* (adapted from Richardson and Zaki [1954]).

layer is $2R_d$, which is also taken to be the distance separating two consecutive planes.

In Fig. 7.15 the area of each hexagon is

$$A_h = 2\sqrt{3}R_d^2 \tag{7.51}$$

Therefore the volume of each prism is

$$V_h = 2R_d A_h = 4\sqrt{3}R_d^3 \tag{7.52}$$

The volume V_p of each particle is

$$V_p = \frac{4}{3}\pi r_p^3 \tag{7.53}$$

Thus the volume of particles per unit volume of fluid is[j]

$$\phi_{vf} = \frac{\frac{4}{3}\pi r_p^3}{4\sqrt{3}R_d^3} = \frac{\pi}{3\sqrt{3}}\left(\frac{r_p}{R_d}\right)^3 \tag{7.54}$$

---

[j] Particles are conveniently treated as flocs even though, as mentioned, the analysis of Richardson and Zaki [1954] does not refer to flocs.

and therefore

$$\frac{r_p}{R_d} = 1.183 \sqrt[3]{\phi_{vf}} \qquad (7.55)$$

In Configuration *II* particles in consecutive hexagonal layers touch each other, thus offering the least resistance to vertical flow of fluid (Fig. 7.15b). Because in *II* the layers are closer than in *I*, for the same volume fraction ϕ_{vf} the spacing between particles in each layer must be greater in *II*. This means that the flow in *II* will experience milder velocity gradients, hence resistance.

In Configuration *II* the volume of each prism is

$$V_h = 2r_p A_h = 4\sqrt{3}R_d^2 r_p \qquad (7.56)$$

and

$$\frac{r_p}{R_d} = 1.286\sqrt{\phi_{vf}} \qquad (7.57)$$

In order to simplify the treatment we will replace the hexagon by a circle of the same area and equivalent radius R_{cl}

$$\pi R_{cl}^2 = 2\sqrt{3}R_d^2 \qquad (7.58)$$

or

$$R_{cl} = 1.05R_d \qquad (7.59)$$

Accordingly, in Configuration *I*

$$\frac{r_p}{R_{cl}} = \frac{1.183}{1.05}\sqrt[3]{\phi_{vf}} = 1.126\sqrt[3]{\phi_{vf}} \qquad (7.60)$$

and in Configuration *II*

$$\frac{r_p}{R_{cl}} = \frac{1.286}{1.05}\sqrt{\phi_{vf}} = 1.225\sqrt{\phi_{vf}} \qquad (7.61)$$

We will further assume that the particles are stationary and the fluid is moving upward with a velocity W_y. This is equivalent to particles falling at the rate w_s in a quiescent fluid. Let r be the radius of an elemental fluid ring around the particle (Fig. 7.16). In any radial cross-section of this ring the height dl is along a streamline, and thickness dn is along an equipotential line orthogonal to the streamline. The coordinates of the center of the cross-section are (r, h). At that point the streamline has a velocity u and is inclined at angle θ relative to the vertical. The vertical line of symmetry between two particles in the same layer is at a distance R_d from the center of the particle.

Since the fluid ring is in equilibrium along a streamline, the shear force must balance the gradient of pressure force driving the flow, *i.e.*,

$$\frac{\partial}{\partial n}\left(2\pi rdl \cdot \tau\right)dn = \frac{\partial}{\partial l}\left(2\pi rdn \cdot p\right)dl \qquad (7.62)$$

where p is the dynamic pressure. The viscous shear stress τ is

$$\tau = \eta \frac{\partial u}{\partial n} \qquad (7.63)$$

Substituting Eq. 7.63 into Eq. 7.62 and rearranging we obtain

$$2\pi\eta\frac{\partial}{\partial n}\left(dl \cdot r\frac{\partial u}{\partial n}\right)dn = 2\pi\frac{\partial}{\partial l}\left(dn \cdot rp\right)dl \qquad (7.64)$$

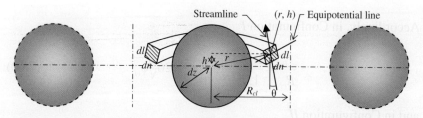

Figure 7.16. Elemental ring of fluid around a particle (from Richardson and Zaki [1954]).

which can be expanded to

$$\eta\left\{dl\cdot\frac{\partial}{\partial n}\left(r\frac{\partial u}{\partial n}\right)dn + r\frac{\partial u}{\partial n}\cdot\frac{\partial}{\partial n}(dl)dn\right\}$$

$$= dn\frac{\partial}{\partial l}(rp)dl + rp\frac{\partial}{\partial l}(dn)dl \qquad (7.65)$$

Next, we will make use of the following geometric relationships (Fig. 7.17) relative to the rates of deformation of streamlines and equipotential lines

$$\frac{\partial}{\partial n}(dl)dn = \frac{\partial\theta}{\partial l}dl\cdot dn \qquad (7.66)$$

and

$$\frac{\partial}{\partial l}(dn)dl = \frac{\partial\theta}{\partial n}dn\cdot dl \qquad (7.67)$$

Therefore

$$\eta\left\{dl\cdot\frac{\partial}{\partial n}\left(r\frac{\partial u}{\partial n}\right)dn + r\frac{\partial u}{\partial n}\cdot\frac{\partial\theta}{\partial l}dl\cdot dn\right\}$$

$$= dn\frac{\partial}{\partial l}(rp)dl + rp\frac{\partial\theta}{\partial n}dn\cdot dl \qquad (7.68)$$

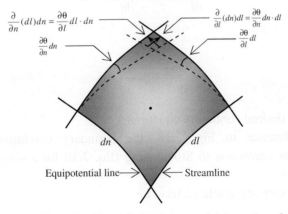

Figure 7.17. Geometric relationships defining the rates of deformation of streamlines and equipotential lines (from Richardson and Zaki [1954]).

or

$$\eta\left\{\frac{\partial}{\partial n}\left(r\frac{\partial u}{\partial n}\right)+r\frac{\partial u}{\partial n}\cdot\frac{\partial \theta}{\partial l}\right\}=\frac{\partial}{\partial l}(rp)+rp\frac{\partial \theta}{\partial n} \qquad (7.69)$$

which is the desired equation of motion.

Now, with respect to continuity of flow we note that the differential discharge dQ through the ring is

$$dQ = 2\pi ru \cdot dn \qquad (7.70)$$

As dQ between two streamlines is constant under the assumed steady-state condition, by definition

$$\frac{\partial(dQ)}{\partial l}=\frac{\partial}{\partial l}(2\pi ru)dn=0 \qquad (7.71)$$

or

$$\frac{\partial(ru)}{\partial l}dn+ru\frac{\partial}{\partial l}(dn)=0 \qquad (7.72)$$

Using Eq. 7.67

$$\frac{\partial(ru)}{\partial l}dn+ru\frac{\partial \theta}{\partial n}dn=0 \qquad (7.73)$$

or

$$\frac{\partial(ru)}{\partial l}+ru\frac{\partial \theta}{\partial n}=0 \qquad (7.74)$$

which is the desired continuity equation.

With reference to Fig. 7.18, the boundary conditions for the calculation of correction to Stokes drag (Eq. 7.10 for a single particle) are as follows.

Zero velocity at particle surface, *i.e.*,

$$u = 0 \quad \text{at} \quad r=a_r=\sqrt{r_p^2-h^2} \qquad (7.75)$$

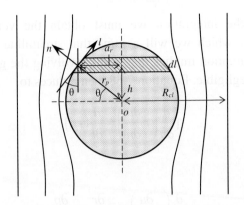

Figure 7.18. Elemental ring on particle surface.

and symmetrical flow between adjacent particles, *i.e.*,

$$\frac{\partial u}{\partial n} = 0 \quad \text{at} \quad r = R_{cl} \tag{7.76}$$

The vertical viscous drag force on the elemental ring (Fig. 7.17) is

$$dF_D = 2\pi a_r \, dl \, \eta \left(\frac{\partial u}{\partial n}\right)_{r=a_r} \cos\theta \tag{7.77}$$

where a_r is the ring radius. We note that

$$dl \cos\theta = dh \tag{7.78}$$

Therefore

$$dF_D = 2\pi a_r \eta \left(\frac{\partial u}{\partial n}\right)_{r=a_r} dh \tag{7.79}$$

Hence the total force is obtained from

$$F_D = \int_{-r_p}^{r_p} 2\pi a_r \eta \left(\frac{\partial u}{\partial n}\right)_{r=a_r} dh \tag{7.80}$$

To carry out the integration we must obtain the velocity gradient $(\partial u / \partial n)_{r=a_r}$, for which we will assume that streamline curvatures are small. This assumption implies that terms involving the gradients $\partial \theta / \partial n$ and $\partial \theta / \partial l$ are negligible. Equation 7.68 then reduces to

$$\eta \frac{\partial}{\partial n}\left(r \frac{\partial u}{\partial n} \right) = \frac{\partial}{\partial l}(rp) \qquad (7.81)$$

or

$$\eta \frac{\partial}{\partial n}\left(r \frac{\partial u}{\partial n} \right) = p \frac{\partial r}{\partial l} + r \frac{\partial p}{\partial l} \qquad (7.82)$$

However, $\partial r / \partial l = \sin \theta \approx 0$. Therefore

$$\eta \frac{\partial}{\partial n}\left(r \frac{\partial u}{\partial n} \right) = r \frac{\partial p}{\partial l} \qquad (7.83)$$

Since the pressure $p = f^{n}(r, h)$, by using the chain rule for differentiation we obtain

$$\frac{\partial p}{\partial l} = \frac{\partial p}{\partial h} \frac{\partial h}{\partial l} + \frac{\partial p}{\partial r} \frac{\partial r}{\partial l} \approx \frac{\partial p}{\partial h} \qquad (7.84)$$

because $\partial h / \partial l = \cos \theta \approx 1$ and $\partial r / \partial l = \sin \theta \approx 0$. Therefore Eq. 7.83 becomes

$$\eta \frac{\partial}{\partial n}\left(r \frac{\partial u}{\partial n} \right) = r \frac{\partial p}{\partial h} \qquad (7.85)$$

Thus by integration

$$r \frac{\partial u}{\partial n} = \frac{1}{\eta} \frac{\partial p}{\partial h} \frac{r^2}{2} + C' \qquad (7.86)$$

where the constant of integration C' can be evaluated by using the boundary condition, Eq. 7.76, to yield

$$\frac{\partial u}{\partial n} = \frac{1}{2\eta} \frac{\partial p}{\partial h} \left(r - \frac{R_{cl}^2}{r} \right) \tag{7.87}$$

As $\partial r / \partial n = \cos \theta \approx 1$, we can restate the above relationship as

$$\frac{\partial u}{\partial r} = \frac{1}{2\eta} \frac{\partial p}{\partial h} \left(r - \frac{R_{cl}^2}{r} \right) \tag{7.88}$$

Then from Eq. 7.80

$$F_D = \int_{-r_p}^{r_p} \pi a_r \frac{\partial p}{\partial h} \left(a_r - \frac{R_{cl}^2}{a_r} \right) dh \tag{7.89}$$

or, after rearrangement

$$F_D = \pi \int_{-r_p}^{r_p} R_{cl}^2 \left\{ \frac{\partial p}{\partial h} \left[\left(\frac{a_r}{R_{cl}} \right)^2 - 1 \right] \right\} dh \tag{7.90}$$

To carry out this integration it is essential to determine the pressure gradient $\partial p / \partial h$. To that end, integration of Eq. 7.83 along with the boundary condition Eq. 7.75 yields

$$u = \frac{1}{2\eta} \frac{\partial p}{\partial h} \left(\frac{r^2 - a_r^2}{2} - R_{cl}^2 \ln \frac{r}{a_r} \right) \tag{7.91}$$

Next, the mean flow velocity W_v is obtained by integrating u over the total area of the cylinder of radius R_{cl} as

$$\pi R_{cl}^2 W_v = \int_{a_r}^{R_{cl}} u \cos \theta \cdot 2\pi r \, dr \tag{7.92}$$

Setting $\cos\theta = 1$ and using Eq. 7.91 for u we obtain

$$\pi R_{cl}^2 W_v = \int_{a_r}^{R_{cl}} \frac{1}{2\eta} \frac{\partial p}{\partial h} \left(\frac{r^2 - a_r^2}{2} - R_{cl}^2 \ln \frac{r}{a_r} \right) 2\pi r \, dr \tag{7.93}$$

which can be solved for the pressure gradient

$$\frac{\partial p}{\partial h} = \frac{\eta W_v}{\frac{3}{8} R_{cl}^2 \left\{ 1 + \frac{1}{3} \left(\frac{a_r}{R_{cl}} \right)^4 - \frac{4}{3} \left(\frac{a_r}{R_{cl}} \right)^2 - \frac{4}{3} \ln \frac{R_{cl}}{a_r} \right\}} \tag{7.94}$$

With substitution of the above equation into Eq. 7.90 the following solution is obtained for the drag force

$$F_D = 3\pi\eta\, W_v d_f K_r \tag{7.95}$$

in which the floc diameter $d_f = d_p = 2r_p$ is conveniently introduced and the correction factor $K_r = K_w^{-1}$ for Stokes drag (for which $K_r = 1$) is given by

$$K_r = \frac{8}{9} \int_0^1 \left\{ \frac{\left(\frac{a_r}{R_{cl}} \right)^2 - 1}{1 + \frac{1}{3} \left(\frac{a_r}{R_{cl}} \right)^4 - \frac{4}{3} \left(\frac{a_r}{R_{cl}} \right)^2 - \frac{4}{3} \ln \left(\frac{a_r}{R_{cl}} \right)} \right\} d\left(\frac{h}{r_p} \right) \tag{7.96}$$

From geometry we have

$$\frac{a_r}{r_p} = \sqrt{1 - \left(\frac{h}{r_p} \right)^2} \tag{7.97}$$

Therefore

$$\frac{a_r}{R_{cl}} = \frac{r_p}{R_{cl}} \sqrt{1 - \left(\frac{h}{r_p} \right)^2} \tag{7.98}$$

For Configuration *I*, from Eq. 7.60

$$\frac{a_r}{R_{cl}} = 1.126 \sqrt{1 - \left(\frac{h}{r_p} \right)^2} \cdot \sqrt[3]{\phi_{vf}} \tag{7.99}$$

and for Configuration *II*, from Eq. 7.61

$$\frac{a_r}{R_{cl}} = 1.225\sqrt{1-\left(\frac{h}{r_p}\right)^2} \cdot \sqrt{\phi_{vf}} \tag{7.100}$$

Thus for both arrangements K_r (Eq. 7.96) depends on the floc volume fraction ϕ_{vf}.

Example 7.3: The following experimental data are provided by Richardson and Zaki [1954] for hindered settling of a sediment suspension:

ϕ_{vf}	K_r
0.05	1.271
0.10	1.634
0.15	2.128
0.20	2.818
0.25	3.806
0.30	5.248
0.35	7.413
0.40	10.73
0.45	15.98
0.50	25.12
0.55	40.83
0.606	75.86

Compare these data with calculations for Configurations *I* and *II*.

In Fig. 7.19, the experimental data yield the best-fit equation

$$K_r = (1-\phi_{vf})^{-4.65} \tag{7.101}$$

For Configuration *I*, K_r is obtained from Eqs. 7.96 and 7.99, and for Configuration *II* from Eqs. 7.96 and 7.100. Configuration *I* in general yields higher than measured values, whereas *II* gives consistently lower values. Thus the data correspond to a configuration of particles that is intermediate between *I* (Fig. 7.15a) and *II* (Fig. 7.15b).

7.2.4 *Empirical settling velocity and flux equations*

Ranges of free settling (Eq. 7.20), flocculation settling (Eq. 7.47) and hindered settling (Eq. 7.49) are schematically shown in Fig. 7.13. Several empirical equations relating the settling velocity to concentration in the flocculation and hindered settling ranges are in commonly use.

As we noted (Fig. 7.2), in the free settling range the upper limit concentration C_f varies approximately between 0.05 to 0.3 kg m^{-3}. The settling flux $F_s = w_s C$ is proportional to C when it is less than C_f. In the

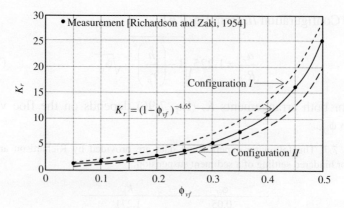

Figure 7.19. Dependence of Stokes drag correction factor on floc volume fraction.

C_f to C_s (flocculation + hindered settling) range Wolanski *et al.* [1989] proposed the expression

$$w_s = a_w \frac{C^{n_w}}{\left(C^2 + b_w^2\right)^{m_w}}$$
(7.102)

where a_w is the settling velocity scaling coefficient, n_w is the flocculation settling exponent, b_w is the hindered settling coefficient and m_w is the hindered settling exponent.[k] Values of these coefficients are given in Table 7.4 based on best-fit application of Eq. 7.102 to data on settling

---

[k] Whitehouse *et al.* [2000a] recommend a formula for w_s (m s^{-1}), which, like Eq. 7.102, covers both low and high concentrations:

$$w_s = \frac{\nu}{d_f}\left[\sqrt{107.33 + 1.049(1 - \phi_{vf})^{4.7} d_*^3} - 10.36\right]$$

$$d_* = \left[\left(\frac{\rho_f}{\rho_w} - 1\right)\frac{g}{\nu^2}\right]^{1/3} d_f$$
(F7.5)

where ν is the kinematic viscosity of fluid (water) in m^2 s^{-1}, ρ_f is floc density and d_f is the floc size in meters. Equation F7.5 is derived for settling velocity data from the Severn estuary (UK). The definition of d_* is analogous to Eq. 6.126. Note that in order to use Eq. F7.5 the flocs volume fraction ϕ_{vf} must be known. Strom and Keyvani [2011] proposed an equation relating w_s with d_f. By using the assumption of fractal flocs defined by the dimension D_f, Eq. F7.5 was generalized by Ferguson and Church [2004].

Table 7.4. Coefficients in Eq. 7.102 for calculation of w_s (m s^{-1}) from C (kg m^{-3}).

Investigator(s)	Sediment source	a_w	b_w	m_w	n_w
Krone [1962]	San Francisco Bay, California	0.048	25.00	1.00	0.40
Owen [1970]	Severn River (UK); salinity 2	0.140	17.00	1.40	1.10
Owen [1970]	Severn River (UK); 8	0.110	11.00	1.53	1.50
Owen [1970]	Severn River (UK); 17	0.160	15.00	1.15	0.50
Owen [1970]	Severn River (UK); 32	0.100	10.00	1.30	1.00
Owen [1970]	Severn River (UK); 48	0.080	9.50	1.34	1.00
Huang *et al.* [1980]	Changjiang (China)	0.012	1.70	2.80	2.20
Thorn [1981]	Severn River (UK)	0.010	2.00	1.46	2.10
Burt and Stevenson [1983]	Thames River (UK); 1981 sample	0.170	3.00	1.90	1.65
Burt and Stevenson [1983]	Thames River (UK); 1982 sample	0.060	2.00	1.90	1.50
Nichols [1984/85]	James River, Virginia	0.039	3.80	1.32	1.52
Odd and Rodger [1986]	Severn River (UK)	0.080	6.50	1.35	1.42
Lott [1986]	Commercial kaolinite	0.010	3.00	1.60	1.30
Ross [1988]	Tampa Bay, Florida	0.001	1.80	1.40	2.10
Hwang [1989]	Lake Okeechobee, Florida; 40% organic matter; particle size 10 μm	0.080	3.50	1.88	1.65
Hwang [1989]	Lake Okeechobee, Florida; 40% organic matter; particle size 15 μm	0.027	5.50	1.60	1.00
Hwang [1989]	Lake Okeechobee, Florida; 40% organic matter; particle size 7 μm	0.090	4.50	1.85	1.80
Costa [1989]	Hangzhou Bay (China)	0.100	6.20	1.60	1.20
Wolanski *et al.* [1991]	Cleveland Bay (Australia); lab tests	0.070	1.30	2.50	2.80
Wolanski *et al.* [1991]	Cleveland Bay (Australia); field test	0.200	1.40	2.25	2.45
Jiang [1999]	Jiaojiang (China); neap tide	0.045	6.00	1.51	1.50
Jiang [1999]	Jiaojiang (China); spring tide	0.230	10.00	1.80	1.50
Marván [2001]	Ortega River, Florida	0.160	4.50	1.95	1.70
Ganju [2001]	Loxahatchee River, Florida	0.190	5.80	1.80	1.80
So [2009]	Lake Apopka, Florida; 62% organic matter	0.8	2.7	2.8	1.00

velocity (m s^{-1}) and concentration (kg m^{-3}) reported in the cited literature.

In the flocculation settling range we have $C \ll b_w$, and Eq. 7.102 is approximated as

$$w_s = a_w b_w^{-2m_w} C^{n_w} \tag{7.103}$$

which is identical to Eq. 7.47 with $a'_w = a_w b_w^{-2m_w}$. In the hindered settling range $C \gg b_w$, and Eq. 7.102 reduces to

$$w_s = a_w C^{n_w - 2m_w} \tag{7.104}$$

which indicates decreasing w_s with increasing C, provided $n_w - 2m_w < 0$.

The following characteristic quantities are derived from Eq. 7.102:

Concentration C_f:

$$C_f = \left(\frac{w_{sf}}{a_w b_w^{-2m_w}} \right)^{1/n_w} \tag{7.105}$$

where w_{sf} is the free settling velocity when $C \le C_f$.

Peak velocity w_{sm}:

$$w_{sm} = a_w b_w^{n_w - 2m_w} \frac{\left(\dfrac{2m_w}{n_w} - 1 \right)^{m_w \frac{n_w}{2}}}{\left(\dfrac{2m_w}{n_w} \right)^{m_w}} \tag{7.106}$$

Concentration C_h:

$$C_h = \frac{b_w}{\sqrt{\dfrac{2m_w}{n_w} - 1}} \tag{7.107}$$

Settling flux F_s:

$$F_s = a_w \frac{C^{n_w+1}}{\left(C^2 + b_w^2\right)^{m_w}} \qquad (7.108)$$

Peak settling flux F_{sm}:

$$F_{sm} = a_w b_w^{n_w - 2m_w + 1} \frac{\left(\dfrac{2m_w}{n_w + 1} - 1\right)^{m_w - \frac{n_w+1}{2}}}{\left(\dfrac{2m_w}{n_w + 1}\right)^{m_w}} \qquad (7.109)$$

Concentration C_m corresponding to F_{sm}:

$$C_m = \frac{b_w}{\sqrt{\dfrac{2m_w}{n_w + 1} - 1}} \qquad (7.110)$$

Mean values and standard deviations of a_w, b_w, n_w and m_w in Table 7.5 reveal trends when the normalized standard deviation, *i.e.*, standard deviation divided by the corresponding mean, is examined. Coefficient a_w shows wide variability manifested as a high value (0.72) of the normalized standard deviation. This variability reflects the dependence of the settling behavior of flocs on sediment composition and the flow environment. For Cleveland Bay (Australia), a_w ranges from a low 0.07 in laboratory tests to 0.2 derived from concentration profiles in the field. This range is due to the difference between quiescent settling in the laboratory and settling under natural shear flow. Faster settling flocs were formed in the field by turbulence-induced inter-particle collisions.

Table 7.5. Statistical measures of coefficients in Eq. 7.102.

Measure[a]	a_w	b_w	m_w	n_w
Mean	0.092	6.67	1.68	1.53
Standard deviation	0.066	5.78	0.43	0.56
Standard deviation/Mean	0.72	0.87	0.26	0.37

[a] Excluding organic-rich muck from Lake Apopka tested by So [2009].

The coefficient b_w is a measure of the onset of hindered settling as it depends on C_h in Eq. 7.107 and C_m in Eq. 7.110. The variability in b_w reflected by its high (0.87) normalized standard deviation is due to the sensitivity of C_h and C_m to the sediment and the fluid.

The value 1.53 for n_w is reasonably close to the "theoretical" value of 1.33 in the flocculation settling range (Eq. 7.46). As we have noted this dependence of the settling velocity on C^{n_w} highlights the role of aggregation in the settling process.

The mean value of exponent $n_w - 2m_w$ in Eq. 7.104 is −1.83 (Table 7.4). The rate of hindered settling is similar for different slurries, as suggested by the somewhat low values of the normalized standard deviations of n_w (0.37) and m_w (0.26). This similarity is brought about by damping of turbulence at high concentrations and the control of settling by dewatering rate that does not vary substantially from slurry to slurry.

An example of settling velocities of sediment from the Jiaojiang estuary in China, and Eq. 7.102, is shown in Fig. 7.20. Data spread is endemic to the methods of measurement and estimation of the settling velocity, as well as the fact that the settling velocity is sensitive to flow shear rate not accounted for in the equation. The overall trend for the Amazon estuary in Fig. 7.21a is similar. In Fig. 7.20, laboratory-derived settling velocities have been augmented by values from measured

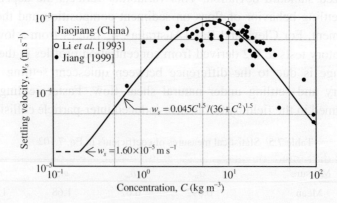

Figure 7.20. Variation of settling velocity with concentration for the Jiaojiang estuary (China) sediment (revised from Jiang [1999]).

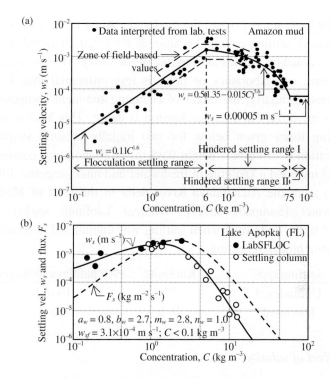

Figure 7.21. Variation of settling velocity with concentration for: (a) Amazon estuary (Brazil) mud (from Vinzon and Mehta [2003]); (b) muck from Lake Apopka, Florida [So, 2009]. In Fig. 7.21b the data points are for the velocity. The flux curve is obtained from Eq. 7.108.

concentration profiles at a field site [Kineke, 1993]. The free settling range has been ignored and hindered settling is divided into two ranges. As we will note further in Chapter 8, in the second range starting at $C = 75$ kg m^{-3}, the settling velocity is influenced by channelized upward (capillary) flows in the slurry.

Settling velocities in Fig. 7.21b are of organic-rich (62% organic content) dark mud, or muck, at the bottom of Lake Apopka in Florida. Data in the flocculation settling range were obtained by testing water samples drawn from the lake on board a vessel. The tests were conducted in a settling column with a camera system called LabSFLOC [Manning, 2006]. The advantage of using this arrangement is evident; the flocs were likely to have been considerably closer to their natural state then, than later in the laboratory. Data in the hindered settling range were obtained in a laboratory settling column because, as we noted, hindered settling is largely governed by the dewatering rate of the slurry. This means that the environment, *i.e.,* laboratory or field, is less influential than in dilute suspensions. Similarly, the effect of changing sediment composition is often less effective when the suspension is dense. This is illustrated by the settling times given below for two initially uniform suspensions (concentration 87.2 kg m^{-3}, height 0.35 m) in a laboratory column. The sediments included a kaolinite in freshwater and mud (smectite, illite and kaolinite) from the Atchafalaya River delta in the Gulf of Mexico in native water [Sampath, 2009]. Whereas kaolinite settled slowly compared to the delta mud when settling was unhindered, the rates were equal during hindered settling.

Settling type	Kaolinite	Atchafalaya Delta mud
Unhindered	2 h	10 min
Hindered	16 h	16 h

7.2.5 *Effect of salinity*

From Stokes law (Eq. 7.15) it is evident that the effects of changing salinity on floc diameter, floc density and to some extent fluid viscosity collectively influence the settling velocity.

In Chapter 4 we briefly examined the effect of salinity on the floc settling velocity of San Francisco Bay mud. Tests using mud from the Avonmouth estuary in England are shown in Fig. 7.22 [Owen, 1970]. This mud consisted of 55% particles in the clay size range and 45% in the silt range. Clay minerals were illite, kaolinite, montmorillonite and chlorite. With respect to mineral composition the Bay mud and the Avonmouth mud were somewhat similar. Yet the results in Fig. 7.22 differ from those in Fig. 4.14. Unlike Bay mud, Avonmouth mud was measurably influenced at salinities well above 10°/$_{oo}$. For each initial sediment concentration, there is a generally consistent trend of gradual rise in the settling velocity with salinity, attainment of a peak value at some salinity, and a relatively rapid decrease with a further increase in salinity. Above 4 kg m^{-3}, concentration hinders settling. The flatness of

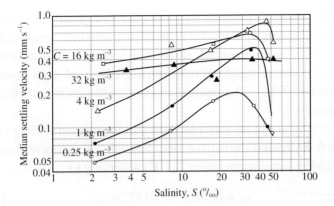

Figure 7.22. Effect of salinity on the settling velocity of mud from Avonmouth, England (after Owen [1970]).

the curve at 32 kg m^{-3} suggests that hindrance likely suppressed the effect of salinity. The decrease in w_s at high salinities and concentrations of 16 kg m^{-3} or less is a significant feature absent in the Bay mud data. Such a reduction in w_s can be brought about by a decrease in the fall diameter and floc density, or an increase in the flatness of floc shape. These changes therefore allude to the role of salinity in altering the internal structure of the floc. Another likely and possibly very significant factor is the biological makeup of flocs. High salinities can chemically alter organic matter and influence microbial communities such as cyanobacteria producing mucus binding the particles. As mentioned in Chapter 4, microbes likely govern the properties and transport behavior of Bay flocs.

7.2.6 *Effect of temperature*

Decreasing temperature increases the settling flux of suspended matter. Partly due to the demand placed on equipment to maintain a constant fluid temperature, only a limited number of studies appear to have been carried out on the effect of temperature. Owen [1972] conducted tests on an estuarine mud in a settling column, whereas the tests of Lau [1994] were in an annular rotating flume using kaolinite. In a set of latter tests, distilled water ranging in temperature from 5 to 26°C was used, and the flow speed was held constant to produce a bed shear stress of about 0.2 Pa. We observe in Fig. 7.23a that at a constant temperature the (depth–mean) concentration of suspended sediment fell at first, then achieved a constant residual value characteristic of data from such flumes (Section 7.3.3). The residual concentration significantly increased with temperature.

Jiang [1999] used Lau's data to show that the median settling velocity of kaolinite decreased during the first 100 min of sediment deposition with increasing water temperature T_θ ($°C$) according to

$$w_s = (1.776 - 0.0518T_\theta)w_{s15} \qquad (7.111)$$

where w_{s15} is the value of w_s at $T_\theta = 15°C$ (Fig. 7.23b). A likely reason is that with increasing temperature the thermal activity of clay micelles increases repulsion between particles and leads to lighter and smaller flocs. It is worth noting however that the effect of temperature on the settling velocity in a flume can be more complex than in a settling column, because in the flume bed erosion, if present, would also increase with temperature (Chapter 9). Thus Eq. 7.111 should be used with caution.

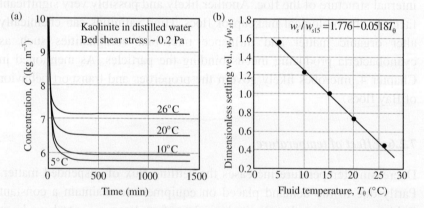

Figure 7.23. (a) Variation of depth–mean suspended sediment concentration with time and temperature during deposition of kaolinite in an annular rotating flume (from Lau [1994]); (b) Dimensionless settling velocity of kaolinite flocs as a function of fluid temperature based on the data of Lau [1994] (from Jiang [1999]).

7.2.7 Effect of shear rate

Among the effects of various factors on the settling behavior of flocs, the effect of turbulence has led to numerous laboratory and field devices for capturing and measuring the settling velocity of suspended flocs as a function of the turbulent shear rate. In the strongly tidal Thames River (England), Owen [1971] measured settling velocities under two salinity ranges, $6-10^o/_{oo}$ and $26-32^o/_{oo}$. The tubular apparatus used in that study has since been named after Owen. As no significant effect was detected in the Thames, it was concluded that the effect of shear rate subsumed any likely effect of salinity. Unfortunately, difficulties in interpreting field data can arise from the effect of stress-history on the flocs, which may pass through variable fields of salinity and shear prior to the point of measurement.

The Owen tube has since been modified for easier and partly automated handling (e.g., Cornelisse [1996]). The bottle (or tube) attached to a vane (Fig. 7.24a) is lowered into the water column where it collects the suspension while in a horizontal position. Then it is raised out of water and made to stand vertically to serve as a settling column (Fig. 7.24b). The settling velocity is determined from measurements of concentration as the material settles, employing the bottom-withdrawal method [van Rijn and Nienhuis, 1985].

The settling velocity of dredged sediment in Cleveland Bay (Australia) was obtained by two methods [Wolanski et al., 1992]. The first involved in situ measurements of the vertical profiles of concentration at different times during the disposal of sediment from a hopper dredger. In the second method the same sediment was tested in a laboratory column in which vertically oscillating rings nearly flush with the inner wall of the column were used to produce flow shear (Fig. 7.24c inset). The depth–mean shear rate could be controlled by setting the amplitude and the frequency of oscillation. Equation 6.108 was solved numerically in conjunction with: (1) an estimate of the (constant) diffusion coefficient D_{sz}, (2) initial condition of constant concentration at the start of the test, (3) no net sediment flux condition at the surface, and (4) no resuspension condition at the bed. Plots thus obtained are shown in Fig. 7.24c. Two effects are noteworthy — differences in laboratory

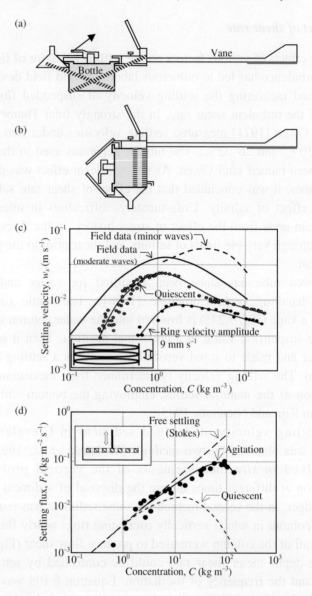

Figure 7.24. (a) Modified Owen tube in horizontal position for sample collection; (b) tube in the vertical position serving as a settling column [Cornelisse, 1996]; (c) settling velocity data based on field and laboratory tests on sediment from Cleveland Bay, Australia (from Wolanski *et al.* [1992]); (d) settling velocity data based on laboratory tests on sediment from the Gironde estuary, France (from Gratiot *et al.* [2005]).

observations depending on the state of the fluid, and differences between field and laboratory environments. Field data indicate higher settling velocities when the weather was calmer (minor wave activity) than under higher wave agitation (with a swell period of about 15 s and a sea period of 3.5 s).[1] Retardation of settling under waves was attributed to the formation of a lutocline which contained large, more slowly settling aggregates than when the conditions were calmer. A similar behavior was recorded in the laboratory tests, in which settling was more hindered when the rings were oscillated than in the still condition. Such differences between field and laboratory indicate the effects of flow on the aggregation processes, and underscore the importance of making *in situ* observations as we noted in Chapter 5.

The tests of Gratiot *et al.* [2005] summarized in Fig. 7.24d were carried out in the laboratory using sediment from the Gironde estuary (France). Agitation was induced in the suspension in a cylinder of square cross-section with a grid made from seven (1.5 cm × 1.5 cm) square bars oscillated vertically by a connecting rod. The stroke was 4.5 cm and the frequency was varied between 3 and 6 Hz. The dashed line in the plot derived from calculated settling flux assuming Stokes law forms a convenient baseline relative to which settling in the flocculation and hindered settling ranges can be assessed. Disaggregation due to agitation resulted in smaller flocs than under quiescent conditions. This meant lower settling velocities and delayed onset of hindered settling with respect to concentration. As a result the position of the peak settling flux shifted rightward from concentration $C_m = 15$–20 kg m^{-3} to 70–80 kg m^{-3}, a four to five-fold increase in C_m possibly implying a substantial reduction in floc size.

Returning to Fig. 7.24c, the effect of shear rate on settling is observed to be important over the entire range of concentration, although above ~10 kg m^{-3} there is a trend of convergence of the curves. This trend is due to the effect of high concentration which damps turbulence and

---

[1] The wind speed at the time of moderate waves was close to 10 m s^{-1} or about 19 knots (1 m s^{-1} =1.944 knot), and under calmer conditions about 4 m s^{-1} (~8 knots). The Beaufort wind scale and the associated international scale of sea description (Appendix F) are useful for a rough estimation of wave height. At 19 knots the wave height would be about 1 m, and at 4 knots less than 0.3 m.

enhances the role of dewatering on the settling behavior. The settling mass is less influenced by the floc structure, and hence sediment composition, than in the flocculation settling range. Schematically, one would expect the actual trends of $w_s(C)$ to appear as shown by the dotted curves in Fig. 7.25, each for sediment of given composition. Teeter [2001a] assumed a simpler model with "linear" trends in the flocculation settling range ($C_f < C < C_h$), in which the exponent $n_w(d_f)$ decreases with increasing floc diameter d_f, as the effect of cohesion also decreases with increasing size of primary particles. Any likely effect of shear rate, although not apparent from the plot, is embodied in n_w, which varies with turbulence. In the hindered settling range all sediments would "ideally" follow the same curve.

The settling velocity in the flocculation settling range is given by

$$w_s(d_f) = w_{sm}\left(\frac{C}{C_h}\right)^{n_w(d_f)} \qquad (7.112)$$

Experimental investigations suggest the following relationship between w_s and a representative flow shear rate $\bar{G}$ (Hz or s^{-1})

$$w_s = \left(\frac{1 + \lambda_1 \bar{G}}{1 + \lambda_2 \bar{G}^2}\right) w_{s|\bar{G}=0} \qquad (7.113)$$

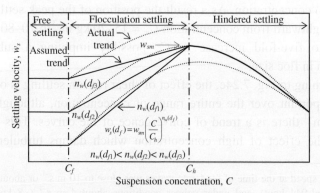

Figure 7.25. Schematic diagram showing the dependence of settling velocity on concentration and floc size.

in which $w_{s|\bar{G}=0}$ must be obtained from a concentration dependent settling velocity function such as Eq. 7.47, and λ_1 and λ_2 are sediment-specific coefficients [van Leussen, 1994; Malcherek and Zielke, 1996; Teisson, 1997]. The shear rate can be estimated from $\bar{G} = v / \lambda_0^2$, where the Kolmogorov eddy length scale λ_0 is defined in Eq. 2.89 and v is the kinematic viscosity of the fluid (water). Equations 7.112 and 7.113 may be combined to yield

$$w_s = w_{sm} \left(\frac{C}{C_h}\right)^{n_w(d_f)} \left[\left(\frac{1+\lambda_1\bar{G}}{1+\lambda_2\bar{G}^2}\right) e^{-\lambda_3 \frac{C}{C_f}} + 1\right] \tag{7.114}$$

where λ_3 is another constant. Since the velocity w_{sm} is also deduced from data, this equation has four fitting coefficients. A set of values of λ_1, λ_2 and λ_3 is 320 s, 75 s^2 and 0.8, respectively [Teeter, 2001a; Letter, 2009].

Figure 7.26 shows a comparison between Eq. 7.114 and settling velocities of a natural mud over the range of concentration, $C_f = 60$ mg L^{-1} to $C_h = 660$ mg L^{-1}. With increasing (initial) concentration the effect of shear rate is gradually subsumed by concentration, which eventually becomes practically the sole determinant of settling velocity.

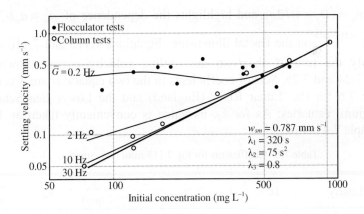

Figure 7.26. Settling velocity of natural sediment varying with shear rate and initial concentration of suspension (adapted from Teeter [2001a]).

Following a mechanistic approach, Winterwerp *et al.* [2006] started with the simple dynamic balance (Eq. 4.120) along with general forms of the growth and breakup functions f_G and f_f in terms of the fractal dimension D_f (see Eqs. 4.121 and 4.125). After further analysis the final expression was obtained as

$$w_s = \left[k_c \frac{C^{1/(2q_b)}}{\tau^{3/8}} - k_a \left(k_c \frac{C^{1/(2q_b)}}{\tau^{3/8}} - k_d d_{f0} \right) e^{-\frac{k_b \tau^{9/8} h}{D_f}} \right]^{D_f - 1} \quad (7.115)$$

where C is the suspended sediment concentration, d_{f0} is the initial ($t = 0$) particle diameter, τ is the local shear stress, h is the water depth and k_a, k_b, k_c, k_d and q_b are sediment-dependent constants. Note that the dimensions of k_b and k_c depend on q_b and D_f.

The second term on the right hand side of Eq. 7.115 accounts for hindered settling. When it is small or nil, *i.e.*, when either τ or h is large, we obtain

$$w_s = k_e \tau^{-\frac{3(D_f - 1)}{8}} C^{n_w} \quad (7.116)$$

where $k_e = k_c^{D_f - 1}$. Equation 7.116 is analogous to Eqs. 7.47 and 7.103 with $n_w = (D_f - 1)/2q_b$, and highlights the dependence of $a_w' = a_w b_w^{-2m_w} = k_c \tau^{-3(D_f - 1)/8}$ on the fractal dimension. Equations 7.115 and 7.116 also identify the role of shear on the settling velocity curves shown in Figs. 7.24 and 7.26. Sets of coefficients in the two equations are given in Table 7.6 for the Tamar River (England) and the Lower Sea Scheldt (Belgium) estuaries. As for k_d, its value is conveniently taken as 1 in Example 7.4 below.

Table 7.6. Coefficients for Eq. 7.115 from two estuaries.

Estuary	D_f	k_a	k_b	k_c	d_{f0} (μm)	n_w	h (m)
Tamar (England)	2.20	1	5	0.007	10	0.20	3
Scheldt (Belgium)	2.15	1	5	0.007	10	0.44	15

Source: Winterwerp *et al.* [2006].

As we saw in Chapter 4, Krone [1962] used heuristic arguments based on the kinetics of flocculation to show that $n_w = 1.33$. This value can be expected to decrease with decreasing cohesion and approach zero when cohesion is negligible. Settling velocity data on the Tamar mud yielded $n_w = 0.20$.[m]

Example 7.4: Plot Eq. 7.115 in terms of settling velocity as a function of shear stress for $C = 0.05$ kg m^{-3} and 6 kg m^{-3}. Use coefficients for the Tamar from Table 7.6 and assume $k_d = 1$.

The plots are shown in Fig. 7.27. Also included are data from the Tamar obtained within approximately the same range of suspended sediment concentration.

Local shear stress values for the Tamar were obtained by Manning and Dyer [2002] from

$$\tau = 0.19 \left[\frac{1}{2} \left(\overline{u'^2} + \overline{v'^2} + \overline{w'^2} \right) \right] \qquad (7.117)$$

where the term within brackets is the turbulent kinetic energy (Eq. 2.4), and the constant 0.19 is specific to the site. An important feature is a threshold (peak) shear stress τ_m representing the boundary between the dominance of floc growth at lower stresses and breakup at higher stresses. It is easily shown (Exercise 7.3) that τ_m is independent of concentration.

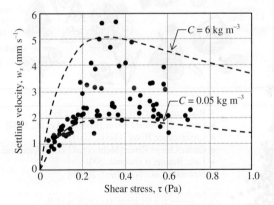

Figure 7.27. Settling velocity versus local shear stress data from the Tamar estuary (England) and Eq. 7.115 (adapted from Winterwerp *et al.*, [2006]).

---

[m] Winterwerp *et al.* [2006] contended that $n_w = 0.20$ was realistic, and that higher values (Table 7.4) reported for other sites suggested measurement error.

7.2.8 *Effect of organic matter*

Particle diameter and material density of clayey sediments are altered by natural organic matter (NOM) including biopolymeric binding, which in turn influences the settling velocity. As we noted in Chapters 3 and 4, biopolymeric binding can generate flocs that may be larger (and often flatter) but typically less dense than flocs held together by electrochemical cohesion (Fig. 7.28a). Little appears to be known quantitatively about the relationship between floc size and its organic content.

The effective density of solids containing mineral as well as organic matter is conveniently calculated by assuming that the mixture is homogeneous, even though aggregates may not be uniform in composition. For sediments from estuaries and lakes in peninsular Florida, wet bulk density ρ (Fig. 7.28b) and the dry bulk density ρ_D

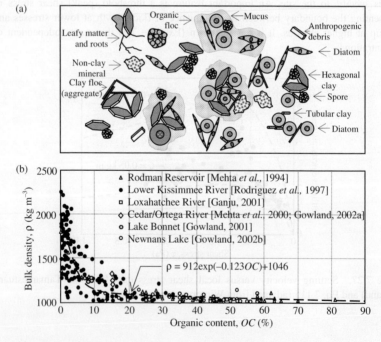

Figure 7.28. (a) Heterogeneity of suspended matter consisting of natural organic and inorganic materials; (b) variation of wet bulk density with organic content (based on Gowland *et al.* [2007]).

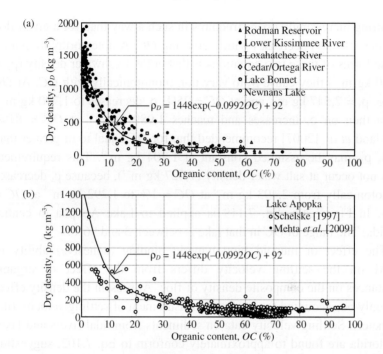

Figure 7.29. Variation of dry bulk density with organic content (a) Data from lakes and rivers in peninsular Florida and Eq. 7.119; (b) Eq. 7.119 applied to data [Schelske, 1997; Mehta *et al.*, 2009] from Lake Apopka in central Florida.

(Fig. 7.29) are plotted as functions of organic content (OC) obtained from loss on ignition tests (Chapter 5). These relationships are applicable to soft beds. Collection was carried out by grab-samplers and push-corers, with core lengths varying approximately between 0.1 m and 1.5 m. Mean trends are as follows:

$$\rho = 912 e^{-0.123OC} + 1046 \qquad (7.118)$$

$$\rho_D = 1448 e^{-0.09992OC} + 92 \qquad (7.119)$$

The effective particle-density is back-calculated from $\rho_s = \rho_w \rho_D / (\rho_D + \rho_w - \rho)$, where ρ_w is water density.

Equations 7.118 and 7.119, which are applicable in the measured range of OC from 1% to 82%, indicate a rapid decline in ρ and ρ_D as OC increases from about 1 to 25%. This trend is the outcome of flocs acquiring increasingly open structures in such a way that, for a given floc volume, its mass decreases with increasing OC. A noteworthy limitation of the forms of these two equations is that at the freshwater density ($\rho_w =$ 1,000 kg m^{-3}) they cause ρ_s to vary non-monotonically with OC. At OC = 1%, ρ_s = 2,547 kg m^{-3} and at OC = 25%, ρ_s decreases to 1,700 kg m^{-3}. From then on ρ_s increases and reaches 1,989 kg m^{-3} at OC = 81%. Gowland *et al.* [2007] recommended that for OC equal to or greater than 25%, ρ_D may be assumed constant at 1,704 kg m^{-3}. This requirement does not occur at salt water density (1,027 kg m^{-3}), because ρ_s decreases monotonically from 2,493 kg m^{-3} at OC = 1% to 1,292 kg m^{-3} at OC = 81%. In Fig. 7.29b, Eq. 7.119 is applied to Lake Apopka in central Florida. The range of OC in that lake is between 3 and 96%.

The effect of mucoid biopolymeric binding (adhesive) ability of NOM on the settling velocity differs from the effect of organic substances on the composite density of floc. At present the density effect can only be accounted for by direct measurement of settling velocity (not diameter). Settling velocity data on sediments from tidal rivers and lakes in Florida are found to approximately conform to Eq. 7.102, suggesting that out of the four coefficients in that equation, a_w is the one most influenced by NOM. Accordingly, we may conveniently ascribe this influence exclusively to a_w, and consider b_w, m_w and n_w to be independent of OC. For prescribed values of these three coefficients, Table 7.7 gives a_w and the Stokes diameter calculated from the free-settling velocity of mineralogically similar sediments with varying OC. These data yield the

Table 7.7. Velocity scaling coefficient a_w, free-settling velocity and derived Stokes diameter of flocs.

Source in Florida	Mean organic content, OC (%)	Scaling coefficient a_w	Free-settling velocity (m s^{-1})	Stokes diameter (µm)
Low OC mud	2	0.20	2.60×10^{-5}	4.9
Loxahatchee River	15	0.19	1.80×10^{-5}	5.1
Ortega River	28	0.16	1.65×10^{-5}	5.2
Lake Okeechobee	38	0.09	0.93×10^{-5}	4.1
Lake Okeechobee	40	0.08	0.78×10^{-5}	3.9
Lake Okeechobee	43	0.03	0.31×10^{-5}	2.5

Sources: Marván [2001]; Ganju [2001].

following expression for a_w

$$a_w = a_{OC0} + a_{OC1}OC + a_{OC2}OC^2 + a_{OC3}OC^3 + a_{OC4}OC^4 \quad (7.120)$$

where $a_{OC0} = 0.2$ (for the inorganic matter), $a_{OC1} = 6.6 \times 10^{-4}$, $a_{OC2} = -1.7 \times 10^{-4}$, $a_{OC3} = 7.1 \times 10^{-6}$ and $a_{OC4} = -1.3 \times 10^{-7}$. These values are compatible with concentration C in units of kg m^{-3} and settling velocity w_s in m s^{-1}.

Table 7.7 indicates a weak trend of decreasing floc diameter and settling velocity with increasing OC. In other words, the aggregates become both smaller and lighter as the effect of binding increases while electrochemical cohesion decreases. It is believed that the flocs associated with Eq. 7.120 were formed mainly by orthokinetic flocculation in shear flow. As we have noted, biopolymeric binding can also produce very large aggregates, *e.g.,* marine snow (Chapter 4), mainly by perikinetic flocculation. However, very large aggregates were not formed in the tests associated with Eq. 7.120.

In general there are three reasons to expect that any empirical relationship between organic content and floc size would be very approximate. Firstly, floc diameter derived from settling velocity data is subject to statistical uncertainties in the floc density. These arise from measurements of sediment concentration and mineral density of primary particles [Fettweis, 2008].

Secondly, imaging techniques, unless sufficiently elaborate, are prone to errors in estimating the diameter of typically odd-shaped organic-rich aggregate networks, as well as small flocs in general [Lintern and Sills, 2006]. Thirdly, if mucus is absent, or is present in insufficient amounts, weakly aggregated organic (planktonic) particles may be small. Their slow settling rate is sensitive to effects of return flow and diffusion [Huisman and Sommeijer, 2002]. In summary, Eq. 7.120 can be used only after carefully establishing its validity in any application.

7.2.9 *Effect of oscillating flow*

Waves and wind-driven boundary-layer current close to the water surface can drive aggregation processes especially if the waves are breaking. Also, in the wave boundary layer very close to the bottom, shear-induced vorticity plays a key role in floc growth-breakup dynamics, and wave-induced shear stress erodes the bed. The notion of entrapment of suspended particles within orderly oscillations of the fluid parcel has been used to qualitatively explain the effect of waves in reducing the settling velocity of particles [Nielsen, 1992; McFetridge, 1985].

To calculate the terminal settling velocity of a particle (without differentiating between flocs and primary units) falling under oscillating flow at an angular frequency σ, let w_r be the difference between the settling velocity of the oscillating particle w_{so} and the vertical velocity of the oscillating fluid w_o, *i.e.*,

$$w_r = w_{so} - w_o \tag{7.121}$$

Now we let

$$w_o = \hat{w}_o \sin \sigma t \tag{7.122}$$

where $\hat{w}_o$ is the amplitude of $w_o(t)$, and

$$w_{so} = \overline{w}_{so} + \hat{w}_{so} \sin(\sigma t - \delta_{so}) \tag{7.123}$$

where $\overline{w}_{so}$ is the mean value of $w_{so}(t)$, δ_{so} is the phase difference between water and particle motions and $\hat{w}_{so}$ is the amplitude of particle velocity oscillation. Substitution of Eqs. 7.122 and 7.123 into Eq. 7.121 and after manipulation we obtain

$$w_r = \overline{w}_s + \alpha_{so} \hat{w}_o \sin(\sigma t - \delta'_{so}) \tag{7.124}$$

where

$$\alpha_{so} = \sqrt{1 - 2\frac{\hat{w}_{so}}{\hat{w}_o}\cos\delta_{so} + \left(\frac{\hat{w}_{so}}{\hat{w}_o}\right)^2} \tag{7.125}$$

and the phase difference

$$\delta'_{so} = \tan^{-1}\left(\frac{\hat{w}_{so}\sin\delta_{so}}{\hat{w}_o - \hat{w}_{so}\cos\delta_{so}}\right) \tag{7.126}$$

The force balance of Eq. 7.13 can be restated as

$$\frac{1}{2}\int_0^{2\pi} C_{Dw}\frac{\pi d_p^2}{4}\rho w_r |w_r| d(\sigma t) = \frac{1}{6}\pi d_p^3 \Delta\rho g \qquad (7.127)$$

where the left hand side is integrated over the wave period to yield the mean drag force that can be equated to the time-independent buoyancy force on the right-hand side. The drag coefficient C_{Dw} is applicable to the oscillating particle and is generally different from C_D defined for steady flows in Chapters 2 and 6.

For $\delta'_{so} \to 0$, representing iso-kinetic flow of particles ($< \sim 100$ μm), Hwang [1985] showed that

$$\alpha_{so} \to 1 - \frac{\hat{w}_{so}}{\hat{w}_o} \approx 1 - \frac{\rho}{\rho_s}\frac{(1+C_M)}{\left(1+C_M\frac{\rho}{\rho_s}\right)} \qquad (7.128)$$

where C_M is the added mass coefficient, whose value for a sphere is 0.5 [Dean and Dalrymple, 1991]. Thus we may take

$$w_r = \overline{w}_s + \alpha_{so}\hat{w}_o \sin\sigma t \qquad (7.129)$$

with

$$\alpha_{so} = 1 - \frac{\rho}{\rho_s}\frac{(1+C_M)}{\left(1+C_M\frac{\rho}{\rho_s}\right)} \qquad (7.130)$$

In order to relate $\hat{w}_{so}$ to the Stokes settling velocity w_s in quiescent water we observe that

$$\frac{1}{2}\int_0^{2\pi} C_{Dw}\frac{\pi d_p^2}{4}\rho w_r |w_r| d(\sigma t) = \frac{1}{6}\pi d_p^3 \Delta\rho g = \frac{1}{2}C_D\frac{\pi d_p^2}{4}\rho w_s^2 \qquad (7.131)$$

Therefore,

$$\int_0^{2\pi} C_{Dw}w_r |w_r| d(\sigma t) = C_D w_s^2 \qquad (7.132)$$

or

$$\int_0^{2\pi} C_{Dw}(\overline{w}_s + \alpha_{so}\hat{w}_o \sin\sigma t)\left|\overline{w}_s + \alpha_{so}\hat{w}_o \sin\sigma t\right| d(\sigma t) = C_D w_s^2 \qquad (7.133)$$

which can be restated as

$$\int_0^{2\pi} \frac{C_{Dw}}{C_D}\left(\frac{\overline{w}_s}{w_s} + \alpha_{so}\frac{\hat{w}_o}{w_s} \sin\sigma t\right)\left|\frac{\overline{w}_s}{w_s} + \alpha_{so}\frac{\hat{w}_o}{w_s} \sin\sigma t\right| d(\sigma t) - 2\pi = 0 \qquad (7.134)$$

For C_D we can use the expressions given in Eq. F7.2 and for C_{Dw} the same expressions with $Re_p = \left|w_r\right|d_p/\nu$. In Fig. 7.30, the solution of Eq. 7.134 in terms of the variation of $\overline{w}_{so}/w_s$ with $\hat{w}_o/w_s$ is compared with the data of Ho [1964]. Observe that when $\hat{w}_o = 0$, *i.e.*, no oscillation, $\overline{w}_{so} = w_s$ is recovered.

7.3 Settling and Deposition in Flow

7.3.1 Quiescent settling and flow effect

As we noted (Fig. 7.24), settling of flocs is substantially influenced by the flowing carrier fluid. As a result, settling velocity measured in still (or rapidly oscillating) water is not accurate enough for calculating the settling flux in flow, even though such an assumption is often made for simplicity of analysis. Another common assumption is that settling is

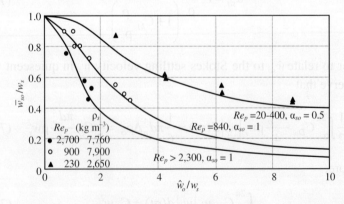

Figure 7.30. Effect of oscillating flow on Stokes settling velocity (from Hwang [1985]).

reasonably well represented by the median floc size or median settling velocity. Although these assumptions are at times acceptable in many engineering applications, it is essential to keep in mind the effect of particle heterogeneity on deposition in flow. As a starting point we will consider particles (flocs) of uniform size.

7.3.2 Settling and deposition of uniform sediment

Consider water flowing at a uniform depth h (Fig. 7.31). The turbulent velocity profile is $u(z)$, τ_b is the (turbulence–mean) bed shear stress and $C(t)$ is the spatially uniform, turbulence–mean suspended sediment concentration at time t. Let the mean mass of each floc be $m_{pf}(t)$, and let n_v be the number of flocs per unit volume of suspension (Fig. 7.32). The probability of deposition (floc sticking to the bed) depends on whether the floc survives the high flow shear rate near the bed, or breaks up and its components are re-entrained. Thus we may define w_s' as an effective floc settling velocity such that for bed area A, mass balance in the suspension while deposition is taking place is

$$\frac{d}{dt}(m_{pf}n_v hA) = -m_{pf}n_v w_s' A \qquad (7.135)$$

Let m_{pf0} and n_{v0} be the initial values of m_{pf} and n_v, respectively. Due to deposition while aggregation processes are active, m_{pf0} will decrease according to

$$m_{pf}(t) = m_{pf0}\frac{n_v(t)}{n_{v0}} \qquad (7.136)$$

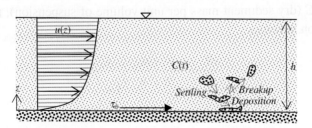

Figure 7.31. Depositing suspended sediment in a channel.

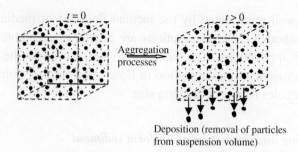

Figure 7.32. A cube of suspension containing flocs and primary particles.

where $n_v(t)/n_{v0}$ is the average number of particles per depositing floc. Note that Eq. 7.136 does not represent mass balance. By considering the flow in the immediate neighborhood of the particle to be laminar, Krone [1962] assumed that n_v/n_{v0} can be obtained from Eq. 4.94, which describes flocculation under non-turbulent conditions in a settling column. Combining Eq. 4.94 with Eq. 7.136 yields

$$m_{pf}(t) = m_{pf0} \frac{1}{\left(1 + \dfrac{t}{t_c}\right)} \qquad (7.137)$$

Introducing Eq. 7.137 into 7.135 results in

$$\frac{d(m_{pf}n_v)}{dt} = -\frac{w_s'}{h}\left(\frac{n_{v0}}{n_v}\right)\frac{m_{pf}n_v}{1 + \dfrac{t}{t_c}} \qquad (7.138)$$

where n_{v0}/n_v is the average number of initial particles in a floc. Since $m_{pf}n_v = C$ (dry sediment mass per unit volume of suspension), the above expression becomes

$$\frac{dC}{dt} = -\frac{w_s'}{h}\left(\frac{n_{v0}}{n_v}\right)\frac{C}{1 + \dfrac{t}{t_c}} \qquad (7.139)$$

As the floc size is limited by flow shear (Chapter 4), we may assume that n_{v0}/n_v remains unchanged. Then, starting with suspended sediment concentration C_0 at $t = 0$, we may integrate Eq. 7.139

$$\int_{C_0}^{C} \frac{dC}{C} = -\frac{w_s'}{h}\left(\frac{n_{v0}}{n_v}\right)\int_0^t \frac{1}{1+\dfrac{t}{t_c}}dt \qquad (7.140)$$

and obtain

$$\ln\frac{C}{C_0} = -\frac{w_s'}{h}\left(\frac{n_{v0}}{n_v}\right)t_c \ln\left(1+\frac{t}{t_c}\right) \qquad (7.141)$$

Similar to the treatment of flocculation-settling under quiescent conditions, two cases of Eq. 7.141 are noteworthy. In the first case, at low concentrations $t_c \gg t$, *i.e.*, aggregation is slow compared to the time of fall. In other words settling is in the free range, *i.e.*, the settling velocity is independent of concentration. Thus, $\ln[1+(t/t_c)] \approx t/t_c$, which leads to

$$\ln\frac{C}{C_0} = -\frac{w_s'}{h}\left(\frac{n_{v0}}{n_v}\right)t \qquad (7.142)$$

or

$$\frac{C}{C_0} = e^{-\frac{w_s'}{h}\left(\frac{n_{v0}}{n_v}\right)t} \qquad (7.143)$$

The quantity n_{v0}/n_v can be absorbed into the effective settling velocity w_s' without loss of generality. Krone [1962] noted that w_s' must be smaller than or equal to the actual settling velocity w_s depending on the probability of deposition $p_d[0,1]$, *i.e.*, $w_s' = p_d w_s$.

From experiments it was found that there was no deposition when the bed shear stress τ_b was equal to or exceeded a characteristic value τ_{cd}, and deposition was at its maximum rate when τ_b was nil (*i.e.*, quiescent settling). Krone [1962] accordingly defined p_d in terms of its (assumed linear) dependence on τ_b. Thus,

$$w'_s = p_d w_s = \left(1 - \frac{\tau_b}{\tau_{cd}}\right) w_s; \quad \tau_b < \tau_{cd}$$

$$w'_s = 0; \quad \tau_b \geq \tau_{cd} \tag{7.144}$$

A practical advantage of using these formulas is that p_d is obtained from measured time–mean quantities τ_b and τ_{cd}, without invoking the stochastic properties of turbulence.[n,o] Equation 7.143 now becomes

$$\frac{C(t)}{C_0} = e^{-\left(1 - \frac{\tau_b}{\tau_{cd}}\right)\frac{w_s}{h}t} \tag{7.145}$$

According to this exponential decay law, C/C_0 approaches zero at large times as long as $\tau_b < \tau_{cd}$. In contrast, for all $\tau_b \geq \tau_{cd}$, $C = C_0$ as there is no deposition at high bed shear stresses.

The upper limit of concentration $C = C_f$, below which Eq. 7.145 holds, has been found to vary over a somewhat different range in quiescent settling (0.05 to 0.3 kg m^{-3} in Fig. 7.2). Figure 7.33 shows data on the time-rate of change of C from three flume experiments with different sediments and flow conditions. In this semi-logarithmic plot

---

[n] Equation 7.144 can be formally expressed as

$$w'_s = p_d w_s = w_s \cdot H\left(1 - \frac{\tau_b}{\tau_{cd}}\right) \tag{F7.6}$$

where $H(x)$ is the heavyside function of any variable x such that $H(x > 0) = 1$ and $H(x \leq 1) = 0$.

[o] The efficiency with which flocs deposit influences the near-bed reference suspended sediment concentration C_a defined in Chapter 6. Teeter [1986] proposed

$$C_a = \left(1 + \frac{Pe_w}{1.25 + 4.75 p_d^{2.5}}\right)C_m \tag{F7.7}$$

which depends on p_d. Here, C_m is the depth–mean suspended sediment concentration and $Pe_w = hw_s / \overline{\varepsilon}_m$ is a fall Péclet number closely related to the Rouse number R_n (Chapter 6). Both p_d and Pe_w together define the efficiency of deposition although the two quantities are to an extent inter-dependent. Selecting the eddy diffusivity $\overline{\varepsilon}_m = 0.067u_*h$ from Eq. 2.66 we find that $Pe_w = 5.97 R_n$. For very slow-settling particles in strongly turbulent flows ($Pe_w \ll 1$), from Eq. F7.7 we obtain $C_a \approx C_m$.

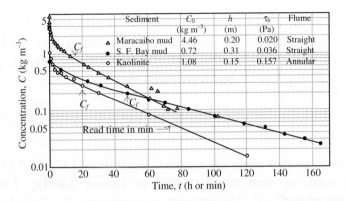

Figure 7.33. Variation of suspended sediment concentration with time in flume tests (from Mehta [1973]).

Eq. 7.145 (representing free-settling) would appear as a sloping straight line [Krone, 1993; Shrestha and Orlob, 1996]. The onset of this line at concentration C_f varies with the sediment as indicated in Table 7.8.

In the second case, Eq. 7.141 can be simplified when $t_c \ll t$, *i.e.*, aggregation is rapid (when $C > C_f$). Since $1 + (t/t_c) \approx t/t_c$ we obtain

$$\ln \frac{C}{C_0} = -\frac{w_s'}{h}\left(\frac{n_{v0}}{n_v}\right)\ln\left(\frac{t}{t_c}\right) \tag{7.146}$$

As before, conveniently absorbing n_{v0}/n_v into w_s' and noting that t_c is a constant for a given sediment yields

$$\ln \frac{C}{C_0} = -\frac{w_s'}{h}\ln t + k' \tag{7.147}$$

which indicates a log-log relationship between C and t. Data in Fig. 7.33 for concentrations greater than C_f can be shown to follow Eq. 7.147, in

Table 7.8. Concentration limits for applicability of Eq. 7.145.

Sediment (in saltwater)	Time at which free settling begins (h)	C_f (kg m^{-3})
Kaolinite	0.33	0.260
Maracaibo (Venezuela) mud	15	0.800
San Francisco Bay mud	50	0.200

which k' is an empirical constant. Krone [1962] noted that, as C approaches C_h marking the onset of hindered settling (Fig. 7.13), the ratio w_s'/h retains only an empirical meaning. Thus, in general, in the range $C > C_f$ Eq. 7.147 is best represented as

$$\ln \frac{C}{C_0} = -k'' \ln t + k' \qquad (7.148)$$

where k'' is another empirical constant. Equations 7.147 and 7.148 are seldom used, as most modelers tend to assume that Eq. 7.145 is applicable when $C > C_f$. This assumption requires that Eq. 7.143 be used instead of Eq. 7.144, with $w_s'(n_{v0}/n_v) \equiv w_s'$ obtained by calibration against site-specific data on concentration variation with time. Although this approach is practical, it is at variance with the physics of aggregation processes at high concentrations, which does not lead to Eq. 7.143 (or Eq. 7.144).

7.3.3 *Deposition of non-uniform sediment*

7.3.3.1 *Rate of deposition and steady-state concentration*

According to Eq. 7.145, when suspended fine sediment at initial concentration C_0 settles in steady flow, $C(t)$ can either decrease monotonically to zero (when $\tau_b < \tau_{cd}$), or not decrease at all ($\tau_b > \tau_{cd}$). In reality there are multiple rates of decrease. This is observed in Fig. 7.34 during the deposition of a kaolinite mixed in distilled water in an annular flume. In this state kaolinite occurs as multi-size flocs and unaggregated (silt-sized) particles represented by the upper tail of the size distribution. The final steady-state concentration C_{fe} depends on τ_b. For a constant C_0, this dependence can be attributed to heterogeneity in sediment size, density and shear strength. This inference is supported by the plot of C_{fe}/C_0 against τ_b in Fig. 7.35, which indicates that for a given τ_b, C_{fe} is a constant fraction of C_0. In other words, at a given bed shear stress the same "depositable" fraction of the initial sediment settles out

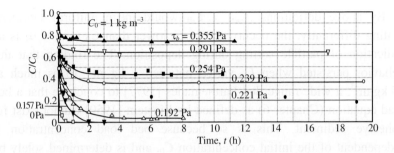

Figure 7.34. Change in suspended sediment concentration with time during deposition of kaolinite in an annular flume filled to a depth of 15 cm (from Mehta [1973]).

[Partheniades, 1962; Rosillon and Volkenborn, 1964; Partheniades *et al.*, 1966; Etter *et al.*, 1968].

The characteristic (or critical) stress for deposition τ_{cd} in Fig. 7.35 is not single-valued as would be expected for a uniform sediment (ideally represented as a step-function), but ranges between τ_{cd1} and τ_{cdM}. Thus C_{fe}/C_0 is zero when $\tau_b < \tau_{cd1}$ and rises to unity when $\tau_b \geq \tau_{cdM}$. For the kaolinite tested the value of τ_{cd1} is 0.18 Pa. When τ_b is equal to or below this value, $C_{fe} = 0$. For τ_b equal to or greater than $\tau_{cdM} = 1.2$ Pa, $C_{fe} = C_0$. When $\tau_{cd1} < \tau_b < \tau_{cdM}$, a fraction C_{fe}/C_0 dependent on τ_b stays in suspension at steady-state, and the remainder, represented by $C_0 - C_f$, settles out. Thus the range τ_{cd1} to τ_{cdM} is a measure of non-uniformity of the suspended sediment; the larger the range the more heterogeneous (multi-size) is the sediment.

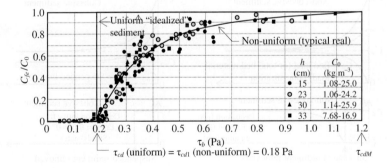

Figure 7.35. Dependence of steady-state suspension concentration on bed shear stress for ideally) uniform sediment (ideally a step-function) and data for kaolinite from an annular flume (adapted from Mehta [1973]).

By above definitions, $\tau_{cdM} = \tau_{cd1}$ would mean uniform sediment. Stating differently, the occurrence of C_f lower than C_0 (and > 0) is an indication of particle sorting during deposition. The finding that this behavior persisted when C_0 varied between 1 kg m^{-3} to as much as 25 kg m^{-3}, a wide range, led Partheniades [1977] to conclude that a bed load equation (Chapter 6) as defined by Einstein [1950] cannot exist for cohesive sediment. This is so because bed load concentration is independent of the initial concentration C_0, and is determined solely by τ_b or, more generally, by flow discharge (see subsequent general discussion on this subject). When the initial concentration exceeded about 25 kg m^{-3}, at a given bed shear stress the ratio C_{fe}/C_0 was found to be lower than predicted from the curve in Fig. 7.35. Assuming that curve is a reasonable mean representation of the data (points), reduction in C_{fe}/C_0 was attributed to the damping of turbulence at high concentrations [Mehta, 1973].

Figure 7.35 indicates that τ_b is not the sole determinant of the sediment-carrying capacity of flow represented by C_{fe}, since it varies with C_0. This inference points to concentration versus time relationships shown in Figs. 7.36a, b. In each case the flow (and hence τ_b) is assumed constant. When the sediment is non-cohesive, starting with any concentration, which can be zero, C_{0A} or C_{0B}, the final concentration is the same equilibrium value C_{fe} representing bed load (or total bed material

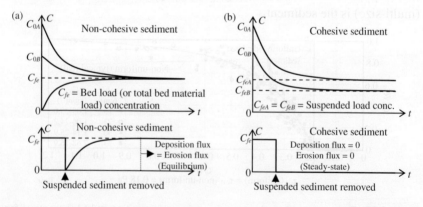

Figure 7.36. Schematic representations of the time-variation of suspended sediment concentration for: (a) Non-cohesive sediment; (b) cohesive sediment.

load). If the suspension at equilibrium were to be replaced by clear water, the concentration would eventually become C_{fe}, because equilibrium implies equality of deposition and erosion fluxes. When the sediment is a mixture of cohesive flocs and non-cohesive fine particles, starting with C_{0A} the final steady-state concentration will be C_{feA}, while with C_{0B} the final value will be C_{feB} such that $C_{feA}/C_{0A} = C_{feB}/C_{0B}$. The constancy of the ratio of initial to final concentration at constant τ_b (Fig. 7.35) implies that while the sediment deposits no erosion occurs. Therefore, in a test in which all suspended matter is removed from flow, there would be no further erosion, and water would remain clear.[p]

These inferences about fine (and typically multi-size) sediment are based on two experiments on kaolinite suspended in water in an annular flume. In the first experiment [Partheniades *et al.,* 1968], initially suspended kaolinite was allowed to deposit at a constant flow until, after several hours, a residual concentration remained in suspension. At that time the suspension was replaced slowly at a constant rate with clear water, without stopping flow in the flume until practically all suspended matter was removed. It was found that, as the flume continued to rotate beyond this time, erosion was minor.

In another experiment [Parchure, 1984], a bed of kaolinite under initially clear water was eroded in an annular flume at a constant bed shear stress until the suspension achieved a practically constant concentration of 3.85 kg m^{-3}. This suspension was replaced with clear water over a period of 4h without changing the flow. A nearly exponential fall in concentration during this time observed in Fig. 7.37 implies removal of suspension by dilution at a constant rate in the absence of significant erosion. By the end removal the concentration had decreased to 0.03 kg m^{-3}. The experiment was continued further for 24 h, at the end of which the concentration increased to only 0.1 kg m^{-3}.

---

[p] This description related to Fig. 7.36 is valid only as long as concentration and bed shear stress are treated as turbulence–mean quantities ignoring the role of the fluctuations, which is considered later.

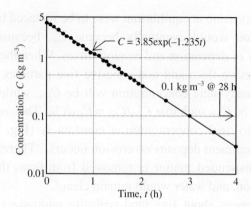

Figure 7.37. Decrease in suspended kaolinite concentration due to dilution in an annular flume (adapted from Parchure and Mehta [1985]).

This small rise was attributed to minor "floc erosion", which continued due to turbulent fluctuations of the bed shear stress.[q]

As we will note in Chapter 9, a uniformly dense bed of flocs will erode at a constant rate until all erodible bed material is entrained, or until damping of turbulence due to the dense suspension decreases the erosion flux. In contrast, a stratified bed in which bed density increases with depth will erode at a decreasing rate until the bed scours to a depth at which the bed shear stress equals the bed shear strength.[r] In this case the final concentration C_{fe} is the result of sediment starvation due to lack of continued supply of particles from the bed.

Example 7.5: In an annular flume in which the volume of suspension is 0.3 m³, it is desired to reduce the concentration from 5 kg m^{-3} to 0.05 kg m^{-3} in 4 h. Determine the volumetric inflow rate of clear water.

---

[q] Floc erosion depends on the spatial variability of floc shear strength over a unit bed surface area containing statistically large number of flocs. As we will note further in Chapter 9, floc erosion flux depends on the frequency distributions of fluctuations in the bed shear stress and the bed floc shear strength. Another influence on floc erosion is that when a floc is detached from its neighbors, elastic rebound (swelling) tends to change the density and shear strength of neighboring flocs.

[r] This condition can be used to experimentally deduce the variation of the erosion shear strength with depth in a stratified bed by carrying out layer-by-layer erosion of the bed (Chapter 9).

At steady-state the rate of withdrawal of water Q equals the rate of inflow. Given suspension volume V_t, we have

$$V_t \frac{dC}{dt} = -CQ \qquad (7.149)$$

which upon integration starting with $C = C_{fe}$ at $t = 0$ yields

$$C = C_{fe} e^{-\frac{Q}{V_t} t} \qquad (7.150)$$

from which $Q = \left[0.3\ln(5/0.05)\right]/4 = 0.345 \text{ m}^3 \text{ h}^{-1}$. If sediment were to be added to the suspension by bed erosion at a mass rate E_r, Eq. 7.149 would become

$$V_t \frac{dC}{dt} = -CQ + E_r \qquad (7.151)$$

which would mean that the dilution law of Eq. 7.150 would not be applicable. (Note that the erosion flux ε is equal to E_r divided by bed area.) The significance of Eq. 7.151 is examined later in this chapter and again in Chapter 9.

The concentration curves in Fig. 7.34 can be approximated by the following empirical equation relating dimensionless concentration C^* to dimensionless time $t_d^* (= t/t_{50})$

$$C^* = \frac{1}{\sqrt{2\pi}} \int_{-\infty}^{t_d^*} e^{-\frac{\omega^2}{2}} d\omega = \frac{1}{2} + \frac{1}{2} erf\left(\frac{t_d^*}{\sqrt{2}}\right)$$

$$C^* = \frac{C_0 - C}{C_0 - C_{fe}}; \quad t_d^* = \frac{1}{\sigma_C} \log\left(\frac{t}{t_{50}}\right) \qquad (7.152)$$

in which σ_C is the standard deviation of the log-normal distribution, t_{50} is the time when $C^* = 0.5$ and $\tau_b^* = \tau_b / \tau_{cd1}$ is a dimensionless bed shear stress (Fig. 7.38). A plot of this equation and data from different tests are shown in Fig. 7.38.

Based on a large number of tests, trends in the variation of σ_C and t_{50} with τ_b^* were qualitatively explained to shed light on the settling behavior of kaolinite as well as natural cohesive sediments [Mehta, 1973]. The effect of τ_b^* on t_{50} was more pronounced in terms of the underpinning physical and physicochemical processes than on σ_C. In

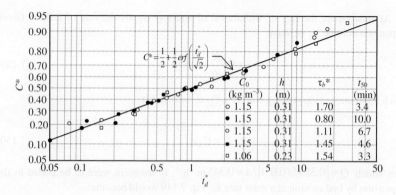

Figure 7.38. Variation of dimensionless concentration C^* with dimensionless time t_d^* for a suspension of kaolinite (from Mehta [1973]).

Fig. 7.39 some results on the variation of $\log t_{50}$ with τ_b^* for kaolinite in distilled water are plotted. Since the range (1.05 to 1.61 kg m^{-3}) of initial concentration C_0 is narrow, the three curves mainly reflect the effect of changing water depth h on the deposition flux. Consider $\log t_{50}$ at $\tau_b^*=1$ as a characteristic value. Since increasing t_{50} would mean decreasing deposition, the data suggest that as h was increased from 0.15 to 0.30 m, characteristically radially-directed secondary currents in the annular flume [Visser *et al.*, 1992] increasingly retarded deposition.

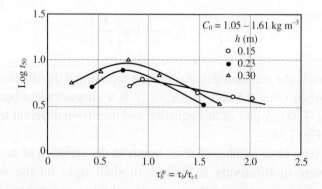

Figure 7.39. Variation of $\log t_{50}$ (t_{50} in min) with dimensionless bed shear stress τ_b^* for suspensions of kaolinite (after Mehta [1973]).

7.3.3.2 *Extension of deposition rate equation*

The application of Eq. 7.145 can be extended to multi-size floc population based on the following considerations: (1) The settling velocity w_s, the initial concentration C_0, the characteristic stress τ_{cd} and the deposition probability p_d are divided into classes $i = 1$ to MM; (2) two relationships are assumed — one between w_{si} and C_{0i}, and another between w_{si} and τ_{ci}; (3) concentration $C(t)$ is obtained by adding the concentrations C_i deduced from a relationship of the form of Eq. 7.145 for each class.

We thus obtain

$$\frac{C}{C_0} = \frac{1}{C_0}\sum_{i=1}^{MM} C_i = \frac{1}{C_0}\sum_{i=1}^{MM} C_{0i}\, e^{-p_{di}\frac{w_{si}}{h}t} \qquad (7.153)$$

in which, following Krone [1962]

$$p_{di} = \begin{cases} 1 - \dfrac{\tau_b}{\tau_{cdi}} & \tau_b < \tau_{cdi} \\[2mm] 0 & \tau_b \geq \tau_{cdi} \end{cases} \qquad (7.154)$$

Figure 7.40 is a schematic drawing of the frequency distribution function $\Phi(w_{si})$ of the settling velocity w_{si}. In that connection we will assume two relationships between w_{si}, C_{0i} and τ_{cdi}. The first is

$$\Phi(w_{si}) = \Phi(C_{0i}) \qquad (7.155)$$

which implies a linear dependence of the settling velocity on concentration ($n_w = 1$ in Eq. 7.47). By relating w_{si} to the respective initial

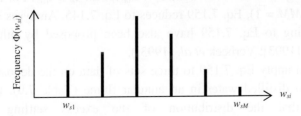

Figure 7.40. Schematic description of discretized frequency distribution of settling velocity.

class concentration C_{0i}, Eq. 7.155 accounts for aggregation processes (but not hindered settling). The second relationship is

$$\tau_{cdi} = k_e w_{si}^{m_e} \qquad (7.156)$$

in which coefficients k_e and m_e depend on the distribution of τ_{cdi} [Mehta and Lott, 1987].

Equation 7.156 incorporates the effect of the characteristic shear stress τ_{cdi} on the settling velocity w_{si}. It permits the reproduction of experimental data on the variation of (depth–mean) concentration with time in the annular flume.

From Eqs. 7.155 and 7.156 we obtain

$$C_{0i} = \Phi(C_{0i}) \cdot C_0 = \Phi(w_{si}) \cdot C_0 \qquad (7.157)$$

and

$$\tau_{cdi} = \tau_{cd1} \left(\frac{w_{si}}{w_{s1}} \right)^{m_e} = \tau_{cd1} \left(\frac{w_{si}}{w_{s1}} \right)^{\frac{\ln(\tau_{cdM}/\tau_{cd1})}{\ln(w_{sM}/w_{s1})}} \qquad (7.158)$$

where m_e has been expressed in terms of $\tau_{cd1}, \tau_{cdM}, w_{s1}$ and w_{sM}. Combining Eqs. 7.153, 7.157 and 7.158 yields the desired expression for concentration decrease with time

$$\frac{C}{C_0} = \frac{1}{C_0} \sum_{i=1}^{MM} C_i = \sum_{i=1}^{MM} \Phi(w_{si}) \, e^{\left[-\left[1 - \frac{\tau_b}{\tau_{cd1}} \left(\frac{w_{s1}}{w_{si}} \right)^{\frac{\ln(\tau_{cdM}/\tau_{cd1})}{\ln(w_{sM}/w_{s1})}} \right] \frac{w_{si}}{h} t \right]} \qquad (7.159)$$

which is subject to the condition, $C_i = C_{0i}$ for all $\tau_b \geq \tau_{cdi}$. For a uniform sediment ($MM = 1$), Eq. 7.159 reduces to Eq. 7.145. Analyses similar to those leading to Eq. 7.159 have also been proposed by others (*e.g.,* Ockenden [1993]; Verbeek *et al.* [1993]).

We will apply Eq. 7.159 to three sets of data on the deposition of a kaolinite in distilled water in an annular flume (Table 7.9). It will be assumed that the distribution of the excess settling velocity $w_{sei} = w_{si} - w_{s1}$ shown in Fig. 7.41 (with $MM = 14$) is applicable. These data were obtained by Yeh [1979] in settling column tests. From annular

Table 7.9. Kaolinite deposition — test parameters.

Run no.	C_0 (kg m^{-3})	τ_b (Pa)	w_{s1} (m s^{-1})	τ_{cd1} (Pa)	m_e
1	1.126	0.333	6.66×10^{-5}	0.084	0.49
2	1.120	0.223	6.66×10^{-5}	0.041	0.62
3	0.968	0.126	1.00×10^{-5}	0.500	0.14

Source: Mehta and Lott [1987].

flume tests τ_{cdM} was found to be 1.2 Pa (Fig. 7.35), and w_{s1} values in Table 7.9 were obtained by calibration of Eq. 7.159. These quiescent settling velocities are in reasonable agreement with values obtained in separate tests on the deposition of kaolinite in a flume under turbulent flow [Dixit *et al.*, 1982].

Concentration curves from Eq. 7.159 are plotted along with measurements (points) in Fig. 7.42. The lower value of the ratio C_{fe}/C_0 (= 0.45) in Run 2 compared to Run 1 (0.78) implies that residual particles in suspension in Run 2 could remain as such at a lower bed shear stress than in Run 1. Due to the size spread of kaolinite spanning clay and silt diameters (Fig. 3.1), the suspended matter contained both flocs and unaggregated silt-size particles. Therefore residual suspended particles in Run 2 would be smaller than those in Run 1. Also, residual particles would be smaller than those at the beginning of the test. In a separate deposition experiment in an annular flume, this inference was

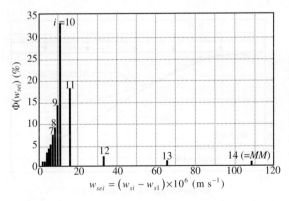

Figure 7.41. Excess settling velocity histogram for kaolinite flocs based on settling column tests by Yeh [1979].

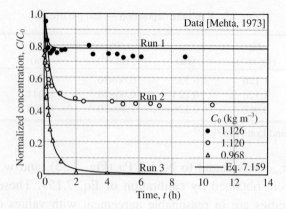

Figure 7.42. Variation of normalized concentration with time: curves based on Eq. 7.159 and data points for deposition of kaolinite (from Mehta and Lott [1987]).

confirmed (Fig. 7.43). The median size of dispersed sediment decreased from about 1 μm initially to 0.2 μm at the end of the test. The size distribution remained essentially intact in the silt size range, but became finer in the clay range.

Interpretation of processes that led to fining of suspended matter is more complex than can be explained by Eq. 7.159. We may infer that silt particles, which were largely unaggregated, were too small to counter turbulence induced lift and deposit on the bed. In the clay range, flocs containing larger clay particles were able to deposit, while smaller ones remained in suspension.

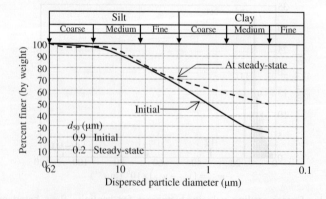

Figure 7.43. Size distribution of dispersed kaolinite in suspension — initial and at the end of the test (from Partheniades *et al.* [1966]).

7.3.4 Significance of bed sediment exchange

7.3.4.1 Exclusive versus simultaneous bed exchange

As described earlier, erosion and deposition processes for clay flocs differ from the same processes for sand and cohesionless silt. For a bed floc the turbulent hydrodynamic force (due to lift and drag) required to erode the floc must just-exceed resistance due to the submerged weight of floc and cohesive (and perhaps adhesive) bonding with the neighbors. Deposition of suspended floc occurs when its submerged weight just-exceeds the hydrodynamic force. When these descriptions for erosion and deposition are considered with the realistic condition that the hydrodynamic fluctuations are finite (or bounded), as opposed to the unbounded normal distribution (Chapter 6), the requirement for equilibrium of erosion and deposition number fluxes is transgressed and bed load transport as defined by Einstein does not occur [Partheniades 1965, 1977, 2009].[s]

Exceptions to absence of bed load are found when cohesive mud occurs as durable balls, pebbles or pellets (e.g., Jacinto and Le Hir [2001]). For their transport as bed load, these particles must be present in statistically significant numbers in a unit area of the bed. Another instance is that of pelletized fecal particles. They tend to be considerably denser than flocs, even when the primary particle size distribution is nearly identical to that of flocs [Edelvang and Austen, 1997]. Thus, the settling velocities of fecal pellets are generally much higher than those of flocs. Furthermore, since pellet density does not decrease with increasing diameter, modeling the transport of pellets must be done apart from the transport of flocs. When pellets occur in large numbers their transport can be expected to be as both bed load and suspended load. This is not a well understood topic particularly because moving mud pellets may reduce in diameter due to abrasion.

Based on conditions for erosion and deposition as mentioned, Partheniades [1965, 1977, 2009] inferred that, under given flow,

---

[s] Mehta [1973] elaborated on this argument based on the bounded (i.e., with finite tails) distribution function of Braswell and Manders [1970].

cohesive particles can either deposit or erode — but never both at the same time. This argument for exclusive deposition or erosion is partly based on laboratory experiments on the deposition of kaolinite. If a floc is strong enough to settle through the high shear layer near the bed and become attached, then it will be strong enough to withstand the bed shear stress (unless this stress increases and exceeds the erosion bed shear strength). If so, there would be no erosion at the flow condition at which deposition occurs. This scenario is reasonable as long as the particles are all of the same size. Partheniades assumed that the sediment (kaolinite) tested was entirely flocculated, with all flocs having the same composition and diameter.

The Partheniades "exclusive model" is based entirely on the observed time-variation of suspended sediment concentration in flumes at mid-depth. Movements of fine particles near the bed surface could not be visually detected due to turbidity. As we noted earlier, supportive evidence for the model included deposition experiments in which the sediment-carrying capacity of flow was found to be independent of flow velocity. As also mentioned, evidence was presented from an erosion experiment in which, once a constant concentration was reached after several hours, the remaining suspension was replaced by clear water without arresting the flow. It was shown that no significant erosion occurred thereafter.

The exclusive model is at variance with the occurrence of simultaneous erosion and deposition in laboratory studies inferred by others (*e.g.,* Krone [1962] and Lick [1982]). Near-bed sediment exchange (sketched in Fig. 7.44) was developed in an attempt to bridge

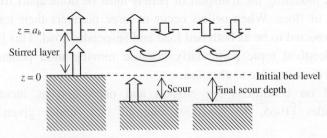

Figure 7.44. Schematic description of sediment fluxes (arrows) in the stirred layer (from Cervantes *et al.* [1995]).

the exclusive model and the simultaneous exchange model of flocs at the bed. The layer extending from the bed (at $z = 0$) to a notional height a_h may be idealized as a thin "stirred layer" within which erosion starting at $t = 0$ sets up a convective cell of diffusive and settling fluxes of flocs. Under a constant bed shear stress these fluxes will eventually approach equality. The bed, assumed to be stratified with respect to density and shear strength (*i.e.*, a bed in which both density and shear strength increase with depth), will practically cease to erode at some depth of scour where the shear stress equals the erosion shear strength (Chapter 9). The outcome will be a steady-state value of suspended sediment concentration. Thus, by applying bed shear stress-based equations for exclusive erosion and deposition (*e.g.*, Eqs. 7.145 and 9.26) at $z = a_h$, simultaneous exchange can be simulated. Flume experiments on deposition, in which two-way exchange between bed and suspension likely occurred, have been described using this concept [McAnally, 1999].

Unfortunately, as we will see next the stirred layer concept does not unequivocally explain the result of radioactive tracer tests of Krone [1959] on the deposition of mud from San Francisco Bay in a flume.

7.3.4.2 *Experiment with tagged particles*

In general, radioisotope-based tracing is not believed to interfere with flocculation in a significant way as long as only a small fraction of the total number of particles per unit volume is tagged. However, it appears to be important to tag the actual sediment, because using a different material may lead to unrepresentative results on floc transport. Spencer *et al.* [2010] mixed different proportions of flocs of a montmorillonite tagged with Holmium (^{67}Ho), a natural element, with mud flocs from Tamar River estuary (UK). The resulting flocs were examined in a settling column. Before mixing, the two sediments had similar flocs. However, while the resulting macroflocs (> 160 μm) were comparable with macroflocs of natural mud, the microflocs (< 160 μm) were smaller and settled at a slower rate. Overall, differences between the original mud flocs and the mixed flocs were substantial.[t]

Based on emission spectra, decay rate, cost and ease of handling, Krone [1957] concluded that the radioactive isotopes ^{198}Au (gold) and ^{46}Sc (scandium) were suitable as tracers for San Francisco Bay mud. In a flume test [Krone, 1959] a small fraction of the initially suspended material was labeled with colloidal ^{198}Au, which has a half-life of 2.697 d. Figure 7.45 shows (depth–mean) suspended sediment concentrations after the flow velocity was reduced to permit deposition. The rate of deposition of labeled sediment was higher than that of the total sediment.

Evaluation of whether erosion occurred during this deposition dominated experiment in which sampling was limited to the suspension is not straightforward, and makes it necessary to assume that floc size, density and shear strength were not altered by gold labeling. Under this assumption the likely process of particle exchange between the suspension and the bed is idealized in Fig. 7.46. Suspended sediment at time t_0 consists of 15 particles, of which 6 (*i.e.*, 40%) are tagged with

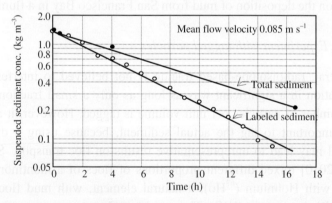

Figure 7.45. Variation of suspended sediment concentration during deposition of San Francisco Bay mud (from Krone [1959]).

---

[t] Methods for detection of natural radioactive tracers associated with sediments (see Table 8.1) offer considerable advantage in terms of the extent of spatial coverage in the field. This technique permits mapping bottom cohesionless and cohesive sediments without manipulation of their properties [van de Graaf *et al.*, 2006].

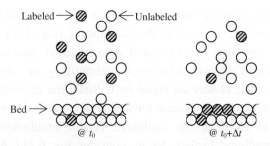

Figure 7.46. Exchange of labeled and unlabeled particles (from Mehta [1991]).

gold. We may assume that thereafter the tagged particles will comprise 40% of the remaining initial suspension. At the end of a period Δt, 12 particles are in suspension of which only two are tagged. This means that 5 of the original 15 particles remain in suspension. If Δt were conveniently taken as 1 s, the deposition of tagged sediment would be $15 - 5 = 10$ particles s^{-1}. The additional 7 particles ($12 - 5$) in suspension would have been eroded from the bed. The net rate of deposition for the total concentration would be $15 - 12 = 3$ s^{-1}, lower than the deposition rate of the tagged sediment mixture. This indicates that bed erosion occurred even while the dominant trend was one of deposition.

7.3.4.3 *Bed exchange paradigms*

Mechanics of particle exchange at the bed must be described accurately particularly when modeling fine sediment loads in unsteady flows. However, basic discrepancies in understanding remain unresolved. For tide-induced sediment transport in a marina basin, Hayter [1983] assumed mutually exclusive erosion and deposition. For a similar application in a navigation channel, Teeter [2001b] argued for deposition exclusive of erosion, along with modeling sediment sorting by size. Lick [1982, 2009] postulated that it was necessary to consider simultaneous exchange to explain flume results on the movement of suspended flocs. For modeling tidal variation of suspended sediment in the Jiaojiang estuary, China, Jiang [1999] modeled simultaneous exchange. Likewise, for modeling wave-induced resuspension in a flume, Maa [1986] made the same assumption. Sanford and Halka [1993] used a numerical modeling approach based on a single particle size to explain the behavior of tidally suspended sediment in the Chesapeake Bay. They showed that in order to reproduce the measured concentration time-series it was essential to permit deposition to occur continuously, *i.e.,* the probability of deposition $p_d[0, 1]$ was set equal to 1. Since the depositional flux is always present, erosion is evident only when it exceeds deposition.

In order to examine the exclusive versus simultaneous erosion–deposition paradigms further, let us consider Eq. 6.111 for the time-averaged balance of suspended sediment mass in the vertical (z) direction. This equation can be restated as

$$\frac{\partial C}{\partial t} - w_s \frac{\partial C}{\partial z} - \frac{\partial}{\partial z}\left(D_z \frac{\partial C}{\partial z}\right) = 0 \qquad (7.160)$$

in which D_z is the turbulent mass diffusion coefficient. As we noted in Chapter 6, for solving Eq. 7.160 the usual boundary condition at the water surface ($z = h$) is that the net sediment flux across the surface must be nil. The solution is then dependent on the condition at the bed ($z = 0$)

$$-w_s C\big|_{z=0} - D_z \frac{\partial C}{\partial z}\bigg|_{z=0} = \varepsilon - \delta_D \qquad (7.161)$$

where ε is the erosion flux and δ_D is the deposition flux, both in units of dry sediment mass per unit bed area and time. Whether there is only erosion, only deposition, or no exchange depends on the bed shear strength (with respect to erosion) τ_s and the characteristic stress for deposition τ_{cd}, recognizing that due to cohesion, $\tau_s > \tau_{cd}$. This is so because, as we will note in Chapter 9, τ_s is defined by the cohesive properties of bed flocs while τ_{cd} is characterized by the properties of falling flocs.

The exclusive paradigm makes it necessary to sub-divide flood and ebb phases of the tidal cycle into periods (Fig. 7.47a). Starting from

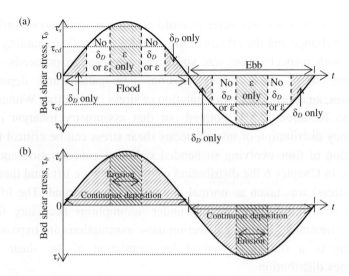

Figure 7.47. Periods of erosion (ε), deposition (δ_D) and no bed exchange during a tidal cycle: (a) Exclusive paradigm; (b) simultaneous (continuous deposition) paradigm.

slack water, the first period in which only deposition can occur corresponds to the duration when the bed shear stress τ_b is less than τ_{cd}. In the second period τ_b is between τ_{cd} and τ_s (which must be exceeded by τ_b for erosion to occur). In this period there is neither erosion nor deposition. In the third period when τ_b is greater than τ_s, there can be erosion but no deposition. The reverse sequence follows as the bed shear stress begins to decrease past its peak at the strength of (tidal) flow. Field evidence supporting this sequence of exclusive processes has not been found.

In contrast with the exclusive paradigm, simultaneous exchange admits the possibility of continuous deposition and a period when both erosion and deposition can occur at the same time (Figure 7.47b). To model this, two important considerations are required.

Firstly, the diameter and other properties of cohesive particles are always multi-class due to the initial size spread and aggregation processes. Secondly, it is essential to define the two stress-related variables, τ_{cd} and τ_s, in terms of the multi-class representation [Letter, 2009].

In Chapter 9, we will refer to multi-class representation of variables in bed exchange and the effects of introducing stochastic variability in τ_{cd} and τ_s with respect to the size spectra as aggregation proceeds. Since turbulent shear stress drives erosion, aggregation and deposition processes, an appropriate representation of stress is essential. Winterwerp and van Kesteren [2004] pointed out that asymmetry inherent in the frequency distribution of instantaneous shear stress can be critical to the prediction of time-evolving suspended sediment concentration and size spectra. In Chapter 6 the distribution of hydrodynamic lift (and therefore shear stress) was taken as normal and also as non-normal. The lift was related to the turbulent velocity under assumptions necessary for an analytic treatment. In the next section these assumptions are bypassed by resorting to a statistical method for simulation of the shear stress frequency distribution.

7.3.5 Shear stress distribution

The instantaneous turbulent bed shear stress $\tau_b(t)$ is expressed as the sum

$$\tau_b = \overline{\tau}_b + \tau_b' \tag{7.162}$$

where $\overline{\tau}_b$ and τ_b' are the time–mean and the fluctuating components, respectively. In terms of velocity u_b ($= \overline{u}_b + u_b'$) characteristically at the top of the particle (Fig. 6.5), or near the bed in general, the quadratic law for bed shear stress analogous to Eq. 6.18 is

$$\tau_b = C_D \rho \frac{u_b^2}{2} = C_D \rho \frac{(\overline{u}_b + u_b')^2}{2} \tag{7.163}$$

For a zero mean velocity (*i.e.*, $\overline{u}_b = 0$), as would occur when for instance a grid oscillating in water at a suitably high frequency produces turbulence, from Eqs. 7.162 and 7.163 we obtain

$$\tau_b' = C_D \rho \frac{u_b'^2}{2} \tag{7.164}$$

in which u_b' is represented by the normal probability density function (pdf). Due to non-linearity in Eq. 7.164, normal pdf of u_b' will result in a skewed (non-normal) pdf of τ_b'.

Referring to Eq. 7.163, let us denote the pdfs of u_b and τ_b by $\Phi(u_b)$ and $\Phi(\tau_b)$, respectively. Sharma [1973] used a hot-wire sensor to measure $\Phi(\tau_b)$ in an air duct with smooth walls and turbulent flow (air Reynolds number = 1.4×10^5). Turbulence intensities were determined 4.9 cm above the duct bottom and reported as hot-wire-voltage E_v. Figure 7.48 shows a pdf with a slight bias towards the high end and an asymmetric peak. This is due to turbulent bursting which in general produces coherent and strongly shearing eddies, and is the basis of the quadratic relationship between velocity and shear stress in turbulent flows. Similar observations in water have been made by, among others, Obi *et al.* [1996].

The normal pdf of $\Phi(u_b)$ is

$$\Phi(u_b) = \frac{1}{\sigma_u \sqrt{2\pi}} e^{-\left(\frac{u_b - \bar{u}_b}{2\sigma_u^2}\right)^2} \tag{7.165}$$

where σ_u is the standard deviation of u_b. For a random value of u_b, the corresponding τ_b is obtained from Eq. 7.163. After selecting $C_D = 0.005$, $\rho = 1{,}027$ kg m^{-3} and $\bar{u}_b = 0.5$ m s^{-1} as characteristic values, Monte

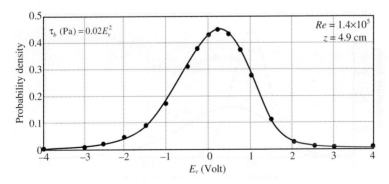

Figure 7.48. Probability density function based on hot-wire-sensor voltage measurements. Bed shear stress can be calculated using the relationship given in the plot. Data points are from Sharma [1973].

Carlo simulation (using Eq. 7.163) was carried out using 10^7 random values of u_b and 200 partitions of both u_b and τ_b. The lower and upper limits of the normalized fluctuation $(u_b - \overline{u}_b)/2\sigma_u^2 = u_b'/2\sigma_u^2$ were chosen as 0 and 1 to represent adequately small and large values, respectively, of this fluctuation.

In Fig. 7.49, simulated $\Phi(u_b)$ is identical to Eq. 7.165, whereas $\Phi(\tau_b)$ is biased towards high stresses. This latter distribution has a mean $\overline{\tau}_b = 0.650$ Pa, standard deviation $\sigma_\tau = 0.232$ Pa, coefficient of skewness (the third moment of pdf about the mean divided by σ_τ^3) equal to 0.535 and the coefficient of kurtosis (the fourth moment of pdf about the mean divided by σ_τ^4) equal to 3.372. For comparison, normally distributed $\Phi(\tau_b)$ would have a skewness of 0 and a kurtosis of 3. Also, the computed $\Phi(\tau_b)$ is much less peaked than normal distribution.

7.4 Exercises

7.1   Under the assumption that $w \ll w_s(t)$, the unsteady form of Eq. 7.13 is

$$\frac{4}{3}\pi\left(\frac{d_p}{2}\right)^3 \rho_s \frac{dw_s(t)}{dt} + \frac{1}{2}C_D \frac{\pi d_p^2}{4}\rho w_s(t)^2 - \frac{1}{6}\pi d_p^3 g \Delta\rho = 0 \qquad \text{(E7.1)}$$

A particle in a fluid falls from rest at time $t = 0$.

(1)   Solve Eq. E7.1 for $w_s(t)$.

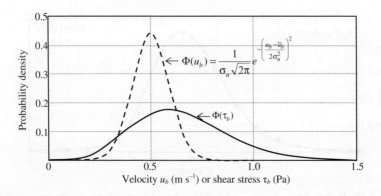

Figure 7.49. Effect of normally distributed velocity on shear stress. The area under each curve is 1 (from Letter [2009]).

(2) The distance of fall $z' = 1 - z$ from the starting point varies with time according to [Bird *et al.*, 1960]

$$z' = \frac{1}{A'} \ln\left[\cosh\left(\sqrt{A'B'}\, t \right) \right] \tag{E7.2}$$

Evaluate the coefficients A' and B'.

7.2 Plot and compare the relationship between the drag coefficient C_D and the Reynolds number Re_p in the range of 0.001 to 10 from Eq. 7.17 with the same relationship from Eq. 7.19. Also include Eq. 7.18 with

$$X = \frac{3}{16} Re_p - \frac{19}{1280} Re_p^2 + \frac{71}{20\,480} Re_p^3 \cdots \tag{E7.3}$$

which was proposed by Goldstein [1929] (for $Re_p \le 2$). By selecting only the first term on the right hand side of Eq. E7.3, the resulting expression becomes identical to one derived by Oseen [1927].

7.3 (1) Using Eq. 7.115, calculate peak settling velocities with coefficients for the Scheldt in Table 7.6. Select concentrations $C = 0.05$, 0.1, 0.3, 1, 5 and 6 kg m^{-3} and assume $k_d = 1$. Can you detect any difference in the effect of cohesion between Tamar (Fig. 7.27) and Scheldt sediments?

(2) Calculate the shear stress τ_m at the peak settling velocity.

7.4 Near-bottom suspended sediment concentration C was measured daily over a period of several months inside a small marina basin. The frequency of occurrence $\varphi(C)$ of concentration is plotted as a histogram in Fig. E7.1. The respective settling velocities (m s^{-1}) can be obtained from Eq. 7.47 with $a_w' = 2.85 \times 10^{-5}$ and $n_w = 1.33$. Determine the annual rate of shoaling in the basin, assuming bed dry density to be 800 kg m^{-3}. Hint: Pro-rate the settling flux corresponding to each concentration by the respective frequency of occurrence. For each frequency bar in the plot, select the mean of the concentration range as the value for that bar, *e.g.*, 12.5 mg L^{-1} for the first bar, *etc.* For use of Eq. 7.47, convert concentrations to kg m^{-3}.

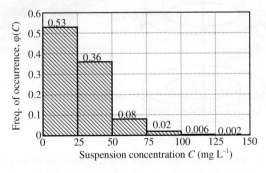

Figure E7.1. Histogram of suspended sediment concentration inside a marina basin.

7.5   The schematic drawing of a counter-rotating annular flume (CRAF) is shown in Fig. E7.2a. The channel's inner radius is r_{ia} and outer radius r_{oa}. It is filled with sediment suspension to the desired height h. If the channel alone were rotated at an angular speed ω_o, it would produce a radially outward secondary current at the bottom. However, there would be no current in the tangential (main flow) direction, since at steady-state water would move at the same speed as the channel (solid-body rotation). To produce a tangential current, the ring is lowered into the channel so that its base is flush with the water surface leaving very narrow annular gaps between the ring's edges and the channel walls. Motion of the ring produces shear at the water surface and a tangential boundary-layer current (Fig. 7.2b). The ring also produces a radially inward secondary current at the bottom. To balance the opposing secondary currents from the channel and the ring, the latter is rotated counter to the channel. The ring speed is adjusted until the secondary currents balance, leaving only tangential flow. Thus CRAF produces an "endless" current in which flocs are not disrupted by pumps and return flow pipes, as they do in conventional flumes. McAnally and Mehta [2002] used a (numerical) model of aggregation processes to show the substantial influence of the return-flow centrifugal pump on the deposition flux of suspended cohesive sediment in the experiments of Krone [1962]. For simulation of the hydrodynamics of annular flumes based on Large Eddy Simulation modeling see Schweim *et al.* [2000], see also Graham *et. al.* (1992).

(1)   Determine the radially directed pressure head due to centrifugal force in the channel. (The ring is rotated to oppose this head.)

(2)   If the ring-induced shear stress at the water surface is τ_{br}, estimate the shear stress on the channel bed. State your assumptions.

(3)   Determine the shape of the velocity profile under an assumed two-dimensional flow structure, *i.e.*, ignore wall effects and the radius of curvature of the channel.

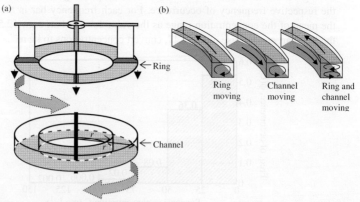

Figure E7.2. Schematic drawing of: (a) CRAF (adapted from Mehta [1973]; Stuck [1996]); (b) secondary flows produced by ring and channel (from Schweim *et al.* [2000]). See also Graham *et al.* [1992].

7.6 Settling velocities deduced from field and laboratory tests on suspended fine sediment in an estuary are given in Table E7.1. Based on concentration, the settling regime can be conveniently divided into free settling, flocculation settling and hindered settling sub-ranges. Determine best-fit equations for these sub-ranges. Plot the data and the equations together.

7.7 Krone [1962] conducted an experiment on the deposition of suspended cohesive flocs from San Francisco Bay in a flume in which the depth of water flowing at a constant velocity was 0.3 m. For this exercise consider that the initial suspended sediment concentration of 2.5 kg m^{-3} was reduced to 0.195 kg m^{-3} after 280 h of deposition (Fig. E7.3 and Table E7.2). This reduced value may be taken as the residual concentration $C_{fe} = C_{fe1}$. At 280 h a wire-mesh grid was inserted into the flume perpendicular to flow and spanning the flume cross-section. As a result, over the next 100 h the concentration fell further to 0.00005 kg m^{-3} ($=C_{fe2}$) nearly clarifying the water column.

(1) Plot pre-grid and post-grid data separately as $[C(t)-C_{fe}]/[C_0-C_{fe}]$ against time t (with a starting time of 0 in each case), and estimate the settling velocity from each curve assuming the validity of Eq. 7.145 (with $p_d = 1$ obtained by setting $\tau_b = 0$).

Table E7.1. Concentration — settling velocity data.

Concentration (kg m^{-3})	Settling velocity (m s^{-1})	Concentration (kg m^{-3})	Settling velocity (m s^{-1})
Free settling		Hindered settling	
0.1	1.00×10^{-5}	8	1.00×10^{-3}
0.2	1.00×10^{-5}	9	9.65×10^{-4}
Flocculation settling		10	9.00×10^{-4}
0.2	1.00×10^{-5}	20	5.40×10^{-4}
0.3	1.70×10^{-5}	30	3.07×10^{-4}
0.4	2.50×10^{-5}	40	1.80×10^{-4}
0.5	3.50×10^{-5}	50	1.15×10^{-4}
0.6	4.50×10^{-5}	60	6.30×10^{-5}
0.7	5.25×10^{-5}	70	4.00×10^{-5}
0.8	6.30×10^{-5}	80	2.40×10^{-5}
0.9	7.30×10^{-5}	90	1.50×10^{-5}
1	8.50×10^{-5}	100	1.00×10^{-5}
2	1.90×10^{-4}		
3	3.00×10^{-4}		
4	4.30×10^{-4}		
5	6.00×10^{-4}		
6	7.50×10^{-4}		
7	9.00×10^{-4}		
8	1.00×10^{-3}		

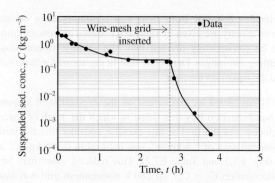

Figure E7.3. Flume experiment in which a wire-mesh grid was inserted to enhance the deposition flux (adapted from Krone [1962]).

Table E7.2. Grid-test data.

t (h)	C (kg m^{-3})
Pre-grid	
0	2.50
10	2.00
20	1.90
35	1.05
45	1.00
70	0.65
120	0.40
130	0.50
175	0.25
220	0.23
235	0.22
270	0.21
280	0.20
Post-grid	
0	0.20
10	0.050
60	0.0025
100	0.0004

(2) Use Stokes law (Eq. 7.15) along with the following relationship between floc excess density $\Delta\rho_f$ (kg m^{-3}) and floc diameter d_f (m)

$$\Delta\rho_f = \frac{\alpha_\Delta}{d_f^{\beta_\Delta}} \qquad (E7.4)$$

Estimate pre-grid and post-grid floc sizes d_f as well as the respective floc densities ρ_f.[u] For the present exercise select $\alpha_\Delta = 0.0006$, $\beta_\Delta = 1$, $\nu = 1.15\times10^{-6}$ m^2 s^{-1} and $\rho_w = 1,025$ kg m^{-3}.

(3) Based on values of d_f and ρ_f comment on the effect of the grid. Discuss in what way the flocs may be peculiar to the flume with a high-shear rate centrifugal pump in the return flow pipe.

7.8 Given the data in Table E7.3 on floc size and settling velocity of organic-rich muck from Lake Apopka in central Florida, plot floc excess density $\Delta\rho_f$ against floc size and determine if a relationship such as Eq. E7.4 exists. Use Eq. 7.19 to calculate the drag coefficient, assuming a particle density of 1,710 kg m^{-3} and water density 1,000 kg m^{-3}. Briefly comment on your answer.

7.9 The settling velocity equation of Strom and Keyvani [2011] is as follows:

$$w_s = \frac{g(s-1)d_f^{D_f-1}}{\alpha_{s1}\nu_w d_p^{D_f-3} + \alpha_{s2}\sqrt{g(s-1)d_f^{D_f}d_p^{D_f-3}}} \qquad (E7.5)$$

Plot w_s against d_f using the following values: $\rho_w = 1,027$ kg m^{-3}, $\rho_s = 2,670$ kg m^{-3}, $s = \rho_s/\rho_w$, $\nu_w = 1.15\times10^{-6}$ m^2 s^{-1}, $d_p = 5$ μm, $\alpha_{s1} = 18$, $\alpha_{s2} = 0.548$, and $D_f = 1.5, 2,$ 2.5 and 3. Vary d_f between 10 and 1,000 μm.

---

[u] Equation E7.4 is based on the experimental work of Kranck and Milligan [1992] on San Francisco Bay mud, with $\alpha_\Delta = 0.035$ and $\beta_\Delta = 1.09$ applicable in the range of d_f from 100 to 1000 μm. This equation has the same form as that obtained by Farrow and Warren [1989] for 1–4 μm glass spheres flocculated by a chemical agent. The trend is also consistent with floc data from the Dollard estuary in the Netherlands, for which $\Delta\rho_f$ was found to decrease from about 600 kg m^{-3} at $d_f = 100$ μm to 100 kg m^{-3} at 350 μm [van der Lee, 2000]. The coefficient α_Δ is sensitive to the sediment, floc size range and whether flocs are formed naturally or in the laboratory. The exponent β_Δ appears to be consistently close to 1. For instance, flocs from the Chesapeake Bay examined by Gibbs [1985] conform to $\beta_\Delta = 0.97$.

Table E7.3. Floc sizes and settling velocities for Lake Apopka muck.

Floc size (μm)	Settling velocity (mm s^{-1})
250	0.554
231	0.832
129	0.517
182	1.079
206	0.870
235	0.830
294	0.785
115	0.977
114	0.800
92	0.669
202	0.659
89	0.642

Chapter 8

Sedimentation and Bed Formation

8.1 Chapter Overview

A bed of particles is formed from sedimentation and consolidation processes defined in Chapter 6. These processes are reviewed for water-saturated soils in the submerged environment. This excludes intertidal mudflats where loss of soil water due to drainage and desiccation creates unsaturated conditions (see Exercise E8.1). Also not considered are gassy sediments. Overall, the treatment is limited to one (vertical) dimensional transport.

The chapter begins with the definition of effective normal stress differentiating sedimentation from consolidation. Next, the equation of flow continuity in bed is given. The subsequent section includes a description of the modes of sedimentation. The formation of drainage channels, pockets of inhomogeneous materials and laminas in the bed are mentioned. Also indicated is the role of tracers as identifiers of bed stratification. Physical bases for simple quantitative analysis of sedimentation are then summarized. These cover empirical, Darcy's law-dependent and mass continuity-dependent developments. In the subsequent section an analytic treatment is introduced to highlight the continuity of the mechanics of combined sedimentation and consolidation.

In the context of bed formation by consolidation, the basic theory of bed volume reduction due to a surcharge or overburden is summarized. A useful analytic development of linearized solutions for the self-weight effect is considered next. The definition of bed rebound and rudimentary modeling of bed densification are given at the end of the chapter.

8.2 Effective Stresses and Flow in Porous Beds

8.2.1 *Effective stresses*

Consider a cubic element of dimension a_c within a saturated sediment bed (Fig. 8.1). The horizontal and vertical normal forces due to the weight of soil above the element are F_h and F_v, respectively, and the corresponding geostatic stresses are $\sigma_h = F_h / a_c^2$ and $\sigma_v = F_v / a_c^2$. The respective pore water pressures are p_{wh} and p_{wv}, the unit weight of wet soil is γ_t and the unit weight of water γ_w. For a sufficiently small volume element containing a large number of discrete particles (in water), the following two relationships hold:

$$\sigma_v = z\gamma_t \tag{8.1}$$

and

$$p_{wh} = p_{wv} = p_w = \rho g z_w \tag{8.2}$$

where z_w is the height of the water table. For a fully saturated bed the pore pressure is $p_w = \rho g z$, where ρ is fluid (water) density.

The (horizontal and vertical) effective stresses are each equal to the respective total stress minus the pore pressure, *i.e.*,

$$\sigma_h' = \sigma_h - p_w \tag{8.3}$$

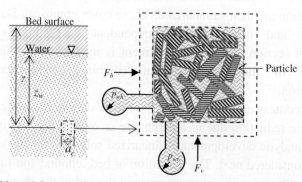

Figure 8.1. Normal forces and stresses in a sediment bed element (adapted from Lambe and Whitman [1979]).

$$\sigma_v' = \sigma_v - p_w \qquad (8.4)$$

The total stress acts over the entire face of the cubic element, whereas pore pressure acts over parts of the face covered by water. Thus the effective stress, which is a notional (mathematical) quantity, is considered to act over the area of the face occupied by particles.[a]

8.2.2 Flow continuity

In order to determine the rate of decrease of soil volume due to compression by stress when sedimentation or consolidation occurs, the stress is related to the rate of volume change of the porous deposit. In that regard consider viscous flow through an elemental volume (Fig. 8.2) of a porous medium.

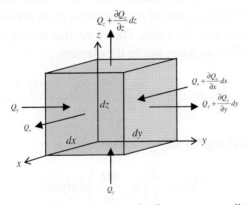

Figure 8.2. Elemental volume of soil as a porous medium.

---

[a] For clay-sand mixtures, Merckelbach and Kranenburg [2004a] give the following expression relating σ_v' to the particle's fractal dimension D_f

$$\sigma_v' = K_\sigma \left(\frac{\phi_c}{1-\phi_s} \right)^{\frac{2}{3-D_f}} + K_{\sigma 0} \qquad (F8.1)$$

where K_σ and $K_{\sigma 0}$ are empirical constants, ϕ_c is the clay volume fraction (of total solids volume) and ϕ_s is the sand volume fraction. When $K_{\sigma 0}$ is small, Eq. F8.1 becomes a simple power-law relating σ_v' to K_σ. For the Ems–Dollard estuary (The Netherlands) mud $K_\sigma = 1.1 \times 10^8$ kPa was obtained.

Assuming steady-state, the total water flow rate or discharge Q in the volume is the sum of its three spatial components, *i.e.,*

$$Q = Q_x + Q_y + Q_z \qquad (8.5)$$

In saturated beds the pressure at any point is represented by the total head $H_t = (p/\rho g) + H$, where p is the flow-induced water pressure and H is the hydrostatic head. Therefore the gradient of H_t in the vertical direction yields the corresponding water discharge Q_z from Darcy's law (Eq. 2.98 with k_p replaced by k_{pz} denoting permeability in the z-direction)

$$Q_z = w_w dxdy = k_{pz} \left(-\frac{\partial H_t}{\partial z} \right) dxdy \qquad (8.6)$$

where Q_z is positive into the element, and w_w is water velocity or Darcy velocity. Note that the so-called seepage velocity is obtained by dividing Q_z by the area occupied by voids. Therefore, this area is a fraction of the total area $dxdy$. For discharge out of the element,

$$Q_z + \frac{\partial Q_z}{\partial z} dz = \left(k_{pz} + \frac{\partial k_{pz}}{\partial z} \right) \left(-\frac{\partial H_t}{\partial z} - \frac{\partial^2 H_t}{\partial z^2} \right) dxdy \qquad (8.7)$$

Thus, the net discharge in the vertical direction is

$$\Delta Q_z = \frac{\partial}{\partial z} \left(k_{pz} \frac{\partial H_t}{\partial z} \right) dxdydz \qquad (8.8)$$

Now the total net discharge is

$$\Delta Q = \Delta Q_x + \Delta Q_y + \Delta Q_z \qquad (8.9)$$

in which the individual net discharges in the x- and y-directions, ΔQ_x and ΔQ_y, respectively, are obtained from terms analogous to Eq. 8.8. Thus

$$\Delta Q = \left[\frac{\partial}{\partial x} \left(k_{px} \frac{\partial H_t}{\partial x} \right) + \frac{\partial}{\partial y} \left(k_{py} \frac{\partial H_t}{\partial y} \right) + \frac{\partial}{\partial z} \left(k_{pz} \frac{\partial H_t}{\partial z} \right) \right] dxdydz \qquad (8.10)$$

We recognize that the volume of water V_w in the element is

$$V_w = \frac{e_v}{1+e_v} dxdydz \qquad (8.11)$$

where e_v is the void ratio. Since the volume of solids in the element, $V_s = dxdydz/(1+e_v)$, is constant, we can express ΔQ as

$$\Delta Q = \frac{DV_w}{Dt} = \frac{1}{1+e_v} \frac{De_v}{Dt} dxdydz \qquad (8.12)$$

where D/Dt denotes the total rate (*i.e.*, sum of temporal and spatial rates) of change of any quantity. Therefore by equating the net discharges ΔQ from Eqs. 8.10 and 8.12 we obtain, after rearrangement,

$$\frac{1}{1+e_v} \frac{De_v}{Dt} = \frac{\partial}{\partial x}\left(k_{px} \frac{\partial H_t}{\partial x} \right) + \frac{\partial}{\partial y}\left(k_{py} \frac{\partial H_t}{\partial y} \right) + \frac{\partial}{\partial z}\left(k_{pz} \frac{\partial H_t}{\partial z} \right) \qquad (8.13)$$

which is the desired form of equation of continuity. This equation along with the definition of effective normal stress is the basis for describing consolidation. In the next section we will review sedimentation of suspensions, in which there are no permanent effective stresses.

8.3 Sedimentation and Related Processes

8.3.1 *General description of settling and consolidation*

Due to low submerged weight, flocculated sediment settles more slowly than coarse or non-cohesive silty sediment of the same size as flocs, and generally undergoes three stages (Fig. 8.3a).

Flocculation stage: In this stage aggregation processes, in particular floc growth, are important. Aggregation is driven by inter-particle collisions due to differential settling and Brownian motion when turbulence is absent. If the initial particles are primary or dispersed and conducive to cohesion, in this stage formation of flocs by

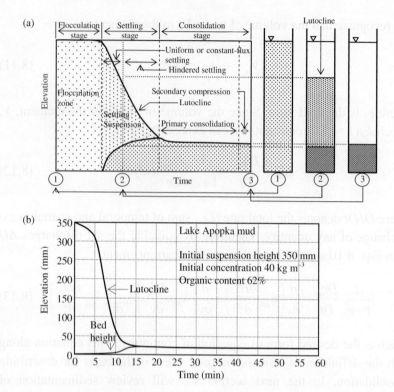

Figure 8.3. (a) Schematic drawing of sedimentation stages: flocculation, settling, and consolidation (adapted from Imai [1981]); (b) lutocline fall and rise of bed height during the settling of a suspension of Lake Apopka (FL) mud (from Sampath [2009]).

coagulation will occur. While for some slurries, *e.g.,* those of kaolinite, the flocculation stage may be only tens of seconds long, for some bentonite slurries it may last effectively for minutes. Even though in the initial phase of this stage settling is minor, substantial physicochemical changes may occur at the microscopic level. The onset of a rapid fall in the interface at the end of the stage, when weighty flocs are formed, shows a smooth transition into the next stage.

Settling stage: Two phases of this stage can be identified. During the initial phase the rate of fall of the interface is uniform because hindrance against fall is negligible. This is the constant-flux settling phase. In the second phase the fall of the interface, a lutocline, is hindered and the rate of fall decreases with time. Below the lutocline a second interface defining the bed rises above the bottom.

The settling stage is accompanied by a significant change in the structure of newly deposited flocs. Krone [1963] described this change in terms of the order-of-aggregation concept (Chapter 4). When aggregates of order arbitrarily designated as n'' deposit they form an initial fluffy, *i.e.,* open-structured and weak, layer of aggregates of order $n''+1$. When this weak layer exceeds a thickness of about 2–3 cm, overburden crushes the $n''+1$ order aggregates back to n'' order aggregates, which then break further into aggregates of order $n''-1$, then $n''-2$, and so on.

Flocs at the bottom of the deposit can be zero-order (tightly packed and strong), or higher if the compressive (crushing) load is not sufficiently high. Since zero-order flocs do not break easily, the maximum consolidated-bed density is defined by the packing arrangement of these flocs. If and when zero-order flocs are crushed by very high loads a dense, shear resistant, bed will result, such as the glacially over-consolidated soils in the Great Lakes region and northward.[b] In Chapter 9 the dependence of bed shear strength of density is examined.

The transition from settling to consolidation occurs at a time t^* when the lutocline meets the rising bed height. In Fig. 8.3b, which shows experimental data for mud (muck) from Lake Apopka in central Florida, t^* is about 15 min when the two curves merge into one. Beyond this time, due to consolidation the newly formed bed surface falls slowly until there is no further reduction in the height of deposit. Whereas the settling stage may last from minutes to hours, full consolidation can take hours to as much as years. In some cases it may take decades or even longer for a hard bed to form. As long as dewatering continues the process is called primary consolidation. Once dewatering ends, internal rearrangements of particles may continue due to physicochemical changes in the material properties accompanied by a lowering of the internal energy of the

---

[b] Present interest is mainly on normally consolidated sediment beds, *i.e.,* beds formed by consolidation under existing stress. For these beds the over-consolidation ratio (OCR), *i.e.,* the ratio of maximum to existing stress is 1. Glacially compressed sediments have high OCR values.

system. These changes can be due to diagenesis of labile organic matter or leaching of inorganic molecules and is known as secondary compression or consolidation.

Settling in the hindered mode and consolidation are both governed by the rate at which pore water seepage (and dewatering) occur. As mentioned, within the settling suspension there is no significant effective normal stress, *i.e.*, pore pressure is practically equal to total hydrostatic pressure. In a consolidating bed, pore pressure is less than total pressure because part of the total weight of the slurry is supported by the particle matrix. The difference in pressure, equal to the effective normal stress, represents load supported by particles (as opposed to water).

Figure 8.4a shows an instantaneous density profile for an estuarine silty-clay from Combwich (UK) in a column 4.75 h after the start of the settling test. The initially uniform suspension had a density of 1,090 kg m^{-3}. Total pressure and pore pressure profiles are also shown. The elevation separating the settling suspension from the developing (and consolidating) bed is at about 55 cm where a lutocline is formed. In Fig. 8.4b, isopycnal lines highlight the time-dependent transition from suspension to bed [Sills and Elder, 1986].

In a general sense there is an association between the time-scale of consolidation, the state of material and its density (Table 8.1). When the rate of deposition in the natural environment is low to moderate, *e.g.*, shoaling is less than ~0.1 m during an event, and the sediment is largely inorganic, consolidation can lead to a stable deposit in about 1–2 weeks. Since this period matches the spring-neap tidal variation, consolidation plays a critical role in governing estuarine sediment accumulation and budget. Sediment freshly deposited at spring tide may not be fully scoured during the subsequent spring tide due to hardening of the deposit.

The formation of a hard bed often includes significant contribution from secondary consolidation. When humic (organic) matter is present in a large proportion, *e.g.*, greater than 10–30 percent, the rate of settling of the bed may measurably slow down. Figure 8.5 shows the consolidation of highly organic (45–50%) bottom muck from Newnans Lake in north-central Florida. The two circles denote *in situ* density data, and the dashed line is derived from the linear consolidation equation of Been and Sills [1981] (Section 8.4.2), starting with an initial "deposit"

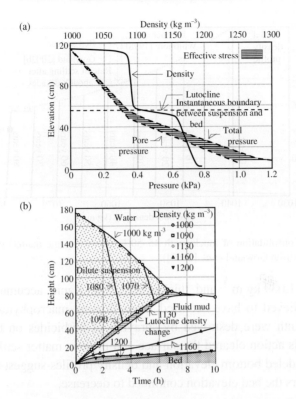

Figure 8.4. Settling of Combwich (UK) mud, a silty-clay: (a) Instantaneous density, total pressure and pore pressure profiles; (b) Development of a bed from suspension (adapted from Sills and Elder [1986]).

Table 8.1. Approximate association between consolidation, material state and bed density.

Degree of consolidation	State	Bulk density, ρ (kg m^{-3})
New deposit (1–3 days)	Fluid-like deposit (Newtonian)	1,005–1,100
Weakly consolidated (1–2 weeks)	Soft deposit (pseudoplastic)	1,100–1,200
Moderately consolidated (1–3 months)	Soft deposit (pseudoplastic)	1,200–1,300
Well consolidated (1–2 years)	Solid (Bingham)	1,300–1,400
Highly consolidated mud (10–20 years)	Solid	1,400–1,500
Hard mud (> 100–200 years)	Solid	>1,500

Source: Modified from van Rijn [1993].

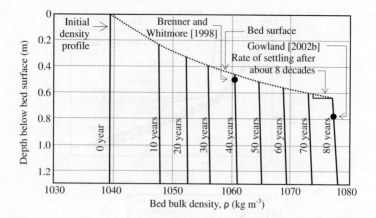

Figure 8.5. Consolidation of muck (with 45–50% natural organic matter) in Newnans Lake, Florida (from Gowland *et al.* [2007]).

density of $1,039 \text{ kg m}^{-3}$ and thickness 1.3 m. Rapid accumulation of muck is believed to have occurred when floating macrophytes such as water hyacinth were destroyed by spraying of herbicides on the water surface. This action cleared the surface and organic matter settled to the bottom. Modeled bottom elevation and density profiles suggest that even after 80 years the bed elevation continued to decrease.

8.3.2 *Non-homogeneity due to drainage channels and layering*

Referring to Fig. 7.13, when the concentration C is lower than, but sufficiently close to, its space-filling value C_s, capillary drainage channels promoting upward transport of pore water tend to develop in dense fluid mud or very soft beds. Even though no significant effective stress (representing permanent particle–particle contact) may exist, pore water transport can be simulated by seepage flow models for predicting the rate of settling of mud mass and increase in density. In biologically active zones drainage channels can occur in the top 5 to 20 cm of the bottom, although, due to local non-homogeneities, they may differ from channels in abiotic beds in structure and water-conducting ability. Figure 8.6a shows a drainage (pore) channel inside a settling slurry of Lake Apopka muck in a laboratory column, and Fig. 8.6b is a cross-sectional

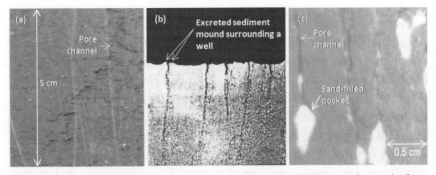

Figure 8.6. (a) Drainage (pore) channel in a settling slurry of Lake Apopka muck (from Mehta *et al.* [2009]); (b) Natural drainage wells in mud (from Migniot [1968]); (c) Capillary channels in a clay–sand slurry with pockets of fine sand at the lower termini (from Dankers [2006]).

view of natural drainage wells formed by burrowing animals. Such channels and wells can also accumulate other material and increase bed non-homogeneity. The mound formed around the well entrance by sediment and waste excreted by the animal tends to increase the hydraulic roughness of the bed.

In general drainage is affected by layers separated by their permeability [King, 2007]. Heterogeneous sediment allowed to settle in cone-shaped columns has been shown to generate remarkable laminae amounting to stratification. Even more complex patterns occur if water current is present [Berthault, 1986; 1988]. The occurrence of such self-organization remains to be explored in the natural environment. Dankers [2006] attempted to examine the settling behavior of mixtures of clayey sediment and fine sand. The sand preferentially settled through drainage channels, forming pockets of accumulation at the lower ends of channels characterized by substantial inhomogeneities (Fig. 8.6c). Modeling the bed as a homogeneous mixture yielded unsatisfactory results on the rate of settling.

A common assumption in hydrodynamic modeling is that the effects of surface non-homogeneities due to preferred paths for upward water movement are small in scale compared to typical dimensions of modeled area. Although this assumption is convenient, it averages out complex interactive relationships between flow, sediment and biology [Neumeier *et al.*, 2006].

8.3.3 *Sedimentation and use of tracers*

An advantage of averaging out small-scale bed features is that it allows the modeler to express erosion and deposition mass fluxes on a spatial mean basis for calculating net changes in bed elevation over large areas. Sediment exchange deduced from these fluxes over discrete time-steps helps in the identification of mechanisms by which heterogeneity in sediment composition and consolidation leads to stratification.

Due to time-lines associated with different aliquots of sediment deposition (per unit time) in the natural environment, it becomes essential to track each aliquot as it undergoes (typically) an increase in density due to settling or consolidation, or erodes over some time-interval. For given sedimentary material, variables that determine the rate of density increase and the ultimate density include the initial density of the aliquot, its thickness and whether its compression is due to its own weight (self-weight), or due to overburden.

As shown schematically in Fig. 8.7a, at time $t = 0$ aliquot A of sediment having a density ρ_A deposits onto a bed of density ρ_{bed}. During a (short) time-increment Δt, A's density increases under self-weight to $\rho_A + \Delta \rho_A$, while an additional aliquot B of density ρ_B deposits on top of A. Within the following time interval, B will undergo the same or nearly the same increase in density (to $\rho_B + \Delta \rho_B$) as A, due to self-weight during the previous time interval. However, A will now dewater and compress due to self-weight *and* overburden (from B). During the third time-increment A will attain the density ($\rho_{bed} = \rho_A + 3\Delta \rho_A$) of the initial bed, while the density of B will increase to $\rho_B + 2\Delta \rho_B$. Within the same interval the upper portion of B is eroded due to a current. Finally, in the fourth interval the density of B will increase to $\rho_B + 3\Delta \rho_B$.

Multiple particle sizes in the bed introduce complexities in the above description. A question important to modeling concerns the initial distribution of sediment size on the bed, which must be consistent with the current velocity patterns and the variation of bed shear stress [Blom, 2008]. A method, which to a degree is an extension of the protocol in Fig. 8.7a, is exemplified in Fig. 8.7b for a sequential approach to model the redistribution of sediment sizes by erosion. At some initial time the

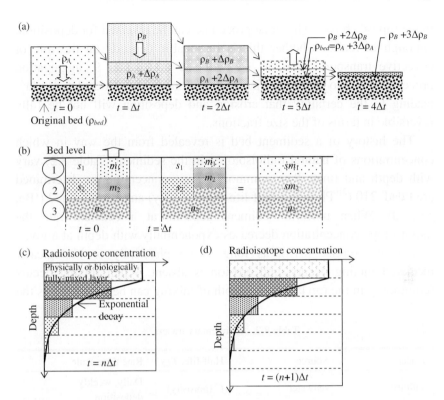

Figure 8.7. (a) Schematic drawing of time-discretized processes of settling and consolidation with deposition and erosion; (b) effect of multiple sizes on bed sediment redistribution; (c) variation of radioisotope concentration with depth at a given time; (d) variation of radioisotope concentration with depth at a later time (n is an arbitrary integer).

bed is divided into sublayers. The top layer is a 50:50 (by weight) mixture of sand s_1 and clay m_1. Similarly the next layer below is a 50:50 mixture of sand s_2 and clay m_2. The third layer consists of uniform sediment u_3. Starting with time $t = 0$ the bed shear stress is only sufficient to erode the clay fraction while sand remains in place. Unlike Fig. 8.7a, the change in bed level due to erosion is ignored, which means that as part of box m_1 erodes, box m_2 takes the vacant space created by m_1 moving upward and box u_3 occupies the vacancy left behind by box 2. The outcome is boxes at time $t = \Delta t$ with new concentrations sm_1 and sm_2

in the top two boxes. The same procedure can be reversed for deposition, although in reality whether this is reasonable depends on the role of advective transport. Also, as we will see in Chapter 9, aggregation processes tend to shift the size distribution of the (flocculated) clay, leading to the certainty that erosion and deposition will not be fully reversible in terms of the size fractions.

The history of a sediment bed is revealed from the way in which concentrations of natural radioisotopes in the sediment (Table 8.2) vary with depth and time. Two common isotope concentrations determined are Lead–210 (^{210}Pb with a half-life $T_{1/2}$ = 22.3 y) and Beryllium–7 (^{7}Be, 53.3 d). When uniform sediment deposits at a constant rate the radioisotope concentration decreases exponentially with depth at a given time (see Exercise 8.13), as long as no physically or biologically mediated mixing occurs, and erosion is absent. When mixing occurs (commonly in the top layer) the depth of mixing can be identified as the

Table 8.2. Sedimentary tracers.

Tracer	Source	Half-life, $T_{1/2}$	Rate time-scale
Salinity	seawater	f^n(porosity)	Daily, weekly deposition
Thorium–234 (^{234}Th)	U–Th decay series	24.1 d	Weekly deposition, bioturbation
Beryllium–7 (7Be)	cosmogenic	53.3 d	Terrestrial sediment deposition
Chlorophyll-a	phytoplankton	f^n(temp)	Weekly deposition, bioturbation
Radium–224/226 (^{224}Ra/^{226}Ra)	U–Th decay series	3.6 d/1600 y	Decadal accumulation
Cesium–137 (^{137}Cs)	anthropogenic	30 y	Decadal accumulation
Lead–210 (^{210}Pb)	U–Th decay series	22.3 y	Centennial accumulation, deep sea bioturbation
Carbon–14 (^{14}C)	cosmogenic	5730 y	Millennial accumulation

Source: John Jaeger personal communication [2009].

thickness over which the decay trend is exponential. In the event that mixing is complete, the concentration of the isotope at the top will be constant (Fig. 8.7c). As deposition continues the entire profile shifts upward at the same rate as the rate of deposition. Deviations from this profile arise due to variable deposition rate, variable sediment composition, erosion and mixing.

Returning to Fig. 8.7, long-term bed elevation changes over a large study area require bookkeeping of grid-by-grid calculations. [Jiang, 1999; Jiang and Mehta, 1999]. In general, for prediction of suspended fine sediment loads, calculations of changes in the density and thickness of the depositing aliquots is essential because density determines the deposit's shear strength against erosion and therefore the bed's erodibility potential [Verreet and Berlamont, 1988]. The relationship between the height of gradually depositing suspension and density during hindered settling is described later in terms of mass continuity by the method of Kynch [1952]. In their numerical model of cohesive sediment transport, Ariathurai *et al.* [1977] included a simple empirical method for hindered settling combined with consolidation.[c] This method (next section) is based on the experimental results of Bosworth [1956] for calculating the increase in the density of the deposited aliquot. An argument for using this rough-estimate approach is that the dependence of bed shear strength on density is empirical and approximate (Chapter 9). Therefore, a more accurate method based on the mechanics of hindered settling and consolidation may not yield a better estimate of the erosion potential of the settling/consolidation layer (*e.g.*, at $t = 3\Delta t$ in Fig. 8.7a).

8.3.4 *Empirical treatment of sedimentation*

From sedimentation of flocculated suspensions in a cylindrical column at initial concentrations in the hindered settling range, the following

---

[c] The two-dimensional (vertical) bed model was based on the work of Ariathurai [1974]. "Layered bed" schematization was implemented to account for the variation of cohesive bed properties with depth.

empirical relationship was proposed between the instantaneous volume
of suspension $V_t(t)$ and the final volume V_{tf}

$$V_t(t) = V_{tf} \left(1 + \frac{t_v}{t} \right) \qquad (8.14)$$

where t_v is the duration required for the volume V_t to reach the value $2V_{tf}$
[Bosworth, 1956]. Figure 8.8 shows plots of V_t against t^{-1} from tests
using mud from San Francisco Bay in water at 23°/$_{oo}$ salinity. In each test
the first phase of sedimentation lasting between 100 and 200 min was
followed by a change to a second, slower, settling phase with a steeper
line slope. The respective lines permit determination of the final volumes
V_{tf1} (obtained by extrapolating the first phase line to the ordinate) and
V_{tf2}.

In Fig. 8.9, V_{tf1} and V_{tf2} are plotted against concentrations C_0 at the
start of the corresponding phases. Linear trends may be construed to
mean that within the range of concentrations for each phase the floc
structure remained independent of time. In particular the fractal
dimension of the flocs (Chapter 4) can be taken as constant. One would
expect the data curves (dashed) to pass through the origin, since at zero
concentration there can be no volume. Therefore, non-negligible

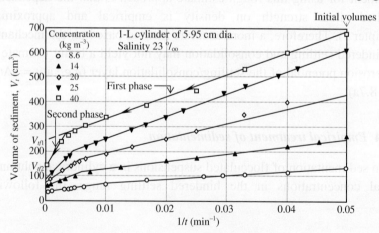

Figure 8.8. Phase changes associated with settling of San Francisco Bay mud (from
Krone [1962]). Note that each plot begins at the right end of the abscissa
($t^{-1} = 0.05$ min^{-1}).

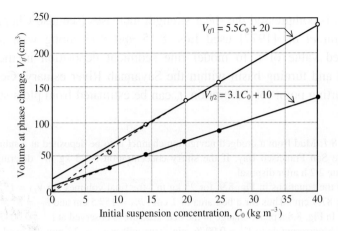

Figure 8.9. Variation of phase change volumes with initial concentration (from Krone [1962]).

intercepts are attributed to loosely packed flocs close to the surface of the suspension compared to the bulk [Krone, 1962].

Since the base area of the column is constant, Eq. 8.14 can also be stated as

$$h(t) = h_f \left(1 + \frac{t_v}{t} \right) \qquad (8.15)$$

where h_f is the final height h of the deposit. Thus, the settling rate w_s of the interface is given by

$$w_s = -\frac{dh}{dt} = \frac{t_v h_f}{t^2} \qquad (8.16)$$

and the final concentration

$$C_{fd} = C_0 \frac{h_f}{h_0} \qquad (8.17)$$

where h_0 is the initial height of the deposit. Knowing C_{fd}, the corresponding wet bulk density is obtained from $\rho = (C_{fd}/\rho_s)(\rho_s - \rho_w) + \rho_w$, where ρ_s is the particle density and ρ_w is water density. As the soil skeleton is compressed the shear strength increases (see also Chapter 9). Ariathurai *et al.* [1977] used Eqs. 8.15 and 8.17 along with a site-calibrated value of t_v to model fine sediment deposition in the ship channel and turning basin within the Savannah River estuary, Georgia. As a starting point for calibration, t_v can be estimated from plots such as Fig. 8.8.

Example 8.1: Mud from a dredged navigation channel is to be deposited at a submerged site in the San Francisco Bay. If the slurry concentration is 25 kg m^{-3} determine the settling rate 2.2 h after disposal.

From the equations in Fig. 8.9, for 25 kg m^{-3} the final volumes are $V_{f1} = 157.5$ cm^3 and $V_{f2} = 87.5$ cm^3. Thus for a bed area of 1 cm^2, $h_{f1} = 157.5$ cm and $h_{f2} = 87.5$ cm are obtained. In Fig. 8.8 a phase change in the bed structure is observed at $t^{-1} = 0.0043$ min^{-1}. Since 2.2 h corresponds to $t^{-1} = 0.0076$ min^{-1}, we will use $t_v = 37$ min obtained for the first phase from the line in Fig. 8.8. Thus, from Eq. 8.16 the settling rate $w_s = (37 \times 157.5)/(132)^2 = 0.333$ cm min^{-1}.

8.3.5 *Sedimentation by seepage flow*

The seepage velocity is a measure of the rate of dewatering and therefore hindered settling. Relying on similarities between consolidation and hindered settling, the hindered settling velocity can be modeled by assuming the applicability of Darcy's law, even though a permanent and continuous porous medium is not formed [Sanchez and Levacher, 2007]. Since the aim is to determine the settling velocity representing sedimentation, this development can as well be part of Chapter 7. However, its inclusion here is meant to emphasize that the process is dependent on seepage.

In Eq. 2.98, the total pressure head H_t can be stated in terms of its components as

$$H_t = \frac{p_w}{\rho g} + H = \frac{p_{ws}}{\rho g} + \frac{p_{we}}{\rho g} + H \qquad (8.18)$$

where p_{ws} is the steady-state pore water pressure and p_{we} is the excess pore water pressure. Since the hydrostatic head H is independent of z, Eq. 2.98 yields

$$w_s = -\frac{K_p}{\eta}\frac{\partial p_{we}}{\partial z} \tag{8.19}$$

where the Darcy velocity w_w has been assumed equal to, and therefore replaced by, the particle settling velocity w_s.[d] We will further assume that the pore channels can be represented by a bundle of equivalent capillary tubes of radius r_c. Then, based on the Hagen–Poiseuille equation for viscous flow (Eq. 5.120), the specific permeability K_p is given by

$$K_p = \alpha_{cp} r_c^2 \left(1 - \phi_{vf}\right) \tag{8.20}$$

in which the shape factor α_{cp} relates the actual flow pathways to idealized tubes and ϕ_{vf} is the floc volume fraction. Considering settling flocs of diameter d_f to be uniformly distributed, it can be shown that

$$r_c = 0.389\beta_{cp}d_f\left(\frac{\pi}{\phi_{vf}}\right)^{1/3}\left[1 - 1.285\left(\frac{\phi_{vf}}{\pi}\right)^{1/3}\right] \tag{8.21}$$

where β_{cp} accounts for irregularities in the floc matrix [Li and Mehta, 1998]. Combining Eqs. 8.19, 8.20 and 8.21 yields the following expression for the settling velocity

$$w_s = \kappa_{cp}d_f^2\frac{g(\rho_s - \rho_w)}{\eta}\left(\frac{\pi}{\phi_{vf}}\right)^{2/3}\left[1 - 1.285\left(\frac{\phi_{vf}}{\pi}\right)^{1/3}\right]\left(1 - \phi_{vf}\right)\phi_{vf} \tag{8.22}$$

where $\kappa_{cp} = 0.151\alpha_{cp}\beta_{cp}$. For freely falling spheres, the term $\kappa_{cp}d_f^2 g(\rho_s - \rho_w)/\eta$ with $\kappa_{cp} = 1/18$ is the Stokes settling velocity (Eq. 7.15). Thus the remaining terms amount to a correction for hindered settling modeled on capillary flow. The floc volume fraction can be obtained from $\phi_{vf} = [(\rho_s - \rho_w)/(\rho_f - \rho_w)]\phi_v = K_{sf}\phi_v$. Assuming K_{sf} to be invariant (for given sediment and ambient fluid) would permit the

---

[d] An assumption in Eq. 2.98 is that the solids are stationary. If they were moving with velocity w_s, momentum balance would require that w_w be replaced by $w_w - w_s$. If now we consider that only the solids are moving, w_w would be nil, and considering w_s positive downward, Eq. 8.19 would ensue.

estimation of ϕ_{vf} from the solids volume fraction ϕ_v. With this assumption, values of $\kappa_{cp}d_f^2$ and K_{sf} from settling experiments in saline water are given in Table 8.3.

8.3.6 Sedimentation from continuity

The theory of sedimentation attributed to Kynch [1952] relies on mass continuity of suspended sediment. It provides an understanding of the mechanism by which a vertical profile of concentration develops during hindered settling under quiescent conditions. An assumption is that the velocity of the solid constituent depends only on the local concentration of particles. This assumption permits us to bypass the use of momentum balance. Another supposition is that the particles attain their final concentration as soon as they reach the bottom. In other words, there is no provision for a change in particle concentration due to compression as flocs are crushed by overburden. Thus the theory as such is not applicable to cohesive flocs. Yet, due to its simplicity, it has been used for estimating the sedimentation rate of flocs (*e.g.*, Sanchez and Levacher [2007]). When the concentration is very small the theory is reduced to Stokes law (Eq. 7.15).

The starting condition is one of uniform suspension of concentration C_0, which is lower than C_m (Fig. 7.13), and therefore hindered settling does not begin immediately. The way in which a denser, hindered-settling, suspension develops at the bottom is examined graphically by the method of characteristics for solving hyperbolic partial differential equations (*e.g.*, Stoker [1992]).

Table 8.3. Coefficient values for Eq. 8.22.

Sediment	Salinity	$\dfrac{\kappa_{cp}d_f^2}{(m^2)}$	K_{sf}
Montmorillonite	0.3	7.47×10^{-12}	38.0
Kaolinite	3.4	1.39×10^{-11}	10.5
Changjiang mud (China)	3.4	2.11×10^{-11}	7.5

Source: Li and Mehta [1998].

Once settling begins at $t \geq 0$, the dilute suspension interface S (Fig. 8.10) falls at a rate[e] w_s, whereas the dense suspension interface P rises at a rate w_u. The P-interface represents a primary lutocline and the S-interface a secondary lutocline. Mass conservation applicable to settling sediment is obtained from Eq. 6.111 by ignoring the diffusion term, leaving

$$\frac{\partial C}{\partial t} + \frac{\partial F_s}{\partial z} = 0 \qquad (8.23)$$

where $F_s = w_s(C) \cdot C$ is the settling flux. The sign of w_s is reversed relative to Eq. 6.108 such that the flux F_s is now positive downward.[f] Using the chain rule from differentiation we may express Eq. 8.23 as

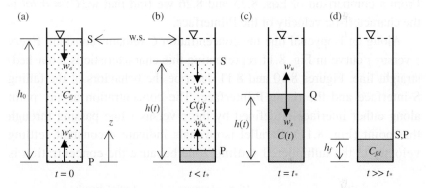

Figure 8.10. (a) Initially uniform dilute suspended sediment in a quiescent column; (b) S-interface falling at a rate w_s and an isopycnal characteristic defining the P-interface rising at a rate w_u; (c) isopycnal characteristic intersecting the S-interface at time t^* and elevation Q; (d) final deposit.

---

[e] Toorman and Berlamont [1993] prefer to call $w_s = -dh/dt$ the settling *rate* in difference to settling *velocity*, which is characteristically associated with individual particles.

[f] In numerical integration of Eq. 8.23 applied to settling of compressible flocs, it is easy to replace the bottom boundary condition of constant final concentration in Kynch's analysis with a condition derived by recognizing that at the bottom the velocities of both constituents of the mixture are zero. See Concha and Bustos [1985] for this extension of the theory.

$$\frac{\partial C}{\partial t} + \frac{\partial F_s}{\partial C}\frac{\partial C}{\partial z} = 0 \tag{8.24}$$

We further define a settling rate $w_u(C) = \partial F_s/\partial C$ and note that

$$\frac{\partial C}{\partial t} + w_u \frac{\partial C}{\partial z} = \frac{DC}{Dt} = 0 \tag{8.25}$$

i.e., the total rate of change of concentration, DC/Dt, is zero. This derivative in the z-t space is also expressed as

$$\frac{DC}{Dt} = \frac{\partial C}{\partial t} + \frac{\partial z}{\partial t}\frac{\partial C}{\partial z} \tag{8.26}$$

From a comparison of Eqs. 8.25 and 8.26 we find that $w_u(C) = \partial z/\partial t$ is the characteristic velocity of the P-interface.

Along an isopycnal line the concentration C remains constant. Every z versus t curve in Fig. 8.11 representing this characteristic is a (dashed) straight line. Figures 8.10 and 8.11 describe the behaviors of the falling S-interface and the rising P-interface. The concentration at any point along either interface is defined by the z versus t line passing through that point (Fig. 8.11). Parallel isopycnals indicate a constant settling velocity in the unhindered settling zone because the concentration is

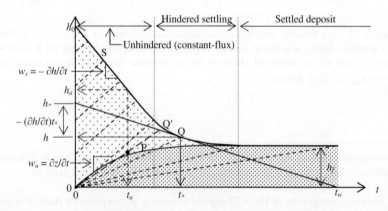

Figure 8.11. Schematic description of the method of characteristics for quiescent settling. Each z-t isopycnal is represented as a dashed line. Initial concentration must be specified at all points over the height of the water column along the $t = 0$ ordinate.

constant (initial value) within this zone, and because concentration associated with any isopycnal is specified at its point of origin at the ordinate. As a result, isopycnals emanate from the $t = 0$ ordinate at all points along the height of the water column over which the initial (constant) concentration is specified. Below the P-interface fluid mud zone occurs due to hindered settling. Within this zone the isopycnals are not parallel because concentration changes with depth and time and the settling velocity decreases with time. All isopycnals start from the origin $(0, 0)$ where the bottom-most initial concentration is specified.

Once the S- and P-interfaces meet settling is practically over and a deposit is formed. However, this does not provide any indication of the state of the deposit, *i.e.*, whether it is a suspension, or a bed with effective stresses. As a practical approach it is often reasonable to assume that the deposit has the consistency of a soft bed.

Any characteristic (z-t line) starting at $t = 0$ from the ordinate intersects the falling S-interface at point Q and time t_*. At Q we draw a tangent which intersects the ordinate at elevation h_*. Given total sediment mass m_t in suspension and the cross-sectional area A of the settling column, mass balance reads

$$\frac{m_t}{A} = w_u C t_* + w_s C t_* = C(w_u + w_s)t_* \qquad (8.27)$$

Now, since

$$\frac{m_t}{A} = C_0 h_0; \quad w_u = \frac{h}{t_*}; \quad w_s = -\frac{dh}{dt} \qquad (8.28)$$

we obtain

$$C_0 h_0 = C\left(h - t_* \frac{dh}{dt}\right) = C h_* \qquad (8.29)$$

Let us look at the settling zones[g] in Fig. 8.11.

---

[g] The flocculation settling zone of Fig. 8.3a is excluded because the settling flux is minor.

In the unhindered (free) settling zone:

$$w_s = -\frac{dh}{dt}; \quad C_0 = \text{constant} \tag{8.30}$$

Therefore this is also the zone of constant settling flux $w_s C_0$. In the hindered settling zone:

$$w_s = -\frac{dh}{dt}; \quad C(t) = \frac{C_0 h_0}{h_*} \tag{8.31}$$

Finally, in the settled deposit:

$$w_s = -\frac{dh}{dt} = 0; \quad C_{fd} = \frac{C_0 h_0}{h_f}; \quad h = h_f = \text{constant} \tag{8.32}$$

where C_{fd} is the final concentration of the deposit. These relationships can be used to calculate any concentration profile $C(t_a)$ of height h_a. In Fig. 8.11 the characteristic of an arbitrarily selected point (dot) at time t_a intersects the S-interface at point Q', where a tangent is drawn and the above method used to determine $C(t_a)$.

Example 8.2: The following suspension heights were obtained at different times in a settling test [Kynch, 1952]. Assume that the initial concentration is 5 kg m^{-3} and plot the concentration profiles at 5, 15 and 40 hours.

Time (h)	z (cm)	Time (h)	z (cm)
0	25.0	10	13.4
1	22.0	15	11.6
2	20.5	20	10.5
3	19.2	25	9.9
4	18.0	30	9.7
5	17.0	35	9.6
6	16.0	40	9.5
8	14.6	50	9.5

The above data (dots) are shown in Fig. 8.12. We can identify the area 0AB (where 0 denotes zero hour) as the zone of unhindered (or free or constant-flux) settling in which the fall of the interface from A to B occurs at a uniform rate. The area 0BC represents hindered settling, and 0CD50 is the settled deposit. Thus within 0AB all the characteristics (z versus t lines) emanating from the origin are parallel to 0B. Beyond 40 h there is no further change in the concentration profile because the interface is no longer

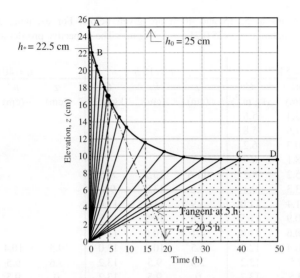

Figure 8.12. Falling interface of a settling suspension plotted against time.

falling (according to this analysis, which is not applied to consolidation). At the boundary point B the surface concentration begins to increase from its initial value of $C_0 = 5$ kg m^{-3}. The concentration at the boundary point C is slightly lower than at all later times. Given $h_0 = 0.25$ m, we find $h_f = 0.095$ m and $C_{fd} = 13.2$ kg m^{-3}. Derived settling rates w_s and heights h_* at different times are given below.

Time, t_* (h)	h_* (cm)	t_w (h)	w_s (m s^{-1})
0	25.0	7.5	9.3×10^{-6}
1	23.2	15	4.3×10^{-6}
2	23.1	17.5	3.7×10^{-6}
3	22.9	19	3.3×10^{-6}
4	22.6	19.5	3.2×10^{-6}
5	22.5	20.8	3.0×10^{-6}
6	21.6	23	2.6×10^{-6}
8	20.3	28	2.0×10^{-6}
10	18.1	36	1.4×10^{-6}
15	15.2	60.5	7.0×10^{-7}
20	13.3	95	3.9×10^{-7}
25	11.4	185	1.7×10^{-7}
30	10.9	275	1.1×10^{-7}
35	10.6	375	7.9×10^{-8}
40	10.4	515	5.6×10^{-8}
50	9.5	-	0

For example, at 5 h a tangent through the data point (dot) gives $h_* = 22.5$ cm $= 0.225$ m. This tangent intersects the time axis at $t_w = 20.5$ h. Therefore $w_s = 0.225/(20.5\times3600) = 3.0\times10^{-6}$ m s^{-1}.

Concentrations at 5, 15 and 40 h are given below. For example, at 5 h and $z = 13.0$ cm we obtain $C = 5 \times 25.0/21.6 = 5.8$ kg m^{-3}. The respective profiles are shown in Fig. 8.13.

	$t_* = 5$ h			$t_* = 15$ h			$t_* = 40$ h	
z (cm)	h_* (cm)	C (kg m^{-3})	z (cm)	h_* (cm)	C (kg m^{-3})	z (cm)	h^* (cm)	C (kg m^{-3})
17.0	22.4	5.6						
13.0	21.6	5.8						
9.0	20.3	6.2						
6.8	18.1	6.9						
3.9	15.2	8.2	11.6	15.2	8.2			
2.7	13.3	9.4	7.8	13.3	9.4			
2.0	11.4	11.0	5.9	11.4	11.0			
1.6	10.9	11.5	4.8	10.9	11.5			
1.4	10.6	11.8	4.1	10.6	11.8			
1.2	10.4	12.0	3.5	10.4	12.0	9.5	10.4	12.0
0.9	9.5	13.2	2.8	9.5	13.2	7.6	9.5	13.2
0	9.5	13.2	0	9.5	13.2	0	9.5	13.2

Figure 8.14 shows a high-frequency (uncalibrated) acoustic record of suspended fine sediment 8 min after disposal of dredged material from a dredger in Townsville Harbour, Australia. Optical backscatter sensors were used to record vertical profiles of settling sediment concentration in 10 m deep water. Two sets of measurements were made, the first during disposal when the sea was calm, and the second when the sea was "moderate". After the first few minutes, wave-induced oscillations and turbulence seemingly reduced the rate of fall of the lutocline compared to the no-waves case. For each case, in Fig. 8.15, somewhat idealized profiles of suspended sediment concentrations at different times are plotted [Wolanski et al., 1991]. Since settling as well as deposition took place

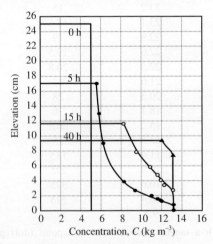

Figure 8.13. Calculated concentration profiles during settling of sediment suspension.

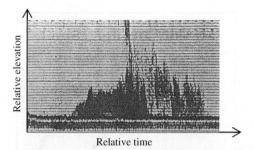

Relative time

Figure 8.14. Non-calibrated acoustic tracking of disposal of muddy sediment in Townsville Harbour, Australia. The depth of water was 10 m (from Wolanski *et al.* [1991]).

the total suspended mass was not conserved. The method of Kynch may be applied to obtain rough estimates of how much sediment was deposited at a given time, and permit an estimation of the effect of wave agitation (Exercise 8.6).

8.3.7 *Hindered settling analogy with consolidation*

As we have noted, hindered settling can be treated by recognizing that, as in consolidation, the seepage velocity controls the rate of fall of the lutoclinal interface. Accordingly, hindered settling has been combined with consolidation in a semi-empirical analysis to obtain a unified expression for the settling rate w_s. In Fig. 8.16, w_s values derived from the rate of fall of the upper interface of suspended fine sediment is

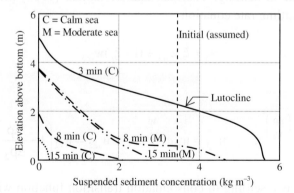

Suspended sediment concentration (kg m^{-3})

Figure 8.15. Profiles of suspended sediment concentration following dredged material disposal during: (a) calm sea; (b) moderate sea. Times are relative to the onset of disposal (idealized interpretation based on Wolanski *et al.* [1992]).

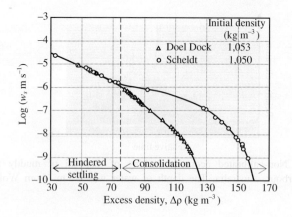

Figure 8.16. Settling rate versus excess density data (from Toorman and Berlamont [1993]).

plotted against the excess density $\Delta\rho = [1-(\rho_w/\rho_s)]C$, which is similar to plotting w_s against concentration C. These data on estuarine muds from Doel Dock (Port of Antwerp) and the Scheldt estuary in Belgium were obtained in a settling column. In the hindered settling range, also called first mode of settling, data from the two sources show self-similar behaviors. However, at about 75 kg m^{-3}, which approximately marks the onset of a bed due to space-filling, the settling rates begin to diverge as the role of effective stress cannot be ignored during consolidation, the second mode of settling. Toorman and Berlamont [1993] proposed the following settling rate equation

$$w_s = \xi_t w_{se1} + (1-\xi_t)w_{se2} \qquad (8.33)$$

where

$$\xi_t = e^{-\left(\frac{\Delta\rho}{\Delta\rho_t}\right)^n} \; ; \qquad w_{se1} = w_{s1}e^{-\frac{\Delta\rho}{\Delta\rho_1}} \; ; \qquad w_{se2} = w_{s2}\left(1-\frac{\Delta\rho}{\Delta\rho_2}\right)^3 \qquad (8.34)$$

in which ξ_t is a settling-to-consolidation transition function with $n > 10$, $\Delta\rho_t$ is the excess density at transition, $\Delta\rho_1$ is a characteristic excess density for the first mode, $\Delta\rho_2$ is the bed excess density corresponding to maximum packing in the second mode, and w_{s1} and w_{s2} are the respective mode settling rates.

Since bulk parameterization is conveniently achieved by using concentration rather than excess density, the following equations are modified forms of Eqs. 8.33 and 8.34, respectively,

$$w_s = \xi_{tc} w_{sc1} e^{-\frac{C}{C_{sc1}}} + \left(1 - \xi_{tc}\right) w_{sc2} \left(1 - \frac{C}{C_{sc2}}\right)^m \quad (8.35)$$

where

$$\xi_{tc} = e^{-\left(\frac{C}{C_t}\right)^n} \quad (8.36)$$

in which the meanings of the four coefficients w_{sc1}, w_{sc2} C_{sc1} and C_{sc2} are analogous to the quantities in Eqs. 8.33 and 8.34 [Jiang, 1999]. These coefficients as well as m and n are given in Table 8.4 for data from different sources. Equation 8.22 is applicable to generally similar settling conditions, and is valid for the first mode of settling only. However,

Table 8.4. Rate parameters for hindered settling combined with consolidation.

Investigator(s)	Sediment source	w_{sc1} (m s^{-1})	C_{s1} (kg m^{-3})	w_{sc2} (m s^{-1})	C_{s2} (kg m^{-3})	C_t (kg m^{-3})	m	n
Burt and Parker [1984]; Jiang [1999]	Estuarine mud, UK	--[a]	--[a]	4.2×10^{-6}	680	15	6	15
Toorman and Berlamont [1993]	Doel Dock, Belgium	5.0×10^{-4}	20	7.0×10^{-6}	205	160	3	13
Jiang [1999]	Jiaojiang estuary, China	1.0×10^{-4}	31	6.0×10^{-6}	350	210	5	15
Marván [2001]	Ortega River, Florida	6.0×10^{-5}	15	3.0×10^{-6}	1000	83	6	18

[a] First mode absent.

Sources: Jiang [1999], Jiang and Mehta [1999], Marván [2001].

Eqs. 8.22 and 8.35 show a similar trend of variation of the settling rate over the full range of excess density $\Delta\rho$, as in Fig. 8.16 (see Exercise 8.4).

8.4 Bed Development

8.4.1 *Consolidation and related processes*

Physics-based analysis of consolidation begins with Terzaghi in 1925 and its three-dimensional extension by Biot [1941], formally identifying the role of soil compression. In Table 8.5, consolidation is defined in the context of dewatering, compaction and subsidence, which are related but distinct processes.

The treatment of consolidation in this chapter is brief in comparison with its coverage in soil mechanics [Lambe and Whitman, 1979; Keedwell, 1984; Winterwerp and van Kesteren, 2004; Bo, 2008]. This is because in fine sediment transport consolidation is viewed mainly as adjunct to the mechanics of erosion. Resistance to erosion increases

Table 8.5. Consolidation, dewatering, compaction and subsidence.

Process	Definition
Consolidation	Consolidation occurs when pressure acting on a saturated soil increases due to overburden either by placement of sediment or application of an external load. As a result water is expelled and particles are forced to be closer to each other. Self-weight consolidation does not require overburden as compression occurs due to the soil's own weight.
Dewatering	Consolidation of saturated soil causes dewatering as the soil matrix compresses. Dewatering also occurs due to lowering of ground water table.
Compaction	Compaction is the process by which soil porosity decreases due to overburden. Compaction occurs when for instance a roller is used to pave soil. For saturated soils compaction is synonymous with consolidation.
Subsidence	Subsidence amounts to lowering of the ground fluid level due to pumping of ground water, oil or gas. Consolidation also causes subsidence.

with volumetric compression of the bed as water is expelled during consolidation. Additional factors include effects of chemical and biochemical coatings and physicochemical transformations within or at the surface of the bed. These factors modulate the dependence of erosion resistance on bed density, at times quite significantly. Thus, physical consolidation is not the only determinant of erosion resistance.

8.4.2 *Consolidation by overburden*

Consider vertical, one-dimensional, primary consolidation of a bed due to overburden, *i.e.,* an additional load (surcharge) over the deposit. Expulsion of water occurs through a compressible particle matrix in which the effective normal stress is non-trivial. It is assumed that the strains are small and the weight of the deposit does not contribute to compression.[h] Under these conditions, the continuity equation (Eq. 8.13) is reduced to

$$\frac{1}{1+e_v}\frac{\partial e_v}{\partial t} = k_p \frac{\partial^2 H_t}{\partial z^2} \tag{8.37}$$

where any spatial change in the void ratio e_v is assumed to be negligible, and the vertical permeability k_p is independent of z. Due to compression, as the effective stress σ_v' increases the void ratio e_v decreases. This relationship is in general non-linear and such that at zero effective stress the void ratio is large and at large effective stresses the void ratio approaches zero. It is expressed in terms of the rate of change of void ratio with effective stress

$$\frac{\partial e_v}{\partial \sigma_v'} = -a_v \tag{8.38}$$

where the coefficient of compressibility a_v (m s^2 kg^{-1}), which can vary with σ_v' (Pa), is a scale for the efficiency with which the effective stress

---

[h] The complete analysis including the self-weight effect on the rate of consolidation permits the treatment of large strains in the bed, which are not considered here.

compresses the bed by reducing the void ratio. The larger the value of a_v the greater is the increment in void ratio reduction (note the negative sign) due to an incremental increase in the effective stress. The typical units of a_v are $m^2 \ kg^{-1}$. However, in the present analysis a_v has been divided by g in order to preserve the unit of stress in Pascal.

Equations 8.37 and 8.38 are combined to yield

$$\frac{\partial \sigma_v'}{\partial t} = -\frac{k_p(1+e_v)}{a_v}\frac{\partial^2 H_t}{\partial z^2}$$

(8.39)

From Eq. 8.18, the second derivative of the total pressure head H_t can be represented in terms of the excess pore water pressure p_{we}. This is so because the hydrostatic head H is independent of z and the steady-state pore pressure p_{ws} varies linearly with depth; hence its second derivative with respect to z is zero. The coefficient of consolidation c_{cv} is now defined as

$$c_{cv} = \frac{k_p(1+e_v)}{\rho g a_v}$$

(8.40)

which indicates an inverse relationship between c_{cv} and a_v. In general, c_{cv} increases with increasing permeability as the sediment changes from clay to sand.

Equation 8.39 can now be expressed as

$$\frac{\partial \sigma_v'}{\partial t} = -c_{cv}\frac{\partial^2 p_{we}}{\partial z^2}$$

(8.41)

Then, since

$$p_w = p_{ws} + p_{we}$$

(8.42)

and from Eq. 8.4

$$\sigma_v' = \sigma_v - p_{ws} - p_{we}$$

(8.43)

the time-rate of change of effective stress is obtained as

$$\frac{\partial \sigma_v'}{\partial t} = \frac{\partial \sigma_v}{\partial t} - \frac{\partial p_{we}}{\partial t} \qquad (8.44)$$

Substitution of Eq. 8.41 into the above equation yields [Terzaghi, 1943]

$$\frac{\partial p_{we}}{\partial t} - \frac{\partial \sigma_v}{\partial t} = c_{cv} \frac{\partial^2 p_{we}}{\partial z^2} \qquad (8.45)$$

When the total stress is independent of time, *i.e.*, $\partial \sigma_v / \partial t = 0$, Eq. 8.45 leads to the following equation for the excess pore pressure p_{we}

$$\frac{\partial p_{we}}{\partial t} = c_{cv} \frac{\partial^2 p_{we}}{\partial z^2} \qquad (8.46)$$

This expression, which describes the diffusion of p_{we}, is applicable when a change in the height of the bed is a minor fraction of the total height. It can be made conveniently dimensionless by introducing the following non-dimensional variables:

$$\tilde{p}_{we} = \frac{p_{we}}{p_{we0}}; \quad \tilde{z} = \frac{z}{h_m}; \quad \tilde{t} = \frac{t}{h_m^2 / c_{cv}} \qquad (8.47)$$

where h_m is conveniently taken as one-half the thickness of the consolidating deposit. Introduction of these quantities in to Eq. 8.46 yields

$$\frac{\partial \tilde{p}_{we}}{\partial \tilde{t}} = \frac{\partial^2 \tilde{p}_{we}}{\partial \tilde{z}^2} \qquad (8.48)$$

For illustration we will examine the solution of Eq. 8.48 for the dissipation of excess pore-pressure in a porous medium subject to the following initial and boundary conditions, respectively:

At time $t = 0$:

$$\tilde{p}_{we} = 1 \quad \text{for} \quad 0 \le \tilde{z} \le 2 \qquad (8.49)$$

i.e., at the beginning, the excess pore pressure is constant throughout the deposit.

At all times $t > 0$:

$$\tilde{p}_{we} = 0 \quad \text{at} \quad \tilde{z} = 0 \text{ and } 2 \qquad (8.50)$$

i.e., at the top and bottom boundaries of the bed the excess pore pressure is entirely (and instantly) dissipated according to this model. The general (series) solution for equations of the form of Eq. 8.48 (along with Eqs. 8.49 and 8.50) is

$$\tilde{p}_{we} = \sum_{n=0}^{n \to \infty} \frac{2}{m} \sin(m\tilde{z}) e^{-m^2 \tilde{t}} ; \quad m = \frac{\pi}{2}(2n+1) \qquad (8.51)$$

which is plotted in Fig. 8.17.

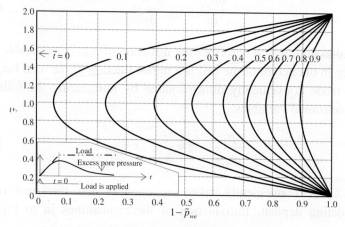

Figure 8.17. Dimensionless plot of variation of excess pore pressure with time and one-half thickness of a porous medium (adapted from Lambe and Whitman [1979]).

Example 8.3: A 3m thick muck-like soil having a liquid limit $W_L = 46\%$ overlies a coarse sandy substrate in a bay. In order to construct a submerged marine habitat, a 5m thick layer of fine sand is disposed on top of the muck. This addition induces a pressure of 10 tons m^{-2} on the muck. Determine: (1) excess pore pressure, (2) pore pressure, and (3) effective normal stress 0.6 m below the muck surface, 6 months after disposal. Use the following data to estimate the coefficient of consolidation c_{cv}:

W_L (%)	c_{cv} (m² yr⁻¹)
30[a]	3.78
60	0.946
100	0.315

[a] Upper limit, undrained remolded soil [Lambe and Whitman, 1979].

Since coarse and fine sands both are considerably more permeable than muck, we can assume that the time required to dissipate excess pore pressure in these media is negligible compared to muck. We obtain the following non-dimensional quantities: $\tilde{z} = 5.6 - 5/1.5 = 0.4$ and $\tilde{t} = 1.588 \times 0.5/(1.5)^2 = 0.3529$. From Fig. 8.17 we get $1 - \tilde{p}_{we} = 0.68$. Therefore, $p_{we} = 10(1 - 0.68) = 3.2$ t m^{-2}. The pore pressure is obtained as $p_w = p_{ws} + p_{we} = 5.6 \times 1 + 3.2 = 8.8$ t m^{-2}. The effective stress is $\sigma'_v = (\sigma'_v)_0 + \Delta\sigma'_v$, where the quantity $(\sigma'_v)_0$ is the magnitude of σ'_v induced by overburden. This stress is obtained from $(\sigma'_v)_0 = (\gamma_{sand} - \gamma_{water}) \times 5 + (\gamma_{clay} - \gamma_{water}) \times 0.6$. Taking $\gamma_{sand} = 2.65$ t m^{-3}, $\gamma_{clay} = 2.65$ t m^{-3} and $\gamma_{water} = 1.00$ t m^{-3} yields $(\sigma'_v)_0 = 9.24$ t m^{-2}. Also, $\Delta\sigma'_v = 10 \times 0.68 = 6.80$ t m^{-2}. Therefore, $\sigma'_v = 9.24 + 6.80 = 16.04$ t m^{-2}.

8.4.3 Self-weight effect

To account for strains due to the self-weight of sediment on the rate of consolidation, it is essential to modify Eq. 8.45. For one-dimensional vertical consolidation the continuity equation (Eq. 8.13) can be written as[i]

---

[i] This equation is obtained as follows. Equation 8.13 for continuity of pore water in the z-direction (only) can be re-derived for an elemental volume as

$$\frac{\partial}{\partial t}(\rho n_r) + \frac{\partial}{\partial z}(\rho n_r w_w) = 0 \qquad (F8.2)$$

where ρ is the fluid (water) density, n_r is the porosity and w_w is the water seepage velocity. Similarly, the conservation of the solid mass moving downward at velocity w_s is

$$\frac{\partial}{\partial t}\left[\rho_s(1 - n_r)\right] + \frac{\partial}{\partial z}\left[\rho_s(1 - n_r)w_s\right] = 0 \qquad (F8.3)$$

where ρ_s is the particle density. By combining Eqs. F8.2 and F8.3 we obtain

$$\frac{De_v}{Dt} + (1 + e_v)\frac{\partial}{\partial z}\left[n_r(w_w - w_s)\right] = 0 \qquad (F8.4)$$

in which the void ratio $e_v = n_r/(1 - n_r)$, and the total derivative of e_v is

$$\frac{De_v}{Dt} = \frac{\partial e_v}{\partial t} + w_s\frac{\partial e_v}{\partial z} \qquad (F8.5)$$

Next, the momentum balance for (vertically) flowing water is

$$k_p\frac{\partial p_{we}}{\partial z} = -n_r(w_s - w_w)\rho g \qquad (F8.6)$$

Substitution of Eq. F8.6 into Eq. F8.4 yields Eq. 8.52.

$$\frac{1}{1+e_v}\frac{De_v}{Dt} = \frac{\partial}{\partial z}\left(-\frac{k_p}{\rho g}\frac{\partial p_{we}}{\partial z}\right) \tag{8.52}$$

A difficulty in solving this equation is that the height of the slurry decreases with time, *i.e.*, this decrease is part of the solution and therefore cannot be ignored [Papanicolaou and Diplas, 1999]. Accordingly, Eq. 8.52 is transformed by using the following relationship between Eulerian coordinates z, t and Lagrangian coordinates Z, t:

$$\frac{\partial Z}{\partial z} = \frac{1+e_{v0}}{1+e_v} \tag{8.53}$$

where e_{v0} is the initial void ratio [Toorman, 1996].

With this transformation Eq. 8.52 becomes

$$\frac{1}{1+e_{v0}}\frac{De_v}{Dt} = \frac{\partial}{\partial Z}\left[-\frac{k_p}{\rho g}\frac{(1+e_{v0})}{(1+e_v)}\frac{\partial p_{we}}{\partial Z}\right] \tag{8.54}$$

which tracks e_v in the Lagrangian framework. Now, by definition the excess pore pressure is

$$p_{we} = \sigma_v - p_w - \sigma_v' = \sigma_v + g(\rho_s - \rho)Z - \sigma_v' \tag{8.55}$$

Thus we obtain the term involving the gradient $\partial p_{we}/\partial Z$ with the help of Eq. 8.53 to be

$$\frac{1+e_{v0}}{1+e_v}\frac{\partial p_{we}}{\partial Z} = g(\rho_s - \rho) - \frac{1+e_{v0}}{1+e_v}\frac{\partial \sigma_v'}{\partial Z} \tag{8.56}$$

Substitution of $\partial p_{we}/\partial Z$ from Eq. 8.56 into Eq. 8.54 yields

$$\frac{1}{1+e_{v0}}\frac{De_v}{Dt} = \frac{\partial}{\partial Z}\left\{-\frac{k_p}{\rho g}\left[g(\rho_s - \rho) - \frac{1+e_{v0}}{1+e_v}\frac{\partial \sigma_v'}{\partial Z}\right]\right\} \tag{8.57}$$

This equation is further transformed by representing it in material coordinates ξ, t, which are related to Z, t by

$$\xi(Z,t) = \int_0^Z \frac{dZ'}{\left[1 + e_v\left(Z',t\right)\right]} = \int_0^Z \phi_v(Z',t)dz' \qquad (8.58)$$

in which the prime denotes a dummy variable. The advantage of using material coordinates is that in spite of the shrinking height of the slurry, the material height of the solid mass ξ_0 remains unchanged.[j] The resulting form of Eq. 8.57 is

$$\frac{\partial e_v}{\partial t} + \frac{\partial}{\partial \xi}\left[\frac{k_p}{\rho g\left(1 + e_v\right)}\frac{d\sigma_v'}{de_v}\frac{\partial e_v}{\partial \xi}\right] + (\rho_s - \rho)\frac{d}{de_v}\left[\frac{k_p}{\rho(1 + e_v)}\right]\frac{\partial e_v}{\partial \xi} = 0 \qquad (8.59)$$

This expression was stated by Gibson *et al.* [1967]. The third term on the l.h.s. represents the effect of the buoyant weight of the solid mass on its own consolidation. De Boer *et al.* [2006] solved Eq. 8.59 by using effective stress-void ratio and permeability-void ratio relationships from experiments of Merckelbach and Kranenburg [2004b]. The time required for consolidation was shown to be sensitive to the fractal dimension D_f of flocs (Eq. 4.53).

8.4.4 *Linearized solutions for void ratio*

Two relationships required to close Eq. 8.59 and solve for the void ratio e_v include one between the effective stress σ_v' and e_v (Eq. 8.38), and another between the permeability k_p and e_v. The latter relationship defines the rate at which dewatering occurs through the pore spaces whose hydraulic conductivity is represented by the void ratio. Also, initial and boundary conditions with respect to e_v must be prescribed. Since the general problem requires a numerical solution, as an approximation we will select a linearized approach, which results in an

---

[j] The material height, also called Gibson height, at any time t_1 is given by

$$\xi_0 = \int_0^{h(t_1)} \phi_v(z,t_1)dz = \int_0^{h(t_1)} \frac{1}{1 + e_v(z,t_1)}dz \qquad (F8.7)$$

where $h(t_1)$ is the height of the bed at t_1 and $\phi_v(z,t_1)$ is the solids volume fraction. Since the total mass of sediment in the cylindrical column does not change, ξ_0 is independent of t_1.

approximate small-strain analytic solution of Eq. 8.59 for the variation of the void ratio with depth and time. The first (assumed linear) relationship is

$$\sigma'_v = a_e - b_e e_v \qquad (8.60)$$

where a_e and b_e are bed-dependent constants. The coefficient of compressibility, assumed to be independent of σ'_v, is $a_v = -\partial e_v / \partial \sigma'_v = b_e^{-1}$. The second (assumed linear) relationship is

$$k_p = \rho g k_0 (1 + e_v) \qquad (8.61)$$

where k_0 is a constant.[k] Thus, the consolidation coefficient c_{cv} defined by Eq. 8.40 is related to b_e by

$$c_{cv} = \frac{k_0 (1 + e_v)^2}{a_v} \approx k_0 b_e \qquad (8.62)$$

according to which c_{cv} is independent of e_v [Lee and Sills, 1981].

Due to the choice of k_p based on Eq. 8.61, the second term in Eq. 8.59 vanishes, yielding a diffusion equation for e_v analogous to Eq. 8.48 for the excess pore pressure[l]

$$\frac{\partial e_v}{\partial \tilde{t}} = \frac{\partial^2 e_v}{\partial \tilde{\xi}^2} \qquad (8.63)$$

---

[k] Based on overburden (loads) applied to very soft soil slurries, Bo [2008] reported the following non-linear relationships

$$\sigma'_v = e^{6.887 - 2.205 e_v + 0.157 e_v^2} \qquad (F8.8)$$

and

$$k_p = e^{3.17(e_v - 8.291)} \qquad (F8.9)$$

in which the unit of σ'_v is kPa and k_p is in m s^{-1}.

[l] Substitution of Eq. 8.60 into Eq. 8.55 yields $p_{we} = \sigma_v - p_w - a_e + b_e e_v$. This *assumed* relationship between p_{we} and e_v makes Eq. 8.48 equivalent to Eq. 8.63.

where $\tilde{\xi} = \xi/\xi_0$, and $\tilde{t} = c_{cv} t/\xi_0^2$.[m] As before we will prescribe a constant initial void ratio e_{v0} over the entire settling sediment mass, *i.e.*,

$$e_v(\tilde{\xi}, 0) = e_{v0} \tag{8.64}$$

Since at the top of the bed ($\xi = \xi_0$) the effective stress is zero (in the absence of overburden), the void ratio does not change, *i.e.*,

$$e_v(\tilde{\xi}_0, t) = e_{v0} \tag{8.65}$$

At the bottom ($\xi = 0$), where there is no transport, the constant gradient of void ratio is specified as

$$\frac{\partial e_v(0, t)}{\partial \tilde{\xi}} = \alpha_v \tag{8.66}$$

where $\alpha_v = g(\rho_s - \rho_w)a_v$. Using Laplace transform, the solution of Eq. 8.63 along with Eqs. 8.64, 8.65 and 8.66 is obtained as

$$e_v(\tilde{\xi}, \tilde{t}) = e_{v0} - \alpha_v \xi_0 \left[1 - \tilde{\xi} - 2 \sum_{n=0}^{n \to \infty} \frac{\cos m\tilde{\xi}}{m^2} e^{-m^2 \tilde{t}} \right]; \quad m = \frac{\pi}{2}(2n+1) \tag{8.67}$$

Experimental observations on the consolidation of clayey deposits do not support the characterization of the surface void ratio as being independent of time. The effect of decreasing void ratio within the settling sediment mass tends to reach the top of the deposit. In order to realistically simulate this behavior of e_v, Been and Sills [1981], for instance, assumed an imaginary overburden, with its height dependent on the difference between the initial void ratio e_{v0} and the ultimate (actual) void ratio e_{vf} at the surface. Based on this assumption the modified Gibson height ξ_{0m}, which is the sum of the original height and the imaginary overburden height, becomes

$$\xi_{0m} = \frac{e_{v0} - e_{vf}}{\alpha_v} + \xi_0 \tag{8.68}$$

---

[m] See Exercise E8.12 for a numerical scheme to solve Eq. 8.63.

Replacing ξ_0 by ξ_{0m} in Eq. 8.67 results in

$$e_v\left(\tilde{\xi},\tilde{t}\right) = e_{v0} - \alpha_v \xi_{0m}\left[1 - \tilde{\xi} - 2\sum_{n=0}^{n\to\infty}\frac{\cos m\tilde{\xi}}{m^2}e^{-m^2\tilde{t}}\right]; \quad m = \frac{\pi}{2}(2n+1) \quad (8.69)$$

along with $\tilde{\xi} = \xi/\xi_{0m}$ and $\tilde{t} = c_{cv}t/\xi_{0m}^2$. In Fig. 8.18, the behavior of void ratio is schematically shown in accordance with Eq. 8.69. The corresponding excess pore pressure is given by

$$p_e\left(\tilde{\xi},\tilde{t}\right) = 2g\left(\rho_s - \rho_w\right)\xi_{0m}\sum_{n=0}^{n\to\infty}\frac{\cos m\tilde{\xi}}{m^2}e^{-m^2\tilde{t}} \quad (8.70)$$

Since this excess pore pressure includes a component due to the imaginary overburden at $\xi = \xi_0$, when calculating the excess pore pressure profile, this component must be subtracted from all values.

From the void ratio profile $e_v(\tilde{\xi},\tilde{t})$, the bulk density profile $\rho(\tilde{\xi},\tilde{t})$ is obtained as

$$\rho(\tilde{\xi},\tilde{t}) = \frac{\rho_s - \rho_w}{1 + e_v(\tilde{\xi},\tilde{t})} + \rho_w \quad (8.71)$$

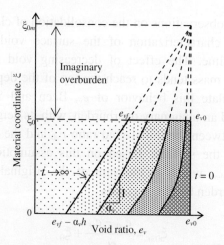

Figure 8.18. Imaginary overburden of material height $\xi_{0m} - \xi_0$ at the top of deposit and change in the void ratio profile with time from the onset of self-weight consolidation (adapted from Been and Sills [1981]).

where ρ_w is the water density. The instantaneous mean density ρ_m over height ξ_0 is found by integrating the above profile

$$\rho_m\left(t = \frac{\xi_0^2 \tilde{t}}{c_{cv}}\right) = \int_0^1 \rho(\tilde{\xi}, \tilde{t}) d\tilde{\xi} \qquad (8.72)$$

Given the initial height of suspension h_0 and density ρ_0, the instantaneous height $h(t)$ is then calculated from

$$h(t) = \frac{\rho_0}{\rho_m} h_0 \qquad (8.73)$$

where $\rho_m(t)$ now refers to density averaged over $h(t)$.

8.4.5 Consolidation tests

To examine Eqs. 8.69 and 8.70, tests were carried out in a 2 m high clear acrylic column (Fig. 8.19), in which the pore pressure was measured by flush-mounted sensors at seven elevations and an additional sensor at the bottom. The total (normal) stress was also measured at the bottom, and

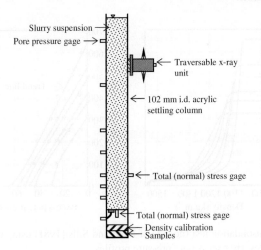

Figure 8.19. Acrylic column with a traversable x-ray unit for density profiling [Been and Sills, 1981].

continuous density profiles were obtained by a vertically traversable x-ray unit. Mud used from an estuarine site at Combwich (UK) was wet-sieved to remove particles larger than 75 μm. Figures 8.20a, b show density and excess pore pressure profiles measured at different times in a test in which the height of an initially uniform slurry with a density of 1,142 kg m^{-3} was 643 mm [Been and Sills, 1981].

For Combwich mud, variation of the void ratio e_v with effective stress σ'_v deduced from measurements is shown in Fig. 8.21a. For comparison, the (non-linear) curve corresponding to Eq. F8.8 is also shown. The absence of an identifiable linear relationship (Eq. 8.60) applicable to the mud data would imply a limitation in using the linear theory. In the plot, at low and high effective stresses there is practically no dependence of void ratio on effective stress. As expected, this leads to deviations in linear theory predictions of the void ratio either at the top of the column (high void ratios) or at the bottom (low void ratios), depending on the choice of best-fit values of coefficients a_e and b_e.

Based on sediment from Canaveral Harbor in Florida (Fig. 8.21b), Govindaraju *et al.* [1999] showed that shearing forces acting at the top of the bed was another reason why the linear approach (Eq. 8.61) must be used with caution.

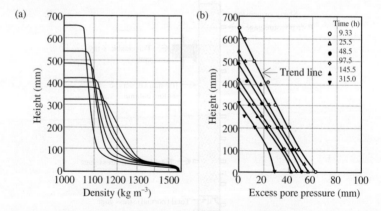

Figure 8.20. Consolidation test (no. 11) of Been and Sills [1981] using Combwich mud: (a) Density profiles; (b) excess pore pressure profiles.

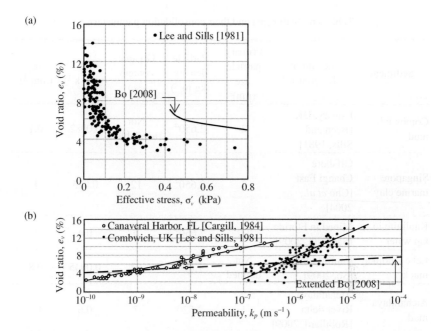

Figure 8.21. Relationships deduced from consolidation tests of Been and Sills [1981] using Combwich mud and Eqs. F8.8 and F8.9 of Bo [2008]: (a) Void ratio and effective stress; (b) void ratio and permeability. For comparison purposes, Eq. F8.9 has been extended beyond the recommended limits.

In Fig. 8.21b, the dependence of void ratio on permeability is approximately log-linear, similar to Eq. F8.9. This observation together with Fig. 8.21a suggests that for application of the linear theory to soft mud consolidation, it is necessary to select piece-wise linear relationships e_v–σ'_v and e_v–k_p.

For the linearized Eqs. 8.69 and 8.70 two coefficients, c_{cv} and a_v, are required. As seen from Table 8.6, these coefficients vary widely depending on bed composition. Note that, following Eq. 8.66, a_v is conveniently represented in terms of $\alpha_v = g(\rho_s - \rho_w)a_v$.

Example 8.4: For Combwich mud assume the following parametric values: $\rho_w = 1{,}000$ kg m^{-3}, $\rho_s = 2650$ kg m^{-3}, $e_i = 10.9$, $e_f = 6.5$, $c_{cv} = 8.20 \times 10^{-9}$ m^2 s^{-1}, $\alpha_v = 100$ m^{-1} and $\xi_0 = 0.0375$ m. The measured fall of bed height is recorded (data points) in Fig. 8.22. For the times indicated in the inset, obtain profiles of wet bulk density, void ratio and normalized excess pressure.

Table 8.6. Sediments and their consolidation parameters.

Sediment	Location/ Reference	Primary particle size (μm)	Particle density (kg m^{-3})	Organic content (%)	c_{cv} (mm^2 s^{-1})	α_v (mm^{-1})
Combwich mud	Estuary, UK [Been and Sills, 1981]	2	2,650^a	Not available	0.0082	0.1
Singapore marine clay	Offshore Changi East [Chu *et al.*, 2004]	3	2,650^a	12	0.025	1
Kaolinite in tap water	Mined in Florida [Sampath, 2009]	1.2	2,650^a	0	0.4	1
Lake Apopka muck	Central Florida [So, 2009]	Not defined	1,690	62	0.0045	0.8
Atchafalaya mud	Atchafalaya River delta [Robillard, 2009]	1	2,520	7	0.6	1.5

^{a} Nominal value.
Source: Sampath [2009].

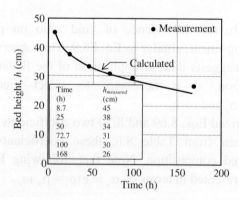

Figure 8.22. Measured (points) and calculated (curve) settling of mud–water interface in test no. 11 of Been and Sills [1981] using Combwich mud.

Sample calculations at 8.7 h are given in Table 8.7 for $t = 8.7$ h, *i.e.*, $\tilde{t} = 0.0387$. The void ratio e_v is calculated from Eq. 8.69 and $\phi_v = 1/(1+e_v)$. Knowing ϕ_v the wet bulk density ρ is calculated next from Eq. 8.71. The excess pore pressure is then obtained from Eq. 8.70. Imaginary excess pore pressure at the top is subtracted from all the values. The resulting pressure is normalized to yield $\tilde{p}_e$ by division with $g(\rho_s-\rho_w)\xi_0$. The actual elevation z of the bed is obtained by first calculating $dz \approx \Delta z = \Delta\xi / \overline{\phi}_v$, where Δ is the differential distance between two consecutive values of ξ, and $\overline{\phi}_v$ is the solids volume fraction averaged over $\Delta\xi$. Finally, the elevation z is obtained by summing cumulatively from the bottom. In Fig. 8.22, bed heights are compared with the calculated curve obtained by interpolating between the calculated heights. The density, void ratio and normalized excess pore pressure profiles are shown in Fig. 8.23a, 8.23b and 8.23c, respectively.

8.4.6 *Bed density profile: empirical representation*

From the above analysis it is evident that the relationship describing the variation of bed density with depth and time is not amenable to simple parameterization, as a variety of factors influence the density profile. These include particle composition (minerals, density, size, shape, organic matter, and biochemical bonding), fluid chemistry, history of deposition (vertical heterogeneity in sediment properties) and the degree of consolidation [Maa and Lee, 2002; McAnally *et al.*, 2007]. Some laboratory tests on the settling/consolidation of thin beds of cohesive sediments suggest that approximate empirical equations can be

Table 8.7. Calculated density, void ratio and excess pressure at $t = 8.7$ h.

ξ (m)	$\tilde{\xi}$	e_v	ϕ_v	ρ (kg m^{-3})	$\tilde{p}_e$	dz (m)	z (m)
0.0375	0.460	10.81	0.0847	1139.74	0	0.02949	0.421
0.0350	0.429	10.78	0.0849	1140.07	0.059	0.02941	0.392
0.0325	0.399	10.75	0.0851	1140.47	0.117	0.02931	0.362
0.0300	0.368	10.70	0.0854	1140.98	0.172	0.02920	0.333
0.0275	0.337	10.65	0.0858	1141.60	0.225	0.02906	0.304
0.0250	0.307	10.59	0.0863	1142.35	0.276	0.02889	0.275
0.0225	0.276	10.52	0.0868	1143.26	0.323	0.02868	0.246
0.0200	0.245	10.43	0.0875	1144.35	0.366	0.02845	0.217
0.0175	0.215	10.33	0.0883	1145.65	0.405	0.06152	0.189
0.0120	0.147	10.04	0.0905	1149.40	0.476	0.05436	0.127
0.0070	0.086	9.71	0.0934	1154.11	0.520	0.06367	0.073
0.0009	0.011	9.18	0.0982	1162.07	0.542	0.00916	0.009

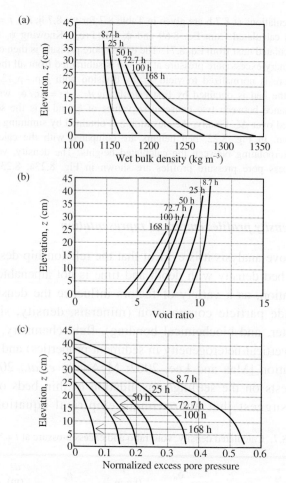

Figure 8.23. Calculated density, void ratio and normalized excess pressure profiles.

obtained for density variation with depth. Scale effects in laboratory-based profiles must be borne in mind, particularly when the amount of organic matter is high.

In Fig. 8.24, dimensionless profiles of bed density ρ_D (dry sediment mass per unit volume of saturated bed) are shown for a bed of kaolinite formed by deposition in a flume. With increasing duration T_c of self-

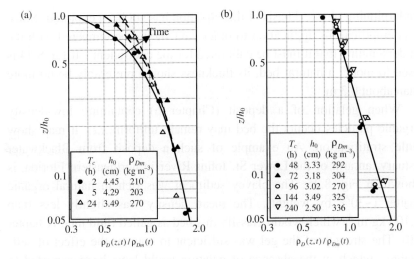

Figure 8.24. Dimensionless bed dry density against normalized bed elevation above flume bottom. Data for kaolinite in tap water: (a) $T_c < 48$ h; (b) $T_c \geq 48$ h (from Dixit [1982]).

weight consolidation, the profile seemingly became increasingly "saturated" from the bottom up (Fig. 8.24a).

As consolidation progressed the profile became stabilized. After a time T_c of about 48 h, the profiles approached self-similarity (Fig. 8.24b), with shape approximated by the dimensionless relationship

$$\frac{\rho_D(z)}{\rho_{Dm}} = \zeta_D \left(1 - \frac{z}{h_0} \right)^{-\xi_D} \tag{8.74}$$

where ρ_{Dm} is the dry density averaged over bed height $h_0(T_c)$. Values of coefficients ζ_D and ξ_D in Eq. 8.74 are 0.794 and 0.288, respectively.[n] The

---

[n] Migniot [1968] proposed the following empirical equation for bed dry density (kg m^{-3}):

$$\rho_D(z') = \rho_{D0} + \hat{n}z' \tag{F8.11}$$

in which z'(m) is the distance measured downward from the bed surface where the dry density is ρ_{D0}. The sediment dependent coefficient $\hat{n}$ was found to range between 50 and 80. The use of this equation is limited to a bed of no more than 2 m in thickness. A maximum of 0.3 m would seem more realistic.

underpinning physical basis of this equation has not been explored. Since it is empirical, its application to other data should be tempered with the understanding that it must be calibrated for the coefficients. If Eq. 8.74 is used to model a natural bed, its thickness should preferably be no more than about 0.3 m.

When gelation of a deposit (Chapter 6) containing low-density organic particles occurs the bed may remain uniform, *i.e.*, it may show little stratification. An example of such a deposit from Blackwater estuary, an arm of the Lower St. Johns River in northeastern Florida, is shown in Fig. 8.25. The clayey sediment was rich in natural organic matter (~30% by weight). The mean density was slightly less than 1,100 kg m^{-3}, which would classify this bed as gelled fluid mud (Chapter 10). The strength of the gel was sufficient to counter the effect of self-weight, which in the absence of gelation could have been expected to result in a stratified density profile. In other words, the low but uniform density suggests that gelation can produce a bed at a density at which a non-gelled flocculated bed would be stratified.

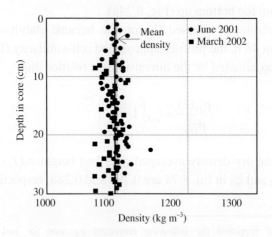

Figure 8.25. Density profiles (June 2001 and March 2002) from a coring site in Blackwater estuary, north-central Florida (from Jaeger *et al.* [2004]).

8.4.7 Bed rebound

When a clayey bed is unloaded, *i.e.*, the existing overburden is removed, the compressed bed partially rebounds, or swells, due to its elastic property. Swelling, the reverse of compression, amounts to an increase in the void ratio e_v due to a decrease in the effective stress σ'_v. It is characterized by the Swell Index C_{SI} defined as

$$C_{SI} = -\frac{\partial e_v}{\partial(\log \sigma'_v)} \qquad (8.75)$$

Empirically, C_{SI} is found to vary in an approximate way with the soil's liquid limit W_L as observed in Fig. 8.26, which includes data for mineral clays (Lambe and Whitman [1979]). As we will see in Chapter 9, the erodibility of beds rich in organic matter also may be influenced by elastic rebound of embedded organic flocs.

8.5 Exercises

8.1 Following the definitions in Chapter 5, the wet bulk density ρ of an unsaturated bed is related to its void ratio e_v by

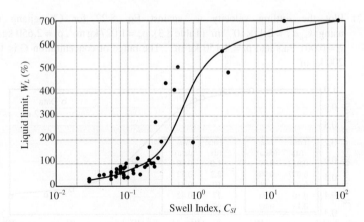

Figure 8.26. Variation of Swell Index (for clays) with the liquid limit (data from Lambe and Whitman [1979]).

$$\rho = \left(\frac{s + e_v S_s}{1 + e_v} \right) \rho_w \qquad (E8.1)$$

where s is the specific gravity of the soil, S_s is the degree of saturation and ρ_w is the density of water. For such a bed

$$e_v = \frac{sW}{S_s} \qquad (E8.2)$$

in which W is the water content. You are given $s = 2.65$, $\rho_w = 1{,}027$ kg m^{-3} and $W = 133\%$. Plot the variation of ρ with S_s ranging from 50% to 100%.

8.2   For a porous medium derive the conservation of flow (Eq. F8.1), and the conservation of solids (Eq. F8.2). Use the differential elemental volume approach.

8.3   Determine the best value of the characteristic time t_v in Eq. 8.15 by applying it to settling data for Doel Dock mud in Fig. E8.1.

8.4   (1) Obtain Eq. 8.21 by noting that the number of flocs per unit volume of water, N, is given by

$$N = \frac{K_{sf} \phi_{vf}}{\frac{\pi}{6} d_f^3} \qquad (E8.3)$$

and the distance between the centers of adjacent flocs is $N^{-1/3}$. Consider a simple cubic lattice, *i.e.*, with a floc occupying each corner of the cube.

(2) Plot the settling velocity w_s against Eq. 8.22 for Changjiang mud using: $k_{cp} d_f^2 = 2.11 \times 10^{-11}$ m^2 (Table 8.3), $\rho_w = 1{,}027$ kg m^{-3}, $\rho_s = 2{,}650$ kg m^{-3}, $\eta = 0.001$ Pa.s and $\rho_f = 1{,}150$ kg m^{-3}. The range of concentration C is 10 to 200 kg m^{-3}.

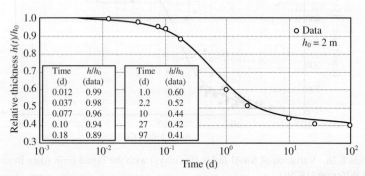

Figure E8.1. Settling interface of Doel Dock mud, Antwerp, Belgium (from Toorman and Berlamont [1993]).

8.5 In Fig. E8.2 data are plotted on the height of mud-water interface as a function of time from a 0.1 m diameter settling column (at 20°C). Out of the eleven estuarine sediments four — Mahury, Douala, Owendo and Escaut — are non-European. Several of these muds had a high content of kaolinite and very fine particle size (Table 9.2). The European muds were mixtures of several clays (illite, smectite, kaolinite and chlorite) and quartz (sand). Assume that for all sediments the initial height h_0 of the suspension is 1 m (at a concentration of 200 kg m^{-3}).

(1) Estimate the interface settling velocity *in the unhindered (constant-flux) zone* for the Mahury (French Guiana) and the Gironde (France) samples, assuming the slurry to be a suspension (as opposed to a bed) at all times. (Why is this assumption necessary?) As a starting point, replot the data in Table E8.1 for the first 20 days for each sample using a linear scale for the time axis.

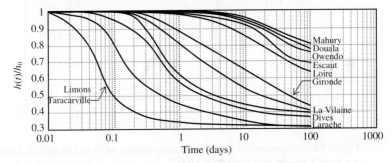

Figure E8.2. Variation of settling suspension height with time for natural muds. The time axis is logarithmic (from Migniot and Hamm [1990]).

Table E8.1. Interface settling of two marine sediments.

Time	h/h_0	
(d)	Mahury	Gironde
0.1	1.0	1.0
0.2	0.98	1.0
0.5	0.94	1.0
0.7	0.90	1.0
1	0.88	1.0
2	0.81	0.99
5	0.72	0.98
7	0.68	0.97
10	0.65	0.94
20	0.58	0.91
50	0.49	0.84
70	0.46	0.82
100	0.43	0.80

(2) Suggest why at 100 h the Mahury mud was compressed more than the Gironde mud?

8.6   For mud from Doel Dock, measured profiles of sediment concentration at 2.7 h, 22 h and 339 h are shown in Fig. E8.3 [Toorman and Berlamont, 1993]. Assume that this sediment is throughout in the suspended state.

(1) Based on the settling interface data given in Fig. E8.1, apply the method of Kynch to predict the concentration profiles at the same three times. Explain the differences between measured (Table E8.2) and predicted profiles in terms of mass fluxes.

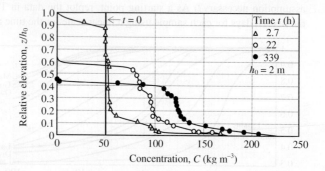

Figure E8.3. Concentration profiles during settling of Doel Dock mud. Data points are from Toorman and Berlamont [1993].

Table E8.2. Concentration profiles during settling of Doel Dock mud.

2.7 h		22 h		339 h	
z/h_0	C (kg m^{-3})	z/h_0	C (kg m^{-3})	z/h_0	C (kg m^{-3})
0.9137	24.0	0.5598	54.0	0.4293	2.5
0.8729	47.5	0.5311	77.5	0.4221	63.0
0.8177	47.5	0.4904	82.0	0.4125	89.0
0.7602	47.5	0.4354	85.0	0.3837	111.5
0.6571	47.5	0.3804	84.5	0.3429	114.0
0.5899	50.5	0.3278	95.0	0.3213	117.0
0.3741	51.5	0.2823	95.0	0.2974	121.5
0.2710	53.5	0.2153	98.5	0.2710	122.0
0.2182	53.5	0.1627	99.0	0.2446	123.0
0.1583	59.0	0.1364	105.5	0.2206	125.0
0.1055	82.5	0.1029	112.0	0.1871	126.5
0.0767	95.5	0.0790	120.5	0.1655	128.0
0.0504	96.5	0.0526	140.5	0.1343	131.5
0.0048	168.0	0.0359	142.0	0.1055	136.0

Table E8.2. (Con't.)

0.0239	147.5	0.0815	153.0
0.0168	161.5	0.0456	170.5
		0.0384	179.0
		0.0240	192.5
		0.0144	207.0

Source: Toorman and Berlamont [1993].

(2) Repeat the Kynch application, this time using the data in Table E8.3, which are similar to those from Townsville Harbour [Wolanski *et al.*, 1991]. From the results, comment on the effect of waves on the settling and deposition of dredged material in the harbor.

(3) With data from Table 8.4 for the Jiaojiang and the Ortega River, plot the settling rate w_s against the excess density $\Delta\rho$ using Eq. 8.35. Assume $\rho_w = 1,027$ kg m^{-3} in both cases, $\rho_s = 2,650$ kg m^{-3} for the Jiaojiang, and 2,300 kg m^{-3} for Ortega.

8.7 Suspended sediment concentration profiles (below mid-tide level) from a river are given in Table E8.4 and plotted in Fig. E8.4. Assume that the current velocity during the data collection period was practically zero.

Table E8.3. SSC profiles during harbor sediment disposal in calm sea (C) and moderate sea (M).

Elevation (m)	Suspended sediment concentration (kg m^{-3})					
	0 min (C, M)	3 min (C)	8 min (C)	15 min (C)	8 min (M)	1 (M)
0	3.5	5.6	2.1	0.3	4.6	2.9
0.5	3.5	5.5	1.0	0.2	3.0	2.2
1.0	3.5	5.4	0.4	0	1.9	1.7
1.5	3.5	4.9	0.2	0	1.5	1.4
2.0	3.5	4.0	0	0	1.1	1.0
2.5	3.5	3.0	0	0	0.8	0.7
3.0	3.5	1.9	0	0	0.4	0.4
3.5	3,5	1.0	0	0	0.1	0.1
4.0	3.5	0.5	0	0	0	0
4.5	3.5	0.2	0	0	0	0
5.0	3.5	0	0	0	0	0
5.5	3.5	0	0	0	0	0
6.0	3.5	0	0	0	0	0

Table E8.4. Concentration profile data from a tidal river.

30 min before slack		10 min before slack		10 min after slack		40 min after slack	
SSC (kg m^{-3})	Depth (m)	SSC (kg m^{-3})	Depth (m)	SSC (kg m^{-3})	Depth (m)	SSC (kg m^{-3})	Depth (m)
0	−0.98	0	−2.20	0	−3.33	0.105	−3.92
1.15	−1.05	1.15	−2.42	1.04	−3.62	0.42	−4.08
2.50	−1.07	2.50	−2.70	2.40	−4.15	0.73	−4.40
5.00	−1.12	3.85	−2.95	3.65	−4.53	1.25	−4.80
7.40	−1.22	4.90	−3.08	4.90	−4.78	1.88	−5.20
8.75	−1.35	6.15	−3.25	6.15	−4.98	2.50	−5.42
9.90	−1.48	7.40	−3.35	7.40	−5.13	3.75	−5.72
11.15	−1.68	8.65	−3.45	8.75	−5.27	5.00	−5.88
12.29	−2.00	9.90	−3.52	10.00	−5.33	6.25	−5.98
13.44	−2.58	12.40	−3.65	12.50	−5.47	7.340	−6.07
14.69	−3.43	14.90	−3.77	15.00	−5.53	8.75	−6.13
16.04	−4.80	17.40	−3.93	17.50	−5.60	10.00	−6.17
17.40	−6.42	18.75	−4.05	19.90	−5.67	12.40	−6.23
18.33	−7.60	20.00	−4.17	24.90	−5.75	17.30	−6.30
18.75	−7.98	21.15	−4.30	29.890	−5.83	22.71	−6.38
		22.40	−4.53	34.90	−5.93	27.40	−6.43
		23.75	−4.95	37.50	−6.02	32.50	−6.53
		25.00	−5.58	38.85	−6.07	37.50	−6.58
		26.04	−6.40	40.00	−6.10	42.50	−6.67
		26.77	−7.20	41.35	−6.18	47.50	−6.72
		27.19	−7.57	42.40	−6.25	52.50	−6.78
		27.71	−7.98	43.85	−6.37	57.40	−6.85
				45.63	−6.58	62.50	−6.93
				47.50	−6.92	67.50	−7.03
				48.65	−7.20	68.85	−7.08
				50.00	−7.58	70.00	−7.18
				51.56	−7.97	71.25	−7.37
						72.61	−7.58
						73.44	−7.80
						74.06	−7.97

(4) Using the method of Kynch, estimate the time t_0 when the initial profile can be assumed to have been (hypothetically) uniform as shown. Hint: This is done by finding the value of t_0 by trial-and-error such that the measured profiles (at times relative to t_0) best match the calculated profiles.

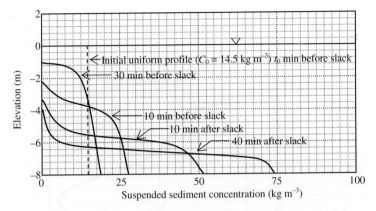

Figure E8.4. Suspended sediment concentration profiles from a tidal river.

(5) Plot the calculated concentration profile at the time of slack water.

8.8   Density profiles from a laboratory column at four different times are given in Table E8.5 for mud from the Atchafalaya River delta along the Gulf of Mexico coast (AD mud, Table 5.5). The particle density is 2,520 kg m^{-3}, water density (nominal value) 1,000 kg m^{-3}, initial void ratio 0.49 and final void ratio 75% of the initial value. Compare these profiles with the corresponding profiles using the self-weight consolidation theory with $c_{cv} = 3.17 \times 10^{-7}$ m^2 s^{-1} and $a_v = 0.5205$ (in conventional units of m^2 kg^{-1}). Based on the comparison, indicate to what extent (and why) the theory is applicable to this experiment.

Table E8.5. Density profiles from a consolidation test using AD mud.

Elevation (cm)	Density (kg m^{-3})			
	10 min	60 min	90 min	1440 min
130	1037	1015	1014	1010
105	1037	1034	1017	1012
80	1041	1038	1032	1012
55	1040	1043	1043	1035
30	1040	1045	1046	1054
15	1044	1057	1064	1078
5	1048	1093	1099	1099

Source: Sampath [2009].

8.9   Bottom density profiles in Fig. E8.5 and Table E8.6 at two locations inside a marina basin show similar profiles in the top 0.15 m. However, the trends deviate substantially below about 0.2 m. Provide a possible explanation (with a theoretical basis) of these trends.

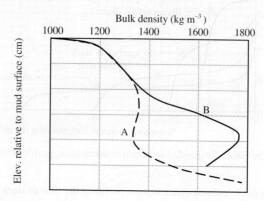

Figure E8.5.   Bottom density profiles from a marina basin (from Srivastava [1983]).

Table E8.6.   Marina density profile data.

Depth (m)	Prof. A (kg m^{-3})	Prof. B (kg m^{-3})
0	1000	1000
-0.02	1160	1160
-0.05	1220	1220
-0.1	1280	1280
-0.15	1320	1320
-0.2	1350	1380
-0.25	1360	1460
-0.3	1355	1620
-0.35	1340	1740
-0.38	1335	1765
-0.4	1340	1765
-0.45	1380	1700
-0.5	1465	1635
-0.55	1700	
-0.56	1785	

8.10   The time-variation of the depth-mean dry density of an initially 6 cm thick deposit in a flume is shown in Fig. E8.6 and given in Table E8.7. The starting normalized density $\rho_D / \rho_{D\infty}$ is 0.42 kg m^{-3}, where $\rho_{D\infty} = 1,280$ kg m^{-3}.

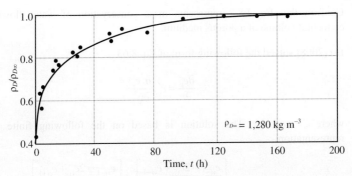

Figure E8.6. Time-variation of the depth-mean dry density of an initially 6 cm thick deposit in a flume.

Table E8.7. Time-density data.

t (h)	ρ_D / ρ_∞
0	0.42
3	0.63
4	0.57
5	0.66
12	0.74
15	0.79
16	0.77
25	0.825
28	0.81
31	0.85
51	0.92
52	0.88
60	0.945
76	0.925
101	0.98
128	0.99
152	0.995
172	0.9999

(1) Determine the best exponential fit equation.

(2) Find the time to reach $\rho_D / \rho_{D\infty} = 0.99$.

(3) Why would an exponential trend be reasonable?

8.11   Confirm that Eq. F8.2 represents mass conservation of water flowing in an elemental volume of a porous medium.

8.12   Bo [2008] solved the following form of Eq. 8.63

$$\frac{\partial e_v}{\partial t} = c'_{cv} \frac{\partial^2 e_v}{\partial \xi^2} \tag{E8.4}$$

where $c'_{cv} = c_{cv}/\xi_0$. The solution is based on the following finite difference representation:

$$\frac{e_{vi,t+\Delta t} - e_{vi,t}}{\Delta t} = c'_{cv} \frac{\left[\dfrac{e_{vi-1} - 2e_{vi} + e_{vi+1}}{(\Delta z)^2}\right]_t + \left[\dfrac{e_{vi-1} - 2e_{vi} + e_{vi+1}}{(\Delta z)^2}\right]_{t+\Delta t}}{2} \tag{E8.5}$$

$$e_{vi,t+\Delta t} = \frac{(e_{vi-1} + e_{vi+1})_{t+\Delta t} + e_{vi,t}}{2\left(1 + \dfrac{1}{\beta_{cv}}\right)} \tag{E8.6}$$

Complete the scheme including initial and boundary conditions, and develop an algorithm to obtain void ratio profiles such as those in Fig. 8.20. Include imaginary overburden correction.

8.13   Consider the unsteady (vertical) transport–reaction equation for the concentration, or activity, C_{At} of a sedimentary tracer in the bed

$$\frac{\partial C_{At}}{\partial t} = \frac{\partial}{\partial z}\left(D_{At} \frac{\partial C_{At}}{\partial z} - w_{At} C_{At}\right) - \lambda_{At} C_{At} \tag{E.8.7}$$

where D_{At} is the diffusion coefficient due to biological activity, w_{At} is the tracer burial velocity representing sediment deposition and λ_{At} is the (assumed first-order) reaction constant which determines the rate at which the tracer dissipates.[o]

---

[o] The diffusion coefficient D_{At} in the bed is smaller than D_m in water (free of sediment) and varies with tortuosity characterizing convoluted capillary pathways that the ions and molecules must follow. In a bed containing particles or flocs of diameter d_f, for an incremental pathway length ΔL_p ($\gg d_f$) the tortuosity $\theta_T = \Delta L_p/\Delta z$ reduces D_m according to

$$D_{At} = \frac{D_m}{\theta_T^2} \tag{F8.12}$$

If ΔL_p and Δz are coincident, *i.e.*, both are 90° with respect to the horizontal plane, then $\theta_T = 1$. If the distribution of capillary angles is random, for a statistically large number of

(1) Assuming that D_{At} and w_{At} are both independent of depth, the general *steady-state* solution is

$$C_{At}(z) = A_{At}e^{a_{At}z} + B_{At}e^{b_{At}z} \tag{E8.8}$$

where A_{At} and B_{At} are arbitrary constants of integration. Define a characteristic Péclet number $Pe = w_{At}h_b/D_{At}$, where h_b is the bed thickness over which Eq. E8.8 is applicable, and a Damkohler number $Da = \lambda_{At}h_b^2/D_{At}$. Letting $\hat{z} = z/h_b$ we may restate Eq. 8.8 as

$$C_{At}(\hat{z}) = A_{At}e^{\hat{a}_{At}\hat{z}} + B_{At}e^{\hat{b}_{At}\hat{z}} \tag{E8.9}$$

Show that $\hat{a}_{At}$ and $\hat{b}_{At}$ are functions of Pe and Da.

(2) Show that a steady-state solution of Eq. 8.8 for constant D_{At} and w_{At} is

$$C_{At} = C_{At0}e^{-\frac{w_{At} + \sqrt{w_{At}^2 + 4\lambda_{At}D_{At}}}{2D_{At}}z} \tag{E8.10}$$

where C_{At0} is the (constant) value of C_{At} at the bed surface. Assume that h_b is large, which means that the bottom boundary condition $dC_{At}/dz|_{z\to\infty} = 0$ can be considered to apply.

(3) Assume the following expression [Boudreau, 1994] for the burial velocity (cm y^{-1})

$$w_{At} = 3.3 \times 10^{-0.875 - 0.000435h} \tag{E8.11}$$

where h (m) is the water depth. Take diffusion coefficient (cm^2 y^{-1}) as

$$D_{At} = 15.7w_{At}^{0.69} \tag{E8.12}$$

These two expressions are *severe* approximations and may be used only for rough estimations of w_{At} and D_{At}. Write an algorithm to solve the steady-state equation and plot the profile of C_{At} for $h = 4.4$ m, given $C_{At0} = 10$ dpm g^{-1} of [234]Th used as a tracer (half-life $T_{1/2} = 24.1$ d $= 0.066$ y). Note that for radioactive elements $\lambda_{At} = \ln 2/T_{1/2}$.

capillaries per unit bed volume the mean capillary angle would be 45°, and $\theta_T = \sqrt{2}$ [Boudreau, 1997].

Chapter 9

Erosion and Entrainment

9.1 Chapter Overview

This chapter begins with a general description of fine sediment erosion by shear flow. Bed surface erosion is reviewed with reference to three analyses — stochastic, rate-process and empirical. This is followed by a description of the bed shear strength and its relationship with bed density, yield stress, degree of consolidation, salinity and organic matter. Other erosion modes including mass erosion and fluid mud entrainment are next reviewed. As a corollary of the erosion process, two contrasting paradigms introduced in Chapter 7 concerning exclusive erosion or deposition versus simultaneous exchange of bed material, are further explored. This is followed by, as a somewhat separate topic, a brief description of the effects of small quantities of clayey sediment on the erosion of sand; effects which can also be viewed as an extension of the erosion behavior of sand summarized in Chapter 6. For a fuller coverage of fine sediment erosion the last section is on elements of the design of channels that are stable with respect to bed scour.

9.2 Modes of Erosion

Among the unit processes related to fine sediment transport mentioned in Chapter 6, erosion has occupied the greatest attention of hydraulic engineers. Just as cemented sand erodes in a different way than loose sand, erosion of fine sediment depends on the properties of the eroding bed. Partly as a result no general theory of erosion exists, and it has become necessary to identify modes of erosion and treat them individually. An outcome is that the representation of erosion in

536

modeling is sometimes unsatisfactory when bed composition and state are variable, or when two or more erosion modes occur simultaneously at comparable intensities. The usual approach is to evaluate each mode separately and calculate the total erosion as the sum of the modes, ignoring inter-mode interactions. Since these interactions are not well understood, the summation approach is typically assumed to be adequate. Accuracy in modeling erosion is dependent in part on calibration, and explains why a variety of erosion flux equations may perform equally well, although perhaps none quite satisfactorily, in model simulations.

Figure 9.1a shows a severely eroded steep-sided tidal flat at Peterstone Wentooge in the Severn River estuary, UK. The flat consists of over-consolidated Holocene clay layers, or rhythmites [Allen, 2004], topped by thin laminas of peat (*i.e.*, accumulated partially decayed vegetative material), which provides protection where the peat remains. The mean thickness of each clay layer is about 5 mm. Illite is the dominant mineral with lesser amounts of smectite. The clay layers are overconsolidated because this is an eroding foreshore from which as little as 1–2 m of overburden of similar material has been naturally removed. Although visually tidal rhythmites may show a well-organized and seemingly uniform banding pattern, detailed spectral analysis of the bands typically reveals wide ranging periodicities associated with variability in local astronomical tide and storms [Stupples and Plater, 2007].

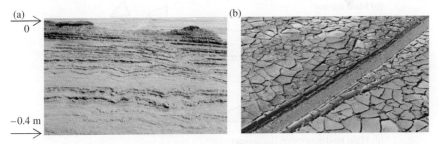

Figure 9.1. (a) View of an eroded mud flat in the Severn River estuary, UK. Vertical distance between the top and bottom layers is about 0.4 m (courtesy Robert Kirby); (b) Desiccated layer of mud over sandy Cassino Beach, Brazil. The approximately 15-cm wide automobile tire track provides a scale of the mosaic mud-sheets (courtesy Lauro Calliari).

Figure 9.1b shows a desiccated layer of mud over a sandy substrate at Cassino Beach (State of Rio Grande do Sul) in Brazil. Mud is deposited on the beach during events of high water discharge and its southerly course from the entrance to Lagoa dos Patos bay at the north end of the beach [Vinzon *et al.*, 2009]. The deposit is gradually washed away by waves.

Different states of bottom mud in the natural environment (*e.g.*, Figs. 9.1a, b) underscore the need to simplify the description of erosion into recognizable and treatable component processes. As schematized in Fig. 9.2, the erosion of fine-sediment bed by current or waves can be roughly categorized into modes including surface erosion, mass erosion, fluid mud entrainment and fluid mud generation by waves. Surface erosion by current occurs when the bed abrades at the surface, by detachment of flocs or small particles. As the overburden is removed by erosion the flocs at the freshly exposed surface swell, which amounts to drainage [Winterwerp *et al.*, 2012]. The outcome of surface erosion is that the bed surface is gradually lowered. In mass erosion the undrained bed below the surface layer fails episodically at some depth where

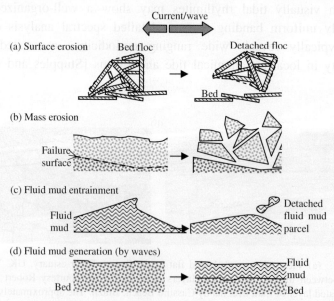

Figure 9.2. Modes of erosion: (a) Surface erosion; (b) Mass erosion; (c) Entrainment of fluid mud; (d) Generation of fluid mud by waves.

fracture occurs and the displaced clasts are transported downstream. The term "entrainment" (of fluid mud) is borrowed from the more extensively studied process of transfer of saltwater into freshwater. When applied to fine sediment this term implies erosion (due to current or waves), because aqueous parcels of particles lift out of fluid mud and mix with the upper fluid.

Two terms associated with erosion are noteworthy. "Resuspension" is used to mean the erosion of sediment recently deposited by currents or waves in which the eroding bed retains detectable signatures of deposition. "Scour" is used in the context of what is called general scour, *e.g.*, lowering of the bed level due to contraction of channel width by, for example, bridge abutments or jetties at tidal entrances. A second use is in reference to local scour around embedded structures such as bridge piles. Generation of fluid mud is another mode of resuspension in which a phase change from bed to suspension occurs.

This classification of erosion is meant to assist in understanding a complex set of mechanisms through simplified analytic and empirical treatments. The first three modes of erosion are reviewed in this chapter. Fluid mud generation by waves is mentioned in Chapter 10.

9.3 Surface Erosion

9.3.1 *Erosion formulation*

In this mode a floc at the bed surface is detached from its neighbors and erodes when cohesive/adhesive bonds (*e.g.*, at points 1, 2 and 3, Fig. 9.2a) connecting the floc to its neighbors are ruptured by fluid stress. In hydraulic engineering efforts have been made to relate the cumulative erosion flux (accounting for a statistically large number of flocs) to easily measurable bulk variables (Chapter 5). This means that formalizing the complex microscale processes is bypassed in favor of a much less accurate but conveniently integrated approach.

For steady or quasi-steady (*e.g.*, tidal) flows numerous formulas relating surface erosion flux to the bed shear stress have been proposed [Mehta *et al.*, 1982; Taki, 2001; Piedra–Cueva and Mory, 2001]. These formulas are useful as long as the concentration of sediment suspended by erosion or advected from elsewhere does not exceed the limit C_m (Fig. 7.13). At concentrations greater than C_m, settling of suspended sediment is hindered and a layer of fluid mud is formed over the bed. The mechanism by which flowing fluid mud erodes the bed below is not modeled well by simple, stress-based formula used for surface erosion, because damping of turbulence occurs within fluid mud. The result is that the bed shear stress often changes substantially from its value in clear or nearly clear flows. A correction for the damping effect on the bed shear stress (or the friction velocity) has been shown to improve model predictions of erosion flux [Sheng and Villaret, 1989; Hsu *et al.*, 2007].[a]

Some investigators have eschewed the use of bed shear stress in favor of more direct measures of boundary layer turbulence. In their observations in Long Island Sound, Bedford *et al.* [1987] found that the turbulent kinetic energy (TKE) was a better descriptor of surface erosion flux than the Reynolds stress representing the turbulent component of

---

[a] Among experimental evaluations of the effect of fine suspended sediment concentration on the friction velocity, based on their field annular flume assembly, the Sea Carousel, Amos *et al.* [1997; 2003] proposed

$$\frac{u_{*ssc}}{u_*} = 1 - 0.036 \log(\rho_s \phi_v) \qquad (F9.1)$$

In this relationship u_* (cm s^{-1}) is the sediment-free friction velocity, u_{*ssc} is its actual value at solids volume fraction $\phi_v = C/\rho_s$, ρ_s (mg L^{-1}) is the particle density and C is the suspended sediment concentration in mg L^{-1}. Thus the ratio u_{*ssc}/u_* represents drag reduction. Equation F9.1 is applicable to comparatively low concentrations from nil to about 1,000 mg L^{-1}, *i.e.*, about 1 kg m^{-3}. From experiments using a kaolinite in a flow-recirculating flume, Li and Gust [2000] proposed

$$\frac{u_{*ssc}}{u_*} = -62.28\phi_v + 9.171 \times 10^{-6} Re + 0.425 \qquad (F9.2)$$

where Re is the flume-flow Reynolds number. Both fresh and seawater were used, and the concentration ranged between 0.1 and 8 kg m^{-3}. In the field $Re = u_m \delta_b / \nu$ is recommended, where u_m is the mean current velocity in the near-bed boundary layer of height δ_b and ν is the kinematic viscosity of water. To calculate u_{*ssc}, u_* is obtained from the mean current using Eq. 2.59.

bed shear stress (Chapter 2). However, a firm conclusion regarding the use of TKE versus bed shear stress does not appear to be available.

Surface erosion has been examined by at least three approaches: (1) stochastic, (2) chemical rate-process analogy, and (3) empirical. They are summarized here.

9.3.2 *Erosion as a stochastic process*

9.3.2.1 *Model of Partheniades*

In a comprehensive study of the surface erosion of cohesive sediment, Partheniades [1962; 1965] tested beds of mud from the Mare Island Strait in San Francisco Bay in an 18 m long, flow-recirculating flume. In each of three series of tests the flow velocity, and therefore the bed shear stress, was increased in steps, with each step lasting up to 200 h. Series I and II beds were prepared by placing the natural sediment at the bottom. In Series III the bed was produced by fully suspending sediment at a high velocity and allowing the material to deposit at a reduced velocity. On a wire-mesh screen of the type used in sieve size analysis, small weights were placed to enable the screen to penetrate the bed (Fig. 9.3a). From the extent of penetration the variation of the load resistance of bed as a function of depth was estimated. Series I and II beds had high penetration resistance, *i.e.,* bed strength,

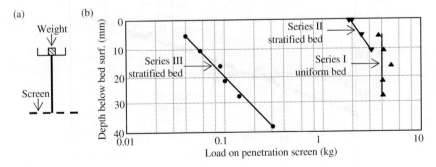

Figure 9.3. (a) Wire-mesh screen penetrometer; (b) penetration resistance in Series I, II and III of Partheniades [1962].

with Series I having uniform strength. Series III bed was the weakest and most stratified (Fig. 9.3b).

Results of two test runs are shown in Fig. 9.4. In Run 3 from Series I the dense and uniform bed was eroded at a shear stress of 0.492 Pa, and in Run 30 from Series III the weak and stratified bed was eroded at about the same shear stress (0.435 Pa). In each case the initial suspended sediment concentration was taken as the value at the end of the previous run.

Series I bed eroded at a constant rate for 150 h, *i.e.*, the concentration increased linearly with time. In other words the erosion flux was constant. In the Series III run, which was initially close to that of Series I, the flux decreased steadily thereafter and became negligible after about 40 h.

Partheniades concluded that once surface flocs eroded, the scoured material did not redeposit, as this would be the only explanation for the constant erosion flux in the Series I run. Such an inference would be commensurate with the variation of depositing sediment concentration with time as mentioned in reference to Fig. 7.36. In contrast, in Series III the bed shear strength increased with continued erosion, which in turn reduced the erosion flux. Reduction in erosion continued until the bed was scoured to a depth at which the bed shear stress (0.435 Pa) became equal to the bed shear strength, and erosion stopped. Thus it was inferred

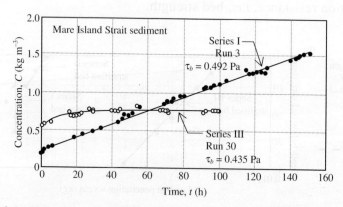

Figure 9.4. Variation of suspended sediment concentration with time during Series I (Run 3) and Series III (Run 30) in cohesive bed erosion tests of Partheniades [1962].

that the decrease in erosion flux was due to bed densification and hardening with depth rather than by redeposition of eroded sediment. Partheniades further noted that there was a paradigmal difference in the transport behavior of cohesive sediments in contrast to cohesionless sediments of uniform size. As mentioned in Chapter 7, in the latter case the erosion flux ultimately decreases to zero when the number flux of eroding particles becomes equal to the number flux of depositing particles under non-silting-non-scouring, or live-bed, equilibrium. Partheniades developed an expression for the erosion flux by using a stochastic description of floc transport in turbulent flows as follows.

In partial analogy with the bed load analysis of Einstein [1950] (Chapter 6), let p_e be the probability of cohesive particle surface erosion. The term "particle" refers to a unit that may exist as a single solid or as a floc. Define t_b as the characteristic time required to detach a particle from the bed. Then, the number flux of eroding particles (number per unit bed area and unit time) N_e is given by

$$N_e = \frac{p_e}{\alpha_2 d_p^2 t_b} \tag{9.1}$$

where d_p is the particle diameter and α_2 is an area shape factor. Since the weight of each particle is $\alpha_1 \gamma_s d_p^3$, where α_1 is a volume shape factor and $\gamma_s = g\rho_s$ is the unit weight of the particle, the erosion flux (dry sediment mass per unit area and unit time) is given as

$$\varepsilon = \frac{\alpha_1 \gamma_s d_p}{\alpha_2 t_b} p_e \tag{9.2}$$

In order to use this expression for calculating ε, the probability p_e is evaluated by noting that erosion occurs when the instantaneous hydrodynamic lift force F_l leads to detachment of the particle from the bed where it is anchored by cohesion (inclusive of biochemical adhesion) to the neighboring particles. Referring to the combined resistance to erosion due to cohesion and buoyant weight as F_C, p_e is the probability that

$$\frac{F_l}{F_C} \geq 1 \tag{9.3}$$

The lift force F_l can be represented as the sum of its time–mean value $\bar{F}_l$ and turbulence-induced fluctuation F_l'. Thus, from Eq. 6.69

$$F_l = \bar{F}_l \left(1 + \eta_l\right) \tag{9.4}$$

in which η_l is the lift fluctuation made dimensionless by $\bar{F}_l$.

The lift force fluctuation F_l' can be positive when the velocity fluctuation that induces it is in the (turbulence-mean) flow direction, or negative when the velocity fluctuation is opposed to flow. Thus in analogy with Eq. 6.70

$$\left|\frac{F_l}{F_C}\right| = \frac{\bar{F}_l}{F_C}\left|1 + \eta_l\right| \geq 1 \tag{9.5}$$

Therefore

$$\left|1 + \eta_l\right| \geq \frac{F_C}{\bar{F}_l} \tag{9.6}$$

Introducing $\eta^* = \eta_l / \sigma_L$, where σ_L is the standard deviation of η_l with a commonly adopted value of 0.5 based on measurements (Chapter 6), we obtain

$$\left|\frac{1}{\sigma_L} + \eta^*\right| \geq \frac{F_C}{\bar{F}_l \sigma_L} \tag{9.7}$$

Squaring both sides

$$\left(\frac{1}{\sigma_L} + \eta^*\right)^2 \geq \left(\frac{F_C}{\bar{F}_l \sigma_L}\right)^2 \tag{9.8}$$

The exact limit $\eta^* = \eta^*_{lim}$ at which incipient erosion occurs is given by the equality

$$\left(\frac{1}{\sigma_L}+\eta^*_{lim}\right)^2=\left(\frac{F_C}{\overline{F}_l\sigma_L}\right)^2 \qquad (9.9)$$

Therefore

$$\eta^*_{lim}=\pm\frac{F_C}{\overline{F}_l\sigma_L}-\frac{1}{\sigma_L} \qquad (9.10)$$

which is analogous to Eq. 6.75. These limiting (positive and negative) values of η^*_{lim} are identified in Fig. 9.5.

As noted in Chapter 6, the probability p_e is now obtained as the sum of the two end areas (*i.e.*, not ignoring the typically small area A_l) representing erosion identified within the frequency distribution of η^*

$$p_e=A_r+A_l=1-\frac{1}{\sqrt{2\pi}}\int_{-\frac{F_C}{\overline{F}_l\sigma_L}-\frac{1}{\sigma_L}}^{+\frac{F_C}{\overline{F}_l\sigma_L}-\frac{1}{\sigma_L}}e^{-\frac{\omega^2}{2}}\,d\omega \qquad (9.11)$$

where ω is a dummy variable. Recognizing that hydrodynamic lift as well as bed shear stress result from the same velocity fluctuation at the top of the particle and are therefore proportional to each other, we will

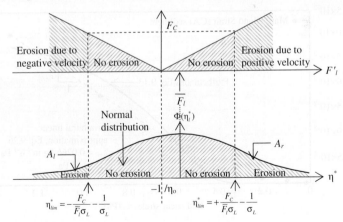

Figure 9.5. Erosion-defining limits and areas of erosion and no erosion specified by particle submerged weight combined with cohesion and hydrodynamic lift force.

set $\overline{F}_L = k_E \tau_b$ where, ignoring the overbar, τ_b now denotes the time–mean value of the instantaneous bed shear stress and k_E is a representative area parameter. Substitution of the resulting expression for p_e into Eq. 9.2 yields the erosion flux

$$\varepsilon = \frac{\alpha_2 \gamma_s d_p}{\alpha_1 t_b} \left(1 - \frac{1}{\sqrt{2\pi}} \int\limits_{\frac{F_C}{k_E \tau_b \sigma_L} - \frac{1}{\sigma_L}}^{+\frac{F_C}{k_E \tau_b \sigma_L} - \frac{1}{\sigma_L}} e^{-\frac{\omega^2}{2}} d\omega \right) \qquad (9.12)$$

Example 9.1: Find the best-fit coefficients for Eq. 9.12 using the erosion data of Partheniades [1962] given below (and in Fig. 9.6).

τ_b (Pa)	ε (kg m^{-2} h^{-1})
0.110	0.0005
0.215	0.0010
0.334	0.0018
0.492	0.0036
0.538	0.0078
0.934	0.0180
1.330	0.0304

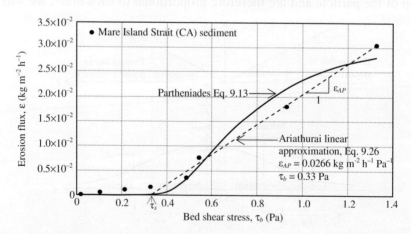

Figure 9.6. Erosion flux data of Partheniades [1962] and analytic representation.

With $\sigma_L = 0.50$, Eq. 9.12 can be rewritten as

$$\varepsilon = k_{E1}\left(1 - \frac{1}{\sqrt{2\pi}}\int_{\frac{k_{E2}}{\tau_b} - 2}^{\frac{k_{E2}}{\tau_b} - 2} e^{-\frac{\omega^2}{2}} d\omega\right) = k_{E1}\left\{1 - \frac{1}{2}\left[\text{erf}\left(\frac{1}{\sqrt{2}}\omega\right)_{-\frac{k_{E2}}{\tau_b} - 2}^{\frac{k_{E2}}{\tau_b} - 2}\right]\right\} \quad (9.13)$$

This two-coefficient equation with best-fit values $k_{E1} = 0.0297$ kg m^{-2} h^{-1} and $k_{E2} = 1.4021$ is plotted in Fig. 9.6. The simulated non-linear trend mimics the data.

9.3.2.2 *Reinterpretation of Partheniades model*

Equation 9.12 has three noteworthy restrictions. The first is that treating the bed shear stress as a stochastic variable while assuming that the bed shear strength is a mean-value variable prohibits erosion at localized sites where the bed shear strength is higher than the instantaneous bed shear stress. The second restriction arises from the fact that, as we saw in Chapter 7, the probability density function of the bed shear stress is non-normal (and skewed) because it is obtained from the pdf of the randomly fluctuating (and therefore normal-distributed) turbulent current velocity u_b. Finally, the third restriction is predicated on the consideration that since the maximum value of the term within parentheses in Eq. 9.12, *i.e.*, the probability of erosion, is 1, the erosion flux can be only as high as $\alpha_2 \gamma_s d_p / \alpha_1 t_b$. This is so because the time t_b required to detach the particle from the bed is assumed to be independent of flow, which is unrealistic because one would expect t_b to decrease as the flow increases [van Prooijen and Winterwerp, 2010; Letter, 2009]. Here we will summarize the significance of relaxing these restrictions starting with a more general definition of the probability of erosion p_e.

Let $\varphi(\tau_s)$ be the pdf of bed shear strength τ_s which varies over the (unit) bed area. The bed shear stress τ_b and τ_s are both assumed to be statistically independent, and erosion is considered to occur where $\tau_b > \tau_s$. As τ_b increases a larger fraction of the bed can be expected to erode until the instantaneous minimum shear stress in $\varphi(\tau_b)$ over the bed is greater than the instantaneous maximum shear strength in $\varphi(\tau_s)$ anywhere on the bed. Then the area of erosion will be the entire bed surface, with the probability of erosion equal to 1. If the bed shear stress is drastically reduced, the probability of erosion will approach zero.

As seen in Fig. 9.7, the lower tail of $\varphi(\tau_b)$ represents the condition when the instantaneous value of τ_b is negative and $|\tau_b|$ exceeds the time–mean value $\overline{\tau}_b$. The effect of an instantaneous negative bed shear stress can be evaluated by folding the $\varphi(\tau_s)$ distribution over the ordinate at the origin. Accordingly, as before (Eq. 9.11), p_e is the sum of the contribution A_r from the positive values of τ_b and (a much smaller and at times negligible contribution) A_l from the negative values of τ_b. Thus the probability of erosion is

$$p_e = A_r + A_l = \int_0^\infty \left\{ \int_0^{\tau_b} \varphi(\tau_s)\, d\tau_s \right\} \varphi(\tau_b)\, d\tau_b + \int_{-\infty}^0 \left\{ \int_{\tau_b}^0 \varphi(-\tau_s)\, d\tau_s \right\} \varphi(\tau_b)\, d\tau_b$$

(9.14)

This expression can be also written as

$$p_e = A_r + A_l = \int_0^\infty \int_0^\infty \varphi(\tau_b)\varphi(\tau_s) H\left(\tau_b - \tau_s\right) d\tau_s\, d\tau_b$$

$$+ \int_{-\infty}^0 \int_{-\infty}^0 \varphi(\tau_b)\varphi(-\tau_s) H\left(\tau_b - \tau_s\right) d\tau_s\, d\tau_b$$

(9.15)

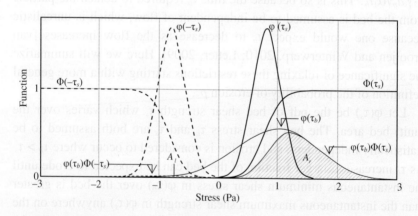

Figure 9.7. Probability of erosion (sum of areas A_r and A_l) when shear stress and shear strength are both represented by their probability density functions.

where the heavyside function $H(\theta) = 0$ if $\theta < 0$, 1 if $\theta > 0$ and 0.5 if $\theta = 0$.

A convenient form of Eq. 9.14 results when it is recognized that the integral over the shear strength within the brackets is the cumulative distribution function (cdf), Φ. Thus

$$p_e = A_r + A_l = \int_0^\infty \Phi(\tau_s)\varphi(\tau_b)d\tau_b + \int_{-\infty}^0 \Phi(-\tau_s)\varphi(\tau_b)d\tau_b \quad (9.16)$$

In Fig. 9.8, plots corresponding to Fig. 9.7 are given for a single shear strength τ_s represented by its mean value $\overline{\tau}_s$ so that the cdf of τ_s is a heavyside step (Dirac delta) function. This amounts to setting $\varphi(\tau_s) = \varphi(-\tau_s) = 1$ in Eq. 9.15 and $\tau_s = \overline{\tau}_s$ for the heavyside function. We then obtain

$$p_e = A_r + A_l = \int_0^\infty \varphi(\tau_b)H(\tau_b - \overline{\tau}_s)d\tau_b + \int_{-\infty}^0 \varphi(\tau_b)\left[1 - H(\tau_b - \overline{\tau}_s)\right]d\tau_b$$
$$(9.17)$$

In order to obtain the erosion flux we will represent the characteristic time t_b by the ratio of a length scale and the friction velocity u_* [van

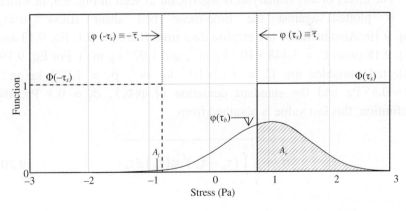

Figure 9.8. Probability of erosion (sum of areas A_r and A_l) when the shear stress is represented by its distribution, and the shear strength is constant, equal to its spatial mean value.

Prooijen and Winterwerp, 2010; Letter, 2009]. In other words t_b is inversely proportional to $\sqrt{\tau_b / \rho_w}$. Then, from Eq. 9.2 the erosion flux can be restated as

$$\varepsilon = C_\varepsilon\, u_* \, p_e(\tau_b) = C_\varepsilon \sqrt{\frac{\tau_b}{\rho_w}}\, p_e(\tau_b) \tag{9.18}$$

where C_ε is a characteristic concentration. To represent Eq. 9.18 in terms of a single, stress-dependent parameter we may move $\sqrt{\tau_b}$ inside the integral defining p_e, which would mean that this integral would no longer represent the probability of erosion. At the same time, dividing the integral by $\sqrt{\rho_w}$ would yield an erosion velocity w_e.[b] Thus, by making use of Eq. 9.16 we obtain

$$\varepsilon = C_\varepsilon w_e = \frac{C_\varepsilon}{\sqrt{\rho_w}} \left[\begin{array}{l} \displaystyle\int_0^\infty\!\!\int_0^\infty \tau_b^{1/2}\varphi(\tau_b)\varphi(\tau_s)H(\tau_b - \tau_s)\,d\tau_s\,d\tau_b \\[2mm] \displaystyle + \int_{-\infty}^0\!\!\int_{-\infty}^0 \tau_b^{1/2}\varphi(\tau_b)\varphi(\tau_s)\left[1 - H(\tau_b - \tau_s)\right]d\tau_s\,d\tau_b \end{array} \right] \tag{9.19}$$

The effect of this reanalysis is significant as seen in Fig. 9.9, in which ε is plotted against the time-mean bed shear stress $\overline{\tau}_b$ using Eq. 9.19. Also included are erosion data from Example 9.1, Eq. 9.13 and Eq. 9.18 (with $C_\varepsilon = 3.448 \times 10^{-1}$ kg m^{-3}, and 1,027 kg m^{-3}). For Eq. 9.19, selected variables are $C_\varepsilon = 1.12 \times 10^{-1}$ kg m^{-3}, $\rho_w = 1{,}027$ kg m^{-3}, $\overline{\tau}_s = 0.55$ Pa and the standard deviation of $\varphi(\tau_b)$, $\sigma_\tau = 0.3$ Pa. By definition, this last value is obtained from

$$\sigma_\tau = \sqrt{\int_{-\infty}^{\infty} \left(\tau_b - \overline{\tau}_b\right)^2 \varphi(\tau_b)\,d\tau_b} \tag{9.20}$$

---

[b] The symbol w_e has also been used elsewhere to denote fluid mud entrainment velocity.

where

$$\overline{\tau}_b = \int_{-\infty}^{\infty} \tau_b \ \varphi(\tau_b) \, d\tau_b \qquad (9.21)$$

In Fig. 9.9, Eq. 9.13 approaches a maximum erosion flux at high shear stresses. Equation 9.18 is not limited in this way; however, as plotted, it does not mimic the data points.

Equation 9.19, which is also free from the erosion-limiting feature of Eq. 9.18, is calibrated to follow the data.

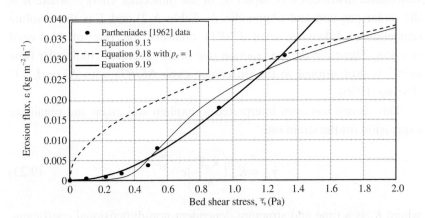

Figure 9.9. Erosion flux against bed shear stress plot including the original and revised Partheniades equations (from Letter [2009]).

9.3.3 *Erosion as a rate process*

An interpretation of the rate process theory for chemical reactions has been used to provide a qualitative explanation of surface erosion. The outcome is a relationship between the erosion flux and fluid temperature.

When two molecules A and BC react to form products AB and C, an intermediate unstable energy-activated molecular complex ABC is formed. This process is expressed as

$$A + BC \rightleftarrows ABC \rightleftarrows AB + C \qquad (9.22)$$

The arrows imply that the process is reversible such that AB reacts with C to reproduce A and BC. However, conditions for the forward reaction are more favorable than the backward reaction because the total energy of $AB+C$ is less than that of $A+BC$ by an amount ΔE (Fig. 9.10). In other words, $AB+C$ represents a more stable state of matter than $A+BC$. The total energy of the activated complex ABC is the highest of the three states. E' is the activation energy for the forward reaction and E'' for the backward reaction. Successful intermolecular collisions are required for the reaction to proceed in either direction. It can be shown that the rate of forward reaction is proportional to $\exp(-E'/RT)$ assuming the Maxwell–Boltzmann distribution (Chapter 4) of the molecular energy, where R is the molar gas constant (Appendix A, Table A.2) and T is the absolute temperature. Likewise, $\exp(-E''/RT)$ controls the rate of backward reaction. Therefore the net reaction rate of the so-called "molecular flow units" is proportional to the difference $\exp(-\Delta E/RT)$ with $\Delta E = E''-E'$ [Eyring, 1936].

This theory has been interpreted to explain soil creep and obtain an expression for the strain rate

$$\dot{\gamma}_{sc} = K_{sc}\left(\frac{k_B T}{h_p}\right)e^{\left(-\frac{\Delta E}{RT}\right)} \tag{9.23}$$

where K_c is a time and structure-dependent non-dimensional coefficient, k_B is the Boltzmann constant and h_p is Planck's constant (Appendix A,

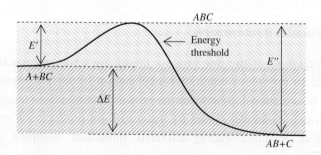

Figure 9.10. Energy levels of reactants A and BC, products AB and C, and activated complex ABC.

Table A.2). It has been shown that ΔE can be notionally considered to be proportional to the shear stress τ causing creep [Mitchell *et al.*, 1968].

For the erosion of flocs at the bed surface, in Eq. 9.23 we may replace $\dot{\gamma}_{sc}$ resulting from the causative quantity ΔE by erosion flux ε due to stress τ, or more generally the excess bed shear stress $\tau_b - \tau_s$, where τ_s is the bed shear strength [Paaswell, 1973]. Thus we may consider bed surface erosion to be a rate-process in which a particle is dislodged when there is a "successful" application of bed shear stress causing the inter-particle bonds to rupture. This interpretation described by the Boltzmann distribution of particle energy is consistent with the probabilistic nature of erosion in turbulent flow.

Kelly and Gularte [1981] explained their experiments on the surface erosion of flocculated clayey beds at different fluid temperatures by stating Eq. 9.23 in the form

$$\frac{\varepsilon}{T} = e^{\left(\Psi_r - \frac{\Lambda_r}{T}\right)} \tag{9.24}$$

where T is the absolute temperature ($^\circ$K), $\Psi_r = \ln(K_{sc} k_B / h_p)$ and $\Lambda_r = \Delta E / R$. From this expression we obtain

$$\log\left(\frac{\varepsilon}{T}\right) = \frac{\Psi_r}{2.303} - \frac{\Lambda_r}{2.303}\left(\frac{1}{T}\right) \tag{9.25}$$

which indicates that $\log(\varepsilon/T)$ varies linearly with $1/T$. Equation 9.25, also called the Arrhenius equation, was tested for the erosion illitic clay called grundite. The apparatus was a horizontally placed flow-recirculating tube (*i.e.*, a covered channel) with controllable fluid temperature [Kelly *et al.*, 1979].

Figure 9.11 shows the result for a test in which the flow velocity was maintained at 0.18 m s^{-1} and the temperature was increased in 5°C steps about every 40 min. In general an increase in temperature only marginally affects the van der Waals attractive force at the particle surface, but the inter-particle repulsive force increases due to thermal agitation. As a result, particle–particle bonds rupture more easily at higher temperatures, thereby leading to enhanced erosion [Lau, 1994].

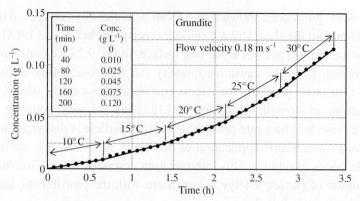

Figure 9.11. Erosion of grundite in a recirculating tube (from Kelly *et al.* [1979]).

As expected, this process is consistent with the commonly observed increase in the rate of chemical reactions with temperature.

In Fig. 9.12a log (ε/T) is plotted against $1/T$, where T is in °K ($= $ °C+273). Also included are data of Christensen and Das [1974] on grundite and a kaolinite, both eroded in a brass tube. The lines are remarkably parallel to each other, suggesting that Λ_r, which is proportional to line slope, is a characteristic constant of erosion when interpreted as a rate-process. Additional tests were carried out using grundite at slightly higher water contents (W), and by systematically varying the fluid salinity (NaCl concentration). The results shown in Fig. 9.12b corroborate the dependence of ε/T on T. However, unlike Fig. 9.12a, data points for different salinities seem to be almost affine, suggesting that a single line suffices to capture the data trends at all salinities. The coefficients $\Psi_r = 34.7$ and $\Lambda_r = 10,145$ for the line in Fig. 9.12b are consistent with the units of ε in g m^{-2} s and T in °K.

9.3.4 *Semi-empirical erosion flux equations*

9.3.4.1 *Uniform beds*

The choice of an unbounded frequency distribution of the bed shear stress in Eq. 9.12 implies the likelihood of erosion, however small, at any value of the time–mean bed shear stress. Carefully conducted laboratory

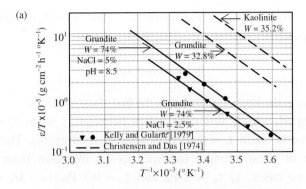

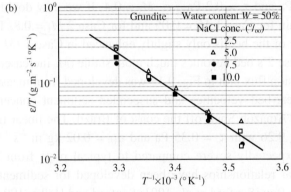

Figure 9.12. Variation of ε/T with T^{-1} from erosion tests using: (a) grundite and kaolinite (from Kelly *et al.* [1979]); (b) grundite (from Kelly and Gularte [1981]).

tests on cohesive beds with delicate surface floc structures have indicated the occurrence of sediment entrainment, referred to as floc erosion, even at low time-mean bed shear stresses (*e.g.,* Parchure and Mehta [1985]; Kuijper *et al.* [1989]). A realistic interpretation is that erosion occurs because flocs with practically no shear strength are exposed at the bed surface as the test continues. As we will note later, by assuming the shear stress to be bounded by a maximum value (in or against the mean flow direction), the criterion of bed shear strength equal to or greater than the maximum shear stress is used to design a stable channel with non-erodible bed.

For uniform beds Ariathurai [1974] assumed a straight line approximating the non-linear trend of Eq. 9.12 (Fig. 9.6) and attributed the linear expression to Partheniades. This so-called Ariathurai–Partheniades equation is

$$\varepsilon = \varepsilon_{AP}(\tau_b - \tau_s) \qquad (9.26)$$

in which ε_{AP} is the line slope and τ_s is the bed shear strength.

Defining $M_\tau = \tau_s/\tau_b$, this ratio varies between 1, when $\tau_s = \tau_b$ (no erosion), and 0, when $\tau_s = 0$ (bed lacking shear strength). The relative importance of shear strength increases as M_τ increases from 0 to 1. Consider two cases: 1) $\tau_b = 8$ Pa and $\tau_s = 0.2$ Pa, *i.e.*, $M_\tau = 0.025$; 2) $\tau_b = 0.5$ Pa and $\tau_s = 0.2$ Pa, *i.e.*, $M_\tau = 0.4$. If we now double τ_s to 0.4 Pa, we obtain for Case 1 $M_\tau = 0.05$ and for Case 2 $M_\tau = 0.8$. Thus while the stability of the bed is only slightly enhanced in Case 1 (M_τ increased to 0.05), Case 2's bed becomes much more stable (M_τ increased to 0.8).

The erosion flux plot in Fig. 9.13 was developed from measured time-series of near-bed velocities and suspended sediment concentration at sites on the Amazon Shelf off the coast of Brazil. The linear mean trend follows Eq. 9.26 with $\tau_s = 0.35$ Pa and $\varepsilon_{AP} = 0.02$ kg m^{-2} s^{-1} Pa^{-1}. Data scatter is expectedly severe compared to typical plots from laboratory tests. Similar relationships have been developed for sediment from the Chesapeake Bay [Sanford *et al.*, 1991; Sanford and Halka, 1993].

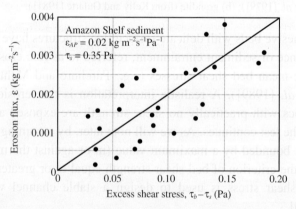

Figure 9.13. Erosion flux versus excess bed shear stress based on the analysis by Vinzon [1998] of Amazon Shelf sediment data (dots) collected by Kineke [1993] (revised from Vinzon and Mehta [2003]).

Equation 9.26 can be restated in the dimensionless form as

$$\frac{\varepsilon}{\varepsilon_M} = \left(\frac{\tau_b - \tau_s}{\tau_s}\right) \qquad (9.27)$$

in which $\varepsilon_M = \varepsilon_{AP}\tau_s$ is the value of ε when $\tau_b = 2\tau_s$ [Kandiah, 1974]. Power-law versions of Eq. 9.27 have been proposed, *e.g.*,

$$\varepsilon = \frac{\varepsilon_{sc}}{\tilde{T}_\varepsilon^2}\left[\frac{1}{2}\left(\frac{\tau_b}{\tau_s} - 1\right) + \frac{1}{2}\left|\frac{\tau_b}{\tau_s} - 1\right|\right]^m \qquad (9.28)$$

where ε_{sc} is an erosion flux constant, $\tilde{T}_\varepsilon$ is a dimensionless constant representing the effect of consolidation on bed shear strength and the exponent $m \leq 1$. The constant $\tilde{T}_\varepsilon = T_\varepsilon / T_{\varepsilon 0}$, where T_ε is the duration of consolidation and $T_{\varepsilon 0}$ is a normalizing parameter usually taken as 1 day [Sheng and Chen, 1992]. If m is assumed to be 1, a reasonable choice, only two coefficients, $\varepsilon_{sc} / \tilde{T}_\varepsilon^2$ and τ_s, would have to be determined by calibration. Selecting $\varepsilon_{sc} = 10^3$ kg m^{-2} d^{-1}, $T_\varepsilon = 40$ d, $\tau_s = 0.2$ Pa, $m = 0.9$ and $\tau_b = 2$ Pa, we obtain $\varepsilon = 4.52$ kg m^{-2} d^{-1}.

Surface erosion has been treated by Winterwerp *et al.* [2012] in terms of Mohr–Coulomb failure of the drained layer of flocs exposed at the bed surface. Referring to Eq. 9.26, the erosion shear strength τ_s is equated to the yield strength τ_y obtained from Eq. 5.15, and $\varepsilon_{AP} = c_{cv}\phi_v\rho_D / 10 d_{50}c_u$. In this expression c_{cv} is the coefficient of consolidation (Eq. 8.40), ϕ_v is the solids volume fraction, ρ_D is the dry density of the bed, d_{50} is the median diameter of the primary particles and c_u is the drained shear strength.

As an alternative to Eq. 5.15, it is convenient to use a power-law expression for τ_s (in Eq. 9.26) in terms of the Plasticity Index *PI* (Eq. 3.3) obtained by Smerdon and Beasely [1959]. This expression is: $\tau_s(\text{Pa}) = \alpha_p PI^{\beta_P}$ with *PI* in percent and coefficients α_P, β_P dependent on the sediment. Consider the following nominal values [see also Winterwerp *et al.*, 2012; Jacobs, 2011]: $c_{cv} = 2 \times 10^{-8}$ m^2 s^{-1}, $\sigma'_v = 2$ Pa, $\varphi_{fr} = 15°$, $\phi_v = 0.15$, $\rho_D = 397.5$ kg m^{-3}, $d_{50} = 7$ μm $= 7 \times 10^{-6}$ m, $c_u = 1$ Pa, $\alpha_P = 0.16$ and $\beta_P = 0.84$. Assume a bed shear stress value of $\tau_b = 5.2$ Pa. We are interested in calculating the erosion flux ε. First we obtain

$\varepsilon_{AP} = 1.07 \times 10^{-2}$ s m^{-1}. Using the *PI* based formula yields $\tau_s = 2.535$ Pa. Finally, from Eq. 9.26, $\varepsilon = 4.4 \times 10^{-2}$ kg m^{-2} s^{-1}.

9.3.4.2 *Non-uniform beds*

For beds stratified with respect to the shear strength τ_s, erosion flux formulas which account for the variation of τ_s with depth have been developed. Although these formulas differ from Eq. 9.27, in most the erosion flux varies with the excess shear stress $\tau_b - \tau_s$. This similarity as well as experience from modeling suggests that Eq. 9.27 can be used for stratified beds with reasonable accuracy by allowing ε_M and/or τ_s to vary with depth, *i.e.*, by replacing the quantities ε_M and τ_s by $\varepsilon_M(z)$ and $\tau_s(z)$, respectively, where z is the vertical coordinate [Ariathurai *et al.*, 1977; Hayter, 1983; Parchure, 1984; Piedra–Cueva and Mory, 2001].

Consider the following extension of Eq. 9.26

$$\varepsilon(z,t) = \varepsilon_{AP}(z)[\tau_b(t) - \tau_s(z)] \qquad (9.29)$$

in which the time-dependence of erosion flux arises from the long-term unsteady (gradually varied) bed shear stress $\tau_b(t)$ such as due to tide. Select

$$\varepsilon_{AP} = \beta_v \rho_D = \beta_v \rho_s \phi_v \qquad (9.30)$$

where ρ_D is the dry density at the bed surface, ρ_s is the particle density, ϕ_v is the solids volume fraction and β_v is a proportionality constant [Sanford and Maa, 2001]. After rearrangement Eq. 9.29 becomes

$$w_e(z,t) = \frac{\varepsilon(z,t)}{\rho_s \phi_v} = \beta_v [\tau_b(t) - \tau_s(z)] \qquad (9.31)$$

where w_e is the erosion velocity. The time-rate of change of w_e is

$$\frac{dw_e}{dt} = \beta_v \left(\frac{d\tau_b}{dt} - \frac{d\tau_s}{dz}\frac{dz}{dt} \right) \qquad (9.32)$$

Let $\gamma_v = d\tau_s/dz$ denote the depth-gradient of bed shear strength and note that $w_e = dz/dt$ is the erosion velocity. Thus Eq. 9.32 becomes

$$\frac{dw_e}{dt} + \beta_v \gamma_v w_e = \beta_v \frac{d\tau_b}{dt} \qquad (9.33)$$

The solution of this equation for w_e depends on the behavior of the bed shear stress and the bed dependent coefficients β_v and γ_v. For a time-independent bed shear stress, *i.e.*, $d\tau_b/dt = 0$, and constant values of these coefficients we obtain

$$w_e = \beta_v (\tau_b - \tau_{s0}) e^{-\beta_v \gamma_v t} \qquad (9.34)$$

where τ_{s0} is the shear strength at the surface when the bed shear stress is first applied. The trend of a gradual decrease in the erosion velocity with time can arise from four causes:

(1) With continued increase in the suspended sediment concentration the deposition flux would increase until no change in concentration occurs. In this case w_e would represent the *net* erosion velocity.

(2) At a stratified bed, increasing scour would expose increasingly resistant bed layers and decrease w_e. This is a type of physicochemical bed armoring effect (Fig. 9.14).

(3) During erosion overcrowding of heavier or strongly bonded particles may occur at the exposed bed surface. In this armoring effect, similar to the effect in coarse-grain beds, erodible particles are increasingly blanketed by large or weighty non-erodible particles and w_e is reduced.

(4) Armoring can also be due to the formation of a biopolymeric film at the bed surface.

The above causes resulting in a gradual decrease in w_e may operate singly or in combination. As a consequence, coefficients β_v and γ_v in Eq. 9.34 require careful calibration using site-specific sediment.

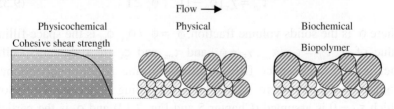

Fig. 9.14. Physicochemical, physical and biochemical armoring of bed.

9.4 Factors Influencing Bed Shear Strength

9.4.1 *Bed shear strength and density*

9.4.1.1 *General equation*

Shear strength representing resistance to bed failure is commonly related to a measure of sediment packing such as the wet bulk density, the dry mass density, solids volume fraction or the water content. The approximate nature of such a relationship must be borne in mind in applications, as several factors contribute to uncertainty as follows:

(1) The top bed layer with relatively high permeability may be in the drained condition. It can experience surface erosion if hard, or plastic yield if soft. The lower layer will be undrained and may undergo mass erosion. The relative magnitudes of surface and mass erosion fluxes will depend on the state of drainage.

(2) In soft beds the shear strength also depends on the dynamic state of mud characterized by the rate of shear. In Fig. 9.15a the plastic yield stress is plotted against shear rate in mud at different water contents (W). These results were obtained from rotational rheometry tests on a kaolinite in freshwater. Even at a constant value of W the yield stress increases by as much as an order of magnitude as the shear rate is increased from 0.01 to 10 s^{-1}.

(3) A relationship between shear strength and any measure of density is dimensionally inconsistent and entirely empirical.

Notwithstanding these limitations, given τ_{sh} as a general symbol representing shear strength, operational relationships of the following form have been proposed

$$\tilde{\tau}_{sh} = \chi_\tau \left(\tilde{\phi}_v - 1 \right)^{\xi_\tau} ; \; \tilde{\phi}_v \geq 1 \tag{9.35}$$

where ϕ_v is the solids volume fraction, $\tilde{\phi}_v = \phi_v / \phi_{vs}$, ϕ_{vs} is the space-filling value of ϕ_v, $\tilde{\tau}_{sh} = \tau_{sh} / \tau_{sm}$, $\chi_\tau = \phi_{vs}^{\xi_\tau}$ and τ_{sm} and ξ_τ are sediment-specific coefficients. The space-filling solids volume fraction is obtained from $\phi_{vs} = C_s / \rho_s$, where C_s is the space-filling concentration below which $\tau_{sh} = 0$ is assumed (Chapter 5 and Fig. 7.13) and ρ_s is the particle mineral density.

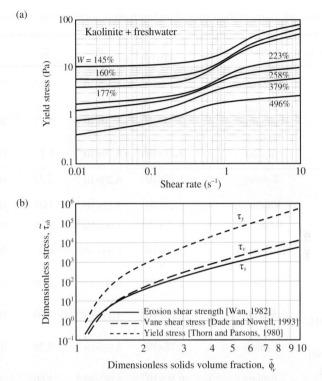

Figure 9.15. (a) Rheometric data on the variation of yield stress with water content and shear rate (adapted from Tsuruya and Nakano [1987]); (b) Examples (from Table 9.1) of variation of dimensionless measures of shear strength with solids volume fraction.

Common measures of τ_{sh} are the plastic yield stress (or strength) τ_y (Chapter 5), the vane shear strength τ_v and the bed surface erosion shear strength τ_s (Table 9.1). An example of each (for different sediments) is shown in Fig. 9.15b. The three dimensionless shear strengths show similar trends with dimensionless volume fraction.

9.4.1.2 *Plastic yield stress and density*

As we noted in Chapter 5, the Bingham plastic yield stress τ_B is a special case of the (upper-Bingham) plastic yield stress τ_y. Depending on the method of measurement, τ_y can be a remolded quantity (such as from rotational testing) or practically undisturbed (from oscillatory testing). The difference between the two values can be significant, but in the general

Table 9.1. Measures of shear strength as functions of solids volume fraction.

Investigator(s)	Sediment	τ_{sh}	τ_{sm} (Pa)	χ_τ	ξ_τ	ϕ_{vs} [a]
Krone [1963]	Estuary	τ_y	4.66×10^1	4.50×10^{-6}	2.55	0.008
Migniot [1968]	Marine, estuarine, fluvial,	τ_y	2.53×10^4	5.79×10^{-6}	5.1^b	0.094
	mined	τ_y	1.21×10^6	5.79×10^{-6}	5.1^b	0.094
Owen [1970]	Estuary	τ_y	1.11×10^3	6.20×10^{-4}	2.33	0.042
Wan [1982]	Kaolinite	τ_y	1.28×10^3	5.12×10^{-4}	3.00	0.080
	Bentonite	τ_y	4.15×10^5	1.00×10^{-9}	3.00	0.001
Mehta *et al.* [2009]	Lake muck	τ_y	3.99×10^6	5.03×10^{-9}	4.4	0.013
Hwang [1989]	Lake	τ_v	2.00×10^1	5.60×10^{-2}	1.00	0.056
Dade and Nowell [1991]	Marine	τ_v	8.00×10^2	3.43×10^{-4}	3.00	0.070
Thorn and Parsons [1980]	Estuary	τ_s	3.75×10^1	5.93×10^{-5}	2.28	0.014
Kusuda *et al.* [1984]	Estuary	τ_s	6.50×10^0	4.06×10^{-3}	1.60	0.032
Villaret and Paulic [1986]	Bay	τ_s	1.65×10^0	1.00×10^{-1}	1.00	0.100
Black [1991]	Estuary	τ_s	1.88×10^0	9.16×10^{-3}	2.30	0.130
Berlamont *et al.* [1993]	Marine	τ_s	5.41×10^0	2.96×10^{-2}	0.90	0.020
Whitehouse *et al.* [2000a]	Estuaries^c	τ_s	4.22×10^1	2.44×10^{-2}	1.24	0.050
Vinzon and Mehta [2003]	Shelf	τ_s	1.30×10^0	2.32×10^{-3}	1.57	0.021

[a] Values of $\phi_{vs} = C_s/\rho_s$ are nominal. The nominal value of ρ_s is 2,650 kg m^{-3} except for Mehta *et al.* [2009] (1,690 kg m^{-3}) and Hwang [1989] (2,140 kg m^{-3}).
[b] Migniot [1968] quoted the range to be 4 to 5.
[c] Mean values based on 14 estuarine sediments reported by Whitehouse *et al.* [2000a].

description given here this distinction has not been emphasized (however, see Chapters 5 and 11).

Figure 9.16a includes data on the dependence of τ_y on bottom sediment concentration C (= $\rho_s \phi_v$) from twelve sediments, among which powdered limestone contained no clay minerals and Lake Apopka mud had 62% organic matter (thus falling within the generic classification of muck). All the lines have an exponent of 5 (except no. 12 with exponent 4.4, which is reasonably close to 5). This similarity of slopes indicates a common physical basis related to the density and structure of inter-

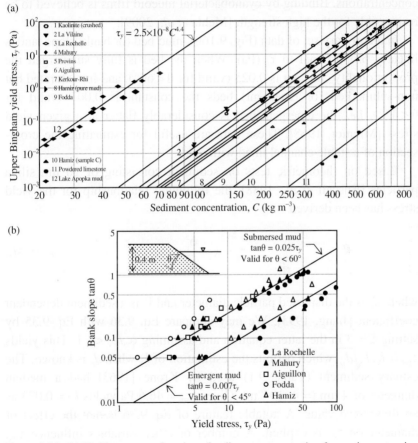

Figure 9.16. (a) Yield stress as a function of sediment concentration for marine, estuarine and fluvial sediment samples; (b) dependence of bed slope on yield stress (revised from Migniot [1968]; Mehta *et al.* [2009]).

particle bonds [Sanchez and Levacher, 2007]. For the clayey sediments, τ_{sm} ranges between 2.53×10^4 (Hamiz sample C) to 1.21×10^6 (crushed kaolinite). This wide range points to the site-specificity of sediment properties and the need for calibration of Eq. 9.35 for every untested sample.

Table 9.2 gives the origin, mineral content and median (dispersed) size of several largely inorganic sediments inclusive of those in Fig. 9.16a. The largely organic Lake Apopka's mud in the same plot has a yield strength range comparable to the other sediments, but at lower concentrations. Binding by cyanobacterial mucoid films is believed to be responsible for the high strength [Mehta *et al.*, 2009].

In a corollary set of data (Fig. 9.16b), mud bed (or bank) slope $\tan\theta$ is linearly proportional to τ_y (Pa). When the bed is fully submerged the proportionality constant is 0.025 (valid for $\theta < 60°$), and for saturated but non-submerged (or emergent) beds the constant is 0.007 (valid for $\theta < 45°$), *i.e.*, the emergent bed is considerably flatter at a given yield stress [Migniot, 1968]. These data are useful for estimating the cross-section of a stable channel lined with mud.

Based on rheometric tests on sediment consisting of 26 μm silica particles [Buscall *et al.*, 1988], the following relationship for the yield stress has been derived

$$\tau_y = k_y \frac{\phi_v^3}{\sqrt{d_p}} \qquad (9.36)$$

where d_p is the dispersed particle diameter and k_y is a sediment dependent coefficient [Jiang, 1993]. We may compare Eq. 9.36 with Eq. 9.35 by setting $\xi = 3$ in the latter equation and assuming $\phi_{vs}/\phi_v \ll 1$. This yields $\tau_{sm} = k_y / \sqrt{d_p}$, which permits the estimation of k_y when d_p is known. The estuary sediment (Table 9.1) used by Krone [1963] had a median diameter of 4 μm ($= 4 \times 10^{-6}$ m). Given $\tau_{sm} = 46.6$ Pa yields $k_y = 0.093$ as an illustrative value. A notable feature of Eq. 9.36 is that the effect of diameter on τ_{sm} is explicit. A number of other variables influence τ_{sm}, many associated with biota and fluid chemistry. At times these influences can be significant enough to overshadow the effect of the diameter. Therefore, as a rule, τ_{sm} must be obtained from direct measurement of the yield stress of specific sediment.

Table 9.2. Origin, minerals and size of selected European and non-European sediments.

Sediment	Origin	Minerals	Dispersed size (μm)
Kaolinite (crushed)	Processed clay	Kaolinite	2.7
La Rochelle	Marine	Illite, kaolinite	–
Mahury, Guinea	Estuarine	Kaolinite, illite	–
Provins (clay)	Mining/quarry residue	Kaolinite	–
Aiguillon Bay	Marine	Kaolinite, illite	0.9
Kerkour–Rhi Morocco	Mining/quarry residue	Montmorillonite, illite, phosphates	0.4
Hamiz, Algeria (pure mud)	Fluvial	Illite, kaolinite, montmorillonite	–
Oued Fodda, Algeria	Fluvial	Kaolinite, illite, montmorillonite, chlorite	3.2
Hamiz, Algeria (sample C)	Fluvial	Clay minerals: illite, kaolinite, montmorillonite	–
Powdered limestone	Mining/quarry residue	$CaCO_3$	–
Croix-de-Vie 19	Estuarine	–	2.0
La Vie	Fluvial	–	3.5
Etang Vaine	Lacustrine	–	5
Durance (silt)	After screening	Detrital powder, quartz	9
La Vilaine	Estuarine	Illite, kaolinite, traces of chlorite	–

Source: Migniot [1968].

9.4.1.3 *Vane shear strength and density*

The vane shear strength τ_v is a measure of remolded resistance to internal shear and is used in geotechnical evaluations of cohesive soil consistency, *i.e.,* stiffness (Table 9.3). Stiffness is typically characterized by values of shear strength in kilopascals. For an assessment of consistency with respect to bed stability against erosion, τ_v values of sewer sediments have been correlated with the water content [Wotherspoon and Ashley, 1992].

Figure 9.17 shows the dependence of τ_v on the wet bulk density of sediment derived from push-coring in the muddy zone of Lake Okeechobee (Florida). With regard to Eq. 9.35, the low value of $\xi = 1$ corresponding to the mean trend-line reflects the organic-rich (40% by weight; see Table 5.5) composition of this sediment. The vane shear strength of this mud does not increase with density nearly as rapidly as in clayey sediments. This is so because intervention from organic matter reduces the scope of electrochemical bonding between the clay mineral particles. Moreover, the elastic behavior of light-weight aggregates means that the compressive stress required for self-weight consolidation is not high enough to densify the bed. The line intersects the horizontal axis at a density of 1,065 kg m^{-3}, which is approximately the space-filling

Table 9.3. Soil consistency classification based on vane shear strength.

Soil consistency	Identification	Vane shear strength, τ_v (kPa)
Very soft	Easily molded by fingers	0–80
Soft	Molded by fingers with some pressure	80–140
Firm	Very difficult to mold. Can be penetrated by a hand-spade	140–210
Stiff	Requires a hand pick for excavation	210–350
Very stiff	Requires a power tool for excavation	350–750

Source: Annandale [1995].

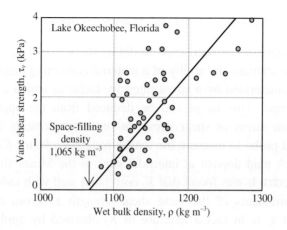

Figure 9.17. Variation of vane shear strength with wet bulk density for bottom sediment from Lake Okeechobee in Florida (revised from Hwang [1989]).

density below which the bottom did not have measurable shear strength. Using a particle density of 2,140 kg m^{-3} (Table 5.5) and water density of 1,000 kg m^{-3}, the space-filling solids volume fraction ϕ_{vs} = (1,065–1,000)/(2,140–1,000) = 0.057. The corresponding space-filling sediment (dry mass basis) concentration C_s would be 0.057×2140 = 122 kg m^{-3}. This value may be contrasted with that for a sediment with a mineral density of 2,650 kg m^{-3} and wet bulk density at the onset of shear strength equal to 1,150 kg m^{-3}, which would yield ϕ_{vs} = 0.09 and C_s = 241 kg m^{-3}. Thus we observe that organic-rich sediment can occur as fluid mud at lower concentrations than mineral mud.

Most studies on the erosion of submerged soils in the marine environment are limited to "very soft" to "soft" cohesive materials (Table 9.3). This is so because wave- and current-induced bed shear stresses are usually not large enough to erode stiffer soils, at least over "short" durations. On the other hand, in rivers with high stream velocities, even firm soils such as a stiff blue-clay can erode significantly over years [Jiang *et al.*, 2004]. In such a situation τ_v is a convenient although approximate [Lee, 1985] parameter for estimating the erosion flux when the bed shear stress τ_b exceeds τ_v.

9.4.1.4 Fracture toughness and penetrometer resistance

Fracture toughness has been used to define the strength of soft marine muds. It characterizes the ability of a material containing cracks to resist fracture parameterized by a stress intensity factor at which a thin crack begins to grow. This factor can be deduced from the application of normal tensile stress or shear stress. *In-situ* measurements were made using a field probe to measure the tensile roughness factor K_I in the top 1-meter thick mud deposit at intertidal sites in the Minas Basin (Nova Scotia, Canada). It was found that K_I correlated well with independently made measurements of the vane shear strength τ_v. From this it was inferred that τ_v is in fact a measure of K_I obtained by applying shear stress (as opposed to tensile stress) (Johnson *et al.* [2011]).

Another measure of mud strength is bed resistance recorded by a penetrometer. The single probe nuclear density cone-penetrometer is a gravity probe which, when dropped from the vessel, records both the density of mud penetrated and also penetration resistance, sleeve friction and the pore pressure [Karthikeyan *et al.*, 2007].

9.4.1.5 Erosion shear strength and density

The shear strength τ_s is a measure of the drained resistance of flocs to erosion at the bed surface. An example of data on τ_s is shown in Fig. 9.18 based on laboratory tests using mud from the Chikugo River estuary in Kyushu, Japan [Kusuda *et al.*, 1984].

The variability in coefficients applicable to τ_s (Table 9.1) can be attributed to differences in mineral composition and pore fluid chemistry. The shear strength is measurably affected by minor changes in the composition of pore water. For example, as we noted in Chapter 4 (Fig. 4.13), a smectite could be altered between the dispersed and the flocculated states merely by changing the pH of the pore fluid either by holding the Sodium Adsorption Ratio constant, or by keeping the total cation concentration at a constant level. A dispersed clay bed will erode with considerably greater ease than a flocculated bed of the same clay. This underscores the need to use native water in laboratory

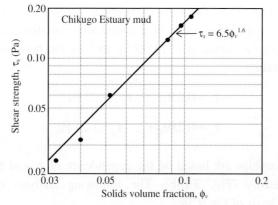

Figure 9.18. Variation of bed shear strength with solids volume fraction for mud from Chikugo estuary in Kyushu, Japan (from Kusuda *et al.* [1984]).

erosion tests. An alternative is to use reconstituted eroding fluid based on the main ionic constituents of native water [Parchure, 1984]. However, this approach cannot account for likely effects of trace constituents in the fluid. Such issues make a strong case for *in situ* testing of bed erodibility in the natural environment [Maa *et al.*, 1993; Ravisangar *et al.*, 2001]. Unfortunately, equipment that is cumbersome to deploy in the field places a practical limit on its use. The upshot is that smaller, onboard devices which provide approximate *ex situ* answers on erodibility but can cover a large study area have gained some popularity, *e.g.*, a portable device designed by Tsai and Lick [1986] for erosion, and by Manning [2006] for the settling velocity. The Gust Microcosm [Gust and Muller, 1997] is a chamber with flowing water for accurate calibration of stress measuring sensors.

9.4.2 *Shear strength from yield stress*

From viscometric measurements of the (upper-Bingham) yield stress τ_y (see Fig. 5.13a) and the critical friction velocity u_{*c} for erosion of seven different sediments, the following relationships were obtained by Migniot [1968]

$$u_{*c} = 0.016\sqrt{\tau_y}; \quad \tau_y > 1.27\,\text{Pa}$$
$$u_{*c} = 0.017\tau_y^{1/4}; \quad \tau_y \leq 1.27\,\text{Pa}$$

(9.37)

As seen in Fig. 9.19, these two empirical expressions are piecewise (log-log scale) approximations of a continuous variation of τ_y with u_{*c}. Since $u_{*_c} = \sqrt{\tau_s / \rho}$, taking $\rho = 1,000$ kg m^{-3} as the nominal density of water we obtain

$$\tau_s = 0.256\tau_y; \quad \tau_y > 1.27 \, \text{Pa}$$
$$\tau_s = 0.289\sqrt{\tau_y}; \quad \tau_y \le 1.27 \, \text{Pa}$$

(9.38)

These relationships are based on the dependence of τ_y and τ_s on bottom sediment density (Fig. 9.15b). The following analysis explains the mechanistic basis of Eq. 9.38.

From Eq. 6.8 we may conveniently define an entrainment parameter for cohesive sediment as

$$\theta_{co} = \frac{\tau_s}{g(\rho_f - \rho_w)d_f}$$

(9.39)

where ρ_f is the floc density and d_f is the floc diameter. Given the analogy with sand transport underpinning this expression, in the strict sense it is applicable only to weakly cohesive sediments [Dade *et al.*, 1992].

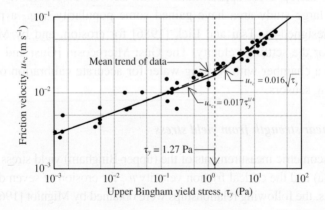

Figure 9.19. Experimental relationships between yield stress and critical friction velocity for erosion (adapted from Migniot [1968]).

The relevant particle-based fall Reynolds number is

$$Re_p = \frac{w_s d_f}{\nu} = \frac{g(\rho_f - \rho_w)d_f^3}{\rho_w \nu^2} \tag{9.40}$$

in which w_s is proportional to the Stokes settling velocity (from Eq. 7.15 ignoring the constant 18).

In analogy with the treatment in Chapter 6, force balance is considered for the hydrodynamic drag F_d, lift F_l, buoyant weight F_g and cohesion (including adhesion) F_c (Fig. 9.20) at the point of incipient detachment and transport of a particle. This balance ultimately yields

$$\theta_{co} = \frac{1}{Re_p}\left(\frac{5X}{\tan\varphi_{ap}}\frac{\sqrt{\frac{4}{X}-3}-1}{1-X}\right)^2 \tag{9.41}$$

$$X = Re_p\left(1+\frac{F_c}{F_g}\right)\frac{\pi}{4800\beta_e}\tan\varphi_{ap} \tag{9.42}$$

where φ_{ap} is the packing angle and β_e is the drag force on a particle divided by the drag force on a sphere of the same volume. Although φ_{ap} is notionally akin to the angle of repose of cohesionless grains, it is

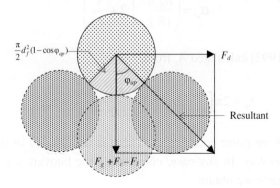

Figure 9.20. Forces and contact-surface area of a bed surface particle cohesively bonded to its neighbors (adapted from Dade *et al.* [1992]).

empirically dependent on the primary particle shape as well as the mode of inter-particle bonding in a floc. Referring to Eq. F7.1 in which a_R is the particle aspect ratio, for particles in random orientation β_e depends on a_R as follows.

For oblate particles ($a_R < 1$, *e.g.*, discs and blades):

$$\beta_e = \frac{\sqrt{1 - a_R^2}}{a_R^{1/3} \cos^{-1} a_R} \tag{9.43}$$

For prolate particles ($a_R > 1$, *e.g.*, rods):

$$\beta_e = \frac{\sqrt{a_R^2 - 1}}{a_R^{1/3} \ln\left(a_R + \sqrt{a_R^2 - 1}\right)} \tag{9.44}$$

Next we may conveniently relate F_c to the yield stress τ_y according to

$$F_c = \tau_y \alpha_e A_s \tag{9.45}$$

where A_s is the mean contact surface area between particles, assumed to be the hatched area in Fig. 9.20, and α_e is a shape factor which adjusts the ratio of particle surface area-to-volume as a function of flatness. For cylinders and spheres α_e can be estimated from

$$\alpha_e = \left(\frac{a_R}{18}\right)^{1/3}\left(2 + \frac{1}{a_R}\right) \tag{9.46}$$

Dade *et al.* [1992] calculated A_s from

$$A_s = 2\pi \int_0^{\varphi_a} r \sin\theta \, r d\theta = \frac{\pi d_f^2}{2}(1 - \cos\varphi_{ap}) \tag{9.47}$$

in which r, θ are polar coordinates. There is uncertainty in the A_s value defined in this way. In any case, considering the buoyant weight F_g of a spherical particle we obtain

$$\frac{F_c}{F_g} = 3\alpha_e (1 - \cos\varphi_{ap}) \frac{\tau_y}{g(\rho_f - \rho_w)d_f} \tag{9.48}$$

In Fig. 9.21, τ_s from Eq. 9.39 is plotted against τ_y for selected values of a_R and the particle diameter. The value $a_R = 1$ represents spherical particles and $a_R = 0.1$ platy particles. The quantities ρ_f and φ_{ap} are held constant at 2,650 kg m^{-3} (nominally for unaggregated silty particles such that $\rho_f \approx \rho_s$ and $d_f \approx d_p$) and 65°, respectively. The shaded area represents bounds of the experimental data of Migniot [1968] (from which Eq. 9.38 is obtained) and of Ohtsubo and Muraoka [1986].

Example 9.2: Determine τ_s given the following parameters: $d_f = 20$ μm, $\tau_y = 1$ Pa, $\rho_f = 2,000$ kg m^{-3}, $\rho_w = 1,027$ kg m^{-3}, $\varphi_{ap} = 55°$ and $\nu = 10^{-6}$ m^2 s^{-1}. Assume the particles are spherical.

For spherical particles $\beta_e = 1$. From Eq. 9.46, $\alpha_e = 1.145$. Applying Eq. 9.48 we obtain $F_c/F_g = 7.67$. From Eq. 9.40, $Re_p = 0.074$ and from Eq. 9.42, X = 0.0006. With these values, Eq. 9.41 yields $\theta_{co} = 0.39$. Therefore, Eq. 9.39 can be restated as

$$\tau_s = \theta_{co} g(\rho_f - \rho_w)d_f \tag{9.49}$$

which yields $\tau_s = 0.074$ Pa.

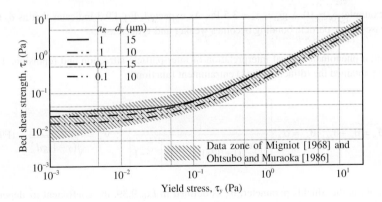

Figure 9.21. Bed erosion shear strength as a function of yield stress (from Dade *et al.* [1992]).

9.4.3 Shear strength and consolidation

9.4.3.1 Shear strength variation and mud scale

Given that τ_s is an indirect measure of the strength of flocs at the bed surface, its value depends on the order of aggregation of these flocs.[c, d] At the bed surface the density corresponds to the space-filling concentration C_s. For this concentration equal to, say, 120 kg m^{-3} the bulk density of a bed of mineral sediment would be about 1,075 kg m^{-3}. In a bed prepared by allowing initially suspended sediment to deposit, the bulk density changes with depth due to two reasons. The first is segregation due to differences in the settling velocities of the falling

---

[c] Only a few analyses seem to have been made to relate τ_s to parameters characterizing floc cohesion. For mixtures of illitic clay and water in which pore fluid chemistry was changed systematically, Kandiah [1974] used the Gouy–Chapman equation (Eq. 4.15) to estimate the double layer thickness $\delta_d = \kappa_D^{-1}$. By eroding the mixture in an annular rotating cylinder apparatus, the following equation for τ_s (Pa) was empirically derived for a homoionic illitic soil:

$$\tau_s = 14.20781 - 1.43766\delta_d + 0.0367\delta_d^2; \quad \delta_d > 5\text{Å} \qquad (F9.3)$$

For example: $\delta_d = 10$ Å gives $\tau_s = 3.5$ Pa. As expected, Eq. F9.3 indicates that as δ_d (Å) increases, inter-particle bonding becomes weaker and τ_s decreases.

[d] From a phenomenological interpretation of inter-particle attraction and repulsion, Taki [2001] obtained the (dimensionless) entrainment function

$$\theta_{ec} = \theta_{es} + \theta_{co}; \quad \theta_{co} = \alpha_E \left\{ \frac{1}{\left[\left(\dfrac{\pi}{6} \right)(1 + 32 s W_L \tilde{v}_m^{-0.4}) \right]^{1/3} - 1} \right\}^2; \quad \tilde{v}_m = \frac{v_m}{\sqrt{(s-1)gd_p^3}} \qquad (F9.4)$$

where θ_{es} is the Shields parameter, θ_{co} is defined in Eq. 9.39, the coefficient α_E depends on sediment composition, s is the specific weight of particles, W_L is the liquid limit, d_p is the particle (not floc) diameter and v_m is the kinematic viscosity of mud. According to this model, in a sand-clay mixture cohesion enhances the critical shear stress for sand. As d_p decreases the dimensionless viscosity $\tilde{v}_m$ increases. In turn, the entrainment function θ_{co} increases with $\tilde{v}_m$. Consider the following values: $d_p = 10$ μm, $s = 2.65$, $g = 9.81$ m s^{-2}, $W_L = 180\%$, $v_m = 0.001$ m^2 s^{-1}, $\alpha_E = 0.5$ and $\theta_{es} = 0.065$. Equation F9.4 yields $\theta_{co} = 0.020$ and $\theta_{es} = 0.085$.

particles. When the sediment is multi-sized, particle size and bulk density can both vary with depth. The second reason is self-weight consolidation of the freshly deposited flocs. In this case the rate of increase of density with depth is usually highest just below the bed surface and gradually decreases with depth until the internal overburden, *i.e.,* self-weight, is insufficient for further compression of the particle matrix. The multi-size effect is not easily isolated from the self-weight effect unless additional experiments on the effect of size gradation on consolidation are carried out.

A cautious rule of thumb in the natural environment is that beds formed by deposition increase in density until it reaches ~1,350 kg m^{-3} at depths on the order of 0.2 to 0.5 m, depending on the age and composition of the deposit (Fig. 9.22). Increasing density with depth is accompanied by decreasing order of aggregation and increasing shear strength. In practical terms the 1-to-10 Lowe scale (modified from Lowe [2007]) in Table 9.4 is conceptually similar to the Beaufort scale of sea state (Appendix F). For typical engineering problems involving erosion in coastal and estuarine environments, the Lowe scale does not exceed about 3 or 4.

9.4.3.2 *Measurement of soft-bed erosion shear strength*

Referring to Fig. 9.3, penetrometer resistance, a measure of the compressive strength of the bed matrix, increases with depth in bed prepared by the deposition of initially suspended sediment slurry. Since

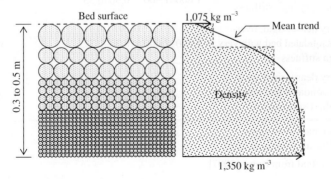

Figure 9.22. Schematic drawing of the relationship between floc structure and density of a self-weight consolidating (or fully consolidated) bed formed by deposition.

Table 9.4. Modified Lowe scale for clay-water mixtures.

Lowe scale	Mixture description	Nominal density, ρ (kg m^{-3})	Nominal solids volume fraction, ϕ_v	Nominal yield stress, $\tau_y{}^a$ (Pa)	Nominal vane shear strength, τ_v (kPa)
0	Flocs in suspension or slurry state with negligible yield stress	1,060–1,100	0.01–0.03	0–0.1	0
1	Flocs in fluid to fluid-like state with negligible to low yield stress	1,100–1,200	0.03–0.07	0.1–10	0–20
2	Compliant mud with moderate yield stress	1,200–1,300	0.07–0.10	10–100	20–50
3	Moderately dense, nearly non-compliant mud with moderate to high yield stress	1,300–1,400	0.10–0.14	100–400	50–80
4	Moderately dense and firm bed with high yield stress	1,400–1,500	0.14–0.18	400–1,000	80–120
5 (mid)	Dense and very firm bed with very high yield stress	1,500–1,600	0.18–0.22	1,000–3,000	120–160
6	Dense and firm bed with moderate to high stiffness	1,600–1,700	0.22–0.25	$-^b$	160–200
7	Dense and firm bed with high stiffness	1,700–1,800	0.25–0.30	$-^b$	200–280
8	Very dense and firm bed with very high stiffness	1,800–1,900	0.30–0.33	$-^b$	280–350
9	Very dense, rigid, over-consolidated bed with very high stiffness	1,900–2,000	0.33–0.37	$-^b$	350–500
10	Very dense, rigid, over-consolidated bed with highest stiffness	2,100	0.37–0.40	$-^b$	500–750

[a] Upper Bingham yield stress.
[b] Too stiff for measurement of τ_y.
Source: Modified from Lowe [2007].

the erosion shear strength of such beds is characteristically low, Parchure [1984], Kuijper *et al.* [1989], and others used a laboratory procedure schematized in Fig. 9.23 originally credited to Thorn and Parsons [1980]. It involves layer-by-layer stripping of the bed by erosion. The bed shear stress is increased in steps (τ_{bi}, $i = 1, 2, 3 ...$), which permits the test to be run at each step of duration T_E to enable scouring to occur up to a depth at which the erosion flux decreases to a negligible value and the (depth–mean) suspended sediment concentration $C(T_E)$ becomes practically constant. The incremental depth of bed scour is

$$\Delta z = h \frac{\Delta C}{\overline{\rho}_{D\Delta z}} \qquad (9.50)$$

where h is the water depth, ΔC is the increase in the concentration C during time T_E and $\overline{\rho}_{D\Delta z}$ is the bed dry density taken as the mean value over the incremental depth of scour Δz. At depth Δz the bed shear stress can be taken to be equal to the bed shear strength, *i.e.*, $\tau_{bi} = \tau_{si}$.

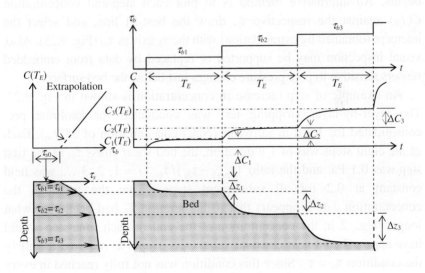

Figure 9.23. Schematic depiction of a method to estimate the erosion shear strength profile of a soft deposited bed.

The same procedure is followed after the bed shear stress is increased to the next step and the shear strength at the new depth of scour is estimated. In this way, the profile of variation of the shear strength with depth is constructed. Note that the determination of each Δz requires the corresponding value of $\overline{\rho}_{D\Delta z}$, which in turn is obtained by averaging the measured density profile $\rho_D(z)$ over Δz. For this purpose the bed density profile must be determined in a separate set of tests. In these the bed is, for instance, cored using a miniature laboratory corer at different times and the profile of density in the core is measured [Parchure, 1984]. As the density varies with depth, Δz is calculated by trial-and-error, starting with an assumed initial value of Δz. When the density is a gradually-varied function of depth, a single iteration for Δz may be sufficient to yield its value consistent with the corresponding $\overline{\rho}_{D\Delta z}$.

The shear strength τ_{s0} at the bed surface corresponding to the space-filling concentration (or density) cannot be determined by this method because for soft beds it is difficult to measure, *e.g.*, by the conventional visual inspection of the bed surface, the shear stress at which erosion begins. An alternative method is to plot each step-end concentration $C(T_E)$ against the respective τ_b, draw the best-fit line, and select the intercept (obtained by extrapolation) with the τ_b axis as τ_{s0} (Fig. 9.23). Also, visual inspection may be supported or replaced by data from embedded sensors sensitive to pore pressure changes just below the bed surface.

An example of step-increase in concentration is shown in Fig. 9.24. The layer-by-layer stripping test was conducted for kaolinite pre-consolidated for 1.7 d in saltwater (NaCl concentration of $35°/_{oo}$). Each of the eight steps was of 1 h duration, the bed shear stress τ_{b1} of the first step was 0.1 Pa, and the ratio $(\tau_{bi+1} - \tau_{bi})/\tau_{bi}$; $i = 1, 2, 3....$, was held constant at 0.2 for all subsequent steps. From the trend in the concentration data it appears that if the duration T_E had been somewhat longer, *e.g.*, 2 h, the concentration at the end of each time-step would have approached a near constant value (zero slope) as required to satisfy the condition $\tau_{bi} = \tau_{si}$. Since this condition was not fully reached in every case, error is present in the estimation of depth in the bed at which a given shear strength occurred.[e] The dry density profile of this bed (Fig. 9.25) was obtained by coring as noted.

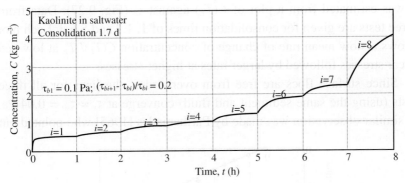

Figure 9.24. Variation of depth-mean concentration with time in a flume erosion test. The sediment was kaolinite in saltwater (from Parchure and Mehta [1985]).

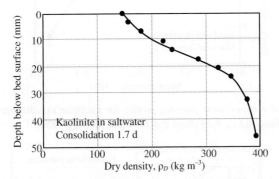

Figure 9.25. Bed density profile for kaolinite in saltwater (from Parchure and Mehta [1985]).

---

[e] Curve-fitting using a single set of data on the variation of suspended sediment concentration with time at a constant shear stress led Krone [1999] to propose

$$C = \frac{\dfrac{1}{\Pi_0} t}{\dfrac{\Pi_0}{\Pi_1} t + 1}$$

(F9.5)

where for a given sediment the coefficients Π_0 and Π_1 depend on the bed shear stress. A useful property of this function is that after a long time C approaches $1/\Pi_0$. Also, at $t = 0$, $dC/dt = 1/\Pi_1$.

The shear strength profile derived from eight steps is shown in Fig. 9.26. The surface value of the shear strength corresponding to point 0 was estimated from a plot of $C(T_E)$ against τ_b (Fig. 9.27). Data from three tests are given for consolidation times of 1, 1.7 and 5.6 d. Each plot shows a slow mean rate of change of concentration $C(T_E)/T_E$ at low bed shear stresses followed by larger rates at higher stresses.

Since surface flocs are free from overburden, the lines for all three tests (using the same sediment and fluid) converge at $\tau_s = \tau_{s0} = 0.04$ Pa. A similar observation was made by Partheniades [1965] who pointed out

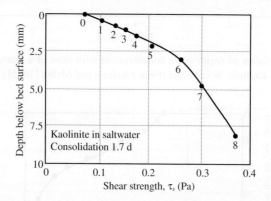

Figure 9.26. Bed erosion shear strength profile for kaolinite in saltwater based on the method schematized in Fig. 9.23 (from Parchure and Mehta [1985]).

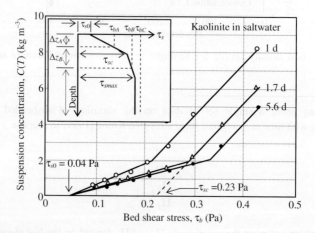

Figure 9.27. Method to estimate characteristic bed erosion shear strengths τ_{s0} and τ_{sc} (adapted from Parchure and Mehta [1985]).

that independence from overburden implies that at the surface erosion occurs by breakup of inter-particle bonds whose strength is determined by electrochemical forces. Therefore, it was argued, for describing surface erosion of flocs in the drained condition (because they are exposed to ambient water and not constrained by surrounding solids), this breakup process cannot be correlated with geotechnical indices such as the Atterberg limits. They represent soil bulk properties encompassing orders of magnitude greater dynamic scales than those assigned to electrochemical forces. However, mass erosion of deeper layers, which remain undrained, has been correlated to geotechnical indices. Winterwerp *et al.* [2012] have argued that surface erosion may also be predictable.

Below the surface each profile may be conveniently divided into three zones, a top zone of high gradient in τ_s, a middle zone of weaker gradient and a lower zone of practically uniform τ_s. When a shear stress τ_{bA} is applied the depth of scour Δz_A is small (Fig. 9.27 inset). When the shear stress is τ_{bB} the scour depth Δz_B is larger. The shear strength τ_{sc}, which separates these two zones of depths Δz_A and Δz_B is represented by a break in the $C(T_E)$ versus τ_b plot (Fig. 9.27). For example, for the test with 1.7 d consolidation time, $\tau_{sc} = 0.23$ Pa. When the shear stress τ_{bC} exceeds τ_{smax}, erosion continues indefinitely as long as the flow (velocity, depth and turbulence) remains unchanged. Due to its uniformity, a placed (*i.e.*, set manually and remolded as opposed to deposited from suspension and stratified) bed that is sufficiently dense can be expected to show a trend of continued erosion at a constant rate. Series III run of Partheniades [1962] in Fig. 9.4 shows such a behavior. In contrast, Series I run corresponds to a shear stress (acting on a deposited bed) whose value is less than τ_{smax}.

The shear strength profiles shown in Fig. 9.28 are for beds formed by deposition of suspended kaolinite in saltwater and consolidated for durations ranging from 1 to 10 d. As the aggregate order did not change at the surface in the absence of overburden, the shear strength remained independent of consolidation.

Below the bed surface, increasing shear strength reflects increasing floc strength and decreasing aggregate order. Given that the time-variation of erosion flux depends on the shear strength profile, which in turn varies with the density profile, it is feasible to empirically correlate the erosion flux directly with the bed density [Jepsen *et al.*, 1997].

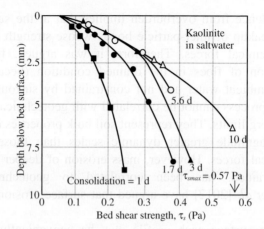

Figure 9.28. Shear strength profiles for a bed of kaolinite in saltwater (from Parchure and Mehta [1985]).

In Fig. 9.29, depth–mean values of shear strength obtained from the profiles in Fig. 9.28 are plotted against the consolidation time. Averaging for each mean value was carried out over the top bed layer of rapidly changing shear strength, with layer thickness ranging between 1.6 and 4.0 mm.

Observe (Fig. 9.29) that the shear strength increased almost three-fold during the first week, then approached a constant value of 0.37 Pa. As we have noted earlier, this weekly time-scale of change of bed resistance is comparable to the spring-neap variation of tide in the estuary. Although the similarity of the two time-scales is fortuitous, it has significant implications. An outcome is that sediment deposited during spring tide is

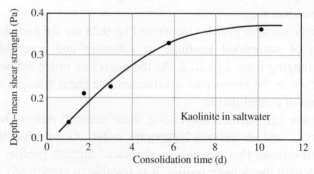

Figure 9.29. Variation of depth–mean shear strength with consolidation of kaolinite in saltwater (from Parchure and Mehta [1985]).

prone to retention as a bed because the tidal range decreases for approximately a week following its highest value in spring. The bed also becomes increasingly resistant to erosion due to consolidation during that week.

In thin beds of thicknesses on the order of a few tens of millimeters, self-weight consolidation is not significant enough to increase the shear strength beyond τ_{smax} without an imposed surcharge (overburden). Even then, strong zero-order flocs are likely to remain intact, and only excessive compressive stresses, such as under glaciation, can crush them into smaller clay units. In Fig. 9.28, τ_{smax} is about 0.57 Pa. To determine τ_{smax} directly, erosion tests represented by results in Fig. 9.30 using kaolinite in freshwater were carried out. At a depth of 20 mm, the 1-day and 8-day profiles converge to a shear strength value of 0.59 Pa, which appears to be τ_{smax} for that bed.

A feature of interest in the 1-day consolidation profile in Fig. 9.30 is the distinct gradient of τ_s at a depth of about 15 mm separating a thin upper layer of weak flocs from the stronger floc layer below. This is the depth below which the weight of the upper layer of flocs of a given order of aggregation was able to crush the flocs to a lower order resulting in a step-increase of ~0.09 Pa in the shear strength. The minimum thickness Δz_m of the upper layer required to break down the layer beneath may be estimated as follows.

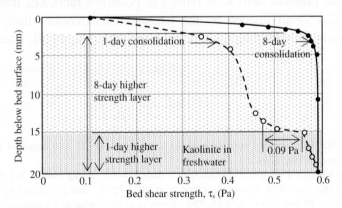

Figure 9.30. Shear strength profiles from two long-duration erosion tests using kaolinite in freshwater (from Parchure and Mehta [1985]).

The downward (normal) stress at depth Δz_m is

$$\sigma_v = \frac{\phi_v}{\phi_{vf}} (\rho_s - \rho_w) g \Delta z_m \qquad (9.51)$$

where ϕ_v is the volume fraction of the primary particles and ϕ_{vf} is the volume fraction of the flocs of a given order in the overburden [Krone, 1963]. From Eq. 4.42 we have

$$\frac{\phi_v}{\phi_{vf}} = \frac{2.5}{\rho_s k_f} \qquad (9.52)$$

Substitution into Eq. 9.51 yields

$$\sigma_v = \frac{2.5 g \Delta z_m}{k_f} \left(1 - \frac{\rho_w}{\rho_s} \right) \qquad (9.53)$$

For cohesive materials σ_v varies with the yield stress τ_y according to

$$\sigma_v = \frac{2 \tau_y}{(1 - K_s)} \qquad (9.54)$$

where the pressure ratio K_s is related to Poisson's ratio, *i.e.*, transverse contraction strain divided by the longitudinal extension strain in the direction of the stretching force [Lambe and Whitman, 1979]. Therefore

$$\Delta z_m = \frac{0.8 k_f \tau_y}{g(1 - K_s)\left(1 - \frac{\rho_w}{\rho_s} \right)} \qquad (9.55)$$

Example 9.3: Estimate Δz_m and compare the answer with the 15 mm height of the low-strength layer in Fig. 9.30.

From Chapter 4 we select characteristic values $k_f = 10^{-2}$ m^3 kg^{-1} and $\tau_y = 1$ Pa. Also, $K_s = 0.93$ [Krone, 1963]. Therefore, given $\rho_w = 1,025$ kg m^{-3} and $\rho_s = 2,650$ kg m^{-3}, we obtain $\Delta z_m = 19$ mm from Eq. 9.55. The closeness of this value to 15 mm supports the inference that the lower step-structure in the variation of shear strength with depth in Fig. 9.30 is the outcome of the collapse of floc structure at a threshold overburden.

9.4.4 *Shear strength and salinity*

The effect of salinity on the bed erosion shear strength is complicated by a variety of factors that influence the electric double layer of the clay micelle including the type of salt and its concentration in the pore fluid. In nature the amplitude and frequency of variation of salinity of the eroding fluid are commonly considerably higher than in the pore fluid. When freshets occur, a high gradient of salinity may develop between the salty fluid in the surface flocs and the fresher ambient water. If such a gradient persists for several hours (*e.g.*, as in Fig. 4.15), a reduction in pore-water salinity due to osmosis may occur. This in turn will influence the inter-particle bonds and the stability of the top bed layer (Chapter 5).

Figure 9.31 shows the effect of increasing salinity from 0.5 to 10°/$_{oo}$ on the shear strength of a lake mud. At the bed surface the strength increased from 0.13 to 0.27 Pa. This for instance is consistent with the influence of salinity on the floc structure of San Francisco Bay mud (Chapter 4). The maximum shear strength (τ_{smax}) was about 0.67 Pa.

In Fig. 9.32 the depth–mean shear strength is plotted against salinity. Averaging has been carried out over the top layer of variable shear strength, with layer thickness ranging from about 0.12 to 1 mm. As the salinity increased from 0.5 to 2°/$_{oo}$, the shear strength nearly doubled. The rate of shear strength increase was slower as the salinity increased to 10°/$_{oo}$. As would be expected, this trend is consistent with the

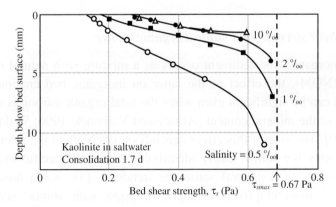

Figure 9.31. Bed erosion shear strength profiles of a lake mud obtained from erosion tests (from Parchure and Mehta [1985]).

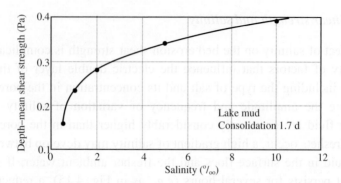

Figure 9.32. Variation of depth–mean bed erosion shear strength of lake mud with salinity (from Parchure and Mehta [1985]).

effect of salinity on the settling velocity of flocs. In many cases (but not all as we noted in Chapter 7) a rule of thumb is that in coastal areas where the salinity is in excess of about $10°/_{oo}$, its effect on floc properties need not be modeled. Flocs can be sensitive to changes in salinity when the salinity is less than about $10°/_{oo}$. Unfortunately, without knowing the dynamic response of the flocs to salinity changes it is risky to model a response. It is safer to assume that one is dealing with stable flocs at all salinities. Such a bias is favored by the fact that even in low salinity waters biotic secretions are efficient in binding individual particles into agglomerates that qualitatively tend to behave like salt-flocculated particles.

9.4.5 *Shear strength and natural organic matter*

When inorganic fine sediment occurs as a mixture with natural organic matter (NOM), the effect of the latter on inorganic bed erosion shear strength can be significant even when the total organic carbon is a small fraction of the mineral content [Ashley and Verbanck, 1996; Widdows *et al.*, 2000]. The size of this effect depends on the biota in the system and may or may not follow easily identifiable trends in specific instances. Nevertheless, in a general sense the trends in Fig. 9.33a have been postulated from empirical evidence. Starting with abiotic sediment, colonization by macrofauna that promote bioturbation (Figs. 9.33b, c) would mean an effective decrease in the bed shear strength from τ_{sA} to

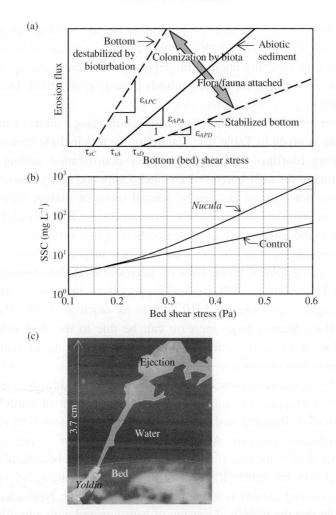

Figure 9.33. (a) Schematic drawing of the effect of colonization by biota and extant flora/fauna on erosion flux; (b) effect of bioturbation by *Nucula* on suspended sediment concentration; (c) sediment ejection by *Yoldia* (adapted from Davis [1991; 1993]).

τ_{sC} and an increase in the erosion flux constant (Eq. 9.26) from ε_{APA} to ε_{APC}. In other words bottom stability would decrease. Microphytobenthos present in the bed secrete extracellular polymeric substances (EPS) such as mucopolysaccharides (mucus). EPS increases bottom stability by binding particles together, which effectively increases the bed shear strength from τ_{sA} to τ_{sD} and decreases the erosion flux constant from ε_{APA}

to ε_{APD}. The effect of thin mats of diatoms[f] at the bed surface falls within this category. The effect of EPS on bed erodibility appears to be limited to water depths typically about 3 m, below which there is no significant penetration of sunlight for photosynthesis [Borsje *et al.*, 2008; Dickhudt *et al.*, 2009].

A general list of bed stabilizing and destabilizing factors related to aquatic life is given in Table 9.5. Stabilizing factors include secretion of EPS-forming biofilms, bed compaction or densification arising from water drainage through burrowed channels, networking by filamentous biota, protection against erosion by animal tubes or diatom mats, bio-filtration and bio-deposition. To this list compiled by Black *et al.* [2002] we may add the indirect effect of desiccation of bed surface exposed to sun or atmospheric heat. In this case the bed may be stabilized even under a *reduced* role of biota. Data on intertidal mud flats compiled by Houwing [1999] from other studies suggested an increase in the bed shear strength from order of 1×10^{-1} Pa to as much as 7×10^{-1} Pa over exposed flats. Such a large increase can be due to the dual effect of suppression of biota by oxidation as well as hardening of inorganic matter due to loss of water.

Destabilization can occur due to detachment of the biofilm from the particle by entrapped gas such as oxygen, pelletization of particles by feces formation, foraging and grazing by feeders at the sediment surface as well as burrow cleaning. A change in the bed surface texture for any reason will change the bed roughness and therefore the bed shear stress. The effect can be either drag enhancement due to increased surface roughness or drag reduction when the surface smoothens. Neumeier *et al.* [2006] note that the pattern of erosion of beds covered with a biofilm can be quite patchy due to selective rupturing of the surface by fluid stress. This may make it difficult to define erosion per unit bed surface area, particularly when this area is small, as in a small flume in which the bed is tested for its erodibility.

Figure 9.33b shows the bioturbation effect of *Nucula annulata* bivalve mollusks (clams) derived from laboratory tests using sediment

---

[f] Diatoms, a form of algae, are relatively simple aquatic organisms that convert inorganic substances into organic matter by photosynthesis. Most diatoms are unicellular.

Table 9.5. Stabilizing and destabilizing biotic influences on natural cohesive sediment.

Stabilizing	Destabilizing
EPS Secretion: Extracellular polymeric substances (EPS) produced by benthic microalgae (microphytobenthos) enhance the effect of electrochemical cohesion, promote floc growth and deposition	*Blistering*: Trapping of oxygen bubbles in biofilms increases the buoyancy of the biofilm to such an extent that it pulls away from the sediment
Compaction (densification): Burrowing macrofauna increase sediment density and bed strength	*Pelletization*: The formation of feces can enhance erodibility
Increased Drainage: Burrows and channel formation promote dewatering and compaction/densification	*Grazing*: Organisms feeding on intertidal flats cause physical disturbance and resuspension of sediment and reduce the stabilizing influence of microbes
Network Effect: Filamentous biota extend branching through the sediment matrix binding sediment particles together	*Burrow Cleaning*: Some animals clean tubes they inhabit in the sediment and give rise to a localized benthic flux of sediment
Flow Effect and Armoring: Plants and animal tubes form dense fields that induce skimming flow in the overlying water, protecting the bed from erosion; armoring can also occur from other causes such as microbial mats	*Boundary Layer Effect*: Burrows, tubes and tracking of the sediment surface increases bed roughness and therefore bed shear stress
Bio-filtration and Bio-deposition: Feeding by organisms removes sediment from suspension and leads to deposition	
Boundary Layer Effect: Smoothing of the sediment surface reduces bed shear stress by decreasing bed roughness	
Desiccation: Periodic drying of sediment at low water levels can have a significant effect on the surface bed shear strength	

Source: Revised from Black *et al.* [2002].

and water from Long Island Sound [Davis, 1993]. When the flow-induced bed shear stress was low there was little effect of the clams on the suspended sediment concentration, which however increased more rapidly than control (*Nucula* absent) when the bed shear stress exceeded about 0.2 Pa. Ejection of sediment by *Yoldia limatula*, another bivalve mollusk, is photographed in Fig. 9.33c.

Since colonization of sediment by microbes and benthic microalgae (microphytobenthos) depends on particle size, their effect on the shear strength τ_s varies with particle diameter d_p. In Fig. 9.34a the sediment ranges between silt (10 μm) and sand (250 μm). There is as much as an order of magnitude spread in the shear strength contingent upon the absence (low shear resistance) or presence (high shear resistance) of

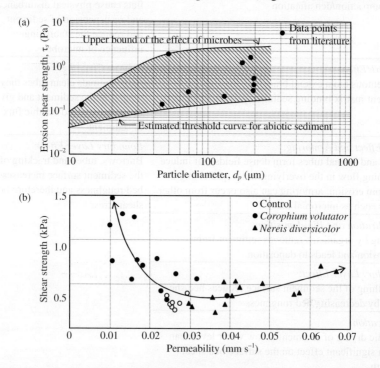

Figure 9.34. Effect of biota on the relationships among bulk parameters representing bed properties: (a) Variation of bed erosion shear strength with particle size based on studies on the influence of microbial mats relative to abiotic sediment (from Black *et al.* [2002]); (b) effects of invertebrates on the relationship between (penetrometer) shear strength and permeability (from Meadows and Tait [1989]).

microbes [Black *et al.*, 2002]. Other bulk measures support such observations. In Fig. 9.34b the bed shear strength is plotted against the (coefficient of) bed permeability k_p (see Table 2.9 for the range of values). The shear strength was obtained by a fall-cone penetrometer [Meadows and Tait, 1989], and the permeability was measured with a falling-head permeameter [Kirkham, 2005]. Sediment was mud from the Clyde estuary in UK. The sediment used as control was free of invertebrates. The same mud was also used with two invertebrate species — *Corophium volutator* and *Nereis diversicolor*. These species showed different effects on control; whereas *C. volutator* caused k_p to decrease and shear strength to increase, *N. diversicolor* increased both k_p and the shear strength. This qualitative difference in the effects of invertebrates underscores the diversity in bulk *physical* properties of bed sediment on biota, and the difficulty in developing predictive equations for these properties.

As we noted in Chapter 7, NOM tends to reduce the density and the settling velocity of clayey flocs. The change in the dry density ρ_D of bed sediment with NOM makes it convenient to relate τ_s to ρ_D as a surrogate for NOM. In Fig. 9.35, the data are from several submerged sites in peninsular Florida. The mean trend is described by the relationship

$$\tau_s = 0.05 \rho_D^{0.17} \tag{9.56}$$

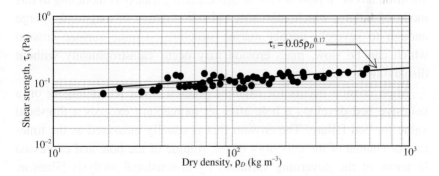

Figure 9.35. Dependence of bed erosion shear strength on bed dry density for organic-rich sediment from peninsular Florida (from Gowland *et al.* [2007]).

which is similar to Eq. 9.36 because $\rho_D = \phi_v \rho_s$. In comparison with the range of exponent values between 0.9 and 2.30 in Table 9.1, the value 0.17 indicates a weak dependence of shear strength on density. This observation along with very low (< 0.2 Pa) values of shear strength indicates low erosion thresholds of comparatively light (low submerged weight) and, in the present case, the weakly inter-bonded structure of organic-rich flocs or floccules.

The weak dependence of the erosion shear strength on bed density appears to be related to the likely presence of an organic-rich layer of fluff at the bed surface. This layer, with a thickness of perhaps a few floc diameters, consists of flocs released from the bed layer below by swelling as the top layer erodes away. Thus the resistance of the fluff flocs to erosion does not wholly depend on bed density.

9.5   Mass Erosion

9.5.1   *Stiff beds*

In mass erosion the top layer of the undrained bed fails and sediment clasts separate from the parent bed (Fig. 9.2). The exact mode of erosion depends on the forcing, *i.e.*, tangential or normal. In general, the failure plane is determined by the resultant of both forces. Surprisingly little attention has been paid in coastal and estuarine transport modeling to this mode of failure, which can cause episodic erosion of magnitudes large enough to severely impact marine structures. Unlike surface erosion which can be slow, mass erosion may occur catastrophically without displaying any early signs.

Mass erosion occurs for example when a water jet impinging on a cohesive bed develops a hole after jet momentum exceeds a threshold value for bed failure. The erosion flux is usually measured as the time-rate of change of the cube-root of the volume of the hole and expressed in terms of the governing variables by dimensional analysis [Hanson, 1990; Hollick, 1976; Moore and Masch, 1962].

In Fig. 9.36, experimental data on eroded mass is plotted against applied bed shear stress for a clay sample from Grande Baleine River in

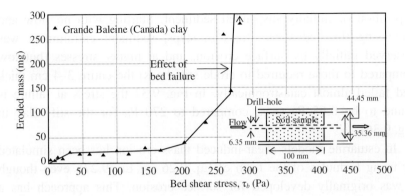

Figure 9.36. Bed failure effect in a plot of eroded mass versus time for Grande Baleine (Canada) clay eroded in a drill-hole apparatus (adapted from Rohan *et al.* [1986]).

Quebec, Canada, using a drill-hole apparatus (inset). This device is designed for testing the erodibility of intact and stiff clays such as an over-consolidated glacial till. A 100 mm long specimen is trimmed to a diameter of about 37 mm and lowered into a metal tube. A mechanical lathe is used to drill a hole of about 6.4 mm diameter, which exposes a total surface area of 20 cm^2 for erosion. A variable pressure-head pump is then used to force flow through the hole. At the high end the velocity can be in excess of 10 m s^{-1}, generating a bed shear stress of several hundred Pascals. The flow is increased in steps and run for a fixed period of time. In the test shown the sample failed catastrophically at about 270 Pa corresponding to a flow velocity of about 8.5 m s^{-1}. In contrast, a sample of clay from the St. Barnabe area (Quebec) did not show any erosion up to 400 Pa, pointing to the substantial site-specificity of bed response [Rohan *et al.*, 1986].

9.5.2 *Weak beds*

Although mass erosion is often considered to occur when stiff beds are subjected to high stresses, weak beds may also mass erode at low stresses. In flume tests, Hwang [1989] reported the failure of beds of low-density and high-porosity fine sediment from Lake Okeechobee in Florida containing 40% by weight of NOM (Table 5.5). The density was fairly uniform even though each bed tested was prepared from the deposition of initially suspended sediment. Starting with no flow and increasing the bed shear stress in steps, sediment entrainment was observed initially by surface erosion, and at higher stresses (but low compared to those required to erode stiff beds) the entire 3–4 cm thick bed failed almost catastrophically. In Fig. 9.37 the stress at failure is found to be 0.65 Pa (*e.g.*, compared to 270 Pa for the stiff clay in Fig. 9.36).

In estuarine models shear-induced mass erosion has been simulated by using formulas of the form of Eq. 9.26 (or Eq. 9.27), even though it was originally developed for surface erosion. This approach has a partial basis in soil mechanics [Winterwerp *et. al.*, 2012], and is convenient because it attempts to represent a complex process in terms of a simple two-parameter equation [Ariathurai *et al.*, 1977]. For data of the type in Fig. 9.27, failure can be considered to be shear-induced and to occur when the (plastic) yield strength τ_y of the bed is exceeded. Thus Eq. 9.27 may be restated as

$$\frac{\varepsilon}{\varepsilon_M} = \left(\frac{\tau_b - \tau_y}{\tau_y} \right) \tag{9.57}$$

As would be expected, the coefficients τ_s and ε_M for surface erosion of Lake Okeechobee sediment are lower than τ_y and ε_M for mass erosion of the same material (Fig. 9.38). For example, for a bed of wet bulk density

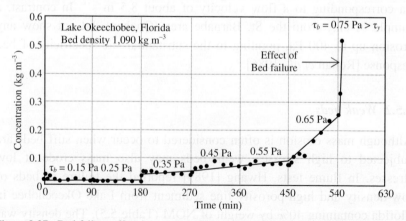

Figure 9.37. Bed failure effect in concentration versus time plot for a bed of organic-rich sediment from Lake Okeechobee, Florida (from Hwang [1989]).

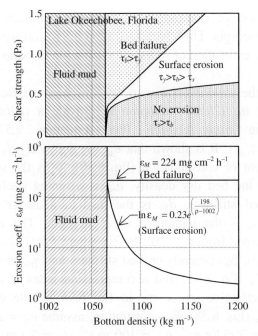

Figure 9.38. Shear-induced mass and surface erosion parameters for bottom sediment (muck) from Lake Okeechobee, Florida (from Hwang [1989]).

ρ = 1,150 kg m^{-3}, τ_s and ε_M for surface erosion are 0.60 Pa and 2.4 mg cm^{-2} h^{-1} respectively. The corresponding values for mass erosion are τ_y = 1.3 Pa and ε_M = 224 mg cm^{-2} h^{-1}.

Recognizing the rapidity with which mass erosion of recently formed beds occurs in the tidal environment, in their model for cohesive sediment transport Ariathurai *et al.* [1977] introduced the following expression for the erosion flux amounting to catastrophic failure of a bed layer of thickness Δz_b in a very short time Δt (perhaps on the order of a minute)

$$\varepsilon = \rho \frac{\Delta z_b}{\Delta t} \tag{9.58}$$

where ρ is the wet bulk density of the bed. This model is applicable when the bed shear stress exceeds the yield stress. The thickness Δz_b (on the order of centimeters) was determined by calibration against data on the time-variation of suspended sediment concentration in Savannah River estuary, Georgia. This characterization of erosion was shown to be essential for the estimation of sediment shoaling in the ship turning basin in the harbor.

Following the experimental work of Smerdon and Beasley [1959], Winterwerp *et al.* [2012] attempted to provide a rational basis for calculating the mass erosion flux. Equation 9.57 is recast as $\varepsilon = \varepsilon'_M (\tau_b - \tau_{sme})$. It is shown that $\varepsilon'_M = c_{cv} \phi_v \rho_D / \alpha_{me} d_{50} s_u$, where c_{cv} is the coefficient of consolidation (Eq. 8.40), ϕ_v is the solids volume fraction, ρ_D is the bed dry density, α_{me} is a tuning coefficient which depends on bed stiffness, d_{50} is the median diameter of dispersed particles and s_u is the undrained shear strength. The erosion shear strength is obtained from $\tau_{sme} = \beta_{me} PI^{\gamma_{me}}$, where *PI* (%) is the Plasticity Index (Eq. 3.33), β_{me} depends on bed stiffness and γ_{me} is scaled by the Swell Index C_{SI} (Eq. 8.75) and a compression index C_{CI} obtained by replacing the effective stress σ'_v in Eq. 8.75 by the coefficient of compression a_v (Eq. 8.38). For illustration, assume $c_{cv} = 2 \times 10^{-9}$ m^2 s^{-1}, $\phi_v = 0.30$, $\rho_D = 795$ kg m^{-3}, $\alpha_{me} = 100$ m s^{-2}, $d_{50} = 7$ μm, $s_u = 80$ kPa, $\beta_{me} = 0.16$ and $\gamma_{me} = 0.84$. These values yield $\varepsilon'_M = 8.5 \times 10^{-6}$s m^{-1} and $\varepsilon = 6.86 \times 10^{-5}$ kg m^{-2} s^{-1}. These calculations are approximations of the fuller analysis by Winterwerp *et al.* [2012].

9.5.3 *Effect of flow acceleration*

The effect of sudden flow acceleration or deceleration on erosion is embodied in Eq. 9.33, in which a non-negligible rate of change of bed shear stress $d\tau_b/dt$ yields a different solution than a monotonic change in the erosion velocity (or erosion flux) predicted by Eq. 9.34.

Figure 9.39a shows data from experiments by Thimakorn [1980] and Cervantes [1987] on the erosion of deposited natural muds in flumes (Table 9.6). Starting with clear water, in each test spikes in suspended sediment concentration are observed during the initial phase of erosion. The concentration $C_m(t)$ is the mean value over height $z = z_a$ (reference near-bed elevation) to h (mean surface elevation), and C_{mf} is the eventual steady (or nearly steady) value of C_m (Fig. 9.39b). Spikes in Fig. 9.39a suggest that each is an erosion pulse due to a ramped input of bed shear

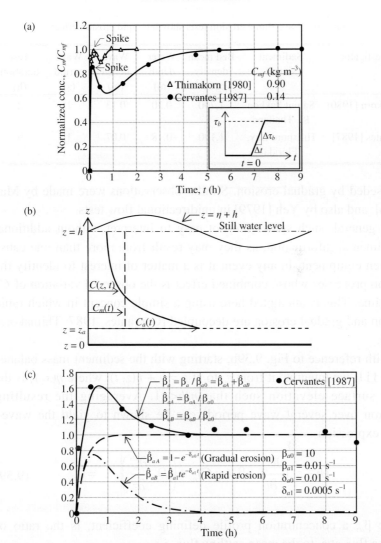

Figure 9.39. (a) Suspended sediment concentration profile under waves; (b) depth–mean concentration variation with time; (c) time-dependence of near-bed erosion parameters.

stress at the onset of wave action (inset) and consequent failure and entrainment of the top loose layer of the bed. This rapid erosion is far more episodic that the gradual surface erosion commonly observed in erosion tests (*e.g.*, Fig. 9.24). Eventually the pulse decays and is

Table 9.6. Wave flume erosion tests showing rapid erosion effect.

Investigator	Sediment source	Bed mean density (kg m^{-3})	Water depth (m)	Wave height (m)	Wave period (s)	Test duration (h)
Thimakorn [1980]	Samut Sakhon R., Thailand	1,700	0.30	0.13	1	2
Cervantes [1987]	Hillsboro Bay, Florida	1,100	0.18	0.07	1	9

superseded by gradual erosion. Similar observations were made by Maa [1986], and also by Yeh [1979] in unidirectional flow tests.

In general, such spikes are not easy to interpret without additional experimental information, as they may result from more than one cause in given equipment. In any event it is a matter of interest to identify the erosion processes whose combined effect is the observed variation of C_m with time. This is attempted here using a simple concept in which rapid erosion and gradual erosion are decoupled [Cervantes, 1987; Thimakorn, 1980].

With reference to Fig. 9.39b, starting with the sediment mass balance Eq. 6.111, we integrate it from z_a to at $z = h + \eta(z, t)$, where $\eta(x, t)$ is the wave surface elevation such that $\eta/h \ll 1$. Averaging the resulting equation over several wave periods can be shown to yield the wave–mean expression

$$(h - z_a)\frac{\partial C_m}{\partial t} + \beta_a (w_s C)_m + \left(D_{sz}\frac{\partial C}{\partial z} \right)\bigg|_{z = z_a} = 0 \qquad (9.59)$$

where β_a, a concentration profile defining coefficient, is the ratio of settling flux at z_a to the mean settling flux, i.e.,

$$\beta_a = \frac{(w_s C)\big|_{z = z_a}}{(w_s C)_m} \qquad (9.60)$$

After a long time ($t \to \infty$), Eq. 9.59 is subject to the conditions $\partial C_m / \partial t = 0$, $C_m = C_{mf}$ and $\beta_a = \beta_{a0}$. With these conditions and considering w_s to be constant Eq. 9.59 reduces to

$$(h - z_a) \frac{\partial C_m}{\partial t} = \left(\beta_a C_m - \beta_{a0} C_{mf} \right) w_s \qquad (9.61)$$

in which β_a will be conveniently represented as the sum

$$\beta_a = \beta_{aA} + \beta_{aB} \qquad (9.62)$$

where β_{aA} is for gradual erosion and β_{aB} for rapid erosion (Fig. 9.39). We will let

$$\beta_{aA} = \beta_{a0} \left(1 - e^{-\delta_{a0} t} \right); \quad \beta_{aB} = \beta_{a1} t e^{-\delta_{a1} t} \qquad (9.63)$$

in which, at steady state, as $t \to \infty$, $\beta_{aA} \to 1$ and $\beta_{aB} \to 0$. In Fig. 9.39c measured concentration profiles during the test of Cervantes (Table 9.6) were used to evaluate β_{a0}, β_{a1} and δ_{a0} and δ_{a1}. These coefficients determine the concentration $C_m(t)$ based on Eq. 9.61.

In general, depending on forcing by current or waves, sediment type and bed density, these coefficients are likely to vary with the bed shear stress τ_b. In addition, β_{a0} can be expected to vary with the ramping rate $\Delta \tau_b / \Delta t$ of the bed shear stress.

9.6 Fluid Mud Entrainment

9.6.1 *Moving and stationary fluid mud*

As mentioned in Chapter 6, entrainment and mixing are distinct processes; fluid mud entrainment is the process by which particle flux occurs from fluid mud into water (or a dilute suspension). The outcome is lowering of the top surface of fluid mud due to scour. In mixing, water flux occurs from the upper fluid into fluid mud with the result that fluid mud is diluted and its surface rises (dilation). Mixing is briefly reviewed in Chapter 10. As pointed out by Bruens [2003], two cases of entrainment are of interest. Figure 9.40a is a rendering of the common case of turbulent flow such as due to tide entraining stationary (or nearly stationary) fluid mud. Figure 9.40b is the opposite case wherein turbulent fluid mud layer is transported downslope as a gravity underflow (Chapter 12), while the water above is still (or nearly still). Water is sucked

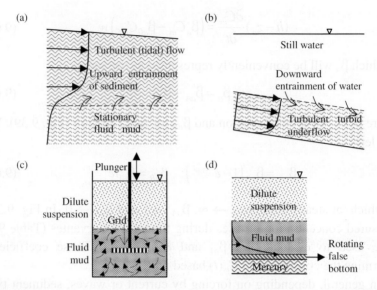

Figure 9.40. Cases of upward and downward entrainment: (a) upward entrainment of sediment from stationary fluid mud to turbulent flow; (b) downward entrainment of water from still water into turbulent underflow; (c) turbulence in fluid mud due to oscillating grid inside a cylindrical tank; (d) turbulence in fluid mud due to rotating false bottom in an annular flume.

downward into flowing mud where the pressure is lower than above the mud due to the Bernoulli effect (Chapter 2). Figure 9.40c describes turbulence due to an oscillating grid in fluid mud inside a cylindrical tank, and in Fig. 9.40d turbulence is produced in fluid mud by a rotating false bottom floating over mercury inside an annular flume.

With regard to the process in Fig. 9.40a, entrainment of stationary fluid mud into a steady turbulent flow can be conveniently thought of in terms of two simultaneous component sub-processes. In the first component, TKE from the mean flow is converted into the potential energy of the fluid mud layer which is diluted by mixing and rises to a height dependent on the equilibrium between energy supplied to maintain the layer at that height and energy removed by viscous loss. The rise may not be significant if fluid mud density is high, and there may be no downward entrainment of water (Fig. 9.40b). In the second sub-process billows occur at the interface, and once it is destabilized sediment mixes with the water layer above fluid mud (Fig. 9.41a). Thus at steady-state the entrainment velocity w_e is determined by the rate at which transfer of sediment from fluid mud to water occurs.

Billows at the interface can be in two modes. One is at the dominant frequency with which the water surface oscillates in the forced mode. If more than one dominant frequency is present in the wave spectrum, billows at various frequencies may also occur due to inter-wave interactions. In contrast to the forced mode, in the free mode the interface oscillates at the buoyancy (or Brunt–Väisälä) frequency. Free oscillations occur for example due to persistent wind over a practically unconfined body of water. Interfacial instability and mixing are reviewed in Chapter 10.

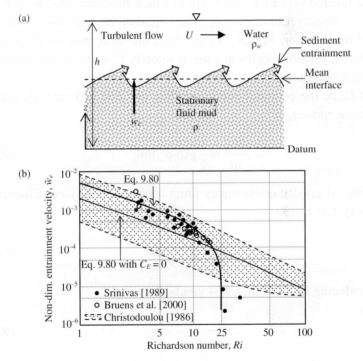

Figure 9.41. Entrainment of fluid mud: (a) Definition sketch for turbulent upper (water) layer and stationary lower (fluid mud) layer; (b) variation of non-dimensional fluid mud entrainment velocity with Richardson number. The data zone of Christodoulou [1986] is for a (sediment-free) saline water layer.

Fluid mud entrainment velocity under a steady shear flow is considered next, and entrainment by waves is briefly reviewed in Chapter 11. Several studies may be consulted for details [Jiang and Mehta, 1992; Scarlatos and Mehta, 1993; Uittenbogaard, 1995; Kranenburg and Winterwerp, 1997; Winterwerp and Kranenburg, 1997].

9.6.2 *Entrainment of stationary fluid mud*

Entrainment of stationary fluid mud is described under the premise of a balance between production of TKE, buoyancy work required to maintain fluid mud in raised condition, and viscous energy loss. Consider the two-layered system of Fig. 9.41a in which fluid mud of density ρ lies below a water layer of density ρ_w. The height of the suspension is h. It is desired to determine velocity w_e at which the fluid mud interface rises. Boundary-layer flow occurs at a mean velocity U, and supplies energy to raise the lower fluid.

Let E_p be the potential energy of flow per unit bed area. Its rate of change, *i.e.,* power, is

$$\frac{dE_p}{dt} = \frac{1}{2} g(\rho-\rho_w)hw_e \qquad (9.64)$$

The reduced gravity or buoyancy jump across the interface (Chapter 2) for this system is

$$g' = g\frac{\rho-\rho_w}{\rho} \qquad (9.65)$$

Introducing Eq. 9.65 into 9.64 yields

$$w_e = \frac{2}{\rho g'h}\frac{dE_p}{dt} \qquad (9.66)$$

Accordingly, a dimensionless entrainment velocity $\tilde{w}_e = w_e/U$ can be defined as

$$\tilde{w}_e = \frac{2}{\rho g'hU}\frac{dE_p}{dt} \qquad (9.67)$$

To calculate $\tilde{w}_e$ it is necessary to evaluate dE_p/dt. The non-stationary, turbulent energy balance in its simplified form for a horizontally homogeneous boundary layer above fluid mud is readily obtained from Eq. 2.94 as

$$\frac{\partial}{\partial t}\left(\frac{\overline{q^2}}{2}\right) = -\frac{1}{\rho}\overline{\rho u'w'}\frac{\partial U}{\partial z} - \frac{g}{\rho}\overline{\rho'w'} + \varepsilon_t \qquad (9.68)$$

where $q^2/2$ is the TKE (per unit mass), u' is the instantaneous turbulent fluctuation of the longitudinal velocity U, w' is the velocity fluctuation in the z-direction, ρ' is the fluctuation of density ρ and ε_t is rate of energy loss per unit mass. Overbars imply turbulence averaging and $-\overline{\rho u'w'}$ is the Reynolds stress. Equation 9.68 states that the temporal rate of change of TKE arises from three processes. These include shear production (first term on the r.h.s.), buoyancy work involving conversion of kinetic to potential energy (second term), and viscous energy loss (third term). Based on the commonly invoked phenomenological argument that shear production and viscous dissipation are proportional to U^3, Eq. 9.68 is reduced to

$$\frac{\partial}{\partial t}\left(\frac{\overline{q^2}}{2}\right) = A_E\frac{U^3}{h} - \frac{g}{\rho}\overline{\rho'w'} \qquad (9.69)$$

where A_E is a proportionality coefficient. Since the ratio h/w_e is the characteristic time-scale of entrainment and $q^2/2$ is scaled by U^2, we have

$$\frac{\partial}{\partial t}\left(\frac{\overline{q^2}}{2}\right) = B_E\frac{U^2}{h/w_e} \qquad (9.70)$$

where B_E is a scaling constant [Zilitinkevich, 1975]. Therefore, Eq. 9.69 becomes

$$B_E\frac{w_e U^2}{h} = A_E\frac{U^3}{h} - \frac{g}{\rho}\overline{\rho'w'} \qquad (9.71)$$

Three noteworthy contributors to the term $-g\overline{\rho'w'}/\rho$ are as follows [Mehta and Srinivas, 1993]:

1. Buoyancy entrainment flux E_e, which by definition is

$$E_e = -g'w_e \tag{9.72}$$

2. Flux E_s due to effects of settling, cohesion (including electrochemical and biological effects) and the difference between the viscosities of mud and water

$$E_s = g'w_{st} \tag{9.73}$$

where the representative velocity w_{st} is

$$w_{st} = -C_E \frac{g'h}{U} \tag{9.74}$$

in which C_E is an empirical coefficient.

3. Buoyancy flux due to molecular diffusion

$$E_d = D_E g'\sqrt{\frac{D_b U}{h}} \tag{9.75}$$

where D_b is the diffusion coefficient, h/U represents the eddy time-scale and D_E is a proportionality constant. These three quantities are substituted into Eq. 9.71 and solved for the non-dimensional entrainment velocity $\tilde{w}_e$ to yield

$$\tilde{w}_e = \frac{w_e}{U} = \frac{2}{\rho g'hU}\frac{dE_p}{dt} = \frac{A_E + D_E\dfrac{Ri}{\sqrt{Pe}} - C_E Ri^2}{B_E + Ri} \tag{9.76}$$

in which ρ is the fluid mud density. The definition of $\tilde{w}_e$ is inserted from Eq. 9.67, from which dE_p/dt can be obtained intrinsically. Also,

$$Pe = \frac{Uh}{D_b} \tag{9.77}$$

is a Péclet number representing the role of advection relative to diffusion. The global or bulk Richardson number Ri is defined as

$$Ri = \frac{g'h}{U^2} \qquad (9.78)$$

This form of Ri follows from Eq. 2.16 because, essentially, $Ri = 1/F'^2$ (Chapter 2).

When $C_E = 0$, such as when the lower layer is a sediment-free saline solution, Eq. 9.76 reduces to the expression for entrainment of salt. For the present purpose we will conveniently invoke the condition of local equilibrium, *i.e.*, set $\partial(q^2/2)/\partial t$ to zero, which means that $B_E = 0$. Thus,

$$\tilde{w}_e = \frac{A_E}{Ri} + \frac{D_E}{\sqrt{Pe}} - C_E Ri \qquad (9.79)$$

Based on saltwater entrainment experiments, the local equilibrium assumption is found to be acceptable but not ideal [Atkinson, 1988]. For most cases one may set the term $D_E/\sqrt{Pe}$ to zero because molecular diffusion is of negligible consequence under typically turbulent conditions that lead to fluid mud entrainment. Accordingly, Eq. 9.79 is simplified to

$$\tilde{w}_e = \frac{A_E}{Ri} - C_E Ri \qquad (9.80)$$

In Fig. 9.41b $\tilde{w}_e$ is plotted against Ri along with data on the entrainment of clayey fluid muds tested in laboratory flumes. The data of Srinivas [1989] were obtained in a "race-track" flume in which two so-called disc pumps, each consisting of vertically stacked horizontal metallic discs were rotated by a central shaft to generate water motion above a stationary fluid mud layer. The two stacks were spun in the opposite directions to produce sufficiently high flow speeds in the flume.

The data of Bruens *et al.* [2000] were obtained in an annular rotating channel. The coefficients A_E and C_E for these data have been adjusted to highlight the self-similarity of the trend between $\tilde{w}_e$ and Ri. The selected values are $A_E = 5.2 \times 10^{-3}$ and $C_E = 1.6 \times 10^{-5}$. Since the second term on the right hand side of Eq. 9.80 arises from the effect of settling, C_E depends on the settling velocity. Setting $C_E = 0$ leads to inverse dependence of $\tilde{w}_e$

on Ri, as in saltwater entrainment. The resulting expression is compared with data on the entrainment of saltwater into fresh water [Christodoulou, 1986]. Due to settling, at relatively high values of Ri sediment entrains less readily than saltwater. Factors responsible for the rapid drop in the entrainment velocity at high Ri include particle weight, flow energy loss due to turbulent viscosity and wall friction in the flume. As noted later wall friction may be substantial in typical laboratory experiments. In addition, increased resistance to entrainment due to stratification of the fluid mud layer, which is usually difficult to eliminate entirely in laboratory tests, tends to retard the entrainment velocity.

Example 9.4: The wet bulk density of a fluid mud layer at the bottom of a 5 m deep stream is 1,017 kg m^{-3}. The water is brackish with a density of 1,015 kg m^{-3}. The sediment is high in NOM with an effective particle density of 1,500 kg m^{-3}. Values of the coefficients in Eq. 9.80 are $A_E = 0.005$ and $C_E = 2\times10^{-4}$. If the mean current velocity is 0.15 m s^{-1}, calculate the entrainment velocity, the depth of scour of the fluid mud layer in 30 minutes, and the critical Richardson number for entrainment.

We will assume that the entrained sediment fully mixes over the water column. Thus, with $h = 5$ m, Eq. 9.78 yields $Ri = 4.3$. Then from Eq. 9.80, $\tilde{w}_e = w_e/U = 0.0003$, and $w_e = 0.000046$ m s^{-1}.

The dry mass concentration of bottom sediment is $[(1017-1015)/(1500-1015)]$ $\times1500 = 6.2$ kg m^{-3}. Therefore the entrainment flux is $0.000045\times6.2 = 0.00028$ kg m^{-2}s^{-1}. In one-half hour the depth of scour would be $0.000045\times0.5\times3600 = 0.082$ m $= 8.2$ cm.

At the critical Richardson number Ri_c there is no entrainment. Setting $\tilde{w}_e = 0$ in Eq. 9.80 yields $Ri_c = \sqrt{A_E/C_E}$. Thus, $Ri_c = 5$.

For a fluid mud underflow of thickness z_f moving downslope at a mean velocity u_m, the characteristic Richardson number may be defined as

$$Ri_b = \frac{g'z_f}{(\Delta u)^2} \qquad (9.81)$$

where $g' = g(\rho_{fm} - \rho_w)/\rho_{fm}$, ρ_{fm} is the fluid mud (wet bulk) density and Δu is the difference between u_m and the velocity of the overlying fluid (water). For the simple case of still water we have $\Delta u = u_m$. Then the entrainment flux ε (dry sediment mass per unit bed area and time) is obtained from

$$\varepsilon = \frac{\alpha_{en} u_m C_{fm}}{\left(1 + 63 Ri_b^2\right)^{3/4}} \qquad (9.82)$$

where $C_{fm} = \rho_s (\rho_{fm} - \rho_w) / (\rho_s - \rho_w)$ is the fluid mud concentration and the entrainment coefficient α_{en} is of order 1. This expression is modified from its original form proposed by Odd and Rodger [1986]. We may consider a 0.15 m thick layer of fluid mud of 1,150 kg m^{-3} wet bulk density flowing downslope at 0.15 m s^{-1}. Assume still seawater (density 1,025 kg m^{-3}) and particle density $\rho_s = 2,650$ kg m^{-3}. Thus we obtain $g' = 1.07$ m s^{-2}, $Ri_b = 7.1$, $C_{fm} = 204$ kg m^{-3} and, finally, $\varepsilon = 0.072$ kg m^{-2} s^{-1}. Whitehouse *et al.* [2000a] have expressed concern regarding the applicability of Eq. 9.82 to the natural environment, indicating the need for careful calibration with respect to α_{en}. Furthermore, as we will note next, the effect of downward entrainment of water into flowing fluid mud must be considered separately.

9.6.3 *Entrainment of flowing fluid mud*

As mentioned in Chapter 6, entrainment of (typically turbulent) fluid mud into still water requires that the upward entrainment velocity of sediment be separated from the downward velocity associated with water entrainment into fluid mud. Separation is achieved from mass balances for the solids and the fluid including the respective upward and downward fluxes. Experiments have been carried out in cylindrical tanks in which a plunger oscillating at a desired frequency (and amplitude) produces turbulence by a horizontal grid (Fig. 9.40c). The rate of mixing depends on the rate of input of mechanical energy of the plunger (converted to TKE).

In some experiments stratification was initially induced by using saltwater for the lower layer and fresh water for the upper one. These experiments are a subset of a broader range of tests, with some using fine-grained sediment suspensions, that are not merely restricted to agitation of the lower layer. Such experiments are meant mainly to examine conditions under which a pycnocline is destabilized by turbulent kinetic energy at zero mean shear [Wolanski and Brush, 1975; Hopfinger and Linden, 1982; Wolanski *et al.*, 1989].

Very few experiments appear to have been conducted in which the lower layer of suspended sediment is actuated to move horizontally and the resulting mixing and entrainment resolved separately. In that context, the experiments of Bruens [2003] in an annular flume are noteworthy. In that setup the ring was removed leaving the water surface free. Close to the channel bed a false bottom was permitted to rotate, generating turbulent shear flow in the fluid mud layer (Fig. 9.40d). By filling the gap between the false bottom and the channel bed with mercury, the gap was effectively prevented from accumulating sediment from the fluid mud above the false bottom. Shear flow produced by the false bottom acting as the bed led to both mixing and entrainment, and it was found that the dependence of non-dimensional entrainment velocity $\tilde{w}_e$ on the Richardson number Ri was qualitatively similar to that shown in Fig. 9.41b for stationary fluid mud below turbulent flow of water. At low values of Ri, $\tilde{w}_e$ was found to vary with -1 power of Ri. It was argued that in the absence of sidewall friction the power would be -0.5, indicating a strong wall effect.

A difficulty associated with experiments on fluid mud entrainment has been the significant influence of any objects that might disturb the flow stream (*e.g.,* Mehta and Srinivas [1993]). Bruens [2003] reported that vertical supports (thin rods) for the false bottom contaminated the data on entrainment at the interface between fluid mud and water.

9.7 Modeling Erosion with Deposition

9.7.1 *Exchange paradigms*

In Chapter 7 we discussed the shortfalls in adopting the model of exclusive bed exchange, *i.e.,* the inability of the model to permit erosion and deposition to occur simultaneously. For simultaneous bed exchange it is necessary that the critical shear stress for deposition τ_{cd} (Eq. 7.144) and the erosion shear strength τ_s introduced earlier in this chapter be permitted to vary with the diameter of particles including strongly bound flocs, weakly held aggregates and even temporarily unbound individual particles. This variation is inherent in two experimentally-based multi-size class relationships. The first, for the critical shear stress for deposition τ_{cdi} of a suspended class-i particle of diameter d_i, is

$$\tau_{cdi} = \tau_{cd1} \left(\frac{d_i}{d_1} \right)^{\xi_e} \tag{9.83}$$

where τ_{cd1} is the value of τ_{cdi} for the smallest particle diameter (d_1) in the population and the exponent ξ_e depends on sediment composition. This relationship is a restatement of Eq. 7.158 and is obtained by replacing the settling velocity in that equation by the diameter determined from Stokes law (Eq. 7.15). This means that $\xi_e = 2m_e$, where m_e is the exponent in Eq. 7.158. Equation 9.83 implies that at low values of the bed shear stress τ_b only small particles can stay suspended. Larger particles will have sufficient masses and settling velocities to overcome turbulence and reach the bed. As τ_b increases, the threshold diameter separating eroding particles from depositing ones also increases. At τ_b values less than τ_{cdi}, a class-i particle bound to its bed neighbors will resist erosion, whereas at τ_b greater than τ_{cdi} turbulence will prevent a suspended particle of the same diameter from deposition.

The second relationship is for bed particles, for which the shear strength τ_{si} is based on Eq. 9.38 along with Eqs. 4.60 and 4.63:

$$\tau_{si} = 0.256 B_F \left(\frac{\Delta \rho_s}{\rho_w} \right)^{2/(3-D_f)} \left(\frac{d_1}{d_i} \right)^2 \; ; \quad \tau_{si} > 0.326 \, \text{Pa} \quad \text{(Upper segment, } U\text{)}$$

$$\tau_{si} = 0.289 \sqrt{B_F} \left(\frac{\Delta \rho_s}{\rho_w} \right)^{1/(3-D_f)} \frac{d_1}{d_i} \; ; \quad \tau_{si} \leq 0.326 \, \text{Pa} \quad \text{(Lower segment, } L\text{)}$$

$$\tag{9.84}$$

where ρ_w is the density of water, $\Delta \rho_s = \rho_s - \rho_w$, ρ_s is the material density of the particle, D_f is its fractal dimension and coefficient B_F depends on the sediment.

According to Eq. 9.84 the class-*i* shear strength τ_{si} decreases as d_i increases, which is consistent with the observation that as d_i increases the floc becomes weaker and its ability to remain attached to the bed decreases. While this may seem contrary to the trend of Eq. 9.83 for suspended particles, we note that when a depositing floc impacts the bed it tends to break up into components that are characteristically denser and more cohesive than the parent (in accordance with Eq. 4.71, with $\psi = 1$). The need to recognize this difference between a depositing floc and an eroding one was pointed out by Krone [1963], and is basic to the understanding how the simultaneous exchange paradigm is realized.

Figure 9.42 includes plots of Eqs. 9.83 and 9.84 using representative values of variables and coefficients from Table 9.7. At the point where the two segments *U* and *L* intersect the shear stress is 0.326 Pa, and the diameter d_{**} is

$$d_{**} = \left[0.8858\sqrt{A_s} \left(\frac{\Delta\rho_s}{\rho_w} \right)^{\frac{1}{3-D_f}} \right] d_1 \tag{9.85}$$

In the present case $d_{**} = 1.43$ μm. Depending on the variables, Eq. 9.83 can intersect the upper or the lower segment at a "junction" diameter d_* obtained by recognizing that at the point of intersection $\tau_{cdi} = \tau_{si}$. The two respective diameters are

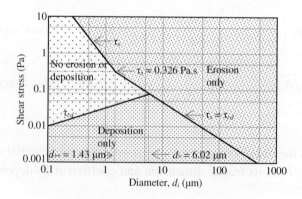

Figure 9.42. Fine sediment exchange zones.

Table 9.7. Deposition and deposition parameters for kaolinite.

τ_{cd1} (Pa)	0.03
d_1 (μm)	0.1
ξ_e	0.5
$\Delta\rho_s$ (kg m^{-3})	1,650
ρ_w (kg m^{-3})	1,000
B_f	75
D_f	2.2

Source: Letter [2009], Letter and Mehta [2011].

$$d_{*U} = \left\{ \left[0.256 A_s \left(\frac{\Delta\rho_s}{\rho_w} \right)^{\frac{2}{3-D_f}} \frac{1}{\tau_{cd1}} \right]^{\frac{1}{\xi_e+2}} \right\} d_1 \tag{9.86}$$

and

$$d_{*L} = \left[\frac{0.289\sqrt{A_s} \left(\frac{\Delta\rho_s}{\rho_w} \right)^{\frac{1}{3-D_f}}}{\tau_{cd1}} \right]^{\frac{1}{\xi_e+1}} d_1 \tag{9.87}$$

In the present case d_{*L} = 6.02 μm from Eq. 9.87. If Eq. 9.83 were to intersect the upper segment, d_{*U} would be obtained from Eq. 9.86.

Due to natural variability in τ_{cdi} and τ_{si}, d_* is not a precise quantity. However, experimental data [Bagnold, 1966; Mantz, 1977] suggest $d_* \approx 10$ μm (Chapter 6). The significance of this diameter is that for $d_i < d_*$ cohesion increases with decreasing d_i, as manifested by the difference $\tau_{si} - \tau_{cdi}$. This anomaly arises because on one hand, for deposition the submerged weight of the suspended particle must exceed the hydrodynamic lift force (which in turn is proportional to the drag force). On the other hand, resistance to the erosion of a bed particle of the same diameter is made up of (submerged) weight *as well as* cohesion. When $d_i > d_*$, electrochemical cohesion rapidly decreases with increasing diameter, and in the present idealized model it is assumed to be nil. Therefore, notwithstanding adhesion or binding due to natural exopolymers forming very large aggregates (say 1,000 μm or more) as in marine snow (Fig. 4.29), the τ_{cdi} and the τ_{si} curves are represented by a single composite curve.

In Fig. 9.43 an attempt has been made to pictorially illustrate bed sediment exchange. For uniform sediment, at a given bed shear stress τ_b the particles might erode from the bed, or deposit from suspension (exclusive mode) (Fig. 9.43a). If particle density is permitted to vary there can be simultaneous erosion and deposition depending on the density distribution (Fig. 9.43b). Effects of size and density (and shear strength) add more scenarios of bed exchange. For example, since both density and shear strength tend to decrease with increasing size, small, dense and cohesive flocs may deposit while larger, lighter and less cohesive flocs are picked up from the bed (Fig. 9.43c).

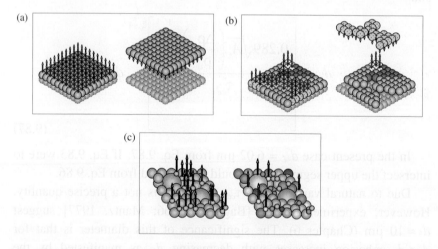

Figure 9.43. Pictorial depiction of bed exchange: (a) Particles have uniform properties; there can be either erosion or deposition but not both at the same time; (b) particle density is variable (spheres of different shades and patterns). There can be simultaneous erosion and deposition; (c) size and density (and shear strength) are variable (spheres of different shades and sizes). Bed exchange (only pickup is shown) becomes more selective than in (b).

9.7.2 *Effects of simultaneous exchange*

The sediment exchange zones in Fig. 9.42 can be identified for deposition-dominated flume tests such as those using kaolinite reported by Mehta and Partheniades [1975]. These tests are revisited to highlight the "invisible" role of erosion during net deposition. Then the significance of a kaolinite erosion-and-dilution test of Parchure and Mehta [1985] (Fig. 7.37) is summarized.

The deposition tests were carried out in the counter-rotating annular flume (CRAF) using kaolinite clay in 0.31 m deep water (Chapter 7). The kaolinite had a broad distribution of dispersed particle size from 0.1 to 40 μm and a median size of 1 μm. These heterogeneous particles collectively occur in states ranging from fully flocculated aggregates to primary mineral units. The states need not be static however, as turbulence-induced collisions and aggregation processes lead to shifting floc size distribution as the experiment proceeds. Thus for modeling the experimental trend it is essential to include mechanisms that describe inter-particle collisions, floc growth and floc breakup, together with erosion and deposition. Successful collisions result in a net increase or decrease in the floc diameter and related changes in density and shear strength depending on turbulence.

The methodology for multi-class aggregation modeling by numerical simulations of collision outcomes is summarized in Chapter 4. Each i-class erosion flux ε_i was modeled by Eq. 9.19. The characteristic concentration C_ε was replaced by the product $f_{bi} C_\varepsilon$, where f_{bi} is the fraction of the i-diameter particles in the bed. Similar to the analysis given after Eq. 9.19, $C_\varepsilon = 2.39 \times 10^{-4}$ kg m^{-3} was selected and f_{bi} values were optimized such that the time-variation of the overall erosion flux was in agreement with kaolinite erosion test data of Parchure [1984] analyzed by Letter [2009]. The deposition flux by class was taken as the product of the class settling velocity w_{si} and class concentration C_i. The settling velocity w_{si}, the stress τ_{cd1} and the exponent ξ_e in Eq. 9.83 were obtained from values for kaolinite reported in Mehta and Lott [1987] and in Table 9.7.

Erosion and deposition were driven by bed shear stress τ_b represented as the sum of a mean value $\bar{\tau}_b$ and a turbulent fluctuation τ_b'. As described in Chapter 7, Eq. 7.164 was used to generate the probability density function (pdf) of τ'_b from the corresponding normal (Gaussian) pdf of the velocity fluctuation u'_b.

Modeling requires tracking at each time-step every class diameter d_i, density ρ_i, shear strength τ_{si} and suspended sediment concentration C_i This concentration was obtained from mass balance for each class after accounting for mass gain and loss due to aggregation processes, erosion and deposition. The total concentration $C(z,t)$ at every modeled elevation z and time t was taken as the sum of all respective C_i values. For the present analysis the (one-dimensional vertical) model had five cells in the vertical covering the water depth. The final outcome was the variation of depth–mean concentration $C(t)$ in the flume as the experiment proceeded. The simulation began with assumed distributions of bed and suspended sediment particle sizes and concentrations by class. Computations showed that in simulations lasting for hours these initial conditions did not influence the outcomes in any significant way [Letter, 2009].

In each of the four deposition runs, kaolinite was initially suspended at a concentration of 1 kg m^{-3}, and then permitted to deposit at a constant bed shear stress $\bar{\tau}_b$. In Fig. 9.44, simulated depth–mean concentration curves are plotted along with the respective data (points) for experimental runs at $\bar{\tau}_b$ = 0.25, 0.40, 0.60 and 0.85 Pa. In each case the concentration fell and approached a residual value which increased with the bed shear stress. For illustration of sediment exchange, zones for the lowest bed shear stress of 0.25 Pa are shown in Fig. 9.45 using the following values of the parameters in Table 9.7: d_1 = 0.1 μm, τ_{cd1} = 0.03 Pa, ξ_e = 0.8, B_F = 1,800, ρ_w = 1,000 kg m^{-3}, ρ_s = 2,650 kg m^{-3} and D_f = 2.2. The selected range of diameters was 0.1 to 1,000 μm, without any effort to segregate well-formed flocs from other particles. The diameter 1,000 μm was meant to represent a large macrofloc.

The deposition, erosion and no-exchange zones in Fig. 9.45 highlight the fact that in general the overall trend of deposition in the flume is a composite of all three exchange modes. Deposition can occur over a size range of about 1 to 9 μm, whereas above 9 μm there can be erosion only. If $\bar{\tau}_b$ were increased the deposition zone would decrease rapidly and the zones of erosion as well as no-exchange would expand. This is seen in

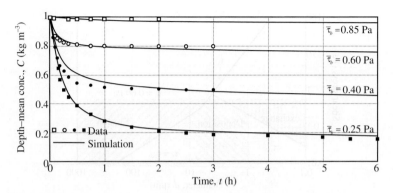

Figure 9.44. Measured (points) and simulated (curves) suspended sediment concentration of kaolinite in four deposition tests at different mean bed shear stresses (from Letter and Mehta [2011]).

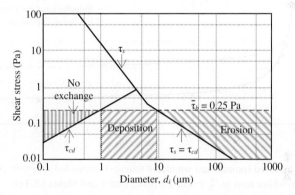

Figure 9.45. Bed exchange zones for shear stress $\overline{\tau}_b = 0.25$ Pa (from Letter and Mehta [2011]).

Fig. 9.46; at 0.60 Pa the size range for deposition has shrunk to 4–5 μm, whereas erosion is prevalent over a wide size range. Above about 0.9 Pa no deposition would occur. However, while Figs. 9.45 and 9.46 identify the exchange zones, they do not indicate the magnitudes of erosion and deposition fluxes controlling the net rate of decrease of concentration, which were calculated from mass balance by class and its cumulative value every time step.

In Fig. 9.44 the simulated overall concentration mimics the data and approaches its expected residual value in each run. Referring to the run

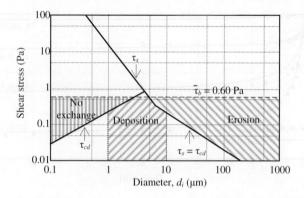

Figure 9.46. Bed exchange zones for shear stress $\overline{\tau}_b$ = 0.60 Pa (from Letter and Mehta [2011]).

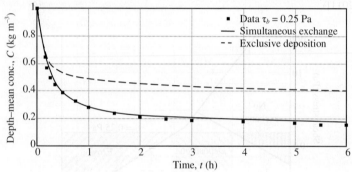

Figure 9.47. Measured (points) and simulated (curves) suspended sediment concentration of kaolinite deposition tests at $\overline{\tau}_b$ = 0.25 Pa (from Letter and Mehta [2011]).

in Fig. 9.47 corresponding to Fig. 9.45 ($\overline{\tau}_b$ = 0.25 Pa), the residual concentration is the outcome of equilibrium between the total erosion and deposition fluxes, with contributions from the selected classes. This is not the case when the exclusive paradigm based solely on deposition is examined. Even though both paradigms can predict the trend of decreasing concentration, the underpinning difference between two mechanisms is pronounced because in the exclusive case continued deposition results in finer and finer suspended sediment while no "new" sediment is added from the bed. An outcome is that aggregation slows down and eventually ceases to produce depositable flocs. The residual concentration represents a steady-state condition consisting of non-depositable very fine suspended particles. In simultaneous exchange mixing of the suspended matter with eroded material brings about more rapid aggregation than in the exclusive case, as manifested by the higher rate of reduction in concentration to its final, lower value, and agreement with measurements. In other words exchange makes deposition more rapid due to the efficiency with which depositable flocs are formed. Furthermore, the residual concentration accounts for non-depositable particles as well as those added and removed by erosion and deposition, respectively.

9.7.3 *Effects of probabilistic representation*

Probabilistic representation (PR) of the bed shear stress implies that as it fluctuates from instant to instant relative to its mean value, the bed exchange zones also change. Thus the description of mean–value representation (MVR) in Fig. 9.42 (as well as Figs. 9.45 and 9.46) must be revised with respect to the effect of time-varying turbulent bed shear stress. As an illustration, in Fig. 9.48 the mean shear stress is 0.25 Pa and the upper and lower boundaries represent ± 1 standard deviation (σ_τ) relative to the mean value. The critical shear stress for deposition τ_{cdi} and the shear strength τ_{si} are taken as mean values by class. Probabilistic representation of the shear stress introduces variability in the boundaries of the bed exchange zones, making them "fuzzy". The variability has the same sense as when the mean value is varied, *e.g.*, between 0.25 Pa (Fig. 9.45) and 0.60 Pa (Fig. 9.46). Thus, while the MVR bed exchange zones remain fixed, in PR bed shear stress fluctuations induce variability in the area of each zone between a minimum and a maximum value. If the shear stress range was increased to, say, ± 3 standard deviations, the effects of variability in τ_b on the bed exchange zones would become more pronounced because the boundaries would be fuzzier. In general such a probabilistic scenario is more realistic than the assignment of fixed values of the governing erosion–deposition variables.

Ultimately, all three variables, τ_b, τ_{cdi} and τ_{si}, must be represented in terms of their pdfs. The latter two can be conveniently taken as normally distributed about their respective mean values. We will consider their collective effect in two cases. The first will be for the variation of

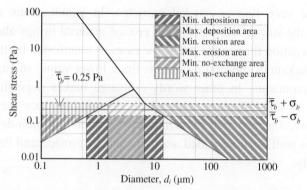

Figure 9.48. Minimum and maximum bed exchange zones in probabilistic representation of the bed shear stress τ_b (Pa) = 0.25 ± 1σ_τ (from Letter and Mehta [2011]).

suspended sediment concentration in the erosion–dilution test of Parchure and Mehta [1985]. The second will be for the likely evolution of suspended particle size distribution relying on the field data of Sanford and Halka [1993].

The dilution test using kaolinite carried out in the CRAF by Parchure and Mehta [1985], and briefly mentioned in Chapter 7, illustrates the basic difference between MVR and PR mechanics. In Fig. 9.49 concentrations from Fig. 7.37 are reproduced until the end of the experiment. Beginning at hour–120 when the concentration was 3.85 kg m^{-3} (5 hours after the start of the experiment), dilution was initiated by extraction of the suspended matter (and synchronous replacement by clear water). In the first 2 hours the simulations closely follow the data and conform to the pure dilution law (Eq. 7.150). During the remaining 2 hours, MVR continues to follow the same law, and after the end of the extraction phase the concentration remains constant at 0.036 kg m^{-3}. In contrast, PR generates a small erosion flux manifested by the slight positive deviation of the (dashed) curve from Eq. 7.150. More importantly, PR indicates erosion after dilution, increasing the concentration to 0.1 kg m^{-3} at hour–148 in agreement with the data. Although the erosion flux between hour–124 and hour–148 is low compared to the flux at the inception of the test, it underscores the importance of probabilistic treatment over the mean–value approach which produced no erosion.

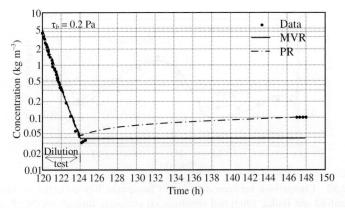

Figure 9.49. Dilution test and simulations.

9.7.4 *Chesapeake Bay data*

Sanford and Halka [1993] made field deployments to estimate the erosion and deposition of placed dredged material at selected disposal sites in the Chesapeake Bay. Shear stresses were estimated and suspended sediment concentrations calculated over single semi-diurnal tidal cycles. Although suspended floc size distributions were not reported, it is instructive to examine the likely outcomes based on the MVR and PR treatments.

In Fig. 9.50, tidal variation of the mean–value bed shear stress and the depth–mean concentration derived from measurements are plotted for a test carried out on January 5, 1989. During this 12-hour period the water depth varied between 3.4 to 4.1 m. Also plotted are the simulated concentration time-series based on MVR and PR. In PR the three variables τ_b, τ_{cdi} and τ_{si} were represented by their respective pdfs. The same values of the parameters for erosion and deposition were used in both cases: $d_1 = 0.1$ μm, $\tau_{cd1} = 0.02$ Pa, $\xi_e = 0.5$, $B_F = 100$, $\rho_w = 1,025$ kg m^{-3}, $\rho_s = 2,650$ kg m^{-3} and $D_f = 2.2$ [Letter, 2009]. No effort was made to manipulate them to improve agreement with measurements. This is one reason why neither approach can be judged to be better than the other. In addition, complexities in the natural processes and use of one-dimensional vertical modeling are likely causes of discrepancies between

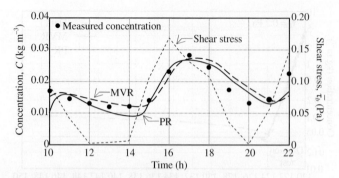

Figure 9.50. Comparison between data from Chesapeake Bay collected on January 5, 1989 [Sanford and Halka, 1993] and simultaneous exchange simulations (MVR and PR) (from Letter [2009]).

measurements and simulations. Overall the simulations are able to reproduce the tidally oscillating behavior of measured concentration, which permits us to tentatively examine the floc size spectra.

As noted by Lick [1982], a shift in the size spectrum of suspended flocs during erosion is a characteristic dynamic feature of cohesive sediment transport. In their deposition tests, Lau and Krishnappan [1994] recorded a shift towards larger flocs as the concentration decreased and approached a residual value. In general, although concentration reduction occurred at all sizes, there was a greater reduction in the mass of larger flocs due to deposition. Since smaller flocs did not deposit rapidly, their loss was mainly due to collisional aggregation which caused them to be captured by the larger flocs.

Simulated size spectra relying on the Chesapeake Bay data are shown in Fig. 9.51 in terms of the volume concentration C_v (ppm) plotted against diameter d_i (μm). Starting with an initial hypothetical distribution of particle size at 10 hours, a shift towards larger flocs is achieved at 17 hours. Note that these curves do not correspond to the distribution of floc mass (not plotted) because large flocs tend to be less dense than smaller ones.

The concentration of particles smaller than about 0.05 μm declined while the concentration of those greater than about 10 μm rose. The size

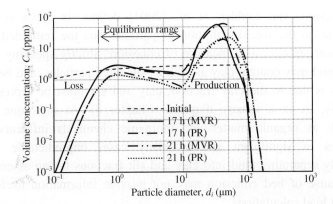

Figure 9.51. Simulated evolution of Chesapeake Bay floc size distribution on January 5, 1989 starting with an assumed initial distribution. Both simulations (MVR and PR) include simultaneous exchange (from Letter [2009]).

range of 0.05 to 10 μm was comparatively unaffected. This suggests that it is an equilibrium range through which the particles transited as the volume concentration of the larger particles increased and smaller ones decreased. The computations showed that the rate of change of the spectrum was high in the beginning when the overall mass concentration was high and aggregation rapid. Between 17 and 21 hours concentration at all sizes declined as aggregation slowed even as deposition continued. The difference between MVR and PR treatments is evident; since PR covers a wider range of sizes the spectrum is more smeared than MVR. Macroflocs on the order of 200 μm are produced when probabilistic variability in the critical shear stress for deposition and floc shear strength is taken into account.

9.8 Erosion Behavior of Sand-Mud (Clay) Mixtures

9.8.1 *Erosion modes*

Marine sediment occurring as pure clay, silt or sand is rare. Even minute quantities of one material within another, such as clay particles in sand, can alter the fabric and the cohesive strength of the bed. Sorbed and dissolved inorganic and organic chemicals can have a similar effect on bed cohesion. van Ledden *et al.* [2004] define six types of beds depending on the size distribution of particles and the remolded (*e.g.,* vane) shear strength as a measure of cohesion. Strictly speaking, since natural mud may contain clay as well as silt-sized particles, and because cohesion is due to the clay fraction, clay and silt must be represented separately, as in a ternary diagram (Chapter 3). At the same time, inasmuch as organic matter and inorganic chemicals influence bed properties in significant and in many cases unpredictable ways, physically separating mud into clay and silt fractions for an assessment of the cause of bed cohesion may not provide information useful for sediment load calculations.

The erosion behavior of mixtures of sand and clay-sized sediment remains poorly understood from experimental as well as interpretational points of view [Flemming and Delafontaine, 2000]. A sand grain surrounded by cohesive flocs at the bed surface can result in local scour of flocs due to high shear stresses surrounding the grain. In contrast, flocs below the surface are protected from scour. This variable behavior of fine particles in sand has significant implications for benthic life. It was noted by de Jonge and van den Bergs [1987] that sand-silt mixtures induced selective resuspension of diatom species. While silt by itself introduced no significant preference, when sand was present the species attached to sand fell into two groups, with the better attached species more resistant to detachment and eroding at a higher current velocity than the other group.

With the addition of mud or clay to sand, three modes of incipient erosion can be identified [Whitehouse *et al.,* 2000a]: (1) mud erosion as suspended load; (2) mixed mud-sand erosion; and (3) sand erosion mainly as bed load. The size of the eroding flocs may decrease with increasing amounts of sand. Raudkivi [1998] observed that mud takes over as the erosion controlling sediment once it exceeds 5–10% by weight. This is due to the space-filling effect of clay particles within the sand matrix as described in the next section.

One can consider the effects of changing mode of erosion either with respect to addition of sand to mud (or clay) or mud to sand. In Fig. 9.52

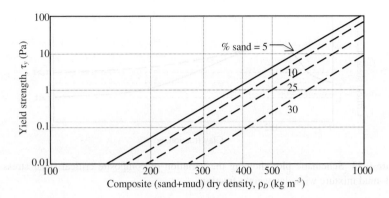

Figure 9.52. Effect of adding sand on the yield strength of mud/sand mixture (from Migniot [1968]).

[Migniot, 1968], we observe that the addition of sand to mud even in small quantities, *e.g.*, 5% by weight, measurably decreased the yield strength of the mixture. Increasing loss of the yield strength of the bed with addition of sand up to 30% must be viewed in conjunction with the general observation that flocculated sediment can erode more easily than, say, medium sand. However, this need not be so if the sand is very fine and the fine-grained fraction is highly cohesive.

The schematic plot in Fig. 9.53 highlights the complexity of the dependence of the critical shear stress τ_c of the mixture on mud weight fraction ψ. The addition of mud to pure sand initially increases τ_c from τ_{co} to its peak τ_{cmax} at a value of ψ estimated to be on the order of 0.2 [Whitehouse *et al.*, 2000a; Lyle and Smerdon, 1965]. As ψ increases further the critical shear stress decreases until at $\psi = 1$ (pure mud) it represents the erosion shear strength τ_s. Depending on sand size and mud cohesion, τ_s may be less than or greater than τ_{co}.

In as much as the definition of a composite critical shear stress of the sand-clay mixture is somewhat notional, erosion fluxes of coarse and fine fractions are often modeled as independent processes with book-keeping incorporated in the model code to track the eroded (and deposited) masses. Although this approach is expedient and perhaps necessary, it is physically untenable.

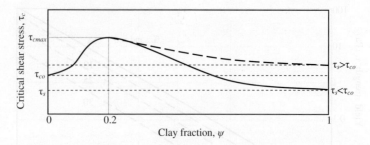

Figure 9.53. Schematic plot showing the variation of composite critical shear stress of sand–mud mixture with clay weight fraction.

9.8.2 *Effect of clay on sand*

It appears that the effect of adding sand to clay on bed erosion has received more engineering attention than the effect of adding clay to sand. As we noted in reference to Fig. 9.53, laboratory data show that the critical shear stress τ_c for erosion of quartz sand (not the mixture) varies non-linearly with increasing weight fraction ψ of clay added to sand. The role of clay at low values of ψ is of particular interest because, up to values of about 0.15 to 0.20, measurable effects of clay particles on sand erosion have been reported [Nalluri and Alvarez, 1992; Torfs, 1995; Huygens and Verhoeven, 1996; Panagiotopoulos *et al.*, 1997; Raudkivi, 1998; Sharif, 2002]. In some experiments the critical shear stress τ_{co} of pure sand ($\psi = 0$) has been found to decrease to τ_{cm} at $\psi = \psi_m$, followed by an increase (Fig. 9.54). At a $\psi = \psi_r$ the τ_c curve intersects the line of constant τ_{co} such that when $\psi > \psi_r$ ($>\psi_m$), $\tau_c > \tau_{co}$. Alternatively, τ_c may initially remain equal to τ_{co}, which would imply that the clay may have a retarding effect on the subsequent rise in τ_c with ψ.

Flume-based observations of Torfs [1995] related to Fig 9.54 are given in Table 9.8. In these tests, erosion of 0.23 mm (median diameter) sand was examined by adding a kaolinite and, in a separate test, a montmorillonite. Due to the addition of kaolinite, τ_c apparently decreased from $\tau_{co} = 0.35$ Pa to $\tau_{cm} = 0.31$ Pa, amounting to an 11% drop at the minimum value of $\psi = \psi_m = 0.02$. The stress τ_c then increased, surpassed τ_{co} at the crossover value $\psi_r = 0.04$, and continued to increase with

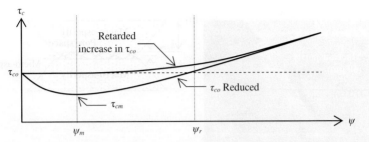

Figure 9.54. Effect of clay fraction on critical shear stress for erosion of sand.

Table 9.8. Critical stress parameters from Torfs [1995].

Sand, d_{50} (mm)	τ_{co} (Pa)	Clay	ψ_m	τ_{cm} (Pa)	$1 - \dfrac{\tau_{cm}}{\tau_{co}}$ (%)	ψ_r
0.23	0.35	Kaolinite	0.02	0.31	11	0.04
0.23	0.35	Montmorillonite	0.03	0.33	6	0.07

further increase in ψ. The respective values for montmorillonite were τ_{cm} = 0.33 Pa (6% drop), ψ_m = 0.04 and ψ_r = 0.07.

We may examine this trend of the $\tau_c(\psi)$ curve under the premise that interstitial clay particles alter grain-to-grain friction and interlocking due to micro-asperities at the surface of (quartz) sand particles, and that the nature of lubrication depends on ψ. Disregarding the likely fractal succession of surface asperities, we may consider that the height of asperity typically ranges from $k_{s.macro}$ of $O(10^2)$ μm for macro-asperities to $k_{s.micro}$ of $O(10^{-1})$ μm for micro-asperities on top of macro-asperities (Figs. 9.55a, b).

With reference to the schematic drawing of Fig. 9.56, we will consider the process of reduction in the critical stress in the x-z plane, treating sand grains as cylinders of unit thickness in the third dimension.

Consider the upper sand grain to be set in motion at a small velocity U by turbulence-induced instantaneous specific load σ_{v1} at the exposed bed plane. Assume that the short duration of the turbulent fluctuation over which this grain-induced load acts is long enough for the forces

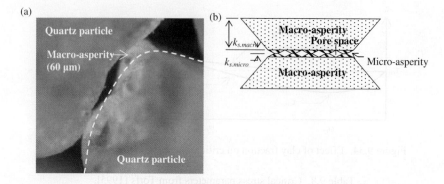

Figure 9.55. (a) 60 μm macro-asperities of a 1.2-mm (1,200-μm) diameter quartz sand particle; (b) sketch showing macro- and micro-asperities (from Barry *et al.* [2006]).

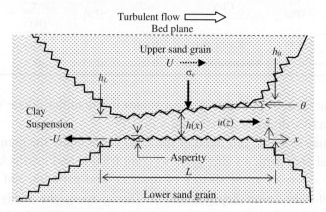

Figure 9.56. Two-dimensional representation of the flow of a fine-sediment suspension as a pore fluid (from Barry *et al.* [2006]).

to be in quasi-equilibrium. Due to generally non-parallel orientation of the upper and the lower grain surfaces separated by a viscous clay suspension of small thickness, *i.e.*, $h(x)/L \ll 1$, the resulting pressure gradient over the pore channel of length L will cause the suspension to flow momentarily. If we choose the upper grain to be stationary and the lower one to translate at a velocity $-U$, for small values of the angle θ (denoting deviation from parallel orientation) the problem becomes analogous to one of slider-bearing type lubrication due to the viscosity of suspension. This effect, which is treated in standard texts on fluid flow (*e.g.*, Liggett [1994]), yields the following Couette–Poiseuille flow velocity profile in the pore fluid

$$u(z) = -U + \left(\frac{4U}{h} + \frac{6Q}{h^2} \right) z + \left(\frac{3U}{h^2} + \frac{6Q}{h^3} \right) z^2 \qquad (9.86)$$

where Q is the discharge considered steady over a sufficiently short duration. In the derivation of Eq. 9.86 it is assumed that grain asperities are of negligible height, *i.e.*, the grains have effectively smooth surfaces. Accordingly, the total shear force (per unit thickness) is obtained from

$$F_{d1} = \int_0^L \eta \frac{du}{dz} \bigg|_{z=0} dx = \frac{\eta}{(k_u - 1)h_L} \left[4 \ln k_u - 6 \left(\frac{k_u - 1}{k_u + 1} \right) \right] UL = K_U UL \quad (9.87)$$

where η is the suspension viscosity, $k_u = h_L/h_0$ and K_U is made up of constants η, h_0 and h_L. Thus the coefficient of kinetic friction f_k is obtained from

$$f_k = \frac{F_{d1}}{\sigma_{v1}} = K_U \left(\frac{\eta U}{\sigma_{v1}} \right) \qquad (9.88)$$

This expression is known as Petroff's law for thick-film lubrication (dashed line in Fig. 9.57).

The dimensionless term $\eta U/\sigma_{v1}$ is a measure of the viscous force relative to specific load. With decreasing value of this term, f_k decreases until the distance h of separation between grains approaches $O(10^{-1} - 10^0)$ μm, the nominal size of clay particle sheets. Then, f_k increases with a further decrease in $\eta U/\sigma_{v1}$ in the zone of thin-film lubrication. Finally, there is a rapid rise in f_k due to boundary lubrication as the film becomes so thin that its properties are no longer those of the bulk. Lubrication is taken over by abrasion/interlocking due to clay particles and sand asperities, $O(10^{-1})$ μm. In accordance with this quasi-hydrodynamic behavior we may extend Eq. 9.88 to yield the so-called Stribeck function

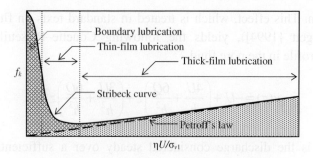

Figure 9.57. Petroff's law and Stribeck function.

$$f_k = K_U \left(\frac{\eta U}{\sigma_{vl}} \right) + K_U' e^{-K_U'' \left(\frac{\eta U}{\sigma_{vl}} \right)}$$

(9.89)

in which K_U' and K_U'' characterize thin-film and boundary layer lubrication respectively [Szeri, 1998]. Equation 9.88 is recovered when $K_U' = 0$.

In Eq. 9.89 we may take the viscosity η of clay suspension to be $O(10^{-2})$ Pa.s, and the instantaneous specific load σ_{vl} of $O(10^{-2}–10^{-1})$ N m^{-1} s^{-1} based on measurements of the hydrodynamic lift fluctuations due to flow-induced turbulence over a sand grain (Chapter 6). The drag coefficient is obtained from $C_D = \tau_c / \rho u_c^2$, where ρ is the (clay) suspension density (nominally 1,000 kg m^{-3}) and u_c is the critical velocity for sand grain erosion, which we will assume remains practically constant. Since the behavior of C_D can be considered to reflect the behavior of f_k, we recognize the analogy between the behaviors of f_k versus $\eta U / \sigma_{vl}$, and τ_c versus ψ. In the latter case we can infer that the retarded-increase curve (Fig. 9.54) is merely a special case of the other, more general, stress reduction curve. Thus the equivalent Stribeck function for the critical shear stress can be stated as

$$\tau_c = \kappa \psi + \tau_{co} e^{-\chi \psi}$$

(9.90)

in which the coefficients κ and χ are specific to the sand/clay mixture. They are determined by clay particle size and shape, distribution of electrical charge on its surface, sorbed impurities in the form of stray ions and the ionic composition of fluid.

In a laboratory flume, beds of 0.83 mm quartz sand mixed with 1 μm kaolinite were subjected to increasing water velocity until incipient sand grain movement was achieved [Barry, 2003]. In one series of tests water was fresh, whereas in a second series the clay in pore water was pre-equilibrated with a $3°/_{oo}$ solution of sodium chloride. In a third set, the same sand was mixed with a clay mixture (50% kaolinite, 35%, attapulgite and 15% bentonite). Results from these tests in Fig. 9.58 mimic the trend of Eq. 9.90. The parameters ψ_m, ψ_r and τ_{cm} vary with clay type, with ψ_m ranging from about 0.04 to 0.06 and ψ_r from 0.06 to ~0.15. The maximum reduction in τ_{cm} occurred when kaolinite in fresh water was used (0.70 Pa to 0.26 Pa, or 63%). This amount of stress reduction appears to be unusually high and suggests the need for further experimentation to ascertain whether the test conditions were exceptional.

The lubricating effect of kaolinite is determined by the orientation of individual (plate-like) particles. As we noted in Chapter 4, in freshwater the particles are stacked like decks of cards (face-to-face orientation), whereas in saltwater their fractal arrangement is more like a card-house (edge-to-face orientation). It can be shown that if sand grains are hexagonally close-packed, within the fresh pore water of the resulting face-centered unit cube there would be around 20 layers of parallel-

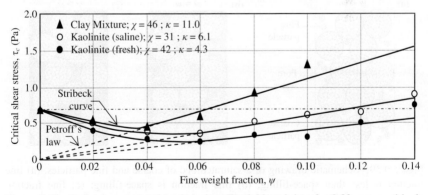

Figure 9.58. Stribeck curve analogy based on Eq. 9.90 and data for 0.83 mm sand beds (from Barry *et al.* [2006]).

oriented dispersed clay particles which would provide lubrication between sand grains at $\psi = \psi_m$. In saltwater at the same clay weight fraction the flocs would contain semi-randomly oriented particles that are less effective in reducing the critical shear stress. An important general observation is that ψ_m marks the transition between thin-film and thick-film lubrication.

The fraction ψ_r appears to be the pore space-filling volume fraction. In a hexagonally packed cubic matrix in which the net number of grains is 4, they touch each other along the diagonal. Hence the cube edge length is $2\sqrt{2}r_p$, where r_p is the grain radius. Consider a bed of 0.83 mm diameter sand in water of density $\rho = 1{,}000$ kg m^{-3}. We will assume that the inter-granular spacing at $\psi = \psi_m$ is equal to the height of the asperities. Selecting a representative value $\psi_m = 0.06$ and the asperity height to be 20 μm, the corresponding spherical shell volume of 0.182 mm^3 can be easily shown to be equivalent to clay porosity $n_r = 0.50$. This approximate calculation suggests that with this value of n_r at $\psi_r = 0.13$, the fraction of pore space occupied by the clay matrix would be 0.96. As this value is close to unity, it essentially amounts to space-filling by clay (Fig. 9.59). It follows that the increase in τ_c above τ_{co} with a further increase in ψ would mean that the effect of the clay fraction increasingly influences the erodibility of sand for $\psi > \psi_r$.

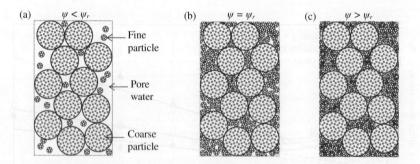

Figure 9.59. Schematic drawing of a saturated bed of coarse and fine particles: (a) fine fraction is less than space-filling; (b) fine fraction is space-filling; (c) fine fraction exceeds its space-filling value (adapted from Torfs *et al.* [2001]).

Once erosion of a mixture commences, sorting of bed sediment varies with mixture makeup characterized by size and type of matter. Law *et al.* [2008] eroded sediment in cores from Golfe du Lion (France) layer by layer employing the Gust Microcosm [Gust and Muller, 1997] mentioned earlier. It is a circular apparatus consisting of housing with a rotating shear plate, a removable lid, and water input and output connections. A critically important feature of the apparatus is that, by controlling both the rotation rate of the shear plate and the rate at which water was pumped through the device, a uniform shear stress can be applied across the sediment surface.

The UMCES-Gust Erosion Microcosm System (U-GEMS) is an advanced version of the Gust Microcosm. Improvements include, among other features, a dual core design for test replication, use of a positive displacement pump instead of two centrifugal pumps in the original device, and an integrated semi-automated control and analysis software [Dickhudt *et al.*, 2011].

Based on erosion tests in U-GEMS, size-specific "mobility plots" of the dispersed sediment were developed by dividing the proportion of each particle size in suspension (after peptization) at each shear stress by its proportion in the sediment before erosion. If all sizes that make up the bottom sediment are eroded equally from the bed, then mobility would be 1 for all sizes. Values < 1 indicate that the suspended sediment is enriched in the size class, and values > 1 indicate that the size class is enriched in the bed. The plots showed that in non-cohesive sandy silt beds, fine particles (clays and fine silts) were eroded preferentially at low shear stresses. With increasing bottom stress, progressively larger particles were eroded. In cohesive silts, preferential erosion of the finer sizes no longer occurred, with all sizes up to medium silts eroding at approximately the same rate. Effectively, sandy silt could be winnowed of its fine fraction during erosion but cohesive silts could not be sorted.

9.9 Stable Channels

9.9.1 Critical shear stress

In muddy natural channels (e.g., Fig. 9.60) that are stable over periods of long durations, the net rate of erosion or deposition is characteristically minor compared to the gross sediment load. It follows that practically stable channels with low dredging maintenance can be designed, provided the critical shear stress for erosion (or erosion shear strength) is known.

For stable channels with cohesionless beds the critical shear stress is a well-defined quantity (Chapter 6). In contrast the dependence of the erosion flux of cohesive sediment on bed density brings about the need for a careful interpretation of data on erosion in order to identify a characteristic critical shear stress.

In general, the cohesive erosion flux ε increases with the bed shear stress τ_b in a non-linear way (Eq. 9.28). In Fig. 9.61a the typical variation of ε with τ_b is divided into two assumed linear segments. At low shear stresses the erosion flux is minor because the rate of increase in bed shear strength with depth is high (Fig. 9.61b). At higher shear stresses the rate of increase in the shear strength is slower and the flux increases more rapidly with shear stress. The two linear segments have slopes ε_{AP1} and ε_{AP2}. Three variables have been defined along the shear stress axis—the surface shear strength τ_{so}, a characteristic critical shear stress τ_{cr} obtained by extrapolating the line of slope ε_{AP2}, and a shear stress τ_c marking the point of intersection of the two straight lines. The last value corresponds to an erosion flux ε_{ch} and scour depth z_{ch}. This depth delineates the transition between a high-porosity, high-order of aggregation upper bed and a low-porosity, low-order of aggregation lower bed. Since ε_{ch} is usually small compared to the erosion flux at higher shear stresses, for channels in which there is to be practically no significant bed erosion, either the shear stress τ_{cr} or τ_c can serve as the design value for the threshold of erosion. The latter is preferable

(a)

(b)

Figure 9.60. Tidal sloughs at mudflats: (a) Western coast of South Korea (courtesy Korea Ocean Research and Development Institute, Ansan, photograph taken in 2007); (b) Turnagain Arm estuary, Alaska (2011). The erosion scarp is about 0.6 m high.

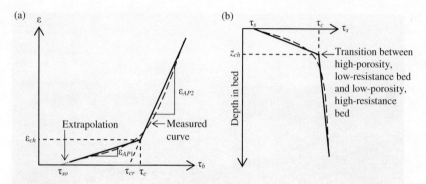

Figure 9.61. Schematic drawing of: (a) Erosion flux against bed shear stress with Type I and Type II segments; (b) bed shear strength profile.

because it yields a more conservative estimate than the former in regard to bed stability. Figure 9.62 is an example of a two-segment erosion curve obtained in the laboratory (Test no. 13 in Table 9.9).

In order to differentiate between two-segment erosion from a single segment trend characterized by τ_{cr}, we will refer to the former as consisting of Type I and Type II linear segments and the latter as consisting of Type II only. The small erosion flux in the Type I range of bed shear stress (τ_b) is of minor consequence to stable channel performance (but may have substantial significance relative to benthic

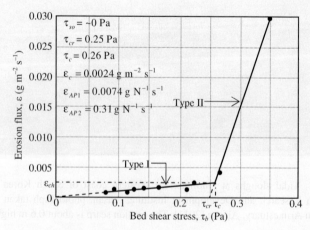

Figure 9.62. Erosion flux versus bed shear stress data of Gularte [1978].

Table 9.9. Summary of selected laboratory erosion tests.

Test no.	Source	Sediment	Water content (%)	Eroding fluid salinity	Pore fluid	Fluid temp. (°C)
1	Partheniades [1962]	San Francisco Bay (Mare Island) mud	110	33	Native (bay) water	–
2			112	33	Native (bay) water	–
3	Espey [1963]	Taylor marl	47	~0	~0	–
4	Christensen and Das [1974]	Kaolinite	35	~0	~0	13
5		Grundite	33	~0	~0	13
6	Raudkivi and Hutchison [1974]	Kaolinite	51	0.01 M NaNO$_3$	0.01 M NaNO$_3$	21
7						32
8			28	0	0.005 N NaCl, SAR = 1.1	–
9	Arulanandan et al. [1975]	Yolo loam	28	0	0.005 N NaCl, SAR = 1.6	–
10			30	0	0.005 N NaCl, SAR = 10.7	–
11	Gularte et al. [1977]	Thames River (CN)	~170	31	Native water	–
12	Gularte [1978]	Grundite	60	2.5	2.5	–
13				10	10	–
14	Ariathurai and Arulanandan [1978]	Illite, silica flour	36	0	Total salt^a conc. 20 meq L^{-1}	9.5
15						18
16						23
17						42
18	Lee et al. [2004]	Sheboygan River (WI)	50	0	–	–
19	Papanicolaou et al. [2007]	Union Flat Creek (WA)	55	0	pH = 8.2	17–18

^{a} No sodium chloride was present.
Sources: Hunt [1981]; Hunt and Mehta [1985].

colonization for example). Type II erosion is initiated when τ_b exceeds τ_{cr}. Ignoring Type I erosion, *i.e.*, taking $\varepsilon_{AP1} = 0$, and setting $\varepsilon_{AP\,2} = \varepsilon_{AP}$ yields Eq. 9.26. Examples of Type II erosion are shown in Fig. 9.58 (Test nos. 14 and 16 in Table 9.9). Table 9.10 gives characteristic parameters from nine out of the nineteen erosion tests in Table 9.8 that are Type I. The remaining tests yielded Type II erosion (Table 9.11) [Mehta and Partheniades, 1982].

The variation of erosion flux with bed shear stress in the laboratory depends on the manner in which the bed is prepared. In a flume, when an adequately dense sediment slurry is manually placed a uniform bed should result. Yet in reality it is difficult to eliminate Type I erosion (*e.g.*, Fig. 9.62) because a thin, very soft, surface layer of flocs is usually formed even when the remainder of the bed is harder. Causes of formation include osmosis due to differences between pore water and eroding water chemistry as well as elastic rebound or swelling of surface flocs especially if the bed has been compacted mechanically during preparation.

In contrast to a flume, in a rotating cylinder apparatus used in test nos. 3, 8, 9, 10, 14, 15, 16 and 17 (Table 9.9), the cylindrical bed hung in water consisted only of soil that did not slough. Any soft layer while the fluid is at rest would have fallen off the vertical surface. The outcome is

Table 9.10. Parameters for two-segment erosion.

Test no.	τ_{so} (Pa)	τ_{cr} (Pa)	τ_c (Pa)	ε_c (g m^{-2} s^{-1})	ε_{AP1} (g N^{-1} s^{-1})	ε_{AP2} (g N^{-1} s^{-1})
1	0.072	0.34	0.50	0.0011	0.0023	0.0065
2	–	1.24	1.20	0.0020	0.0018	0.011
3	14.9[a]	60.0[a]	77.3[a]	8.36[a]	0.13	0.51
4	–	0.47	0.58	0.056	0.094	0.55
5	–	0.76	0.85	0.0060	0.0043	0.063
11	0.078	0.27	0.47	0.022	0.056	0.11
12	0.014	0.061	0.062	0.0015	0.030	0.97
13	~0	0.25	0.26	0.0024	0.0074	0.31
18[b]	0.12	0.51	0.58	0.025	0.054	0.313

[a] Values as reported.
[b] Reported [Lee *et al.*, 2004] values of ε_c, ε_{AP1} and ε_{AP2} have been divided by 1,000.

Table 9.11. Parameters for single-segment erosion.

Test no.	τ_{cr} (Pa)	ε_{AP} (g N^{-1} s^{-1})
6	0.22	0.13
7	0.24	0.080
8	0.10	5.3
9	0.062	21
10	0.006	147
14	2.7	0.51
15	2.4	0.69
16	2.2	1.13
17	1.2	3.47
19	4.16	0.034

Type II erosion only (*e.g.*, Fig. 9.63). As described later, in test no. 18 an undisturbed core was eroded in a different type of apparatus.

A rotating cylinder apparatus (Simulator of Erosion Rate Function or *SERF* developed by Jiang *et al.* [2004]) shown in Fig. 9.64 was a modified version of a similar device of Arulanandan *et al.* [1975]. *SERF* had the advantage of employing small clay samples that typically require a high-flow-speed facility for erosion tests. The cylindrical sample had to be of strength sufficient to remain intact under its own weight without significant deformation. After carefully molding a sample of 76 mm diameter around a shaft-and-disc mandrel, it was suspended inside an acrylic outer cylinder of 102 mm diameter from a load-cell and

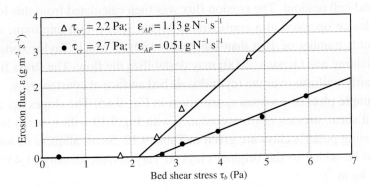

Figure 9.63. Erosion flux versus bed shear stress data of Ariathurai and Arulanandan [1978] for two clayey beds.

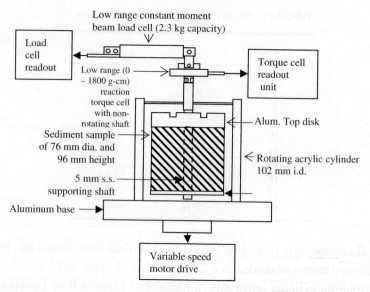

Figure 9.64. Schematic drawing of Simulator of Erosion Rate Function (SERF) (from Jiang *et al.* [2004]).

submerged in water. Thus water was confined to a 26 mm annular gap and acted as the eroding fluid. The outer cylinder was spun at a selected *rpm* (1,600 maximum), which yielded the desired shear stress on the sample from the torque measured by a strain-gage cell. After a suitable duration (up to 15 min), spinning was stopped and the loss of suspended sample mass due to erosion at the cylindrical surface was obtained from the load-cell readout. The erosion flux was then calculated from this loss. The device operated at a considerably higher *rpm* (~800 minimum at a Reynolds number of 10^5) than the *rpm* (~150) at which Taylor vortices [Schlichting and Gersten, 2000] may destabilize the flow. The stress field was fairly uniform over the (cylindrical) bed surface at constant *rpm*.

Sample plots of ε versus τ_b are given in Fig. 9.65. The slopes ε_{AP} and critical shear stresses τ_{cr} (in accordance with Eq. 9.19 with $\tau_s = \tau_{cr}$) from tests using potter's clays are given in Table 9.12. Small aliquots of water had to be added to the original clays to vary sample density (1,435 to 1,963 kg m^{-3}).

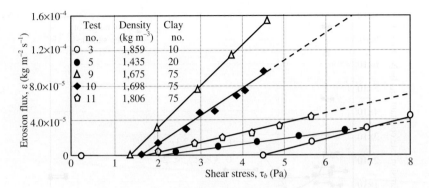

Figure 9.65. Example plots of erosion flux versus bed shear stress (from Jiang *et al.* [2004]).

Table 9.12. Erosion parameters obtained in the SERF.

Test no.	Clay number	Pore water salinity	Bed density (kg m^{-3})	Erosion parameters ε_{AP} (kg N^{-1} s^{-1})	τ_{cr} (Pa)
1	10	0	1,665	5.0×10^{-5}	3.1
2	10	0	1,710	1.9×10^{-5}	3.4
3	10	0	1,859	1.2×10^{-5}	4.5
4	10	0	1,928	1.0×10^{-5}	4.7
5	20	0	1,435	3.9×10^{-5}	1.9
6	20	0	1,537	1.2×10^{-5}	2.8
7	20	0	1,721	7.7×10^{-6}	3.0
8	20	0	1,905	7.1×10^{-6}	3.0
9	75	0	1,675	4.8×10^{-5}	1.4
10	75	0	1,698	3.2×10^{-5}	1.7
11	75	0	1,806	1.1×10^{-5}	1.8
12	75	0	1,963	9.9×10^{-6}	1.9
13	10	35	1,928	1.7×10^{-5}	0.6
14	20	35	1,894	4.4×10^{-5}	1.6
15	75	35	1,940	5.1×10^{-5}	3.3

For a given clay bed, as the density increases τ_{cr} increases while ε_{AP} decreases, as for example in test nos. 9, 10, 11 and 12 for clay no. 75 (Table 9.12) in freshwater as the eroding fluid. Erosion tests carried out by Arulanandan *et al.* [1980] in a flume (Fig. 9.66) as well as in a rotating cylinder apparatus (Fig. 9.67) showed similar trends of decreasing

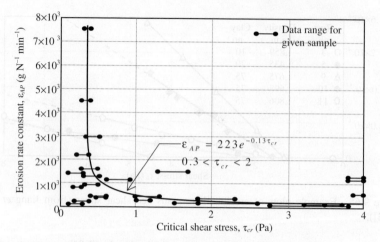

Figure 9.66. Erosion flux constant plotted against critical bed shear stress for undisturbed soils tested in a flume with distilled water as eroding fluid (adapted from Arulanandan *et al.* [1980]).

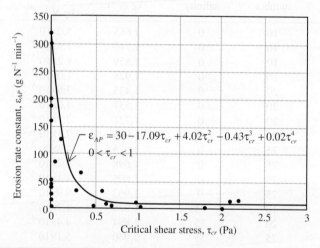

Figure 9.67. Erosion flux constant plotted against critical bed shear stress for remolded soils tested in a rotating cylinder apparatus with distilled water as eroding fluid (adapted from Arulanandan *et al.* [1980]).

ε_{AP} with increasing τ_{cr} for a large number of soil samples from locations in the US.

In regard to Eq. 9.26, laboratory erosion test data from several sources appear to generally conform to the relationship

$$\varepsilon_{AP} = \varepsilon_{APo} \, e^{-\chi_s \tau_s^{\lambda_s}} \qquad (9.91)$$

in which the coefficient χ_s was found to vary from 1.345 to 10.582[g] (when the unit of ε_{AP} is g N^{-1} s^{-1} and τ_s is in Pa), while the exponent λ_s varied within the narrow range of 0.252 to 0.386 with a mean value of 0.353. It was determined that $\varepsilon_{APo} = 200$ g N^{-1} s^{-1} was a reasonable value applicable to all data. Unresolved effects of bed properties and fluid chemistry are believed to be mainly manifested in χ_s. It is likely that the test geometry and the flow field had significant effects on ε_{AP} and τ_{cr} [Mehta and Parchure, 2000].[h]

With reference to test no. 18 in Tables 9.9 and 9.10, a class of open- as well as closed-conduit flumes has been used to measure the erosion of undisturbed sediment cores [McNeil *et al.*, 1996; Roberts *et al.*, 1998;

---

[g] As χ_s increases ε_{AP} decreases, *i.e.*, the erosion flux decreases. The value 1.345 corresponds to light-weight (low ρ_D) organic-rich sediment, whereas 10.582 is applicable to densely packed (high ρ_D) clayey beds. This experimental trend contradicts Eq. 9.30 in which, for analytical purposes ε_{AP} is assumed to vary linearly with ρ_D.

[h] The erosion shear strength of soft beds has also been expressed in other ways. For example, Dickhudt *et al.* [2009] state

$$\tau_s = \chi_{es} m_C^{\lambda_{es}} + \tau_{s0} \qquad (F9.6)$$

where m_C is the sediment dry mass concentration multiplied by unit bed height (kg m^{-3} m^{+1} $\equiv$ kg m^{-2}), χ_{es} and λ_{es} are site-specific coefficients, and τ_{s0} (Pa) is the erosion shear strength at the bed surface. Five sets of data, each from a different estuarine environment, correspond to the coefficient ranges in Table F9.1.

Table F9.1. Coefficients for Eq. F9.6.

Data set	a_{es}	b_{es}	τ_{s0} (Pa)
1	0.833	0.790	0.115
2	0.590	0.366	0.010
3	0.835	0.508	0.010
4	0.531	0.646	0.017
5	0.243	0.754	0.030

Source: Dickhudt *et al.* [2009].

The measured range of m_C was 0 to 2.5 kg m^{-2}.

Ravens and Gschwend, 1999; Briaud *et al.*, 2001; Trammel, 2004]. The core is extruded into the flume from the bottom, and as its top end erodes extrusion is continued by forcing the core upward at the lower end with a piston (Fig. 9.68). Thus an advantage of this method is that actual bed scour is measured, as opposed to bed erosion interpreted from the concentration of eroded sediment [Jones and Gailani, 2007]. Typically the core diameter is 7.62 cm corresponding to a standard Shelby tube as the coring device. A core of square or rectangular cross-section is also used. The core density profile reflects the effects of stratification due to composition, consolidation, erosion, deposition and biotic presence (Fig. 9.69). Use of an x-ray or gamma-ray transmission densitometer permits the measurement of core density profile without disturbing the sediment.

In some of the devices an automated imaging method with servo-control is used to force the eroding core upward at the desired rate. The open flume of Lee *et al.* [2004] uses laser light-scattering to illuminate the sediment-water interface (Fig. 9.70). A charge coupled device (CCD) camera records the image, and a detection algorithm identifies the interface as the edge of a white band on the image. A servo-control program determines the difference between the interface and the flume bottom and triggers a precision stepper-motor to extrude the core when the difference exceeds a small value, *e.g.*, 0.1 mm. An example of the relationship for a core from the Sheboygan River is shown in Fig. 9.71.

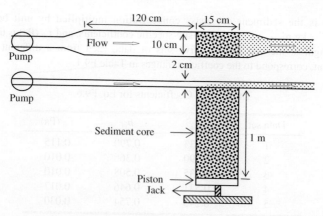

Figure 9.68. Schematic drawing of a closed-conduit flume (Sedflume) for measuring the erosion of undisturbed bottom-cored sediment (from McNeil *et al.* [1996]).

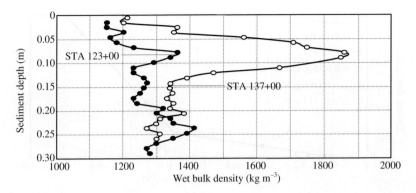

Figure 9.69. Density profiles from sediment cores collected in Sheboygan River, WI (adapted from Lee *et al.* [2004]).

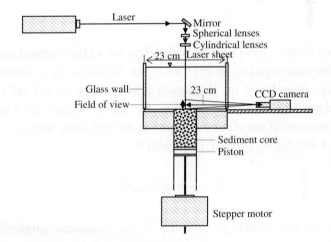

Figure 9.70. Essential features of the method for imaging the surface of eroding core (from Lee *et al.* [2004]).

9.9.2 *Channel cross-section*

For channels in sedimentary equilibrium, or "in regime", empirical equations for cross-sectional dimensions and current velocity have been developed from streams in the US, India and elsewhere. The regime concept, meant to provide a descriptive framework for these equations, is given in works on sediment transport (*e.g.,* Blench [1960]; Simon and

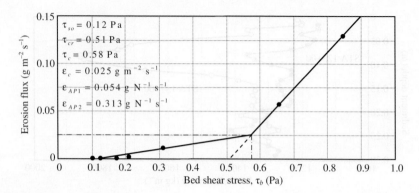

Figure 9.71. Variation of erosion flux with bed shear stress for a core from the Sheboygan River, WI. Note that the flux values in Lee *et al.* [2004] are 10^3-fold larger and may indicate a plotting error.

Albertson [1960]; Graf [1971]). With reference to the simplest case of a rectangular cross-section in Fig. 9.72, the following is a summary of regime equations based on Henderson [1966]. They do not refer to the critical shear stress because they are for stable streams in which, on average, the critical shear stress is equal to the bed shear stress.

Area A divided by depth h, i.e., width B:

$$B = \frac{A}{h} = 0.9 K_1 \sqrt{Q} \tag{9.92}$$

where Q is the discharge and K_1 is an empirical proportionality coefficient.

Hydraulic radius R_h:

$$R_h = \frac{A}{P_w} \tag{9.93}$$

where P_w is the wetted perimeter:

$$P_w = B + 2h \tag{9.94}$$

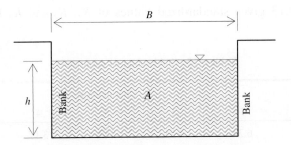

Figure 9.72. Channel of rectangular cross-section.

Depth h ($R_h < 2.13$ m):

$$h = 1.21 K_2 Q^{0.36} \qquad (9.95)$$

Depth h ($R_h \geq 2.13$ m):

$$h = 0.61 + 0.93 K_2 Q^{0.36} \qquad (9.96)$$

where K_2 is an empirical discharge coefficient. The mean velocity u_m is obtained from the Chézy discharge equation (Eq. 2.32) using

$$\frac{C_z^2}{g} = \frac{u_m^2}{ghS_f} = K_3 \left(\frac{u_m B}{v} \right)^{0.37} \qquad (9.97)$$

where S_f is the slope of the energy grade line (bed slope in uniform flow), v is the kinematic viscosity of water and K_3 is another discharge coefficient.

The coefficients K_1, K_2 and K_3 vary with the bed and the banks. Five channel types are as follows [Simon and Albertson, 1960]:

(1) Sand bed and banks
(2) Sand bed and cohesive banks
(3) Cohesive bed and banks
(4) Coarse sediment (not limited to sand)
(5) Same as 2, but with heavy sediment load (2 to 8 kg m^{-3})

Table 9.13 gives standardized values of K_1, K_2 and K_3 for the five channels.

Table 9.13. Stable channel parameters.

	Channel type				
	1	2	3	4	5
K_1	6.34	4.71	3.98	3.17	3.08
K_2	0.57	0.48	0.41	0.25	0.37
K_3	0.33	0.54	0.87	–	–

Source: Revised from Henderson [1966].

9.9.3 *Channel distribution*

Branching stable channels formed by erosion in muddy wetland areas illustrate a seemingly "pseudo-living" system in which the morphology of the channel network may be conditioned by life requirements of the vegetation. Figure 9.73 shows a small part of the mangrove coast bordering the Atlantic Ocean in the Brazilian states of Maranão and Pará. The mangroves are connected to the sea by well-developed systems of channel networks, one of which is enclosed within the dotted rectangle. A question is whether such networks conform to scaling relationships germane to their function as feeder and drainage conduits for the mangroves.

In biological systems allometric, *i.e.,* growth related, scaling follows the simple power-law

$$Y = Y_0 m_r^{\beta} \tag{9.98}$$

where Y is some observable variable, m_r is the mass of organism and Y_0 is the value of Y when the exponent $\beta = 1$. It is almost always a multiple of $\frac{1}{4}$, *i.e.,* $\beta = n/4$, where n is a positive or negative integer. An example

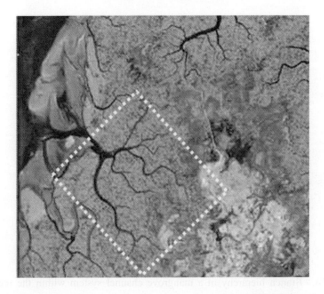

Figure 9.73. Shore segment in the states of Maranão and Pará, Brazil showing multiple branching levels of mangrove-serving channels (inset) (adapted from Ab'Sáber and Holmquist [2001]).

is the basal metabolic rate of mammals and birds. Over four orders of magnitude of the biomass the metabolic rate follows Eq. 9.98 remarkably well with $n = -3$. Branched trees also conform to the same equation [West and Brown, 2004; Arnot and Gobas, 2004].

In Fig. 9.74, channels in the system within the dotted rectangle (Fig. 9.73) have been categorized by numbers starting with 1 for the primary channel. Numbers 11 and 12 denote the first level of branches, 111, 112, 113 and 114, the second level branches of channel 11 and so on. Data on channel width (using an arbitrary unit) from three systems are plotted in Fig. 9.75. The widths show remarkably affine behavior with $n = -2$. A seeming inference is that the mangrove distribution symbiotically interacts with the channels in producing an optimal feeder system to serve the vegetation.

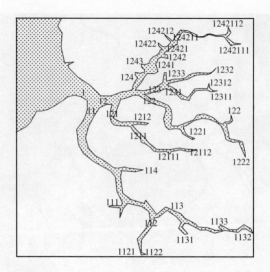

Figure 9.74. Branch hierarchy in a mangrove channel system within the rectangle in Fig. 9.73.

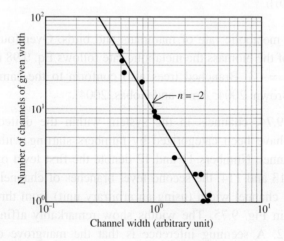

Figure 9.75. Number of channels of a given width plotted against channel width from Fig. 9.74. All widths are relative to the width of the primary channel.

9.10 Exercises

9.1   Plot the variation of drag reduction u_{*ssc}/u_* with concentration ranging from 0 to 4 kg m^{-3}. You are given the following data: water density 1,000 kg m^{-3}, particle density 2,650 kg m^{-3}, water dynamic viscosity 0.001 Pa.s, flow depth 0.3 m and flow velocity 0.3 m s^{-1}. For the dependence of dynamic viscosity on concentration use Eq. 5.33 with $\alpha_r = 5$ and $\beta_r = 0.5$. Is the dependence of fluid density on concentration also important?

9.2   Using the parametric values given in Example 9.2 in the text, plot the shear strength τ_s against the aspect ratio a_R ranging from 0.1 to 1. State the reason for the resulting trend.

9.3   Calculate the erosion flux using Eq. 9.19 with $C_\varepsilon = 2.39 \times 10^{-4}$ kg m^{-3} and $\rho_w = 1,027$ kg m^{-3}. You are given the following distributions (for a general sketch see Fig. 9.9):

$$\varphi(\tau_b) = \begin{cases} 0.5; & 0 \le \overline{\tau}_b \le 2 \\ 0; & 2 < \overline{\tau}_b \end{cases} \qquad (E9.1)$$

$$\varphi(\tau_s) = \begin{cases} 2; & 0.7 \le \overline{\tau}_s \le 1.2 \\ 0; & \overline{\tau}_s < 0.7; \overline{\tau}_s > 1.2 \end{cases} \qquad (E9.2)$$

9.4   Figure E9.1 shows suspension concentration variation with time in a large cylindrical flask using a spindle stirrer over a bed consisting of a mixture of kaolinite and bentonite clays in saltwater.

The depth of water was 27.5 cm. Starting from 0 *rpm* the stirrer speed was increased in six two-hourly increments, and the suspension concentration was measured at mid-depth. Assume that this concentration represents its depth–mean

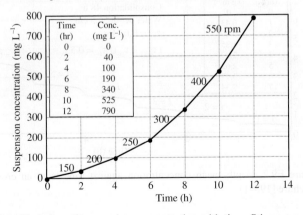

Time (hr)	Conc. (mg L^{-1})
0	0
2	40
4	100
6	190
8	340
10	525
12	790

Figure E9.1. Variation of suspension concentration with time. Stirrer *rpm* indicated was increased every two hours.

value. Derive an expression for the erosion flux, considering each of the six segments to be linear. Use the following formula to convert *rpm* to bed shear stress τ_b (Pa):

$$\tau_b = 0.0138 \times rpm \qquad (E9.3)$$

9.5   Results from an erosion test in a flume with a deposited bed of kaolinite in tap water are shown in Fig. E9.2. The water depth was 56.5 cm and the bed was consolidated for 48 h before eroding it "strip-by-strip". Assume that the bed density profile is given by Eq. 8.74 (and Fig. 8.25). Determine the shear strength profile analogous to Fig. 9.26.

9.6   Variations of suspended sediment concentration with time are plotted in Fig. E9.3 from three erosion tests in a CRAF (counter-rotating annular flume). In test KT a bed of kaolinite was eroded. In KN nutrients (organic matter + inorganic nutrients + vitamins) were added to the bed. Finally, in KM the bed from KN was allowed to remain in water under artificial sunlight for several days in order for biotic (algae and microbes) growth to cover the bed. In all tests the fluid was tap water with a depth of 15 cm, and the bed shear stress was held constant at 0.60 Pa. Estimate the effects of nutrients and biota on the erosion flux of kaolinite.

9.7   Beds of an illitic-clay were eroded in a *SERF*-like apparatus filled with distilled water at different temperatures. Plots of the variation of erosion flux with bed shear stress are shown in Fig. E9.4. Decide if these data agree with the Arrhenius relationship (Eq. 9.25). Use values given in Table E9.1. [Hint: Plot ε against the excess bed shear stress $\tau_b - \tau_s$, then plot Eq. 9.25 for an arbitrarily chosen fixed value of the excess bed shear stress.]

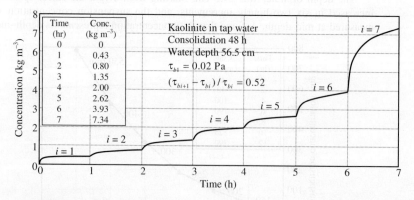

Figure E9.2. Depth–mean concentration against time from an erosion test (from Parchure [1984]).

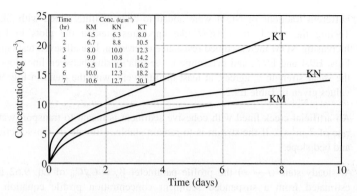

Figure E9.3. Variation of suspended sediment concentration with time in three erosion tests with inorganic and organic-rich beds to assess the effect of biota on erodibility (from Parchure [1984]).

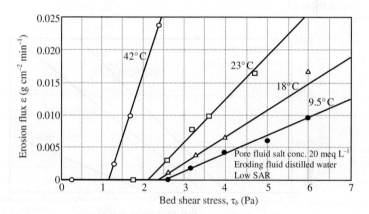

Figure E9.4. Variation of erosion flux with bed shear stress and fluid temperature (from Ariathurai and Arulanandan [1978]).

Table E9.1. Erosion flux values from Fig. E9.4.

Temp. (°C)	τ_b (Pa)	
	$\varepsilon = 0$ (g cm^{-2} min^{-1})	$\varepsilon = 0.010$ (g cm^{-2} min^{-1})
9.5	2.52	6.10
18	2.35	4.80
23	2.10	3.60
42	1.15	1.63

9.8   Different fractions of NOM were added to beds of illite mixed with silica flour. Testing these beds in a *SERF*-like apparatus yielded the plots in Fig. E9.5. Increasing NOM increased bed resistance to erosion. Use all seven test curves from Figs. E9.4 and E9.5, and determine if the relationship between line slope ε_{AP} and the shear strength τ_s agrees, at least qualitatively, with the trend in Fig. 9.67. Use values given in Table E9.2.

9.9   An artificial creek lined with cohesive sediment is meant to transport water at the rate of 7 m^3 s^{-1}. If the stream is to remain stable, estimate its cross-sectional area and bed slope.

9.10   At steady state ($t \to \infty$) the profile parameter $\beta_a = C_d/C_m$ of Eq. 9.62 is readily estimated from a suspended sediment concentration profile equation such as Eq. 6.119.

(1)   From that equation derive the expression for β_a.

(2)   Calculate its value, given $h = 0.8$ m, $w_s = 1\times10^{-5}$ m s^{-1}, $z_a = 0.05$ m and $u_* = 0.05$ m s^{-1}.

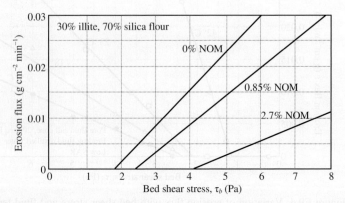

Figure E9.5. Variation of erosion flux with bed shear stress and NOM (from Kandiah [1974]).

Table E9.2. Erosion flux values from Fig. E9.5.

NOM (%)	τ_b (Pa)	
	$\varepsilon = 0$ (g cm^{-2} min^{-1})	$\varepsilon = 0.010$ (g cm^{-2} min^{-1})
0	1.90	3.36
0.85	2.50	4.25
2.7	4.10	7.60

Chapter 10

Fluid Mud Properties and Behavior

10.1 Chapter Overview

Physical properties of fluid mud and its behavior in tidal as well as wave dominated regimes are reviewed in this chapter. The definition of fluid mud, its significance in the marine environment and an empirical approach to identify its thickness in practical applications are summarized. Some noteworthy features of fluid mud behavior are reviewed, the first being the potential for its production by liquefaction. Next, conditions for mixing of fluid mud and water are mentioned, followed by mechanisms that cause the fluid mud-water interface to become unstable. These instabilities can lead to sediment entrainment into the water column. Entrainment flux due to current is reviewed in Chapter 9 and due to waves in Chapter 11. The energetic condition which causes the lutocline to rise is mentioned in this chapter.

The definition of saturation concentration which may result in turbulence collapse is introduced. Next, the effect of dense suspension on the Prandtl mixing length, a key bulk variable governing entrainment in turbulent flow, is covered in some detail as a special topic. Illustrative field observations on fluid mud transport in the tidal and wave environments conclude the chapter.

10.2 Nature and Significance of Fluid Mud

Very soft mud is commonly described as a fluid based on a subjective assessment of the slurry with a thick soup-like appearance. As a rule this mud cannot be molded into a shape except when it gels and develops the consistency of yogurt. Fluid mud has a higher viscosity than water, as

653

well as some elasticity due to close inter-particle spacing. Such materials can be loosely called Maxwell fluids because they approximately conform to that rheological description of fluids with high viscosity and low but often measurable elastic modulus (Chapter 5). A rule-of-thumb proposed by Migniot [1968] is that for a cohesive bed to behave as a fluid when subjected to waves, the bed's plastic yield stress τ_y must be below about 20 Pa.[a] This transition is defined by the space-filling concentration and the dynamic state of the bed determined by waves and the rate of viscous energy loss.

The contribution of fluid mud-related sediment load as a fraction of total suspended sediment load can be overwhelming. For instance a 10 mm thick fluid mud layer at a concentration 10 kg m^{-3} is equivalent to a 1 m thick suspension at a concentration of 0.1 kg m^{-3}. In other words, even very thin layers of fluid mud can carry significant quantities of sediment compared to dilute suspensions. At many coastal locations, current and wave-induced forces strong enough to resuspend sufficient sediment to form fluid mud occur only during storms. This mud, which settles out to form an immobile bed when the forcing subsides, often remains unrecorded because thin layers are difficult to detect by commonly deployed sensors. Also, sensors are only occasionally left in place for sufficiently long periods to cover storms. An outcome is that sedimentation due to fluid mud in episodic environments is sometimes identified only by indirect inferences based on other measurements and, partly as a result, its contribution is often estimated incorrectly.

A case in point is the Camachee Cove Marina basin in northern Florida. In the 1980s, this semi-enclosed basin with a single entrance had a water surface area of 33,370 m^2 and a mean depth of 3 m below mean sea level. Sediment entered the basin during flood tide (range 1.6 m). From bathymetric surveys it was found that during a 2.5 year period starting in 1980, the mean rate of sediment deposition in the basin was 3,375 kg per (semi-diurnal) tide. On the other hand, measurements of suspended load entering with tide during calm weather yielded a 30%

---

[a] When bed rigidity is sufficiently reduced by wave action, the water depth effectively increases by an amount approximately equal to the thickness of the resulting liquefied mud. As a result the wave length increases and the pattern of wave refraction in the shallow sea is altered [Rodriguez, 2000].

lower rate of deposition. It was inferred that the difference consisted of material that had entered as a dense near-bed suspension during storms. In such events tidal flood current in the entrance may remain practically unaffected by the storm; however, waves stir up bottom sediment outside the entrance. Since as we noted in Chapter 7 the fine sediment-carrying capacity of flow can be high with concentrations on the order of 10–20 kg m^{-3} (which as we will see later is in the fluid mud range), the stirred-up sediment is easily transported into the basin [Srivastava, 1983].

As mentioned in Chapter 6, the fluid state of mud is transitional and is sustained only as long as sufficient flow energy is available to prevent it from reverting to the (solid) bed state that is much more stable. Fluid mud is only rarely found in a true state of equilibrium, *i.e.*, a state in which its thickness and mass remain constant for long periods.

The processes by which fluid mud is formed and reworked by current and waves are schematically shown in Fig. 10.1. These are as follows: (1) Beginning with a well-mixed suspension at a high current velocity, fluid mud is formed below a less dense suspension when the velocity decreases sufficiently and settling occurs, partially stratifying the water

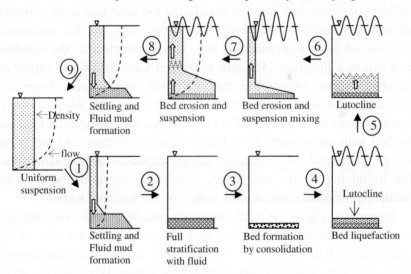

Figure 10.1. Schematic description of sequential response of fine sediment to current and waves.

column. (2) The flow stops, water is clarified and fluid mud dewaters during hindered settling. The settling slurry is further stratified. (3) Due to continued dewatering a consolidated bed is formed, often within hours. If the deposit is thick (*e.g.,* a meter or more) and dewatering slow, the fluid state of mud may persist for longer periods, especially if sufficient flow energy is supplied.

At the Amazon shelf and vicinity fluid mud persists over patches that are hundreds of square kilometers in area. Low rates of consolidation of low-permeability mud as well as continual replacement of mud lost downstream by fresh mud are two important causes.

As we noted in Chapter 7, when sediment contains significant fractions of organic matter, even thin layers of fluid mud may not dewater due to low particle density and buoyant weight. At Lake Okeechobee in Florida where the bottom muck contains about 40% organic matter by weight (Table 5.5), a fluid mud layer of 0.10 to 0.15 m thickness persists under mild to moderate wave action over a ~200 km^2 central zone of the lake. Numerous other shallow lakes in Florida are similarly laden with persistent "fluid muck" [Kirby *et al.,* 1994].

Continuing with the sequence in Fig. 10.1 we note that: (4) Waves of sufficient intensity and duration liquefy the top-most bed layer, possibly without significantly decreasing the density due to the undrained condition of the low-permeability bed. The top interface of the liquefied mud is now a lutocline. (5) Under continued wave action and supply of turbulent kinetic energy the lutocline rises, mixing and downward entrainment of water occur at the interface and the density of the liquefied layer decreases. (6) Wave action increases sufficiently to erode the bed with entrainment of sediment from the bottom into the lower water column and the suspension density increases. At the same time, due to mixing of the two layers and sediment entrainment at the oscillating (unstable) interface the water column becomes turbid.

Figure 10.2 shows the relationship between Secchi disc visibility and suspended sediment concentration at a site in the coastal waters of the Great Barrier Reef along the east coast of Australia. Visibility changes slowly when the volume concentration C_v (Chapter 3) is less than about 0.4 cm^3 m^{-3}, but decreases rapidly at higher concentrations. Assuming a particle density of 2,650 kg m^{-3}, we obtain 1 mg L^{-1} as the transition

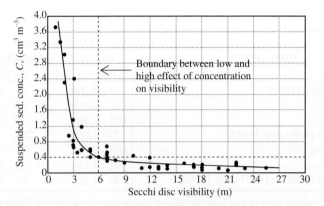

Figure 10.2. Volume concentration of suspended sediment as a function of Secchi disc visibility (in meters of water depth) applicable to the coastal waters of the Great Barrier Reef, Australia (adapted from Wolanski *et al.* [1981] and Wolanski [2007]).

concentration with a visibility of 6 m. At $C_v = 3.2 \text{ cm}^3 \text{ m}^{-3}$ (= 8.5 mg L^{-1}) the visibility would drop to about 1.5 m. Needless to say fluid mud is entirely opaque to visible light (shortest wave length $\sim 10^{-7}$ m), although (for instance) not to gamma rays ($\sim 10^{-11}$ m) passed through the fluid mud sample to measure its density [PIANC, 2008].

The last three components of the sequence in Fig. 10.1 are: (7) Increasing current velocity causes greater erosion and mixing. (8) Wave motion ceases and the thickness of the bottom suspension layer increases due to settling. (9) Current increases further and a uniform suspension ensues [Kineke and Sternberg, 1995; Gowland *et al.*, 2007].

10.3 Estimation of Fluid Mud Thickness

Although fluid mud properties depend on physicochemical, biological and hydrodynamic factors defining its present state and stress-history, it is convenient to broadly characterize fluid mud by the wet bulk density (or concentration) range. The lower-limit density or concentration corresponds to the upper elevation of the fluid mud layer and the upper-limit density to the lower elevation. Variable ranges reported in different investigations (Table 10.1) can be ascribed to three causes: sediment composition, fluid composition (*e.g.,* saltwater versus fresh or brackish water) and criteria selected to identify fluid mud.

Table 10.1. Fluid mud density and concentration ranges.

Investigator(s)	Parameter ranges	
	Density (kg m^{-3})	Concentration (kg m^{-3})
Inglis and Allen [1957]	1,030 – 1,300	10 – 480
Krone [1962]	1,010 – 1,110[a]	10 – 170
Nichols [1985]	1,003 – 1,200	3 – 320
Kendrick and Derbyshire [1985]	1,120 – 1,250[a]	200 – 400
Hwang [1989]	1,002 – 1,065[b]	4.4 – 120
Mean range	1,033 – 1,185	4.6 – 300
Mean range without data of Hwang [1989]	1,040 – 1,215	56 – 340

[a] Conversion between density and concentration based on particle density of 2,650 kg m^{-3}.
[b] Conversion between density and concentration based on particle density of 2,140 kg m^{-3}.

Sediment and fluid compositions as well as stress-history determine the structure of fluid mud flocs. The importance of floc structure is highlighted by the dependence of the space-filling concentration C_s, i.e., the upper limit (at the lower elevation) of the fluid mud concentration range, on the floc fractal dimension D_f (Eq. 4.53). In Fig. 10.3, for the Ems River in the Netherlands, C_s varies with D_f and the energy dissipation rate $\overline{G}$ representing the effect of turbulent flow shear. The dissipation rate is obtained from $\overline{G} = v/\lambda_0^2$, where λ_0 is the Kolmogorov eddy length scale

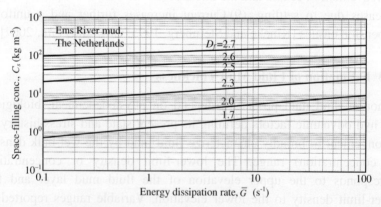

Figure 10.3. Relationship between space-filling concentration (defining the upper limit of fluid mud), floc fractal dimension and energy dissipation rate (from Winterwerp [2002]).

(Eq. 2.89) and v is the kinematic viscosity of the fluid (water). A noteworthy feature of the plot is that it emphasizes the importance of not only floc structure but also the dynamic state of mud on the transition between fluid mud and the cohesive bed. For flocs of fixed fractal dimension C_s increases with $\bar{G}$. The more the agitation the higher would be the concentration required to form a continuous particle matrix (*i.e.*, a bed). For a given $\bar{G}$, C_s increases with D_f, because floc porosity increases concurrently with D_f [Winterwerp, 2002]. Typically, C_s values are commensurate with high values of D_f (2.7 and 2.6). Low values of D_f (2.0 and less) represent flat and (often) large flocs. Such flocs are promoted by organic matter such as mucus, which also reduces floc weight in water. Accordingly, the effect of changing sediment composition, especially with respect to organic matter, is to an extent implicit in Fig. 10.3.

Low densities reported by Hwang [1989] in Table 10.1 reflect the presence of the high (~40% by weight) fraction of low-density organic matter in Lake Okeechobee. The mean particle density was 2,140 kg m^{-3}, which is 19% lower than 2,650 kg m^{-3} typical of abiotic muds (Chapter 3). The density of native water, which affects fluid mud's wet bulk density, was close to that of freshwater, whereas, for example, in the flume tests of Krone [1962], saltwater was used. The density range of Kendrick and Derbyshire [1985] was partly based on the horizontal mobility of estuarine mud in which current speed and density were measured *in situ*. Fluid mud was defined as bottom sediment in motion. In contrast, Krone deduced fluid mud densities from hindered settling of San Francisco Bay mud in the absence of horizontal flow.

The mean density range from all five sets of values in Table 10.1 is 1,033 to 1,185 kg m^{-3}. When the organic-rich sediment of Hwang is excluded the range becomes 1,040 to 1,215 kg m^{-3}. On average the density range of fluid mud can be conveniently chosen as 1,050 to 1,200 kg m^{-3}, recognizing that in specific instances, particularly when organic matter is present, the range may span lower densities. As mentioned this is also indicated in Fig. 10.3 by the low values of the space-filling concentration at low fractal dimensions and energy dissipation.

660   *An Introduction to Hydraulics of Fine Sediment Transport*

It is feasible to assign an operational meaning to the density range of fluid mud in terms of sediment transport processes in preference to density alone — a static measure of mud state. We may delineate the upper and lower elevations z_A and z_B of fluid mud as in Fig. 10.4. The upper level corresponds to the peak settling flux F_{sm} (from Fig. 7.13). At this peak we have the condition

$$\left.\frac{\partial F_s}{\partial \rho}\right|_{F_s = F_{sm}} = \left.\frac{\partial F_s}{\partial C}\right|_{F_s = F_{sm}} = 0 \tag{10.1}$$

This level also corresponds to the primary lutocline characterized by hindered settling below. The lutocline is a density or concentration discontinuity because at that level we have

$$\left.\frac{\partial F_s}{\partial C}\right|_{F_s = F_{sm}} = \left.\frac{\partial F_s}{\partial z}\frac{\partial z}{\partial C}\right|_{F_s = F_{sm}} = \left.\frac{\partial F_s / \partial z}{\partial C / \partial z}\right|_{F_s = F_{sm}} = 0 \tag{10.2}$$

Therefore, on account of Eq. 10.1

$$\left.\frac{\partial C}{\partial z}\right|_{F_s = F_{sm}} = \left.\frac{\partial \rho}{\partial z}\right|_{F_s = F_{sm}} \to \infty \tag{10.3}$$

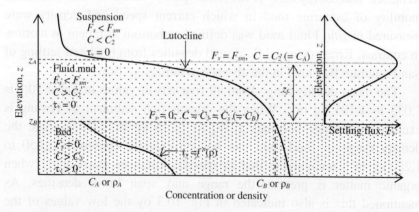

Figure 10.4. Schematic definition of fluid mud layer. The settling flux curve and symbols correspond to Fig. 7.13.

In reality a finite gradient exists over a small vertical distance called the lutocline shear layer (Fig. 6.22). The greater the density difference (and therefore buoyancy jump) across the interface the sharper will be the lutocline boundary. In turbulent flows the interface can be perturbed by oscillating instabilities which, when intense, result in mixing and entrainment of sediment and water between the dense suspension below the lutocline and a typically dilute suspension above. Entrainment is suggested in Fig. 10.5, which is a raw, ~1-minute long time-record derived from an acoustic-suspended-solids-monitor (ASSM) operated at 0.5 MHz with a sampling interval of 0.6 s. This record was obtained in the Jiaojiang estuary near Hainan in China. The heights of the oscillations are on the order of 0.1 m.

A convenient definition of the elevation z_B is one below which mud has a measurable yield strength τ_y and is not a suspension. At z_B, where the flow-induced shear stress τ is equal to or just less than τ_y, the flow velocity is nil (Fig. 10.6a). Due to the dependence of yield strength on bed density (Chapter 9), τ_y typically increases with depth below z_B. If at z_B, τ is greater than τ_y, and the mud has, say, Bingham rheology, *i.e.*, $\tau_y = \tau_B$, the Bingham yield stress, viscoplastic plug flow with a uniform velocity core will occur down to the elevation z_B' (Fig. 10.6b). When the bed is practically homogeneous or weakly stratified the thickness of plug flow may not be negligible. As we have noted in Chapter 9, in the shallow natural environment, mud density commonly increases rapidly in the first ~0.3 m due to consolidation, with a slower rate of increase up to

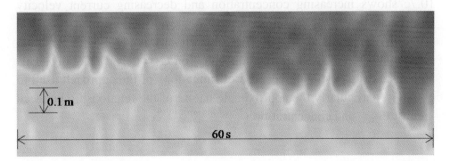

Figure 10.5. ASSM record showing a lutocline modulated by instabilities and sediment entrainment. The image is from Jiaojiang estuary in China (from Jiang [1999] and Jiang and Mehta [2000]).

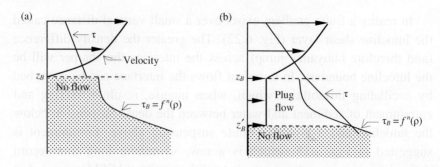

Figure 10.6. (a) Coincidence of zero velocity plane and onset of yield stress; (b) occurrence of plug flow.

a maximum density at some depth. Accordingly, one can anticipate the maximum thickness of plug flow to be no more than about 0.3 m. Mobile layers of such a thickness can be expected in tidal channels with strong flows or in densely turbid gravity slides. Elsewhere the thickness may be smaller. Notwithstanding exceptions, we may consider 0.3 m as the rule-of-thumb upper limit of the expected thickness of fluid mud in, say, meso-tidal environments.[b] In macro to hyper-tidal areas the thickness may be as much as a meter.

Figure 10.7 illustrates an approximate method to estimate the thickness $z_f = z_A - z_B$ of the fluid mud layer. In Fig. 10.7a the concentration profile indicates a well-defined lutocline. Its characteristic concentration C_A is 20 kg m^{-3} and the corresponding elevation $z_A = -5.4$ m is determined from the settling flux plot in Fig. 10.7b. Figure 10.7c shows increasing concentration and decreasing current velocity with depth below the lutocline. The lower level of fluid mud occurs at -0.16 m, *i.e.*, $z_B = -5.56$ m, where the velocity is zero. This elevation corresponds to the space-filling concentration $C_B = C_s = 260$ kg m^{-3}. This value is a rough estimate because of limitations of the approach used to obtain it, particularly the approximate nature of the relationship between yield stress τ_y and concentration C.

---

[b] Conventional typology for the tidal range is: micro-tidal < 2 m, meso-tidal between 2 and 4 m, macro-tidal between 4 and 6 m, and hyper-tidal > 6 m.

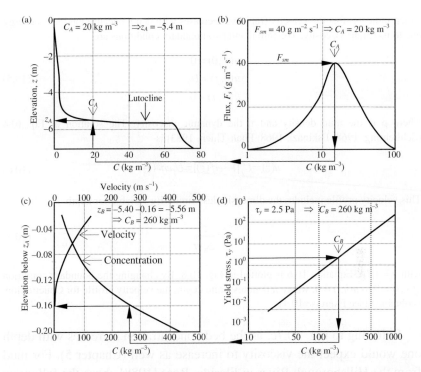

Figure 10.7. Estimation of fluid mud layer thickness. Plots are based on estuary measurements by Odd and Rodger [1986] and Kendrick and Derbyshire [1985].

From Fig. 10.7d we determine that at -0.16 m, where the bed surface occurs, τ_y would be 2.5 Pa. For plug flow the fluid stress τ would have to exceed this value. Thus the thickness of the fluid mud layer z_f is 0.16 m.

Example 10.1: Show that the velocity profile in fluid mud in Fig. 10.7c is consistent with momentum diffusion in accordance with Raleigh flow.

In its simplest form the well known Raleigh flow problem (or Stokes first problem) can be formulated as locally horizontal flow with longitudinal and lateral (x and y directions, respectively) uniformity. We would like to calculate the flow velocity $u(z,t)$, where z is the vertical coordinate. Let $z' = 1 - z$ such that z' is positive downwards with its horizontal datum at the lutocline (Fig. 10.7). At time $t = 0$ the interface is set in horizontal motion at an instantaneously achieved constant velocity U_L. We are interested in knowing the manner in which the boundary layer develops below the lutocline. The momentum balance is

$$\frac{\partial u}{\partial t} = -\frac{1}{\rho}\frac{\partial}{\partial z'}\left(\eta\frac{\partial u}{\partial z'}\right) \tag{10.4}$$

which indicates that the velocity (rapidly) decreases with depth because mud loses energy due to its (high) viscosity. The initial and two boundary conditions are

$$u(z',0)=0$$
$$u(0,t)=U_L \qquad\qquad (10.5)$$
$$u(\infty,t)=0$$

where ρ is the mud density and η the dynamic viscosity. The solution of Eq. 10.4 [Schlichting, 1968; Eskinazi, 1968; Phan-Thien, 1983] is

$$u(\hat{z})=U_L[1-erf(\hat{z})]=U_L erfc(\hat{z}) \qquad\qquad (10.6)$$

This velocity profile includes the dimensionless similarity variable for depth

$$\hat{z}=\frac{z'}{2\sqrt{vt}} \qquad\qquad (10.7)$$

with $v=\eta/\rho$. Equation 10.6 is plotted in Fig. 10.8. By changing the boundary condition $u(\infty,t)=0$ to $u(h,t)=0$, where h denotes finite depth, the velocity profile for Couette flow is obtained (see Exercise 10.4).

Referring to Fig. 10.7c, as the bed concentration increases with depth one would expect the viscosity to increase as well (Chapter 5). For mud from the Hillsborough River in Florida, Ross [1988] chose the following linear expression for the dynamic viscosity (Pa.s) as a function of concentration (kg m^{-3})

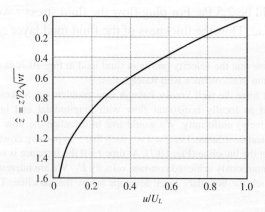

Figure 10.8. Rayleigh flow in fluid mud.

$$\eta = \eta_w(1+0.2C) \qquad (10.8)$$

along with fluid mud density ρ given by

$$\rho = \rho_w + 0.61C \qquad (10.9)$$

Let water viscosity $\eta_w = 0.001$ Pa.s and water density $\rho_w = 1,030$ kg m^{-3}. We will change the mud-water interface condition, based on a constant fluid velocity, to the more realistic constant stress

$$\eta \frac{\partial u}{\partial z'}\bigg|_{z'=0} = \tau_L \qquad (10.10)$$

where τ_L is the instantaneously applied constant fluid stress (corresponding to U_L) at the lutocline. Its value will be chosen as 0.02 Pa [Ross, 1988].

Equation 10.4 together with the first and the third conditions in Eq. 10.5 (and Eqs. 10.8, 10.9, 10.10) were solved using a finite difference approximation of Eq. 10.4. A spatial increment of 0.02 m was chosen for the incremental depth $\Delta z'$ and 4 s for time-step Δt. The resulting boundary layer development is shown in Fig. 10.9a. At 20 min, after application of the stress τ_L, the velocity profile reached a

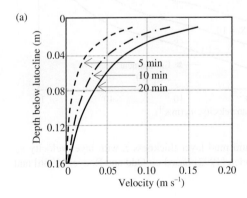
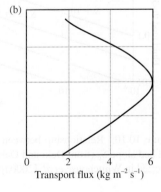

Figure 10.9. Fluid mud velocity profiles and dry sediment flux (from Ross and Mehta [1989]).

configuration for which the dry sediment mass flux $u(z')C(z')$ is plotted against depth in Fig. 10.9b. The peak flux of 6 kg m^{-2} s^{-1} occurs about 0.08 m below the lutocline (and not at the surface).

For a more accurate estimation of the thickness z_f the momentum equation must include advection terms (in Eq. 10.4) and fluid mud must be treated as a dense liquid apart from water. For gravity underflow of fluid mud at assumed steady-state, z_f may be estimated from Fig. 10.10 knowing the mean velocity u_m of the layer and bed slope S_o. This plot is based on a numerical solution of fluid momentum and sediment mass balances [Whitehouse *et al.*, 2000a and Chapter 12]. It is limited to fluid mud having a density of 1,075 kg m^{-3} and a viscosity of 0.7 Pa.s. When the shear stress τ_b at the bottom of fluid mud is less than 0.1 Pa, the underflow is unstable and prone to deposition. Similarly, when the Richardson number Ri_b defined in Eq. 9.81 is less than 10 (in order of magnitude agreement with the value of 20 in Fig. 9.41), sediment entrainment occurs at the interface. The time-rate of loss of sediment mass per unit bed area is given by Eq. 9.82.

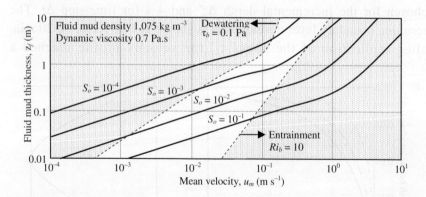

Figure 10.10. Relationship between fluid mud layer thickness z_f with layer velocity u_m and bed slope S_o (from Delo and Ockenden [1992] based on Odd and Rodger [1986] and reported in Whitehouse *et al.* [2000a]).

10.4 Fluid Mud by Liquefaction

10.4.1 *Bed failure by waves: drained and undrained conditions*

Failure of bed by liquefaction due to waves occurs mainly by two mechanisms. They are: (1) buildup of pore water pressure, and (2) upward water pressure in the bed during the passage of a wave trough. The first mechanism is called residual liquefaction and the second momentary liquefaction. In either case the model for evaluating the criterion of liquefaction depends on the ratio of two time scales. One is the wave period, which sets the time of wave loading and therefore pore pressure buildup. The other is the time for pore water flow, which dissipates the pore pressure. The ratio of these scales determines if strain-softening of the bed and failure will occur under undrained condition represented by low value of the ratio, or strain-hardening under drained condition characterized by a high ratio [van Kessel and Kranenburg, 1998; Sumer and Fredsøe, 2002].

For a bed of given permeability, the longer the wave the greater is the drainage. For undrained response we may consider the bed to be a viscoelastic continuum. However, when drainage is significant, energy losses in the solid and the fluid phases are considered separately, as in a poroelastic medium.

Once the bed matrix at rest is disturbed by waves the depth at which liquefaction begins depends on cohesion, floc structure, bed thickness and degree of consolidation. When the bed is thin and unconsolidated or partially consolidated, mud displacement can be high near the hard bottom due to the dynamic pressure gradient. In this case liquefaction may be initiated near the bottom and if so it will proceed upward. In a thick or fully consolidated bed, wave-induced displacement will be large at the top and smaller or possibly nil at the bottom. Liquefaction initiated at the top will proceed downward up to a depth at which the hardness of the bed and lack of wave effect prevent further advance. A stable density gradient, *i.e.*, one in which the density is low at the top and increases downward, would make it easier for liquefaction to start at the top. In fact, bed stratification (which is often significant) is believed to be the main reason for top-down liquefaction observed in many laboratory studies [Chou, 1989].

We will limit the review to top-down residual liquefaction under undrained condition as would occur when deposited clayey beds are subjected to short-period waves.[c]

10.4.2 *Fluid mud thickness*

Persistence of fluid mud in shallow seas and lakes is the outcome of sea state determined by "steady" wind-waves and availability of mud. At Lake Okeechobee in Florida where water depths are 2–5 m, fluid mud thickness characteristically ranges between 0.05 and 0.20 m depending on the water depth. Thicker and persistent layers are found along the coasts of Korea and Suriname/Guyana, the Amazon shelf and the southwestern coast of India [Wells, 1983; Augustinus, 1987; Eisma *et al.*, 1991; Kirby *et al.*, 1994; Mathew and Baba, 1995; Kineke and Sternberg, 1995; Li and Mehta, 2001].

Consider a water layer and a bed with fluid mud in-between (Fig. 10.11). As we noted in Chapter 8, in water and in fluid mud (which is a suspension, not a homogeneous liquid) the total normal stress, *i.e.*, the hydrostatic pressure σ_v, increases linearly with depth because the (water-containing) pores form a continuous network, which permits the transmission of pressure through the water phase. At the bed surface the concentration is at its space-filling value C_s. In the bed σ_v is the sum of effective normal stress σ'_v and the pore pressure p_w. The pore pressure consists of two components, namely hydrostatic pore pressure p_{ws} and excess pore pressure p_{we}.

---

[c] In geotechnical analysis, the criterion for soil liquefaction under wave loading is obtained by determining the soil's (load-controlled) cyclic triaxial strength at the point of sample failure. The sample is maintained in the undrained condition. Test results suggest that marine mud samples with density in excess of 1,300 kg m^{-3} may not liquefy under loading equivalent to that due to wave action in the prototype environment. This threshold density appears to be realistic for the natural environment, although in general the triaxial test does not fully replicate the movement of pore water due to a progressive wave.

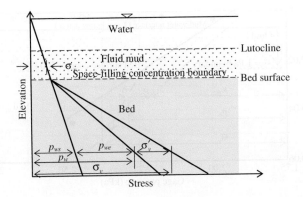

Figure 10.11. Instantaneous stress profiles in a layered water-mud system.

When wave motion is initiated over a silty sediment bed, measurements of pore pressure oscillations indicate that a momentary, resonant amplification of the pressure amplitude may accompany a sudden increase in the wave-mean pore pressure p_w in localized bed cavities [Foda *et al.*, 1991]. Flume experiments in which clayey beds are subjected to waves suggest a similar occurrence in selective pore spaces [Feng, 1992]. Under continued wave-induced motion in a bed in which the inter-particle bonds are broken easily, the effective normal stress σ_v' gradually diminishes and may ultimately become negligible. At that time the bed is considered to be liquefied. This may occur even when the concentration is equal to or exceeds the (static value of) space-filling concentration C_s.

In experimental observations it is useful to set a small but measurable value of the effective stress as the boundary separating fluid mud from the bed. Measurements supportive of the stress profiles in Fig. 10.11 were carried out in a wave flume using mud from Hillsborough Bay in Florida [Cervantes, 1987; Ross, 1988]. The water depth was 31 cm and mud thickness 12.4 cm. The bed, pre-consolidated for 72 h, was subjected to 6 cm high waves at a frequency of 1 Hz. Total and pore pressure time-series were obtained from gages flush-mounted on a sidewall of the flume. An example of the profiles of wave-mean stresses 1.5 h after test initiation is shown in Fig. 10.12. The bed had eroded to a depth of 7 mm below the initial level (datum). The boundary between fluid mud and the bed was set at effective stress $\sigma_v' = 1$ Pa, a small value.

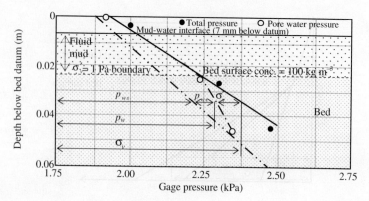

Figure 10.12. Total and pore pressure profiles in Hillsborough Bay mud bed 1.5 h after initiation of waves in a flume (from Ross *et al.* [1987]).

At this boundary, 0.023 m below the datum, the concentration was 100 kg m^{-3}. This is likely to be a non-static value of the space-filling concentration C_s. It would be lower than the respective static value, and consistent with the bed's dynamic condition characterized by the high rate of wave energy loss (*e.g.*, Fig. 10.3). The thickness of the fluid mud layer was 0.016 m.

Starting with a bed at rest, the increase in pore pressure at a given depth is schematically shown in Fig. 10.13. Liquefaction occurs at time t^* equal to a number N_L multiplied by the wave period T, and is defined by the condition

$$\sigma_v'(t^*) = \sigma_v(t^*) - p_w(t^*) = \sigma_v(t^*) - [p_{ws}(t^*) + p_{we}(t^*)] = 0 \quad (10.11)$$

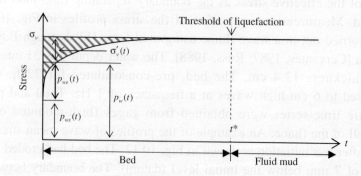

Figure 10.13. Liquefaction condition for a cohesive bed subjected to waves.

Experimental data on the liquefaction of non-cohesive beds have suggested the following empirical expression for N_L:

$$N_L = \left(\alpha_L \frac{\tau^0}{\sigma_v'} \right)^{\beta_L} \tag{10.12}$$

where τ^0 is the amplitude of the oscillatory stress, and α_L and β_L are sediment-specific coefficients dependent on the relative density of the soil D_r (Table 5.2). For example, for $D_r = 0.54$ the coefficients are $\alpha_L = 4$ and $\beta_L = -6$. The negative value of β_L is consistent with the observation that N_L increases with the initial effective normal stress σ_v' [Sumer and Fredsøe, 2002]. Assuming the applicability of Eq. 10.12 to cohesive beds, as an example, let $\tau^0 = 5$ Pa be the stress amplitude due to a 6 s wave. If $\sigma_v' = 50$ Pa characterizes the initial state of the bed, we obtain $t^* = 1,465$ s $= 24.4$ min, a likely laboratory value.

In Fig. 10.14a, we observe the growth of the fluid mud layer due to liquefaction of Hillsborough Bay mud. As noted, the boundary between the fluid mud layer and the bed was established by tracking the level corresponding to effective stress $\sigma_v' = 1$ Pa. Figure 10.14b shows density profiles obtained by coring the bed at different times during the test. The locus of the 1-Pa level in Fig. 10.14a indicates that at the beginning of the test the top 2.2 cm thick layer was fluid mud. At the end of the two-

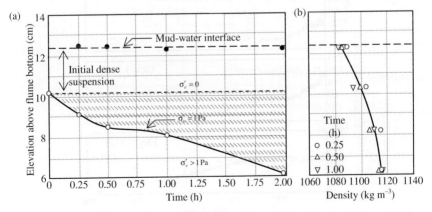

Figure 10.14. Fluid mud layer growth in a laboratory flume using mud from Hillsboro Bay, Florida: (a) Time-tracking the 1 Pa effective stress boundary; (b) density profiles (adapted from Ross [1988]).

hour test this layer grew to a thickness of 6.3 cm, thus adding 4.1 cm by liquefaction. However, there was little systematic change in the bed density, suggesting that drainage or dilution were not significant.

When there is no measurable change in bed density during liquefaction, once wave action ceases, inter-particle contacts are rapidly reestablished and fluid mud reverts to the bed state with concentration at least as high as the static space-filling value. However, if and when dilution of fluid mud occurs as the lutocline rises due to upward entrainment of sediment and downward entrainment of water, dewatering must occur before the bed is reformed. The time of reformation of the bed depends on sediment composition, thickness of the fluid mud layer and the extent of dilution.

The growth rate of the fluid mud layer was examined in a wave flume with a bed of a 1:1 mixture of kaolinite and attapulgite clays (AK bed) at a wet bulk density of 1,170 kg m^{-3}. Mud thickness was 0.16 m and the water depth was 0.19 cm. The bed was allowed to consolidate for 85 h before initiating waves of 4 cm height and 1 s period [Feng, 1992]. A sample output is shown in Fig. 10.15. The effective stress σ'_v is plotted against time at three elevations. The time t^* required for σ'_v to vanish increased with depth. At depths of 4, 6.5 and 8.5 cm, the respective times for practically zero effective stress were 170, 305 and 480 min, as

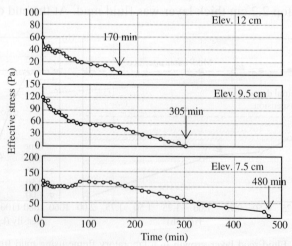

Figure 10.15. Time-variation of effective normal stress in a clay bed and the formation of fluid mud (adapted from Feng [1992]).

liquefaction progressed downward. The rate of liquefaction decreased with penetration of wave energy. From the elevation of 12 cm down to 9.5 cm the rate was 0.0185 cm min^{-1}. It decreased to 0.0114 cm min^{-1} between 9 cm and 7 cm.

Studies on the generation of fluid mud summarized in Table 10.2 suggest that comparatively narrow ranges of wave properties and bed thicknesses were selected presumably due to limitations of laboratory test dimensions. These results therefore do not permit a full assessment of the effects of the governing variables, and are also likely to be influenced by scale effects. However, they do provide background information for setting up more comprehensive tests. We will review an informative experiment by van Kessel and Kranenburg [1998] in a later section.

The instantaneous thickness $z_f(t)$ of the fluid mud layer can be generally expressed as

$$z_f(t) = z_{fe} \cdot f^n(t) \qquad (10.13)$$

where z_{fe} is the equilibrium value of z_f such that $f^n(t) \rightarrow 1$ as $t \rightarrow \infty$. The time to reach equilibrium can vary widely; when it is long the test duration may be insufficient to reach z_{fe}. This is apparent in Fig. 10.16a showing z_{fe} versus time data derived from Fig. 10.14a for the Hillsborough Bay mud.

Table 10.2. Selected fluid mud generation experiments in flumes

Source	Mud	Water depth (cm)	Bed thick. (cm)	Mud density (kg m^{-3})	Wave ampl. (cm)	Wave freq. (Hz)	Mud viscosity (Pa.s)	Mud rigidity (Pa)	Fluid mud thick. (cm)
Ross [1988]	Tampa Bay, Florida	31–32	12–13	1,080	3.1–3.6	1.0–1.1	25.0	100	5.0–6.3
Lindenberg et al. [1989]	Kaolin	25	5	1,300	2.4–3.6	0.4–0.7	3.0	5	1.0–2.5
Feng [1992]	AK^a	18–20	15–17	1,170	1.9–4.0	1.0	6.1	295	2.0–3.5

[a] 1:1 (by weight) clayey mixture of an attapulgite and a kaolinite.

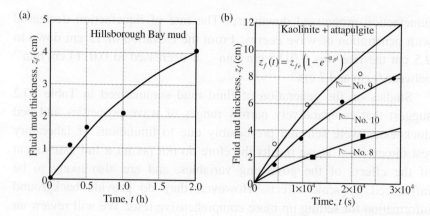

Figure 10.16. Growth of fluid mud thickness under sustained wave motion in flume tests using: (a) Mud from Hillsborough Bay, Florida; (b) a 1:1 mixture of kaolinite and attapulgite (AK).

Data such as those in Fig. 10.15 have been used to plot the growth of the fluid mud layer in Fig. 10.16b for the clay bed. The time dependence of fluid mud thickness is approximated by the relationship

$$z_f(t) = z_{fe}\left(1 - e^{-\alpha_{fl}t}\right) \qquad (10.14)$$

in which $\alpha_{fl} = 3\times10^{-5}$ s^{-1}, and values of z_{fe} are given in Table 10.3 along with the test conditions. Since equilibrium fluid mud thicknesses were not achieved in the tests, z_{fe} must be treated as an empirical best-fit coefficient applicable over the test duration only (3×10^4 s). The estimated z_{fe} values are likely to be higher than actual, as they are artifacts of imprecise curve-fitting.

Table 10.3. Fluid mud production in beds of a 1:1 mixture of kaolinite and attapulgite

Test no.	Wave amplitude[a] (cm)	Pre-test bed consolidation time (h)	z_{fe} (m)
8	2.0	240	0.07
9	2.8	65	0.20
10	4.0	85	0.14

[a] Wave period was 1 s.
Source: Feng [1992].

The z_{fe} values indicate a complex dependence of fluid mud thickness on the wave height and the pre-test bed consolidation time. Test no. 8, with a well-consolidated (240 h) bed subjected to the smallest (2 cm amplitude) waves, expectedly showed the lowest rate of fluid mud formation. In no. 9 the bed was less consolidated (65 h) and was subjected to higher (2.8 cm) waves. These two factors collectively led to a higher rate of growth of the fluid mud layer compared to no. 8. In no. 10 the waves were increased to 4 cm, but presumably because the bed was more consolidated (85 h) than in no. 9, layer growth was slower.

10.4.3 *Effect of liquefaction on shear modulus*

In situ rheometric measurements of changes in the bed as liquefaction proceeds allow one to identify the transition from a bed to fluid mud. As mentioned in Chapter 5, this transition can be detected by passing a high-frequency shear-wave of amplitude τ^0 and measuring the strain amplitude γ^0 and phase lag δ_s (Eqs. 5.72 and 5.73, respectively). Let us consider the Kelvin–Voigt element for which $G' = G$ (Eq. 5.81), the elastic shear or rigidity modulus. We may state

$$G(t) = \left[\frac{C_{sh}(t)}{C_{sh}(0)} \right]^2 G(0) \qquad (10.15)$$

where $G(0)$ is the initial value of $G(t)$ and $C_{sh}(0)$ is the initial shear wave celerity $C_{sh}(t)$. The factor $[C_{sh}(t)/C_{sh}(0)]^2$ is the reduction in bed rigidity at any time. Equation 10.15 is applicable to a constant density (ρ) medium subjected to waves.

The time-dependent behavior of G was examined in a wave flume with AK bed (Table 5.5). The bed thickness was 0.16 m and the water depth 0.19 m. A miniature shear-wave transducer (Fig. 10.17) was inserted in the bed which was subjected to water waves of 0.02 m amplitude at 1 Hz frequency. The celerity of a 1.8 kHz shear-wave was measured 0.028 m above the flume bottom by flexibly tethering the transducer. It consisted of a piezo-ceramic bimorph element bonded to a steel plate as the generating or detecting surface. Shear-wave is produced

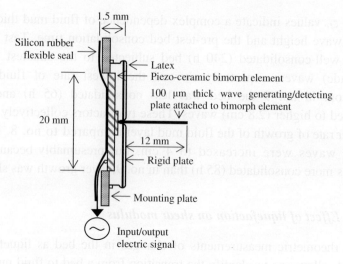

Figure 10.17. Shear-wave transducer (from Mehta *et al.* [1995]).

by the piezoelectric effect[d], for which an AC voltage is applied to the ceramic element. The resulting deformation of the element oscillates the attached plate in its own plane, generating a shear-wave [Williams and Williams, 1992; Mehta *et al.*, 1995].

The oscillating plate simultaneously produces two sets of shear waves, one on each side. They travel in opposite directions along paths of unequal lengths x_1 (= 13.0 mm) and x_2 (= 13.1 mm), and are reflected by two respective wave detecting plates at the ends of the paths (Fig. 10.18). The motion of these plates induced by wave arrival produces electrical output signals by the inverse piezoelectric effect. Knowing the virtual gap Δx (= 0.1 mm) between the two paths and the phase difference $\Delta x/C_{sh}$ between wave arrivals at the two plates permits the calculation of C_{sh} (Fig. 10.19).

Figure 10.20 shows the result from a test in which water wave motion was initiated over a clay bed pre-consolidated for 20 h and having a density of 1,170 kg m^{-3} (and solids volume fraction ϕ_v nominally equal to 0.1). The relative celerity $C_{sh}(t)/C_{sh}(0)$, where $C_{sh}(0)$ was 2 m s^{-1},

---

[d] When certain crystalline materials are mechanically stressed electric charge accumulates on the surface, which is manifested as a piezo-electric current. This process is reversible and can be used to generate deformation in the material by passing a current.

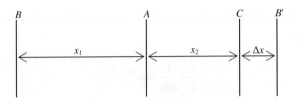

Figure 10.18. Virtual gap geometry consisting of wave-generating plate A and detecting plates B and C. The virtual gap is the difference Δx between plate C and virtual plate B'.

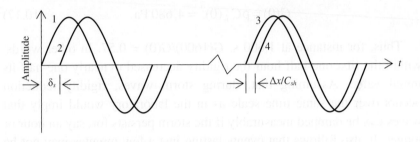

Figure 10.19. Shear-wave celerity and phase lag using the virtual gap geometry. Curve 1: stress at plate A (calculated), curve 2: displacement of plate A (measured), curve 3: stress at plate C (measured) and curve 4: stress at plate B (measured).

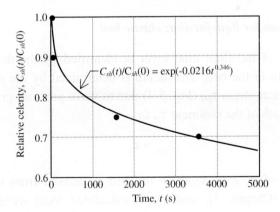

Figure 10.20. Relative shear wave celerity measured by the shear-wave transducer (from Mehta *et al.* [1995]).

rapidly decreased with time (seconds), as approximated by the empirical relationship

$$\frac{C_{sh}(t)}{C_{sh}(0)} = e^{-\alpha_{sw}t^{\beta_{sw}}}$$ (10.16)

with coefficients $\alpha_{sw} = 0.0216$ and $\beta_{sw} = 0.346$. The initial rigidity is obtained from

$$G(0) = \rho C_{sh}(0)^2 = 4,680 \, \text{Pa}$$ (10.17)

Thus, for instance at 1,600 s, $G(1600)/G(0) = 0.57$. In other words, within the first one-half hour the rigidity decreased to nearly one-half its initial value. Assuming that, during storm waves, rigidity reduction occurs over the same time scale as in the laboratory would imply that waves can be damped measurably if the storm persists for, say an hour or longer. It also follows that events lasting just a few minutes may not be effective in wave damping, as storm intensity and duration together determine the post-storm energy absorption by mud (Chapter 5).

In the following two sections we will examine criteria for the equilibrium thickness of liquefied bed represented as continua.

10.4.4 *Criterion for liquefaction: elastic bed*

The thickness of the fluid mud layer is identified by the depth of the plane of failure in the bed. This depth is determined by the equality of the deviator shear stress τ_{qm} (Fig. 5.4) due to wave pressure gradient and the yield strength of the sediment τ_y, *i.e.*,

$$\tau_{qm} = \tau_y$$ (10.18)

The variation of τ_y with depth in the bed is derived from rheometric measurement (Chapter 5), and τ_{qm} is calculated from wave and bed properties [van Kessel and Kranenburg, 1998]. The deviator stress can also be obtained from a triaxial test, although it is uncertain if this would be consistent with a bed stressed by water waves. In any event, here we will determine τ_{qm} driven by wave pressure at the bed surface.[e]

Referring to Fig. 10.21, liquefaction of a mildly sloping elastic bed occurs by plastic yield. The liquefied layer is detected by its gravity-slide, which leaves the bed behind (Fig. 10.22). For simplicity the long wave assumption is invoked and pore-water flow is assumed to be negligible, *i.e.*, the bed is considered to remain undrained. The governing equations for bed displacement are

$$G\frac{\partial^2 x_s}{\partial z^2} = \frac{\partial p}{\partial x} \tag{10.19}$$

$$G\frac{\partial^2 z_s}{\partial z^2} = \frac{\partial p}{\partial z} \tag{10.20}$$

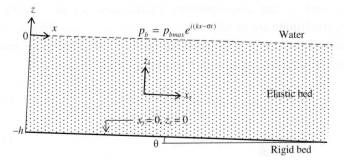

Figure 10.21. A mildly inclined elastic bed subjected to wave-induced pressure.

---

[e] For long waves the horizontal pressure gradient is the dominant cause of liquefaction, and for dense beds cohesion may prevent or retard liquefaction. Madsen [1974] proposed the failure criterion

$$-\left(\frac{\partial p_{we}}{\partial x}\right)_{cr} d_p = s_u \tag{F10.1}$$

where p_{we} is the excess pore pressure, d_p is the particle or floc diameter and s_u is the undrained shear strength (*e.g.*, Table 5.3). Bed failure is considered to occur when the pressure difference across the particle equals or just exceeds the yield strength. All inter-particle bonds are broken at that depth and above that level the bed is instantaneously liquefied.

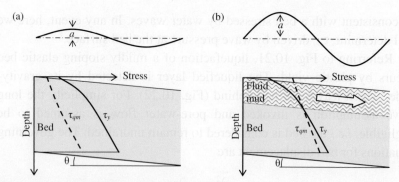

Figure 10.22. Wave effect on a mildly inclined fine sediment bed: (a) No liquefaction at low wave amplitude; (b) liquefaction and gravity-slide of the upper bed under a high wave.

$$\frac{\partial x_s}{\partial x} + \frac{\partial z_s}{\partial z} = 0 \tag{10.21}$$

where x_s and z_s are displacements [Yamamoto *et al.*, 1978]. The boundary conditions are as follows:

At the bed surface $(z = 0)$ the shear stress τ_{xz} is taken to be zero and the surface wave pressure p_b to be harmonic, *i.e.*,

$$\tau_{xz} = \frac{\partial x_s}{\partial x} + \frac{\partial z_s}{\partial z} = 0 \tag{10.22}$$

$$p_b = p_{bmax} e^{i(kx-\sigma t)} \tag{10.23}$$

where p_{bmax} is the pressure amplitude, k is the wave number and σ is the angular frequency. At the rigid bottom $(z = -h)$ the fluid is at rest, *i.e.*,

$$x_s = 0; \quad z_s = 0 \tag{10.24}$$

Equations 10.19 through 10.24 are solved[f] to obtain

$$x_s = i \frac{p_{bmax}}{Gk} \left(b_{s0} + b_{s1} e^{kz} + b_{s2} e^{-kz} \right) e^{i(kx-\sigma t)} \tag{10.25}$$

---

[f] An assumption in the original analysis by Yamamoto *et al.* [1978] is that the so-called stiffness ratio G/K_e' is negligible, where K_e' is the elastic bulk modulus of the fluid, a measure of its compressibility. For water, K_e' is 2.15×10^9 Pa. For muds that can be liquefied, G is lower than 10^4 Pa, and G/K_e' is on the order of 10^{-5}.

$$z_s = i\frac{p_{bmax}}{Gk}\left(b_{00} + b_{s0}kz + b_{s1}e^{kz} - b_{s2}e^{-kz}\right)e^{i(kx-\sigma t)} \qquad (10.26)$$

$$p_b = p_{bmax}\left(b_{s1}e^{kz} + b_{s2}e^{-kz}\right)e^{i(kx-\sigma t)} \qquad (10.27)$$

where

$$b_{s0} = -\left(\frac{e^{kh} + c_{s1}e^{-kh}}{c_{s1}+1}\right) \qquad (10.28)$$

$$b_{s1} = \frac{1}{c_{s1}+1} \qquad (10.29)$$

$$b_{s2} = \frac{c_{s1}}{c_{s1}+1} \qquad (10.30)$$

$$b_{00} = 2\left(\frac{c_{s1}-1}{c_{s1}+1}\right) \qquad (10.31)$$

$$c_{s1} = \frac{\frac{1}{2}e^{kh}(1-kh)-1}{\frac{1}{2}e^{-kh}(1+kh)-1} \qquad (10.32)$$

The peak deviator normal stress is $\sigma_{qm} = (\sigma_1 - \sigma_3)/2$ (Fig. 5.4). The corresponding peak deviator shear stress is

$$\tau_{qm} = \sqrt{\tau_{xz}^2 + \sigma_{qm}^2} \qquad (10.33)$$

where, for the present analysis, the shear stress τ_{xz} is taken to be

$$\tau_{xz} = ip_{bmax}\left(b_{00} + b_{s0}kz + 2b_{s1}e^{kz} - 2b_{s2}e^{-kz}\right)e^{i(kx-\sigma t)} + \tau \qquad (10.34)$$

in which τ is the gravity-induced shear stress along the slope. For the assumed mild slope condition we have $\tau = gz(\rho_s - \rho_w)\sin\theta$. The principal normal stresses are

$$\sigma_1 = -p_{bmax}\left(2b_{s0} + b_{s1}e^{kz} + b_{s2}e^{-kz}\right)e^{i(kx-\sigma t)} + \sigma_{1s} \qquad (10.35)$$

and

$$\sigma_2 = -p_{bmax}\left(2b_{s0} + 3b_{s1}e^{kz} + 3b_{s2}e^{-kz}\right)e^{i(kx-\sigma t)} + \sigma_{2s} \qquad (10.36)$$

where σ_{1s} and σ_{2s} are gravity-induced normal stresses. For simplicity we will assume that these two stresses are equal at the time of liquefaction.

Depth-profiles of calculated deviator shear stress and measured yield stress are shown in Fig. 10.23. The data were derived from tests using a 2-week consolidated bed of kaolinite in a flume inclined by 0.05 radians and subjected to 1.65 s waves. The yield stress profile was obtained *in situ* based on resistance against a rod used as a penetrometer. The wave amplitude was increased in steps from 4 mm to 55 mm. Each step was run until the bed approached its final condition of either no liquefaction or fluid mud formation down to equilibrium depth.

When the wave amplitude was 5 mm (Fig. 10.23), liquefaction could not occur as the deviator stress τ_{qm} was less than the yield stress τ_y over

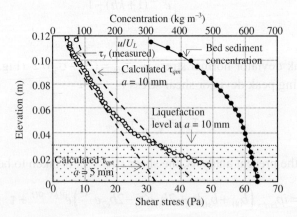

Figure 10.23. Conditions for liquefaction of a kaolinite bed in a wave flume (adapted from van Kessel and Kranenburg [1998]).

the entire bed. However, at the amplitude of 10 mm the top 9 cm of bed was liquefied because τ_{qm} was greater than τ_y over this thickness.

10.4.5 *Criterion for liquefaction: viscoelastic bed*

Consider short-period progressive waves over soft viscoelastic mud. A floc of mass m_{pf} (Fig. 10.24a) is acted on by four forces (Fig. 10.24b) as follows:

(1) Buoyancy force

$$m_{pf}\, g' = m_{pf}\, g\, \frac{\rho(z') - \rho_w}{\rho_w} \qquad (10.37)$$

where $g'(z')$ is the depth-dependent reduced gravity (with $\rho \approx \rho_w$ in the denominator), $\rho(z')$ is the mud density varying with depth $z' = -z$ and ρ_w is the water density.

(2) Normal components F_{p1} to F_{p4} and tangential components f_{p1} to f_{p4} of the inter-particle contact forces.

(3) Inertial force L_{pi}, due to bed oscillation.

(4) Cohesion as cumulative resistive force F_C.

We will assume that the normal and shear contact forces surrounding a floc are randomly directed. This assumption permits us to further assume that these forces have no tangible net effect on bed liquefaction. Although the bed can withstand a certain amount of compressive force, it cannot resist strong stretching forces. Thus, when the upward inertia

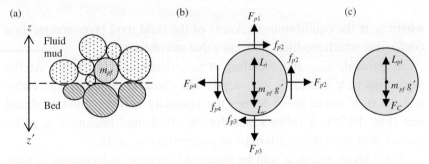

Figure 10.24. (a) Flocs represented as integral particles in bed and in fluid mud; (b) forces on a particle; (c) vertical forces on a particle.

force equals or just exceeds the sum of buoyant weight of the particle and cohesion, the bed is critically stretched, inter-particle connections are broken and the effective normal stress vanishes. We may accordingly select the criterion for liquefaction down to any depth in the bed to be the condition (Fig. 10.24c)

$$L_{pimax} = m_{pf} g' + F_C \qquad (10.38)$$

where L_{pimax} is the amplitude of the upward force at depth z'. By considering bed properties to vary in the vertical direction only, the above equality, which is based on forces on a single particle at depth z', may be reinterpreted as the condition causing the separation of the entire fluid mud layer from the bed at depth z'. Also, as the right hand side of the expression represents a downward pull against lift induced by the wave, we may combine particle weight and cohesion as a single term proportional to g'.

Now let $z_s(z',t)$ be the wave-induced vertical displacement of the bed particle. The corresponding acceleration is given by

$$\ddot{z}_s(z',t) = \frac{d^2 z_s}{dt^2} = \sigma^2 z_s \qquad (10.39)$$

where σ is the wave angular frequency. Given $z_{smax}(z')$ as the amplitude of $z_s(z',t)$ we will restate Eq. 10.38 as

$$\sigma^2 z_{smax}(z_{fe}) = \alpha_{bc} g'(z_{fe}) \qquad (10.40)$$

where z_{fe} is the equilibrium thickness of the fluid mud layer and α_{bc} is a coefficient which modifies buoyancy due to cohesion.

In general, z_{smax} and therefore $\sigma^2 z_{smax}$ decrease with depth. At the same time $\alpha_{bc} g'$ increases with depth. The plane where the $\sigma^2 z_{smax}$ curve and the $\alpha_{bc} g'$ curve cross defines the boundary between fluid mud and bed (Fig. 10.25). It follows that for the fluid mud thickness z_{fe} to be greater than zero, $\sigma^2 z_{smax}(0)$ must be greater than $\alpha_{bc} g'(0)$.

For a given bed $\alpha_{bc} g'$ can be assumed to remain independent of time over the duration of wave motion. Wave height, period and water depth determine $z_{smax}(0)$, and the profile $z_{smax}(z' > 0)$ depends on bed properties

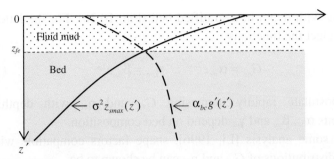

Figure 10.25. Equilibrium thickness of fluid mud over a viscoelastic bed.

[de Wit, 1995]. Accordingly, z_{fe} can be obtained by an approximate method as follows.

We may relate the amplitude of displacement $z_{smax}(z')$ to $z_{smax}(0)$ by

$$z_{smax}(z') = \Psi_{sh}(z') \cdot z_{smax}(0) \qquad (10.41)$$

where Ψ_{sh} is a depth-wise shape function. Equation 10.40 then becomes

$$\sigma^2 \Psi_{sh}(z_{fe}) \cdot z_{smax}(0) = \alpha_{bc} g'(z_{fe}) \qquad (10.42)$$

which can be solved for z_{fe}. The shape function depends on the bed properties and their variation with depth. Since cyclic stretching and compression of mud are of interest, internal shearing may be characterized by viscoelastic parameters from extensional rheology (Chapter 5) according to

$$\tau = G_{ne}\gamma + \eta_{ne}\dot{\gamma} \qquad (10.43)$$

where G_{ne} and η_{ne} are the extensional modulus of elasticity and loss modulus, respectively (Chapter 5). These coefficients are approximately related to the shear modulus G and viscosity η by

$$G_{ne} = \frac{3}{2}G; \quad \eta_{ne} = \frac{3}{2}\eta \qquad (10.44)$$

which permits the determination of G_{ne} and η_{ne} from shear rheometry [Li, 1996]. For illustration we will consider two forms of dependence of these coefficients on the depth, namely uniform in a constant-density bed, and increasing with depth in a stratified bed (Fig. 10.26). For the latter case we will select the functions

$$G_{ne} = \alpha_{ne} e^{\gamma_{ne} z'}; \quad \eta_{ne} = \beta_{ne} e^{\gamma_{ne} z'} \qquad (10.45)$$

which postulate rapidly increasing G_{ne} and η_{ne} with depth. The coefficients α_{ne}, β_{ne} and γ_{ne} depend on bed composition.

After some analysis [Li, 1996], shape factors compatible with the selected distributions of G_{ne} and η_{ne} can be shown to be

$$\Psi_{sh} = 1 - \frac{z'}{h} \qquad (10.46)$$

for the uniform bed and

$$\Psi_{sh} = \frac{e^{-\gamma_{ne} z'} - e^{-\gamma_{ne} h}}{1 - e^{-\gamma_{ne} h}} \qquad (10.47)$$

for the stratified bed. For large depths ($h \to \infty$) Eq. 10.47 reduces to the exponential decay function

$$\Psi_{sh} = e^{-\gamma_{ne} z'} \qquad (10.48)$$

We will now consider bed response to wave-induced pressure at the bed surface. The bed has two layers (1 and 2) from the start, and both are

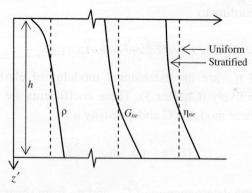

Figure 10.26. Vertical distributions of bed density, extensional elastic modulus and extensional loss modulus.

represented by the Kelvin–Voigt (KV) element. However, by selecting different sets of the coefficients it is permissible to treat the upper layer as soft and fluid-like compared to the lower, relatively harder, layer (Fig. 10.27). Mechanically the two layers are equivalent spring-mass-damper harmonic oscillators in series with two degrees of freedom (as opposed to one degree for a single oscillator). The displacement z_s of each layer is dependent on the equivalent mass-related term m_e, elastic spring constant-related term k_e and viscous damping coefficient-related term c_e. These three parameters depend on the bed properties as follows:

$$k_e = \frac{G_{ne}}{h}; \quad c_e = \frac{\eta_{ne}}{h}; \quad m_e = \int_0^h \rho(z')\Psi_{sh}dz' \quad (10.49)$$

where Ψ_{sh} is given by Eq. 10.46 and h can be h_1 or h_2 . The upper layer surface is oscillated by dynamic pressure $p_b(t)$ having an amplitude p_{bmax} and angular frequency σ (Eq. 10.23).

For this bed representation the equations of motion are

$$m_{e1}\frac{\partial^2 z_{s1}}{\partial t^2} + c_{e1}\left(\frac{\partial z_{s1}}{\partial t} - \frac{\partial z_{s2}}{\partial t}\right) + k_{e1}(z_{s1} - z_{s2}) = p_{bmax}e^{i\sigma t} \quad (10.50)$$

and

$$c_{e1}\left(\frac{\partial z_{s1}}{\partial t} - \frac{\partial z_{s2}}{\partial t}\right) + k_{e1}(z_{s1} - z_{s2}) = m_{e2}\frac{\partial^2 z_{s2}}{\partial t^2} + c_{e2}\frac{\partial z_{s2}}{\partial t} + k_{e2}z_{z2} \quad (10.51)$$

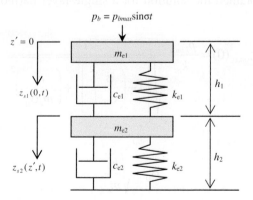

Figure 10.27. Soft bed layers of mud represented as mechanical analogs.

The displacements are specified as

$$z_{s1} = z_{smax1}e^{i(\sigma t - \delta_{s1})}; \quad z_{s2} = z_{smax2}e^{i(\sigma t - \delta_{s2})} \tag{10.52}$$

where z_{smax1} and z_{smax2} are the amplitudes of z_{s1} and z_{s2}, respectively, and δ_{s1}, δ_{s2} are the corresponding phase shifts relative to pressure $p_b(t)$. Substitution of displacements from Eq. 10.51 into Eqs. 10.49 and 10.50 leads to the following expressions for the amplitudes z_{smax1} and z_{smax2}

$$z_{smax1} = P_{bmax}\sqrt{\frac{(k_{e1} + k_{e2} - m_{e2}\sigma^2)^2 + \sigma^2(c_{e1} + c_{e2})^2}{Z_{s1}^2 + Z_{s2}^2}} \tag{10.53}$$

$$z_{smax2} = P_{bmax}\sqrt{\frac{k_{e1}^2 + \sigma^2 c_{e1}^2}{Z_{s1}^2 + Z_{s2}^2}} \tag{10.54}$$

where

$$Z_{s1} = c_{e2}\sigma(k_{e1} - m_{e1}\sigma^2) + c_{e1}[k_{e2} - (m_{e1} + m_{e2})]\sigma^3 \tag{10.55}$$

$$Z_{s2} = (k_{e2} - m_{e2}\sigma^2)(k_{e1} - m_{e1}\sigma^2) - (k_{e1}m_{e1} + c_{e1}c_{e2})\sigma^2 \tag{10.56}$$

Two limiting cases must be mentioned. In one, setting $k_{e1} = 0$ would change the upper layer from a KV solid into a Newtonian fluid. In the second limit, with $c_{e1} \to \infty$ or $k_{e1} \to \infty$ and $m_{e1} = 0$ the upper layer ceases to exist and we obtained the solution for a single-layer harmonic oscillator, *i.e.,*

$$z_{smax}(0) = \frac{P_{bmax}}{k_e} \frac{1}{\sqrt{\left[1 - \left(\dfrac{\sigma}{\sigma_0}\right)^2\right]^2 + \left(\dfrac{c_e}{m_e\sigma_0}\right)^2}} \tag{10.57}$$

where

$$\sigma_0 = \sqrt{\frac{k_e}{m_e}} \tag{10.58}$$

is the resonant frequency of the oscillator.

Example 10.2: Consider a soft fine-sediment bed without fluid mud at the beginning. The water depth is 19 cm, mud thickness is 16 cm, water density is 1,000 kg m^{-3}, uniform mud density is 1,170 kg m^{-3}, wave amplitudes are 2, 3 and 5 cm, wave angular frequency range is 1 to 10 rad s^{-1}, shear modulus of elasticity is 356 Pa and the viscous coefficient is 9 Pa.s. Assume $\alpha_{bc} = 1$. For each wave amplitude plot the variation of fluid mud thickness with frequency (0.5 to 2 Hz).

From Eq. 10.44 we obtain $G_{ne} = 534$ Pa and $\eta_{ne} = 13.5$ Pa.s. Thus, $k_e = 3,337.5$ N m^{-3} and $c_e = 84.4$ N.s m^{-3}. The shape function is

$$\Psi_{sh}(z') = 1 - 6.25z' \qquad (10.59)$$

From Eq. 10.49 we obtain $m_e = 93.6$ kg m^{-2}, and from Eq. 10.58 $\sigma_0 = 5.97$ rad s^{-1}. From Eq. 2.133 the pressure amplitude p_{bmax} is a function of frequency σ by way of the wave number k (Eq. 2.119). Note that in Eq. 2.133 we set $\cosh(h+z) = 1$ at the water bottom. From Eq. 10.57, $z_{smax}(0)$ is expressed as a function of σ and substituted into Eq. 10.42 along with $\Psi_{sh}(z_f) = 1 - 6.25 z_f$ and g' = 1.67 m s^{-1}. The ensuing expression is solved for z_{fe} in the frequency range of 0.5 to 2 Hz and amplitude range of 2 to 5 cm.

Results in Fig. 10.28 indicate that for a given frequency, the thickness of the liquefied layer increases with increasing wave amplitude, which also determines the frequency range over which liquefaction occurs. For given amplitude, at very low frequencies wave pressure is insufficient for liquefaction. Likewise at very high frequencies liquefaction does not occur because the pressure effect does not reach the bottom. Within the frequency range over which liquefaction does occur, fluid mud thickness reaches a maximum at the characteristic frequency $\sigma_0 = 5.97$ rad s^{-1}, or 0.95 Hz.

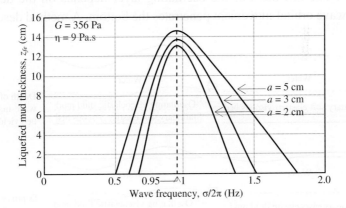

Figure 10.28. Variation of liquefied mud thickness with wave frequency and amplitude.

10.4.6 *Dynamic thickness of fluid mud*

The thickness of liquefied mud layer by wave motion is practically coincident with the oscillating layer that contributes to wave energy loss and damping. On the other hand, if the initial bottom is fluid mud, the thickness of the layer which oscillates with the waves (and absorbs energy) is *a priori* unknown. In wave prediction models (Chapter 11) this thickness is a required input. Its value can be extracted by back-calculation if wave damping is measured, which would however mean that the wave model would not be entirely useful for predicting wave damping under different conditions. Commonly therefore a value of the fluid mud layer participating in wave-mud interaction is assumed as a tuning parameter.

Thus far we have considered methods to estimate the liquefied mud thickness when the bottom is a soft solid (Fig. 10.29a). The thickness of the oscillating layer depends on the depth up to which waves cause the bed to yield. A corollary case is one of bottom mud that is initially at rest, or, more realistically, sustained in the liquefied state by (mild) wave action, with wet bulk density low enough ($< \sim 1{,}200$ kg m^{-3}) to be called fluid mud. Our interest is in the equilibrium thickness z_{fe} of this mud (Fig. 10.29b). The thickness of the oscillating layer depends on the depth to which wave orbital movement penetrates the bottom. A method described

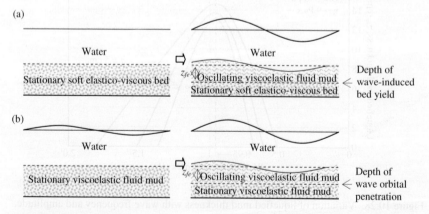

Figure 10.29. Wave effects on bottom mud: (a) Phase change from stationary soft solid (bed) to oscillating fluid mud; (b) transition from stationary elastico-viscous fluid mud to oscillating and fully liquefied viscoelastic mud.

by Robillard [2009] yields an approximate value. It was developed for mud from the Atchafalaya submersed delta along the Gulf of Mexico coast of Louisiana (Table 5.5).

Referring to Fig. 10.30, water with mean depth h is inviscid and shallow relative to the wave length. At rest the mud layer is in a fluid-like elastico-viscous state. When waves occur and the strain rate attains a value $\dot{\gamma}_{liq}$, the bottom changes to a viscoelastic fluid mud due to a reduction in the elastic component of the viscosity η'', as schematized in Fig. 5.43. Thus the equilibrium thickness z_{fe} is defined by the plane at which the induced shear rate $\dot{\gamma}(z')$, where $z' = z + h$, is equal to $\dot{\gamma}_{liq}$. Since $\dot{\gamma}_{liq}$ is an empirical coefficient, it is necessary to obtain an expression for $\dot{\gamma}(z')$ in terms of measurable quantities.

Mud motion is produced by the horizontal velocity $u_b(x,t)$ at the water bottom ($z = -h$) and is obtained from Eq. 2.124 assuming inviscid flow. Within mud, viscous energy loss is high and the rotational velocity u_r is the main contributor to motion. This velocity is deduced from the simplified momentum balance

$$\frac{\partial u_r}{\partial t} = \frac{\left| \eta^* \right|}{\rho} \frac{\partial^2 u_r}{\partial z^2} \qquad (10.60)$$

in which the absolute value of the complex viscosity η^* is given by Eq. 5.144 for the delta mud. To obtain u_r, Eq. 10.60 is integrated along with the specification that u_r must match $u_b(x,t)$ at the mud surface ($z = -h$).

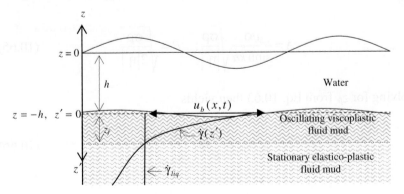

Figure 10.30. Definition sketch for the estimation of dynamic fluid mud layer thickness.

The resulting expression is

$$u_r\left(x,z',t\right)=\frac{a\sigma}{\sinh kh}e^{-\sqrt{\frac{\sigma\rho}{2|\eta^*|}}z'}\cos\left(kx-\sigma t+\sqrt{\frac{\sigma\rho}{2|\eta^*|}}z'\right) \quad (10.61)$$

Therefore

$$\dot\gamma(x,z',t)=\frac{\partial u_r}{\partial z'}=$$

$$\frac{a\sigma}{\sinh kh}\sqrt{\frac{\sigma\rho}{|\eta^*|}}e^{-\sqrt{\frac{\sigma\rho}{2|\eta^*|}}z'}\left[\sin\left(kx-\sigma t+\sqrt{\frac{\sigma\rho}{2|\eta^*|}}z'+\frac{\pi}{4}\right)\right] \quad (10.62)$$

At $z'=z_{fe}$ the amplitude of this harmonic function is equated to $\dot\gamma_{liq}$, *i.e.*,

$$\frac{a\sigma}{\sinh kh}\sqrt{\frac{\sigma\rho}{|\eta^*|}}e^{-\sqrt{\frac{\sigma\rho}{2|\eta^*|}}z_{fe}}=\dot\gamma_{liq} \quad (10.63)$$

in which $|\eta^*|$ is evaluated from Eq. 5.144 at $\dot\gamma=\dot\gamma_{liq}$, *i.e.*,

$$|\eta^*|=\frac{\eta'_\infty+\eta'_{lev}}{1+\sin\delta_{lev}} \quad (10.64)$$

Let

$$A=\frac{a\sigma}{\sinh kh}\sqrt{\frac{\sigma\rho}{|\eta^*|}};\quad B=\sqrt{\frac{\sigma\rho}{2|\eta^*|}} \quad (10.65)$$

Solving for z_{fe} from Eq. 10.63 then yields

$$z_{fe}=\frac{1}{B}\ln\left(\frac{A}{\dot\gamma_{liq}}\right) \quad (10.66)$$

Example 10.3: Estimate the equilibrium thickness z_{fe} of a marine mud based on the following laboratory data: $a = 0.03$ m, $\sigma = 4$ rad s^{-1}, $h = 0.19$ m, $\rho = 1,200$ kg m^{-3}, $\eta'_\infty = 0.07$ Pa.s, $\eta'_{lev} = 0.94$ Pa.s, $\sin\delta_{lev} = 0.15$ and $\dot{\gamma}_{liq}$ 2.25 (s^{-1}). Assume waves are in shallow water.

In shallow water the wave number $k = 2.93$ m^{-1}. Then, from Eq. 10.64 we obtain $|\eta^*| = 0.878$ Pa.s. Next, with $A = 15.1$ and $B = 52.3$ from Eq. 10.65, Eq. 10.66 yields $z_{fe} = 0.036$ m $= 3.6$ cm.

10.5 Suspension Mixing

10.5.1 *Mixing criterion*

We have considered some of the mechanisms associated with scenarios 1 through 4 in Fig. 10.1. Relative to scenarios 5, 6 and 7 involving mixing and destratification, energy must be supplied by waves, and for scenario 9 by a current. In stratified flows, vertical mixing is retarded because work must be done to raise heavy fluid parcels containing particles and (for volumetric continuity) lower the relatively lighter parcels. Also, sufficient energy must be available to overcome particle settling, and to balance energy loss when inter-particle collisions occur, especially when the suspension is dense. For dilute suspensions considered below we will ignore particle related effects in order to emphasize the main factors responsible for mixing.

Consider a stratified water column with two fluid layers (Fig. 10.31) of depths h_1 and h_2, densities ρ_1 and ρ_2 with $\rho_2 > \rho_1$, *i.e.*, in a stable condition, and depth–mean velocities u_1 and u_2, respectively. Mixing can occur as long as the two velocities are unequal. Once mixing is complete the density of the fluid will be ρ and velocity u.

Treating the simple case of two layers of the same height, *i.e.*, $h_1 = h_2$, the center of mass of the two layers together would be at elevation $(3\rho_1 + \rho_2)h_1 / 2(\rho_1 + \rho_2)$, whereas in the fully-mixed fluid the center of mass would be at $h/2$, where $h = h_1 + h_2 = 2h_1$. Therefore, since $\rho_2 > \rho_1$, the following condition must be satisfied

$$\frac{(3\rho_1 + \rho_2)h_1}{2(\rho_1 + \rho_2)} < \frac{h}{2} \qquad (10.67)$$

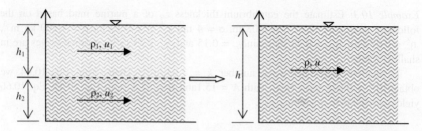

Figure 10.31. Fully-stratified and fully-mixed water column.

In other words, work would be required to raise the potential energy per unit volume (*PE*) of the system to mix the fluids. The excess potential energy ΔPE (per unit volume) is equal to the *PE* of the mixed system minus that of the stratified system. Given $h_1 = h_2$ we obtain

$$\Delta PE = \frac{1}{8}(\rho_2 - \rho_1)gh^2 \qquad (10.68)$$

This increase in *PE* must be equal to the excess kinetic energy per unit volume ΔKE extracted from flow. By invoking the Boussinesq approximation (*e.g.*, Tennekes and Lumley [1972]), *i.e.*,

$$\rho_1 \approx \rho_2 \approx \rho \qquad (10.69)$$

we obtain

$$\Delta KE = \frac{1}{8}\rho(u_1 - u_2)^2 h \qquad (10.70)$$

Thermodynamically, mixing can occur only if ΔKE is greater than ΔPE. In other words

$$\frac{(\rho_2 - \rho_1)gh}{\rho(u_1 - u_2)^2} = \frac{g'h}{(u_1 - u_2)^2} < 1 \qquad (10.71)$$

i.e.,

$$Ri_0 = \frac{g'h}{(u_1 - u_2)^2} < 1 \qquad (10.72)$$

where g' is the reduced gravity and Ri_0 is a bulk Richardson number. When this criterion is not met molecular mixing due to thermal energy is limited to the region close to the mud-water interface. When the criterion is satisfied, interfacial undulations such as those seen at the top of the lutocline in Fig. 10.1 can appear.

In a system of two layers separated by a sharp interface such as those in Fig. 10.31, the vertical gradients of velocity and density are theoretically infinite. In contrast a common case is one of continuously stratified flow with gradually varied gradients in the interfacial region (Fig. 10.32). Let us consider the energy balance for two neighboring (one above another) fluid parcels. They have equal volumes and are at heights z and $z+dz$. If they are interchanged, the change in the potential energy per unit volume corresponding to work needed to overcome gravity is

$$\Delta PE = -g \frac{d\rho}{dz} \qquad (10.73)$$

For conservation of horizontal momentum the particle at z must acquire a new velocity $u+\Theta(du/dz)dz$ and the particle at $z+dz$ will have its new velocity $u+(1-\Theta)(du/dz)dz$, where Θ is some number between 0 (*i.e.*, no change in velocity) and 1 (*i.e.*, maximum change in velocity). The kinetic energy change in the basic flow as it releases this energy is

$$\Delta KE = \frac{1}{2}\rho u^2 + \frac{1}{2}\left(\rho + \frac{d\rho}{dz}dz\right)\left(u + \frac{du}{dz}dz\right)^2$$

$$-\frac{1}{2}\rho\left(u + \Theta\frac{du}{dz}dz\right)^2 - \frac{1}{2}\left(\rho + \frac{d\rho}{dz}dz\right)\left[u + (1-\Theta)\frac{du}{dz}dz\right]^2 \qquad (10.74)$$

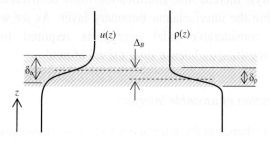

Figure 10.32. Density and velocity profiles in continuously-stratified flow.

or

$$\Delta KE = \Theta(1-\Theta)\rho\left(\frac{du}{dz}dz\right)^2 + u\frac{du}{dz}dz \cdot \frac{d\rho}{dz}dz \qquad (10.75)$$

Since the maximum value of $\Theta(1-\Theta) = 1/4$ occurs at $\Theta = 0.5$, we may restate the above equality as

$$\Delta KE \le \frac{1}{4}\rho\left(\frac{du}{dz}dz\right)^2 + u\frac{du}{dz}dz \cdot \frac{d\rho}{dz}dz \qquad (10.76)$$

Since for an interchange of fluid parcels ΔKE must be greater than ΔPE, we may replace ΔKE in Eq. 10.76 by ΔPE from Eq. 10.73 to yield

$$-g\frac{d\rho}{dz} < \frac{1}{4}\rho\left(\frac{du}{dz}dz\right)^2 + u\frac{du}{dz}dz \cdot \frac{d\rho}{dz}dz \qquad (10.77)$$

which reduces to the well-known condition

$$Ri_g = -\frac{g(\partial\rho/\partial z)}{\rho(\partial u/\partial z)^2} < \frac{1}{4} \qquad (10.78)$$

This inequality indicates that for mixing the gradient Richardson number Ri_g must be less than $1/4$. The Eq. 10.78 criterion is qualitatively analogous to Eq. 10.72. For a thin fluid column of thickness Δz we may express Ri_g in Eq. 10.78 by Ri_0. This would indicate that mixing a continuously stratified fluid requires greater kinetic energy than one that is fully-stratified, given the same buoyancy force (represented by reduced gravity). Incremental stratification must be overcome at every elevation within the interfacialcur boundary layer. As we will see later, in estuaries considerable tidal energy is required to break up continuously stratified and dense suspensions of mud.

10.5.2 *Generation of unstable interface*

Whether a disturbance at the interface between water and suspended mud dies out or grows depends on the densities and velocities of the two layers. Instability at a weakly stratified parallel shear flow leads to a warping of the interface which under certain conditions takes the form of Kelvin–Helmholtz (KH) billows. In general the degree of stratification can be characterized by layer thicknesses δ_u representing the velocity defect, the density jump δ_ρ and the offset Δ_B between the velocity and the density gradients (Fig. 10.32). The condition for the primary wave mode to be KH is $\delta_u \approx \delta_\rho$, $\Delta_B \approx 0$ and a sufficiently low gradient Richardson number Ri_g.

Other noteworthy effects include the interaction between a compliant, weakly rigid but not significantly viscous cohesive bed and the so-called Benjamin–Feir instabilities within frequency side-bands bracketing the water wave frequency. These interactions can result in substantial wave damping, although their significance in the natural mud environment remains unknown because high damping also leads to a rapid die-out of the side-bands [Foda, 1989].

The treatment of KH waves in inviscid flows, although approximate, is instructive for understanding the condition actuating interfacial warping. Referring to Fig. 10.33, consider the flows of two incompressible, inviscid fluids of densities ρ_1 and ρ_2 and velocities u_1 and u_2, respectively. Both fluids are effectively semi-infinite in thickness. It is assumed that $\eta(x,t)$ is small compared to the wave length, which permits linearization of the relevant equations [Lamb, 1945].

The velocity potentials ϕ_1 and ϕ_2 in the two layers must satisfy equations of flow continuity

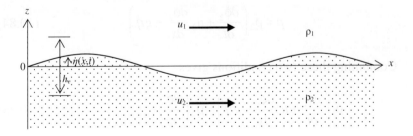

Figure 10.33. Definition sketch of flow producing KH instability.

$$\nabla^2 \phi_1 = 0$$
$$\nabla^2 \phi_2 = 0 \tag{10.79}$$

For which the respective velocity potentials are

$$\phi_1 = u_1 x + \phi_1'$$
$$\phi_2 = u_2 x + \phi_2' \tag{10.80}$$

where $u_1 x$ and $u_2 x$ are the steady current components and ϕ_1' and ϕ_2' the wave-induced components of ϕ_1 and ϕ_2, respectively. At the interface (at $z = 0$ in this linearized treatment) the kinematic and the dynamic conditions are invoked to obtain the interfacial wave celerity C_{wi}. The kinematic conditions for the two fluids at the interface are

$$\frac{\partial \eta}{\partial t} + u_1 \frac{\partial \eta}{\partial x} = -\frac{\partial \phi_1}{\partial z} \tag{10.81}$$

$$\frac{\partial \eta}{\partial t} + u_2 \frac{\partial \eta}{\partial x} = -\frac{\partial \phi_2}{\partial z} \tag{10.82}$$

where $\eta(x,t)$ is the interfacial elevation relative to the still water surface. The dynamic pressure $p(x,t)$ at the interface is obtained for the two fluids in the linearized forms as

$$p = \rho_1 \left(\frac{\partial \phi_1}{\partial t} + u_1 \frac{\partial \phi_1}{\partial x} - g\eta \right) \tag{10.83}$$

$$p = \rho_2 \left(\frac{\partial \phi_2}{\partial t} + u_2 \frac{\partial \phi_2}{\partial x} - g\eta \right) \tag{10.84}$$

Combining these two conditions gives

$$\rho_1 \left(\frac{\partial \phi_1}{\partial t} + u_1 \frac{\partial \phi_1}{\partial x} - g\eta \right) = \rho_2 \left(\frac{\partial \phi_2}{\partial t} + u_2 \frac{\partial \phi_2}{\partial x} - g\eta \right) \tag{10.85}$$

With both fluids of unlimited depths the following harmonic functions are chosen for $\phi_1'(x, z, t)$ and $\phi_2'(x, z, t)$, respectively

$$\phi_1'(x,z,t) = A_1 e^{kz} e^{i(kx-\sigma t)} = A_1 e^{kz+i(kx-\sigma t)} \tag{10.86}$$

$$\phi_2'(x,z,t) = A_2 e^{-kz} e^{i(kx-\sigma t)} = A_2 e^{-kz+i(kx-\sigma t)} \tag{10.87}$$

$$\eta(x,t) = ae^{i(kx-\sigma t)} \tag{10.88}$$

where k is the wave number, σ is the wave angular frequency and a is the amplitude of the interfacial wave. The velocity potential amplitudes $A_1 e^{kz}$ and $A_2 e^{-kz}$ rapidly die away with distance from the interface, and the effect of the interfacial oscillation is confined to the region in the vicinity of the interface. Coefficients A_1 and A_2 are obtained by substitution of the above three expressions into by Eqs. 10.81, 10.82 and 10.85. These steps lead to the wave dispersion equation

$$\frac{\sigma}{k} = \frac{\rho_1 u_1 + \rho_2 u_2}{\rho_1 + \rho_2} \pm \sqrt{\frac{g}{k}\frac{\rho_2 - \rho_1}{\rho_1 + \rho_2} - \frac{\rho_1 \rho_2}{(\rho_1 + \rho_2)^2}(u_2 - u_1)^2} \tag{10.89}$$

in which the first term on the right hand side is the mean velocity of the system. The second term is the interfacial wave celerity (relative to a frame of reference moving with the mean velocity)

$$C_{wi} = \sqrt{\frac{g}{k}\frac{\rho_2 - \rho_1}{\rho_1 + \rho_2} - \frac{\rho_1 \rho_2}{(\rho_1 + \rho_2)^2}(u_2 - u_1)^2} \tag{10.90}$$

In the special case of $\rho_2 \gg \rho_1$, *e.g.*, when the upper layer is air and the lower one is water, setting $\rho_1 = 0$ we recover $C_{wi} = \sqrt{g/k}$, the celerity of a linear deep-water surface wave (Eq. 2.123). Another feature of Eq. 10.90 is that when

$$\frac{g(\rho_2 - \rho_1)}{k_{st}\rho_1(u_2 - u_1)^2} = \frac{g'}{k_{st}(u_2 - u_1)^2} = Ri_{0L} < \frac{\rho_2}{\rho_1 + \rho_2} \approx \frac{1}{2}(\text{when } \rho_1 \approx \rho_2) \tag{10.91}$$

the celerity becomes an imaginary number. Here k_{st} is the threshold wave number (and $L_{st} = 2\pi/k_{st}$, the wave length) characterizing interfacial stability. Instability occurs if $k > k_{st}$ or $Ri_{0L} < 1/2$, where Ri_{0L} is a bulk Richardson number based on k_{st}. Qualitatively this condition is similar to one for complete mixing in Eq. 10.76, because both depend on the relative roles of the inertia force associated with kinetic energy production and buoyancy force with the potential energy. In both cases stability implies a low ratio of buoyancy to inertia and instability a high ratio.

10.5.3 *Observations of interfacial disturbance*

10.5.3.1 *Laboratory setting*

Equation 10.91 defines the threshold of theoretically unbounded growth. The analysis does not indicate the rate of growth of the resulting KH billows, or the rate of mixing of the two fluids. Inclusion of the non-linear terms in the governing equations and boundary conditions coupled with numerical modeling may enhance the analysis; however, the basis for validation of results would remain in experimental evidence.

Several methods have been used in the laboratory to examine the effects of disturbances at the mud-water interface, relying on procedures developed for haloclinic and thermoclinic boundaries.

Using a simple lock-exchange flow setup in a Plexiglas channel (Fig. 10.34), turbid flows of dense suspensions of kaolinite (200 kg m^{-3} and higher) were produced [Scarlatos and Wilder, 1990]. In each test, initially the suspension was separated from sediment-free saltwater in the channel by a closed gate. It was then raised rapidly to permit the resulting

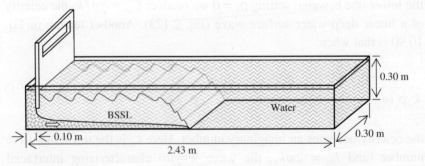

Figure 10.34. Schematic drawing of interfacial instabilities in lock-exchange flow setup (from Scarlatos and Mehta [1993]).

turbid underflow to move into the channel. Transient undulations occurred at the secondary lutocline boundary of the so-called benthic suspended sediment layer (BSSL) between clear water and the dense underflow. In BSSL the mean sediment concentration was on the order of 0.5 kg m^{-3}. The wave lengths of the undulations ranged between 0.05 and 0.2 m and amplitudes between 0.005 and 0.025 m. The velocity difference between the BSSL and water had an average value of 0.15 m s^{-1}.

Example 10.4: From tests in a lock-exchange flow setup the following values were derived: $\rho_1 = 1{,}000$ kg m^{-3}, $\rho_2 = 1{,}020$ kg m^{-3} and $u_2 - u_1 = 0.12$ m s^{-1}. Determine if the critical condition for the growth of the undulations was met, given $k_{st} = 40$ m^{-1}.

From Eq. 10.91 we obtain $Ri_{0L} = g(\rho_2 - \rho_1)/k_c \rho_1 (u_2 - u_1)^2 = 0.34$, which is less than 0.5. Therefore one would expect the undulations to grow.

In another experimental setup, a small Plexiglas channel initially containing a BSSL of kaolinite (1 kg m^{-3}) in saltwater below a layer of sediment-free saltwater was placed in a wind tunnel [Scarlatos and Zhang, 1991]. Wind-induced circulation produced convective mixing by billows at the BSSL-water interface with wave lengths ranging from 0.06 to 0.1 m and amplitudes on the order of 0.01 m. The velocity discontinuity across the interface (Fig. 10.32) was about 0.01 m s^{-1}. The state of the interface was defined by the Wedderburn number

$$Wd = \frac{Ri_*}{a_R'}; \quad Ri_* = \frac{2gh_1(\rho_2 - \rho_1)}{(\rho_2 + \rho_1)u_*^2} \tag{10.92}$$

where Ri_* is the bulk Richardson number based on the friction velocity u_* with subscript 1 for the water layer and 2 for the sediment suspension. In the eddy scale parameter $a_R' = \lambda_0'/h_1$, λ_0' is the characteristic dimension of the eddy associated with the undulations. When $Wd > 1$, a sloping interface due to wind stress occurred; however mixing was minor. When $Wd < 1$, KH instabilities occurred along with high mixing at the interface [Thompson and Imberger, 1980].

Example 10.5: In a two-layered flow setup driven by a constant wind, Scarlatos and Zhang [1991] carried out two tests (nos. 4 and 5) with layer densities specified in terms of the condition $2(\rho_2-\rho_1)/(\rho_2+\rho_1) = 0.04$. The wind produced water current with $u_* = 6.3$ mm s^{-1}. The following data were reported:

Test no.	h_1 (cm)	χ_0 (cm)
4	7	34
5	2	52

Determine the state of the interface in each test.

We obtain the following values of the Richardson number and the Wedderburn number:

Test no.	Ri_*	Wd
4	69.1	14.2
5	19.7	0.8

which suggests that KH instabilities would have occurred in test no. 5 only.

In addition to KH, a second mode of interfacial undulations, the Holmboe instability, can be identified under the conditions $\delta_u \gg \delta_\rho$ and $\Delta_B = 0$ (Fig. 10.32). This consists of two trains of growing waves traveling opposite to the mean flow. They eventually result in a series of sharply cusped crests protruding alternatively into each layer, with wisps of fluid erupting from the cusps. In laboratory experiments it is common to find cusps mainly protruding into the upper fluid. In Fig. 10.35, the lower layer is a clayey suspension (1,030 kg m^{-3}) with the interface sheared by freshwater flow above [Holmboe, 1962; Hazel, 1972; Mehta and Srinivas, 1993].

The state of the interface in experiments producing Holmboe like features is often characterized by the bulk Richardson number

$$Ri_{he} = \frac{g(\rho_2 - \rho_1)h_e}{\rho_1(u_2 - u_1)^2} = \frac{g'h_e}{(u_2 - u_1)^2} \qquad (10.93)$$

where h_e is the thickness of the moving layer scaled by the interfacial wave height (Fig. 10.33). When only the top layer of thickness h moves with a depth-mean velocity u, the Richardson number becomes

$$Ri_h = \frac{g'h}{u^2} \qquad (10.94)$$

The condition in Fig. 10.35 is characterized by $Ri_h \approx 12$.

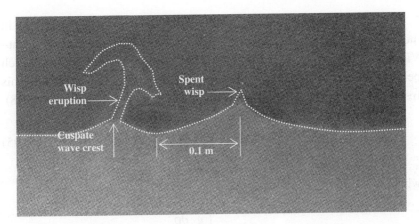

Figure 10.35. Erupting wisp and a spent wisp of sediment at interfacial crests in a flume experiment (from Mehta and Srinivas [1993]).

10.5.3.2 *Field setting*

Large-scale interfacial oscillations have been reported at the top of a gravity-driven turbid underflow out of the Huang He (River) and into the Bohai Gulf, China. The description in Fig. 10.36 was obtained with a 200 kHz Ross fathometer. The suspension in the so-called hyperpycnal plume contained sediment in the silt-size range. Near the surface a lighter (hypopycnal) plume of lower sediment concentration and fresher water moved offshore as a surface jet (see also Chapter 12) of the river discharge. The interfacial wave amplitudes were 1-m and higher, and in the frequency range of 2.5×10^{-3} to 5×10^{-3} Hz [Wright *et al.*, 1988].

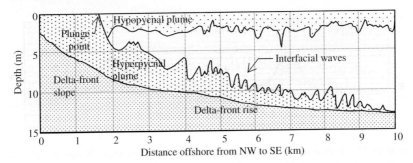

Figure 10.36. Interfacial oscillations at the surface of a dense (hyperpycnal) suspended sediment underflow out of the Huang He in China (from Wright *et al.* [1988]).

A characteristic feature of free undulations at the interface such as in the Huang He plume is that they occur at their natural resonant Brunt–Väisälä or buoyancy frequency f_{0b} [Tennekes and Lumley, 1972], which permits an estimation of the density gradient across the interface. This frequency is associated with the gradient Richardson number (Eq. 10.78)

$$Ri_g = -\frac{f_{0b}^2}{(\partial u / \partial z)^2} \tag{10.95}$$

where

$$f_{0b} = \sqrt{\frac{g}{\rho_w} \frac{\partial \rho}{\partial z}} \tag{10.96}$$

Example 10.6: The mean buoyancy frequency f_{0b} in the Huang He plume was found to be 3.8×10^{-3} s^{-1}. Assume a water density of 1,027 kg m^{-3} and calculate the density gradient across the interface.

From Eq. 10.96 we obtain $\Delta\rho/\Delta z \approx \partial\rho/\partial z = 1.5 \times 10^{-3}$ kg m^{-4}. Taking 2 m as the characteristic vertical dimension of the interfacial oscillation yields $\Delta\rho = 3 \times 10^{-3}$ kg m^{-3}, a low value which confirms that the hyperpycnal plume was dilute.

Waves at the top of a fluid mud layer can interfere with navigation if they are common and have large amplitudes. Well organized, ~2 m high and non-linear waves of cuspate configuration are seen in Fig. 10.37 at the upper surface of fluid mud in the Rotterdam Waterway near the entrance to Europort in the Netherlands. These interfacial waves were

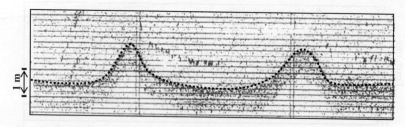

Figure 10.37. Fathometer record at the mouth of the Rotterdam Waterway after the passage of a large vessel. Fluid mud is dredged from this channel regularly to maintain the navigation routes (from van Leussen and van Velzen [1989]).

actuated by the passage of a large vessel with a draft of 15.5 m in 25 m deep water [van Leussen and van Velzen, 1989].

In the Jiaojiang estuary in China, instabilities at the top of the lutocline (Fig. 10.5) were observed at two distinct frequencies ("low" and "high"). In Table 10.4 characteristic parameters for the two sets of waves are derived from three 550 s long time-series of interface elevation [Jiang, 1999]. The mean height of the long internal waves was 0.38 m and its frequency was in the range of 0.07 to 0.11 rad s^{-1}. The buoyancy frequency f_{0b} was calculated as $\sqrt{g'/h_e}$, where and h_e is the effective thickness of the interfacial layer. For example, given $g' = 0.096$ m s^{-2} and $h_e = 1.38$ m yields $f_{0b} = 0.26$ s^{-1}. Overall, the frequency f ranged between 0.20 to 0.26 rad s^{-1}. Given that the definition of h_e is somewhat arbitrary (Fig. 10.33), it can be inferred that the measured and estimated long internal wave frequencies were comparable. In turn this would imply that the long-wave interface was oscillating freely. The short internal wave with a mean height 0.21 m had a frequency range of 1.1–1.2 rad s^{-1}, comparable with the surface wave frequency [Jiang, 1999]. Thus the short interfacial wave was the source of energy for the long interfacial wave.

Table 10.4 gives the celerities C_{wi} of the two sets of interfacial waves, the corresponding wave lengths L and the characteristic KH wave length L_{st} below which the interface would be unstable. These values imply that the long waves were stable but short ones were not. The role of unstable waves in entraining sediment was confirmed by measurement of suspended sediment concentration above and below the interface.

Table 10.4. Low and high-frequency interfacial wave parameters in the Jiaojiang

h_e (m)	g' (m s^{-2})	f_{0b} (s^{-1})	Low-frequency			High-frequency			L_{st} (m)
			f (s^{-1})	C_{wi} (m s^{-1})	L (m)	f (s^{-1})	C_{wi} (m s^{-1})	L (m)	
1.38	0.096	0.26	0.11	0.46	28	1.2	0.08	0.39	2.71
1.50	0.097	0.25	0.09	0.47	32	1.2	0.08	0.44	1.88
2.86	0.119	0.20	0.07	0.54	48	1.1	0.11	0.65	0.06

Source: Jiang and Mehta [2002].

10.5.4 *Effect of stratification on mass diffusion*

Due to the effect of density stratification on the turbulent eddy length scale, mass diffusion in all directions is altered relative to non-stratified or homogeneous flows. We will consider vertical (z-direction) diffusion by Fick's law applied to turbulent mass transport (see Eq. 6.101)

$$F_d = -D_s \frac{\partial C}{\partial z} \qquad (10.97)$$

where F_d is the mass flux and D_s the mass diffusivity. The negative sign ensures that mass transport occurs in the direction of decreasing gradient $\partial C/\partial z$ of suspended sediment concentration C. The diffusivity can be obtained by measuring F_d at a given concentration gradient. Alternatively it can be estimated by resorting to the use of the k-ε equations (Appendix E) for the transport of turbulent kinetic energy (TKE). Even so empiricism cannot be entirely avoided when dealing with dense suspensions in which the effect of concentration on turbulence is not negligible [Hsu *et al.*, 2007]. Presently we will consider D_s to be simply obtained from

$$D_s = D_{so} \Phi_s \qquad (10.98)$$

where

$$\Phi_s = f^n \left(1 + \frac{z}{L_m} \right) \qquad (10.99)$$

Therefore

$$D_s = D_{so} \cdot f^n \left(1 + \frac{z}{L_m} \right) \qquad (10.100)$$

Here D_{so} is the neutral diffusion coefficient applicable to homogenous flows and $\Phi_s (\leq 1)$ is the damping term arising from stratification. It is the mass transport counterpart of Φ_m for momentum transport mentioned in Chapter 2. The quantity L_m is the Monin–Obukov eddy length scale

(Eq. 2.78), which is obtained from the steady-state form of the TKE balance from a simplification of Eq. 9.68 as

$$-\frac{1}{\rho}(\overline{\rho u'w'})\frac{du}{dz} = -\frac{g}{\rho}\overline{\rho'w'} + \varepsilon_t \qquad (10.101)$$

For specification of L_m consider a flow in which production of energy (represented by the term on the l.h.s.) and buoyancy flux (first term of the r.h.s.) required to sustain the particles in suspension are in approximate balance, thus permitting us to ignore ε_t. Accordingly, we will characterize the boundary layer height by an arbitrary near-bed elevation z_b according to

$$-\overline{\rho u'w'}\frac{du}{dz}\bigg|_{z=z_b} = -g\overline{\rho'w'}\bigg|_{z=z_b} \qquad (10.102)$$

Representing the Reynolds stress $-\overline{\rho u'w'}$ in terms of friction velocity u_*, and the shear rate du/dz in terms of log-velocity distribution (Eq. 2.57) yields

$$\rho u_*^2 \frac{u_*}{\kappa z}\bigg|_{z=z_b} = -g\overline{\rho'w'}\bigg|_{z=z_b} \qquad (10.103)$$

Then, conveniently scaling the elevation z by L_m we obtain

$$L_m = \frac{\rho u_*^3}{\kappa g\overline{\rho'w'}} \qquad (10.104)$$

which indicates that L_m is a measure of the ratio of kinetic energy of shear flow to the potential energy of the buoyant suspension. When inertia dominates z/L_m in Eq. 10.99 approaches zero with $\Phi_s \to 1$ and $D_s \to D_{so}$. The greater the degree of stratification the lower will be the value of Φ_s and greater the effect of buoyancy on diffusion. Thus when the flow consists of water above a dense suspension, D_s can significantly differ from D_{so} [Friedrichs *et al.*, 2000].

Empirical expressions relating Φ_s to relevant parameters are deduced from velocity and concentration or density profiles. The following formula based on phenomenological analysis [Rossby and Montgomery, 1935; Munk and Anderson, 1948] is commonly used

$$\Phi_s = \frac{1}{\left(1 + \alpha_{st} Ri_g\right)^{\beta_{st}}} \qquad (10.105)$$

where α_{st} and β_{st} are sediment dependent coefficients and Ri_g is the gradient Richardson number (Eq. 10.78). Some values of these coefficients for diffusion by a current or by waves are given in Table 10.5.

The flux Richardson number Ri_f is the appropriate parameter characterizing the stratifying effect of sediment relative to destratification by shear production and mixing. This number

$$Ri_f = -\frac{g\overline{\rho' w'}}{\rho \overline{u'w'} \dfrac{du}{dz}} \qquad (10.106)$$

quantifies the relative effects of two terms in Eq. 10.102. In the wave-dominated environment with fine suspended sediment, Ri_f can vary significantly with water depth and thereby identify depths at which damping due to stratification would be important (high Ri_f) and where kinetic energy production would dominate (low Ri_f). Safak *et al.* [2010] collected time-series data on hydrodynamic and suspended sediment parameters in 5 m deep water over the submersed delta of the Atchafalaya River along the coast of Louisiana. In this environment tidal currents (~0.3 m s^{-1} at peak) are strong enough to resuspend fine sediment. However, frontal waves (with a significant wave height of

Table 10.5. Turbulence damping related coefficients.

Source/forcing	α_{st}	β_{st}
Ross [1988] /current	4.2	2.0
Ross [1988] /waves	2.0	0.5
Costa [1989] /current	1.0–8.0	2.0
Hwang [1989] /waves	0.5	0.5
Jiang [1999] /current	10	0.5

about 1 m) enhance the turbulent kinetic energy, and within the wave boundary layer TKE production dominates (low Ri_f).[g]

Equation 10.106 indicates a close relationship between Ri_f and L_m. In that regard, in Eq. 10.105, which pertains to damping due to suspended sediment, the Richardson number must be represented by Ri_f. However Ri_g is often used because it is easier to deduce from measurements. It should also be noted that

$$Ri_g = ScRi_f \qquad (10.107)$$

where Sc is the turbulent Schmidt number (Chapter 6). It follows that in Eq. 10.105 the coefficient α_{st} is inclusive of Sc, which is therefore empirically accounted for.

The neutral mass diffusivity D_{so} is related to the momentum diffusivity ε_{mo} (Chapters 2 and 6) by

$$D_{so} = Sc^{-1}\varepsilon_{mo} \qquad (10.108)$$

which makes it convenient to use expressions for ε_{mo} for the current-induced boundary layer or the wave boundary layer, provided Sc is known or estimated. Overall, far fewer sets of data are available for deducing D_{so} than ε_{mo}. In estuarine measurements Sc has been observed to vary over the tide cycle, suggesting that it is not reliable as a constant. In Hangzhou Bay, China, values were reported to be in the range of 0.9 to 2.5, which would mean that the ratio of momentum diffusivity to mass diffusivity straddled the neutral value of 1.0 [Costa and Mehta, 1990].

---

[g] Safak *et al.* [2010] use the unsteady TKE balance for flow in both horizontal directions

$$(1-\phi_{vf})\frac{\partial k}{\partial t} = \varepsilon_m \left[\left(\frac{\partial u}{\partial z}\right)^2 + \left(\frac{\partial v}{\partial z}\right)^2\right] + \frac{\partial}{\partial z}\left[\left(v + \frac{\varepsilon_m}{\sigma_\varepsilon}\right)\frac{\partial(1-\phi_{vf})k}{\partial z}\right]$$

$$- (1-\phi_{vf})\varepsilon_t - (s-1)g\frac{\varepsilon_m}{Sc}\frac{\partial \phi_{vf}}{\partial z} \qquad (F10.2)$$

where ϕ_{vf} is the floc volume fraction, k denotes TKE, u and v are the horizontal velocity components, v is the kinematic viscosity of water, ε_m is the eddy diffusivity, σ_ε is a modeling coefficient, ε_t is the TKE dissipation rate per unit mass, s is the specific weight of particles and Sc is the Schmidt number (Appendix E). Under the assumptions of steady-state, $\phi_{vf} \ll 1$ and $v = 0$, Eq. F10.2 is equivalent to Eq. 10.101.

From field and laboratory observations combined with analytic and (1D vertical) numerical analysis of turbulent velocity and suspended sediment concentration profiles, Winterwerp *et al.* [2009] concluded that the best-fit value of *Sc* was 2. It was shown that the effect of *Sc* on the velocity profiles was not large, but pronounced on the concentration profiles. On the other hand, data from other studies indicate that as the suspended sediment concentration rises to a "high" value, *Sc* has a measurable effect on momentum diffusivity ε_m (the neutral value of which is ε_{mo}). In Fig. 10.38 this effect is evident as *Sc* increases from 0.7 to 2.0 [West and Oduyemi, 1989; Costa, 1989; Winterwerp, 2006].

Mixing of suspended sediment in the absence of wind occurs due to two diffusion mechanisms that can be considered additive. These include diffusion due to tangential components of turbulent velocity fluctuations and centrifugal force arising from the curvature of fluid particle pathlines. Experimental observations have yielded the following expressions for D_{so}:

$$D_{so} = \alpha_{js}\kappa u_* z\left(1-\frac{z}{h}\right) + \beta_{js}u_* h\left(\frac{1-z/h}{0.9}\right)^3 \;;\quad \frac{z}{h} \geq 0.1 \quad (10.109)$$

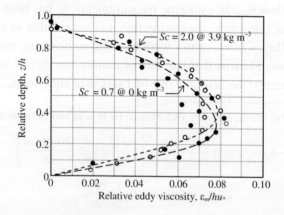

Figure 10.38. Measured dimensionless vertical eddy diffusivity profiles in steady, sediment-laden flows (adapted from Winterwerp [2006]).

$$D_{so} = \alpha_{js} \kappa u_* z \left(1 - \frac{z}{h}\right) + \beta_{js} u_* h \left(\frac{z/h}{0.1}\right)^3 ; \quad \frac{z}{h} \le 0.1 \quad (10.110)$$

Values of the coefficients α_{js} and β_{js} were reported to be 0.038 and 0.98, respectively [Jobson and Sayre, 1970].

Example 10.7: Plot the variation of the mass diffusive flux F_d with concentration gradient $\partial C/\partial z$ ranging from 10 to 1,000 kg m^{-4}. Assume that the velocity gradient $\partial u/\partial z = u_*/\kappa z$ is based on the logarithmic profile. Note also that

$$\frac{\partial \rho}{\partial z} = \left(1 - \frac{\rho_w}{\rho_s}\right)\frac{\partial C}{\partial z} \quad (10.111)$$

You are given the following values: water depth $h = 3$ m, friction velocity $u_* = 0.1$ m s^{-1}, elevation $z = 0.1$ m, Karman constant $\kappa = 0.4$, water density $\rho_w = 1,000$ kg m^{-3}, sediment density $\rho_s = 2,650$ kg m^{-3} and damping coefficients $\alpha_{st} = 4$, $\beta_{st} = 2$.

Figure 10.39 shows the plot obtained from Eqs. 10.97, 10.98 and 10.99. For convenience the absolute value of the concentration gradient is selected. Mass flux increases with the gradient and reaches a maximum at about 260 kg m^{-4}. This behavior reflects the dominant effect of the gradient in enhancing mass flux. With a further increase in the gradient the damping effect of stratification on diffusion dominates and leads to a fall in the flux, which would imply that the steeper the gradient the greater the difficulty in destabilizing the flow by mixing. This effect provides an explanation for the persistence of the (primary) lutocline throughout the tidal cycle even in many macro-hypertidal estuaries such as the Severn (UK) and the Amazon [Kirby, 1986; Kineke, 1993].

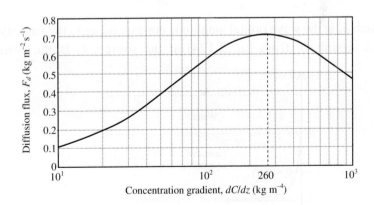

Figure 10.39. Variation of vertical mass diffusive flux with suspended sediment concentration gradient.

10.5.5 *Lutocline rise*

When wind waves transfer their energy to the bottom, stirring of the near-bottom suspension can stretch the lutocline to a higher level h_L. It is a matter of interest to determine the wave height required to maintain the lutocline at that level. The following analysis applies to a dilute suspension because turbulence damping is omitted. Stretching implies dilution if the reduced concentration is not compensated by the addition of new sediment. In the present analysis the risen lutocline is treated only in its final equilibrium state.

Referring to Fig. 10.40, the source of TKE is at some convenient near-bottom level z_b associated with the turbulent boundary layer. The horizontal velocity amplitude at that level is u_b, which decays rapidly with height above the source, as for instance when the source is an oscillating grid crudely representing the effect of a surface wave [Hopfinger and Toly, 1976; Hopfinger and Linden, 1982; Wolanski *et al.*, 1989].

From laboratory experiments, Sleath [1991] reported the following decay law for the velocity amplitude $u(z)$ with elevation z above the source at z_b

$$u(z)=0.16\frac{\sqrt{a_b^3 k_s}}{zT} \qquad (10.112)$$

in which a_b is the wave orbital amplitude corresponding to the velocity u_b ($= a_b C_w k/\sinh kh$ from Eq. 2.124) at the source, k_s is the bottom

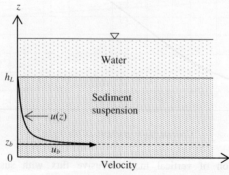

Figure 10.40. Sediment-stratified water column with equilibrium lutocline height h_L.

roughness and T is the wave period. To relate this velocity to the lutocline height h_L we will consider the TKE balance of Eq. 10.101. To proceed, it is necessary to represent the fluctuating quantities in terms of the respective time–mean values based on the mixing length theory (Chapter 2). This analysis goes as follows.

Following conventional approach (Chapter 2) the Reynolds stress $-\rho\overline{u'w'}$ is scaled by u^2. Thus we may restate the kinetic energy production term (without introducing a scaling coefficient) as

$$-\frac{1}{\rho}(\rho\overline{u'w'})\frac{du}{dz} = \frac{du^3}{dz} \tag{10.113}$$

The buoyancy driven flux is evaluated by conveniently equating it to the corresponding settling flux at equilibrium

$$-\frac{g}{\rho}\overline{\rho'w'} = -g'_{se}w_s\phi_v\left(z - h_L\right) \tag{10.114}$$

In this equality, $g'_{se} = g(\rho_s - \rho_w)/\rho_w$ (with ρ_w in the denominator standing for the mean density) and ϕ_v is the solids volume fraction. The flux is maximum at $z = 0$ and reduces to zero at the lutocline ($z = h_L$).

The rate of energy loss is proportional to u^3 and inversely proportional to the mixing length and, therefore, inversely proportional to elevation z, *i.e.*,

$$\varepsilon_t = \alpha_b\frac{u^3}{z} \tag{10.115}$$

where the coefficient α_b depends on the flow field. Substitution of Eqs. 10.113, 10.114 and 10.115 into Eq. 10.101 results in

$$\frac{du^3}{dz} = -g'_{se}w_s\phi_v\left(z - h_L\right) + \alpha_b\frac{u^3}{z} \tag{10.116}$$

The two relevant boundary conditions (Fig. 10.40) for solving this equation are:

$$u(z_b) = u_b; \quad u(h_L) = 0 \tag{10.117}$$

Then, from integration of Eq. 10.116 we obtain

$$h_L = 2.57 \left[\frac{(u_b z_b)^3}{g'_{se} w_s \phi_v} \right]^{1/4} \qquad (10.118)$$

in which the value of α_b (Eq. 10.115) has been taken as 3 following the measurements of E and Hopfinger [1987].[h] The product $u_b z_b$ is a characteristic near-bottom unit discharge obtained from Eq. 10.112 as

$$u_b z_b = 0.16 \frac{\sqrt{a_b^3 k_s}}{T} \qquad (10.119)$$

Equation 10.118 then becomes [Vinzon and Mehta, 1998]

$$h_L = 0.65 \left[\frac{(a_b^3 k_s)^{3/2}}{T^3 g'_{se} w_s \phi_v} \right]^{1/4} \qquad (10.120)$$

Example 10.8: Measurements of waves and suspended sediment concentration at Newnans Lake in north-central Florida yielded the following values: $a_b = 0.035$ m, $T = 1.4$ s and $k_s = 0.1$ m, $\rho_s = 1,700$ kg m^{-3}, $\rho_w = 1,000$ kg m^{-3}, $w_s = 1.15 \times 10^{-4}$ m s^{-1}, and $\phi_v = 4.1 \times 10^{-5}$. Calculate h_L.

From Eq. 10.120 we obtain $h_L = 0.37$ m. A typical concentration profile is shown in Fig. 10.41. An estimate of $h_L = 0.80$ m has been made using equal-area assumption as indicated. At low wind speeds heights as small as 0.20 m were measured. Note that from $\rho = \phi_v (\rho_s - \rho_w) + \rho_w$ we obtain $\rho = 1,029$ kg m^{-3}. At this low density the suspension is too dilute to experience hindered settling, and is referred to as a Benthic Suspended Sediment Layer (BSSL), with h_L defining the height of a secondary lutocline.

10.5.6 *Saturation concentration*

As the suspension concentration increases the intensity of turbulence eventually decreases with the kinetic energy expended to sustain the suspended particles. Additional energy is taken up by inter-particle collisions and against cohesion between closely spaced particles. In their wave flume experiments Lamb *et al.* [2004] reported a contraction

---

[h] E is the first author's last name.

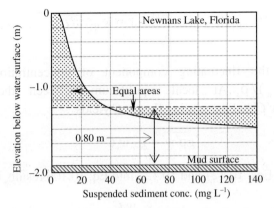

Figure 10.41. Typical suspended sediment profile and estimation of BSSL height in Newnans Lake, Florida (from Jain *et al.* [2007]).

in the height of the sediment-free boundary layer as more sediment was introduced. This was attributed to sediment-induced stratification and limited vertical exchange of momentum. This description applies to fluid mud.

In the muddy environment the ratio $u_* / \sqrt{B_u}$, where B_u is the integral of the buoyancy anomaly $g(\rho_s - \rho_w)C(z,t)/\rho_s$ over height z from 0 to ∞, has been used to represent the effect of stratification on momentum exchange. When this ratio is small stratification rather than the vertical extent of flow limits the scale of turbulence [Trowbridge and Kineke, 1994]. At a critical or saturation concentration C_{cr} turbulence tends to collapse, the flow becomes predominantly viscous and its load-carrying capacity diminishes substantially. Rapid deposition occurs and the suspension becomes subsaturated. As shown below, C_{cr} is proportional to the flux Richardson number Ri_f (Eq. 10.106) representing the consumption of the kinetic energy for sustaining the sediment in suspension.

We will assume steady turbulent flow with a log-velocity profile and parabolically distributed eddy diffusivity (Eq. 2.64). The Reynolds stress representing turbulent shear is given by

$$-\rho \overline{u'w'} = \rho \kappa u_* z \left(1 - \frac{z}{h}\right)\frac{du}{dz} = \rho u_*^2 \left(1 - \frac{z}{h}\right) \qquad (10.121)$$

where h is the water depth. Next we recognize that

$$g\overline{\rho'w'} = g\frac{\rho_s - \rho_w}{\rho_s}\overline{w'C'} = g''_{se}\overline{w'C'} \tag{10.122}$$

where C' is the fluctuating component of concentration C and $g''_{se} = g(\rho_s - \rho_w)/\rho_s$. At local equilibrium between the upward diffusive flux $\overline{w'C'}$ and the settling flux $w_s C$ yields

$$g\overline{\rho'w'} = g''_{se}w_s C \tag{10.123}$$

Substitution of Eqs. 10.121 and 10.123 into Eq. 10.106 results in

$$Ri_f = \frac{g''_{se}w_s C}{\rho_w u_*^2\left(1 - \dfrac{z}{h}\right)\dfrac{du}{dz}} \tag{10.124}$$

in which we have assumed that $\rho \approx \rho_w$ in the denominator. Based on the log-velocity profile we then obtain

$$Ri_f = \frac{g''_{se}F_s}{\dfrac{\rho_w u_*^3}{\kappa h}\left(\dfrac{h}{z} - 1\right)} \tag{10.125}$$

where $F_s = w_s C$ is the settling flux. Substitution of Eq. 7.108 for F_s in the above expression gives

$$Ri_f = \frac{g''_{se}}{\dfrac{\rho_w u_*^3}{\kappa h}\left(\dfrac{h}{z_b} - 1\right)}\frac{a_w C^{n_w+1}}{(b_w^2 + C^2)^{m_w}} \tag{10.126}$$

Figure 10.42 shows the dependence of Ri_f on concentration C evaluated at a characteristic boundary layer elevation $z_b = 0.05h$ using the following values: $a_w = 0.092$, $b_w = 6.67$, $m_w = 1.68$, $n_w = 1.53$, $u_* = 0.02$ m s^{-1}, $\rho_w = 1{,}025$ kg m^{-3}, $\rho_s = 2{,}650$ kg m^{-3}, $h = 2$ m and $\kappa = 0.4$. Above a typical experimental value of the critical Richardson number $Ri_{fc} = 0.2$ [Tennekes and Lumley, 1972], subsaturated turbulent flow becomes supersaturated and viscous.

Since at saturation Ri_f is equal to Ri_{fc}, from Eq. 10.125 we may solve for the saturation concentration to yield

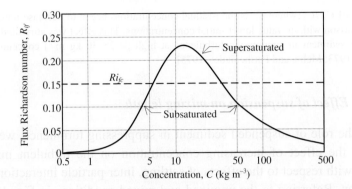

Figure 10.42. Variation of flux Richardson number with concentration (adapted from Winterwerp [2001] and van Maren *et al.* [2009]).

$$C_{cr}(z) = \frac{\rho_w u_*^3 Ri_{fc}}{g_{se}'' w_s}\left(\frac{h}{z}-1\right) \qquad (10.127)$$

In general C_{cr} is in the hindered settling range. For simplified analysis Winterwerp [2001] proposed an approximate expression for the depth–mean saturation concentration

$$\overline{C}_{cr} = K_s \frac{\rho_w u_*^3}{g_{se}'' w_{sref} h} \qquad (10.128)$$

where K_s is a sediment-specific proportionality constant inclusive of Ri_{fc} and w_{sref} is a reference settling velocity. From an evaluation of stratification in a hypothetical channel it was concluded that K_s must be about 0.7. We observe that $\overline{C}_{cr}$ is a strong function of u_* and varies inversely with w_{sref}. These associations imply that for cohesive sediment $\overline{C}_{cr}$ can be expected to vary with the locality.

Example 10.9: Consider the transport of fine sediment in a flume. You are given the following parameters: $u_* = 0.02$ m s^{-1}, $\rho_w = 1,025$ kg m^{-3}, $\rho_s = 2,650$, $w_{sref} = 2\times10^{-5}$ m s^{-1} and $h = 2$ m. Calculate $\overline{C}_{cr}$.

These values yield $g_{se}'' = 6.016$ m s^{-2}. Then, with $K_s = 0.7$, from Eq. 10.128 we obtain $\overline{C}_{cr} = 23.9$ kg m^{-3}. In experiments using a kaolinite in freshwater in an annular flume (CRAF, Chapter 7), parametric values were similar to those selected here. At a constant flow speed and initial suspension concentration C_0 in excess of about 25 kg m^{-3}

the ratio of C_{fe}/C_0 (where C_{fe} is the residual concentration) began to decrease noticeably in comparison with its value lower initial concentrations. This effect was attributed to the effect of sediment in suppressing turbulence at high ($\geq$ 25–30 kg m^{-3}) concentrations [Mehta, 1973; Mehta and Partheniades, 1975].

10.5.7 *Effect of suspension on mixing length*

As to the role of suspended sediment in suppressing turbulence, we will review the effect of increasing concentration on the turbulent mixing length with respect to the effects of settling, inter-particle interaction and cohesion. Referring to the unmixed and mixed conditions in Fig. 10.31, the sum of kinetic, potential and dissipated energy per unit volume of fluid in homogeneous (completely mixed) turbulent flow must be equal to the sum of the same energy components in flow stratified by sediment. In terms of turbulence-mean properties this equality is expressed as

$$\tau_{turbo}\frac{\partial u}{\partial z} = \tau_{turbm}\frac{\partial u}{\partial z} + gF_b + gF_s + \tau_j\frac{\partial u}{\partial z} + \tau_f\frac{\partial u}{\partial z} \qquad (10.129)$$

in which τ_{turbo} is the Reynolds stress in homogeneous flow, τ_{turbm} is the Reynolds stress in non-homogeneous flow, F_b is the buoyancy flux, F_s is the settling flux, τ_j is the shear stress associated with inter-particle interactions, and τ_f is the shear stress associated with cohesion. These terms are evaluated as follows [Jiang, 1999].

In homogeneous flows τ_{turbo} is expressed in terms of the momentum mixing length l_{t0} according to

$$\tau_{turbo} = -\rho_w l_{t0}^2 \left(\frac{\partial u}{\partial z}\right)^2 \qquad (10.130)$$

which is a representation of Eq. 2.67 with the overbar on the velocity u omitted for convenience and the density taken to be that of water. Then the rate of kinetic energy production per unit volume of fluid is

$$\tau_{turbo}\left(\frac{\partial u}{\partial z}\right) = -\rho_w l_{t0}^2 \left(\frac{\partial u}{\partial z}\right)^3 \qquad (10.131)$$

Furthermore

$$l_{to}^2 \left(\frac{\partial u}{\partial z} \right)^3 = \frac{u_*^3}{\kappa z_b} = T_{ke} \tag{10.132}$$

where u_* is the friction velocity, κ is the Karman constant, z_b is a representative near-bottom boundary layer height and T_{ke} symbolizes the rate of TKE production per unit mass of fluid. Therefore

$$\tau_{turbo} \left(\frac{\partial u}{\partial z} \right) = -\rho_w \cdot T_{ke} \tag{10.133}$$

For stratified flows the condition analogous to Eq. 10.131 is

$$\tau_{turbm} \left(\frac{\partial u}{\partial z} \right) = -\rho l_{tm}^2 \left(\frac{\partial u}{\partial z} \right)^3 \tag{10.134}$$

where ρ is the density of the suspension.

The rate of work against buoyancy in stratified flow is

$$gF_b = g l_s l_{tm} \frac{\partial \rho}{\partial z} \left| \frac{\partial u}{\partial z} \right| = gSc l_{tm}^2 \frac{\partial \rho}{\partial z} \left| \frac{\partial u}{\partial z} \right| \tag{10.135}$$

where l_s is the sediment mass mixing length and $Sc = l_s/l_{tm}$ is the turbulent Schmidt number (see also Eq. 10.108).

The rate of work against gravity is

$$gF_s = g \left(1 - \frac{\rho}{\rho_s} \right) w_s C \tag{10.136}$$

where ρ_s is the particle density, w_s is the settling velocity and C is the concentration.

The rate of work against inter-particle interaction is taken to be

$$\tau_j \left(\frac{\partial u}{\partial z} \right) = \left[\gamma_k \eta \left(\frac{\partial u}{\partial z} \right) \right] \left(\frac{\partial u}{\partial z} \right) = \gamma_k \eta \left(\frac{\partial u}{\partial z} \right)^2 \tag{10.137}$$

where η is the fluid dynamic viscosity. The term γ_k accounts for the effect of concentration and sediment composition on τ_j. Given dimensionless concentration $\tilde{C}_m = C_{mf} / C,$ where C_{mf} is a characteristic maximum value of C corresponding to floc-floc contact, γ_k is selected as

$$\gamma_k = \alpha_k \frac{(2\tilde{C}_m^{1/3} - 1)\tilde{C}_m^{1/3}}{(\tilde{C}_m^{1/3} - 1)^2} \qquad (10.138)$$

in which the coefficient α_k depends on sediment composition [Bagnold, 1956; Jiang, 1999].

The mixing length ratio $l_{tm}/l_{to} = \Phi_{tm}(\le 1)$ is a measure of the damping of turbulent eddies. Substitution of Eq. 10.131 through 10.138 into Eq. 10.129 yields the following solution for Φ_{tm}

$$\Phi_{tm} = \frac{l_{tm}}{l_{to}} = \sqrt{\frac{1 - (\lambda_w + \lambda_k + \lambda_f)}{s + Ri_g}} \qquad (10.139)$$

where $s = \rho/\rho_w$ is the specific gravity of the suspension. The other parameters are

$$\lambda_w = g\left(1 - \frac{\rho}{\rho_s}\right)\frac{w_s C}{\rho_w \cdot T_{ke}} \qquad (10.140)$$

which represents the ratio of the potential energy associated with settling to T_{ke},

$$\lambda_k = \frac{\gamma_k \eta (\partial u/\partial z)^2}{\rho_w \cdot T_{ke}} \qquad (10.141)$$

which represents the ratio of viscous energy to T_{ke}, and

$$\lambda_f = \frac{\tau_f (\partial u/\partial z)}{\rho_w \cdot T_{ke}} \qquad (10.142)$$

which is the ratio of cohesion-related energy to T_{ke}.

The quantities λ_w, λ_k and λ_f modify the mixing length by enhancing the influence of the gradient Richardson number Ri_g on Φ_{tm}, with fluid density ρ in Eq. 10.78 conveniently replaced by ρ_w. As expected from the equality in Eq. 10.129, setting $C = 0$, the shear stress $\tau_f = 0$ and the fluid density $\rho = \rho_w$ reduces Eq. 10.139 to $l_{tm} = l_{to}$, *i.e.*, $\Phi_{tm} = 1$ in homogeneous flows.

Example 10.10: Calculate λ_w, λ_k and λ_f, given: $\rho_s = 2{,}650$ kg m^{-3}, $\rho_w = 1{,}012$ kg m^{-3}, $\alpha_k = 0.5$, $C = 25$ kg m^{-3}, $C_{mf} = 145$ kg m^{-3}, $w_s = 1 \times 10^{-3}$ m s^{-1}, $s = 1$, $\eta = 0.1$ Pa.s, $\kappa = 0.4$, $z_b = 0.3$ m, $\tau_f = 0.04$ Pa and $u_* = 0.08$ m s^{-1}. Assume a log-velocity profile. Also estimate the saturation concentration for turbulence collapse in 2 m deep water.

Using the log-velocity representation (Eq. 2.55) we obtain the following expressions for λ_w, λ_k and λ_f, respectively,

$$\lambda_w = g \left(1 - \frac{\rho_w}{\rho_s} \right) \frac{w_s C}{\rho_w} \frac{\kappa z_b}{u_*^3} \tag{10.143}$$

$$\lambda_k = \gamma_k \frac{\eta}{\rho_w \kappa z_b u_*} \tag{10.144}$$

$$\lambda_f = \frac{\tau_f}{\rho_w u_*^2} \tag{10.145}$$

From these we calculate $\lambda_w = 0.035$, $\lambda_k = 0.038$ and $\lambda_f = 0.006$. These values imply that cohesion plays a minor role compared to gravity and inter-particle interaction in reducing the mixing length. From Eq. 10.139

$$\Phi_{tm} = \frac{0.96}{\sqrt{1 + Ri_g}} \tag{10.146}$$

which indicates that l_{tm} is reduced to values less than 96% of l_{to}. From Eq. 10.128 with $K_s = 0.7$ we obtain $\bar{C}_{cr} = 29.9$ kg m^{-3}.

Using values similar to those in the above example, Jiang [1999] obtained the plot shown in Fig. 10.43. The five curves of Φ_{tm} against concentration C correspond to five values of the turbulent kinetic energy production T_{ke}. Damping of the mixing length measurably increases from its homogeneous value corresponding to $\Phi_{tm} = 1$, when C exceeds about 5 kg m^{-3}. Also plotted is the settling flux F_s based on Eq. 7.108 and described by

$$F_s = \frac{0.085 C^{2.5}}{(C^2 + 100)^{1.6}} \tag{10.147}$$

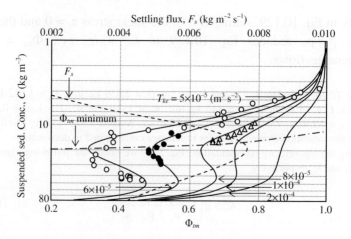

Figure 10.43. Dependence of mixing length on suspended sediment concentration and rate of turbulent kinetic energy production. Data points are for the Jiaojiang estuary, China (adapted from Jiang and Mehta [2000]).

With increasing C, the flux F_s increases at first, reaches a peak at about 18 kg m^{-3}, then decreases due to hindered settling. Since the concentration typically increases with depth, one may consider C as a surrogate for depth in water. A lutocline will occur at the depth of the peak flux (Fig. 10.4), which is also the level at which the mixing length has a minimum due to turbulence damping. As the suspended sediment concentration increases (from top to bottom), inter-particle interaction and cohesion eventually lead to a drastic reduction in Φ_{tm}. Turbulence collapse is implied after the concentration exceeds about 40 kg m^{-3}.

Values of Φ_{tm} from measurements in the Jiaojiang estuary (Fig. 10.43) are consistent with the trend of increase in Φ_{tm} with turbulent kinetic energy production, T_{ke}, a manifestation of which would be the enhancement of interfacial mixing. In Fig. 10.43 the rise in the locus of the elevation of a minimum in Φ_{tm} is a reflection of this behavior. More directly we see this effect in Fig. 10.44, in which mixing between the fluids below and above the lutocline is represented by the mixing index

$$M_I = \frac{(\partial C / \partial z)_{mean}}{(\partial C / \partial z)_{max}}$$

(10.148)

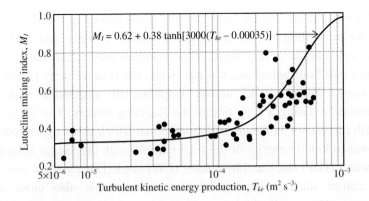

Figure 10.44. Dependence of lutocline mixing on kinetic energy production (adapted from Jiang and Mehta [2000]).

where $(\partial C/\partial z)_{mean}$ is the mean concentration gradient over depth and $(\partial C/\partial z)_{max}$ is the gradient across the lutocline shear layer (Fig. 6.22). For the Jaojiang the mean trend is described by

$$M_I = 0.62 + 0.38 \tanh[3000(T_{ke} - 0.00035)] \qquad (10.149)$$

Mixing becomes increasingly vigorous once T_{ke} exceeds about 0.00013 m^2 s^{-3}. This can be taken as the critical kinetic energy for entrainment of sediment in this estuary.

10.6 Observations of Dense Suspension Behavior and Effects

10.6.1 *Tidal environment*

In the tidal environment two noteworthy causes of fluid mud are: (1) advection of bottom mud as the current accelerates after slack water, and (2) deposition of dense suspended matter as the current decelerates before slack. The spring-neap variation of the tidal range plays a role in determining the contribution from each cause. In the mixed wave-current environment a more complex picture usually emerges depending on the relative strengths and variability in waves and tidal current.

The method to estimate fluid mud thickness shown in Fig. 10.7 holds only when settling is the dominant transport process and turbulent diffusion is comparatively weak. This condition may occur more or less within a half hour before slack water, when the flow velocity is low and decreasing. Figure 10.45 shows isopleths of suspended sediment concentration during a neap tide in the Jiaojiang estuary. The 20 kg m^{-3} isopleth represents the height of the lutocline, which shifts up or down in response to kinetic energy of the tidal current, with the lowest heights at high water (HW) slack and low water (LW) slack. During about one-half hour before slack water the lutocline elevation is also close to a minimum. The small thickness of fluid mud during this period is accompanied by a compression of the isopleths indicating densification of mud. Almost immediately following slack the lutocline begins to rise at a rapid rate as the flow accelerates. This rise is manifested as rapid stretching of the isopleths due to dilution.

Concentration isopleths from the hyper-tidal Severn River estuary in the UK (Fig. 10.46) highlight the effect of boundary layer growth on lutocline rise. The elevation z_L is the upper level of the lutocline shear layer (Fig. 6.22). This elevation is characteristically less distinct than for example z_A in Fig. 10.7 marking the onset of hindered settling. The parameters z_{u99} and z_{u95} are elevations at which the current velocity u is equal to $0.99U$ and $0.95U$, respectively, where U is the maximum value

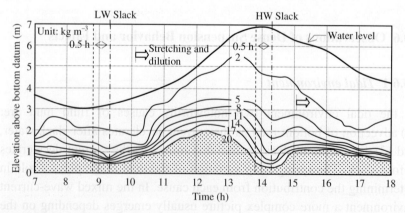

Figure 10.45. Concentration isopleths during neap tide on November 15, 1995, in the Jiaojiang estuary near Hainan, China (adapted from Jiang [1999]).

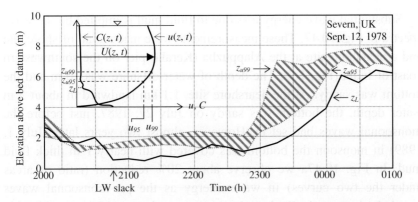

Figure 10.46. Movement of lutocline related to boundary layer height. Measurements are from the Severn estuary, UK (adapted from Parker [1987]).

of u. The thickness $z_{u99}-z_{u95}$ is a measure of boundary layer height, and its variation generally agrees with lutocline elevation z_L as both depend on the current. From about 2015 to 2245 h, which includes LW slack at 2045 h, the lutocline remained close to the bed and the boundary layer was compressed. The layer $z_{u99}-z_{u95}$ was equal to or less than about 1 m. At 2245 h, the lutocline began to rise rapidly in tandem with $z_{u99}-z_{u95}$ until the rate of rise slowed at about 15 minutes past midnight. The layer $z_{u99}-z_{u95}$ increased to as much as 3 m, implying a significant amount of stretching and dilution of the boundary layer.

10.6.2 *Wave environment*

In the wave-dominated environment, fluid mud can form either by liquefaction of the bed or by rapid deposition of suspended sediment. Typically, when wave motion ends fluid mud forms a bed if the concentration is equal to or greater than its space-filling value. When storm waves resuspend large quantities of sediment, its deposition in the post-storm period will produce fluid mud during hindered settling. As this mud dewaters to form a bed, it will absorb wave energy and bring about calmer conditions more rapidly than in an environment devoid of fluid mud.

Substantial wave energy loss due to fluid mud is implied by the wave spectra in Fig. 10.47. These measurements were made at an offshore site and a nearshore site at the Alappuzha (Kerala) pier on the southwestern coast of India (Fig. 1.7). The depth of water offshore was 10 m and the bottom was sandy. At the nearshore site, 1.1 km landward in about 5 m water depth, the bottom was sandy on July 27, 1987, just before the monsoonal waves had set in. However, almost two years later (July 1, 1989) in monsoon the bottom was covered with nearly 1 m thick fluid mud. In Fig. 10.47a we observe about 20% reduction (ratio of areas under the two curves) in wave energy as the pre-monsoonal waves traveled shoreward. During the monsoon there was a considerable increase in wave energy at the offshore site, but due to fluid mud nearly 85% of the energy was lost. As a result the nearshore station recorded lower wave energy under monsoon waves than under calmer, pre-monsoonal waves (Fig. 10.47b). High turbidity at the onset of the monsoon became measurably lower within weeks thereafter. As fluid mud occurs only when the offshore wave energy increases, this is a case of *reduction* in nearshore suspended sediment by an *increase* in offshore wave energy [Mathew, 1992; Mathew *et al.*, 1995; Li and Parchure, 1998].

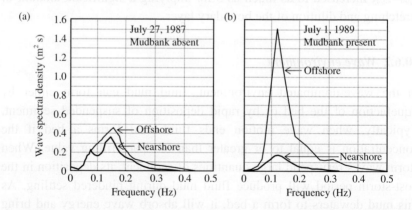

Figure 10.47. Offshore and nearshore wave spectra near Alappuzha in Kerala, India. (a) In the absence of mudbank; (b) when a mudbank was present (from Mathew [1992]).

The shallow and muddy coastal shelf of the contiguous shorelines of French Guiana, Suriname and Guyana is a case of energetic waves losing energy over stretches tens of kilometers long. The main source of mud is the Amazon River. Figure 10.48 illustrates the extent of wave damping in coastal waters of Suriname. From wave measuring station A to station B, a distance of 11 km, the *rms* wave height (H_{rms}) decreased from 0.93 m to 0.32 m, amounting to an 88% loss of energy (proportional to wave height squared). At station C, H_{rms} reduced further to 0.19 m, which meant a 96% loss over the distance of 18 km from A [Wells and Kemp, 1986].

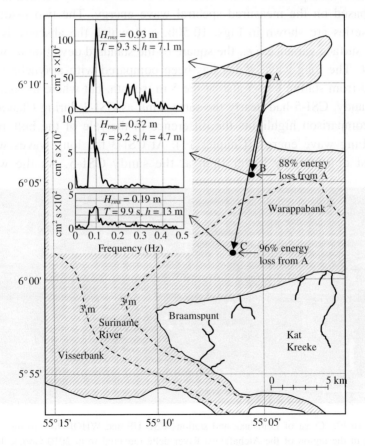

Figure 10.48. Wave spectra measured off Suriname (from Wells and Kemp [1986]).

Sheremet *et al.* [2005] analyzed the time-series of sea state and suspended sediment measurements at Louisiana State University stations CSI-3 and CSI-5 (Fig. 10.49). Figure 10.50a shows wind speed during July 11–18 in 2003 due to the passage of Hurricane Claudette across the Gulf of Mexico. At CSI-3 water level and suspended sediment concentration were measured. This station was at the 5 m isobath on the muddy Terrebonne Shelf near the Atchafalaya Bay in southwestern Louisiana. The wave record was resolved as two time-series, one including long waves with frequencies ≤ 0.2 Hz and the other with shorter waves > 0.2 Hz. Division at 0.2 Hz (*i.e.*, wave period of 5 s) was chosen for interpretation as the data could be conveniently divided in this way based on the measured spectral wave energy. The two resulting time-series are shown in Figs. 10.50b, c in which the abscissa is the water surface variance, *i.e.*, the square of the standard deviation of wave height. The long-wave series has been compared with a synchronous record from station CSI-5 also at the 5 m isobath, but where the bottom was sandy. CSI-5 had nearly the same wind as CSI-3 during Claudette. The comparison highlights the difference in the role of the bottom in absorbing wave energy (Chapter 11). At CSI-3 the long waves were almost consistently lower than at the sandy CSI-5. As the wind

Figure 10.49. Coast of Louisiana and stations CSI, UF and WHOI close to the –5 m isobath in the region of the Atchafalaya River delta (adapted from 2010 Google Earth image).

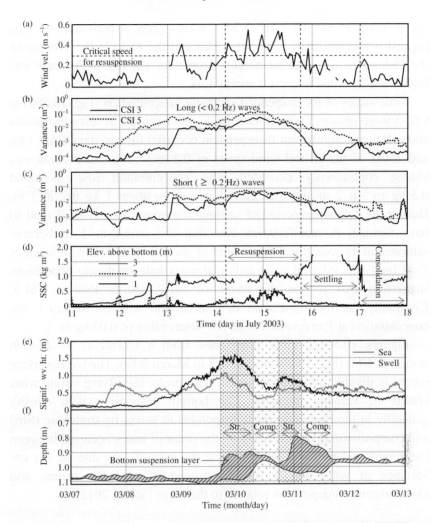

Figure 10.50. Observations near the coast of Louisiana: (a) Time-series of wind during Hurricane Claudette in July 2003; (b) long-wave height variances at CSI-3 and CSI-5; (c) short-wave height variances at CSI-3 and CSI-5; (d) SSC at CSI-3; (e) sea and swell time-series obtained in March 2006 at a UF2 station; (f) bottom mud layer thickness (from Sheremet *et al.* [2005], Sahin [2012]).

speed increased on the 12th of July, waves grew at both stations. The highest waves occurred during the 14th and the 15th. In the evening of the 15th as the wind fell the heights decreased. The rate of decrease was higher at CSI-3 due to greater energy loss.

Time-series of the suspended sediment concentration (Fig. 10.50d) at CSI-3 were obtained with optical backscatter sensors at three elevations above bottom. Significant resuspension occurred for two days (July 14th and 15th), as the critical wind speed of 0.3 m s^{-1} for resuspension was almost continuously exceeded. The concentration rose to about 0.5 kg m^{-3} at 2 m and 3 m elevations and to about 1 kg m^{-3} at 1 m. During the following period of settling one would expect fluid mud to have formed. A manifestation was that the sensor at 1 m became saturated at its operational upper limit of about 2 kg m^{-3}. The presence of fluid mud was inferred from sediment mass balance during deposition, suggesting that it was the cause of wave energy loss at CSI-3. Eventually, as the fluid mud dewatered and was consolidated, the concentration at 1 m reverted to its pre-storm value (< 0.05 kg m^{-3}).

Figures 10.50e, f show time-series from a University of Florida station at UF2 (Fig. 10.49) near CSI-3 in March 2006. The water surface record is divided into sea and swell, with peaks identifying strong wind effects of variable intensity. The bottom suspension with density generally in the fluid mud range responded to waves by stretching (Str.) and compression (Comp.). A lag on the order of hours occurred between the waves and the mud layer. It was the sum of lags due to: (1) co-variance of wind and waves, (2) to-and-fro of tidal current, and (3) resuspension/deposition relative to the wave [Sahin, 2012].

10.6.3 *Rate of fluid mud formation*

As mentioned (see Chapter 6), fluid mud is formed when the rate of sediment accumulation by deposition exceeds the rate at which the deposit dewaters to form a bed. The rate of formation of fluid mud can be related to the settling number,

$$Sw = \frac{w_s}{h\sigma} \tag{10.150}$$

where w_s is the settling velocity, h is the water depth and σ is the wave angular frequency. A high value of Sw would mean rapid formation of fluid mud. Substitution for w_s from Eq. 7.15 (without the factor 18) yields

$$Sw = \frac{g'd_p^2}{h v \sigma} \tag{10.151}$$

where $v = \eta/\rho$ is the kinematic viscosity of water.

In shallow-water Eq. 10.151 reduce to

$$Sw = \frac{g'}{2\pi v\sqrt{g}} \frac{1}{h/L} \frac{d_p^2}{\sqrt{h}} \tag{10.152}$$

where L is the shallow-water wave length. Introducing the threshold of shallow-water condition $h/L = 1/20$ (Chapter 2) yields

$$Sw = \frac{10g'}{\pi v\sqrt{g}} \frac{d_p^2}{\sqrt{h}} \tag{10.153}$$

which indicates that Sw is proportional to $d_p^2/\sqrt{h}$.

In field measurements during material disposal from a hopper dredger at Townsville Harbor in Australia, $d_p^2/\sqrt{h}$ was found to range over two orders of magnitude, as was the rate of formation of fluid mud in the shallow bay [Wolanski *et al.*, 1992].

10.7 Exercises

10.1 Consider the viscous flow of clayey fluid mud layer in a channel. The layer has uniform thickness h, density ρ and Newtonian viscosity η. The width of the channel is $\bar{B}$ and the bed slope angle θ. The flow momentum balance is

$$\frac{d\tau}{dz} = -g(\rho - \rho_w)\sin\theta \tag{E10.1}$$

where τ is the laminar shear stress and ρ_w is the water density.

(1) Derive an expression for dry sediment load.

(2) Given a 0.12 m thick fluid mud layer with wet bulk density of 1,133 kg m^{-3}, saltwater density 1,025 kg m^{-3}, particle density 2,707 kg m^{-3} and viscosity of 1 Pa.s, calculate the discharge (kg per second) of the (wet) mud. The channel is 3.15 m wide with a 1° slope.

10.2 A layer of fluid mud below a layer of water, each of thickness h, are both flowing steadily between two parallel plates (separated by height $2h$) due to pressure drop Δp over distance l (Fig. E10.1). Assume that the two fluids (water of viscosity η_1 and fluid mud of viscosity η_2) are immiscible and Newtonian. The momentum balance

$$\frac{d\tau}{dz} = \frac{\Delta p}{l} \qquad (E10.2)$$

where τ is the shear stress, is applicable to each layer.

(1) Obtain by integration the wall shear stresses τ_{Rt} and τ_{Rb}.

(2) Show that the plane of zero shear ($\tau = 0$) is at distance h_0 from the $z = 0$ axis such that

$$h_0 = \frac{h}{2}\left(\frac{\eta_2 - \eta_1}{\eta_1 + \eta_2}\right) \qquad (E10.3)$$

(3) Determine the velocity distributions $u_1(z)$ and $u_2(z)$.

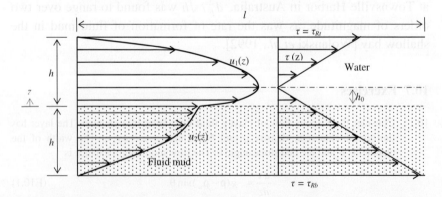

Figure E10.1. Flowing layers of fluid mud and water between two parallel plates (adapted from Bird *et al.* [1960]).

10.3   Consider the development of boundary layer flow in a fluid mud underlayer shown in Fig. E10.2. A convenient definition of boundary layer thickness from Eq. 10.6 is δ_b at which the velocity ratio $u/U_L = 0.01$.

(1)   Show that $\delta_b = 3.64\sqrt{\nu t}$, and determine its value at $t = 10$ s, given $\nu = 0.0001$ m^2 s^{-1}.

(2)   It is assumed that the tilt of the interface is negligible. How would you estimate the tilt if the fluid mud was in a trench of length l starting at the hard soil wall? What additional assumptions would you have to make?

10.4   Couette flow is obtained from the Stokes first problem by introducing a stationary bottom at depth h below the moving water surface. This surface is instantaneously accelerated to the constant speed U_L, while the flow velocity at the bottom is always zero. Equation 10.4 is the governing equation with the initial and boundary conditions

$$u(z',0) = 0$$

$$u(0,t) = U_L \qquad\qquad\qquad\text{(E10.4)}$$

$$u(h,t) = 0$$

The solution using Laplace transform is

$$u(\tilde{z}) = U_L \left\{ \sum_{n=0}^{\infty} erfc[(n+1)\tilde{z} - \tilde{h}] - \sum_{n=0}^{\infty} erfc(n\tilde{z} + \tilde{h}) \right\} \qquad\text{(E10.5)}$$

where $\tilde{z} = (h - z'')/2\sqrt{\nu t}$, $z'' = 0$ at the top and $-h$ at the bottom, and $\tilde{h} = h/2\sqrt{\nu t}$. You are given: $h = 0.2$ m, $U_L = 0.1$ m s^{-1} and $\nu = 0.001$ m^2 s^{-1}. Plot the velocity profile u against z'' for $t = 1, 2, 5$ and 25 s. Note: In Eq. E10.5 select $n = 6$.

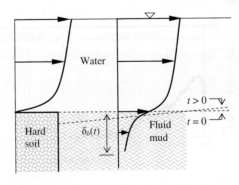

Figure E10.2.   Transient boundary layer flow of a fluid mud underlayer.

10.5 In a flume experiment the rate of liquefaction of a cohesive bed by waves was measured with an array of pore-pressure sensors tracking the loss of effective stress [Feng, 1992]. The water depth was 35 cm, wave amplitude 2.5 cm, wave frequency 1.06 Hz, bed thickness 16.6 cm and bed density 1,180 kg m^{-3}. The test duration was 300 min, and the following variation of the thickness of fluid mud layer z_f (cm) with time (min) was recorded:

$$z_f = 15.5 - 0.047t + 7.11 \times 10^{-5} t^2 \qquad (E10.6)$$

This relationship is valid from 5 min to 300 min after the onset of waves. The rate of wave energy loss in the bed ε_D (Pa m s^{-1}) was found to be

$$\varepsilon_D = 1.52 - 0.0019t - 7.89t^{-1} + 1.40 \times 10^{-6} t^2 \qquad (E10.7)$$

This equation is valid from 70 min to 300 min.

(1) Plot the absolute values of z_f and dz_f / dt against time (5 to 300 min).

(2) Plot the absolute values of ε_D against time (70 to 300 min).

(3) Plot the absolute value of ε_D against the absolute value of dz_f / dt, both in the range of 70 to 300 min. Provide your interpretation of the resulting trend.

10.6 From the flume experiment in Exercise 10.5, the time-variation of the effective normal stress 9.5 cm below the bed surface is plotted in Fig. E10.3. Provide an interpretation of the trend.

10.7 The four-parameter Carreau [1972] model for viscosity as applied to fluid mud is

$$\eta = \eta_\infty + \frac{\eta_0 - \eta_\infty}{[1 + (a_{cm}\dot{\gamma})]^{p_{cm}}} \qquad (E10.8)$$

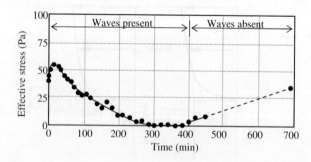

Figure E10.3. Variation of effective normal stress with time at a depth of 9.5 cm below bed surface in a wave flume test (adapted from Feng [1992]).

It is characterized by viscosities η_0 and η_∞, and empirical coefficients a_{cm} and b_{cm}. When the shear rates are low, *i.e.*, $a_{cm}\dot{\gamma} \ll 1$, the model predicts Newtonian behavior with viscosity η_0. At higher values of $a_{cm}\dot{\gamma}$ the relationship between η and $\dot{\gamma}$ is a power-law. Herbich *et al.* [1989] tested several muds from underwater sites in a laboratory viscometer and applied Eq. E10.8. Results for samples with fluid-mud consistency from the Canaveral Barge Canal (FL) are given in Table E10.1.

(1) Based on the six mud samples taken together, determine the dependence of η_0 on the wet bulk density of mud, ρ.

(2) Plot the viscosity η for each sample against the shear rate $\dot{\gamma}$ ranging from 0 to 500 s^{-1}.

(3) Indicate if there is a recognizable trend with respect to mud density.

10.8 Figure E10.4 shows the schematized lateral (east-west) cross-section (width *l*) of an estuary. The longitudinal (north-south) axis is perpendicular to the diagram. The estuary is stratified due to a turbid layer of thickness h_2 supplied by a tributary (not shown). Water layer thickness is h_1. When wind of sufficient speed blows from west to east, the fresh-turbid water interface tilts and causes a turbid outflow over a sill at depth h_0 below surface. The outflow pours into an area where marine life is adversely affected. Based on steady-state force balance in the lateral *l*-direction, derive an expression for the turbid water unit discharge *q* over the sill as a function of the wind-induced stress τ_w at the water surface. The shear stress due to counterflow driving turbid discharge is τ_i. Assume that the tidal range is small compared to the depth of water. Also assume that the height of turbid outflow Δz and the surface and interfacial slopes $d\eta_1/dx$ and $d\eta_2/dx$, respectively, are small. State any other assumptions you make.

Table E10.1. Carreau viscosity equation coefficients

Sample	ρ (kg m^{-3})	η_0 (Pa.s)	η_∞ (Pa.s)	a_{cm}	b_{cm}
FL-11	1,170	4.0	0.009	0.24	0.50
FL-12	1,220	5.0	0.009	0.22	0.50
FL-13	1,230	5.0	0.012	0.19	0.50
FL-14	1,260	5.5	0.014	0.18	0.50
FL-43	1,150	2.5	0.009	0.43	0.46
FL-44	1,100	1.9	0.008	0.44	0.47

Source: Herbich *et al.* [1989].

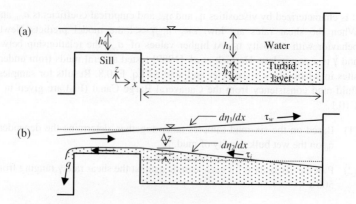

Figure E10.4. Schematic of lateral cross-section of an estuary: (a) No-wind; (b) wind-induced outflow (= inflow) at steady-state.

10.9   Consider the quantities in Table E10.2 obtained under non-storm (fair weather) conditions. The settling velocity is defined by Eq. 7.102, with the same set of coefficient values (means given in Table 7.3) for all three locations. The Darcy–Weisbach friction factor is assumed to be 0.024. Make an assessment of the likelihood of the formation of fluid mud at each site.

10.10 Show that the ratio on the right hand side of Eq. 10.128 resembles the Knapp–Bagnold criterion for auto-suspension (Eq. 12.82).

Table E10.2. Fair weather parameters at three locations[a]

Location	Water depth (m)	Maximum current (m s^{-1})	Concentration[b] (mg L^{-1})
Ortega River, FL	3	0.7	10–30
San Francisco Bay, CA	5	1.1	30–1,000
Severn Estuary, UK	10	2	1,000–10,000

[a] All values are nominal.
[b] 1-meter above bottom.

Chapter 11

Wave–Mud Processes

11.1 Chapter Overview

Interactive processes underpinning the damping of waves as they travel in the muddy coastal sea, a lagoon, or a large lake, are presented in the context of bottom properties. In wave–mud interaction modeling a rheological equation representing the mechanism of energy loss is required. Typically, the relationship between the coefficients in the equation and bed sediment properties are unknown, with the result that it is cumbersome to apply the model to conditions other than those for which it is calibrated. Although this deficiency persists, in this chapter some efforts made to link the mode of energy loss and bed properties are summarized.

In Chapter 2 we reviewed basic features of waves in an inviscid fluid over a rigid bed. Here, several mechanisms responsible for the energy loss over rigid as well as non-rigid, *i.e.*, compliant, beds are summarized, some of which are from Dean and Dalrymple [1991]. The simple case of waves at the surface of a viscous fluid over a rigid bed is considered first. Such a fluid may be a turbid suspension which can absorb wave energy at a higher rate than water due to the comparatively large viscosity of suspension even when it is dilute. Other cases can arise depending on the rheological properties of mud and whether water is treated as an inviscid or a viscous fluid. In compliant beds which can heave due to wave motion, energy loss is due to fluid viscosity and inter-granular friction, one or both of which can be important depending on how the bed

737

interacts with the wave. Following a brief discussion of possible approaches to parameterize the damping mechanisms, wave-induced local erosion and entrainment processes are summarized.

11.2 Wave–Mud Processes and Imprint

Table 11.1 illustrates an imprint of waves on mud offshore of the coast of French Guiana [Migniot, 1968]. In this region the muddy nearshore zone is extensive, as in the neighboring Suriname and Guyana. There is a shoreward increase in the water content W of bottom mud, starting at a distance of 6 km in 5 m of water. This variation in W has been cited as evidence of the fluid mud generating potential of shoaling waves in the sense that waves can prevent or retard the densification of mud close to the bed surface. The nearer the wave approaches the shoreline the greater is the orbital motion at the bottom and the higher the rate of its overturning. As orbital motion penetrates the soft mud (Fig. 11.1a), wave energy loss also increases. As seen in the laboratory data of Fig. 11.1b, wave damping is sensitive to mud viscosity. At a given depth in mud, the orbital amplitude decreases rapidly with increasing viscosity as mud density increases, or water content decreases.

The rate at which wave height decays is measured in units of inverse distance. A remarkably consistent trend is seen in Fig. 11.1c, in which this rate, *i.e.,* the wave damping coefficient, is plotted against water

Table 11.1. Offshore mud water content in French Guiana.

Distance from shoreline (km)	Water depth (m)	Water content, W (%)
0	0	200
2	2	150–175
3	3	125–150
4.5	4	100–125
6	5	100

Source: Migniot [1968].

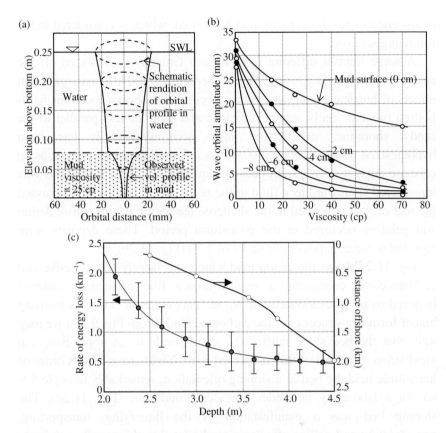

Figure 11.1. Damping of wave motion in marine mud: (a) orbital contraction in mud with a dynamic viscosity of 25 cp; (b) variation of orbital amplitude with depth and viscosity in mud (from Migniot [1968]); (c) variation of rate of wave energy loss with water depth on the submersed Atchafalaya Delta, Louisiana (from Elgar and Raubenheimer [2008]).

depth based on wave data collected from a shore-normal array of gages in the vicinity of the Atchafalaya Delta along the coast of Louisiana (WHOI site in Fig. 10.49). The primary source of this mud is the Mississippi River, from which part of the sediment load is diverted by the Atchafalaya River, a distributary of the Mississippi. Starting at the offshore depth of 4.5 m, as the water became shallower the wave damping coefficient increased as wave energy was absorbed with increasing efficiency. As we noted in Chapters 5 and 10, this efficiency is associated with the dynamical state of mud, which is considered in the following sections.

A wave imprint different from French Guiana was reported off the east coast of Florida. During August–September 2004, when significant effects of hurricane generated waves occurred, at several locations hard-bottom biogenic reefs, which run close and more or less parallel to the sandy shoreline, experienced mud accumulation. It has been hypothesized [Gorham *et al.*, 2008] that soft mud normally present as an offshore deposit (Fig. 11.2a) became liquefied by storm waves and was transported shoreward as a fluid to the reefs. After the fluid mud crossed the reef crest it deposited in the sheltered lee region where consolidation and gelation occurred in the post-storm period. These deposits were reported to have slopes on the order of 1:10 (Fig. 11.2b).

Fig. 11.2c shows the *in situ* mud which was largely non-cohesive and carbonaceous, containing a reddish-brown fine particulate material believed to be of cyanobacterial origin. This material acting as a strong binder formed a veneer over the carbonate mud. From Fig. 9.13b we may infer that the bed slope $\tan\theta = 0.1$ corresponds to an upper-Bingham yield stress $\tau_y = 4$ Pa of the submersed mud. Yield stress of this order of magnitude held the bed in a sloping orientation, remarkably at angle $\theta >$ 45° in a laboratory jar under quiescent conditions (Fig. 11.2d). The sloping bed was a manifestation of the liquefying, transporting, consolidating and gelling effects on mud related to the growth and decay of storm waves. When the binding veneer in the jar was disturbed the underlying sloping carbonate layer failed (Fig. 11.2e). After the binding agent in laboratory samples disintegrated in a few months, a sloping bed could not be sustained.[a]

Wave theories help us understand the ramifications of wave–mud interaction, and in recent years a great deal of relevant literature has developed. The need to consider waves over planar water surface has led to the production of numerical models that allow spectral wave description and time-dependent forcing (*e.g.*, Holthuijsen [2007]). Here

---

[a] Tilting mud in a jar or a cylinder at different times after a deposit is formed from suspension was described by Krone [1957] as a simple way to assess whether the mud is fluid or has gelled.

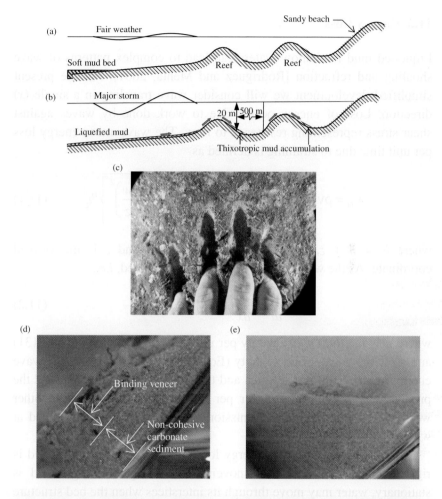

Figure 11.2. (a) Mud deposit over coarse sediment offshore of a reefed shoreline; (b) shoreward transport of mud liquefied during a hurricane strength storm and post-storm retention of hardened mud; (c) mud coated with cyanobacterial film (adapted from Gorham *et al.* [2008]); (d) sediment in a jar forming a stable slope (> 45°); (e) slight shaking of the jar dispersed the veneer and the bed failed.

we will summarize only basic aspects of wave–mud interaction and emphasize analytic treatments, since the main focus is on the relationship between bed properties and energy loss.

11.3 Energy Loss

Liquefied mud in shallow waters can lead to complex patterns of wave shoaling and refraction [Rodriguez and Mehta, 2001]. For the present simplified development we will consider wave traveling in a single (x) direction. Loss of energy occurs due to work done by waves against shear stress representing resistance to flow. The wave–mean energy loss per unit time due to straining is defined as

$$\varepsilon_D = \rho \nu \int_0^h \overline{\left[2\left(\frac{\partial u}{\partial x}\right)^2 + \left(\frac{\partial w}{\partial x} + \frac{\partial u}{\partial z}\right)^2 + 2\left(\frac{\partial w}{\partial z}\right)^2 \right]} dh_z \qquad (11.1)$$

where $h_z = h + z$, h is the local water depth and z is the vertical coordinate. As the wave travels its energy is dissipated, *i.e.*,

$$\frac{dE}{dt} = \frac{d}{dx}\left(EC_g \right) = -\varepsilon_D \qquad (11.2)$$

where E is the total wave energy per unit water surface area (Eq. 2.131) and C_g is the wave group velocity (Eq. 2.132). Therefore ε_D is the wave energy loss per unit surface area and time. As we noted in Chapter 2 the product EC_g is the wave power per unit water surface area. In other words EC_g is the rate of transmission of energy, with C_g the speed at which energy is transmitted.

The mechanism of wave energy loss depends on whether the bed is rigid or heaves under wave movement. Although the rigid bed is stationary, water may move through its interstices when the bed structure is sufficiently porous. In a soft, compliant bed, heaving is due to pressure variation at the bed surface. As a result, additional energy loss mechanisms come into play. Based on particle descriptions in Chapters 2 and 3, 40 μm may be conveniently taken as the approximate threshold diameter below which one can expect a sufficiently soft bed to heave.

For the constant water depth case considered here, Eq. 11.2 reduces to

$$C_g \frac{dE}{dx} = -\varepsilon_D \qquad (11.3)$$

We will take the decrease in wave amplitude $a(x)$ with distance to be of the form

$$a(x) = a_0\, e^{-k_i x} \tag{11.4}$$

where a_0 is the amplitude at an initial position $x = 0$ and k_i is the wave damping coefficient. This simple law follows from the standard harmonic solution for the wave form which satisfies the equation of motion for a progressive wave (see *e.g.*, Chapter 5), and is not universally applicable as we will note later.

Combining Eqs. 11.3 and 11.4 yields

$$k_i = \frac{\varepsilon_D}{2 C_g E} \tag{11.5}$$

Thus, determination of k_i requires ε_D, which in turn depends on the mechanism for energy loss. Relevant mechanisms are mentioned in Table 11.2.

The straining of the bottom to applied stress depends on bottom state (continuum or two-phased particle-water mixture, solid or fluid; Table 11.2). Loss of wave energy due to bed friction occurs in the laminar or

Table 11.2. Bottom sediment constitutive properties and models.

Bed type	Property or medium	Constitutive behavior	Constitutive models
Non-compliant (rigid) solid	Surface roughness	Frictional resistance	Laminar or turbulent boundary-layer
Non-compliant (rigid) porous solid	Two-phase	Percolation	Darcy viscous loss
Fluid	Continuum	Viscous	Newtonian or non-Newtonian viscous
Compliant solid	Continuum	Viscoelastic	Kelvin–Voigt or higher order
Fluid	Continuum	Viscoelastic	Maxwell or higher order
Compliant solid and fluid	Continuum	Viscoplastic	Bingham plastic or higher order
Compliant porous solid	Two-phase	Poroelastic	Coulomb friction and Darcy viscous loss

turbulent boundary layer. It is the only mechanism associated with the bed surface, and varies with surface roughness. All the other mechanisms involve the bulk of the bed.

Percolation occurs below the bed surface, as pore water flows due to wave pressure gradient. Energy loss is expressed in terms of Darcy viscous flow of pore water. The bed is rigid and has two (solid and liquid) phases, although here it will be treated as a porous continuum. In contrast, energy loss in Newtonian and non-Newtonian viscous flows, *e.g.*, fluid mud flow, is in a continuum. For continua with viscoelastic properties, rheological models include the Kelvin–Voigt and the Maxwell elements, and others that are higher-order combinations of these two basic types.

As we noted in Chapter 5, viscoplastic behavior is often modeled as a Bingham plastic, which is a basic model, or using models such as the Casson or the Herschel–Bulkley plastic. Viscoplastic models can be thought of as variants of commonly observed shear-thinning behavior of soft mud.

Poroelastic response of a two-phased bed treated as a porous continuum involves Coulomb friction loss due to inter-particle friction, and Darcy viscous loss in the pore fluid.

11.4 Wave in Viscous Fluid above Rigid Bed

A simple case of the damping of a traveling wave (specified by Eq. 2.118) is one in which a viscous fluid such as a dilute suspension or not too dense fluid mud is bounded by the oscillating free surface and a rigid bed. The fluid is assumed to remain viscous, *i.e.*, non-turbulent. The viscous fluid assumption conveniently simplifies the rheological behavior of real muds, and is popularly incorporated into wave-forecasting numerical models such as SWAN, in which mud viscosity is the only required rheological parameter [Kranenburg *et al.*, 2011]. SWAN does not treat the non-linear dynamics of natural wave spectra (*e.g.*, Kaihatu *et al.* [2007]), which requires that the effect of variable frequency on mud rheology be accounted for (Chapter 5).

Referring to Eq. 2.90 and Fig. 2.21, the relevant (linearized) equations of motion in the x-z plane are

$$\frac{\partial u}{\partial t} = -\frac{1}{\rho}\frac{\partial p}{\partial x} + v\left(\frac{\partial^2 u}{\partial x^2} + \frac{\partial^2 u}{\partial z^2}\right) \tag{11.6}$$

$$\frac{\partial w}{\partial t} = -\frac{1}{\rho}\frac{\partial p}{\partial z} + v\left(\frac{\partial^2 w}{\partial x^2} + \frac{\partial^2 w}{\partial z^2}\right) - g \tag{11.7}$$

For further treatment we will make the x-equation dimensionless by using the following variables

$$\tilde{x} = xk, \quad \tilde{z} = \frac{z}{\delta_v}, \quad \tilde{t} = \sigma t, \quad \tilde{u} = \frac{u}{a\sigma}, \quad \tilde{p} = \frac{p}{\rho ga} \tag{11.8}$$

where $\delta_v = \sqrt{2v/\sigma}$ is the thickness of the laminar (or Stokes) boundary layer in periodic motion and k is the wave number.[b] Thus Eq. 11.6 becomes

$$\frac{\partial \tilde{u}}{\partial \tilde{t}} = -\left(\frac{gk}{\sigma^2}\right)\frac{\partial \tilde{p}}{\partial \tilde{x}} + \left(\frac{vk^2}{\sigma}\right)\frac{\partial^2 \tilde{u}}{\partial \tilde{x}^2} + \left(\frac{v}{\sigma\delta_v^2}\right)\frac{\partial^2 \tilde{u}}{\partial \tilde{z}^2} \tag{11.9}$$

in which the dimensionless terms in parentheses can be restated as

$$\frac{gk}{\sigma^2} = \frac{1}{\left(\dfrac{C_w}{\sqrt{gk^{-1}}}\right)^2}; \quad \frac{vk^2}{\sigma} = \frac{1}{\left(\dfrac{\sigma k^{-2}}{v}\right)}; \quad \frac{v}{\sigma\delta_v^2} = \frac{1}{\left(\dfrac{\sigma\delta_v^2}{v}\right)} \tag{11.10}$$

where $C_w / \sqrt{gk^{-1}}$ is a Froude number and $\sigma k^{-2}/v$ as well as $\sigma\delta_v^2/v$ are Reynolds numbers. Introduce characteristic values: $k = 10^{-1}$ m^{-1}, $\sigma = 1$ rad s^{-1}, $g \approx 10$ m s^{-2} and $v = 10^{-4}$ m^2 s^{-1} for fluid mud. We then obtain $gk/\sigma^2 = 1$ and $vk^2/\sigma = 10^{-6}$, and further, $\delta_v \sim \sqrt{v/\sigma}$ yields $v/\sigma\delta_v^2 = 1$. As a result of the comparatively small value of the middle number

---

[b] For analyses in which it is essential to distinguish the wave number from the wave damping coefficient k_i, the symbol k is used to denote the complex wave number whose real part is k_r and imaginary part is k_i. In the present description the subscript r has been ignored.

we can ignore viscous energy loss associated with the derivative of the non-dimensional horizontal velocity gradient $\partial \tilde{u} / \partial \tilde{x}$, *i.e.*, the second term on the right hand side of Eq. 11.9 [Dean and Dalrymple, 1991].

Consider velocity u to be the sum of an irrotational component u_i representing inviscid flow and a rotational component u_r for viscous flow. At the bottom ($z = -h$) the no-slip condition is $u = u_i + u_r = 0$, *i.e.*, $u_r = -u_i$. At the surface ($z = 0$) we note that $u_r = 0$ because significant rotational effects are mainly confined within the relatively thin bottom boundary layer $\delta_v = \sqrt{2\nu / \sigma}$. For example, for fluid mud the layer thickness is $\sqrt{2 \times 10^{-4} / 1} = 0.014$ m ≈ 14 mm. Increasing the viscosity to, say, 10^{-3} m^2 s^{-1} would result in a boundary layer that is only 45 mm in thickness.

The velocity u_i satisfies the equation of motion for inviscid flow

$$\frac{\partial u_i}{\partial t} = -\frac{1}{\rho} \frac{\partial p}{\partial x} \tag{11.11}$$

and u_r satisfies the approximate (linearized) viscous flow equation

$$\frac{\partial u_r}{\partial t} = -\nu \frac{\partial^2 u_r}{\partial z^2} \tag{11.12}$$

The inviscid solution for u_i is (see Eq. F2.3)

$$u_i = \frac{agk}{\sigma} \frac{\cosh k(h+z)}{\cosh kh} \cos(kx - \sigma t) \tag{11.13}$$

This expression can be written in the complex notation as

$$u_i = \frac{agk}{\sigma} \frac{\cosh k(h+z)}{\cosh kh} e^{i(kx - \sigma t)} \tag{11.14}$$

in which the real part of the complex function

$$e^{i(kx - \sigma t)} = \cos(kx - \sigma t) + i \sin(kx - \sigma t) \tag{11.15}$$

is of present interest. The surface wave (corresponding to Eq. 2.118) is $\eta(x,t) = ae^{i(kx-\sigma t)}$. To determine u_r the method of separation of variables is used along with the no-slip boundary condition at the bed. The result, [Dean and Dalrymple, 1991] is

$$u_r = -\frac{agk}{\sigma\cosh kh} e^{-(1-i)\frac{z+h}{\delta_v}} e^{i(kx-\sigma t)} \qquad (11.16)$$

Therefore

$$u = \frac{agk}{\sigma\cosh kh}\left[\cosh k(h+z) - e^{-(1-i)\frac{z+h}{\delta_v}}\right] e^{i(kx-\sigma t)} \qquad (11.17)$$

Equation 11.17 indicates a phase difference in the velocity relative to water surface that depends on the elevation z.

Example 11.1: Consider a wave of 0.25 m amplitude in 1.5 m deep water. Plot the dimensionless velocity u/u_δ (where u_δ is the value of u at the height of the boundary layer δ_v) against dimensionless elevation $\tilde{z} = (z+h)/\delta_v$ for $k\delta_v = 0.01$ and three values of $\delta_\psi (= kx - \sigma t)$, *i.e.*, $= 0$, $-\pi/4$ and $-\pi/2$. Next, for a wave period of 7 s calculate u in a fine-sediment suspension at an elevation of 0.1 m above bottom. Select $\delta_\psi = -\pi/6$. Note that the velocity u_δ is approximated from

$$u_\delta \approx u_b = \frac{agk}{\sigma\cosh kh} \qquad (11.18)$$

The plot is shown in Fig. 11.3. Due to higher inertia in the upper layer of the oscillating flow relative to the greater effect of viscous drag in the lower layer, the two layers generally move at different speeds and are not in phase. For the selected wave we obtain $\sigma = 0.897$ rad s^{-1}, $k = 0.239$ m^{-1}, $\delta_v = 0.042$ m and $\tilde{z} = 2.388$. Therefore, at $\tilde{z} \approx 2.4$ we get $u/u_b = 0.894$ and $u = 0.55$ m s^{-1}.

The stipulation of zero velocity at the bottom is a noteworthy improvement over the solution based on inviscid and irrotational fluid theory (Chapter 2), and permits the calculation of bed shear stress τ_b arising mainly from the vertical gradient of u_r, *i.e.*,

$$\tau_b \approx \rho v \left.\frac{\partial u_r}{\partial z}\right|_{z=-h} \qquad (11.19)$$

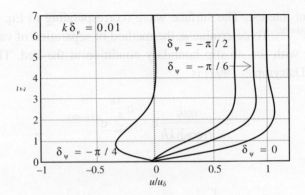

Figure 11.3. Normalized plot of horizontal velocity against elevation in wave-induced viscous boundary layer (based on Dean and Dalrymple [1991]).

Making use of Eq. 11.16 then yields

$$\tau_b \approx \rho \sqrt{\frac{\delta_v}{2}} \frac{agk}{\cosh kh} \cos\left(kx - \sigma t - \frac{\pi}{4}\right) \tag{11.20}$$

which indicates that τ_b lags the water surface consistently by $\pi/4$.

Example 11.2: Calculate the maximum (viscous) bed shear stress due to a wave of 0.1 m amplitude traveling in a 1.5 m thick layer of fluid mud having a density of 1,150 kg m^{-3} and kinematic viscosity 10^{-3} m^2 s^{-1}. The wave period is 7 s.

From Eq. 2.119 we obtain $k = 0.239$ m^{-1} and $C_w = \sigma/k = 3.75$ m s^{-1}. Then, from Eq. 11.20, $\tau_{bmax} = agk\rho\sqrt{\delta_v/2}/\cosh kh = 38.9$ Pa. This large value compared to sediment-free water (6.9 Pa at a viscosity of 10^{-6} m^2 s^{-1}) is due to the high mud viscosity. Thus fluid mud in viscous flow exerts a significantly higher drag at the bottom than clear water.

In Eq. 11.1 the shear rate $\partial u/\partial z$ is the dominant term contributing to wave energy loss. From that term, the work done against shear stress (Eq. 11.20) is given by

$$\varepsilon_D = \rho v \int_0^h \overline{\left(\frac{\partial u_r}{\partial z}\right)^2} dh_z \tag{11.21}$$

where the overbar represents a wave–mean quantity. Substituting for u_r from Eq. 11.16 yields

$$\varepsilon_D = \frac{\nu k\, E}{\delta_\nu \sinh 2kh} \qquad (11.22)$$

Then from Eq. 11.5

$$k_i = \frac{k^2 \delta_\nu}{2(2kh + \sinh 2kh)} \qquad (11.23)$$

and

$$a(x) = a_0\, e^{-\frac{k^2 \delta_\nu}{2(2kh+\sinh 2kh)} x} \qquad (11.24)$$

Example 11.3: A wave of 7 s period is traveling in a 1.5 m deep fine-sediment suspension having a density of 1,150 kg m^{-3} and dynamic viscosity of 1 Pa.s. Calculate the percent reduction in wave amplitude due to viscous damping over a distance of 1 km. Take the viscosity of water to be 10^{-6} m^2 s^{-1}. What would be the percent reduction in water free of sediment?

From Eq. 11.23 $k_i = 8.39 \times 10^{-4}$ m^{-1}, and from Eq. 11.24 $a/a_0 = 0.432$ or a reduction of 56.8%. As in Example 11.2, this high rate of damping arises from the viscosity of the fluid that is 1000-fold greater than that of water, which by itself would yield a reduction of only 2.8%.

Liu and Mei [1989] extended the description to include denser mud with Bingham plastic properties, non-linear surface waves, low to moderate Reynolds number flows and flat as well as sloping beds.

11.5 Shallow-Water Wave in Inviscid Fluid above Viscous Fluid

We will now consider a layer of inviscid fluid above a fluid mud layer. Shallow water wave traveling in this two-layered system was analytically modeled by Gade [1958]. Jiang and Mehta [1992] applied the model to diagnostically analyze wave damping and viscous oscillations of fluid mud in Lake Okeechobee, Florida. It was found that waves at the down-wind end of the lake were only about a quarter as high as they would have been in the absence of fluid mud.

The shallow water condition for waves over a rigid bottom is $kh < \pi/10$ (Chapter 2). Combined with the wave dispersion relationship $\sigma/k = \sqrt{gh}$, the condition becomes

$$\frac{h\sigma^2}{g} < 0.1 \qquad (11.25)$$

Thus, the depth h must be equal to or less than $0.1g/\sigma^2$. When the bottom is compliant the maximum depth can be somewhat smaller. This is so because, due to the deformable lower layer, for given water depth the wave length is greater than over a rigid bottom. Although fluid mud is visually opaque its density is only about 5 to 20% greater than that of freshwater. Thus when fluid mud of thickness h_2 (which may be the same as the liquefied mud thickness z_{fe} in Chapter 10) is present the shallow water criterion can be approximated as $(h_1 + h_2)\sigma^2/g < 0.1$, where h_1 is the depth of water.

Referring to Fig. 11.4, let us consider water (density ρ_1) and fluid mud (ρ_2) layers to be inviscid and viscous, respectively. The vertical coordinate $z' = -z$ is positive downward and the fluid domain, bounded between $z' = 0$ and $h_1 + h_2$, is infinite in the $\pm x$ directions. The assumption of inviscid fluid implies the absence of wave boundary layer above the water–mud interface. The surface and interface variations relative to still water levels are $\eta_1(x,t)$ and $\eta_2(x,t)$. The corresponding amplitudes are assumed to be small enough to conform to the linear theory.

As a result of the shallow water assumption only horizontal water motion is considered. For the upper layer the equation of motion is

$$\frac{\partial u_1}{\partial t} + g\frac{\partial \eta_1}{\partial x} = 0 \qquad (11.26)$$

and flow continuity

$$\frac{\partial(\eta_1 - \eta_2)}{\partial t} + h_1\frac{\partial u_1}{\partial x} = 0 \qquad (11.27)$$

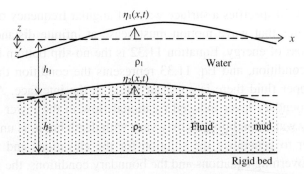

Figure 11.4. Two-layered inviscid-viscous shallow water system subject to a progressive surface wave.

For the lower layer the respective equations are

$$\frac{\partial u_2}{\partial t} + g\widetilde{\Delta\rho}\frac{\partial \eta_2}{\partial x} + (1-\widetilde{\Delta\rho})\frac{\partial \eta_1}{\partial x} = \nu\frac{\partial^2 u_2}{\partial z'^2}$$ (11.28)

and

$$\int_0^{h_i} \frac{\partial u_2}{\partial x} dz' + \frac{\partial \eta_2}{\partial t} = 0$$ (11.29)

where $u_1(x,t)$ and $u_2(x,t)$ are the wave-induced velocities, $\widetilde{\Delta\rho} = (\rho_2 - \rho_1)/\rho_2$ is the normalized density jump and $h_i = h_1 + \eta_2$ is the instantaneous elevation of the water–mud interface. The relevant boundary conditions are

$$\eta_1(0,t) = a_0 \cos \sigma t$$ (11.30)

$$u_1(\infty,t), u_2(\infty,z,t), \eta_1(\infty,t), \eta_2(\infty,t) = 0$$ (11.31)

$$u_2(x, h_1 + h_2, t) = 0$$ (11.32)

and

$$\frac{\partial u_2(x, h_1, t)}{\partial z'} = 0$$ (11.33)

Equation 11.30 specifies a surface wave of angular frequency σ, and Eq. 11.31 requires that wave motion must cease at infinite distance due to complete loss of energy. Equation 11.32 is the no-slip bottom kinematic boundary condition, and Eq. 11.33 represents the condition that due to inviscid upper fluid there is no shear stress at the interface. Also, as a result of boundary layer induced rotational flow in the lower layer and the shallow water condition, u_2 may vary with depth but u_1 is uniform.

In order to generalize the solution for the elevations and velocities from the governing equations and the boundary conditions, the following dimensionless quantities are introduced

$$\tilde{x} = \frac{x}{h_1}, \quad \tilde{z} = \frac{z'}{h_1}, \quad \tilde{t} = \sigma t, \quad \tilde{k} = k h_1, \quad \tilde{u}_1 = \frac{u_1}{\sigma h_1}, \quad \tilde{u}_2 = \frac{u_2}{\sigma h_2},$$

$$\tilde{\eta}_1 = \frac{\eta_1}{h_1}, \quad \tilde{\eta}_2 = \frac{\eta_2}{h_2}, \quad \tilde{h}_2 = \frac{h_2}{h_1}, \quad \tilde{h}_i = \tilde{h}_2 + \tilde{\eta}_2 \tag{11.34}$$

Thus, in the upper layer

$$\frac{\partial \tilde{u}_1}{\partial \tilde{t}} + \frac{1}{F^2} \frac{\partial \tilde{\eta}_1}{\partial \tilde{x}} = 0 \tag{11.35}$$

$$\frac{\partial (\tilde{\eta}_1 - \tilde{\eta}_2)}{\partial \tilde{t}} + \frac{\partial \tilde{u}_1}{\partial \tilde{x}} = 0 \tag{11.36}$$

and in the lower layer

$$\frac{\partial \tilde{u}_2}{\partial \tilde{t}} + \frac{\widetilde{\Delta \rho}}{F^2} \frac{\partial \tilde{\eta}_2}{\partial \tilde{x}} + \frac{1 - \widetilde{\Delta \rho}}{F^2} \frac{\partial \tilde{\eta}_1}{\partial \tilde{x}} = \frac{1}{Re_w} \frac{\partial^2 \tilde{u}_2}{\partial \tilde{z}^2} \tag{11.37}$$

$$\int_0^{\tilde{h}_i} \frac{\partial \tilde{u}_2}{\partial \tilde{x}} d\tilde{z} + \frac{\partial \tilde{\eta}_2}{\partial \tilde{t}} = 0 \tag{11.38}$$

in which the Froude number F and the Reynolds number Re_w are

$$F = \sigma \sqrt{\frac{h_1}{g}}, \quad Re_w = \frac{\sigma h_1^2}{\nu} \tag{11.39}$$

The dimensionless slope terms in Eqs. 11.36 and 11.37 are scaled by $1/F^2$, and the loss term on the right hand side of Eq. 11.37 is scaled by

$1/Re_w$. Based on typical natural conditions, from the magnitude of the multiplier of each term we find that the surface slope term is (expectedly) most important. Also, as a response to the surface wave the velocity gradient in the mud layer can be significant. Compared to the surface, the response of the interface is generally mild and the interface slope is much smaller than the surface slope.

For further examination of the influences of the terms in the equations of motion consider a set of characteristic values: $\sigma = 2\pi$ rad s^{-1}, $h_1 = 1$ m, $\nu = 10^{-2}$ m^2 s^{-1} and $\widehat{\Delta\rho} = 0.1$ yields $F = 2$, and $Re_w = 628$, a low. Using these values, $1/F^2$, $\widehat{\Delta\rho}/F^2$, $(1 - \widehat{\Delta\rho})/F^2$ and $1/Re_w$ would be 0.25, 0.025, 0.225, and 0.0016, respectively. Since low values of Re_w are common, fluid mud flow is often viscous, even when the boundary layer in water is turbulent [Maa and Mehta, 1990]. A strong flow is required to induce turbulence in fluid mud, which, however, may also destabilize it by mixing.

The boundary conditions for Eqs. 11.30 to 11.33 in their dimensionless forms are

$$\tilde{\eta}_1(0, \tilde{t}) = \tilde{a}_0 \cos \tilde{t} \tag{11.40}$$

$$\tilde{u}_1(\infty, \tilde{t}), \tilde{u}_2(\infty, \tilde{z}, \tilde{t}), \tilde{\eta}_1(\infty, \tilde{t}), \tilde{\eta}_2(\infty, \tilde{t}) = 0 \tag{11.41}$$

$$\tilde{u}_2(\tilde{x}, 1 + \tilde{h}_2, \tilde{t}) = 0 \tag{11.42}$$

and

$$\frac{\partial \tilde{u}_2(\tilde{x}, 1, \tilde{t})}{\partial \tilde{z}} = 0 \tag{11.43}$$

where $\tilde{a}_0 = a_0/h_1$ is the dimensionless wave amplitude. Equations 11.35 to 11.39 are solved along with these boundary conditions as follows.

Let us assume the following harmonic solutions for the elevations and velocities

$$\tilde{\eta}_1 = \tilde{a}_0 \, e^{i(\tilde{k}\tilde{x} - \tilde{t})} \tag{11.44}$$

$$\tilde{u}_1 = \tilde{B} e^{i(\tilde{k}\tilde{x} - \tilde{t})} \tag{11.45}$$

$$\tilde{\eta}_2 = \tilde{c}\, e^{i(\tilde{k}\tilde{x}-\tilde{t})} \tag{11.46}$$

$$\tilde{u}_2 = \tilde{D}\cdot \tilde{E}(z) e^{i(\tilde{k}\tilde{x}-\tilde{t})} \tag{11.47}$$

in which $\tilde{a}_0$ (given), $\tilde{B}$ (unknown) and $\tilde{c}$ (unknown) are the amplitudes of $\tilde{\eta}_1$, $\tilde{u}_1$ and $\tilde{\eta}_2$, respectively. The amplitude of $\tilde{u}_2$ is $\tilde{D}\cdot \tilde{E}(z)$, consisting of an unknown quantity $\tilde{D}$ and another unknown $\tilde{E}$ that varies with z (or z'). These quantities must be evaluated by substitution in the dimensionless governing equations and the boundary conditions.

From Eqs. 11.35, 11.44 and 11.45 we have

$$\tilde{B} = \tilde{a}_0\, \frac{\tilde{k}}{F^2} \tag{11.48}$$

Similarly

$$\tilde{c} = \tilde{a}_0\left(1-\frac{\tilde{k}^2}{F^2}\right) \tag{11.49}$$

and

$$\tilde{D}\left(\frac{1}{Re_w}\ddot{\tilde{E}}+i\tilde{E}\right) = i\tilde{a}_0\, \frac{\tilde{k}}{F^2}\left(1-r\frac{\tilde{k}^2}{F^2}\right) \tag{11.50}$$

where

$$\ddot{\tilde{E}} = \frac{\partial^2 \tilde{E}}{\partial \tilde{z}^2} \tag{11.51}$$

Next we set $\tilde{F}$ as

$$\tilde{F} = i\tilde{a}_0\, \frac{\tilde{k}}{F^2}\left(1-\widetilde{\Delta\rho}\,\frac{\tilde{k}^2}{F^2}\right) \tag{11.52}$$

and further let

$$\tilde{D} = Re_w \cdot \tilde{F} \tag{11.53}$$

Thus Eq. 11.50 becomes

$$\ddot{\tilde{E}} + i\,Re_w \cdot \tilde{E} = 1 \tag{11.54}$$

which is readily solved to yield

$$\tilde{E} = -\frac{i}{Re_w} + \tilde{M}_1 \cosh m\tilde{z} + \tilde{M}_2 \sinh m\tilde{z} \tag{11.55}$$

where

$$m = (1-i)\sqrt{\frac{Re_w}{2}} = \sqrt{\frac{Re_w}{i}} \tag{11.56}$$

Based on Eqs. 11.42 and 11.43 we get

$$\tilde{M}_1 = \frac{i}{Re_w} \tag{11.57}$$

$$\tilde{M}_2 = -\tilde{M}_1 \tanh m\tilde{h}_2 \tag{11.58}$$

Therefore, Eq. 11.55 becomes

$$\tilde{E} = -\frac{i}{Re_w}(1 - \cosh m\tilde{z} + \tanh m\tilde{h}_2 \cdot \sinh m\tilde{z}) \tag{11.59}$$

Now, from Eq. 11.38 we have

$$\frac{1}{\tilde{k}} = \frac{\tilde{D}}{\tilde{c}}\int_0^{\tilde{h}_i} \tilde{E}\,d\tilde{z} \tag{11.60}$$

Substituting Eq. 11.59 into Eq. 11.60 and carrying out the integration yields

$$\frac{1}{\tilde{k}} = -\frac{\tilde{D}}{\tilde{c}}\frac{i}{Re_w}\left[\tilde{h}_i + \frac{\tanh m\tilde{h}_2(\cosh m\tilde{h}_i - 1) - \sinh m\tilde{h}_i}{m}\right] \tag{11.61}$$

As $\tilde{\eta}_2 \ll \tilde{h}_2$ we may conveniently introduce the linear approximation

$$\tilde{h}_i \approx \tilde{h}_2 \tag{11.62}$$

Then, by simplifying Eq. 11.61 and substituting for $\tilde{c}$ and $\tilde{D}$ from Eqs. 11.49 and 11.53 we obtain

$$\left(\frac{1}{\tilde{k}}\right)^2 = \frac{\tilde{h}_2}{F^2}\left(\frac{1-\widetilde{\Delta\rho}\dfrac{\tilde{k}^2}{F^2}}{1-\dfrac{\tilde{k}^2}{F^2}}\right)\tilde{\Gamma} \qquad (11.63)$$

where

$$\tilde{\Gamma} = 1 - \frac{\tanh m\tilde{h}_2}{m\tilde{h}_2} \qquad (11.64)$$

Therefore, $\tilde{k}/F$ can be obtained as

$$\frac{\tilde{k}}{F} = \sqrt{\frac{1+\tilde{h}_2\tilde{\Gamma} \pm \sqrt{\left(1+\tilde{h}_2\tilde{\Gamma}\right)^2 - 4\widetilde{\Delta\rho}\tilde{h}_2\tilde{\Gamma}}}{2\widetilde{\Delta\rho}\tilde{h}_2\tilde{\Gamma}}} \qquad (11.65)$$

Thus, in general,

$$\tilde{k} = f^n\left(F, \tilde{h}_2, \widetilde{\Delta\rho}, Re_w\right) \qquad (11.66)$$

There are two solutions for $\tilde{k}$ from Eq. 11.65 corresponding to the positive and the negative signs. The positive sign yields a larger value of the amplitude at the interface relative to the surface, and the negative sign corresponds to the opposite case, which is of present interest because the former solution does not conserve energy.

As $\tilde{B}$, $\tilde{c}$, $\tilde{D}$ and $\tilde{E}$ have been evaluated, $\tilde{\eta}_1$, $\tilde{\eta}_2$, $\tilde{u}_1$, and $\tilde{u}_2$ can be stated as

$$\tilde{\eta}_1 = \tilde{a}_0\, e^{i(\tilde{k}\tilde{x}-\tilde{t})} \qquad (11.67)$$

$$\tilde{\eta}_2 = \tilde{a}_0\left[1-\left(\frac{\tilde{k}}{F}\right)^2\right]e^{i(\tilde{k}\tilde{x}-\tilde{t})} \qquad (11.68)$$

$$\tilde{u}_1 = \tilde{a}_0 \frac{\tilde{k}}{F^2} e^{i(\tilde{k}\tilde{x}-\tilde{t})} \tag{11.69}$$

and

$$\tilde{u}_2 = \tilde{a}_0 \frac{\tilde{k}}{F^2} \left[1 - \widetilde{\Delta\rho} \left(\frac{\tilde{k}}{F} \right)^2 \right] (1 - \cosh m\tilde{z} + \tanh m\tilde{h}_2 \cdot \sinh m\tilde{z}) e^{i(\tilde{k}\tilde{x}-\tilde{t})} \tag{11.70}$$

We observe that $\tilde{\eta}_2$ is reduced relative to $\tilde{\eta}_1$ by the damping factor, $1 - (\tilde{k}/F)^2$. Similarly, the velocity $\tilde{u}_2$ is damped relative to $\tilde{u}_1$. Next, the dimensionless wave number $\tilde{k}$ is separated into its real part $\tilde{k}_r$ and the wave damping coefficient $\tilde{k}_i$ according to

$$\tilde{k} = \tilde{k}_r + i\tilde{k}_i \tag{11.71}$$

where

$$\tilde{k} = k h_1, \tilde{k}_r = k_r h_1, \tilde{k}_i = k_i h_1 \tag{11.72}$$

Now let

$$\tilde{Y} = \left(\frac{\tilde{k}}{F} \right)^2 = Y_R + i Y_I \tag{11.73}$$

and

$$\tilde{h}_2 \tilde{\Gamma} = \tilde{R} + i\tilde{I} \tag{11.74}$$

Therefore, Eq. 11.73 can be written as

$$\tilde{Y} = \frac{1}{2\widetilde{\Delta\rho}\left(\tilde{R}+i\tilde{I}\right)} \left[-\sqrt{\left(1+\tilde{R}+i\tilde{I}\right)^2 - 4\widetilde{\Delta\rho}\left(\tilde{R}+i\tilde{I}\right)} + 1 + \tilde{R} + i\tilde{I} \right] \tag{11.75}$$

From this expression we obtain

$$Y_R =$$

$$\frac{1}{2\widetilde{\Delta\rho}\left(\tilde{R}^2+\tilde{I}^2\right)}\left[\tilde{R}\left(1+\tilde{R}-\cos\frac{\theta}{2}\sqrt{\sqrt{\tilde{p}^2+\tilde{q}^2}}\right)+\tilde{I}\left(\tilde{I}-\sin\frac{\theta}{2}\sqrt{\sqrt{\tilde{p}^2+\tilde{q}^2}}\right)\right]\quad(11.76)$$

and

$$Y_I =$$

$$\frac{1}{2\widetilde{\Delta\rho}\left(\tilde{R}^2+\tilde{I}^2\right)}\left[\tilde{R}\left(\tilde{I}-\sin\frac{\theta}{2}\sqrt{\sqrt{\tilde{p}^2+\tilde{q}^2}}\right)-\tilde{I}\left(1+\tilde{R}-\cos\frac{\theta}{2}\sqrt{\sqrt{\tilde{p}^2+\tilde{q}^2}}\right)\right]\quad(11.77)$$

where

$$\tilde{p}=(1+\tilde{R})^2-4\widetilde{\Delta\rho}\tilde{R}-\tilde{I}^2 \tag{11.78}$$

$$\tilde{q}=2\tilde{I}(1+\tilde{R}-2\widetilde{\Delta\rho}) \tag{11.79}$$

$$\theta=\tan^{-1}\left(\frac{\tilde{q}}{\tilde{p}}\right) \tag{11.80}$$

and

$$\tilde{R}=\tilde{h}_2\left[1-\frac{e^{4\tilde{\chi}}-1+2\sin 2\tilde{\chi}\cdot e^{2\tilde{\chi}}}{2\tilde{\chi}(e^{4\tilde{\chi}}+1+2\cos 2\tilde{\chi}\cdot e^{2\tilde{\chi}})}\right] \tag{11.81}$$

$$\tilde{I}=\tilde{h}_2\left[1-\frac{e^{4\tilde{\chi}}-1-2\sin 2\tilde{\chi}\cdot e^{2\tilde{\chi}}}{2\tilde{\chi}(e^{4\tilde{\chi}}+1+2\cos 2\tilde{\chi}\cdot e^{2\tilde{\chi}})}\right] \tag{11.82}$$

in which

$$\tilde{\chi}=\tilde{h}_2\sqrt{\frac{Re_w}{2}}=h_2\sqrt{\frac{\sigma}{2\nu}}=\frac{h_2}{\delta_\nu} \tag{11.83}$$

In this expression for $\tilde{\chi}$, δ_ν is the Stokes boundary layer thickness defined previously. Now from Eqs. 11.72, 11.73, 11.76 and 11.77 we obtain

$$\frac{\tilde{k}_r}{F} = \frac{\sqrt{gh_1}}{C_w} = \sqrt{\frac{\sqrt{Y_R^2 + Y_I^2} + Y_R}{2}} \qquad (11.84)$$

which is the desired relationship for shallow water celerity C_w. Also,

$$\frac{\tilde{k}_i}{F} = \frac{k_i\sqrt{gh_1}}{\sigma} = \sqrt{\frac{\sqrt{Y_R^2 + Y_I^2} - Y_R}{2}} \qquad (11.85)$$

for the wave damping coefficient. It is readily seen that $\tilde{k}_r/F$ and $\tilde{k}_i/F$ depend on $\tilde{h}_2$, $\tilde{\Delta}\rho$ and $\tilde{\chi}$.

Equation 11.84 is plotted in Fig. 11.5a for a representative value of the normalized density jump $\tilde{\Delta}\rho = (\rho_2 - \rho_1)/\rho_2 = 0.15$ at the interface between fluid mud and water. For each value of $\tilde{h}_2 = h_2/h_1$, the ratio $\sqrt{gh_1}/C_w$ decreases as $\tilde{\chi}$ increases above unity. To explain this trend, consider a system with given values of h_1 and h_2, and subject to a wave of frequency σ. The case $\tilde{\chi} = 0$ implies a rigid bottom as $v \to \infty$; therefore $C_w = \sqrt{gh_1}$ is recovered. Increasing $\tilde{\chi}$ indicates increasing mud softness, which effectively increases the total depth beyond h_1. As $\tilde{\chi} \to \infty$, the lower layer becomes inviscid and $C_w = \sqrt{g(h_1 + h_2)}$. Therefore, the limiting value of the vertical axis is $\sqrt{gh_1}/\sqrt{g(h_1 + h_2)}$. The lower layer becomes water-like when the value of $\tilde{\chi}$ exceeds about 3, and the celerity does not change significantly beyond that value of $\tilde{\chi}$.

Equation 11.85 is plotted in Fig. 11.5b, in which noteworthy features are the maxima in $k_i\sqrt{gh_1}/\sigma$ at $\tilde{\chi} \approx 1$, which indicates a peak in the damping coefficient k_i. In other words wave damping is at its maximum when mud thickness h_2 is of the same order as boundary layer thickness δ_v. As $\tilde{\chi}$ increases above ~ 1, k_i decreases because δ_v decreases. As $\tilde{\chi}$ decreases below ~ 1, k_i again decreases because even though the boundary layer occupies the entire mud thickness, the rate of energy loss decreases along with the depth–mean shear rate (velocity gradient) in mud. The same explanation holds if we keep v constant and vary σ.

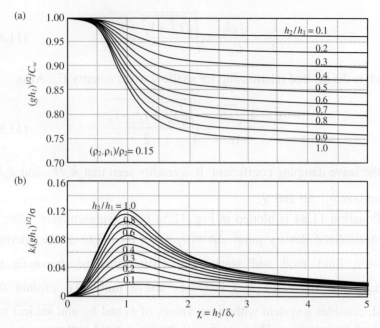

Figure 11.5. Sample results from Gade's [1958] model: (a) dispersion relationship; (b) wave damping coefficient.

In wave flume experiments k_i was determined by Jiang [1993] for beds of clay mixtures: AK composed of a 1:1 (by weight) mixture of an attapulgite and a kaolinite (Table 5.5), and ABK, a 1:1:2 mixture of the same two clays with Wyoming bentonite added to increase cohesion. The bed density for AK and ABK was maintained at 1,200 kg m^{-3} (*i.e.*, $\phi_v = 0.121$). In Fig. 11.6, k_i values are plotted against σ from three sets of tests. The mean trends resemble the curves in Fig. 11.5b. The peak value and location of k_i vary with the sediment and bed thickness. For AK sediment, reducing the thickness by 33% (from 0.18 to 0.12 m) decreased the peak k_i from about 0.40 to 0.35 m^{-1}, *i.e.*, by 13%, and the corresponding frequency increased from 6.5 to 7.0 rad s^{-1}. Changing the sediment to ABK but keeping the same thickness (0.12 m) increased the peak k_i from 0.35 to 0.37 m^{-1}, *i.e.*, by 6%, and shifted the frequency to about 9 rad s^{-1}. These changes can be attributed to the highly viscous

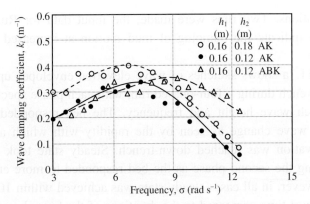

Figure 11.6. Variation of wave damping coefficient with angular frequency for AK and ABK sediments (from Jiang [1993]).

bentonite in ABK. Increasing σ increased the thickness of the boundary layer in mud relative to its actual thickness with the result that energy loss increased. Once the boundary layer thickness theoretically exceeded the mud thickness, further increase in σ decreased the depth–mean shear rate in mud, and energy loss decreased.

Example 11.4: A 10 s wave travels in 1 m deep water over 0.3 m thick fluid mud. (1) If mud viscosity is 2.5×10^{-3} m^2 s^{-1}, calculate the wave speed and the damping coefficient. Assume $\overline{\Delta \rho} = 0.15$. (2) What would be the speed if mud was rigid? (3) What would be the speed if mud was replaced by water?

(1) Using these parameters in Eq. 11.83 we obtain $\tilde{\chi} = 3.36$. Therefore, $Y_R = 0.7917$ and $Y_I = -0.1499$, $\tilde{k}_r = 0.179$, $\tilde{k}_i = 0.0168$ and $C_w = 3.50$ m s^{-1}. (2) For rigid mud $\tilde{\chi} = 0$, and $C_w = \sqrt{9.81 \times 1} = 3.13$ m s^{-1}. (3) If mud was replaced by water the wave speed would increase because $C_w = \sqrt{9.81 \times (1+0.3)} = 3.57$ m s^{-1}.

In flume tests Robillard [2009] measured wave damping over AD mud (Table 5.5) from the Atchafalaya Delta (LA) bordering the Gulf of Mexico. The solids volume fraction of mud (ϕ_v) was 0.125, the water and mud depth were held constant at $h_1 = 0.19$ m and $h_2 = 0.08$ m, respectively, and the trench in which mud was placed was 5 m long. Test parameters for Run 1 are given in Table 11.3 in which $|\dot{\gamma}|$ is the estimated depth–mean shear rate in mud based on visual observations of mud oscillations. Two runs were made; the input data for Run 2 are included to indicate the damping of mud motion at increased density ($\phi_v = 0.214$).

Figure 11.7a shows sketches of wave elevation envelopes up-trench and down-trench during a four-phased run, with each phase specified by the up-trench wave height and frequency. The bed responded within minutes to wave change as seen by the rapidity with which a steady surface elevation was reached down-trench. Steady state took slightly longer during the second phase as the bed responded to more energetic waves. However, in all cases steady state was achieved within 10 min, a generally short time compared to the durations of the four phases meant to represent real storm time-scales.

In order to assess the stress-strain response of the bed to waves, samples of AD mud were exposed to a three-phase, controlled oscillatory strain test in a rheometer for durations comparable to the flume runs [Robillard, 2009]. The behaviors of the loss and storage components η' and η'', respectively, of the complex viscosity η^* (Eq. 5.77) were found to be thixotropic, (Fig. 11.7b). During Phase A the viscosity adapted quickly to a low rate of strain (0.106 s^{-1}) indicated by the near-constant values of η' and η''. In Phase B a substantially higher rate of strain (6.39 s^{-1}) was imposed, which initiated a plastic yield manifested by a switch in the dominant complex viscosity component. Initially the

Table 11.3. Test parameters for two wave flume runs with AD mud.

Run	Volume fraction ϕ_v	Parameters	Phase A	Phase B	Phase C
1	0.125	σ (rad s^{-1})	7.30	6.28	7.30
		$\vert\dot{\gamma}\vert$ s^{-1}	0.11	6.39	0.11
		Duration (min)	42	60	30
2	0.214	σ (rad s^{-1})	7.30	6.28	7.3
		$\vert\dot{\gamma}\vert$ (s^{-1})	0.03	1.73	0.03
		Duration (min)	42	60	30

Source: Robillard [2009].

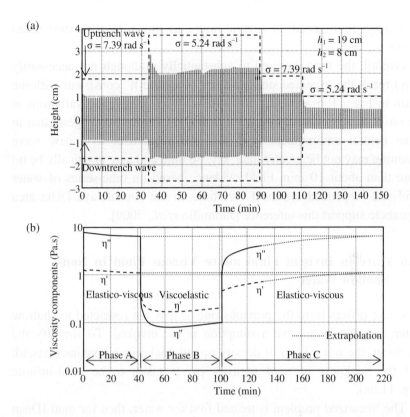

Figure 11.7. Tests on AD mud: (a) sketches of measured up-trench and down-trench wave envelopes in a flume; (b) viscosity components measured in a rheometer (from Robillard [2009]).

mud was in an elastico-viscous state with storage η'' dominanting viscous loss η', but then transformed into viscoelastic state. The mud became equilibrated with the higher shear rate within 25 min. In Phase C, which involved relaxation of strain back to its initial value, the mud reverted quickly to the elastico-viscous state, although it took several hours to return to the Phase A condition. For time-scales longer than order of minutes, thixotropic response of AD mud to increased oscillatory strain (as for instance would occur under a storm) was of secondary significance. On the other hand the response to decreased oscillatory strain lasted for several hours, as shown by the extrapolated curves.

Overall, the time required to substantially (although not necessarily fully) reach the expected steady values of η' and η'' consistent with the strain in Fig. 11.7b explains down-trench wave amplitude variations at the onset of a new wave condition in Fig. 11.7a. It also suggests that in some natural environments the time for adjustment to a new wave condition may not be greater than, say, 30 min, and may practically be no more than about 10 min. Field evidence based on time-series of water level and suspended sediment variations in the Atchafalaya Delta area appears to support this inference [Jaramillo *et al.*, 2009].

11.6 Wave in Inviscid Fluid above Viscous Fluid in Non-Shallow Water

This case differs from the previous one as it is not restricted to shallow water, *i.e.*, the long wave assumption is not invoked. To simplify the treatment, the water layer of density ρ_1 and thickness h remains inviscid, and the thickness of mud (density ρ_2) is taken to be semi-infinite (Fig. 11.8a).

The linearized problem is treated first for water, then for mud [Dean and Dalrymple, 1991]. In water the velocity potential is pre-selected as

$$\sqrt{gh_1}\ \phi_1(x,z,t) = \left[A_1' \cosh k(h+z) + B_1' \sinh k(h+z) \right] e^{i(kx-\sigma t)} \quad (11.86)$$

where A_1' and B_1' are unknowns to be determined and ϕ_1 satisfies the Laplace equation for flow continuity (Eq. 2.107). Applying the dynamic free surface boundary condition (Eq. 2.111) we obtain

$$A_1' \cosh kh + B_1' \sinh kh = i\frac{ga_0}{\sigma} \quad (11.87)$$

Next, the kinematic free surface boundary condition (Eq. 2.109) yields

$$A_1' \sinh kh + B_1' \cosh kh = i\frac{\sigma a_0}{k} \quad (11.88)$$

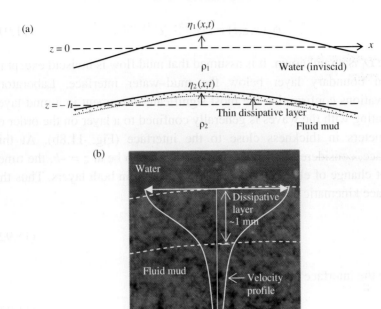

(a) $\eta_1(x,t)$

$z = 0$

ρ_1 Water (inviscid)

$\eta_2(x,t)$

$z = -h$

Thin dissipative layer

ρ_2 Fluid mud

(b) Water

Dissipative layer ~1 mm

Fluid mud

Velocity profile

Figure 11.8. Inviscid water layer and lower fluid with a thin viscous layer subject to a progressive surface wave: (a) Definition sketch; (b) notional dissipative layer at the surface of a flocculated marine mud in laboratory setting.

We may now solve for A_1' and B_1' in terms of the unknown wave number k

$$A_1' = i \frac{a_0 \cosh kh}{\sigma k} (gk - \sigma^2 \tanh kh) \tag{11.89}$$

$$B_1' = i \frac{a_0 \cosh kh}{\sigma k} (\sigma^2 - gk \tanh kh) \tag{11.90}$$

For a rigid bed, application of the bottom boundary condition of Eq. 2.108 specifying zero vertical velocity leads to $B_1' = 0$, and as a result the wave dispersion relationship of Eq. 2.119 is recovered. In the present case the bottom boundary condition is linked to the movement of the compliant mud layer, for which we will select the velocity potential function

$$\phi_2(x,z,t) = D_1' e^{k(h+z)} e^{i(kx-\sigma t)} \tag{11.91}$$

where D_1' is an unknown. It is assumed that mud flow is inviscid except in a thin boundary layer below the mud–water interface. Laboratory observations confirm that the most significant movement in a mud layer oscillating with the waves is generally confined to a layer on the order of millimeters in thickness close to the interface (Fig. 11.8b). At this interface, considered in this linearized problem to be at $z = -h$, the time-rate of change of elevation η_2 must be the same in both layers. Thus the interface kinematic condition is

$$\frac{\partial \eta_2}{\partial t} = -\frac{\partial \phi_1}{\partial z} = -\frac{\partial \phi_2}{\partial z} \tag{11.92}$$

where the interface elevation is

$$\eta_2(x,t) = a_2 e^{i(kx-\sigma t)} \tag{11.93}$$

with a_2 as the amplitude. Substitution of Eqs. 11.86, 11.91 and 11.93 into Eq. 11.92 yields

$$D_1' = B_1' \tag{11.94}$$

and

$$a_2 = -i\frac{kB_1'}{\sigma} \tag{11.95}$$

At the interface the fluid pressure p must be equal on both sides, *i.e.*, $p_1 = p_2$. Thus the corresponding linearized dynamic condition at $z = -h$ is

$$p_1 = \rho_1 \frac{\partial \phi_1}{\partial t} - \rho_1 g \eta_2 = p_2 = \rho_2 \frac{\partial \phi_2}{\partial t} - \rho_2 g \eta_2 \tag{11.96}$$

Substitution of Eqs. 11.86, 11.91 and 11.93 into Eq. 11.96 yields

$$A_1' = \left[\frac{\rho_2}{\rho_1}\left(1 - \frac{gk}{\sigma^2}\right) + \frac{gk}{\sigma^2} \right] B_1' \tag{11.97}$$

Now substituting A_1' and B_1' from Eqs. 11.89 and 11.90, respectively, into the above equation leads to the following two solutions for the wave dispersion relationship

$$\sigma^2 = gk \tag{11.98}$$

and

$$\sigma^2 = gk \left[\frac{\left(\dfrac{\rho_2}{\rho_1} - 1\right) \tanh kh}{\dfrac{\rho_2}{\rho_1} + \tanh kh} \right] \tag{11.99}$$

These relationships are plotted in Fig. 11.9. The two possible wave modes can be identified by the ratio of the amplitudes of η_1 and η_2, *i.e.,*

$$\frac{a_0}{a_2} = e^{kh} \tag{11.100}$$

which indicates a larger surface wave than interfacial wave, and

$$\frac{a_0}{a_2} = -\left(\frac{\rho_2}{\rho_1} - 1\right) e^{-kh} \tag{11.101}$$

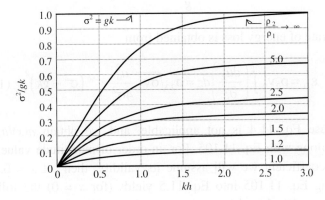

Figure 11.9. Dispersion relationship for waves over a semi-infinitely deep lower fluid (from Dean and Dalrymple [1991]).

according to which the interfacial wave is larger of the two. It can be shown that the first mode implies no effect of the lower layer on the upper one. The second mode, which creates a larger and 180° out-of-phase interfacial wave relative to the surface wave is of present interest. Note that the deep water asymptotes are represented by $\sigma^2 / gk = [(\rho_2 / \rho_1) - 1][(\rho_2 / \rho_1) + 1]$.

To obtain the wave damping coefficient, the velocity $u_2 = -\partial \phi_2 / \partial x$ is modified by adding a component u_2' representing the thin viscous layer. At the interface the sum of these two velocities must be equal to the velocity u_1 associated with the upper layer. Thus at $z = -h$

$$u_1 = -\frac{\partial \phi_1}{\partial x} = u_2 + u_2' = -\frac{\partial \phi_2}{\partial x} + u_2' \qquad (11.102)$$

where u_2' is conveniently taken as a harmonic function with depth-varying amplitude

$$u_2' = F_1' e^{(1-i)\frac{(h+z)}{\delta_v}} e^{i(kx - \sigma t)} \qquad (11.103)$$

in which F_1' is an unknown term. Substitution of Eq. 11.103 along with Eqs. 11.86 and 11.91 into Eq. 11.102 yields

$$F_1' = -\frac{a_0}{g} e^{kh} (\sigma^2 - kh) \qquad (11.104)$$

Then the rate of energy loss is obtained from

$$\varepsilon_D = \rho_2 \nu_2 \int_\infty^{-h} \left(\frac{\partial u_2'}{\partial z} \right) dz = \rho_2 \sqrt{\sigma \nu_2} \frac{a_0^2}{4\sigma^2} e^{2kh} \left(\sigma^2 - gk \right)^2 \qquad (11.105)$$

In this case Eq. 11.4 is not applicable, and we obtain $a(x)/a_0$ from Eq. 11.3 along with Eq. 11.105. For an order of magnitude value of the damping coefficient we will assume its validity, then set $x = 0$. Thus, substituting Eq. 11.105 into Eq. 11.5 yields (for $x = 0$) the following initial (nominal) value of k_i

$$\frac{a(x)}{a_0} = \sqrt{\left[1 - \frac{\rho_2}{\rho_1}\frac{\delta_{v2}}{2\sqrt{2}gC_g\sigma}e^{2kh}(\sigma^2 - gk)^2 x\right]} \qquad (11.106)$$

Example 11.5: A 7 s wave travels in 7 m deep water over fluid mud. If the viscosity of mud is 2.5×10^{-3} m^2 s^{-1}, calculate the wave speed and a/a_0 at a distance of 1,000 m. Assume $\rho_1 = 1,020$ kg m^{-3} and $\rho_2 = 1,200$ kg m^{-3}. Then obtain a representative value of k_i over 1,000 m by assuming the applicability of Eq. 11.4.

From Eq. 11.99 we obtain $C_w = \sigma/k = 7.49$ m s^{-1}, and from Eq. 2.213 $C_g = 6.18$ m s^{-1}. Next, from Eq. 11.106 $a/a_0 = 0.763$. Finally, Eq. 11.4 yields $k_i = 2.71 \times 10^{-4}$ m^{-1} at $x = 1,000$ m.

11.7 Energy Loss and Rheology

11.7.1 *Percolation, bed friction and breaking wave*

11.7.1.1 *Zones of applicability*

When mud is hard enough to be rigid, bed friction, wave breaking and percolation are the three main modes of energy loss. Percolation is important only as long as bed permeability is large enough to permit water movement within the bed due to the dynamic pressure gradient at the bed surface. For percolation and poroelastic effects to be significant the particle size must be about 40 μm or larger. This threshold size is based on an approximate relationship between diameter and permeability. It is also an approximate diameter above which cohesion is largely absent.

Figure 11.10 roughly sorts out sand, silt and clay by permeability k_p and void ratio e_v. In general, sands have medium to high k_p values ($> 10^{-3}$ cm s^{-1}) and e_v less than 1.5. Silts have low k_p values (between 10^{-6} and 10^{-3} cm s^{-1}) and e_v lower than 1.5. Some silts, most compacted clays and mixtures of sand with silt or clay have very low to negligible permeability.

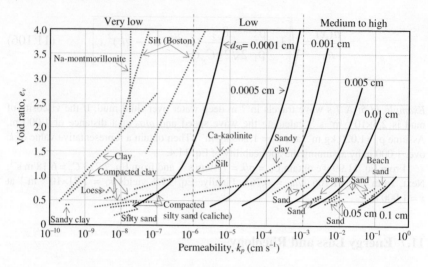

Figure 11.10. Void ratio against permeability for sediments (revised from Lambe and Whitman [1979]). Curves for diameters d_{50} (cm) are based on the Kozeny–Carman equation and the modified Buscall *et al.* [1987] equation.

Loess from Vicksburg, Mississippi, accumulated from fine sediment transported by wind during the Pleistocene, is silty and has low permeability as in a compacted clay. Although the particles are held together by of calcium carbonate in trace quantities, this sediment is not cohesive to the same degree as clays.

Most mixtures also have void ratios less than 0.5. This includes caliche (or calcrete) from Argentina, a sandy clay containing minerals such as sodium nitrate and sodium chloride. Boston silt and sodium-montmorillonite have very low permeabilities but high (> ~1.5) void ratios. High void ratios of smectites (including Na-montmorillonite) are due to their open structure when flocculated. Boston silt may have contained flocculated clays, which would explain its high void ratio. An interesting case is that of Ca-kaolinite, which is far more compact than Na-montmorillonite. Kaolinites form a variety of structures depending on salt concentration; some structures are open and others contain tightly packed particles (Chapter 5).

For beds of particles smaller than about 40 μm, energy losses associated with wave-induced heaving increasingly overwhelm bed friction and percolation as size decreases. Wave breaking also results in significant energy loss.

We will first consider selective expressions for wave damping due to percolation, bed friction and wave breaking over a rigid bed. This is followed by damping over compliant beds whose constitutive behavior can be poroelastic, viscoelastic or viscoplastic.

11.7.1.2 *Percolation*

Energy loss by percolation is determined for a rigid but porous bed of semi-infinite thickness and specific permeability K_p. The bed is below a water layer of given thickness. An analytic solution is obtained for the wave damping coefficient based on Laplace equations for flow continuity in water (Eq. 2.107) and in the porous bed (Eq. 2.95). If in addition it is assumed that water is inviscid everywhere except in the bed and in a thin boundary layer just above the bed (Fig. 11.8), the following expression is obtained [Liu, 1973]

$$k_i = \frac{2k}{2kh + \sinh 2kh} \left(\frac{K_p \sigma}{v} + \frac{k\delta_v}{2} \right) \tag{11.107}$$

where v is the kinematic viscosity of water, $\delta_v = \sqrt{2v/\sigma}$ is the boundary layer thickness and K_p is related to the coefficient of permeability k_p by (Eq. 2.98)

$$K_p = \frac{\eta k_p}{\rho_w g} = \frac{v k_p}{g} \tag{11.108}$$

In Eq. 11.107, setting the boundary layer thickness $\delta_v = \sqrt{2v/\sigma}$ to zero reduces the equation to the expression of Reid and Kajiura [1957] for a porous bottom of semi-infinite thickness.[c] King *et al.* [2009]

---

[c] The Reid and Kajiura [1957] solution can be obtained directly by replacing the zero velocity assumption at $z = -h$ in the rigid bottom case (Fig. 2.21), and equating the vertical velocity w in the water phase with the pore water velocity w_w. The resulting wave dispersion equation is

extended the analysis to a bottom layer of finite thickness; see Exercise 11.4, which is more realistic because a boundless domain over-estimates energy loss.

For permeability, the commonly used Kozeny–Carman equation is

$$k_p = \frac{1}{C_{kc}A_{pv}^2}\frac{\rho g}{\eta}\frac{e_v^3}{(1+e_v)} = \frac{1}{C_{kc}A_{pv}^2}\frac{\rho g}{\eta}\frac{n_r^3}{(1-n_r)^2}$$
$$= \frac{1}{C_{kc}A_{pv}^2}\frac{\rho g}{\eta}\frac{(1-\phi_v)^3}{\phi_v^2}$$

(11.109)

where the coefficient C_{kc} depends on pore geometry and on the ratio of the length of the actual flow path to bed thickness, and A_{pv} is the specific surface area of the particle on volume basis, *i.e.*, surface area per unit volume [Lambe and Whitman, 1979]. For spherical particles $C_{kc} = 5$ [Carrier, 2003] and $A_{pv} = 6/d_p$, where d_p is the diameter to be taken as d_{50} in centimeters. In water, with $\rho = 1{,}000$ kg m^{-3} and $\eta = 0.001$ Pa.s, we obtain $\rho g / \eta C_{kc}A_{pv}^2 \approx 550 d_{50}^2$.

Strictly speaking Eq. 11.109 can be applied to sands and silts but not to clays because cohesion is not included in its formulation. However, for flocculated clay suspensions k_p values have been found to be consistent with this equation by taking $\rho g / C_{kc}A_{pv}^2 \eta = 5.56 \times 10^5 d_{50}^3$, when d_{50} is less than 0.002 cm. This relationship is revised from one by Buscall *et al.* [1987]. Equation 11.109 may be used when $d_{50} \geq 0.001$ cm. Curves for the variation of k_p with e_v are plotted in Fig. 11.10 for d_{50} ranging from 0.0001 cm (1 µm) to 0.1 cm (1 mm). Thus,

$$k_p = \begin{cases} 550 d_{50}^2 \dfrac{e_v^3}{(1+e_v)}; & d_{50} \geq 0.001\,\text{cm} \\[3mm] 5.56 \text{x} 10^5 d_{50}^3 \dfrac{e_v^3}{(1+e_v)}; & d_{50} < 0.001\,\text{cm} \end{cases}$$

(11.110)

$$\sigma^2 - gk\tanh kh = -i\left(\frac{\sigma K_p}{\nu}\right)(gk - \sigma^2\tanh kh)$$

(F11.1)

which reduces to Eq. 2.119 when K_p is zero [Dean and Dalrymple, 1991].

Example 11.6: Consider three sediment beds each with solids volume fraction $\phi_v = 0.6$ (or $n_r = 0.4$) and particle diameters 0.0062 cm (62 μm), 0.0040 cm (40 μm) and 0.0002 cm (2 μm). Select $\rho_w = 1,000$ kg m^{-3} and $v = \eta/\rho_w = 10^{-6}$ m^2 s^{-1} for water. Calculate k_p from Eq. 11.110 and then K_p from Eq. 11.108.

Values of k_p and K_p are as follows:

d_{50} (cm)	k_p (cm s^{-1})	K_p (m^2)
0.0062	2.85×10^{-2}	2.91×10^{-11}
0.0040	1.19×10^{-2}	1.21×10^{-11}
0.0002	6.00×10^{-6}	6.12×10^{-15}

From Fig. 11.10 we note that the range of k_p is from low to medium.

As we noted in Chapter 8, the more compact the sediment the greater the effective stress, and consequently for a given mud as this stress increases the permeability should decrease. This trend is evident in Fig. 11.11, in which K_p varies with the dry density ρ_D of the bed and the effective stress σ_v'. The data are for a "typical mud" [Whitehouse *et al.*, 2000a].

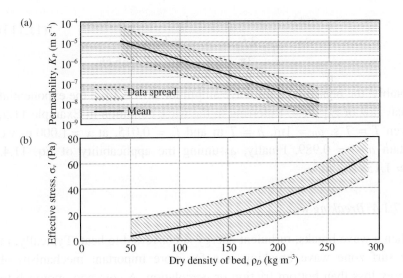

Figure 11.11. Variation of mud dry density with: (a) permeability; (b) effective normal stress. Data of Delo and Ockenden [1992] reported in Whitehouse *et al.* [2000a].

11.7.1.3 *Bed friction in turbulent flow*

In the turbulent boundary layer, the starting point for derivation of the wave damping coefficient k_i is the rate of energy loss due to friction at the rigid bed surface

$$\varepsilon_D = \overline{\tau_b u_{bt}} = \overline{\frac{f_w}{4} \rho u_{bt}^2 |u_{bt}|} \qquad (11.111)$$

where u_{bt} is the wave-induced velocity at the bottom from Eq. 2.124, τ_b is the bed shear stress from Eq. E2.8 and the overbar indicates averaging over the wave period. The wave friction factor f_w can be determined from Eq. E2.9. Using Eqs. 2.124 and 11.111 we obtain

$$\varepsilon_D = \frac{\rho f_w}{6\pi} \left(\frac{a\sigma}{\sinh kh} \right)^3 \qquad (11.112)$$

Equation 11.112 is now substituted into Eq. 11.3 and solved for the wave amplitude to yield

$$\frac{a(x)}{a_0} = \frac{1}{1 + \dfrac{2 f_w}{3\pi} \dfrac{a_0 k^2 x}{(2kh + \sinh 2kh) \sinh kh}} \qquad (11.113)$$

Notably, the behavior of the wave amplitude differs from the exponential decay law of Eq. 11.4. Referring to the wave condition in Example 11.5, given $T = 7$ s, $a_0 = 1$m, $h = 7$ m and $f_w = 0.015$, at $x = 1,000$ m we obtain $a/a_0 = 0.989$. Finally, assuming the applicability of Eq. 11.4, $k_i = 1.13 \times 10^{-5}$ m^{-1}.

11.7.1.4 *Breaking wave*

When a wave breaks, turbulent energy is converted to heat. Typically, in the surf zone wave breaking is a far more important mechanism of energy loss than bottom friction or percolation. A common approach to obtain an effective value of the wave damping coefficient is based on the classical breaking hydraulic bore analogy [Battjes and Janssen, 1978].

Another approach is based on inference concerning the rate of energy loss associated with a particular shape of the equilibrium beach profile. For this, we will use Eq. 12.1, which describes the beach profile as having an overall concave shape. The underpinning basis of this profile is that it is consistent with a uniform loss of turbulent energy per unit volume of water [Dean, 1977; Dean and Dalrymple, 2002]. Denoting this uniform loss as ε_{Dv}, the total energy loss over the water column of local depth h and unit surface area is given by

$$\varepsilon_D = h\varepsilon_{Dv} \qquad (11.114)$$

Therefore, from Eq. 11.2

$$\frac{1}{h}\frac{d(EC_g)}{dx} = -\varepsilon_{Dv} \qquad (11.115)$$

Since nearshore breaking is in shallow water, the group velocity $C_g = \sqrt{gh}$. Then, making use of Eqs. 2.131 and 2.149 we obtain

$$\varepsilon_{Dv} = \frac{5}{16}\rho g^{3/2}\kappa_b^2\sqrt{h}\frac{dh}{dx} \qquad (11.116)$$

Finally, given the beach profile $h(x)$ from Eq. 12.1, the wave damping coefficient is

$$k_i = \frac{5A_{pr}^{3/2}}{6\sqrt{h}} \qquad (11.117)$$

where A_{pr} is the beach profile scaling parameter. As we will see in Chapter 12, this parameter varies systematically with the settling velocity when the sediment is sand or silt. However, at a cohesive beach there does not appear to be a well-defined dependence of A_{pr} on the settling velocity. In addition, Eq. 12.1 is not applicable to convex profiles. Lastly, when energy absorption by mud is high, breaking may be minor and not an important contributor to energy loss.

11.7.2 *Poroelastic bed*

11.7.2.1 *Energy loss*

When waves travel over a porous bed, viscous and pressure forces within the pores are transmitted as effective stresses to the particle matrix, which is consequently deformed. As the matrix possesses rigidity as well as compressibility, two kinds of stress waves, shear and compressional, occur. As a result, wave energy in a heaving bed is lost by skin friction at the solid boundaries and by internal friction (Coulomb damping) at points of contact between particles grinding against each other (Fig. 11.12). Internal friction can be a major contributor to energy loss in mud [Maa and Mehta, 1990].

Depending on assumptions concerning the particle matrix and the pore fluid, four types of poroelastic models have been developed: fully dynamic model, drained-condition model, consolidation model and Coulomb damping model (Table 11.4). The wave dispersion relation is different in each model [Jeng and Lin, 2003].

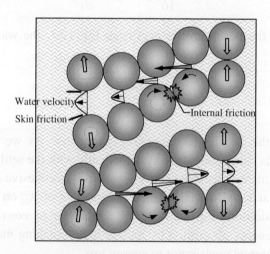

Figure 11.12. Energy loss in a heaving poroelastic bed due to viscous skin friction drag and internal friction between particles.

A comparison of the different models in Table 11.4 suggests that the Coulomb damping approach is appropriate for fine-grained silty sediment [Jeng and Lin, 2003; Jain, 2007]. Energy loss by Coulomb friction was considered by Mindlin and Deresiewicz [1953]. Yamamoto [1983] proposed an analytic framework in which the thickness of the poroelastic bottom was semi-infinite. Lin [2001] and Lee *et al.* [2002] extended the model to a bottom of finite thickness. The latter investigators obtained an analytic solution by including the effects of Coulomb friction and fluid acceleration on soil response. The following experiment-based development is due to Yamamoto and Takahashi [1985].

Table 11.4. Types of poroelastic models.

Model/ properties	Full dynamic	Drained behavior	Consolidation	Coulomb-damping
Investigators	Zienkiewicz *et al.* [1980]	Dean and Dalrymple [1991], Kim *et al.* [2000a]	Yamamoto *et al.* [1978], Madsen [1978], Jeng [1997]	Yamamoto [1983], Lin [2001], Lee *et al.* [2002]
Assumptions	Incompressible and irrotational flow in water; linearized free surface boundary condition; small loading wave amplitude	Rigid porous medium; incompressible pore fluid	Large time-scale; all accelerations negligible	Coulomb damping

Sources: Jain [2007], Jeng and Lin [2003].

The poroelastic constitutive equation relating the local shear stress τ to shear strain γ is given by Eq. 5.74, in which for the present purpose the complex shear modulus G^* will be defined as

$$G^* = G(1 + i\delta_p) \qquad (11.118)$$

where G is the shear modulus of elasticity and δ_p is known as the specific energy loss. The latter quantity is related to the rate of energy loss ε_D by

$$\delta_p = \frac{\sigma \varepsilon_D}{E_m} \qquad (11.119)$$

where the maximum elastic energy E_m is given by

$$E_m = \frac{1}{2} G \gamma^0 \qquad (11.120)$$

in which γ^0 is the strain amplitude.

For applications relying on the results of Yamamoto and Takahashi [1985], G and δ_p are obtained from a series of empirical formulas:

$$G = G_{p0} \frac{1}{1+X} \qquad (11.121)$$

$$\delta_p = \delta_{p0} + \delta_{pm} \frac{\sqrt{Y}}{1+\sqrt{Y}} \qquad (11.122)$$

$$X = \begin{cases} Z/(1-Z) & \text{for } Z < 0.95 \\ 19 & \text{for } Z \geq 0.95 \end{cases} \qquad (11.123)$$

$$Y = \frac{\gamma_G^0}{\gamma_D^0} X; \quad Z = \frac{\tau}{G_{p0} \gamma_G^0} \qquad (11.124)$$

where γ_G^0 and γ_D^0 are relative strain amplitudes, and δ_{p0} and δ_{pm} are the minimum and maximum Coulomb specific-energy losses, respectively. Typical values are $\gamma_G^0 = 0.003$, $\gamma_D^0 = 0.05$, $\delta_{p0} = 0.03$ and $\delta_{pm} = 1$. The shear modulus G_{p0} (Pa) is obtained from

$$G_{p0} = \alpha_p \frac{(\beta_p - n_r)^2}{1 - n_r} \sqrt{\sigma_0'} \qquad (11.125)$$

in which the porosity n_r and coefficients α_p and β_p depend on the bed. For a given bed G_{p0} varies with the square root of the confining effective stress σ_0' in the bed. The following sets of values of α_p and β_p are based on tests using sand and clay.

Sediment	α_p	β_p
Sand (round)	2.19×10^5	2.17
Sand (angular), kaolinite	1.03×10^5	2.97
Bentonite (a smectite)	0.14×10^5	4.40

Source: Yamamoto and Takahashi [1985].

These ranges of α_p and β_p suggest that typical values for silts would be about $\alpha_p = 0.5 \times 10^5$ and $\beta_p = 3.5$. The confining effective stress is obtained from

$$\sigma_0' = \frac{1}{3}(1 + 2K_0)(\rho - \rho_w)gz' \tag{11.126}$$

where K_0 is the coefficient of earth pressure at rest with values ranging between 0.4 and 1, z' is the depth coordinate, ρ_w is the pore fluid (water) density and ρ is the wet bulk density of the bed obtained from Eq. 4.46

$$\rho = (\rho_s - \rho_w)\phi_v + \rho_w = (1 - n_r)\rho_s + n_r\rho_w \tag{11.127}$$

where ρ_s is the particle density and ϕ_v the solids volume fraction. As G_p and δ_p vary with the wave amplitude, the following approximate method has been recommended:

1. Replace τ in Eq. 11.124 by τ_p, where

$$\tau_p = 0.27 \frac{\rho_w ga}{\cosh k_r h} \tag{11.128}$$

where a is the wave amplitude, k_r is the wave number assuming a rigid bottom and h is the water depth.

2. Set $z' = L/6$ in Eq. 11.126, where $L = 2\pi/k_r$.

The above method yields high estimates of the shear modulus of elasticity G and, therefore, for soft muds requires calibration with respect to α_p in Eq. 11.125.

Example 11.7: Select water depth $h = 2$ m, wave period $T = 6$ s, wave amplitude $a = 0.4$ m, $\rho_w = 1,000$ kg m^{-3}, $\rho_s = 2,650$ kg m^{-3}, $\alpha_p = 3$, $\beta_p = 4.4$, $n_r = 0.7$ and $K_0 = 0.7$. Also, $\gamma_{G0} = 0.003$, $\gamma_{D0} = 0.05$, $\delta_{po} = 0.03$ and $\delta_{pm} = 1$. Determine G and δ_p from Eqs. 11.121 and 11.122, respectively.

We obtain $G = 2,986$ Pa and $\delta_p = 0.063$.

In order to obtain the wave damping coefficient, the rate of energy loss is calculated over a bed of semi-infinite thickness, for which

$$\varepsilon_D = \sigma \delta_p \int_0^{-\infty} \frac{1}{2} \tau \gamma dz \qquad (11.129)$$

Thus ε_D depends on the functional forms of shear stress τ and strain γ. We will consider two types of bed responses.

11.7.2.2 *Quasi-static bed response*

In this case τ and γ are allowed to vary only with elevation z in the bed (with datum at the bed surface), *i.e.*, they are independent of time. These quantities are defined as

$$\tau = -p_b k_r z e^{k_r z} \qquad (11.130)$$

$$\gamma = -\frac{p_b}{G} k_r z e^{k_r z} \qquad (11.131)$$

When combined, these two expressions give Hooke's law, $\tau = G\gamma$. The amplitude of bottom fluid pressure p_b is obtained from Eq. 2.133 by setting $z = -h$, *i.e.*,

$$p_b = \frac{a\rho_1 g}{\cosh k_r h} \qquad (11.132)$$

With these definitions, Eq. 11.129 is integrated and the resulting expression for ε_D upon substitution into Eq. 11.5 yields

$$k_i = \frac{\rho_1 g \delta_p}{4G \cosh^2 k_r h \left(1 + \dfrac{2k_r h}{\sinh 2k_r h}\right)} \qquad (11.133)$$

Yamamoto and Takahashi [1985] give 2 rather than 4 as the factor in the denominator. This is the result of a slightly different definition of δ_p in Eq. 11.129 based on Eq. 11.118.

11.7.2.3 *Dynamic bed response*

The quasi-static approach ignores the likelihood of dynamic amplification of bed (heaving), which can greatly increase energy loss by Coulomb friction. To incorporate amplification the analysis requires solving for bed oscillation under wave pressure prescribed by Eq. 11.132. The effect of amplification (resonance) is characterized by the ratio σ/σ_0, in which the resonance frequency σ_0 is given by

$$\sigma_0 = \sqrt{2}\sqrt{\frac{G}{\rho_2}}\, k_r = 1.41 C_{sh} k_r \qquad (11.134)$$

After further analysis,

$$k_i = \frac{\rho_1 g \delta_p}{4G \cosh^2 k_r h \left(1 + \dfrac{2k_r h}{\sinh 2k_r h}\right)} \frac{1}{\left[1 - \left(\dfrac{\sigma}{\sigma_0}\right)^2\right]^2 + \delta_p^2} \qquad (11.135)$$

Equation 11.133 is recovered by setting $\sigma/\sigma_0 = 0$ and $\delta_p^2 \ll 1$.

The damping coefficient k_i from Eq. 11.135 is plotted in Fig. 11.13 against the shear modulus G for beds having representative parametric values: $h = 0.3$ m, $\sigma = 6.28$ rad s^{-1} (*i.e.*, wave period of 1 s), $\rho_1 = 1{,}000$ kg m^{-3}, $\rho_s = 2{,}650$ kg m^{-3} and energy loss parameter $\delta_p = 0.05$

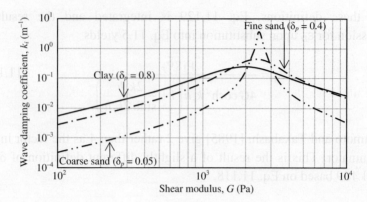

Figure 11.13. Variation of wave damping coefficient with shear modulus in poroelastic beds subject to Coulomb friction.

(coarse sand), 0.4 (fine sand) and 0.8 (clay). Resonance ($\sigma/\sigma_0 \to 1$) is marked by a peak in k_i. In clays the phenomenon is closely related to the microstructure of the particle matrix. It appears that those pore-cavities whose natural frequency matches σ_0 become preferentially susceptible to the onset of liquefaction [Tzang *et al.*, 1992, Chapter 8]. When σ is not close to σ_0, higher Coulomb friction loss occurs in clays than sands. This may be partially explained by the effect of particle size. For a given bed volume and porosity, the number of clay particles is substantially higher than sand. Accordingly, the greater the number of particles, the higher the energy loss.

11.7.3 *Viscoelastic bed*

In Chapter 5 we examined the general characteristics of a viscoelastic material, and introduced simple mechanical analogs describing the oscillatory behavior of mud under waves. The Kelvin–Voigt (KV) viscoelastic model is expressed by Eq. 5.54, while from Eqs. 11.117 and 11.118 the poroelastic model is

$$\tau = G(1 + i\delta_p)\gamma \qquad (11.136)$$

From a comparison of Eq. 5.54 with Eq. 11.136 we note that both represent an elastic response (characterized by G) coupled with energy loss due to viscosity in Eq. 5.54 and Coulomb friction in Eq. 11.136. In Eq. 5.54 the separation of elastic and viscous effects between the bed and fluid phases is not explicit and the material is treated as a continuum. In Eq. 11.136 elasticity and specific loss are mainly associated with the structure of the particle matrix. Even so, the porous medium is a continuum.

The equivalent viscosity $\eta^* = \tau/\dot{\gamma}$ (Eq. 5.76) of the KV model can be determined from Eq. 5.54 along with the oscillatory strain response and stress forcing given by Eqs. 5.72 and 5.73, respectively. This analysis yields

$$\eta^* = \eta_2 + i\frac{G}{\sigma} \qquad (11.137)$$

Similarly, the Maxwell model (Eq. 5.55) yields

$$\eta^* = \frac{\eta_2}{1 + \left(\dfrac{\eta_2\sigma}{G}\right)}\left(1 + i\frac{\eta_2\sigma}{G}\right) \qquad (11.138)$$

Replacing η by η^* in equations of motion (*e.g.*, Appendix B, Eqs. B.7, B.8 and B.9) permits the consideration of the material's elastic property. For a KV solid this approach makes the analysis empirical because a solid is treated as a fluid. In spite of this, several investigators have used the approach. For instance, MacPherson [1980] examined wave damping due to a KV bed of semi-infinite thickness beneath water assumed to be inviscid. An explicit solution for the wave damping coefficient k_i could be obtained only for a bed considered to be "almost rigid", *i.e.*, one in which the interfacial amplitude of oscillation is small compared to surface amplitude. It is found that

$$k_i = \frac{\rho_1 g \eta_2}{4\sigma\left(\eta_2^2 + \dfrac{G^2}{\sigma^2}\right)} \qquad (11.139)$$

11.7.4 *Viscoplastic bed*

11.7.4.1 *Two-layer wave–mud system*

Figure 11.14 shows a general system of Bingham viscoplastic mud flow below inviscid "water" wave with subscripts 1 and 2 for the upper and lower layers, respectively. Present interest is in wave damping, for which we will consider a solitary wave (Mei and Liu [1987] also considered a sinusoidal wave). The choice of this shallow-water wave permits the use at a viscoplastic description, and a simplification of the problem because the horizontal velocity in water is uniform over depth (Chapter 2). The surface elevation is given by

$$\eta_1 = a \left[\sec h^2 \sqrt{\frac{3a}{4h_1^3}} \cdot (x - C_w t) \right] \qquad (11.140)$$

in which $a(x)$ is the wave amplitude, C_w is the celerity (Eq. 2.140), h_1 is the water depth relative to the mean position of the mud–water interface at $z = 0$, and k is the wave number. Because it is solitary, this wave acts on the mud for only a few wave lengths. For an assumed weakly non-linear wave $(a/h_1 << 1)$, k and the depth–mean velocity $u_1(x, t)$ are respectively approximated as

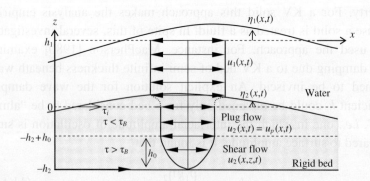

Figure 11.14. Shallow water and viscoplastic system subject to surface water wave.

$$k = \sqrt{\frac{3a}{4h_1^3}} \qquad (11.141)$$

$$u_1 = \frac{g\eta_1}{C_w} \qquad (11.142)$$

The surface elevation relative to mean water level is always positive, as is the velocity (in the positive x direction).

In general, in simple shear flow the constitutive equation for Bingham flow (Eq. 5.27) relating the shear stress τ, the Bingham yield stress τ_B, the viscosity η_2 and the flow shear rate $\partial u_2/\partial z$ is given as

$$\tau = \eta_2 \frac{\partial u_2}{\partial z} = 0; \quad |\tau| \le \tau_B \qquad (11.143)$$

$$\tau = \tau_B \cdot sign\left(\frac{\partial u_2}{\partial z}\right) + \eta_2 \frac{\partial u_2}{\partial z}; \quad |\tau| > \tau_B \qquad (11.144)$$

The multiplier, $sign(\partial u_2/\partial z)$, in which $sign$ can be $+$ or $-$, preserves the direction of τ_B consistent with the sign of $\partial u_2/\partial z$.

We are interested in a mild environment in which the waves are not too high and mud flow is not turbulent. In general, the onset of turbulence in a viscoplastic flow is specified by an effective Reynolds number Re_e defined as

$$\frac{1}{Re_e} = \frac{1}{Re^\eta} + \frac{1}{Re^\tau} \qquad (11.145)$$

$$Re^\eta = \frac{4\rho_2 u_{2m} h_2}{v_2}; \quad Re^\tau = \frac{8\rho_2 u_{2m}^2}{\tau_B} \qquad (11.146)$$

in which $u_{2m}(x)$ is the amplitude of the depth–mean horizontal velocity in the mud of thickness h_2. The Reynolds number Re_e combines the influences of viscosity and yield stress in the same sense as the Hedstrom number $He = 16\rho_2 \tau_B h_2^2 / \eta_2^2$ for Bingham flow of dense slurries (*e.g.*, Wasp *et al.* [1977]). The transition from viscous to turbulent flow is specified by the critical Reynolds number $Re_{ec} = 4\rho_2 u_{2m} h_2 / [v_2 + (\tau_B h_2 / 2u_{2m})]$ in the range of 2,000 to 3,000 [Engelund and Zhaohui, 1984].

The velocity $u_p(x, t)$ represents uniform or plug flow within mud in a layer between the interfacial elevation $\eta_2(x, t)$ and depth (negative elevation) $z = -h_2 + h_0$, where h_0 is the thickness of the shear–flow layer at the bottom. Plug flow exists because within it $\tau \le \tau_B$, which prevents flow shearing and the entire layer moves as a "solid" plug.

Since all vertical displacements are small, the governing equations are simplified. Conservation of mass in water and in mud is given by, respectively,

$$\frac{\partial}{\partial t}(\eta_1 - \eta_2) + h_1 \frac{\partial u_1}{\partial x} = 0 \qquad (11.147)$$

$$\frac{\partial \eta_2}{\partial t} + h_2 \frac{\partial u_2}{\partial x} = 0 \qquad (11.148)$$

The momentum equations for water flow, plug flow and bottom shear layers are, respectively,

$$\rho_1 \frac{\partial u_1}{\partial t} = -\frac{\partial p_1}{\partial x} - \frac{\tau_i}{h_1} \qquad (11.149)$$

$$\rho_2 \frac{\partial u_p}{\partial t} = -\frac{\partial p_2}{\partial x} + \frac{\partial \tau}{\partial z} \qquad (11.150)$$

$$\rho_2 \frac{\partial u_b}{\partial t} = -\frac{\partial p_2}{\partial x} + \eta_2 \frac{\partial^2 u_b}{\partial z^2} \qquad (11.151)$$

where p denotes pressure, τ_i is the shear stress at the mud–water interface and u_b is the value of u_2 in the bottom shear layer in which $\tau > \tau_B$. Whether mixing and entrainment of mud occurs at the interface depends on the ratio τ_i/τ_B, assuming that, for entrainment, τ_B representing the threshold of movement must be exceeded by τ_i. The interface is destabilized and a shear layer forms below the interface when $\tau_i/\tau_B > 1$.

The ratio τ_i/τ_B can be estimated from $f_i \rho_1 g a^2 / 2\tau_B h_1$, where f_i is the interfacial friction factor [Mei and Liu, 1987, Chapter 2]. For example, $h_1 = 3$ m, $a = 0.1$ m, $\rho_1 = 1,000$ kg m^{-3}, $f_i = 0.01$ (for a smooth interface) and $\tau_B = 10$ Pa would yield $\tau_i/\tau_B = 0.016$, which is much smaller than unity. This indicates that for mixing and entrainment the wave must be high and yield strength low. In the mild environment, the term τ_i/h_1 in Eq. 11.148 can be ignored, given that most energy loss occurs within mud.

Water pressure is approximately hydrostatic under the solitary long-wave, both in water and in mud, *i.e.*,

$$p_1(x,z,t) = \rho_1 g(\eta_1 - z + h_1) \tag{11.152}$$

$$p_2(x,z,t) = -\rho_2 g z + \rho_1 g(\eta_1 + h_1) \tag{11.153}$$

The pressures p_1 and p_2 are about equal and permit the substitution of p_1 for p_2. As a result, Eqs. 11.150 and 11.151 for mud reduce to, respectively,

$$\rho_2 \frac{\partial u_p}{\partial t} = \rho_1 \frac{\partial u_1}{\partial t} + \frac{\tau'}{h_2} \tag{11.154}$$

$$\rho_2 \frac{\partial u_b}{\partial t} = \rho_1 \frac{\partial u_1}{\partial t} + \eta_2 \frac{\partial^2 u_b}{\partial z^2} \tag{11.155}$$

In Eq. 11.154, τ' is the shear stress at bottom of the plug flow layer $(z = -h_2 + h_0)$, *i.e.*, $\left| \tau' \right| = \tau_B$. Since in this layer the velocity is independent of z, $d\tau/dz = \tau'/h_2$ is constant.

The required boundary conditions for the mud layer are:
At $z = -h_2 + h_0$ the velocities and stresses must be continuous, *i.e.*,

$$u_b(x, z = -h_2 + h_0, t) = u_p \text{ and } \frac{\partial u_b}{\partial z} = 0 \tag{11.156}$$

At the bottom, $z = -h_2$,

$$u_b = 0 \tag{11.157}$$

For the wave form of Eq. 11.140 (along with Eqs. 11.141 and 11.142), Mei and Liu [1987] outline the procedure to solve for $\eta_2(x, t)$, which is unimportant because $\eta_2 \ll \eta_1$, and $u_p(x, t)$ using Eqs. 11.154 to

11.157. All three variables are gradually-varied functions of x. For u_b the following boundary layer profile is assumed:

$$u_b = u_p \left[2 \left(\frac{z}{h_0} \right) - \left(\frac{z}{h_0} \right)^2 \right] \qquad (11.158)$$

For the system as a whole the following dimensionless yield stress term emerges as a characteristic number

$$\tilde{\tau}_B \tilde{\rho}_2 = \frac{2}{\sqrt{3}} \frac{\tau_B}{\rho_2 g h_2} \left(\frac{h_1}{a} \right)^{3/2} \tilde{\rho}_2 \qquad (11.159)$$

where $\tilde{\rho}_2 = \rho_2 / \rho_1$.

11.7.4.2 *Wave damping*

For estimating wave damping the total rate of wave energy loss ε_D in mud is taken as the sum of three contributions

$$\varepsilon_D = \varepsilon_B + \varepsilon_i + \varepsilon_v \qquad (11.160)$$

in which ε_B is work done by the yield stress and ε_v represents viscous loss. Both quantities are associated with shearing in the mud layer below plug flow. The term ε_i is work associated with interfacial shear stress τ_i. The three energy losses are obtained from

$$\left. \begin{array}{l} \varepsilon_B = \overline{\tau_B |u_p|} \\[2mm] \varepsilon_v = \eta_2 \overline{\int_{-h_2}^{-h_2+h_0} \left(\frac{\partial u_b}{\partial z} \right)^2 dz} \\[2mm] \varepsilon_i = \overline{\frac{1}{2} f_i \rho_1 |u_1 - u_p|^3} \end{array} \right\} \qquad (11.161)$$

where overbars imply integration from $-\infty$ to ∞ for the solitary wave.

For further analysis the following dimensionless wave amplitude, water depth and distance are introduced

$$\tilde{a} = \frac{a(x)}{a_0}; \quad \tilde{h} = \frac{h_1(x)}{h_{10}}; \quad \tilde{x} = k_0 x \tag{11.162}$$

where the amplitude a_0, water depth h_{10} and wave number k_0 (in general for a compliant bed) are specified at an arbitrary initial location $x = 0$. For the solitary wave

$$k_0 = \frac{\sqrt{3}}{4} \sqrt{\frac{a_0}{h_{10}}} \tag{11.163}$$

After time-integrating the wave power per unit water surface area EC_g from $-\infty$ to ∞ in Eq. 11.2, the resulting balance is:

$$\frac{d}{d\tilde{x}}(\tilde{a}\tilde{h})^{3/2} = -\frac{\sqrt{3}}{4}\frac{h_2}{h_{10}}\frac{\tilde{a}^2}{\tilde{h}}\left(\frac{4}{3}\Omega_1 + \frac{f_i}{h_2}\sqrt{a_0 h_{10}}\sqrt{\tilde{a}\tilde{h}}\,\Omega_2 + \frac{4}{3\sqrt{3}}\frac{\delta_{v2}}{h_2}\Omega_3\right) \tag{11.164}$$

in which the thickness of the boundary layer where most of the energy loss occurs is $\delta_{v2} = \sqrt{\eta_2 / \sigma \rho_2}$. The quantities Ω_1, Ω_2 and Ω_3 are functions of $\tilde{\tau}_B \tilde{\rho}_2$. The dimensionless wave amplitude $\tilde{a}$ is obtained by numerically solving Eq. 11.164.

Based on calculations summarized by Liu and Mei [1987], in Fig. 11.15, Ω_1, Ω_2 and Ω_3 are plotted against $\tilde{\tau}_B \tilde{\rho}_2$ representing bed stiffness. The curves are for $a_0 = 1$ m, $h_{10} = 10$ m, $h_2/h_{10} = 1$, $\tilde{\rho}_2 = 1.1$,

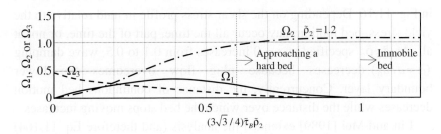

Figure 11.15. Variation of characteristic mud flow parameters Ω_1, Ω_2 and Ω_3 with $\tilde{\tau}_B \tilde{\rho}_2$ (adapted from Mei and Liu [1987]).

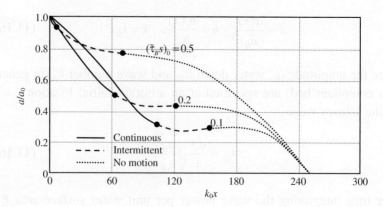

Figure 11.16. Variation of dimensionless wave amplitude $\tilde{a} = a/a_0$ with dimensionless distance $\tilde{x} = k_0 x$ for a viscoplastic mud subjected to a solitary wave. Dots define conditions at which the mud dynamic regime changes (revised from Mei and Liu [1987]).

$f_i \sqrt{a_0 h_{10}} / h_2 = 0.1$ and $\delta_{v2}/h_2 = 0.1$. Values of $\tilde{\tau}_B \tilde{\rho}_2 \leq 0$ imply a suspension with no damping role for the yield stress (Ω_1) and negligible effect of viscosity (Ω_3). When $\tilde{\tau}_B \tilde{\rho}_2 > 0$, Ω_1 and Ω_3 vary with $\tilde{\tau}_B \tilde{\rho}_2$, while Ω_2 depends on $\tilde{\tau}_B$ and separately on $\tilde{\rho}_2$. As the bed becomes stiff ($\tilde{\tau}_B \tilde{\rho}_2 = 1$), interfacial friction (Ω_2) approaches bed friction, while the damping role of yield stress vanishes. Likewise, because plug flow thickness (Fig. 11.14) increases and approaches the bottom, the role of the lower boundary layer becomes insignificant.

Equation 11.164 was solved numerically for $a(x)$ with results shown in Fig. 11.16. Depending on the shear stress profile in mud relative to the yield stress, bed motion may occur all the time, part of the time, or not at all. As $\tilde{\tau}_B \tilde{\rho}_2$ specified at $x = 0$ increases from 0.1 to 0.5, wave damping ($\tilde{a} = a/a_0$) decreases because plug flow overcomes the lower boundary layer. Concurrently, the distance of continuous bed motion decreases while the distance over which the bed stops moving increases.

Liu and Mei [1989] extended the analysis (and therefore Eq. 11.164) to the case where τ_i is large enough to generate a shear layer below the interface. It becomes necessary to account for additional loss of energy in this layer as it can be turbulent.

11.7.5 *Parameters influencing wave damping coefficient*

11.7.5.1 *Wave damping coefficients from analytic models*

In Table 11.5, four expressions for the wave damping coefficient k_i are given. In each case the amplitude ratio $a(x)/a_0$ is obtained from Eq. 11.4. For the last two cases Eq. 11.4 is not used. The relative significance of damping from each mechanism is illustrated by numerical examples.

Example 11.8: Consider a 6 s wave of 1 m amplitude in seawater (density 1,030 kg m^{-3}) of 4.5 m depth traveling at a wave group velocity of 5.14 m s^{-1}. The bed consists of mud having a density of 1,200 kg m^{-3}. Bed surface resistance is characterized by the wave friction factor equal to 0.035, the kinematic viscosity of water is 1×10^{-6} m^2 s^{-1}, mud dynamic viscosity is 2.5 Pa.s, and the shear rigidity modulus is 500 Pa. Assume a specific permeability value of 1.27×10^{-13} m^2, specific energy loss parameter equal to 0.052 and bed porosity 0.5. Calculate a/a_0 and k_i using the equations in Table 11.5.

Table 11.5. Wave damping coefficients over rigid and compliant beds.

Physical basis	k_i or a/a_0	Investigator(s)
Viscous fluid over rigid bed	$k_i = \dfrac{k_r^2 \delta_{v1}}{2(2k_r h + \sinh 2k_r h)}$	Dean and Dalrymple [1991]
Percolation	$k_i = \dfrac{2k_r}{2k_r h + \sinh 2k_r h}\left(\dfrac{K_p \sigma}{\nu} + \dfrac{k_r \delta_{v1}}{2}\right)$	Liu [1973]
Poroelastic bed	$k_i = \dfrac{\rho_1 g \delta_p}{4G\cosh^2 k_r h\left(1 + \dfrac{2k_r h}{\sinh 2k_r h}\right)\left[1 - \left(\dfrac{\sigma}{\sigma_0}\right)^2\right]^2 + \delta_p^2}$	Yamamoto and Takahashi [1985]
Viscoelastic Kelvin–Voigt bed	$k_i = \dfrac{\rho_1 g \eta_2}{4\sigma\left(\eta_2^2 + \dfrac{G^2}{\sigma^2}\right)}$	MacPherson [1980]
Inviscid fluid over viscous fluid of indefinite thickness in non-shallow water	$\dfrac{a(x)}{a_0} = \sqrt{\left[1 - \dfrac{\rho_2}{\rho_1}\dfrac{\delta_{v2}}{2\sqrt{2}gC_g\sigma}e^{2k_r h}(\sigma^2 - gk_r)^2 x\right]}$	Dean and Dalrymple [1991]

Table 11.5 (Con't)

Physical basis	k_i or a/a_0	Investigator(s)
Bed friction in turbulent flow	$\dfrac{a(x)}{a_0} = \dfrac{1}{1+\dfrac{2f_w}{3\pi}\dfrac{a_0 k_r^2 x}{(2k_r h+\sinh 2k_r h)\sinh k_r h}}$	Dean and Dalrymple [1991]

k_r = wave number (*i.e.*, real part of the complex wave number), σ = wave angular frequency, h = water depth, v_1 = fluid viscosity, v_2 = mud viscosity, $\delta_{v1}\,(=\sqrt{2v_1/\sigma})$ = wave boundary layer thickness in water (or viscous fluid with rigid bottom), $\delta_{v2}\,(=\sqrt{2v_2/\sigma})$ = wave boundary layer thickness in mud, ρ_1 = upper fluid (water) density, ρ_2 = mud density, η_2 = mud dynamic viscosity, C_g = wave group velocity, K_p = specific permeability, a = wave amplitude, f_w = wave friction factor, g = acceleration due to gravity, G = shear modulus of elasticity of mud, δ_p = Coulomb specific energy loss, and $\sigma_0 (= k_r\sqrt{2G/\rho_2})$ = poroelastic resonance frequency.

Source: Partially from Jain and Mehta [2009a].

For the last two cases estimate the effective k_i by applying Eq. 11.4 at a distance $x = 1,000$ m.

The calculated values of k_i indicate that for the selected parameters the viscoelastic KV model is the only important contributor to wave damping. Although the poroelastic contribution is low, it can be high if the bed is silty.

Physical system or process	a/a_0	k_i (m^{-1})
Viscous fluid over rigid bed	0.995	5.4×10^{-6}
Inviscid fluid over viscous fluid	0.435	8.3×10^{-4}
Percolation	0.989	1.1×10^{-5}
Bed friction in turbulent flow	0.936	6.6×10^{-5}
Poroelastic bed	0.953	4.8×10^{-5}
Viscoelastic Kelvin–Voigt bed	~0	2.6×10^{-2}

Example 11.9: Consider the parametric values given in Table 11.6 relevant to Newnans Lake in north-central Florida. This lake has a hydraulically very rough ($f_w = 0.1$) mud bottom. (1) Calculate the wave damping coefficients for viscous fluid mud, percolation, bed friction, poroelastic bed and viscoelastic (KV) bed. (2) Determine percent reductions in wave energy over a distance of 5.5 km from lake edge to the center.

Using expressions from Table 11.5 along with derived values of the wave number $k_r = 2.80$ m^{-1} and group velocity $C_g = 0.95$ m s^{-1}, the results are summarized in Table 11.7. Percent energy reduction is defined as

$$\Delta E = 100\frac{(a_0^2 - a^2)}{a_0^2} \tag{11.165}$$

Since the poroelastic and viscoelastic beds are semi-infinitely deep they overestimate energy loss. Due to very low permeability, use of the poroelastic model may be as

Table 11.6. Characteristic lake parameters.

T (s)	a_0 (m)	h_1 (m)	h_2 (m)	G (Pa)	ν $(\text{m}^2\,\text{s}^{-1})$	η_2 (Pa.s)	ρ_1 (kg m^{-3})	ρ_2 (kg m^{-3})	K_p (m^2)	δ_p
1.2	0.05	1.3	0.3	300	1×10^{-6}	2.4	1,000	1,200	1×10^{-10}	0.5

Source: Jain [2007].

Table 11.7. Lake wave damping coefficients and percent reduction in energy.

Quantity	Viscous fluid	Percolation	Bed friction	Poroelastic bed (Coulomb friction)	Viscoelastic bed (KV)
a/a_0	0.98	0.94	1	0.18	0
$k_i\ (\text{m}^{-1})$	3.3×10^{-6}	1.1×10^{-5}	4.9×10^{-8}	3.1×10^{-4}	3.4×10^{-1}
$\Delta E\ (\%)$	4	11	0	97	100

Source: Jain and Mehta [2009a].

qualitative as the viscoelastic representation. Also, given a low shear modulus ($G = 300$ Pa) the assumption of a viscoelastic bed may offer no tangible advantage over a viscous fluid bottom.

We can arrive at a similar conclusion regarding bed behavior in the lake by calculating the characteristic dimensionless parameters given in Table 11.8. Since the normalized boundary layer in water $\delta_{v1}/h_1 = 0.0005 \ll$ normalized boundary layer in mud $\delta_{v2}/h_2 = 0.09$, the water boundary layer is negligibly small and energy loss is mainly within mud. Thus the boundary layer friction effect is also negligible. The value 0.14 of the shear Mach number Ma_G (wave speed in mud divided by speed in water; see Eq. 11.167) indicates that due to energy loss in mud, wave speed in water is about seven-times that in mud. The very low value of the ratio of loss due to mud elasticity to viscosity, η''/η_2 ($= 0.002$), corroborates the inference that the lake mud is fluid-like.

11.7.5.2 *Parametric effects on wave damping*

For soft muds in general, viscoelastic fluid mud and the poroelastic (silty) bed are arguably the two most important bottom descriptions relative to wave damping [Jain and Mehta, 2009a,b; Jain, 2007]. From the pi-theorem it can be shown that for a viscoelastic fluid the dimensionless wave damping coefficient $\tilde{k}_i = k_i h$ can be represented as

$$\tilde{k}_i = k_i h_1 = f^n\left(\frac{\sigma^2}{gk_r}, \frac{h_2}{h_1}, \frac{\delta_{v2}}{h_2}, \frac{\Delta\rho}{\rho_1}\right) \qquad (11.166)$$

Table 11.8. Characteristic dimensionless numbers for Newnans Lake.

Δ_{v1}/h_1	δ_{v2}/h_2	Ma_G	η''/η_2
0.0005	0.09	0.14	0.002

Source: Jain and Mehta [2009a].

where $\Delta\rho = \rho_2 - \rho_1$ is the density difference between mud and water. The dimensionless wave damping coefficient $\tilde{k}_i$ depends on σ^2/gk_r (*i.e.*, on the surface wave dispersion relationship), relative mud thickness h_2/h_1, relative thickness of the Stokes boundary layer δ_{v2}/h_2 in mud, and the buoyancy of the lower layer represented by $\Delta\rho/\rho_1$.

For a poroelastic bed

$$\tilde{k}_i = f^n\left(\frac{\sigma^2}{gk_r}, \frac{h_2}{h_1}, \frac{C_{sh}}{C_{ws}}, Pe\right) \tag{11.167}$$

Note that in C_{sh}/C_{ws} (= Ma_G), $C_{sh} = \sqrt{G/\rho_2}$ is the shear wave velocity (Chapter 5), G is the shear modulus of elasticity (or the rigidity modulus) and $C_{ws} = \sqrt{gh}$ is the shallow water wave celerity. The Péclet number Pe identifies drained (low Pe) versus undrained (high Pe) bed condition [Winterwerp and van Kesteren, 2004; Zienkiewicz *et al.*, 1980, Chapter 8].

These bed states are relevant because under drained condition excess pore pressure generated by wave action is dissipated by pore water flow, while under undrained condition pore pressure does not dissipate. This difference in bed behavior is related to particle size. A large particle is associated with high permeability (k_p) and easy drainage, and a small particle with low k_p, *i.e.*, poor, drainage. Thus the mode of wave energy loss is dependent on Pe, which is the product of a characteristic velocity scale and a length scale divided by the isotropic coefficient of consolidation (Chapter 8). This coefficient in turn varies with k_p. Accordingly, we may conveniently introduce an approximate but easily measurable number $\tilde{W} = w_s/k_p$, the ratio of particle settling velocity w_s to k_p, as a proxy for the degree of drainage. The significance of this number will be apparent in subsequent comments on mechanisms representing wave energy loss.

The shear Mach number Ma_G varies with the solids volume fraction $\phi_v = 1 - n_r$, where n_r is the porosity [$= (\rho_s - \rho_2)/(\rho_s - \rho_1)$] and ρ_s is the particle density. When $\phi_v \leq \phi_{vs}$, where ϕ_{vs} is the space-filling value of ϕ_v below which mud is a fluid and therefore cannot transmit a shear wave, C_{sh} and therefore Ma_G are trivial. In the range of $\phi_v > \phi_{vs}$, C_{sh} rapidly increases with ϕ_{vs}. When waves begin to liquefy mud, C_{sh} decreases even when ϕ_v ($> \phi_{vs}$) remains unchanged. As mentioned in Chapter 9, in a flume experiment on a clayey mud bed at $\phi_v = 0.10$, the onset of steady wave action caused C_{sh} to decrease from 2 to 1.4 m s^{-1} in one hour.

Expressions for the wave damping coefficient k_i in Table 11.5 are restrictive in their applications, especially since none includes interaction between the boundary layers in water and in mud taken together. The earliest development in this regard is the semi-analytic treatment of Dalrymple and Liu [1978] in which the upper fluid and mud are viscous as opposed to mud alone. This analysis has led to several extensions of the approach. Here we will summarize the work of Jain [2007], which permits us to review mud mass transport.

11.8 Fluid Mud: Semi-Analytic Solutions

11.8.1 *Method of analysis*

An idealized system consisting of a viscous fluid (water) above a linear viscoelastic fluid is shown Figure 11.17.

Following the observation in Chapter 5 that fluid-like mud is generally neither a KV solid nor a Maxwell fluid, the Jeffreys model (Eq. 5.58) is selected as a better descriptor of mud rheological behavior:

$$G' = \frac{(\alpha\beta - \xi)\sigma^2}{1 + \alpha^2\sigma^2} \tag{11.168}$$

$$G'' = \frac{(\alpha\xi\sigma^2 + \beta)\sigma}{1 + \alpha^2\sigma^2} \tag{11.169}$$

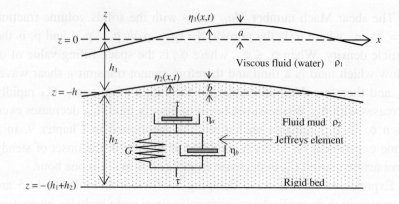

Figure 11.17. Water-wave in a two-layered system with viscous fluids.

$$\eta' = \frac{\left(\alpha\xi\sigma^2 + \beta\right)}{1 + \alpha^2\sigma^2} \qquad (11.170)$$

$$\eta'' = \frac{\left(\alpha\beta - \xi\right)\sigma}{1 + \alpha^2\sigma^2} \qquad (11.171)$$

where the empirical coefficients α, β and ξ are related to the viscosity and shear modulus coefficients of the Jeffreys model. Limited results from rheometric tests of Jiang and Mehta [1995] on clayey sediments partially justify the choice of this model. When the KV component of the element is deactivated, the bottom becomes a Newtonian fluid.

For determination of the wave damping coefficient, solutions for the induced velocities and dynamic pressure of first and second-order accuracy are obtained from the two components of the equation of flow momentum in the x- and z-directions, and flow continuity. In the following development details are omitted but are given in Jain [2007]. The basic boundary value problem is found in Dalrymple and Liu [1978].

The two momentum equations and continuity equation are, respectively,

$$\frac{\partial u_{ji}}{\partial t} = -\frac{1}{\rho_j}\frac{\partial p_{ji}}{\partial x} + \nu_{ej}\left(\frac{\partial^2 u_{ji}}{\partial x^2} + \frac{\partial^2 u_{ji}}{\partial z^2}\right) \tag{11.172}$$

$$\frac{\partial w_{ji}}{\partial t} = -\frac{1}{\rho_j}\frac{\partial p_{ji}}{\partial z} + \nu_{ej}\left(\frac{\partial^2 w_{ji}}{\partial x^2} + \frac{\partial^2 w_{ji}}{\partial z^2}\right) \tag{11.173}$$

$$\frac{\partial u_{ji}}{\partial x} + \frac{\partial w_{ji}}{\partial z} = 0 \tag{11.174}$$

where $u_{ji}(x,z,t)$ and $w_{ji}(x,z,t)$ are the horizontal and vertical components of the wave orbital velocity, respectively, and p_{ji} is the dynamic pressure. Subscripts $j = 1$ and 2 are for the upper and lower layers, and $i = 1$ and 2 denote first and second-order solutions, respectively. The quantity ν_{ej} is an equivalent kinematic viscosity, which for simplicity is considered to be independent of the spatial coordinates in each layer. Since the upper layer is assumed to be Newtonian (and the flow laminar), we may set $\nu_{e1} = \nu_1$, the kinematic viscosity of the upper fluid. For the lower layer, depending on the constitutive model, ν_{e2} can be a complex number, and is treated as a constant for a given set of bed properties and wave parameters.

The dynamic pressure $p_j(x,z,t)$ is given by

$$p_j = p_j^T + \rho_j gz + p_j^0 \tag{11.175}$$

where p_j^T is the total pressure and

$$p_j^0 = \begin{cases} 0 & \text{at } j = 1 \\ (\rho_2 - \rho_1)gh_1 & \text{at } j = 2 \end{cases} \tag{11.176}$$

First-order solutions of these equations along with ten required (surface, boundary and bottom, dynamic and kinematic) boundary conditions are obtained. It is assumed that the surface wave amplitude a is so small that all terms of $O(ak_r)^2$ can be ignored relative to $O(ak_r)$. As a result, linear approximations of the boundary conditions can be used.

Solutions for the wave dispersion relationship, amplitudes $U_{11}(z)$ and $U_{21}(z)$ of horizontal velocities $u_{11}(x,z,t)$ and $u_{21}(x,z,t)$, $W_{11}(z)$ and $W_{21}(z)$ of vertical velocities $w_{11}(x,z,t)$ and $w_{21}(x,z,t)$, and $P(z)$ of pressure $p(x,z,t)$, respectively, are as follows:

$$\sigma = \frac{i}{a_1}\left[A_1 \sinh\left(kh_1\right) + B_1 \cosh\left(kh_1\right) + C_1 + D_1 \exp\left(-\lambda_1 h_1\right) \right] \qquad (11.177)$$

$$U_{11} = i\left\{ \begin{array}{l} A_1 \cosh\left(kz'\right) + B_1 \sinh\left(kz'\right) \\ +\dfrac{\lambda_a}{k}\left[C_1 \exp\left(\lambda_a z\right) - D_1 \exp\left(-\lambda_a z'\right) \right] \end{array} \right\} \qquad (11.178)$$

$$U_{21} = i\left\{ \begin{array}{l} E_1 \cosh\left(kz''\right) + F_1 \sinh\left(kz''\right) \\ +\dfrac{\lambda_b}{k}\left[G_1 \exp\left(\lambda_b z'\right) - H_1 \exp\left(-\lambda_b z''\right) \right] \end{array} \right\} \qquad (11.179)$$

$$W_{11} = A_1 \sinh\left(kz'\right) + B_1 \cosh\left(kz'\right)$$
$$+ C_1 \exp\left(\lambda_a z\right) + D_1 \exp\left(-\lambda_a z'\right) \qquad (11.180)$$

$$W_{21} = E_1 \sinh\left(kz''\right) + F_1 \cosh\left(kz''\right)$$
$$+ G_1 \exp\left(\lambda_b z'\right) + H_1 \exp\left(-\lambda_b z''\right) \qquad (11.181)$$

$$P_{11} = \frac{\rho_1 v_{e1}}{k}\left(k^2 - \lambda_a^2\right)\left[A_1 \cosh\left(kz'\right) + B_1 \sinh\left(kz'\right) \right] \qquad (11.182)$$

$$P_{21} = \frac{\rho_2 v_{e2}}{k}\left(k^2 - \lambda_b^2\right)\left[E_1 \cosh\left(kz''\right) + F_1 \sinh\left(kz''\right) \right] \qquad (11.183)$$

in which $z = n_1 + z$, $z = n_1 + n_2 + z$, and

$$\lambda_a^2 = k^2 - i\frac{\sigma}{v_{e1}}; \quad \lambda_b^2 = k^2 - i\frac{\sigma}{v_{e2}} \qquad (11.184)$$

After the real part of the wave number k_r is calculated from the wave dispersion equation, the wave damping coefficient k_i is obtained from

$k = k_r + ik_i$. Determination of the quantities A_1, B_1, C_1, D_1, E_1, F_1, G_1, H_1, k and b_1 from the boundary conditions requires a solution of ten simultaneous equations in which coefficients A_1 through H_1 vary with mud thickness. The complete analysis is provided in Lowes [1993], and independently in Jain [2007].

The second-order Stokes perturbation method [Holthuijsen, 2007] is next applied to the first-order solution. All variables are expanded as convergent power series of the wave steepness ς. It is equal to twice the wave amplitude divided by the wave length, and is assumed to be much smaller than the normalized mud thickness h_2/h_1, so as to make the series convergent. The wave amplitude is not negligible, and $O(ak_r)^2$ contributions are considered to be important. The criterion for convergence of the solution is based on the Ursell number

$$Ur = \left(\frac{2\pi}{k_r h_1}\right)^3 \left(\frac{\varsigma}{h_1}\right) h_1 \qquad (11.185)$$

Thus, Ur is directly proportional to the ratio of wave steepness to water depth. Convergence is achieved when $Ur < 26$ [Holthuijsen, 2007].

Results for the velocity amplitudes are

$$U_{12} = i \left\{ \begin{array}{l} A_2 \cosh\left(2kz'\right) + B_2 \sinh\left(2kz'\right) \\ + \dfrac{\beta_a}{2k}\left[C_2 \exp\left(\beta_b z\right) - D_2 \exp\left(-\beta_a z'\right)\right] \end{array} \right\} \qquad (11.186)$$

$$U_{22} = i \left\{ \begin{array}{l} E_2 \cosh(2kz'') + F_2 \sinh(2kz'') \\ + \dfrac{\beta_b}{2k}\left[G_2 \exp\left(\beta_b z'\right) - H_2 \exp\left(-\beta_b z''\right)\right] \end{array} \right\} \qquad (11.187)$$

$$W_{12} = A_2 \sinh\left(2kz'\right) + B_2 \cosh\left(2kz'\right) \\ + C_2 \exp\left(\beta_a z\right) + D_2 \exp\left(-\beta_a z'\right) \qquad (11.188)$$

$$W_{22} = E_2 \sinh\left(kz''\right) + F_2 \cosh\left(kz''\right) \\ + G_2 \exp\left(\beta_b z'\right) + H_2 \exp\left(-\beta_b z''\right) \qquad (11.189)$$

in which

$$\beta_a^2 = 4k^2 - 2i\frac{\sigma}{v_{e1}}; \quad \beta_b^2 = 4k^2 - 2i\frac{\sigma}{v_{e2}} \qquad (11.190)$$

Determination of A_2, B_2, C_2, D_2, E_2, F_2, G_2, H_2, a_2 and b_2 from the ten second-order boundary conditions is carried out in Jain [2007]. The final (semi-analytic) expressions for the velocities are obtained by adding the first- and the second-order results. The significance of the second-order effect is examined later in connection with mud mass transport.

11.8.2 Wave damping

In Fig. 11.18, the first-order solution for dimensionless wave number $\tilde{k}_r = k_r\sqrt{gh_1}/\sigma$ is plotted against dimensionless lower layer thickness $\tilde{h}_2 = h_2/\delta_{v2}$. Also shown are two other solutions of the same governing equations and data from laboratory experiments of Gade [1958] (Table 11.9). All solutions are for a Newtonian viscous lower layer. The data, obtained in a flume using kerosene as the upper fluid and a sugar solution as the lower, conform to the shallow water condition ($k_rh_1 < \pi/10$). Although these liquids are not representative of real water–mud systems, they are reasonable surrogates for two-layered Newtonian viscous fluids.

Gade's analytic solution for wave motion summarized in Section 11.5, which considers the upper layer to be inviscid, was extended by Dalrymple and Liu [1978] to cover the intermediate water depth range ($\pi/10 \le k_rh_1 < \pi$) and viscosity in both layers. In one of their derivations leading to what might be called the "thick-layer" model, they simplified the physical system by assuming mud thickness to be much greater than the Stokes boundary layer δ_{v2} ($h_2 \gg 1$). The result from that model is included in Fig. 11.18. Citing evidence for the need to account for the effect of typically found thin marine mud layers, Ng [2000] obtained an analytic solution for the lower layer thickness of order δ_{v2} ($h_2 \approx 1$). An outcome of these assumptions is that the Dalrymple and Liu solution deviates from data (and the Jain–Lowes solution) as h_2 decreases below

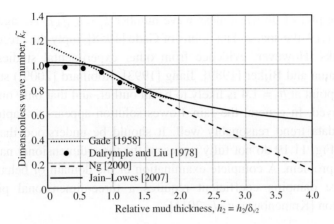

Figure 11.18. Dimensionless wave number versus relative mud thickness (from Jain and Mehta [2009b]).

Table 11.9. Characteristic parameters in laboratory flume and field data.

Source	Upper layer	Lower layer	h_1 (m)	h_2 (m)	a (m)	σ (rad s^{-1})	ρ_2 (kg m^{-3})	ν_2 (m^2 s^{-1})
Gade [1958]	kerosene[a]	sugar solution	0.038	0.010–0.057	0.01	4.45	1504	0.0026
Sakakiyama and Bijker [1989]	water[b]	kaolinite	0.3	0.09	0.016	5.2	1150–1370	0.001–0.015
Jiang [1993]		AK[c]	0.14–0.16	0.12–0.18	0.100–0.125	3.14–6.20	1200	0.015
Jain et al. [2007][d]		muck[e]	0.9	0.3	0.05	5.23	1200	0.001

[a] Density 859.3 kg m^{-3} and $\nu_1 = 2.42 \times 10^{-6}$ m^2 s^{-1}.
[b] Density (nominal) $\rho_1 = 1,000$ kg m^{-3}, kinematic viscosity $\nu_1 = 10^{-6}$ m^2 s^{-1}.
[c] A 1:1 mixture of kaolinite and attapulgite.
[d] Data from Newnans Lake in Florida.
[e] Organic-rich compliant mud.

about 0.5. In the Ng solution, since the lower layer is assumed to be of the order of boundary layer thickness, the wave number does not account for energy loss due to mud thicker than the boundary layer. Therefore the model result approaches measurement only when $\tilde{h}_2$ is less than or about 1.

In the plot of dimensionless wave number $\tilde{k}_i = k_i\sqrt{gh_1}/\sigma$ against h_2 in Fig. 11.19, the sparse data points of Gade [1958] imply the presence of two peaks. However, evidence from other experimental studies (*e.g.*, Sakakiyama and Bijker [1989], Jiang [1993], Robillard [2009]) suggests that the point at $\tilde{h}_2 = 1.4$ is likely to be an outlier, and that only one peak is displayed. In general the Jain–Lowes solution appears to capture the overall data trend reasonably well. It should be understood that plots such as Fig. 11.19 do not fully represent the three-dimensional nature of the real problem. A complete examination of the domain of behavior of the wave damping coefficient requires a three-dimensional pictorial description [Kranenburg *et al.*, 2011].

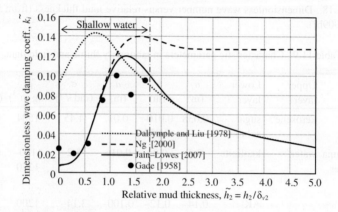

Figure 11.19. Dimensionless wave damping coefficient versus relative mud thickness (from Jain and Mehta [2009b]).

In Fig. 11.19 the Dalrymple and Liu solution predicts the peak in $\tilde{k}_i$ about $0.5\tilde{h}_2$ sooner, and about 20% higher than data trend. The solution diverges from the trend when $\tilde{h}_2$ decreases below about 2.5. This is so because as $\tilde{h}_2$ decreases the boundary layer δ_{v2} in mud begins to occupy an increasing fraction of mud thickness h_2.

The Ng solution, which seemingly exaggerates the peak by about 20%, mimics the data trend at small lower layer thicknesses ($\tilde{h}_2 <\sim 1$), as it accounts for energy loss in the boundary layer. At high values of $\tilde{h}_2$ the solution is restricted by the assumption of independence of the wave length from lower layer thickness. The outcome is over-prediction of wave damping.

The above comparisons pre-suppose that the flow regime is laminar. The relevant wave Reynolds number is $Re_w = u_b^2/\sigma v_1$, where u_b is the horizontal velocity amplitude obtained from U_{11} at $z = -h_1$. The highest value of Re_w based on the data in Table 11.9 is around 1,900, which well exceeds the laminar flow requirement, but does indicate that turbulence was weak. Data from Newnans Lake in Florida highlight the prevalence of laboratory-like conditions in this 27 km^2, shallow body of water [Jain *et al.*, 2007]. As a result of order of magnitude similarities in the parametric values with laboratory results, practically no scaling of data was required for simulating wave–mud interaction in the lake [Jain, 2007].

The Ursell number in the tests of Gade was 258, a high value consistent with the restriction that the perturbation method cannot be applied to the shallow water solution. In the other two sets of laboratory data, the highest value of *Ur* was 3. In Newnans Lake it was less than 1.

11.8.3 *Comparison of wave damping coefficient estimates*

Since a variety of constitutive models have been used to describe the energy dissipating behavior of the lower fluid of the two-phase system, it is useful to examine the effect of changing the equivalent viscosity $v_{e2} = \eta^*/\rho_2$ in Eq. 11.190. In Fig. 11.20 results on the variation of the wave damping coefficient with frequency are shown. Simulated curves are based on η^* derived for assumed behaviors of bottom mud including Newtonian (viscous), Maxwell and Jeffreys. The rheometric data (Table 11.9) were obtained in the controlled-stress mode using AK mud (Table 11.5). Based on test results reported by Jiang [1993], the coefficients of the constitutive equations were taken as $\eta^* = \eta = 20$ Pa.s for the viscous model, $\eta = 20$ Pa.s and $G = 300$ Pa for the Maxwell model (Eq. 5.55), and $\alpha = 1.07$, $\beta = 21$ and $\xi = 21$ for Jeffreys' model (Eqs. 11.168 to 11.171).

Overall, the results from the three models do not differ significantly because the low rigidity ($G = 300$ Pa) did not have a significant influence on the effect of high viscosity.

From Fig. 11.20 we can identify a characteristic frequency σ_f corresponding to the peak value of k_i at which the rate of energy loss in mud is a maximum. For the lower layer as a harmonic oscillator, this frequency is

$$\sigma_f = \frac{\alpha_f}{h_2}\sqrt{\frac{G}{\rho_2}} = 1.73 C_{sh} \frac{1}{h_2} \qquad (11.191)$$

in which the coefficient $\alpha_f = 1.73$ has been derived from the rheological model for axial (as opposed to shear) extension of an element of the lower layer by normal stress, *i.e.*, pressure [Jain, 2007]. This expression

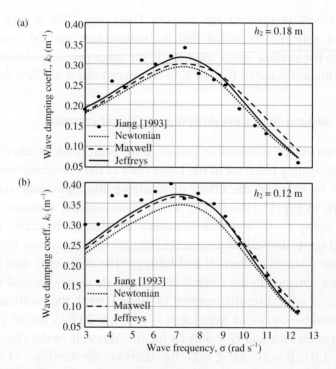

Figure 11.20. Variation of wave damping coefficient with frequency for three constitutive models and data of Jiang [1993]: (a) $h_2 = 0.18$ m; (b) $h_2 = 0.12$ m (from Jain and Mehta [2009b]).

for the resonance frequency is similar to the relationship for a poroelastic bed losing energy by Coulomb friction (Eq. 11.134). Given $h_2 = 0.12$, $\rho_2 = 1,200$ kg m^{-3} and $G = 300$ Pa yields $\sigma_f = 7.2$ rad s^{-1}.

11.8.4 *Mud mass transport*

Mass transport in fluid mud perturbed by waves results from an interaction between moving parcels of water and mud. The significance of second-order effects to motion (Eqs. 11.186 to 11.190) arises in this context. The fluid particle velocity averaged over a wave period, *i.e.*, the mass transport velocity, is given by

$$\overline{U}_m = \overline{u}_S + \overline{u}_E \tag{11.192}$$

where the overbar denotes a time-averaged quantity. The quantity $\overline{u}_S$ is Stokes drift [Dean and Dalrymple, 1991] and $\overline{u}_E$ is the Eulerian streaming velocity. Streaming occurs because at force equilibrium the Reynolds stress representing the momentum flux due to wave velocity fluctuations (Chapter 2), is balanced by a mean stress associated with streaming. Stokes drift is obtained from

$$\overline{u}_S = \overline{\frac{\partial u_E}{\partial x} \int_0^t u_E dt} + \overline{\frac{\partial u_E}{\partial z} \int_0^t w_E dt} = \frac{k}{2\sigma} u_E^2 + \frac{1}{2\sigma} w_E \frac{\partial u_E}{\partial z} \tag{11.193}$$

where $u_E = u_{21} + u_{22}$, and u_{21} and u_{22} are the Eulerian velocities obtained from Eqs. 11.186 to 11.189. Streaming occurs both in the upper layer (u_{E1}) and the lower layer (u_{E2}). These velocities are obtained from a solution of the simplified momentum balances in upper and lower layers

$$\frac{\partial \overline{u_{E1} w_{E1}}}{\partial z} = \nu_1 \frac{\partial^2 \overline{u}_{E1}^2}{\partial z^2} \tag{11.194}$$

$$\frac{\partial \overline{u_{E2} w_{E2}}}{\partial z} = \nu_2 \frac{\partial^2 \overline{u}_{E2}^2}{\partial z^2} \tag{11.195}$$

along with the boundary conditions

$$\overline{u_{E2}} = 0 \quad \text{at} \quad z = -(h_1 + h_2) \tag{11.196}$$

$$\overline{u_{E2}} = \overline{u_{E1}} \text{ and } \eta_1 \frac{\partial \overline{u}_{E1}}{\partial z} = \eta_2 \frac{\partial \overline{u}_{E2}}{\partial z} \quad \text{at} \quad z = -h_1 \tag{11.197}$$

$$\eta_1 \frac{\partial \overline{u}_1}{\partial z} \to 0 \quad \text{outside the boundary layer} \tag{11.198}$$

where $\eta_1 = \rho_1 v_1$ and $\eta_2 = \rho_2 v_2$ are the dynamic viscosities of the upper and lower layers, respectively. A fuller treatment is found in Sakakiyama and Bijker [1989] and Jain [2007].

In Fig. 11.21 the three velocities of Eq. 11.192 are compared with the experimental data of Sakakiyama and Bijker [1989] from a flume test using a bed of kaolinite (Table 11.9 with $\rho_2 = 1{,}150$ kg m^{-3} and $v_2 = 0.001$ m^2 s^{-1}). As evident from Fig. 11.22, the bed was at the threshold of developing a yield stress, and therefore could be approximated as a viscous fluid. The contribution from streaming is of

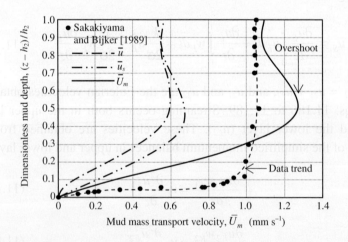

Figure 11.21. Mud mass transport velocity contributions and data of Sakakiyama and Bijker [1983] (from Jain and Mehta [2009b]).

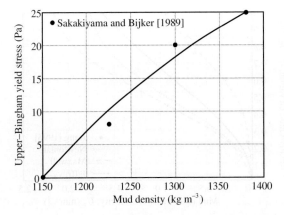

Figure 11.22. Upper–Bingham yield stress plotted as a function of the density of kaolinite bed (derived from data of Sakakiyama and Bijker [1989]) (from Jain and Mehta [2009b]).

the same order of magnitude as Stokes drift, indicating that the latter by itself cannot be taken as a measure of mass transport in this analysis. Although the predicted mass transport velocity is comparable in order of magnitude with the data, the overshoot in simulated $\overline{U}_m$ is absent from the measurements. This suggests that the experimental boundary layer was not fully developed, possibly due to the limited length of the flume.

The role of the constitutive model to simulate mass transport is noteworthy in Fig. 11.23. Results using Newtonian, Maxwell and Jeffreys models have been plotted in conjunction with the data of Jiang [1993] (Table 11.9 with $h_1 = 0.14$ m, $h_2 = 0.17$ m, $a = 0.03$ m, and $\sigma = 6.28$ rad s^{-1}). The analysis of Jiang [1993] has been summarized by Jiang and Mehta [1996]. All three models mimic the shape of the measured velocity profile. Differences in the velocities between the simulations and data are believed to be due to the method of measurement, which was crude. Overall, Jeffreys model shows the closest agreement with the data, with velocities about 40% lower than Maxwell, which is the poorest choice. These tests suggest that mass transport is sensitive to the choice of the lower layer model (Maxwell or Jeffreys).

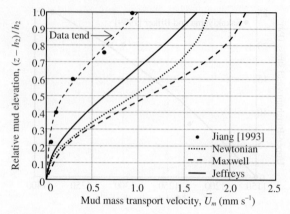

Figure 11.23. Mud mass transport velocity profiles and data of Jiang [1993] (from Jain and Mehta [2009b]).

11.9 Applicability of Energy Loss Mechanisms

In Fig. 11.24 the wave damping coefficient k_i has been plotted against wave frequency for poroelastic and viscoelastic models along with data of Sakakiyama and Bijker [1989] (Table 11.9, with $\rho_2 = 1,150$ and $1,300$ kg m^{-3} along with $v_2 = 0.001$ and 0.01 m^2 s^{-1}, respectively). At the bed density of $1,150$ kg m^{-3}, the bottom material was fluid-like, and at $1,300$ kg m^{-3} a solid matrix occurred with low but measurable yield strength (20 Pa; Fig. 11.22). For viscoelastic representation the Jeffreys model (Chapter 5) was selected. This model combines a KV element and a viscous element (or damper) in series, and therefore represents a viscoelastic fluid. It was applied by assuming its validity over the entire density range. This assumption holds when the bed experiences plastic yield under typical wave loading, which is often the case [Jiang and Mehta, 1995]. Jeffreys is observed to agree reasonably well with measured wave damping coefficients, and indicates that increasing the density to $1,300$ kg m^{-3} would result in as much as five-fold increase in k_i due to viscous energy loss. The poroelastic model, which is not valid in shallow water application, does capture the domain of k_i for the $1,300$ kg m^{-3} bed. It highlights the sensitivity of Coulomb friction to bed porosity. For instance, at the frequency of 6.28 rad s^{-1}, with porosity

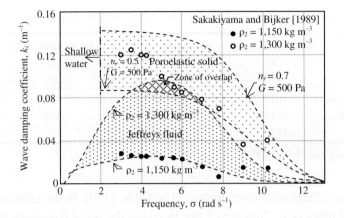

Figure 11.24. Application of viscoelastic fluid and poroelastic solid bed models to wave damping data of Sakakiyama and Bijker [1989].

increasing from 0.5 to 0.7, *i.e.*, by 40%, k_i increases by 80%, from 0.075 to 0.135 m^{-1}.

The overlap in Fig. 11.24 between the domains of viscoelastic and poroelastic models is an artifact of the coefficients used, and underscores difficulties in choosing models for the bed's constitutive behavior. This is so in part because the flow behavior of mud that has undergone cyclic loading differs from the bed's initial behavior.

It is evident that there are practical reasons against imposing strict limits on the applicability of different mechanisms for energy loss. On the other hand, it can be helpful to the modeler to have a descriptive representation of the domains of these mechanisms. The scope of such representation is discussed below.

The bed state can be characterized in terms of the solids volume fraction ϕ_v ($= 1 - n_r$) and a Péclet number $Pe = \sigma d_p / k_p$ [Jain and Mehta, 2009b].[d] The Péclet number increases with increasing wave frequency σ and particle size d_p, and decreases with increasing permeability k_p. Thus Pe also depends on whether the bed is sandy, silty or clayey.

---

[d] Even though the use of ϕ_v is appropriate for non-flocculating sediment ($> \sim 10 - 20$ μm), it is a poor choice for cohesive material, because the bed's constitutive behavior varies with the *floc* volume fraction ϕ_{vf} (Chapter 4). Unfortunately, since ϕ_{vf} is not readily available, ϕ_v is used in its place as an approximate measure of bed state.

Table 11.10. Péclet number for different beds and wave frequencies.

Bed	Nominal diameter, d_p (m)	Permeability, k_p (m s^{-1})	Péclet number, Pe		
			$\sigma = 6$ (rad s^{-1})	$\sigma = 1$ (rad s^{-1})	$\sigma = 0.5$ (rad s^{-1})
Coarse sand	1×10^{-3}	1×10^{-2}	0.6	0.1	0.05
Fine sand	1×10^{-4}	5×10^{-4}	1.2	0.2	0.1
Clay	1×10^{-6}	1×10^{-6}	6	1	0.5

Source: Jain and Mehta [2009b].

These effects on Pe are illustrated in Table 11.10. In the laboratory or in small lakes (*e.g.*, Newnans Lake in which the wave frequency σ is on the order of 6 rad s^{-1} or a period of about 1 s), Pe values are generally high and increase ten-fold (from 0.6 to 6) as the substrate changes from coarse sand to clay. In open sea ($\sigma = 0.5$ rad s^{-1}), Pe values would be comparatively low, even as they increase ten-fold when the substrate changes from coarse to fine. It is also important to note that Pe characterizes the bed as a drained, two-phased medium (even though it is analyzed as a porous continuum) at low values, and an undrained continuum at high values. A corollary is that the viscoelastic model is appropriate for low values of Pe and poroelastic model for higher values.

In Fig. 11.25 the compactness of the bed of given sediment diameter has been defined by ϕ_v (or n_r). The abscissa is specified by the Péclet number with values corresponding to Table 11.10. The wet bulk density ρ is based on a (nominal) particle density of 2,650 kg m^{-3} and water density of 1,000 kg m^{-3}. Domain boundaries are approximate due to the paucity of available information, and also because porosity and permeability depend on particle composition, shape and texture, in addition to size. Another factor contributing to uncertainty in the domain boundaries is due to the thixotropic response of mud to stress, which means that mud response depends on the duration of wave action.

In Fig. 11.25 (oscillatory) viscoelastic and (rotational) viscoplastic behaviors (Chapter 5) occupy much of the cohesive mud region. The less dense viscous fluid region is essentially a subset [Liu and Mei, 1989; van Kessel and Kranenburg, 1998]. When the bed is static (rigid) and the porosity is low, energy loss occurs due to friction at the bed surface. The

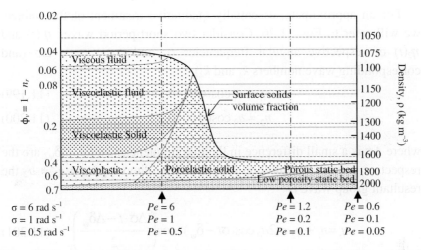

Figure 11.25. Semi-quantitative representation of the domains of constitutive models of mud behavior (revised from Jain and Mehta [2009b]).

domain boundary can vary substantially depending on surface roughness. Porous static beds experience loss due to percolation. Its contribution to energy loss is increasingly subsumed by Coulomb friction as the material becomes finer and the bottom heaves as a poroelastic solid.

Although the choice of the constitutive model varies significantly with ϕ_v and Pe, some approximate trends are evident in Fig. 11.25. Most importantly, the poroelastic response is bracketed between $Pe = \sim0.4$ and ~2. At higher values of Pe one of the viscoelastic/viscoplastic models (including purely viscous fluid model) applies depending on ϕ_v. At lower values of Pe bed friction and percolation are the two important loss mechanisms.

11.10 Long-Wave in Fluid Mud

When water surface variation is represented in terms of a wave energy spectrum (*e.g.*, Fig. 2.18), the fluid mud-water interface is actuated by oscillations resulting in energy loss not accounted for by monochromatic wave models. In Chapter 10 we made a brief reference to the potential effect of side-band instabilities in surface wave damping by fluid mud.

For an approximate, essentially qualitative treatment of this subject we will refer to Fig. 11.26. Consider two short-period waves $\eta_1(t)$ and $\eta_2(t)$ of respective angular frequencies σ_1 and $\sigma_2 = \sigma_1 + \Delta\sigma$ (and corresponding wave numbers k_1 and k_2)

$$\eta_1 = a_1 \cos(\sigma_1 t - \delta_{\psi 1}) \qquad (11.199)$$

$$\eta_2 = a_2 \cos(\sigma_2 t - \delta_{\psi 2}) \qquad (11.200)$$

where $\Delta\sigma$ is a small difference in the two frequencies and $\delta_{\psi 1}$, $\delta_{\psi 2}$ are the respective (arbitrary) time-lags. Conveniently choosing $a_0 = a_1 = a_2$ the resultant group wave is

$$\eta_r = \eta_1 + \eta_2 = 2a_0 \cos(\sigma t - \delta_\psi) \cdot \cos\left(\frac{\Delta\sigma \cdot t - \Delta\delta_\psi}{2}\right) \qquad (11.201)$$

where $\sigma = (\sigma_1 + \sigma_2)/2$, $\delta_\psi = (\delta_{\psi 1} + \delta_{\psi 2})/2$ and $\Delta\delta_\psi = \delta_{\psi 1} - \delta_{\psi 2}$. Thus a wave having the form $2a_0 \cos(\sigma t - \delta_\psi)$ is modulated by $\cos[(\Delta\sigma \cdot t - \Delta\delta_\psi)/2]$ and results in the beat effect well-known in acoustics [Longuet–Higgins and Stewart, 1962; Sharma and Dean, 1979; Dean and Dalrymple, 1991].

Determination of the amplitude of the long wave of frequency $\Delta\sigma = \sigma_L - 0 = \sigma_L$ resulting from the two waves requires that the wave amplitudes be treated as non-negligible, or finite. The long-wave ensues from retention of (non-linear) kinetic head in the dynamic free surface

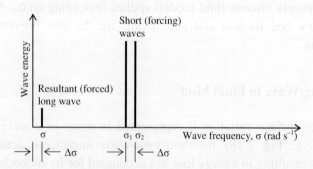

Figure 11.26. Schematic representation of wave energy spectrum showing two short-period wave frequencies and the resultant long wave frequency.

boundary condition (Eq. 2.111 with a quadratic term involving velocities added). The resulting wave is

$$\eta = -\frac{a_0^2 k}{\sinh kh}[1 + \cos(\Delta\sigma \cdot t - \delta_\psi)] \qquad (11.202)$$

where the wave number $k = (k_1 + k_2)/2$.

In shallow waters in which fluid mud persists due to waves induced by steady wind, long-waves forced by short-period waves have been detected. Given that, in general, mud is characteristically rich in nutrients, long-wave oscillations may play an important role in life cycle of benthic biota. A case in point is Lake Okeechobee in Florida where measurements were made of water surface variation and fluid mud motion. The depth of water at the selected site close to the eastern (down-wind) shore of the lake was 1.43 m. A neutrally buoyant accelerometer was embedded at a depth of 20 cm below the mud surface where the density was 1,180 kg m^{-3}. Water level was detected by a pressure gage mounted 0.87 m below surface. Figure 11.27 shows

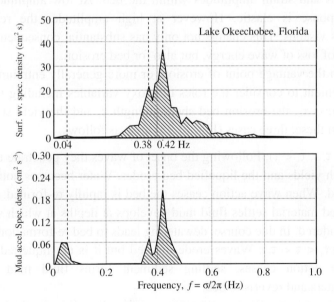

Figure 11.27. Measured surface-wave and mud-acceleration spectra in Lake Okeechobee, Florida (adapted from Jiang and Mehta [1992]).

synchronous spectra of the water surface and horizontal component of mud acceleration. The latter shows three peaks, two corresponding to the short-period waves at frequencies $f = 0.38$ Hz and 0.42 Hz. A long-wave signature occurs at $f \approx 0.42 - 0.38 = 0.04$ Hz. These three frequencies are also discernible in the surface spectrum.

11.11 Local Wave Erosion Processes

11.11.1 *Bed state and behavior*

Wave-induced beach erosion can be described as a profile-wide general scour, or as local scour. General scour and accretion are reviewed in Chapter 12. Local erosion formulas are summarized in this section. As a starting point we will revisit the relationship between bed state and its response introduced in Chapters 5 and 10. Bed response to waves depends on the wave frequency and amplitude which in turn determine the stress and strain amplitudes within the bed. At low amplitudes the bed response is elastic. However at high amplitudes the response becomes viscous. This change not only has substantial consequences for the rate of loss of wave energy, but also for bed erosion.

From the vantage point of erosion, or more generally entrainment, it is convenient to consider the roles of proxy variables including the bed shear stress τ_b, the erosion bed shear strength τ_s and the yield stress τ_y. Based on these, three cases (Fig. 11.28) arise as follows.

Case 1 ($\tau_s < \tau_y < \tau$): Following the onset of waves they penetrate the bed which yields and the liquefied bed erodes. Water becomes noticeably turbid. When wave action ceases the bed is rapidly re-formed. As the eroded material settles fluid mud develops at depths at which settling is hindered. In due course, dewatering leads to bed re-formation.

Case 2 ($\tau_s < \tau < \tau_y$): Waves erode the bed but it is not liquefied. Once wave action ceases, settling sediment forms fluid mud which dewaters and reverts to bed.

Case 3 ($\tau \leq \tau_s < \tau_y$): Wave action is too weak to liquefy or erode the bed.

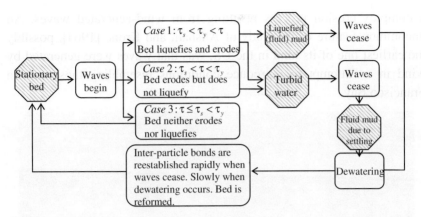

Figure 11.28. Effect of waves of fine-grained sediment erosion (from Jain and Mehta [2009a]).

11.11.2 *Erosion by non-breaking waves*

Although erosion by non-breaking waves mainly occurs in the shallow offshore zone seaward of the surf zone, depending on the bathymetry and sediment other situations can arise. Near a beach in Louisiana shown in Fig. 11.29, shoreward of the sandy surf zone there is another non-breaking zone with a muddy bottom and damped swell activity.

In the present context the muddy bed is considered to be soft but without fluid mud, with wet bulk density equal to or greater than about 1,200 kg m^{-3}. For such a bed, erosion flux equations determined from wave flume tests tend to support the applicability of formulas for steady flows. In general it is found that for waves

$$\varepsilon = \varepsilon_W \left(\frac{\tau_b - \tau_s}{\tau_s} \right)^{\delta_W} \tag{11.203}$$

When $\delta_W = 1$ this equation reduces to Eq. 9.27 for steady flows. Some experimental data (Table 11.11) have yielded values of δ_W close to unity (0.95 to 1.15). The bed shear stress τ_b is the peak value during the wave cycle (although in applications of Eq. 11.203 this distinction relative to the wave–mean bed shear stress is rarely made), τ_s is the bed shear strength and ε_W is the value of ε when $\tau_b = 2\tau_s$ (and $\delta_W = 1$). Most laboratory studies have used monochromatic mechanical waves to avoid a complex erosion process resulting from wind-generated waves. An uncommon feature of the study of Alishahi and Krone [1964], possibly the earliest one of its kind in the US, was that waves were generated by wind in an attempt to reproduce conditions resembling those in San Francisco Bay.

Figure 11.29. Non-breaking swell in the muddy surf zone along the coast of Louisiana (courtesy: Louisiana State University, Baton Rouge).

Table 11.11. Parameters for non-breaking wave-induced erosion equation.

Investigator(s)	Wave generation	Sediment: density $(kg\ m^{-3})$	Parameter ranges[a] a (cm); σ (rad s^{-1}); k (cm^{-1}); h (cm)	Coefficient values		
				ε_W $(g\ m^{-2}\ s^{-1})$	τ_s (Pa)	δ_W
Alishahi and Krone [1964]	Wind	Bay mud: 1,140	$0.9 \le a \le 3.4$ $h = 15$ cm	Test 1: 0.48 Test 2: 11.2	0.29 0.39	1.72 1.15
Thimakorn [1980], Thimakorn [1984]	Mechanical	River mud: 1,210–1,640	$0.16 \le ak \le 1.60$ $3.1 \le \sigma \le 12.6$ $h = 30$ cm	$= u_\delta \delta_v \tau_b / \tau_s{}^b$	0.05– 0.10	1
Maa [1986]	Mechanical	Kaolinite; bay mud: 1,100–1,300	$1.4 \le a \le 3.7$ $3.3 \le \sigma \le 6.3$	Kaolinite: 131 Mud: 30	0–0.17 0–0.20	1.15 0.95
Mimura [1993]	Mechanical	Kaolinite: Water content 160%	$0.6 \le a \le 6.9$ $4.8 \le \sigma \le 8.2$ $h = 10$ cm	0.27	0.15	1.82

[a] a = wave amplitude, σ = wave angular frequency, k = wave number and h = water depth.

[b] u_δ = amplitude of near-bottom velocity (Eq. 11.18), $\delta_v = \sqrt{2\nu/\sigma}$ and ν = kinematic viscosity of water.

An example of Eq. 11.203 is shown in Fig. 11.30 based on a wave flume study using mud from a bay near Cedar Key on the Gulf of Mexico coast of Florida [Maa, 1986]. Wave and sediment characteristics are given in Table 11.11. Also tested in the same study was a kaolinite bed. Its erosion shear strength τ_s was determined by using the method of layer-by-layer stripping of the bed by a series of steps of increasing bed shear stress outlined in Chapter 9, and replacing steady current by waves. It was found that τ_s values were lower than those obtained when the same kaolinite was eroded by steady-current steps. Figure 11.31 shows that in both types of tests τ_s increased with the duration of pre-test bed consolidation. It is conceivable that, with increasing duration beyond 14 days for which τ_s values were not determined, the bed would have hardened by consolidation and gelation. In the "zone of wave softening", waves led to a reduction in τ_s from the upper to the lower curve.

Since the initial state of the deposit also influences the erosion shear strength, the trends suggested by Fig. 11.31 may not be unique. In flume tests using kaolinite, Mimura [1993] measured the erosion flux that was an order of magnitude *lower* under waves than under current, suggesting that the bed had hardened by wave-induced reworking. A reason could be that the initially placed soft deposit underwent accelerated consolidation, rather than liquefaction. This can occur if the initially settled flocs are weak and breakable.

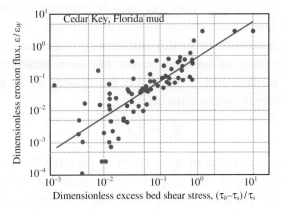

Figure 11.30. Variation of wave erosion flux with dimensionless shear stress for Cedar Key, Florida mud (from Maa [1986], Maa and Mehta [1987]).

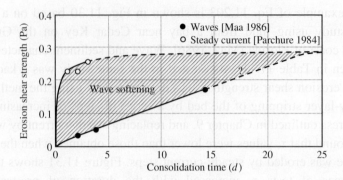

Figure 11.31. Measured and potential effects of waves on erosion shear strength of kaolinite.

11.11.3 Erosion by breaking waves

In bays, rivers and lakes with sufficient wind fetch to generate breaking waves, the rate of bank erosion can vary with the wind. At Cumbarjua Canal in Goa, India, which is lined with muddy sediment, waves due to steady sea breeze tend to erode the banks. However, during periods of low current velocities in the tidal cycle, the bed shear stress is insufficient to erode the submerged bottom. An unexpected outcome is that during such times the suspended sediment concentration in the canal is higher close to the water surface than near the bottom as sediment derived from the eroding banks increases the turbidity of surface water more than the lower water column [Mehta et al., 1983].

At the open coast in the muddy surf zone, bed and bank erosion due to the impact force of breaking waves has been treated as follows. Consider the scenario in Fig. 11.32. Let the velocity $\vec{u}_w$ at the crest of the wave be proportional to wave celerity C_w

$$|\vec{u}_w| = A_w C_w \qquad (11.204)$$

where the absolute value of $\vec{u}_w$ is taken to preserve the positive sign assigned to this velocity. The proportionality constant A_w depends on the breaker type, i.e., plunging, spilling or surging (Chapter 2), and therefore

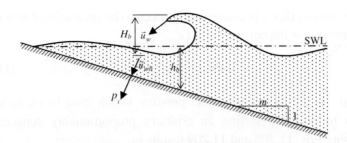

Figure 11.32. Schematic drawing of wave breaking at a muddy beach.

on bed slope m (ignoring effects of wave period). Since breaking of present interest occurs in shallow water, the above expression becomes

$$|\vec{u}_w| = A_w \sqrt{gh_b} \qquad (11.205)$$

where h_b is the water depth at breaking [Azam, 1998; Yamanishi *et al.*, 2001].

The head due to impact pressure p_i as the wave breaks is proportional to the respective velocity head, *i.e.*,

$$\frac{p_i}{\rho g} \propto \frac{\vec{u}_{wb}^2}{2g} \qquad (11.206)$$

Let the fluid particle velocity $\vec{u}_{wb}$ incident on the bed due to breaking be proportional to $\vec{u}_w$, with the proportionality coefficient dependent on, among other factors, the bed slope m. Therefore

$$\frac{p_i}{\rho g} \propto \frac{A_w^2 h_b}{2} = \frac{A_w^2 H_b}{2\kappa_b} \qquad (11.207)$$

where κ_b is the breaking wave index (Eq. 2.149). Finally,

$$p_i = \alpha_{wb} \rho g H_b \qquad (11.208)$$

where α_{wb} is another proportionality coefficient dependent on m.

The erosion flux ε is conveniently taken to be proportional to a power δ_{wb} of the excess impact pressure, *i.e.,*

$$\varepsilon \propto (p_i - p_{icr})^{\delta_{wb}} \qquad (11.209)$$

where p_{icr} is the threshold impact pressure which must be exceeded for erosion to occur. Selecting an arbitrary proportionality constant and combining Eqs. 11.208 and 11.209 results in

$$\varepsilon = \varepsilon_{W0} \left(\frac{H_b - H_{bcr}}{H_{bcr}} \right)^{\delta_{wb}} \qquad (11.210)$$

where H_{bcr} is the critical value of H_b for erosion and ε_{W0} is the value of ε when $H_b = 2H_{bcr}$ and $\delta_{wb} = 1$. In Eq. 11.210, ε depends on the breaker height as opposed to the local wave height in the surf zone. Therefore, ε must be considered to be a representative mean value applicable over the entire surf zone. This can be a reasonable approximation when the surf zone is narrow. However, in a wide surf zone, H_b must be taken as the *local* height and the equation calibrated for ε_{W0}, H_{bcr} and δ_{wb}.

Values of the coefficients in Eq. 11.210 from some studies are given in Table 11.12. In each case the erosion flux was estimated from comparisons of nearshore bottom profiles at different times by calculating the volumetric changes (per unit beach width). From these the corresponding sediment mass changes were obtained by knowing the bottom density. Erosion flux parameters ε_{W0} and H_{bcr} for the glacial till from Lake Erie were calculated from laboratory tests as well as field profiles. Notwithstanding differences in sediment properties among the investigations, the field profiles indicated lower values of ε_{W0} (by as much as three orders of magnitude) and order-of-magnitude higher values of H_{bcr} in comparison with the laboratory data. This suggests that remolding of the field soil samples took place when they were placed in the flume.

Table 11.12. Coefficients for breaking wave-induced erosion equation.

Investigator(s)	Sediment, source of values	Nominal bed density $(kg\ m^{-3})$	ε_{W0} $(kg\ m^{-2}\ s^{-1})$	H_{bcr} (m)	δ_{wb}
Lee [1995], Tarigan [1996]	1:1 (by weight) mixture of a kaolinite and an attapulgite, laboratory tests	1,200–1,300	7.56×10^{-6}	0.027	1
Kemp [1986]	Louisiana coast mud, Based on field profiles	1150–1,300	2.37×10^{-6}	0.087	1
Bishop and Skafel [1992], Bishop *et al.* [1992], Skafel and Bishop [1994]	Lake Erie glacial till, laboratory tests	2,000	1.39×10^{-3}	0.083	1
Nairn [1992]	Lake Ontario glacial till, based on field profiles	2,000	4.18×10^{-5}	0.57	1
Kamphuis [1986]	Lake Erie glacial till, based on field profiles	2,000	7.34×10^{-6}	0.29	1
Gelinas and Quigley [1973]	Lake Erie glacial till, based on field profiles	2,000	5.48×10^{-6} 1.07×10^{-5}	0.23 0.36	1 0.5

Source: Rodriguez [2000].

Figure 11.33 includes erosion fluxes for a consolidated glacial till from the northern shore of Lake Erie between Pointe aux Pins and Long Point, Ontario. Both, a linear regression fit ($\delta_{wb} = 1$) and a power-law fit ($\delta_{wb} = 0.5$) appear reasonable. The power-law is somewhat better (coefficient of determination $r^2 = 0.80$ compared to 0.71 for the linear fit), but possibly unsupported by the scarce data. For all other cases in Table 11.12 $\delta_{wb} = 1$ was found to be reasonable.

11.11.4 *Bluff erosion*

Wave-induced erosion at the toe of bluffs of hard cohesive soils (*e.g.,* Fig. 11.34) such as overconsolidated glacial tills occurs by both pressure and shear forces. Whereas impact due to pressure weakens the soil, shearing erodes the toe. Data compiled from sites in northern US and

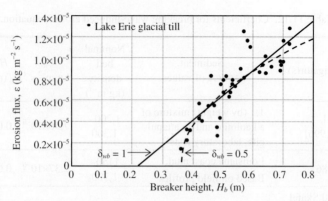

Figure 11.33. Erosion flux within surf zone as a function of breaking wave height for the northern shore of Lake Erie using data of Gelinas and Quigley [1973]. Curves are based on Eq. 11.210 (from Rodriguez and Mehta [2001]).

Figure 11.34. Erosion scarp (2 to 2.5 m in height) at the Ragueneau shoreline of St. Lawrence Estuary in Canada. The top-most layer is the intertidal low marsh silty facies (hardened deposit) and below it, the foreshore bank sandy facies (from Bernatchez and Dubois [2008]).

Canada in Table 11.13 indicate that the rate of erosion measured as bluff recession (in meters per year) approximately varies with the unconfined compressive strength q_u of the soil.[e] In general, the lower the compressive

---

[e] The unconfined compressive strength of a cohesive soil q_u is twice the undrained shear strength s_u (Chapter 5) when the soil condition is characterized by zero value of the internal angle of friction φ_{fr}.

Table 11.13. Bluff erosion rates and unconfined compressive strength.

Investigators	Region	Site	Soil type at bluff toe	Erosion rate $(m\ y^{-1})$	Unconfined compressive strength, q_u $(kN\ m^{-2})$
Quigley *et al.* [1977]	Lake Erie	Patrick Point	Basal till	0.60	400
		Iona Research Site	Silty clay	1.40	350
		Geography Field Station	Silty clay	2.00	290
Pezzeta [1974], Edil and Vallejo [1977]	Lake Michigan (Wisconsin coast)	Kewaunee-B	Clayey silt	5.43	95
		Port Washington	Silty clay	1.60	350
Berg and Collinson [1976], DuMontelle *et al.* [1977]	Lake Michigan (Illinois coast)	Lake Bluff Profile 1	Clayey silt till	0.86	431
		Lake Bluff Profile 3	Clayey silt till	0.55	431
		Lake Bluff Profile 4	Clayey silt till	0.77	431

Sources: Vallejo [1977]; Vallejo [1980].

strength the greater is the rate of recession. Accordingly, without reference to differences in soil types, we may state the following relationships:

q_u (kPa)	Erosion rate $(m\ y^{-1})$
400–431	0.55–0.86
290–350	1.40–2.00
95	5.43

11.11.5 *Fluid mud entrainment*

Entrainment of stationary fluid mud below the water layer by non-breaking waves occurs seaward of the (typically narrow) muddy surf zone when fluid mud is present. Similar to Eq. 9.79 for current-induced entrainment, the wave-induced process depends on the ratio of buoyancy force to inertia force characterized by the Richardson number. The velocity amplitude u_δ (Eq. 11.18) close to the interface between fluid mud and water replaces the unidirectional velocity U. The following relations [Li, 1996] have been developed for the *net* entrainment flux ε (mass per unit bottom area per unit time)

$$\varepsilon = \begin{cases} \alpha_{fm} \rho u_\delta \left(\dfrac{Ri_c^2}{Ri} - Ri \right) - w_s C_a & Ri < Ri_c \\ 0 & Ri \geq Ri_c \end{cases} \quad (11.211)$$

The first term, which includes the sediment dependent coefficient α_{fm}, represents upward movement of sediment and the second term $w_s C_a$ is the flux of sediment settling and merging with fluid mud. In Eq. 11.211, ρ is the density of fluid mud, C_a is the sediment concentration just above the top of fluid mud and Ri is the Richardson number defined by Li [1996] as

$$Ri = \frac{\sqrt{\dfrac{\pi}{2}} g' \delta_v}{(\Delta u)^2} \quad (11.212)$$

where $g' = (\rho - \rho_w) g / \rho_w$, $\delta_v = \sqrt{2v/\sigma}$ is the Stokes boundary layer in mud and Δu is the difference in the wave velocity amplitude across the fluid mud–water interface. In Eq. 11.211, Ri_c is the critical value of Ri below which there is no entrainment.

For typical applications, an order-of-magnitude analysis can be used to justify that typically the velocity in fluid mud can be ignored relative to that in water, thus permitting the replacement of Δu by u_δ [Liu and Mei, 1989]. Therefore, after ignoring the constant $\sqrt{\pi/2} = 1.253$,

$$Ri = \frac{g' \delta_v}{u_\delta^2} \quad (11.213)$$

For simplification it appears permissible to assume $w_s = 0$ in Eq. 11.211. This revised equation is plotted in Fig. 11.35 in terms of the dependence of the dimensionless entrainment flux $\varepsilon / \rho u_\delta$ on Ri. The data points are derived from wave flume tests using a kaolinite, mud from Cedar Key in Florida and from the Changjiang in China [Li, 1996]. The

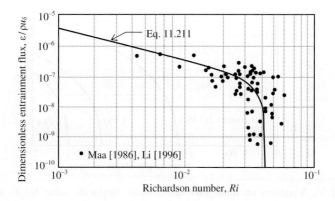

Figure 11.35. Dimensionless wave-induced entrainment flux as a function of Richardson number. Data of Maa [1986] and Li [1996].

coefficients $\alpha_{fm} = 2\times10^{-6}$ and $Ri_c = 0.043$ were selected for both sets of data. The overall bed density range was 1,100 to 1,300 kg m^{-3}. As we noted in Chapter 10, mud at 1,300 kg m^{-3} density may occur in a fluid-like state depending on its composition and as long as wave action of sufficient intensity maintains that state.

The trend of variation of $\varepsilon/\rho u_\delta$ with Ri is qualitatively similar to the trend for entrainment by current (Fig. 9.41). When Ri exceeds about 0.03 the buoyancy force begins to dominate inertia and entrainment flux is rapidly reduced. Once Ri exceeds 0.043 waves are unable to provide sufficient energy for entrainment.

Example 11.10: Plot the variation of fluid mud entrainment flux with wave height H up to 2 m. Select wave period 4 s, water depth 3 m, fluid mud dynamic viscosities $\eta = 0.1$, 1 and 2 Pa.s, $\rho = 1,150$ kg m^{-3}, $\rho_w = 1,027$ kg m^{-3}, $\alpha_{fm} = 2\times10^{-6}$ and $Ri_c = 0.043$. Ignore the settling flux $w_s C_a$.

Equation 11.18 is used to calculate u_δ in which the wave number is obtained from Eq. 2.119 for assumed rigid bottom. The plot based on Eqs. 11.211 and 11.213 is shown in Fig. 11.36. As expected, increasing the viscosity of fluid mud reduces its entrainment. As mentioned previously, flowing fluid mud will increasingly resist entrainment if and when it picks up bed sediment and thereby increases its own viscosity (and density). In practical applications it appears permissible to ignore the settling flux as long as α_{fm} is adjusted by calibration against site-specific conditions.

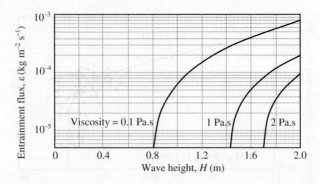

Figure 11.36. Variation of fluid mud entrainment flux with wave height and mud viscosity.

11.12 Suspended Sediment Concentration due to Oscillating Flow

In situ images of suspended fine sediment in estuaries and bays often show particles oscillating due to wave action as they fall. As these particles settle more slowly than in quiescent water, it is a matter of interest to examine the effect of oscillating flow on the suspended sediment concentration profile. For that purpose we will simplify Eq. 6.114 by assuming $f^{n}(C, \bar{G}) = 1$ and $\Phi_s = 1$, which yields the steady mass balance

$$\bar{w}_s C + D_{so} \frac{\partial C}{\partial z} = 0 \qquad (11.214)$$

where $\bar{w}_s$ (which replaces w_{s0}) is the mean component of the oscillating settling velocity. The flow will be assumed to oscillate at angular frequency σ and with a vertical velocity amplitude $\hat{w}_o$ (Eq. 7.122).

Following the wave effect analysis in Section 7.2.9, $\bar{w}_s$ is taken as

$$\bar{w}_s = w_s \left(1 - K_1 \frac{\hat{w}_o}{w_s} \right) \qquad (11.215)$$

where w_s is the Stokes settling velocity in quiescent water and K_1 is an empirical constant [Hwang, 1985].

Let

$$D_{so} = K_2 \hat{w}_o h \tag{11.216}$$

where the constant K_2 defines the scale of mass diffusion. In Exercise 11.6.

The velocity $\hat{w}_o$ is obtained from Eq. 2.125 as

$$\hat{w}_o = a\sigma \frac{\sinh kz}{\sinh kh} \tag{11.217}$$

After substitution of Eqs. 11.215, 11.216 and 11.217 into Eq. 11.214 and integration of the resulting equation from the near-bed elevation z_a (where the concentration is C_a) to an arbitrary elevation z yields

$$\frac{C}{C_a} = \left(\frac{\tanh \dfrac{kz}{2}}{\tanh \dfrac{kz_a}{2}} \right)^{-\frac{w_s \sinh kh}{a\sigma K_2 kh}} e^{\frac{K_1(z-z_a)}{K_2 h}} \tag{11.218}$$

In shallow water this equation reduces to

$$\frac{C}{C_a} = \left(\frac{z}{z_a} \right)^{-\frac{w_s}{a\sigma K_2}} e^{\frac{K_1(z-z_a)}{K_2 h}} \tag{11.219}$$

in which $\exp[K_1(z - z_a)]/K_2 h$ is a concentration enhancing term associated with settling velocity reduction relative to its value in quiescent fluid. The typical range of this term appears to be 1.3 to more than 7 [Hwang, 1985].

11.13 Exercises

11.1   You are given a 10 s wave with an amplitude of 0.05 m in 1 m deep fresh water. The bottom consists of a 0.3 m thick layer of fluid mud with a dynamic viscosity of 0.001 m^2 s^{-1}. Water density is 1,020 kg m^{-3} and mud density 1,200 kg m^{-3}. Determine:

   (1)   the amplitude of mud–water interface velocity,

   (2)   surface wave celerity, and

(3)   wave damping coefficient. Use Eqs. 11.70, 11.84 and 11.85, respectively. Is their use justified for the prescribed wave condition?

11.2   A wave with a period T (= $\sigma/2\pi$) = 7 s is traveling through seawater of depth $h = 4.5$ m, density $\rho_1 = 1{,}027$ kg m^{-3} and dynamic viscosity $\eta_1 = 0.0019$ Pa.s. The mud bed density $\rho_2 = 1{,}200$ kg m^{-3}, dynamic viscosity $\eta_2 = 2.5$ Pa.s, rigidity $G = 895$ Pa, specific permeability $K_p = 3\times10^{-11}$ m^2 and specific energy loss parameter $\delta_p = 0.052$. Calculate the wave damping coefficient using each of the four expressions for k_i in Table 11.5.

11.3   The mechanical analog of the Jeffreys model is shown in Fig. E11.1. Derive the coefficients α, β and ξ in Eqs. 11.168 through 11.171 in terms of shear modulus G and viscosity coefficients η_a and η_b.

11.4   Equation 11.107 is partly based on the following expression of Reid and Kajiura [1957] for the wave damping coefficient k_i, for which the effect of boundary layer on wave damping has not been considered (*i.e.*, $\delta_v = 0$)

$$k_i = \frac{2k}{2kh+\sinh 2kh}\left(\frac{K_p\sigma}{v}\right)$$ (E11.1)

This expression is applicable to surface waves over an infinitely thick porous medium. King *et al.* [2009] presented the following generalized solution for k_i,

$$k_i = \frac{2k}{2kh+\sinh 2kh}\left(\frac{K_{p1}\sigma}{v}\right)C_G$$ (E11.2)

where the bed is composed of one of the following: (1) a semi-infinitely thick porous medium of permeability K_{p1}; (2) a porous medium of thickness h_1 and permeability K_{p1}; or (3) a porous medium of thickness h_1 and permeability K_{p1} over a semi-infinitely thick porous medium of permeability K_{p2}. The generalization coefficient C_G is defined as follows:

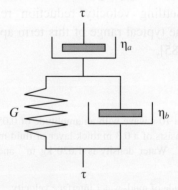

Figure E11.1. Mechanical analog of Jeffreys model representing mud rheology (enlarged from Fig. 11.17).

Bed surface layer thickness (layer 1)	Underlying layer (layer 2)	C_G
Infinite	–	1
h_1	–	$\tanh k_r h_1$
h_1	Infinite	$\dfrac{\tanh k_r h_1 + \dfrac{K_{P2}}{K_{P1}}}{\dfrac{K_{P2}}{K_{P1}} \tanh k_r h_1 + 1}$

For the five cases of waves over porous media sketched in Fig. E11.2, determine the wave damping coefficient. You are given: wave period $T = 4$ s, kinematic viscosity of water $v = 1.17 \times 10^{-6}$ m^2 s^{-1}, bed porosity $n_r = 0.4$ for Sand A (diameter 5 mm) and B (2.5 mm), and 0.5 for clay (2 μm).

11.5   From wave measurements in a 1.65 m deep coastal lagoon you are given the following data: $f_1 = 0.38$ Hz, $f_2 = 0.42$ Hz, $a_1 = 4.8$ cm, $a_2 = 5.0$ cm, $a = 0.6$ cm (measured long-wave amplitude), $\delta_{\psi 1} = 191°$ and $\delta_{\psi 2} = 171°$. Assume a_0 to be the mean of a_1 and a_2. Plot the group wave η_r (Eq. 11.201) and the long-wave η. Briefly comment on the observed trends.

11.6   A 0.30 m high, 7 s wave in 2.5 m water depth is approaching the shoreline.

(1)   Plot the suspended sediment concentration profile for cohesive bottom sediment whose particle settling velocity is 1×10^{-4} m s^{-1}. Measured concentration 0.1 m above bottom is 2.73 kg m^{-3}. In Eq. 11.218 select $K_1 = 10^{-4}$ and $K_2 = 10^{-3}$.

(2)   What would the profile look like if the wave height was doubled?

11.7   In a wave flume experiment a bed of kaolinite was resuspended and concentration profiles were measured at different times as shown in Fig. E11.3. The following

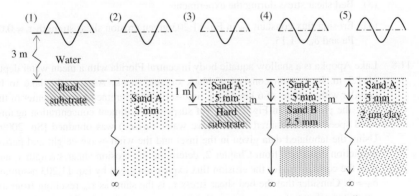

Figure E11.2. Five combinations of waves over porous media.

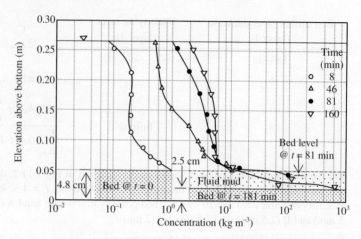

Figure E11.3. Concentration profiles in a wave flume experiment on erosion of kaolinite (adapted from van Leussen [1994]).

parameters are given: water surface elevation (above flume bottom) 0.27 m, wave height 7 cm, wave period 1.5 s and water density 1,027 kg m^{-3}. The initial (at time $t = 0$) bed thickness was 4.8 cm, which decreased to 2.5 cm after 181 min of test run, with a 2.3 cm thick layer of fluid mud generated on top. Assume that the water column was initially free of sediment, the bed density was 1,200 kg m^{-3}, and at 181 min fluid mud concentration was 250 kg m^{-3}. Assume a wave friction factor of 0.04. Use the assumed concentration values in Table E11.1. Determine:

(1) Erosion flux for each time interval (0–8 min, 8–46 min and so on). Then obtain the mean erosion flux and the total mass of sediment eroded per unit bed area (kg m^{-2}) at the end of 181 min.

(2) Total mass of fluid mud per unit bed area.

(3) Bed shear stress during the experiment.

(4) Erosion flux constant ε_W in Eq. 11.203 given erosion shear strength $\tau_s = 0.01$ Pa and $\delta_W = 1.15$.

11.8 Lake Apopka is a shallow aquatic body in central Florida with a mean water depth of 1.3 m. Its bottom consists of black organic muck. Assume the lake to be circular with a radius of 6 km. Using time-series measurements at the center of the lake, the plot given in Fig. E11.4a (of suspended sediment concentration against wind speed U_{10} measured 10 m above water surface) was obtained [So, 2009]. Using the tabulated data given in the inset and the wind–wave height and period prediction equations from Chapter 2, determine the erosion shear strength τ_s and then the constant ε_W for the erosion flux expression given by Eq. 11.203 assuming $\delta_W = 1$. Consider that the bed shear stress τ_b is the same as τ_{cw} resulting from the combined effects of waves and wind-driven current in the lake. Use the following method of Soulsby *et al.* [1993] to estimate τ_{cw}.

$$\tau_{cw} = Y(\tau_{cur} + \tau_{wv}) \tag{E11.3}$$

Table E11.1. Suspended sediment concentration profiles based on interpolated profiles in Fig. E11.3.

Elevation (m)	Suspended sediment concentration (kg m^{-3})			
	8 min	46 min	81 min	160 min
0.025	3.700	12.0	65.0	350.0
0.050	0.980	10.5	60.5	155.0
0.075	0.420	4.00	57.0	57.0
0.100	0.230	2.65	47.0	55.0
0.125	0.175	1.85	40.0	52.5
0.150	0.177	1.03	34.5	52.0
0.175	0.190	0.70	29.0	46.8
0.200	0.197	0.60	22.5	38.0
0.225	0.175	0.55	17.8	29.5
0.250	0.113	0.51	12.0	22.0

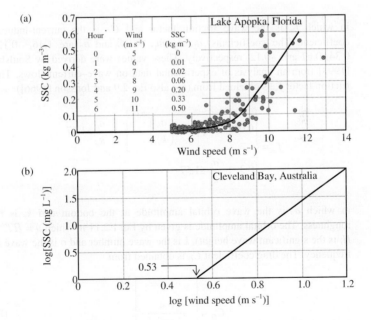

Figure E11.4. Suspended sediment concentration (assumed depth–mean) against wind speed from: (a) Lake Apopka in Florida (from So [2009]); (b) Cleveland Bay, Australia (from Lambrechts *et al.* [2010]).

where τ_{cur} is the current-induced shear stress and τ_{wv} is the wave-induced shear stress. Also,

$$Y = 1 + \hat{a}X^{\hat{m}}(1-X)^{\hat{n}} \tag{E11.4}$$

and

$$X = \frac{\tau_{cur}}{\tau_{cur} + \tau_{wv}} \tag{E11.5}$$

The coefficients in Eq. E11.4 are obtained from

$$\hat{a} = \hat{a}_1 + \hat{a}_2 \log\left(\frac{f_w}{C_D}\right) \tag{E11.6}$$

$$\hat{m} = \hat{m}_1 + \hat{m}_2 \log\left(\frac{f_w}{C_D}\right) \tag{E11.7}$$

$$\hat{n} = \hat{n}_1 + \hat{n}_2 \log\left(\frac{f_w}{C_D}\right) \tag{E11.8}$$

in which f_w is the wave friction factor and C_D is the current-induced drag coefficient. The coefficients $\hat{a}_1$, $\hat{a}_2$, $\hat{m}_1$, $\hat{m}_2$, $\hat{n}_1$ and $\hat{n}_2$ are 1.46, –0.12, 0.47, 0.34, 0.67 and –0.12, respectively. These values were derived by Soulsby *et al.* [1993] from an analysis of experimental data on wave–current flows. The wave friction factor f_w is obtained from (see also Eq. 2.9 and Jonsson [1966])

$$f_w = \begin{cases} 0.00251\exp\left[5.21\left(\dfrac{a_b}{k_s}\right)^{-0.19}\right] & ; \ \dfrac{a_b}{k_s} > 1.57 \\[2ex] 0.3 & ; \ \dfrac{a_b}{k_s} \le 1.57 \end{cases} \tag{E11.9}$$

in which a_b is the wave orbital amplitude at the bottom and k_s is the bed roughness. The orbital amplitude is given by Eq. E2.11 in which $a = H_s/2$ (where H_s is the significant wave height), k is the wave number and σ is the wave angular frequency. The drag coefficient C_D is obtained from

$$C_D = \left[\frac{0.4}{\ln\left(\dfrac{30}{\hat{k}_s}\right) - 1}\right]^2 \tag{E11.10}$$

where $\hat{k}_s = k_s/h$ is the relative bed roughness and h is the water depth. The current-induced shear stress is calculated from

$$\tau_{cur} = C_D \rho u_c^2 \qquad (E11.11)$$

where u_c is the current velocity at a representative elevation above the bottom and ρ is the water density. The wave-induced shear stress is obtained from

$$\tau_{wv} = 0.5 \rho f_w \left(\frac{H_s \sigma}{2 \sinh kh} \right)^2 \qquad (E11.12)$$

Select water density $\rho = 1{,}000$ kg m^{-3} and bed roughness $k_s = 0.006$ m. For calculation of significant wave height and wave period ($= 2\pi/\sigma$) as functions of wind speed U_{10} use Eqs. 2.152, 2.153 and 2.154. Assume the following relationship between u_c (m s^{-1}) and U_{10} (m s^{-1}):

$$u_c = 0.006 U_{10} \qquad (E11.13)$$

11.9 Based on the set of equations in Exercise 11.8 repeat the calculations for Cleveland Bay in Australia by changing the water depth to 4 m and using Fig. E11.4b [Lambrechts *et al.*, 2010]. In this plot the line represents the approximate mean trend of measured SSC with wind speed. Use k_s as the calibration parameter. Compare the constants for the erosion flux Eq. 11.203.

Chapter 12

Sedimentation Phenomena

12.1 Chapter Overview

Some natural and engineered sedimentation phenomena are mentioned in this chapter. There can be net deposition *or* net erosion, or an equilibrium condition in which the two unit processes have equal fluxes. Barring a few exceptions, sedimentation in such cases is examined in terms of analytic developments.

A review of the dynamics of mud shore profiles, basic features of non-cohesive sediment profiles and the well-known Bruun Rule for shoreline recession are summarized. Mud shore profiles are highlighted with reference to similarities and differences relative to non-cohesive, especially sandy, profiles. A simple mud profile equation based on energy conservation is introduced and profile shape is shown to depend on the rheological properties of sediment. Tidal flats and submersed mudbanks are described qualitatively, and likely causes of mudbank formation and dissipation are mentioned. The definition and potential use of a bulk index are introduced for an assessment of the stability of mud shore profile against erosion.

Development of equations for longshore sediment discharge is next given. A simple treatment of the actuation of mud underflows on a sloping bed such as the shelf follows. Sedimentation in entrance channels and small basins including marinas, closed-end canals and dredged traps is the next topic. The treatments are approximate, and meant as a means for first-cut assessments of sedimentation in semi-confined shallow waters. The last topic briefly deals with depth definitions and the utility of the nautical depth concept related to navigation in heavily sedimenting harbors.

12.2 Shore Profiles

12.2.1 *Non-cohesive beach profiles*

Non-cohesive beach profiles, which also include many silty shores, vary greatly in shape depending on sediment composition, offshore bathymetry and waves.[a] A distinct feature of such profiles is the submerged offshore bar (Fig. 12.1), whose volume varies with the wave condition — steep, high-energy waves enlarge the bar during winter months, whereas under low-steepness, low-energy summer waves, bar volume reduces and the profile becomes smoother [Bruun, 1954].

In Fig. 12.1, the depth of closure h_0 marks the seaward limit of the profile at $x = x_0$ such that the movement of non-cohesive sediment is largely confined to the littoral zone shoreward of point x_0, h_0. The depth of closure may change in the alongshore direction as profile shape varies due to erosion, accretion, geomorphic changes or structure-induced impacts, even when waves offshore remain (hypothetically) invariant. In beach engineering it is common to assume a constant depth of closure over a reasonably long stretch of the shoreline bordering more-or-less shore-parallel isobaths (Fig. 12.2a). Over this stretch, when a sufficient

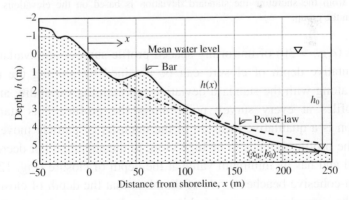

Figure 12.1. Non-cohesive beach profile definitions and power-law approximation.

[a] Some data in Chapter 6 suggest that if silt contains more than about 10% clay minerals, the erosion behavior of the mixture is influenced by cohesion.

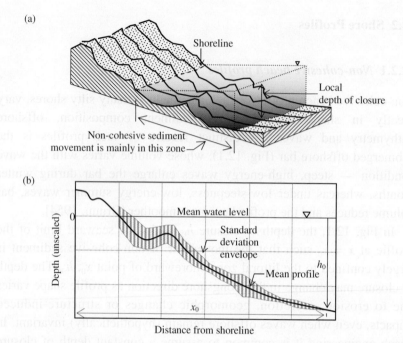

Figure 12.2. Schematic plots showing non-cohesive beach profiles: (a) Profiles and depth of closure; (b) identification of a representative (mean) depth of closure from a minimum in the standard deviation envelope for selected group of equidistant profiles. At any distance from the shoreline the standard deviation is based on the elevations of all profiles in the group.

number (*e.g.*, > 30) of uniformly spaced profile surveys are available, a representative depth of closure can be obtained by plotting the mean profile along with the standard deviation representing variability among the profiles at every survey point along the profile. This standard deviation is a qualitative measure of the intensity of particle movement along the profile. It tends to be high close to the shoreline but decreases seaward and has its minimum value at the depth of closure (Fig. 12.2b). For non-cohesive beaches a rule-of-thumb is that the depth of closure is twice the annual maximum breaking wave height. At that depth the sediment-active profile is considered to account for about 90% of the littoral sediment movement. To reach close to 100% the depth of closure should be taken as 3.5 times the maximum breaking wave height [Bruun and Schwartz, 1985].[b]

Given adequate supply of sediment, non-cohesive beaches tend to assume a shape that is in equilibrium with the wave condition provided it is invariable over the duration to reach equilibrium. For example, in typical laboratory basins full equilibrium may take a day or as much as a week depending on model dimensions. If x is the distance from the shoreline and $h(x)$ is the water depth, examinations of a large number of profiles indicate that in its simplest description h increases with x according to the power-law

$$h = A_{pr} x^{2/3} \qquad (12.1)$$

The bar is not featured in Eq. 12.1, for which other more complex expressions containing more than two adjustable coefficients have been developed [Dean and Dalrymple, 2002].

For given sediment size a representative mean profile scaling parameter is obtained from the empirical relationship

$$A_{pr} = 0.067 w_s^{0.44} \qquad (12.2)$$

in which w_s (cm s^{-1}) is the settling velocity and the unit of A_{pr} is m$^{1/3}$. This relationship is shown in Fig. 12.3 [Dean, 1977].

Example 12.1: Determine the water depth at the seaward end of a 250 m long sandy beach profile. The particle settling velocity is 3 cm s^{-1}.

From Eqs. 12.1 and 12.2 the water depth is 4.3 m, which means that a representative beach slope would be 4.3/250 = 1:58. The smaller the particle the flatter the beach will be. The utility of Eq. 12.2 becomes increasingly subjective as the settling velocity decreases below about 3 cm s^{-1} unless the particles consist of uniform sand without cohesive matter. When the settling velocity is less than about 0.1 cm s^{-1}, Eq. 12.2 is not applicable because, as we will see, the response of cohesive profiles to waves qualitatively differs from that of non-cohesive ones.

---

[b] The depth of closure is a convenient but notional definition, because the annual maximum wave height varies from year to year with the result that a century-averaged annual maximum height is larger than over a decade. In the probabilistic sense 100% closure is never achieved.

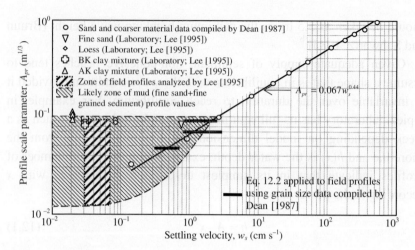

Figure 12.3. Variation of beach profile scaling parameter with settling velocity (adapted from Lee [1995]).

12.2.2 Bruun Rule

Consider a rise a_r of the sea level relative to the bottom having an equilibrium shape (*e.g.*, Eq. 12.1) determined by sea conditions before the rise. For full adjustment of the profile to the new sea level the duration of rise must be sufficiently long. Let x and y denote distance coordinates normal to and along the shoreline, respectively. At a non-cohesive profile the required volume of deposition of sediment per unit beach width (*i.e.*, distance along the shoreline) must be

$$\Delta V_d = \int_0^{x_0}[h(x) + a_r]\,dx - \int_0^{x_0} h(x)\,dx = x_0\,a_r \qquad (12.3)$$

where x_0 is the profile length. For the profile to be in equilibrium with the new sea conditions it is assumed that:
(1) The entire profile is in equilibrium even though profile fluctuations that are short-term compared to the time of sea level rise a_r may occur.
(2) Along the shore where the rule is applied the net volumetric change (per unit width of beach) over the profile due to erosion/deposition is always zero.
(3) Longshore sediment transport, does not vary with alongshore distance and therefore has no effect on the profile.

Under these conditions *only*, the profile (Fig. 12.4a) must rise by an elevation a_r (Fig. 12.4b) and move landward by a distance R_S such that $\Delta V_d = \Delta V_e$, the incremental landward volume (per unit beach length) of sand taken away by erosion (Fig. 12.4c). Since profile uplift and landward translation occur simultaneously, the intermediate profile shown represents a hypothetical stage [Bruun, 1962].

As the profile slope is characteristically mild, *i.e.*, and $h_0/x_0 \ll 1$, it is seen that

$$\Delta V_e \approx \int_0^{x_0} f(x)dx + h_0 R_s - \int_0^{x_0} f(x)dx = h_0 R_s \qquad (12.4)$$

Then, from Eqs. 12.3 and 12.4,

$$R_s = \frac{x_0 a_r}{h_0} \qquad (12.5)$$

which is the rule for estimating R_s. For the above relationship to hold, R_s/x_0 must remain small.

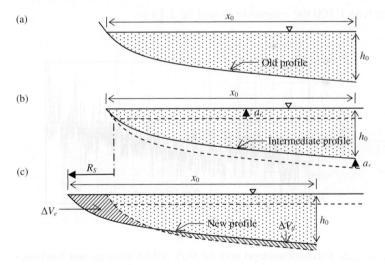

Figure 12.4. Schematic representation of Bruun Rule "steps": (a) original profile; (b) profile uplift due to sea level rise; (c) profile translation landward to equalize the volume of sediment gained with volume lost (per unit width of beach).

The sea level rise value a_r must be accurate because, due to mild profile slope, the error in R_s is x_0/h_0 times greater than the error in a_r. Satellite altimetry based data are superior to conventional tide gage data, but cover only the two most recent decades. Tide gage data recording more than six decades are few and occur mostly in the northern hemisphere [Grinsted *et al.*, 2009].

The trend in monthly mean sea level (*msl*) from the NOS–NOAA tide station no. 9414290 near Golden Gate in San Francisco is shown in Fig. 12.5. The plotted values are relative to the 1983–2001 *msl*. An earthquake in 1906 caused a downward movement of the local datum, which resulted in a break in the record. Thus the long-term linear trend is divided into pre and post-1906 periods. For the 1906 to 1999 post-earthquake period the rate was 2.13 mm y^{-1} with a standard error of 0.14 mm y^{-1}.

Let us consider a hypothetical case for a sandy beach profile of 225 m length ending at the depth of closure of 14 m. Thus the beach slope is 1:16.1, which is steep. Over a period of 10 years we get $a_r = 0.0213$ m, which yields $R_s = 0.342$ m. A flatter beach would recede more; if the slope was 1:100 the recession would be 2.13 m.

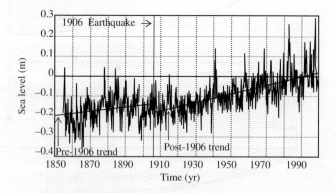

Figure 12.5. Record of sea level from the NOS–NOAA tide gage near San Francisco.

12.2.3 *Cohesive mud shore profiles*

12.2.3.1 *Profile features and nomenclature*

In coastal waters the effects of wind-waves and swell are modulated by tide, whereas in lakes and reservoirs wind-driven current and waves are the transporting agents. In the wave-dominated environment (Fig. 12.6a), the sediment-active nearshore zone is bounded by the shoreline and the terminal depth defined as that depth seaward of which wave action is unable to cause significant changes in the profile. This definition permits transport of suspended sediment across the terminal depth (TD) in the seaward direction but not landward (inset of Fig. 12.6b). The landward profile is sub-divided by the wave breakerline into the surf zone and the offshore zone between the breakerline and the terminal depth. If landward transport is also not permitted, the terminal depth becomes synonymous with the depth of closure (DoC) at sandy beaches.

Coastal cohesive mud deposits often do not show features as prominent as alongshore bars at sandy profiles. Fig. 12.6b is a schematic rendition of a beach profile consisting of both sand and mud from the western part of Mont Saint Michel Bay in Manche, France [Caline, 1994]. It shows distinct submerged bars of coarse to medium sand with largely featureless mud beds in between. The nearly linear profile in Fig. 12.6c is from Sargent Beach in Texas about 96 km southwest of Galveston along the Gulf of Mexico coast. This shoreline has experienced erosion, exposing relict clayey deposits. The bar is small compared to other profiles along the same beach where sand has deposited [Tem Fontaine, Coastal Tech. Inc., personal communication, 2010].

Evidence from large-tide estuaries in Europe indicates that significant rhythmic bedforms may occur even where the sediment is mainly mud. Whitehouse *et al.* [2000b] identify four types of features at intertidal mud flats: (1) channels, creeks and gullies; (2) ridge-and-runnels; (3) ripples and other micro-topography; and (4) cliffs. The ridge-and-runnels systems in the direction of tidal current have attracted considerable attention due to their large dimensions. The genesis of these self-organized and stable systems is not well understood. Mont Saint

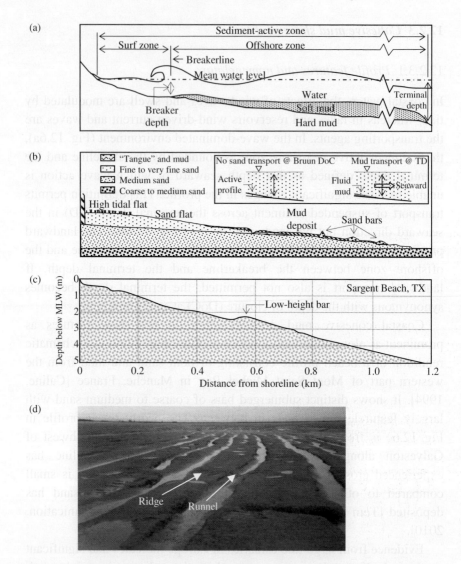

Figure 12.6. (a) Definitions related to the sediment-active nearshore zone; (b) schematic drawing of sedimentary formations in the western part of Mont Saint Michel Bay. "Tangue" refers to coastal sediment found in France (adapted from Caline [1994]); (c) eroding profile exposes old clayey deposit at Sergeant Beach, Texas [courtesy Tem Fontaine]; (d) ridge-and-runnel system at the Brouage mudflat, Marennes–Oléron Bay, France (1997 photograph modified from Whitehouse *et al.* [2000b]).

Michel Bay is an estuary where a ridge-and-runnel sequence has developed due to sand. Figure 12.6d shows a well-developed system at the Brouage mudflat in Marennes–Oléron Bay (France). The widths of the ridges are 1.2–1.5 m and mean height 0.21 m above the runnel. The runnels are 0.30–0.50 m wide. The hydraulic significance of such forms is high bed resistance. While a typical value of the bed roughness k_s (Chapter 2) at mudflats could be, say, 6 mm, at large bedforms k_s values as high as 6 to 9 cm have been reported [Whitehouse *et al.*, 2000b]. Where ridges and runnels occur, as the water recedes with falling tide it drains through the runnels. Drainage current on the order of 0.2 m s^{-1} has been measured. At these speeds measurable sediment flux occurs in the runnel [Le Hir *et al.*, 2000].

Figure 12.7 shows a stratigraphic sequence and nomenclature for a prograding muddy tidal flat made up of a variety of deposits. These descriptions are in part based on mudflats along the southwestern coast of Korea. In general they may include: (1) cross-stratified sandy silts and sandy muds in sub-tidal channel deposits above bedrock; (2) non-stratified, shelly, silty-sand in the low-tidal flat; (3) mottled sandy silts and silts with parallel to lenticular bedding in the mid-tidal flat; and (4) mottled to homogeneous mud in the high-tidal flat. The column on the left is a complete sequence, while the one on the right shows a composite of truncated sequences above a full sequence, with truncations suggesting the effects of local sea level variation [Alexander *et al.*, 1991].

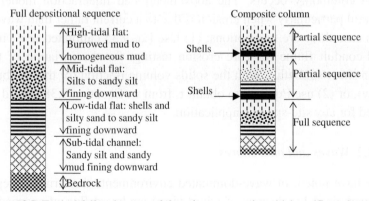

Figure 12.7. Typical full and truncated deposit sequences in a tidal mudflat (adapted from Alexander *et al.* [1991]).

For recording the erosion of a core consisting of a mixed sequence of sediment layering such as that in Fig. 12.7, a closed-conduit flume of the type sketched in Fig. 9.63 is often used. An advantage is that erosion generates a record of scour of thin bed layers of millimeter scale regardless of its composition (which could include clay, silt or sand), as long as the core erodes. This is achieved by high velocities in the duct. Since the protruding portion of the bed must stand on its own, cores with wet bulk density less than about 1,200 kg m^{-3} are not always suitable. Thus the erodibility of very soft or fluid mud cannot be determined in this flume. An option for bottom sediment < ~1,200–1,250 kg m^{-3} is a viscometer or rheometer for characterizing the stress versus rate of strain flow curve as in Fig. 5.13b. Yield strengths (*i.e.*, the upper-Bingham yield stress) obtained from such curves are plotted against the respective mud densities in Fig. 9.16a.

Assessments of the erosion potential of stratified beds from a core depend on the composition and consistency of the stratified profile. So [2009] used the plastic yield stress τ_y (in place of the erosion shear strength τ_s) as the threshold value of the bottom shear stress τ_b at which fluid mud entrainment was postulated to begin. The premise was that the organic-rich, low submerged weight (particle density $\rho_s = 1,690$ kg m^{-3}) muck was "liquefied" when τ_b exceeded τ_y. Core samples of different densities were tested in a viscometer.

When the sediment is clayey but not stiff, (floc-by-floc) surface erosion commonly occurs. The abovementioned liquefaction model is not useful particularly when erosion is due to a current rather than waves. In this case there are two options: (1) Use Eq. 9.35 calibrated (*e.g.*, in a closed-conduit flume for core erosion testing) for the variation of the erosion shear strength τ_s with the solids volume fraction (or the wet bulk density), or (2) use Eq. 9.38 to obtain τ_s from τ_y. Equation 9.38 should be verified for any site-specific application.

12.2.3.2 *Waves and mud shores*

As we have noted, in wave-dominated environments the terminal depth is characterized by absence of (short-term) landward flux of suspended fine sediment, even as seaward flux is not precluded. When a fine sediment beach erodes the material may be transported well seaward of the terminal depth due to the low settling velocity of resuspended matter. The transport distance can be scaled by the settling lag $t_s = s_e / w_s$, where s_e is the height to which bed particle or floc is resuspended and w_s is the settling velocity. For example, given $s_e = 3$ m and a fine sand profile with $w_s = 0.01$ m s^{-1}, we obtain $t_s = 300$ s. An offshore flowing stream such as a rip current with a velocity of 0.5 m s^{-1} would deposit the grains 150 m offshore assuming the depth remains constant. This distance will be equal to or less than the depth of closure for the sandy profile. In contrast, a mud floc with a settling velocity $w_s = 10^{-4}$ m s^{-1} could travel considerably further. Hypothetically, at the same current velocity the offshore distance would be 15 km. More realistically the flocs are carried alongshore by the longshore current. In some situations this current may be too weak to change the beach profile but strong enough to winnow fine suspended matter. When the eroded bed is hard, such as an overconsolidated clay layer, the eroded material will not reform a bed of the same strength.

In Fig. 12.8 the likely sequence of events due to storm waves is shown. Initially, under a calm sea the hardened profile represents mud bed formed following the previous storm (Fig. 12.8a). When storm waves arrive, the bottom shear erodes the profile and water turbidity increases. Where waves break mass erosion may result in high levels of turbidity (Fig. 12.8b), and an erosion scarp may develop. Eventually the submerged bottom may liquefy and form fluid mud. When liquefaction occurs, wave orbitals penetrate the bottom. The shear stress at the new bed surface beneath fluid mud may be high enough to cause some erosion, but the erosion flux may remain low particularly if fluid mud damps the wave and retards breaking. As fluid mud density increases with continued bed scour, the erosion flux may be suppressed and even arrested given sufficiently long storm duration (Fig. 12.8c). Fluid mud will move shoreward or seaward due to mass transport and reconfigure the profile. In the final stage after the storm passes, the shape of the new profile will be determined by the rates of consolidation and gelation (Fig. 12.8d) [Nairn, 1992; Lee and Mehta, 1997].

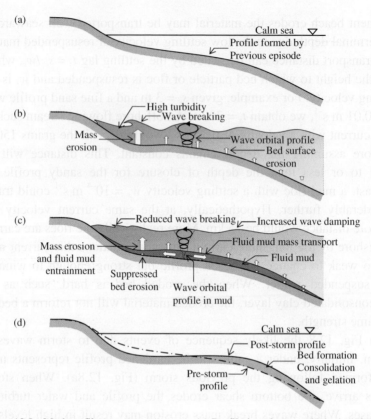

Figure 12.8. Mud profile response to waves: (a) Profile in calm sea; (b) bed surface and mass erosion with turbidity generation; (c) production and transport of fluid mud; (d) reconfigured profile at the end of episode (from Lee and Mehta [1997]).

In order to examine the wave response of beach profiles of different sediments, four 0.9 m wide profiles were subjected to the same set of waves in a large flume [Lee, 1995]. The beds were made of fine quartz sand, loess from Vicksburg in Mississippi, and two clay mixtures: 1:1 attapulgite and kaolinite (AK), and a 1:3 Wyoming bentonite and kaolinite (BK). The median dispersed sizes are given in Table 12.1. In Fig. 12.9a, elevation changes after 44.3 h of wave runtime are plotted relative to the initial profiles (zero elevation line) prepared at a nearly uniform slope of about 1:200. The data are for sand, loess and one (AK) of the two clay mixtures. The sand and the loess were scoured

Table 12.1. Physical characteristics of (mainly) mud profiles.

Name	Median size (μm)	Wave characteristics	Tidal range (m)	Sediment characteristics	Source
Western Louisiana in the vicinity of Cheniere au Tigre	1–5	Non-storm wave height 0.3 m; period 7 s; storm wave height 1 m; period 5–6 s	0.5 (mean)	Illite, smectite and kaolinite; bed density: 1,150–1,300 kg m^{-3}	Morgan *et al.* [1953]; Kemp [1986]
Corte Madera Bay in San Francisco Bay	5	Typical wave height < 0.2 m, period 3 s	1.3 (mean)	> 50% clay (illite, kaolinite and montmorillonite)	Liang and Williams [1993]
West coast of peninsular Malaysia: Selangor, Perak, Kedah	2–15	Non-storm height 0.30 m to 0.75 m, period 5–6 s; storm ht. 1–2 m; period 6–9 s	2–2.5 (mean)	Top 0.1–0.3 m mud 1,200 kg m^{-3}; lower mud 1,500 kg m^{-3}	Malaysian EPU [1986]; Hor [1991]; Tarigan [2002]
West coast of peninsular Malaysia: Pantai Punggur	3–10	Non-storm height 0.30 m to 0.75 m, period 5–6 s; storm ht. 1–2 m; period 6–9 s	2 (mean)	Top 0.1–0.3 m mud 1,200 kg m^{-3}; lower mud 1,500 kg m^{-3}	Tarigan [2002]
Lian Island, Jiangsu coast, P. R. China	2–4	Non-storm height 0.6–1.2 m, period 7–8 s; storm height 4 m	3.4 (mean)	Illite, kaolinite, chlorite; bed density: 1,100 kg m^{-3}	Yu *et al.* [1987a,b]; Rodriguez [2000]
Surabaya coast, Indonesia	5–74	Non-storm heights 0.2–0.8 m, period 3–6 s; storm height 1.7–2.1 m, period 5–8 s	2.4 (spring)	60% clay, 35% silt, 5% sand	Tarigan [1996]
Teluk Waru coast, Indonesia	3–6	Non-storm height < 0.2 m; storm ht. 2 m, period 8 s	1.7 (spring)	49% clay, 47% silt, 4% sand	Tarigan [1996]
Grimsby, Lake Ontario, Canada	0.5	Root-mean-square height 0.9 m	–	Hard glacial till: 46% clay, 33% till, 21% sand and gravel	Bishop and Skafel [1992]
Brouage flat, Marennes–Oléron Bay, France	Clay/silt	Root-mean-square height 0.5 m	5	Entirely mud	Bassoullet *et al.* [2000]

Table 12.1 — Continued

Name	Median size (μm)	Wave characteristics	Tidal range (m)	Sediment characteristics	Source
Skeffling flat, Humber estuary, UK	63 (mean)	Typically small	6	Muddy to sandy	Whitehouse *et al.* [2000b]
Seine estuary flat, Le Havre, France	Clay/silt /sand	Ht. 0.2–3 m	7	Sandy to muddy	Le Hir *et al.* [2000]
Vicksburg loess	18	–	–	Chlorite, illite, kaolinite, montmorillonite, vermiculite, quartz, feldspar, dolomite	Lee [1995]
BK (in lab)	1.1	–	–	Bentonite + kaolinite (1:3)	Lee [1995]
Fine sand (in lab)	90	–	–	Quartz	Lee [1995]
AK in (lab)	1.1	–	–	Attapulgite + kaolinite (1:1)	Lee [1995]

in the first approximately 6 m reach of the flume, and the eroded material deposited in the profile reach between 6 and 14 m.

The elevation changes in Fig. 12.9a suggest an approximate volumetric balance (and therefore mass balance assuming a constant mean bed density) between the eroded and deposited materials. As we have noted volumetric balance is an essential requirement for the applicability of the Bruun Rule. In noteworthy contrast to the sand and loess profiles, the clayey beach showed net erosion over the first 18 m reach. At the same time the eroded material did not deposit between 18 and 23 m. Instead it was transported to the flume sump beyond the 23 m length of the profile.

In analogy with the above laboratory observations, in Fig. 12.9b we see a lack of recovery at a mud beach in western Louisiana [Kemp, 1986]. Thus, for reasons cited, the Bruun Rule may not be applicable to cohesive beaches [Tarrigan *et al.,* 1996].

(a)

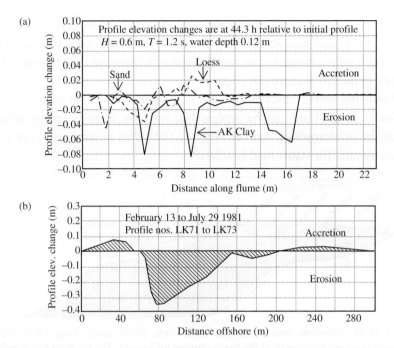

(b)

Figure 12.9. (a) Changes in three profiles under identical wave conditions in flume experiments of Lee [1995]; (b) mud profile changes based on field surveys by Kemp [1986] in western Louisiana.

Table 12.1 includes a brief description of the wave, tide and sedimentary environment in the vicinity of Cheniere au Tigre along the coast of Louisiana, where mud beach profiles were surveyed. This site is west of the mouth of the Atchafalaya River, which supplies cohesive sediment as a westward alongshore stream. Shoreline erosion begins around November and ends typically in March. This is the time of wave activity associated with the southwesterly cold-weather fronts. At neighboring sandy beaches that appear to be in near-equilibrium over the annual period, recovery is essentially complete between March and June. When the imbalance between erosion and deposition in Fig. 12.9b is viewed together with the observation that the mud shores have been stable over time-scales shorter than those associated with sea level changes, it may be inferred that beach dynamics is not dependent on waves alone. In fact there appears to be a relationship between profile dynamics and sediment load in the Atchafalaya mud stream. The dual role of waves and mud stream on profile evolution makes the behavior of this beach more complex than sandy beaches where sediment transport is mainly in the cross-shore direction.

From the above observations we may further conclude that loss of sediment to offshore can be replenished only by a fresh supply from terrestrial or alongshore sources (Fig. 12.10).[c] This requirement appears to be essential for stable mud beaches.[d]

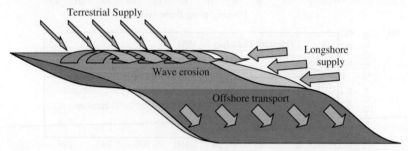

Figure 12.10. Schematic drawing of mud beach where profile shape changes but shore sediment mass per unit length of the beach remains unchanged. The mudbanks of Kerala cannot be explained by this model as it does not account for offshore sources of mud.

---

[c] The Kerala mudbanks seem to be sustained by offshore pools of fluid-like mud "pushed" shoreward by monsoonal waves. There are no active riparian or terrestrial sources of mud in the vicinity [Mathew and Baba, 1995, Chapters 1 and 10]. Such a natural system dependent on an offshore sediment source may be rare.

[d] From an analysis of winter and summer sandy beach profiles, Dean [1973] showed that a threshold wave-steepness (Chapter 2) can be defined. For actual wave steepness that is less than the threshold, the shoreward profile would gain sand from loss at the seaward profile (shoreward of the depth of closure). The result would be a summer profile with a wide beach. When wave steepness exceeds the threshold value the gain and loss zones switch and result in a winter profile with a narrower beach. The evaluation of threshold steepness was predicated on the hypothesis that when the particle settling lag t_s is less than one-half the wave period T, forward orbital motion under the wave crest will carry particles (suspended by the breaking wave) shoreward. This leads to beach buildup. When t_s is greater than $T/2$, the backward orbital movement under the wave trough will transport the particles seaward. Thus for example, for $T = 8$ s, if t_s was less than 4 s the profile would build up. Since for fine sediment t_s is easily an order of magnitude greater than 4 s, the condition favoring shoreward transport *by this mechanism* is not realized.

"Pocket" mud beaches similar to sandy ones where the sediment is prevented from moving alongshore by, say, headlands, may occur where there is no loss or supply of sediment, *i.e.,* the beach is in steady-state rather than in equilibrium. It would follow that where no alongshore or offshore sources of mud exist, any net erosion, say during an unusually severe storm, would cause the shoreline to recede. This is observed in Fig. 12.11, which shows the recession of hard, glacial till (Table 12.1) profile on Lake Ontario between 1980 and 1984. When the till, over-consolidated by glacial surcharge, eroded, the resulting material was transported either offshore or alongshore. The terminal depth of about 6.1 m is easily identified (although based on two profiles only). Assuming that this depth remains unchanged from year to year, all profiles during the four years of recession could be expected to merge at the 6.1 m depth [Davidson–Arnott, 1986; Coakley *et al.*, 1988].

An instance of hard mud profile erosion with no replenishment from natural sources is Mahin Beach about 100 km west of the city of Lagos, Nigeria (Fig. 12.12). At this coast waves are dominated by swell (height 1.4 m, period 14 s) in the wet season originating at storm centers in the southern Atlantic region [Bakker, 2009]. The tide is semi-diurnal with a mean range of 1.5 m and the shore and nearshore material consists mainly of poorly sorted silt with mean size ranging from 20 to 50 µm. Due to submarine canyons that act as sinks for the littoral drift, this region is starved of sediment. In the 1970s a navigation cut was dredged

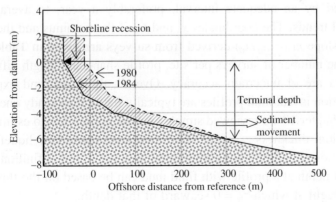

Figure 12.11. Glacial till profiles at Grimsby along Lake Ontario, Canada (from Coakley *et al.* [1988]).

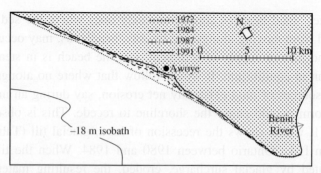

Figure 12.12. Receding mud shoreline in Nigeria (adapted from Eedy *et al.* [1994]).

perpendicular to the coast near the village of Awoye about 20 km west of Benin River in order to connect inland creeks and canals with the ocean. This cut apparently permitted larger waves to penetrate inland, exacerbating erosion of the shoreline near the cut. In addition, saltwater intrusion occurred, which affected vegetation sensitive to brackish water. The ensuing die-back exposed bottom sediment otherwise protected by roots, and considerably increased land loss. Erosion was rapid in the first few years, and a recession on the order of 1.5 to 2.0 km occurred near Awoye between 1972 and 1991 [Eedy *et al.*, 1994; Ibe *et al.*, 1989].

12.2.3.3 *Terminal depth*

Estimation of the terminal depth requires at the minimum two surveys separated by an adequate interval, preferably a year, to average out seasonal trends. The coordinates x_0 and h_0 of the terminus and the mean profile slope m (= h_0/x_0) derived from surveys are given in Table 12.2. Since the number of surveys per site, profile shape and length differ, the m-values are of uncertain accuracy. Overall, the main (and expected) observation is that mud profiles are typically flatter than sandy ones.

In Chapter 10 we reviewed simple analytic approaches to estimate the equilibrium thickness z_{fe} of the fluid mud layer formed by liquefaction. As mentioned earlier a somewhat restrictive but useful definition of the terminal depth at a profile with fluid mud can be based on the threshold wave height at which z_{fe} = 0 seaward of that depth.

Table 12.2. Mud profile dimensions and mean slope.

Location	No. of profiles	Prof. length, x_0 Mean (m)	Prof. length, x_0 Range (m)	End depth, h_0 Mean (m)	End depth, h_0 Range (m)	Profile slope, $m = h_0/x_0$ Mean	Profile slope, $m = h_0/x_0$ Range	Data source
Western LA near Cheniere au Tigre	71	188	21–364	1.06	0.27–2.83	0.0067 (1:149)	0.0021–0.0375	Kemp [1986]
Corte Madera Bay, CA	1	1260	NA^a	1.44	NA	0.00115 (1:870)	NA	Liang and Williams [1993]
Selangor, Malaysia	6	880	615–988	2.98	1.30–4.60	0.0033 (1:303)	0.0020–0.0047	Malaysian EPU [1986]
Perak, Malaysia	13	140	68–184	0.82	0.59–1.01	0.0063 (1:159)	0.0046–0.0111	Lee [1995]
Kedah, Malaysia	3	145	126–168	1.74	1.65–1.82	0.012 (1:83)	0.0109–0.0139	Hor [1991]
Pantai Punggur, Malaysia	1	1100	NA	3.50	NA	0.0032 (1:313)	NA	Tarigan [2002]
Jiangsu, China	2	363	356–370	5.66	5.58–5.74	0.0156 (1:64)	0.0155–0.0157	Wang et al. [1988]
Madura, Indonesia	21	640	268–1006	3.14	2.00–4.50	0.0053 (1:189)	0.0029–0.0123	Tarigan [1996]
Surabaya, Indonesia	18	415	56–1034	2.83	1.00–5.00	0.0211 (1:47)	0.0020–0.0132	Tarigan [1996]
Teluk Waru, Indonesia	4	193	112–388	3	3.00–3.00	0.0198 (1:50)	0.0077–0.0268	Tarigan [1996]
Grimsby, Lake Ontario, Canada	1	500	NA	6.5	NA	0.0130 (1:77)	NA	Coakley et al. [1988]
Brouage flat, Marennes–Oléron Bay, France	NA	4000	NA	5	NA	0.0013 (1:800)	0.0007 (1:1400) Flatter section	Bassoullet et al. [2000]
Skeffling flat, Humber estuary, UK	NA	4000	NA	5.7	NA	0.0014 (1:700)	0.0008 (1:1300) Flatter section	Whitehouse et al. [2000b]
Seine estuary flat, Le Havre, France	NA	400	NA	2.7	NA	0.0067 (1:150)	NA	Le Hir et al. [2000]

^{a} Not available.

Based on Eq. 10.42, h_0 is plotted against the hypothetical deep water wave height H_0 (Chapter 2) in Fig. 12.13 using wave, water depth and mud thicknesses from Lake Okeechobee in Florida. A beach observation from Pantai Punggur in Malaysia near Johor is also included. The following best-fit line is drawn

$$h_0 = 12.7H_0 - 9.7; \text{ valid for } H_0 > 0.8 \text{ m and } h_0 > 0.5 \text{ m} \qquad (12.6)$$

The effect of wave period on h_0 is not considered under the assumption that for estimations of h_0 the effect of wave period is likely to be less significant than height. The value 0.8 m is the threshold height below which no fluid mud is generated. The relationship is not valid when the water depth is 0.5 m or less. In general, the slope and the intercept of the line depend on bed properties. For further application, H_0 may be replaced by the significant (non-breaking) wave height exceeded for a total of 12 hours during the year [Hallermeier, 1978; Rodriguez, 2000]. Equation 12.6 may be used only after site-specific calibration for the slope.

A somewhat different interpretation of the effect of wind–waves on fluid mud is implied in Fig. 12.14, which includes data from a neutrally buoyant accelerometer embedded 0.2 m below the mud surface in Newnans Lake, Florida. The amplitude of wave-induced vertical acceleration in mud increases with wind speed. The threshold speed of 3.2 m s^{-1} marks the onset of mud heave. As noted in Chapter 11,

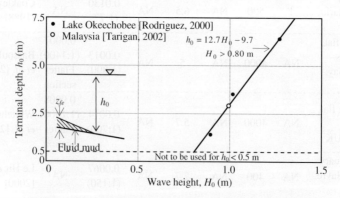

Figure 12.13. Dependence of terminal depth on wave height.

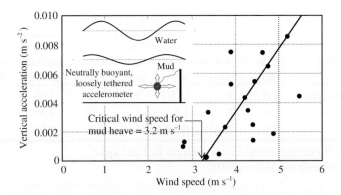

Figure 12.14. Vertical acceleration in mud plotted against wind speed during February–May 2004 in Newnans Lake, Florida (from Jain *et al.* [2007]).

this causes internal friction which significantly enhances the loss of wave energy relative to a non-heaving bed.

12.2.3.4 *Mass transport*

In addition to gravity slide of fluid mud at a beach, its movement can occur by transport over a flat-bed due to an interaction with waves (Chapter 11). A consequence is that pools of mud can form or disperse on the shallow coastal shelf. For mud transport that is co-linear with the wave direction, the mass transport velocity under waves approaching obliquely to the beach can be resolved into a cross-shore and an alongshore component.

For the locations mentioned in Table 12.3, mass transport velocities along the coasts of Suriname, Guyana, Louisiana and Kerala have been estimated (Table 12.4). At the Kerala site offshore of the city of Alappuzha (Alleppey), mass transport is assumed to be shore-normal. Elsewhere the interest has mainly been in the alongshore component which can be slow (and perhaps weakly turbulent) compared to cross-shore transport. In general, a divergence, *i.e.,* advective gradient, in the alongshore mud flux will influence shoreline configuration. The time-scale of this influence ranges from days (India) to weeks (Louisiana) to years (Suriname/Guyana).

Table 12.3. Mud mass transport estimates.

Location	Wave height (m)	Wave period (m)	Littoral zone width (m)	Water depth (m)	Mud thickness (m)	Mud specific gravity[a]	Mud viscosity (Pa.s)
Suriname/Guyana	1.0	9	11.5	3	1.0	1.08	1.3
Louisiana	0.6	5	1.5	1	0.5	1.17	1.6
Kerala	2.0	7	0.5	6	1.5	1.07	0.5

[a] To obtain the corresponding density, multiply by 1,027 kg m^{-3}.
Source: Rodriguez and Mehta [1998].

Table 12.4. Calculated and measured cross-shore and alongshore velocities.

Location	Calculated cross-shore velocity (km d^{-1})	Calculated alongshore velocity (km yr^{-1})	Measured velocity range	Reynolds number	Alongshore velocity due to breaking (km d^{-1})
Suriname	–	1.7	0.5–2.5 (km y^{-1})	88	26
Guyana	–	1.0	0.4–2.0 (km y^{-1})	88	16
Louisiana	–	2.3	0.5–3.0 (km y^{-1})	46	25
Kerala	0.31	–	1.2–2.4 (km d^{-1})	1,170	–

Source: Rodriguez and Mehta [1998].

12.2.3.5 *Wave effect on profile shape*

Depending on waves and bottom sediment properties, mud shore profiles can be "convex-upward" or "concave-upward"; the latter being descriptively similar to sandy beach profiles [Kirby, 1992; Friedrichs, 1993]. By considering wave height to decrease with: (1) distance shoreward due to viscous energy loss within mud, and (2) wave breaking in the surf zone, an equation can be derived for the profile depth $h(x)$ (below mean water level). The simplest approach is to assume that the profile is in equilibrium with the wave field.

The profile equation is based on Eq. 11.2 for steady wave movement over duration sufficiently long to result in an equilibrium profile. More appropriately we may refer to a "target" profile, since true equilibrium may never be achieved when the substrate is cohesive. Equation 11.2 is restated as

$$\frac{d}{dx}(EC_g) = \varepsilon_{De} \qquad (12.7)$$

where ε_{De} is the value of ε_D (representing the wave–mean energy loss per unit bed area and unit time) for the equilibrium cohesive mud profile. The offshore coordinate x is 0 at the shoreline and x_0 at the offshore end of the profile. For further analysis we note that holding the product $h\varepsilon_D$ constant over x_0 is consistent with Eq. 12.1 for sandy profiles (see also Section 11.7.1.3). Lee [1995] proposed that for mud profiles the requirement would be to hold ε_{De} (as opposed to $h\varepsilon_D$) constant. The main cause of wave damping over a sandy profile is the loss of energy due to wave breaking. At a mud profile absorption of energy by fluid mud formed by liquefaction or deposition is the main cause of wave damping. As an approximation we may restate Eq. 11.4 for flat beds as

$$a(x) = a_{x_0} e^{-k_i(x_0 - x)} \qquad (12.8)$$

where a_{x_0} is the incident wave amplitude at $x = x_0$ and k_i is the wave damping coefficient.

Since in shallow water $C_g = \sqrt{gh}$, Eq. 12.7 combined with Eq. 12.8 results in

$$\frac{\rho g^{3/2} a_{x_0}^2}{2} \frac{d}{dx}\left[e^{2k_i(x-x_0)} \sqrt{h} \right] = \varepsilon_{De} \qquad (12.9)$$

where ρ is the fluid (water) density. This equation is integrated from the shoreline (0, 0) to any point x, h

$$\int_{0,0}^{x,h} d\left[e^{2k_i(x-x_0)} \sqrt{h} \right] = \int_0^x \frac{2\varepsilon_{De}}{\rho g^{3/2} a_{x_0}^2} dx \qquad (12.10)$$

which yields

$$e^{2\bar{k}_i(x-x_0)} \sqrt{h} = \frac{2\varepsilon_{De}}{\rho g^{3/2} a_{x_0}^2} x \qquad (12.11)$$

where $\bar{k}_i$ is a representative profile-averaged wave damping coefficient. This expression must satisfy the condition $h = h_0$ at $x = x_0$. Thus,

$$\varepsilon_{De} = \frac{\rho g^{3/2} a_{x_0}^2 \sqrt{h_0}}{2_{x_0}} \qquad (12.12)$$

Combining this equation with Eq. 12.11 yields

$$h(x) = h_0 e^{4\bar{k}_i(x_0-x)} \left(\frac{x}{x_0}\right)^2 \qquad (12.13)$$

A limitation of Eq. 12.13 is that the profile slope dh/dx is zero at the shoreline, a condition usually not met in reality, as effects of breaker-induced erosion and gravity-driven adjustment of the profile typically result in a non-trivial slope [Dean and Dalrymple, 2002]. To account for this feature, Eq. 12.13 is modified to include a nearshore correction term $m_0 x e^{-\beta_0 x}$, where m_0 is the slope at $x = 0$ and β_0 is an empirical exponent. This correction models the offshore extent of the combined influence of shoreline slope and erosion due to wave breaking. Thus,

$$h(x) = h_0 e^{4\bar{k}_i(x_0-x)} \left(\frac{x}{x_0}\right)^2 + m_0 x e^{-\beta_0 x} \qquad (12.14)$$

Equations 12.13 and 12.14 are plotted in Fig. 12.15a along with a profile from the mud coast of Louisiana. The profile is neither strongly erosional (*i.e.*, concave) nor significantly accretionary (convex).[e] While both equations converge at a distance exceeding about 70 m, the absence of the nearshore correction term in Eq. 12.13 prevents the simulation of the trough next to shoreline. Note also that in contrast to this profile, the one from Malaysia in Fig. 12.15b is distinctly concave (erosional) and the one in Fig. 12.16a from Louisiana is convex (accretionary).

---

[e] A concave profile in equilibrium (or target) configuration is neither eroding nor accreting. For the present purpose "erosional" will simply mean concave (or concave-upward) rather than eroding. Similarly, "accretionary" will mean convex (or convex-upward) rather than accreting.

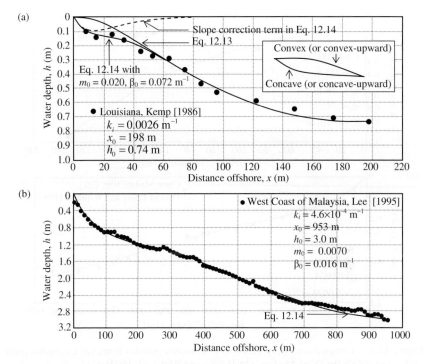

Figure 12.15. Comparison of profile equations with measured profiles from: (a) Louisiana coast obtained by Kemp [1986] (from Lee and Mehta [1997]); (b) the west coast of Malaysia compiled by Lee [1995].

Since the influence of shoreline slope correction typically occurs over a short distance offshore, among variables characterizing profile shape the one most important variable is the damping coefficient k_i (with overbar implied), as it influences the major portion of the profile. This coefficient depends on the erosional or accretionary nature of the environment, sediment composition and bed density. An outcome is that two physically different environments may yield similar values of k_i. Between two nearshore wave gages 3.4 km apart in more or less shore-normal direction in the Gulf of Mexico, nearly 50% reduction in wave energy corresponding to $k_i = 2 \times 10^{-4}$ m^{-1} was recorded by Tubman and Suhayda [1976]. At a profile (Fig. 12.16b) in a very different physical setting near Triangular Marsh at Corte Madera within San Francisco Bay [Liang and Williams, 1993] nearly the same value of k_i was estimated.

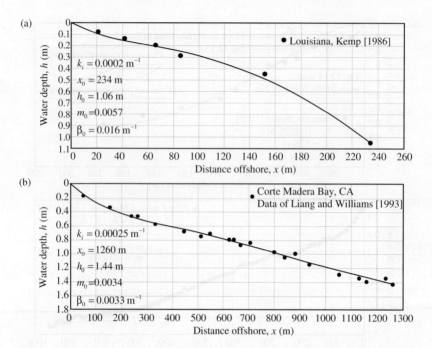

Figure 12.16. Comparison of Eq. 12.14 with measured profiles from: (a) Louisiana coast by Kemp [1986]; (b) Corte Madera, CA [Liang and Williams, 1993].

Example 12.2: Estimate the wave damping coefficient for mud from Mobile Bay, Alabama based on the following rheometric data. Assume that the wave period is 8.9 s and the mud is a Kelvin–Voigt material (Chapter 5).

Mud density, ρ (kg m^{-3})	η' (Pa.s)	η'' (Pa.s)
1,139	1.14×10^{-1}	1.40×10^{2}
1,204	2.51×10^{0}	1.69×10^{3}
1,302	1.32×10^{1}	1.05×10^{4}

Source: Lee and Mehta [1997].

Equation 11.139 can be restated as

$$k_i = \frac{\rho g \eta'}{4\sigma(\eta'^2 + \eta''^2)} \qquad (12.15)$$

Calculated values of k_i given below are plotted against mud (wet bulk) density in Fig. 12.17. The wave damping coefficient changes by two orders of magnitude as a transition occurs in the state of mud, from fluid-like at 1,139 kg m^{-3} with a high potential for wave damping, to a (soft) solid at 1,302 kg m^{-3} with a lower ability to damp waves (Fig. 12.17).

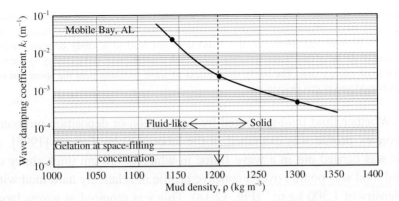

Figure 12.17. Wave damping coefficient against density for mud from Mobile Bay, Alabama.

The space-filling wet bulk density appears to be 1,200 kg m^{-3} (Chapters 5 and 10). Phase change at this density supports the general observation that in the absence of supply of mud from external sources, bed liquefaction would be required for wave damping to significantly increase above that over a solid bottom.

Mud density, ρ (kg m^{-3})	k_i (m^{-1})
1,139	2.3×10^{-2}
1,204	3.7×10^{-3}
1,302	5.4×10^{-4}

The parameters m_0, β_0 and k_i in Eq. 12.14 have been related to profile shape using 96 surveys from Malaysia, China and the US. The mean values and standard deviations are given in Table 12.5. In general, for a given length x_0, high values of k_i are associated with the concave ("erosional") profile, and as k_i decreases and approaches zero the profile becomes convex and ("accretionary"). This trend can be subjectively explained in terms of profile response to waves. A high k_i indicates the presence of fluid mud, which can be carried away. Accordingly, a high k_i is associated with large wave height and bed

Table 12.5. Coefficients in Eq. 12.14 for concave and convex profile shapes.

Profile shape	m_0		β_0 (m^{-1})		$K_i = k_i x_0$	
	Mean	Standard deviation	Mean	Standard deviation	Mean	Standard deviation
Concave (81 profiles)	0.059	0.083	0.046	0.054	0.42	0.13
Convex (15 profiles)	0.026	0.019	0.015	0.0084	0.016	0.027

Source: Lee [1995].

liquefaction (erosion). Conversely, a low k_i implies low wave height and accretionary profile in an environment favoring deposition. Also, decreasing k_i correlates with increasingly rigid bed. However, this trend does not lead to a sandy concave profile because in the latter case wave breaking rather than absorption of wave energy by the bed is the main energy sink. The likely small effect of wave breaking near the shore has not been considered in this description.

Whether fluid mud generated by liquefaction or deposition promotes a convex profile was examined in the laboratory by Tarigan [1996]. A test was carried out in a wave flume in which a ~10 cm thick slurry of fluid mud was poured over a concave profile of relatively hard mud with a density of 1,300 kg m^{-3} (Fig. 12.18). This was required as waves large enough to liquefy mud could not be produced in the flume. The slurry was made of a 1:1 mixture of attapulgite and kaolinite clays (AK mud) at a density of 1,120 kg m^{-3}. Low waves of 6 cm height and 1.2 s period were conducive to retention of fluid mud. During 72 h of wave movement the profile acquired a seemingly convex shape, in part due to seaward transport of a portion of fluid mud from the area of erosion close to the shoreline. This behavior contrasted with a second test employing higher waves (0.14 m height, 1.2 s period and duration 53 h) in which the entire fluid mud layer moved seaward and exposed the original concave profile (Fig. 12.18). An inference from these two tests was that profile change observations were qualitatively not inconsistent with the basis of Eq. 12.14.

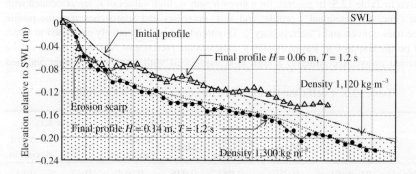

Figure 12.18. Redistribution of placed fluid mud by waves. Results from two tests are included, one with low waves conducive to retention of fluid mud and another with higher, erosional waves (from Tarigan [1996]).

12.2.3.6 *Profile change*

If the wave intensity increases it will erode the profile to form a new shape. Modeling this change can be done by a closed-loop or an open-loop approach. In a closed-loop model the initial profile, not in equilibrium with the more intense waves, will erode to a pre-specified target profile consistent with the new wave condition. An open-loop model is not constrained to converge to a target profile, and reaches a shape that is unknown ahead of time. Thus when the target profile is known, the closed-loop approach offers a "foolproof" morphodynamic method to determine the shapes of the intermediate profiles and the time to reach the final (target) profile. The downside is that the transport process is assumed to follow an assumed path.

Here we will briefly examine profile change leading to the target profile. The governing dynamic equation is formulated on the premise that cross-shore transport varies with the deviation of the wave–mean rate of energy dissipation per unit area ε_D from its equilibrium (target) value ε_{De}, *i.e.*,

$$q_s(x) = K_q \left| \varepsilon_D(x) - \varepsilon_{De} \right|^{n_q} \cdot \text{sign}[\varepsilon_D(x) - \varepsilon_{De}] \tag{12.16}$$

where $q_s(x)$ is the unit volumetric discharge of sediment (Chapter 6) in the cross-shore direction and K_q is a dimensional rate-scaling constant [Dean and Dalrymple, 2002]. The sign term preserves the landward/seaward direction of $q_s(x)$ depending on the difference $\varepsilon_D - \varepsilon_{De}$. At equilibrium $-q_s(x) = 0$, $\varepsilon_D = \varepsilon_{De}$, where the latter is obtained from Eq. 12.12. Using Eq. 12.13 (thereby ignoring the localized effect of wave breaking at the shoreline) and conveniently choosing the linear model ($n_q = 1$), Eq. 12.16 becomes

$$q_s(x) = K_q \left| \varepsilon_D(x) - \varepsilon_{De} \right| \cdot \text{sign}[\varepsilon_D(x) - \varepsilon_{De}] \tag{12.17}$$

or

$$q_s(x) = K_q \frac{\rho g^{3/2} a_{x_0}^2}{2} \left[e^{-2k_i x_0} \frac{d}{dx} \left[e^{2k_i x} \sqrt{h(x)} \right] - \frac{\sqrt{h_0}}{x_0} \right] \cdot \text{sign}[\varepsilon_D(x) - \varepsilon_{De}] \tag{12.18}$$

In order to obtain $q_s(x)$ and $h(x)$, this equation is coupled with the depth–mean form of the volumetric sediment continuity equation

$$\frac{\partial x}{\partial t} = \frac{\partial q_s}{\partial h} \qquad (12.19)$$

For solving the finite-difference forms of Eq. 12.18 and 12.19, Lee [1995] used a double-sweep algorithm treating h, t as the independent variables and x, q_s as dependent ones [Dean and Dalrymple, 2002; Rodriguez, 2000]. As a result, after every iterative time-step the model output was the distance x at which a particular depth h occurs. In other words the bottom contours were directly tracked in time and distance from shoreline.

An example based on profile data from a 33.5 m long and 0.6 m wide flume using AK sediment is shown in Fig. 12.19. The bed density was 1,300 kg m^{-3}, the incident wave height 0.20 m and the period 1.2 s. The initial and final (at 52 h) measured profiles are plotted along with the simulated final profile. In the experiment, eroded sediment had deposited (not shown) seaward of the 18 m distance. Parameters selected for simulation are given in figure inset. Simulation is not valid in the surf zone landward of the breakerline. The initial and final model profiles are smooth and capture the overall trend of erosion represented by the hatched area. The simple model does not permit the simulation of local undulations in the laboratory profile, which may have resulted from gelation of the newly deposited sediment.

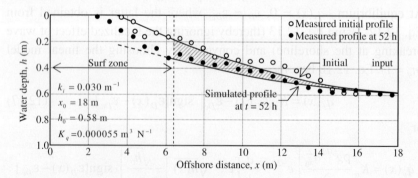

Figure 12.19. Simulation of clayey sediment profile change in a flume (adapted from Lee and Mehta [1997]).

12.2.3.7 *Tide effects*

At tide-dominated mudflats, combinations of longshore and cross-shore components of sediment load, water level change and waves collectively determine accretionary and erosional morphology. An understanding of the sediment load pathways is achieved from modeling water level and currents over mudflats (*e.g.,* Roberts *et al.* [2000], Le Hir *et al.* [2000]). In Fig. 12.20a the mudflat transport regime is characterized by relative magnitudes of flow velocity (u in the x-direction and v in the y-direction) and critical friction velocity u_{*_c} at the threshold of sediment movement.

Starting with an arbitrary uniform slope and deposition dominated transport, accretionary growth lines represent laminas, or rhythmites, of deposits formed under receding tide. As the profile becomes increasingly convex its shape approaches an equilibrium or near-equilibrium configuration. The shore zone with low currents (u and $v < u_{*_c}$), is

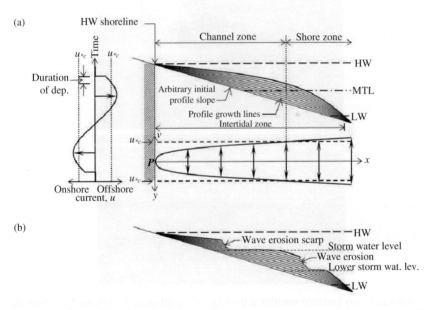

Figure 12.20. (a) Mudflat profile accretion due to cross-shore and longshore tidal currents; (b) erosional terrace development due to storm wave effects at different water levels.

relatively flat compared to the channel zone,[f] in which a steeper profile is molded by deposition balancing erosion due to higher currents in the channel. Point *P* at the intersection of the profile with the shoreline traverses along with the shoreline as water rises and falls. The shape of the profile varies with the phase lag between water level and current.

Loss of sediment from the profile is enhanced by wave episodes, with breaking resulting in erosion scarps. Under the impacts of successive storms at different water levels and insufficient time to rebuild the profile between storms, a terraced structure (Fig. 12.20b) can develop. Figure 12.21a shows such a mudflat at a tidal creek along the western coast of Korea. A close view (Fig. 12.21b) of an erosion scarp at the site reveals mud rhythmites formed by accretion. The thinnest rhythmites are on the order of 2 mm. Similar scarping is evident in Fig. 9.1 in the Severn River (UK) estuary.

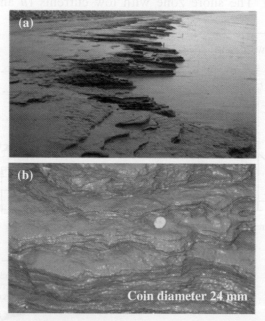

Figure 12.21. (a) Terraced mudflat at the edge of a tidal creek in Incheon Bay along the west coast of Korea; (b) mud rhythmites formed by sequential deposition.

---

[f] The submerged mud mound seaward of low water level is a mudbank. The exposed profile landward of high water level is the shore face.

In Fig. 12.20a, the intertidal zone is the profile between low and high water levels. Thus far we have referred mainly to shore profiles with the assumption that the profile shape does not change in the along shore direction, the shoreline is straight and, as consequence, the profile hypsometry, *i.e.*, the cumulative horizontal bottom area below a given depth, is linear. This is sketched in Fig. 12.22a in which the hypsometry is plotted in terms of depth h normalized by the height of the intertidal zone h_m against horizontal area A below depth h normalized by the maximum area A_m at the base. Examples of actual hypsometric curves from tide-dominated locations are shown in Fig. 12.23.

The effect of tide on the intertidal profile was examined by Friedrichs [1993] under the assumption that a tidal flat is in morphodynamic equilibrium when the maximum (tide-induced) bed shear stress is constant, *i.e.*, uniform, over the length of the profile. The analysis makes use of the maximum tidal current velocity as a surrogate for shear stress. The derivation is independent of sediment size, which means that the analysis does not refer to sediment being either coarse or fine.

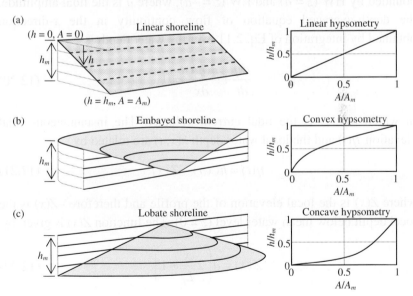

Figure 12.22. Schematic shore profiles and hypsometric plots for: (a) linear shoreline; (b) embayed shoreline; (c) lobate shoreline (adapted from Boon and Byrne [1981]).

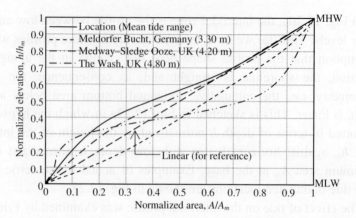

Figure 12.23. Examples of intertidal hypsometric curves in high-tide-range estuaries based on data reported in Dieckmann *et al.* [1987] and Kirby [1992]. Note: Values of h_m and A_m are not available.

The simplest case is of profile and hypsometry at a linear shoreline. Referring to the definition sketch of Fig. 12.24 in which the profile is bounded by HW ($z = a$) and LW ($z = -a$), where a is the tidal amplitude, the depth-averaged equation of flow continuity in the x-direction obtained by integration of Eq. 2.11 over the water depth is

$$\frac{d\eta}{dt} + \frac{\partial(hu)}{\partial x} = 0 \qquad (12.20)$$

in which $u(x, t)$ is the tidal current velocity. The instantaneous tidal elevation $\eta(t)$ and the local water depth $h(x, t)$ are related by

$$\eta(t) = h(x,t) - Z(x) \qquad (12.21)$$

where $Z(x)$ is the local elevation of the profile and therefore $-Z(x)$ is the local depth below mean water level ($z = 0$). The function $Z(x)$ is given by

$$Z(x) = a\left(2\frac{x}{L_I} - 1\right) \qquad (12.22)$$

where L_I is the projection of the profile onto the x-axis.

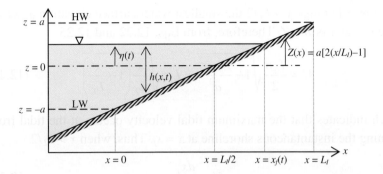

Figure 12.24. Definition sketch of instantaneous tidal water level at a linear shore profile.

Equation 12.20 is integrated to solve for $u(x, t)$ giving

$$u(x,t) = \frac{x_f(t) - x}{h(x,t)} \frac{d\eta(t)}{dt}$$

(12.23)

where x_f is the instantaneous position of the shoreline. Let the tidal variation be specified as

$$\eta(t) = a \sin \sigma t$$

(12.24)

with tidal frequency σ. Thus, from Eqs. 12.23 and 12.24 and profile geometry we obtain

$$u(t) = \frac{L_l \sigma}{2} \sqrt{1 - \frac{\eta^2}{a^2}}$$

(12.25)

which indicates that the maximum velocity U during a tidal cycle occurs when η^2 is a minimum.

In the profile range $x \le L_l/2$, η^2 is a minimum when $\eta = 0$, *i.e.*, the maximum current occurs at mid-tide. We obtain

$$U = U_0 = \frac{L_l \sigma}{2}$$

(12.26)

In the profile range $x > L_I/2$ the smallest water surface elevation at which there is water is $\eta = Z$. Therefore, from Eqs. 12.22 and 12.25

$$U = \frac{L_I \sigma}{2} \sqrt{1 - \frac{Z(x)^2}{a^2}} = L_I \sigma \sqrt{\frac{x}{L_I} \left(1 - \frac{x}{L_I}\right)} \tag{12.27}$$

which indicates that the maximum tidal velocity occurs at the tidal front defining the instantaneous shoreline at $x = x_f$. Thus, when $x > L_I/2$

$$U = \frac{dx_f}{dt} \tag{12.28}$$

In Fig. 12.25 the normalized current velocity U/U_0, where U_0 is the velocity in the range $x \le L_I/2$, is plotted against normalized profile distance x/L_I from LW ($x/L_I = 0$) to HW ($x/L_I = 1$). Since $U = U_0$ is constant when $x \le L_I/2$, it follows that the linear profile is in morphodynamic equilibrium in the seaward half of the intertidal zone. In the landward range of the profile ($x > L_I/2$), U decreases until it is zero at the front. If the velocity is to remain constant the depth must decrease due to net sedimentation.

To obtain the profile shape we will refer to the sketch in Fig. 12.26. Defining $x = L^*$ at $Z = 0$ we observe that the profile in the range $x \le L^*$ is linear, *i.e.,*

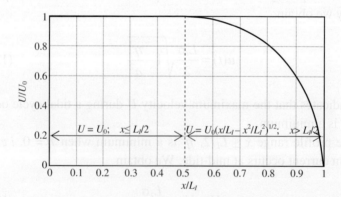

Figure 12.25. Variation of normalized maximum tidal velocity with normalized profile distance from LW shoreline (from Friedrichs and Aubrey [1996]).

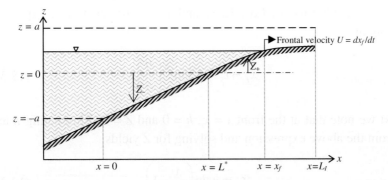

Figure 12.26. Definition sketch of equilibrium profile of a tidal flat at a straight shoreline. Note that Z is equal to Z_+ and Z_- in the positive and negative sense, respectively, with respect to mid-tide level.

$$Z(x) = a\left(\frac{x}{L^*} - 1\right) \tag{12.29}$$

and

$$u = L^*\sigma \tag{12.30}$$

at mid-tide. This means that U is at least as large as this value of u.

On the basis of Eq. 12.28, even for a non-linear profile in range $x > L^*$ it is reasonable to assume that the maximum velocity occurs at the front. It is therefore necessary to know the value of Z required to satisfy

$$U = \frac{dx_f}{dt} = L^*\sigma \tag{12.31}$$

To that end, following a water parcel at the front, from the chain rule we have

$$\frac{dx_f}{d\eta} = \frac{dx_f}{dt}\frac{dt}{d\eta} \tag{12.32}$$

for which $dt/d\eta$ is found from the geometric identity

$$t = \frac{1}{\sigma}\sin^{-1}\left(\frac{\eta}{a}\right) \tag{12.33}$$

Thus by making use of Eqs. 12.31 and 12.33 and integrating Eq. 12.32 we obtain

$$x_f - L^* = L^* \sin^{-1}\left(\frac{\eta}{a}\right) \tag{12.34}$$

Next we note that at the front $x = x_f$, $h = 0$ and $Z = \eta$. Eliminating η and x_f from the above expression and solving for Z yields

$$Z_+ = a \sin\left(\frac{x}{L^*} - 1\right) \tag{12.35}$$

Since $Z_+ = a$ at $x = L_I$, for an equilibrium profile

$$\frac{L^*}{L_I} = \frac{1}{2\pi + 1} \tag{12.36}$$

In Fig. 12.27a the profile at a straight shoreline is observed to be convex (relative to the linear reference line). As expected, the corresponding hypsometry in Fig. 12.27b is identical to the profile. The general similarity between the profile shapes obtained by this analysis and the one for mud profiles described in Section 12.2.3.5 is not surprising. Both are closed-loop assessments describing approach to equilibrium dependent on a velocity-related variable. In the present case it is the maximum tidal velocity, and for mud profiles it is the energy loss rate which in turn is proportional to the cube of the wave-induced velocity. In both cases the difference between the instantaneous value of the variable relative to equilibrium is the driving force morphing the profile.

In order to extend the analysis to a lobate or embayed shoreline (Figs. 12.22b, c), the respective curvatures are defined by widths b_0 and b_L in Fig. 12.28. Since the isobaths are circular arcs, radial symmetry permits the use of the continuity equation in the polar coordinates. For

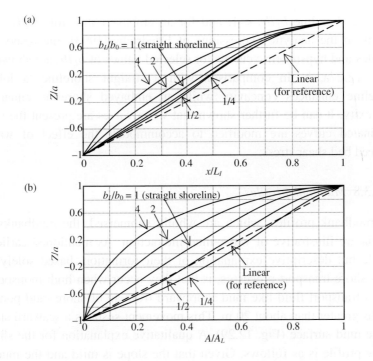

Figure 12.27. Normalized shore geometry: (a) profiles; (b) hypsometry (from Friedrichs and Aubrey [1996]).

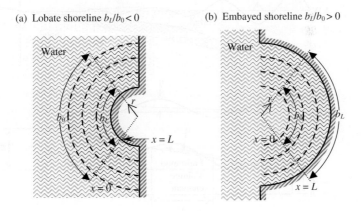

(a) Lobate shoreline $b_L/b_0 < 0$

(b) Embayed shoreline $b_L/b_0 > 0$

Figure 12.28. Idealized depth contours at lobate and embayed shorelines.

a lobate shoreline $b_L/b_0 < 1$, and for an embayed shoreline $b_L/b_0 > 1$. Based on the analytic expressions of Friedrichs [1993], the respective profiles and hypsometric curves for selected values of b_L/b_0 are shown in Figs. 12.27a, b. In comparison with a straight shoreline, a lobate shoreline enhances concavity while an embayed shoreline enhances convexity. It can be further shown that when waves are present the tide-dominated curves are modified to accommodate the effect of wave-induced bed shear stress.

12.2.3.8 *Mudbank profiles*

For mudbank profiles, which are wholly submersed, the mudbanks of Kerala are illustrative of wave–mud interaction. As mentioned earlier a simple but descriptive explanation of their formation, based solely on cross-shore transport processes, is that they occur when high monsoonal waves transport fluid-like mud shoreward from an offshore mud pool in depths greater than about 20 m. This movement sustains a seaward slope of the mud surface (Fig. 12.29). A qualitative explanation for the shape of the profile is as follows. Given that the slope is mild and the mud is

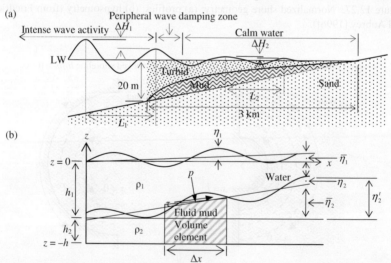

Figure 12.29. Schematic drawings of: (a) a typical monsoonal mudbank profile along the southwest coast of India in Kerala (adapted from Nair [1988]); (b) fluid mud elemental volume.

fluid-like, the net forward thrust due to waves is smaller over wave length L_2 than over L_1 because the decrease in the wave height ΔH_1 over L_1 is greater than ΔH_2 over L_2. Generalizing this over the entire cross-shore length of the mudbank, one could argue that the mudbank shape mimics the wave height (in the negative sense) over the same distance.

In a flume a flat bed of kaolinite (with wet bulk density 1,200 kg m^{-3}) was subjected to waves and the bottom profile measured at different times [Jiang, 1993]. Three profiles shown in Fig. 12.30 indicate that due to net transport of sediment the profile evolved although after 10 hours there appeared to be practically little change. When the waves ceased the bed did not become flat because the mud had hardened due to thixotropy, and its consistency was no longer sufficiently fluid-like for downslope flow in the offshore direction.

When the mudbank persists as a viscous fluid, the equilibrium profile over a mildly sloping bottom can be estimated from a balance between the wave-induced shoreward force and seaward hydrostatic pressure force. By applying the force balance to a volume element of differential length Δx (Fig. 12.29b), profile shape was calculated by Jiang [1993]. For this the equation of motion in the x-direction combined with flow continuity is

$$\frac{\partial u}{\partial t} + \frac{\partial u^2}{\partial x} + \frac{\partial (uw)}{\partial z} = -\frac{1}{\rho_2}\frac{\partial p_t}{\partial x} + \nu_2\left(\frac{\partial^2 u}{\partial x^2} + \frac{\partial^2 u}{\partial z^2}\right) \quad (12.37)$$

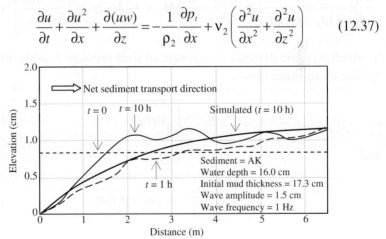

Figure 12.30. Measured and calculated (curve) mud–water interface profiles in a wave flume (from Jiang [1993]).

where u and w are the horizontal and vertical components of water velocity, respectively, and p^t is the total pressure. Let the total depth be $h = h_1 + h_2$ and the instantaneous elevation of the interface $z_i = -h_1 + \eta_2^t$. Equation 12.37 is now integrated over the mud layer from the bottom at $z = -h$ to the surface at $z = z_i$, *i.e.*,

$$\int_{-h}^{z_i} \rho_2 \frac{\partial u}{\partial t} dz + \int_{-h}^{z_i} \rho_2 \frac{\partial u^2}{\partial x} dz + \int_{-h}^{z_i} \rho_2 \frac{\partial (uw)}{\partial z} dz$$

$$= -\int_{-h}^{z_i} \frac{\partial p_t}{\partial x} dz + \int_{-h}^{z_i} \rho_2 \nu_2 \left(\frac{\partial^2 u}{\partial x^2} + \frac{\partial^2 u}{\partial z^2} \right) dz \qquad (12.38)$$

Each term in this equation is evaluated by integration, given the no-slip ($u = 0$) boundary condition at the bottom along with equalities of viscous shear stress τ and pressure p at the interface between fluid mud and water. The mean water surface slope is ignored in comparison with the steeper mean slope of the mud-water interface, *i.e.*, $\partial \bar{\eta}_1 / \partial x \ll \partial \bar{\eta}_2 / \partial x$. Then, after time-averaging the resulting terms over the wave period and solving for the mean interfacial slope yields

$$\frac{\partial \bar{\eta}_2}{\partial x} = \frac{1}{(\rho_2 - \rho_1) g h_2} \left[\overline{p_{z_i} \frac{\partial \eta_2}{\partial x}} - \frac{\partial}{\partial x} \left(M_f + E_m \right) - 2\rho_2 \nu_2 \overline{\frac{\partial u_{z_i}}{\partial x} \frac{\partial \eta_2}{\partial x}} \right] \qquad (12.39)$$

in which p_{z_i} is the dynamic component of total pressure p_t at the interface z_i,. The momentum flux is

$$M_f = \overline{\int_{-h}^{z_i} \rho_2 u_2^2 dz} \qquad (12.40)$$

and

$$E_m = g \left[\left(\frac{\rho_2}{2} - \rho_1 \right) \overline{\eta_1^2} + \rho_1 \overline{\eta_1 \eta_2} \right] \qquad (12.41)$$

Thus E_m includes interaction between the surface water wave and the interfacial mud wave in terms of time-averaged product $\overline{\eta_1 \eta_2}$. By calculating the slope $\partial \bar{\eta}_2 / \partial x$ from Eq. 12.39 and integrating it in the *x*-direction we obtain the desired profile $\overline{\eta}_2(x)$ of the mudbank. All terms involving elevations, velocities and interfacial dynamic pressure were then calculated by a wave–mud interaction model of second-order accuracy similar to that of Jain [2009b] mentioned in Chapter 11. The measured mud profile at 10 hours close to the equilibrium shape is included in Fig. 12.30. The overall curvature of the profile depends on the wave height, period and water depth as well as mud thickness, density and viscosity.

12.2.3.9 *Variability in particle size and turbidity*

Tides and waves together regulate the distribution of particle size at muddy beaches. The winnowing action of flowing water causes the leftover particle size to increase with wave height and current speed. Parameters in Table 12.6 are from the low-tidal energy clayey flats along Bohai Gulf and moderate-energy silty flats of the Jiangsu coast in China. The tidal range and wave height are higher at the Jiangsu coast than at the Bohai Gulf. As a result, although silt and clay are present at the latter flats, the clay fraction has been reduced due to winnowing by waves. Similarly, winnowing mainly by waves is manifested in the Jiangsu coast profiles shown in Fig. 12.31 for which the source of mud is the Huang He (Yellow River). The convex profile A indicates a stable or accretionary state. In contrast, the concave profile B, about 150 km north of the stable profile, indicates severe erosion due to storm waves. Accompanying the steepening and lowering of the profile is coarsening of surface sediment [Bird, 1993].

Table 12.6. Characteristics of muddy tidal flats in China.

Site	Tidal range (m)	Tidal current (m s^{-1})	Wave height (m)	Wave period (s)	Intertidal width (km)	Intertidal slope	Sediment	Source
Bohai Gulf	3	0.2– 0.6	0.1– 0.4	1–4	3–6	0.7	Silty clay	Wang *et al.* [1988]
Jiangsu coast	2–4	0.4– 1.0	0.6– 1.2	4–6	3.3 (erosion) 7.7 (accretion)	1.3 0.2	Silt	Wang *et al.* [1988]

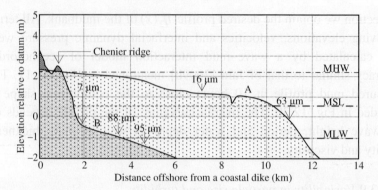

Figure 12.31. Mudflat profiles and sediment size along the Jiangsu coast of China. A: stable or accretionary flat; B: receding flat (adapted from Ren [1992]).

As the squares of tidal amplitude and wave height are proportional to the respective energies (Chapter 2), one would expect a correlation between the squared quantities and particle diameter. In the low tidal energy environment, coarsening of sediment with increasing wave energy is evident in Fig. 12.32. In the present case this could mean that waves erode the fine material which is carried away by the tidal current. When tidal energy is high the sediment appears to be finer, as observed from the bar representing the western coast of Korea. This suggests that in this instance tide supplies fine sediment. When both tide and waves

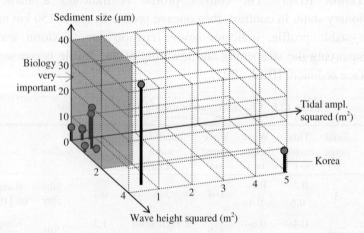

Figure 12.32. Dispersed particle diameter plotted against squares of wave height and tidal amplitude (from Mehta [2002]). In the darkened zone biology may play a dominant role.

are too weak to drive sediment, biological influence on profile stability can become dominant (Chapter 9).

At mudbanks, the relationship between tide, waves and sediment is complicated by the potential for mud mobility, and may explain why data in Table 12.7 yield no obvious relationships between particle diameter, tidal range and wave height. Due to the movement of fluid-like mud at the Kerala mudbanks (Fig. 12.33), sediment sizes measured in different months of the year do not seem to correlate with mudbank location [Narayana *et al.*, 2001]. These plots show the spread of the mud patch without reference to its thickness. Therefore, the effectiveness of the patch in reducing wave energy cannot be inferred.

Based on long-term wave height measurements at an offshore site, Mathew [1992] made inferences related to the dynamic behavior of mudbanks. Although that analysis is limited by the absence of synchronous bottom surveys, it is instructive to mention the seasonal variability of suspended sediment concentration. According to Mathew, when a mudbank appears at the beginning of the monsoon the turbidity decreases as wave energy is absorbed by mud. In a later phase of the

Table 12.7. Characteristics of selected mudbanks.

Location	Tidal range (m)	Waves Max. height (m)	Period (s)	Dimensions Along-shore (km)	Cross-shore (km)	Particle diameter (μm)	Soft mud thickness (m)	Soft mud density (kgm^{-3})	Mud source	Source
Louisiana	0.5	1	5–7	1–5	0.5–3	3–5	0.2–1.5	1,150–1,300	River	Wells [1983], Kemp [1986]
Kerala (India)	0.6	0.3	6–9	2–8	1–3	0.5–3	0.5–2	1,080–1,300	Offshore	Mathew [1992], Shepard [1963]
Suriname	2	1.5	6–9	10–20	10–20	0.5–1	0.5–2	1,030–1,300	River	Wells [1983], Augustinus [1978]
Korea	5–9	2	6–9	1–30	2–30	6–11	0.1–3	1,200–1,300	River	Wells [1983]

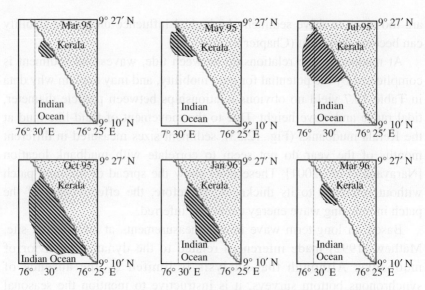

Figure 12.33.   Seasonal variability of Ambalapuzha mudbank (from Narayana *et al.* [2001]).

monsoon, weak currents raise the turbidity by increasing fluid mud entrainment. Suspended sediment concentration profiles in Fig. 12.34 were simulated by a (numerical) solution of Eq. 6.114 for different significant wave heights H_s and depth–mean current speeds u_m [Li and Parchure, 1998]. The panels show profiles before large mudbank formation (November–May; Fig. 12.34a), during formation in early monsoon (June–July; Fig. 12.34b), during fully formed mudbank (July–September; Fig. 12.34c), early dissipating mudbank (September–October; Fig. 12.34d) and finally later stages of dissipation (October–November; Fig. 12.34e). Note the effect of a very small current (0.01 m s^{-1}) in raising the concentration (Fig. 12.34c). Since in some years the mudbanks are, area-wise, small or thin during November–May, under fair weather and milder sea state the concentration is reduced but remains higher than during the time of formed mudbanks when waves are nearly absent in the mudbank area. The sequence is repeated with inter-annual variability as might be expected. It is noteworthy that when mudbanks do return they tend to occur at about the same locations along

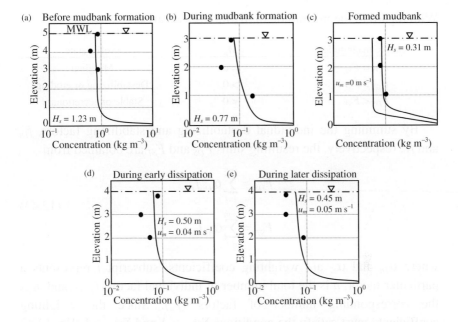

Figure 12.34. Measured (dark circles) and simulated (curves) suspended sediment concentrations: (a) Before mudbank; (b) during mudbank; (c) when mudbank in fully formed; (d) during the early stage of mudbank dissipation; (e) during later stages of mudbank dissipation [Mathew, 1992; Li and Parchure, 1998].

the coast. The refraction pattern of incoming swell has been proposed as a reason for this repeatability [Mathew and Baba, 1995].

12.2.3.10 *Shore profile stability index*

The stability of a shore profile (excluding the submerged mudbank), as defined in relation to whether the profile will accrete sediment, remain unchanged or erode, may be semi-quantitatively characterized by the ratio of a shore stabilizing factor F_{ST} divided by a shore destabilizing factor F_{DS}. A profile stability index M_K can then be specified as

$$M_K = 1 - \frac{F_{ST}}{F_{DS}} \tag{12.42}$$

With regard to the magnitude of M_K three cases arise:

Relative magnitudes	M_K	Profile state
$F_{DS} = F_{ST}$	$= 0$	Marginal
$F_{DS} > F_{ST}$	> 0	Destabilizing or eroding
$F_{DS} < F_{ST}$	< 0	Stable or accretionary

By summing the individual destabilizing and stabilizing factors, f_{Di} and f_{Si}, respectively, the resulting sums F_{DS} and F_{ST} are obtained from

$$F_{DS} = \sum_{i=1}^{i=n} \alpha_{Di} f_{Di}$$

$$F_{ST} = \sum_{i=1}^{i=m} \alpha_{Si} f_{Si}$$

(12.43)

where α_{Di} and α_{Si} are weighting coefficients, subscript i represents a particular factor, n is the total number of individual factors (f_{Di}), and m is the corresponding number of factors (f_{Si}). Since the weighting coefficients must satisfy the conditions $\Sigma\alpha_{Di} = 1$ and $\Sigma\alpha_{Si} = 1$, Eq. 12.42 becomes

$$M_K = 1 - \frac{\displaystyle\sum_{i=1}^{i=n} \alpha_{Si} f_{Si}}{\displaystyle\sum_{i=1}^{i=m} \alpha_{Di} f_{Di}}$$

(12.44)

To calculate M_K from Eq. 12.44 the weighting coefficients must be known. For a broad categorization of stability, the effectiveness of the individual factors contributing to M_K may be considered to assume the following relative numerical values and effects on stability:

Value	Effect
0	None or low
1	Moderate
2	Significant
3	Highly significant

We will further assume that "moderate" effect for any particular factor represents the equilibrium condition ($M_K = 0$), *i.e.*, a non-eroding–non-accreting profile. Thus, in Eq. 12.44 setting $M_K = 0$ and each f_{Di} and

f_{Si} to unity yields $0 = 1 - (\sum \alpha_{Si} / \sum \alpha_{Di})$, which is consistent with the definition of the weighting coefficients. Thus they represent relative magnitudes of the various factors when they individually have moderate effects.

Twenty influential factors relevant to eighteen intertidal mudflats in northwestern Europe were identified by Dyer *et al.* [2000]. The effects of tide, waves and sediment supply on profile shape have been examined by solving the governing equations of flow momentum, continuity and sediment transport in the cross-shore direction [Roberts *et al.*, 2000]. Properties of mudflat surface and sediment distribution at scales much smaller than the profile length are found to affect profile shape. In part this is due to the complex interactions between morphology and biota.

Among the numerous factors defining shore stability relevant to Eq. 12.44, the ones in Table 12.8 are noteworthy. Tides, waves, storm surge and intrusive structures can contribute to erosion. Since storm surge data are not readily available, one may consider waves as surrogates for storm surge. Among biological processes, bioturbation is often the most important destabilizing factor (Fig. 9.33), although biochemical production of gas, *e.g.*, methane, can also destabilize the bottom. Seasonal effects complicate biologically driven systems, which can be dominant in low wave and tidal energy environments (Fig. 12.32).

Sediment supply is a significant factor governing shore stability because, regardless of the erosive force, a profile can remain unchanged or accrete as long as the rate of sediment supply equals or exceeds the rate of depletion. If supply becomes insufficient, the likelihood that a

Table 12.8. Profile destabilizing and stabilizing factors.

Destabilizing factors	Stability influencing factors
1 Waves and storm surge	1 Sediment supply
2 Structures	2 Bottom hardness
3 Tide	3 Structures
4 Biological processes	4 Morphodynamic control
	5 Sediment composition
	6 Vegetative cover
	7 Biological processes

shoreline will remain static is low especially in the long run (except for instance, by hardening with structures). Bottom hardness, as defined by standard measures of soil strength (Chapter 10), can be important in distinguishing between the effects of well-consolidated versus poorly consolidated beds.

Some structures promote stability by sheltering the shore from erosive forces, and their role can be as prominent as that of bottom hardness. Other structures tend to have the opposite effect, namely destabilization by reduction or elimination of sediment supply. Control is exerted by the offshore bathymetry on waves and on alongshore water and sediment transport. Shoreline configuration can be equally significant, as for instance defined by coastal promontories which either cause erosion of the downstream shoreline or sequester sediment by forming pocket beaches. Sediment composition partly determines hardness, although it also varies with consolidation. On the other hand, given two beds of the same density, the less cohesive bed may erode more easily.

Vegetative canopies increase stability, as do certain benthic processes, *e.g.,* surficial mats produced by secretions including mucopolysaccharides. Tide can increase stability through intertidal wetting-and-drying, as desiccation and weathering tend to enhance profile hardness by encrustation. Drying causes a reduction in pore water pressure, which increases the stress between particles and over-consolidates clays. Oxidation may also contribute to over-consolidation.

From inspections of the effects of various factors at muddy coasts the following expression for M_K is introduced

$$M_K = 1 - \frac{0.25 f_{Ssed} + 0.20 f_{Sbh} + 0.15 f_{Sst} + 0.15 f_{Smor} + 0.10 f_{Scom} + 0.10 f_{Sveg} + 0.05 f_{Sbio}}{0.40 f_{Dwv} + 0.30 f_{Dst} + 0.25 f_{Dtc} + 0.05 f_{Dbio}}$$

(12.45)

where the subscripts are: *wv* for waves, *st* for structures, *tc* for tide and tidal currents, *sed* for sediment supply, *bh* for bottom hardness, *mor* for morphology, *com* for sediment composition, *veg* for vegetation and *bio* for biophysicochemical effects [Mehta and Kirby, 1996].

In order to assess the applicability of Eq. 12.45, comprehensive data sets for evaluating the coefficients and factors are required. We will illustrate how this might be accomplished by considering diagnostic examples giving consideration only to the most important factors.

The Gulf of Mexico shore of Louisiana near Cheniere au Tigre undergoes seasonal fluctuations due to variable wave climate, and its mean position is governed by mud supply derived from the Mississippi River. In this usually moderate coastal environment, the mean astronomical tidal range of 0.5 m is modulated by frontal wind-induced oscillations, especially during the winter months. The shore, backed by saltmarshes, has biologically active mudflat morphology with particles in the 1 to 5 μm size range, and is dominated by fluid mud. On the basis of a scale of effects ranging from 0 to ± 3 we will assign f_{Dwv}, f_{Ssed}, f_{Sbh}, f_{Sveg} and f_{Sbio} the value 1, and the remaining factors the value 0. Equation 12.45 then yields $M_K = -0.50$, reflecting the observed seasonal-mean stability of the shoreline, notwithstanding longer term changes due to the rapid rise in the relative sea level in this region (*e.g.,* Gornitz [1995]).

Estimates of M_K for coastal Louisiana and other sites including Kerala, Indonesia, and Suriname are given in Table 12.9. As noted earlier, the Kerala mudbanks are enhanced by monsoonal waves but are much less prominent in fair weather. The shorelines of Indonesia at Teluk Waru, Madura and Surabaya are partly sheltered against waves by neighboring islands, but due to the lack of sufficient sediment supply they display marginal stability. Notwithstanding the example from Kerala, which is somewhat unique in terms of the transient nature of the mudbanks, if a long established muddy coast is presently eroding, it would be natural to look for possible human causes of the altered state of the shoreline. For example, at the Selangor coast of Malaysia, $M_K = 0.67$. This once stable, mangrove-fringed coast has been eroding in recent decades due to reduced sediment supply from rivers resulting from the diversion of water for agricultural use.

Table 12.9. Calculated profile stability index values versus observed stability.

Location	Stability index, M_K	Observed profile stability
Louisiana	–0.5	Generally stable, with seasonal variability
Kerala (monsoon), India	Close to 0	Shore-attached mudbanks occur at specific locations
Kerala (fair weather), India	0.75	Mudbanks are absent
Teluk Waru, Indonesia	Close to 0	Marginally stable environment
Suriname	Close to 0	Mudbanks are stable over short term; translate alongshore over a decadal time scale
Selangor, Malaysia	0.67	Eroding coast
Madura, Indonesia	Close to 0	Marginally stable environment
Surabaya, Indonesia	Close to 0	䂞䂞䂞ginally stable environment

Data sources: Wells [1983], Kemp [1986], Midun and Lee [1989], Eisma *et al.* [1991], Mathew and Baba [1995], Tarigan [1996].

12.3 Nearshore Sediment Transport

12.3.1 *Longshore sediment discharge*

The simulation of mud movement shaping beach profiles by erosion-and-accretion cycles generally requires modeling cross-shore as well as longshore sediment fluxes. To estimate the longshore sediment flux we will use the reference frame and definitions in Fig. 12.35.

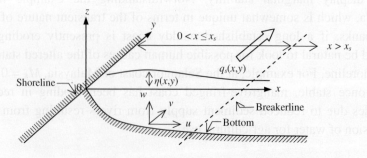

Figure 12.35. Longshore sediment transport definitions.

The wave–mean suspended sediment mass transport rate per unit cross-shore distance $q_s(x,y)$, *i.e.*, unit sediment discharge as dry sediment mass per unit width of the surf zone and unit time, is given by

$$q_s(x,y) = \frac{1}{T}\int_0^T \int_{-h(x,y)}^{\eta(x,y)} C(x,y,z,t)u_{sed}(x,y,z,t)dzdt \qquad (12.46)$$

in which x and y are the cross-shore and alongshore coordinates, respectively, z is the vertical coordinate with reference to mean water level, T is the wave period, η is the local water surface elevation, C is the suspended sediment concentration and u_{sed} is the velocity of the particles.[g] The following approximate analysis is revised from Rodriguez [2000].

For steady, wave–mean transport in the y-direction, Eq. 12.46 can be simplified to obtain

$$q_s(x) = \alpha_u \overline{C}(x)V(x)h(x) \qquad (12.47)$$

where $\overline{C}(x)$ is the depth–mean value of C and $V(x)$ is the depth–mean longshore current velocity.[h] Coefficient α_u arises because the integral in Eq. 12.46 has been replaced by the product of $\overline{C}$ and V, and also because V has been used in place of u_{sed}. To calculate $q_s(x)$, $\overline{C}(x)$ and $V(x)$ (and α_u) must be evaluated.

---

[g] In general, particle velocity differs from water velocity. However, when particle size decreases below 80-100 μm the two velocities may be assumed to be equal. When this occurs, sediment transport is called iso-kinetic [Jobson and Sayre, 1970].

[h] As waves break in the surf zone, some or all the wave energy is converted into heat, but wave momentum is conserved. Longshore current is driven by the alongshore component of the wave momentum flux. A well-rounded review of the mechanics of longshore current is found in Komar [1998].

12.3.2 *Cross-shore suspension concentration variation*

12.3.2.1 *General equation for concentration*

We will ignore the influence of cross-shore advection on resuspension and assume a fixed water level (at $z = 0$). Suspended sediment concentration C at any position along the x-direction is obtained from the local balance between sediment erosion and deposition

$$D_{so} \frac{\partial C}{\partial z} + w_s C = 0 \tag{12.48}$$

which is simplified from Eq. 6.114. Assuming a constant settling velocity w_s this equation is integrated to yield

$$\bar{C} = \frac{C_a D_{so}}{h w_s} \left(1 - e^{-\frac{w_s h}{D_{so}}} \right) \tag{12.49}$$

where C_a is a near-bed reference concentration and $h(x)$ is the local water depth. The diffusion coefficient D_{so} will be obtained from

$$D_{so} = \frac{\alpha_w H^2 \sigma}{6} \tag{12.50}$$

where α_w is an empirical coefficient, $H(x)$ is the wave height and σ is the wave angular frequency [Hwang, 1989]. To solve for the depth–mean concentration $\bar{C}$ it is necessary to determine the reference concentration C_a, for which the surf zone ($0 < x \le x_s$) must be treated separately from the offshore zone ($x > x_s$). For simplicity Eq. 12.50 is selected for both zones with the understanding that in the surf zone H denotes the breaking wave height and in the offshore zone the non-breaking wave height. Due to turbulence in the surf zone the bottom is an erodible bed with no fluid mud above it. In the offshore zone the bottom consists of fluid (or fluid-like) mud.

12.3.2.2 *Surf zone concentration*

In the surf zone at equilibrium, the zero net flux condition at the bottom is represented by the equality between erosion flux ε and settling flux $w_s C_a$, *i.e.*,

$$\varepsilon = w_s\, C_a \qquad (12.51)$$

where ε is given by Eq. 11.210. Therefore

$$C_a = \frac{\varepsilon_{W0}}{w_s}\left(\frac{H_b}{H_{bcr}}-1\right)^{\delta_{wb}} \qquad (12.52)$$

Substitution of Eq. 12.50 and 12.52 into Eq. 12.49 gives

$$\bar{C} = \frac{\varepsilon_{W0}}{w_s}\frac{\alpha_w \kappa_b^2 m\sigma}{6w_s}\left(\frac{H_b}{H_{bcr}}-1\right)^{\delta_{wb}}\left(1-e^{-\frac{6w_s}{\alpha_w \kappa_b^2 m\sigma}\frac{1}{x}}\right)x;\ 0 < x \le x_s \qquad (12.53)$$

in which the breaking wave index κ_b is introduced from Eq. 2.149. For convenience we will make Eq. 12.53 non-dimensional by using the variables

$$\tilde{C} = \frac{\bar{C}}{\left(\dfrac{\varepsilon_{W0}}{w_s}\right)},\ \Delta\tilde{H}_b = \left(\frac{H_b}{H_{bcr}}-1\right)^{\delta_{wb}}\ ;\ \tilde{x}=\frac{x}{x_s},\ \tilde{\sigma}_{sz}=\frac{\alpha_w \kappa_b^2 m x_s}{6w_s}\sigma \qquad (12.54)$$

Thus, Eq. 12.53 becomes

$$\tilde{C} = \left(1-e^{-\frac{1}{\tilde{\sigma}_{sz}\tilde{x}}}\right)\Delta\tilde{H}_b\,\tilde{\sigma}_{sz}\tilde{x};\ 0 < \tilde{x} \le 1 \qquad (12.55)$$

12.3.2.3 *Offshore zone concentration*

In the offshore zone the entrainment flux ε of soft mud is taken as a simplified form of Eq. 11.210

$$\varepsilon = C_e \frac{\rho_m u_\delta}{Ri^n} \qquad (12.56)$$

where C_e and n are sediment-specific constants, ρ_m is the mud density, u_δ is the near-bottom wave velocity amplitude and Ri is the wave Richardson number Eq. 11.213. Following the approach used for the surf zone, the bottom concentration is obtained as [Rodriguez, 2000]

$$C_a = C_e \frac{\rho_m u_\delta}{w_s Ri^n} \qquad (12.57)$$

The unknown C_e is now obtained by requiring that at the breakerline ($x = x_s$) sediment concentration in the offshore zone must match that in the surf zone. Thus, from Eqs. 12.52, 12.57 and 11.213, we obtain

$$\varepsilon_{W0}\left(\frac{H_b}{H_{bcr}}-1\right)^{\delta_{wb}} = C_e \rho_m \left[\frac{\rho_w}{(\rho_m - \rho_w)g\delta_{vb}}\right]^n u_{\delta b}^{2n+1} \qquad (12.58)$$

where $u_{\delta b}$ is the wave-induced bottom velocity amplitude at the breakerline and δ_{vb} is the corresponding wave boundary layer. At the breaking point $u_{\delta b}$ is obtained from Eq. 2.124 applied to shallow water depth h_b and noting that the surface wave amplitude $a = H/2$. Then, solving for C_e from Eq. 12.58 and substitution of the resulting expression into Eq. 12.57, yields

$$C_a = \frac{\varepsilon_{W0}}{w_s}\left(\frac{H_b}{H_{bcr}}-1\right)^{\delta_{wb}} \frac{H^{2n+1}}{\left(\kappa_b h H_b\right)^{(2n+1)/2}} \qquad (12.59)$$

in which it is further assumed that $\delta_{vb}/\delta_v \approx 1$. In this expression the wave height $H(x)$ depends on damping and shoaling, and H_b is its value at the breaker line. Since wave energy is absorbed by mud, conservation of the rate of energy transport (Eq. 2.134) does not hold. On the other hand, we may consider the loss of wave height to be effectively due to the spreading of wave orthogonals (rays). Assuming the loss to be given by Eq. 12.8, we then obtain

$$H(x) = H_0 \sqrt{\frac{g}{2\sigma C_g(x)}} e^{-k_i(x_0 - x)} \tag{12.60}$$

where h_0 is the terminal depth and k_i is the wave damping coefficient. The wave group velocity in the offshore zone is assumed to be reasonably approximated by its shallow water value. Thus, $C_g(x) = \sqrt{gh(x)}$. At the breakerline Eq. 12.60 reduces to

$$H_b = \left[\frac{H_0(g\kappa_b)^{1/4} e^{-k_i(x_0 - x_s)}}{\sqrt{2\sigma}} \right]^{4/5} ; \; x_s = \frac{H_b}{\kappa_b m} \tag{12.61}$$

For the input parameters in Table 12.10, we obtain $H_b = 2.22$ m and $x_s = 493$ m.

Table 12.10. Input for sediment flux calculation.

Parameter	Value	
H_0	2.5	m
T	6	s
x_0	10000	m
m	0.005	
κ_b	0.9	
k_i	0.00001	m^{-1}
w_s	0.0005	m s^{-1}
H_{bcr}	0.3	m
ε_{W0}	0.00005	kg m^{-2} s^{-1}
δ_{wb}	0.5	
α_w	0.2	
n	0.2	
g	9.81	m s^{-2}
C_D	0.022	
α_v	0.9	
α_b	25	deg.
α_u	1	

Substitution of Eqs. 12.50, 12.59 and 12.60 into Eq. 12.49, the depth–mean concentration complementary to Eq. 12.53 is found to be

$$\bar{C} = \frac{\varepsilon_{W0}}{w_s} \frac{\alpha_w \kappa_b^2 m x_s \sigma}{6 w_s} \left(\frac{H_b}{H_{bcr}} - 1 \right)^{\delta_{wb}} \left(\frac{H}{H_b} \right)^{\frac{2n+3}{2}} \left(1 - e^{-\frac{6 w_s}{\alpha_w \kappa_b^2 m x_s \sigma} \left(\frac{1}{\frac{H}{H_b}} \right)} \right); \ x > x_s$$

$$(12.62)$$

Let

$$\tilde{H} = \frac{H}{H_b} \qquad (12.63)$$

Equation 12.62 then assumes the dimensionless form

$$\tilde{C} = \left(1 - e^{-\frac{1}{\tilde{\sigma}_{sz} \tilde{H}}} \right) \Delta \tilde{H}_b \tilde{\sigma}_{sz} \tilde{H}^{\frac{2n+3}{2}}; \ \tilde{x} > 1 \qquad (12.64)$$

12.3.3 *Longshore sediment discharge estimation*

The longshore current $V(x)$ (Eq. 12.47) due to wave-induced excess momentum flux is obtained by using the formulation of Longuet–Higgins [1972] (see also Komar [1998]). The peak velocity occurs within the surf zone and rapidly falls offshore, with the shape of the cross-shore velocity profile dependent on momentum diffusion in the lateral x-direction. Expressions for velocities in the surf zone and the offshore zone are

$$\tilde{V} = \begin{cases} A\tilde{x} + B_1 \tilde{x}^{\gamma_1}; & 0 < \tilde{x} < 1 \\ B_2 \tilde{x}^{\gamma_2}; & 1 < \tilde{x} < \infty \end{cases} \qquad (12.65)$$

where $\tilde{V} = V / V_b$, V_b is the longshore velocity at breakerline obtained by neglecting lateral momentum diffusion, and A, B_1, B_2, γ_1, and γ_2 are constants given by

$$A = \frac{1}{\left(1 - \frac{5\Gamma_v}{2}\right)}, \gamma_1 = -\frac{3}{4} + \sqrt{\frac{9}{16} + \frac{1}{\Gamma_v}}, \gamma_2 = -\frac{3}{4} - \sqrt{\frac{9}{16} + \frac{1}{\Gamma_v}} \qquad (12.66)$$

$$B_1 = \frac{\gamma_2 - 1}{\gamma_1 - \gamma_2} A, B_2 = \frac{\gamma_1 - 1}{\gamma_1 - \gamma_2} A \qquad (12.67)$$

In Eq. 12.66

$$\Gamma_v = \frac{\pi}{2} \frac{m\tilde{\sigma}_H}{\alpha_v C_D} \qquad (12.68)$$

where α_v scales the eddy length relative to the water depth (but is treated here as an empirical coefficient). C_D is the bottom drag coefficient and the dimensionless parameter $\tilde{\sigma}_H$ is

$$\tilde{\sigma}_H = \sqrt{\frac{g}{\kappa_b}} \frac{H_b^{3/2} \sigma}{2\pi} \qquad (12.69)$$

For use in Eq. 12.65 the longshore velocity at the breakerline is obtained from

$$V_b = \frac{5\pi}{8} \frac{\alpha_v m}{C_D} \sqrt{gh_b} \sin \alpha_b \qquad (12.70)$$

where α_b is the breaking wave crest angle with respect to the y-coordinate. For the input parameters in Table 12.10 we obtain $\tilde{\sigma}_H = 1.818$ and $V_b = 0.835$ m s^{-1}.

Now the unit sediment discharge $q_s(x)$ in Eq. 12.47 can be expressed in the non-dimensional form for the two zones as

$$\tilde{q}_s = \alpha_u \tilde{\sigma}_{sz} m\tilde{x}^2 \left(1 - e^{-\frac{1}{\tilde{\sigma}_{sz}\tilde{x}}}\right) \left(A\tilde{x} + B_1\tilde{x}^{\gamma_1}\right); \quad 0 < \tilde{x} < 1 \qquad (12.71)$$

and

$$\tilde{q}_s = \alpha_u \Omega_{s1} m B_2 \frac{e^{-(2n+3)\tilde{k}_i(\tilde{x}_0 - \tilde{x})}}{\tilde{x}^{(6n+5)/4 - \gamma_2}} \left[1 - e^{-\frac{\Omega_{s2}\tilde{x}^{3/2}}{e^{-2\tilde{k}_i(\tilde{x}_0 - \tilde{x})}}} \right] ; \ 1 < x < \tilde{x}_0 \quad (12.72)$$

in which $\tilde{q}_s$, Ω_{s1} and Ω_{s2} are defined as

$$\tilde{q}_s = \frac{q_s}{\left(\dfrac{\varepsilon_{w0}}{w_s} \right) \Delta \tilde{H}_b V_b x_s} ;$$

$$\Omega_{s1} = \frac{\alpha_w \left(H_0 g^{1/4} \right)^{2n+3}}{12 w_s \left(2\sigma\kappa_b H_b \right)^{\frac{2n+1}{2}} \left(m x_s \right)^{\frac{6n+9}{4}}} ; \qquad (12.73)$$

$$\Omega_{s2} = \frac{12 w_s (m x_s)^{3/2}}{\alpha_w \sqrt{g} H_0^2}$$

Since the majority of suspended sediment occurs in the surf zone, in order to assess the applicability of the transport equations, tests were conducted in a wave basin shown in Fig. 12.36. It included a paddle-type wavemaker at one end, mud at the bottom and a flow-return channel to generate longshore current by obliquely incident waves. The wave crest angle was selected to be 20° in order to obtain currents of measurable strength.

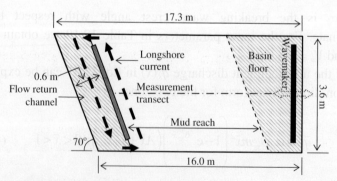

Figure 12.36. Basin setup for wave, alongshore velocity and suspended sediment concentration measurements (from Rodriguez [1997]).

The bottom, mud consisting of AK mud (Table 5.5) with a profile slope of 1:40, had two layers. The bottom layer was comparatively dense with a nominal density of 1,360 kg m^{-3}. The 3-cm thick top layer of softer mud had a nominal density of 1,200 kg m^{-3}. In each test water level and suspended sediment concentration were measured along the centerline transect, and longshore current in the narrow surf zone. Coefficients for Eqs. 12.56 and 12.62 are given in Table 12.11 based on 21 tests for longshore current measurements. Also included is a second set of values representative of the littoral zone off Lian Island near Lianyungang port in China [Yu *et al.*, 1987b; Yu *et al.*, 1994; Rodriguez, 2000].

Table 12.11. Parameters for calculation of sediment concentration and flux.

Parameter/coefficient	Wave basin	Lian Island
H_0 (m)	0.2	1.0
T (s)	1.9	7
x_0 (m)	14	1,500
M	6.7×10^{-2}	2.0×10^{-3}
κ_b	0.9	0.9
k_i (m^{-1})	3.0×10^{-2}	3.0×10^{-4}
w_s (m s^{-1})	3.0×10^{-5}	2.5×10^{-4}
H_{bcr} (m)	0.027	0.18
ε_{W0} (kg m^{-2} s^{-1})	7.56×10^{-6}	3.60×10^{-5}
δ_{wb}	1.0	0.7
α_u	1.0	1.0
α_w	0.2	0.2
N	1	0.2
α_v	0.4	0.4
C_D	0.066	0.015
A	_a	1.147
B_1	_a	−0.797
B_2	_a	0.350
ρ_2 (mud density) (kg m^{-3})	1,200	1,200
η_2 (mud viscosity) (Pa.s)	1.0	1.6

[a] Not required because calculations are for cross-shore transport.
Source: Rodriguez [2000].

In Fig. 12.37, suspended sediment concentrations from three laboratory tests have been compared with values from Eqs. 12.53 and 12.62. In the seaward reach of the surf zone a gradual change in water depth results in practically uniform concentrations, which decrease to zero (assumed) at the boundary. In the offshore zone, the concentration decreases with distance as the effect of waves diminishes with increasing depth.

Measured sediment discharges in the basin's surf zone are compared with calculated values (Eq. 12.47) in Fig. 12.38. Figure 12.39 includes concentration measurements from Lianyungang, and values from Eqs. 12.53 and 12.62. In spite of the three orders of magnitude increase in the length-scale from basin to sea, the general character of the concentration profile is captured by the equations. For an easterly wave approach in deep water at an angle of 45° relative to shoreline, and assuming shore-parallel isobaths as waves travel shoreward, Eqs. 12.46, 12.47 and 12.48 are solved for the unit discharge of sediment at Lianyungang. These equations apply only approximately because the shallow water assumption does not hold over the entire 14,000 m length of the profile. However, transport was negligible at distances greater than about 6,000 m, where the depth was 4 m, which is shallow for 8 s waves.

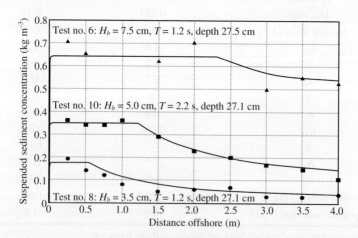

Figure 12.37. Variation of measured and calculated depth–mean suspended sediment concentration with distance from shoreline in test nos. 6, 8 and 10. Water depths are at the toe of the profile (from Rodriguez and Mehta [2000]).

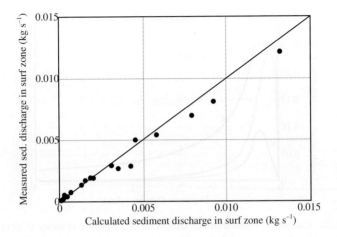

Figure 12.38. Measured versus calculated sediment discharges in the surf zone of the wave basin (from Rodriguez and Mehta [2000]).

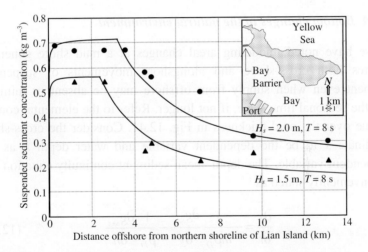

Figure 12.39. Measured and calculated depth–mean concentrations against offshore distance from the northern shoreline of Lian Island. The quantity H_s is the representative significant wave height (from Rodriguez and Mehta [2000]).

In Fig. 12.40 the longshore velocity $V(x)$ (Eq. 12.65), depth–mean concentration $\bar{C}(x)$ (Eqs. 12.53 and 12.62) and the unit suspended sediment discharge $q_s(x)$ (Eqs. 12.71 and 12.72) are plotted using input data from Table 12.10.

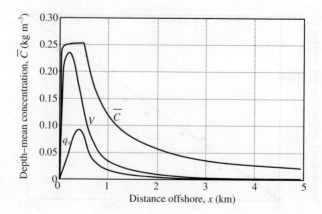

Figure 12.40. Calculated cross-shore distributions of longshore velocity V, depth–mean concentration $\bar{C}$ and unit suspended sediment discharge q_s using input data in Table 12.10.

12.3.4 *Bottom changes in the natural environment*

As we have noted, modeling areal changes at a mud shore generally requires both cross-shore and alongshore movements of suspended sediment even when supply from offshore may be absent or minimal over the seasonal time scale, if not longer. Refer to the elemental control volume over the mud thickness in Fig. 12.41. Consider the cross-shore coordinate x to be the dependent variable and water depth h as the independent variable. The depth–mean sediment continuity equation can be conveniently stated as

$$\frac{\partial x}{\partial t} = -\frac{1}{m\rho_D}\frac{\partial g_{ssy}}{\partial y} - \frac{1}{\rho_D}\frac{\partial g_{ssx}}{\partial h} \tag{12.74}$$

where g_{ssx} and g_{ssy} are the components of suspended sediment mass fluxes per unit length in the x and y directions, respectively, ρ_D is the dry density of the deposit and $m = \partial h/\partial x$ is the local bottom slope. This form of the equation permits direct tracking of isobaths when they are displaced from their initial positions due to erosion or deposition. Thus, interpolation for the identification of any isobath position, as would be

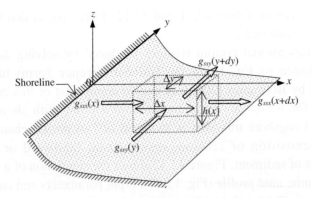

Figure 12.41. Elemental control volume over the full water depth for which sediment mass continuity is established.

essential using a fixed x, y grid, is not required. The output at every time step is a bottom contour "map" dependent on suspended sediment fluxes.

Equation 12.74 reduces to the one-dimensional Eq. 12.19 with $q_s = g_{ssx}/\rho_D$. Based on the analysis of Perlin and Dean [1983] for sandy beaches, the following variant of Eq. 12.18 in the finite-difference form is used

$$g_{ssx} = k_q \overline{C} h u_\delta \left(\frac{\Delta x - \Delta x_e}{\Delta x_e} \right)$$ (12.75)

where k_q is a non-dimensional rate constant, Δx now represents the instantaneous cross-shore distance between two consecutive isobaths and Δx_e is the distance along the target profile. In Eq. 12.75, for any two consecutive contours Δx_e can be calculated from the target profile approximated by Eq. 12.13. The alongshore flux is obtained from Eqs. 12.71 and 12.72.

The simplest way to obtain the bottom velocity amplitude u_δ in Eq. 12.75 is from the Airy linear theory (Chapter 2), which then permits calibration of k_q. However, Rodriguez [2000] calculated u_δ from the boundary layer model of Jiang [1993] similar to Jain–Lowes [Jain, 2009b] mentioned in Chapter 11. In these models mud is moldable and energy-absorbing. Mud density and viscosity required for the calculation of u_δ are given in Tables 12.11 and 12.12. It was found that 0.1 was a reasonable value of k_q.

Two-dimensional bottom changes produced by solving the general mass balance of Eq. 12.74 (in finite-difference form) have been discussed by Rodriquez [2000]. A simple application of this equation is obtained for waves normally incident on a coast with shore-parallel isobaths. Longshore transport is switched off, so profile change, either due to deposition or erosion, results from landward or seaward movement of sediment. Figure 12.42 shows the simulation of a change in a wave-flume mud profile (Fig. 12.19). Input parameters and coefficients are given in Table 12.11. Since in this analysis the initial and final

Table 12.12. Constants for calculation of concentration and sediment flux.

Parameter/coefficient	Grimsby	Mahin
H_0 (m)	1.0	0.7
T (s)	7	8
x_0	700	14,500
m	1.0×10^{-2}	1.3×10^{-3}
κ_b	0.9	0.9
k_i (m^{-1})	7.2×10^{-4}	2.0×10^{-4}
w_s (m s^{-1})	3×10^{-5}	3×10^{-5}
H_{bcr} (m)	0.57	0.10
ε_{W0} (kg m^{-2} s^{-1})	4.18×10^{-5}	5.00×10^{-5}
δ_{wb}	1.0	1.0
α_u	1.0	1.0
α_w	0.2	0.2
n	1	1
α_v	0.4	0.4
C_D	0.015	0.015
A	$-^a$	1.065
B_1	$-^a$	-0.678
B_2	$-^a$	0.388
ρ_2 (mud density) (kg m^{-3})	1,100	1,100
η_2 (mud viscosity) (Pa.s)	0.01	0.1

^{a} Not required because calculations are for cross-shore transport.
Source: Rodriguez [2000], Rodriguez and Mehta [2001].

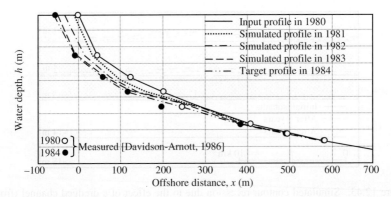

Figure 12.42. Erosion of glacial till profiles at Grimsby, Lake Ontario. Data and numerical simulation (from Rodriguez [2000]).

profiles are known, the output reveals the shape of the profiles in 1981, 82 and 83.

It is instructive to examine the bathymetric change near Awoye mentioned earlier (Fig. 12.12). This can be done by considering shoreline retreat due to the effect of shore-normal waves over a bottom with shore-parallel isobaths disturbed by a dredged cut acting as a sediment sink. The change in the bottom due to increasing deviation from the initial "target" profile, is shown in Fig. 12.43 [Rodriguez and Mehta, 2001]. In order to mimic the observed pattern of bottom change, sediment was withdrawn through the cut at the rate of 800 kg s^{-1}. This exceptionally high rate is a proxy for the rapid rate of erosion that actually occurred. At the end of 7 years, the shoreline had retreated by about 1 km.

12.4 Turbid Gravity Current

Turbidity current in general results from density difference due to spatially varying suspended sediment concentration. On a bed slope over which the bottom stress exceeds the yield stress, turbid underflow can accumulate sediment in navigation channels and reservoirs. Offshore, gravity-induced transport of fluid mud underflow over the sloping shelf transports originally terrestrial sediment eventually to the abyssal

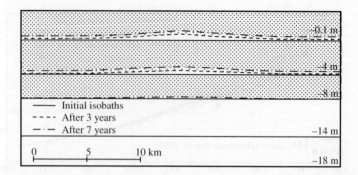

Figure 12.43. Simulated contour recession due to the effect of a dredged channel (from Rodriguez and Mehta [2001]).

seafloor. Under appropriate conditions the current can become self-sustaining and may even accelerate. If this behavior is over an erodible sediment bed, scour can be substantial along a submarine canyon (Fig. 12.44a) and may reach catastrophic proportions [Parker, 1982; Garcia and Parker, 1993]. The material may ultimately accumulate on the seafloor as a submarine fan. Viscoplastic models, such as those based on Eqs. 5.23, 5.24 or 5.27, have been used to model dense turbid underflows [Krone, 1963; Williams, 1986; Mei and Liu, 1987; Toorman, 1996; Huang and Garcia, 1996, 1998, 1999].

When mud from a stream accumulates as a consolidated bed over the shallow shelf slope, wave-induced orbital velocity during storms can bring about liquefaction and actuate turbid underflow. Consider the system in Fig. 12.44b, in which an underflow trapped within the wave-induced boundary layer occurs over a mild slope of angle θ [Einstein, 1941; Ali and Georgiadis, 1991; Friedrichs and Scully, 2007].

The slope is small enough such that the x-z coordinate system can be rotated slightly without significantly affecting the analysis. Thus the thickness of the mud layer h_m is considered to remain unchanged by rotation. The suspended mud concentration is represented by its depth–mean value C_m. At the top of the underflow, u_g is the gravity-induced velocity and u_w is the wave velocity amplitude. We will define a composite velocity u whose absolute value is

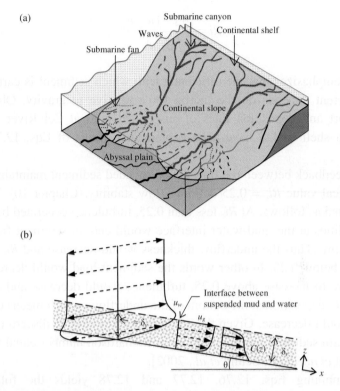

Figure 12.44. Gravity underflow of mud: (a) schematic drawing of a submarine canyon (adapted from Allen [1977]); (b) definition sketch for the effect of waves on the actuation of underflow (from Friedrichs and Scully [2007]).

$$|u| = \sqrt{u_w^2 + u_g^2} \tag{12.76}$$

At equilibrium the down-slope pressure gradient balances up-slope resistance represented by the turbulent bed shear stress τ_b, *i.e.*,

$$gC_m h_m (s-1)\sin\theta = \tau_b s = C_D \rho_s |u| u_g \tag{12.77}$$

where ρ_w is the density of water, ρ_s is the particle mineral density, $s = \rho_s/\rho_w$ and C_D is the bottom drag coefficient (= ~0.003–0.004). The bulk Richardson number for the boundary layer is

$$Ri = \frac{g(s-1)C_m h_m}{\rho s |u|^2} \qquad (12.78)$$

which emphasizes that the efficiency with which sediment is carried by the current depends on the inertia force relative to gravity. Observed transport and deposition rates of suspended mud at Eel River on the western shelf of US have been explained in terms of Eqs. 12.77 and 12.78.

A feedback between the flow and suspended sediment maintains Ri at its critical value $Ri_c = 0.25$ defining flow stability (Chapter 10). This is explained as follows. At Ri_c less than 0.25, turbulence generated by shear instabilities at the mud-water interface would entrain sediment from the underflow. Thus the underflow thickness would decrease and Ri_c would reduce below 0.25. In other words the sediment load would decrease. If Ri_c were to increase above 0.25, turbulence would decrease and lead to sediment deposition at the bottom of the underflow, which means that the load would decrease. Given these behaviors it is easy to discern that the maximum sediment load must occur at Richardson number equal to 0.25 [Wright *et al.*, 2001; Scully *et al.*, 2002].

Combining Eqs. 12.76, 12.77 and 12.78 yields the following expression for the maximum unit sediment (wave-supported) mass load

$$g_{ss} = C_m h_m u_g = \frac{\rho s Ri_c^2 \sin\theta u_w^3}{C_D g(s-1)\left[1-\left(\dfrac{Ri_c \sin\theta}{C_D}\right)^2\right]^{3/2}} \qquad (12.79)$$

from which we obtain the deposition flux

$$F_D = -\frac{1}{h_m}\frac{dg_{ss}}{dx} = g_{ss}\left[-\frac{3}{u_w}\frac{du_w}{dx} - \frac{(1+2\vartheta)}{(1-\vartheta)\sin\theta}\frac{d(\sin\theta)}{dx}\right] \qquad (12.80)$$

where

$$\vartheta = \frac{Ri_c \sin\theta}{C_D} \qquad (12.81)$$

For waves in shallow water, small values of ϑ and uniform bed slope, Eq. 12.80 reduces to

$$F_D = \frac{3}{16}\frac{\rho_s\sqrt{gh}Ri_c^2\sin\theta}{C_D(s-1)}H^3 \qquad (12.82)$$

where H is the wave height and h is the water depth. Thus we observe that F_D increases rapidly with H. On the other hand, with the assignment of Ri as a constant Ri_c, F_D would not be dependent on the amount of sediment in the underflow. Also, for this model to be applicable, the supply of upstream sediment must locally exceed the capacity for down-slope sediment flux [Friedrichs and Scully, 2007]. Assuming this to be so, consider $\rho_w = 1{,}027$ kg m^{-3}, $\rho_s = 2{,}650$ kg m^{-3}, $\theta = 5°$, $h = 2$ m, $H = 0.3$ m and $C_D = 0.004$. From Eq. 12.82 with $Ri_c = 0.25$ we obtain $F_D = 4.46$ kg m^{-1} s^{-1}. Given $C_m = 10$ kg m^{-3} and $h_m = 0.2$ m, Eq. 12.88 with $Ri = Ri_c$ yields $u = 0.22$ m s^{-1}. In general, the bed can be expected to erode at this velocity.

The sustainability and efficiency of sediment transport by turbid gravity current was examined by Bagnold [1956, 1966] based on energy requirements. Following the onset of down-slope turbid flow, whether it is sustained or decays due to deposition is considered from the view point of energy supply. If the rate of production of turbulent energy per unit volume of flowing suspension equals the rate at which energy is lost by boundary layer friction per unit volume, the turbid flow will be a self-sustaining auto-suspension.[i] If the supply cannot keep up with the loss deposition will occur. The threshold for auto-suspension is Eq. 6.41, also called the Knapp–Bagnold (K–B) criterion. The threshold of suspension curve in Fig. 6.7 is based on this criterion.

The K–B criterion has been used to predict the condition of self-acceleration, which however may not be easily reproduced in the laboratory due to limited lengths of flumes [Parker *et al.*, 1986; 1987]. Krishna [2009] used a pseudo-1D (vertical) numerical code in which

---

[i] As we noted in Chapter 5, a quiescent auto-suspension, *i.e.*, non-settling suspension, can be produced by peptizing a flocculated suspension into a colloidal solution which remains as such by Brownian motion alone.

fluid and solids were treated as two separate phases (Chapter 6). Primary variables tested for their effects on the turbidity current were shelf foreset (bed) slope, flow friction velocity and particle size. Initially a pressure driven boundary layer flow was generated by the model structured with ghost boundary grid cells [Hsu *et al.*, 2007]. After a very short period of 150 s the pressure force was switched off and the bed was instantaneously "tilted" by reconfiguring the bottom to have a mild foreset slope (Fig. 12.45).

For a given bed particle size, an "ignitive" condition was defined as one in which the divergence du_m/dx of mean flow velocity u_m in the flow direction (x), and the divergence dF/dx of sediment flux F, are both positive. It was tentatively concluded that in general the threshold condition for ignition as defined did not coincide with the K–B criterion.

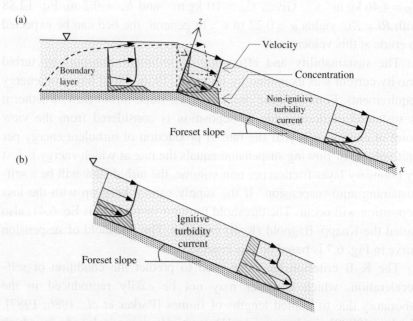

Figure 12.45. Basis for model testing of ignition: (a) Ghost boundary grid induced pressure gradient produces flow. A turbid current is generated when the bed is tilted. Under non-ignitive conditions (characterized by friction velocity, bed slope and particle size) the concentration decreases due to deposition. (b) Conditions are conducive to ignition (from Krishna [2009]).

12.5 Sedimentation in Channels and Small Basins

12.5.1 *Channel and basin geometry*

Predictive assessments of sedimentation in coastal channels and basins are complicated by boundary features such as docks, as well as by uneven bathymetry. Complexities also arise from the three-dimensionality of flow circulation due to tide and wind influenced by stratification. In many instances predictive accuracy of models is also limited by the quality and quantity of data including boundary conditions for flow and sediment transport. In such instances two or three-dimensional numerical models may not yield more useful information than analytic approaches. The following treatments are largely restricted to idealized geometries and conditions. The focus is on the rate of sedimentation rather than patterns of shoaling, which generally require the use of numerical models.[j] Only in some simple cases is an analytic approach to simulate the shoaling pattern feasible. The pattern of deposition rate at a delta near tidal entrances or river mouths is reviewed below.

12.5.2 *Offshore entrance channel*

Consider a channel of length L_c offshore of a harbor entrance (Fig. 12.46) in which deposition is due to interception of (longshore) littoral drift. Given depth–mean suspended sediment concentration $\overline{C}$, the total sediment load Q_s spanning the width of the littoral zone taken to be equal to the width of the channel is

---

[j] In small basins, sedimentation rate may be more important than pattern, especially if depth is maintained by dredging. However, some non-dredging mitigation methods require the identification of locations of heavy sedimentation as well as pattern of particle size variation. A case in point is the use of a suspended-sediment deflecting wall for minimizing sedimentation in basins in which circulation is driven by lateral shear between the main flow and basin water. The shear-driven eddy formed tends to extract suspended matter from the main channel and pulls it into the basin where it accumulates often along the path of navigation [Vollmers, 1976]. Knowing the sediment pathway and the zone of accumulation are essential aspects in wall design [Smith *et al.*, 2001; McNeal and Mehta, 2002].

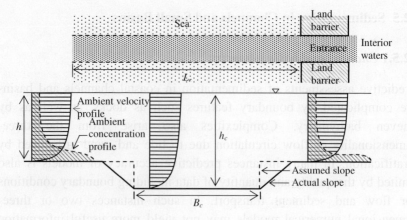

Figure 12.46. Offshore channel at a harbor entrance (adapted from Gole *et al.* [1971]).

$$Q_s = V\bar{C}L_c h\Delta T \qquad (12.83)$$

where V is the mean longshore current velocity, h is the ambient mean water depth and ΔT is the duration of sedimentation.

The sediment load Q_{sP} potentially available for deposition is obtained from mass balance

$$Q_{sP}VhL_c = Q_s w_s B_c L_c \qquad (12.84)$$

where B_c is the channel width and w_s is the sediment settling velocity. Thus,

$$Q_{sP} = \frac{B_c}{h}\frac{w_s}{V}Q_s \qquad (12.85)$$

Let τ_b be the ambient bed shear stress and τ_{bch} the bed shear stress in the channel. Since the channel's depth is greater than its ambient depth, the condition $\tau_{bch} \leq \tau_b$ exists in the channel. As bed shear stress in turbulent flow is proportional to the square of the mean velocity (Eq. 2.31), given the mean velocity in the channel V_{ch}, the actual net sediment load deposited Q_{sN} is obtained from

$$Q_{sN} = K_c \left(1 - \frac{\tau_{bch}}{\tau_b}\right) Q_{sP} = K_c \left(1 - \frac{V_{ch}^2}{V^2}\right) Q_{sP} \qquad (12.86)$$

where K_c is a shoaling coefficient. If the channel is fully filled τ_{bch} will be equal to τ_b and Q_{sN} would be zero. From flow continuity

$$Vh = V_{ch} h_c \qquad (12.87)$$

where h_c is the mean depth in the channel. Thus,

$$Q_{sN} = K_c \frac{B_c w_s}{hV} \left(1 - \frac{h_c^2}{h^2}\right) Q_s \qquad (12.88)$$

This expression permits the calculation of Q_{sN}, provided K_c is known. It can be evaluated for channels where the deposition loads are available. In Table 12.13, data from four ports in India are given. Mumbai, Mormugao and Kochi are on the west coast of peninsular India and Visakhapatnam is on the east coast. The K_c values are similar, which suggests that at these channels the assumed conditions of sedimentation are reasonably met. However, given that they are significantly idealized, the applicability of the approach is limited to similar channels. In such applications $K_c = 0.29$ (the mean of four values) can be used as a trial value.

12.5.3 Deposition pattern near entrances

The spatial distribution of sediment in deltas near river mouths or tidal entrances depends on the role of waves relative to current. When current dominates the delta tends to be elongated in the direction of flow, whereas wave-dominated deltas tend to be crescentic in planform [Devine and Mehta, 1999]. The wave-current field is not only time dependent but is complicated by shoreline geometry, bathymetry and entrance orientation relative to the shoreline. Figure 12.47 shows a turbid plume from the tidal entrance to the Paraiba do Sul River (Brazil). The new ebb plume is observed to mix with remnants of the previous ebb plume.

Table 12.13. Calibration of channel shoaling coefficient K_c.

Parameter	Mumbai	Mormugao	Kochi	Visakha-patnam
Channel length (km)	0.7	7.0	2.1	1.0
Channel width (m)	330	390	180	160
Channel depth (m)	5.8	18.7	12.2	13.4
Ambient water depth (m)	3.4	13.4	8.8	11.6
Ambient alongshore velocity (m s^{-1})	0.6	0.2	0.3	0.1
Sediment concentration (kg m^{-3})	0.18	0.04 (Jun–Sep) 0.01 (Oct–May)	0.13	0.08
Particle diameter (mm)	0.048	0.078	0.120	0.120
Period of sedimentation (months)	12	12	4	12
K_c	0.25	0.30	0.31	0.29

Source: Gole *et al.* [1971].

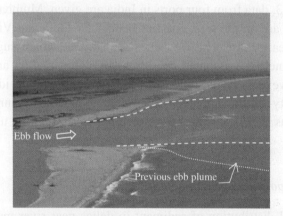

Figure 12.47. Paraiba do Sul River (Brazil) entrance and turbid ebb plume. Shown is one of the two entrances near the town of Atafona (courtesy Prof. Dieter Muehe and Susana Vinzon, Federal University of Rio de Janeiro).

Disregarding wave effects and assuming that water level change along the plume is small, it can be idealized as a shallow-water steady turbulent jet. The rate of spreading of the plume over a generally sloping bottom is determined by jet momentum and bottom resistance. Referring to Fig. 12.48 the depth and half-width at the entrance are h_0 and b_0, respectively, and jet current velocity at the entrance is u_0. The relevant

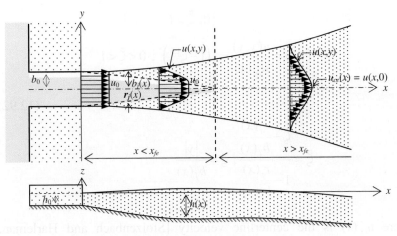

Figure 12.48. Definition sketch of turbulent, shallow-water jet.

steady equations of continuity, momentum and conservation of suspended sediment mass are, respectively,

$$\frac{\partial(hu)}{\partial x} + \frac{\partial(hv)}{\partial y} = 0 \qquad (12.89)$$

$$\frac{\partial(hu^2)}{\partial x} + \frac{\partial(huv)}{\partial y} = \frac{1}{\rho}\left(\tau_{bx} + \frac{\partial F_{yx}}{\partial y}\right) \qquad (12.90)$$

$$\frac{\partial(hCu)}{\partial x} + \frac{\partial(hCv)}{\partial y} = \frac{\partial J_y}{\partial y} + S_r \qquad (12.91)$$

where $\tau_{bx} = -f_c \rho_w u^2 /8$ is the turbulent bed shear stress in the offshore x-direction (the viscous component of stress is ignored), f_c is the Darcy–Weisbach friction factor (Eq. 2.38), F_{xy} is the depth–mean lateral shear force due to turbulence in the x-direction, J_y is depth–mean turbulent diffusive mass flux component in the shore-parallel (y) direction and S_r is a source/sink term accounting for sediment mass flux at the bottom [Özsoy and Ünlüata, 1982].

Following classical development in turbulent jet mechanics the distribution of longitudinal velocity $u(x,y)$ is assumed to be self-similar in terms of a dimensionless distribution function:

$$\tilde{F}(\tilde{\zeta}) = \frac{u(x, y)}{u_{ce}(x)} = \begin{cases} 0; \ \tilde{\zeta} > 1 \\ \left(1 - \tilde{\zeta}^{1.5}\right)^2; \ 0 < \tilde{\zeta} < 1 \\ 1; \ \tilde{\zeta} < 0 \end{cases}$$

(12.92)

$$\tilde{\zeta} = \frac{\zeta - \dfrac{r_j(x)}{b_j(x)}}{1 - \dfrac{r_j(x)}{b_j(x)}}; \ \zeta = \frac{|y|}{b_j(x)}$$

where $u_{ce}(x)$ is the centerline velocity [Stolzenbach and Harleman, 1971].[k] The jet is divided into a near-field zone of flow establishment (ZFE) in which the effect of lateral shear between the jet boundary and ambient water does not reduce the initial jet velocity in the inner core. Thus, in this core of half-width $r_j(x)$ the velocity remains u_0, but in the outer area of half-width $b_j(x) - r_j(x)$, where $b_j(x)$ is the jet half-width, $u(x,y)$ is lower than u_0. Beyond the distance x_{fe} of ZFE, in the zone of established flow (ZEF) the centerline velocity u_{ce} continues to decrease below u_0 as the jet spreads and gradually loses its forward momentum.

For use of Eqs. 12.89 and 12.90, we note that $u(x,y)$ vanishes as y approaches the jet boundary b_j, and F_{yx} also vanishes because the velocity gradient $\partial u(x,y)/\partial y$ approaches zero at the boundary. Due to the Bernoulli effect at the boundary, the jet draws water from outside at an entrainment velocity v_e. Most simply this velocity can be taken to be proportional to the centerline velocity $u_{ce}(x)$, *i.e.*,

$$v_e = \alpha_e u_{ce}(x)$$

(12.93)

where α_e is a water entrainment coefficient. This coefficient is variable; values of 0.036 in ZFE and 0.050 in ZEF have been used [Özsoy and Ünlüata, 1982]. The value 0.21 in ZFE was measured at Sikes Cut in

---

[k] Based on tests in a (1:49 undistorted scale) fixed-bed physical model of Jupiter Inlet in Florida, Purandare and Mehta [1987] confirmed the self-similar character of Eq. 12.92, but found the jet to be narrower and defined by the function $\tilde{F}(\tilde{\zeta}) = \left(1 - \tilde{\zeta}^{1.5}\right)^5$; $0 < \tilde{\zeta} < 1$.

Florida, a tidal entrance connecting Apalachicola Bay to the Gulf of Mexico [Mehta and Zeh, 1980].

For the variation of jet centerline velocity and jet width with distance from the entrance, the resulting forms of Eqs. 12.89 and 12.90 are now integrated across the jet width, *i.e.,*

$$\frac{d}{dx}\left(h\int_{-b_j}^{b_j} u\,dy\right) = 2hv_e \tag{12.94}$$

$$\frac{d}{dx}\left(h\int_{-b_j}^{b_j} u^2\,dy\right) = -\frac{f}{8}\int_{-b_j}^{b_j} u^2\,dy \tag{12.95}$$

For further analysis we will introduce the following dimensionless variables:

$$\tilde{x} = \frac{x}{b_0};\ \tilde{x}_{fe} = \frac{x_{fe}}{b_0};\ \tilde{h}(\tilde{x}) = \frac{h}{h_0};\ \tilde{r}_j(\tilde{x}) = \frac{r_j}{b_0};\ \tilde{b}_j(\tilde{x}) = \frac{b_j}{b_0};\ \tilde{u}_{ce}(\tilde{x}) = \frac{u_{ce}}{u_0} \tag{12.96}$$

Solutions of Eqs. 12.94 and 12.95 are obtained separately in the two zones as follows.

In ZFE ($\tilde{x} < \tilde{x}_{fe}$):

$$\tilde{u}_{ce} = 1;\ \tilde{r}_j = \frac{I_1\tilde{j} - I_2\tilde{s}}{(I_1 - I_2)\tilde{h}};\ \tilde{b}_j = \frac{(1-I_2)\tilde{s} - (1-I_1)\tilde{j}}{(I_1 - I_2)\tilde{h}} \tag{12.97}$$

In ZEF ($\tilde{x} > \tilde{x}_{fe}$):

$$\tilde{u}_{ce} = \frac{\tilde{j}}{\sqrt{\tilde{l}}};\ \tilde{b}_j = \frac{\tilde{l}}{I_2\tilde{h}\tilde{j}} \tag{12.98}$$

where $I_1 = 0.450$ and $I_2 = 0.316$ are numerical constants, and $\tilde{j}(\tilde{x})$, $\tilde{s}(\tilde{x})$ and $\tilde{l}(\tilde{x})$ are obtained from

$$\tilde{j} = e^{-\tilde{f}\int_0^{\tilde{x}}\frac{d\omega}{\tilde{h}(\omega)}} \tag{12.99}$$

$$\tilde{s} = 1 + \alpha_e \int_0^{\tilde{x}} \tilde{h}(\omega) d\omega \qquad (12.100)$$

$$\tilde{l} = \tilde{j}^2(\tilde{x}_{fe}) + \frac{2\alpha_e I_2}{I_1} \int_{\tilde{x}_{fe}}^{\tilde{x}} \tilde{h}(\omega) \tilde{j}(\omega) d\omega \qquad (12.101)$$

in which $\tilde{f} = f_c b_0 / 8h_0$ represents bed resistance and ω is a dummy variable. For an offshore bottom of uniform slope m the depth variation would be

$$\tilde{h} = 1 + \tilde{m}\tilde{x} \qquad (12.102)$$

where $\tilde{m} = mb_0 / h_0$ is the slope parameter.

In the simple case of constant depth ($\tilde{m} = 0$), Eqs. 12.97 through 12.101 are used to obtain the dimensionless jet velocity $\tilde{u}_{ce}(\tilde{x})$ and the width $\tilde{b}_j(\tilde{x})$ (Fig. 12.49). Note that the distance x_{fe} is the value of x when $\tilde{r}_j = 0$. Setting $\tilde{f} = 0$ results in the classical surface jet solutions [Abramovich, 1963; Schlichting, 1968] with linearly expanding jet width; see Eqs. E12.3 and E12.4. Increasing $\tilde{f}$ retards the jet's seaward transport and lateral spreading becomes increasingly rapid.

An application of the equations for jet hydrodynamics is shown in Fig. 12.50 for a flood-flow plume at Sikes Cut transporting water from the Gulf of Mexico into Apalachicola Bay. Water surface variation and current velocities were obtained during a field campaign [Mehta and Zeh, 1980]. Equations 12.89 and 12.90 were modified to include the centerline curvature of the plume resulting from a tidal cross-current and possibly the Coriolis effect. The presence of a small island in the bay deflected the plume from the path it would have been expected to take. The simulated width (dashed line) is influenced by the uneven depth $\tilde{h}(\tilde{x})$ in the bay.

As the jet velocity u_0 at the entrance decreases with distance, suspended sediment deposits in a pattern depending on the rate at which the jet spreads. In a depositional (non-eroding) plume the sediment sink term $S_r = w_s' C$ is obtained from Eq. 7.144 as

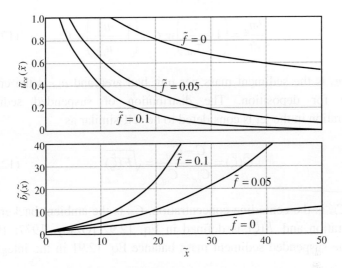

Figure 12.49. Dimensionless variations of jet centerline velocity and width over a horizontal bottom (from Mehta and Özsoy [1978]).

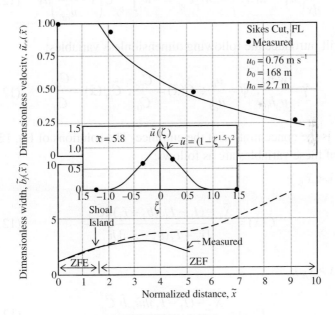

Figure 12.50. Measured and calculated flood-flow plume in Apalachicola Bay near Sikes Cut, Florida (from Mehta and Zeh [1980]).

$$S_r = \frac{dm_a}{dt} = \left(1 - \frac{\tau_b}{\tau_{cd}}\right) w_s C = \left(1 - \frac{u^2}{u_{cd}^2}\right) w_s C \tag{12.103}$$

where m_a is the sediment mass per unit bed area and u_{cd} is the critical velocity for deposition. The distribution of suspended sediment concentration in the jet is considered to be self-similar as

$$\tilde{G}(\tilde{\zeta}) = \frac{C - C_{af}}{C_{ce} - C_{af}} = \sqrt{F(\tilde{\zeta})} \tag{12.104}$$

where C_{ce} is the centerline concentration, C_{af} is the ambient background concentration and $F(\tilde{\zeta})$ is defined in Eq. 12.92 [Özsoy, 1977; 1986]. Next, the suspended sediment mass balance Eq. 12.91 in the integrated form is

$$\frac{d}{dx}\left(h \int_{-b_j}^{b_j} uc\, dy\right) - 2hv_e C_{af} = -w_s \int_{-b_j}^{b_j} C\left(1 - \frac{u^2}{u_{cd}^2}\right) dy \tag{12.105}$$

We will introduce the following dimensionless variables:

$$\tilde{\gamma} = \frac{w_s b_0}{u_0 h_0}; \quad \tilde{\psi} = \frac{u_0}{u_{cd}}; \quad \tilde{C}(\tilde{x}) = \frac{C_{ce}}{C_0}; \quad \tilde{C}_{af}(\tilde{x}) = \frac{C_{af}}{C_0} \tag{12.106}$$

where C_0 is the concentration at the entrance. The solutions of Eq. 12.105 in terms of these variables are as follows:

In ZFE ($\tilde{x} < \tilde{x}_{fe}$):

$$\tilde{C}(\tilde{x}) = \frac{\tilde{X}_1 - (I_1 - I_4)(\tilde{b}_j - \tilde{r}_j)\tilde{h}\tilde{C}_{af}}{[\tilde{r}_j + I_4(\tilde{b}_j - \tilde{r}_j)]\,\tilde{h}} \tag{12.107}$$

In ZEF ($\tilde{x} > \tilde{x}_{fe}$):

$$\tilde{C}(\tilde{x}) = \frac{\tilde{X}_2 - (I_1 - I_4)\tilde{h}\tilde{u}_{ce}\tilde{b}_j\tilde{C}_{af}}{I_4 \tilde{h}\tilde{u}_{ce}\tilde{b}_j} \tag{12.108}$$

where

$$\tilde{X}_1 = \frac{1}{\tilde{P}_1}\left(\int_0^{\tilde{x}} \tilde{P}_1 \tilde{M}_1 d\omega + 1\right) \tag{12.109}$$

$$\tilde{P}_1 = e^{\int_0^{\tilde{x}} \tilde{Q}_1 d\omega} \tag{12.110}$$

$$\tilde{Q}_1 = \frac{\tilde{\gamma}\left[\tilde{r}_j + I_3(\tilde{b}_j - \tilde{r}_j)\right] - \tilde{\gamma}\tilde{\psi}^2\left[\tilde{r}_j + I_5(\tilde{b}_j - \tilde{r}_j)\right]}{\tilde{h}\left[\tilde{r}_j + I_4(\tilde{b}_j - \tilde{r}_j)\right]} \tag{12.111}$$

$$\tilde{M}_1 = \alpha_e \tilde{h}\tilde{C}_{af} + \tilde{\gamma}\left[\tilde{\psi}^2(I_2 - I_5) - (1 - I_3)\right](\tilde{b}_j - \tilde{r}_j)\tilde{C}_{af}$$
$$+ (I_1 - I_4)\tilde{h}(\tilde{b}_j - \tilde{r}_j)\tilde{Q}_1\tilde{C}_{af} \tag{12.112}$$

$$\tilde{X}_2 = \frac{1}{\tilde{P}_2}\left[\int_{\tilde{x}_{fe}}^{\tilde{x}} \tilde{P}_2 \tilde{M}_2 d\omega + X_1(\tilde{x}_{fe})\right] \tag{12.113}$$

$$\tilde{P}_2 = e^{\int_{\tilde{x}_{fe}}^{\tilde{x}} \tilde{Q}_2 d\omega} \tag{12.114}$$

$$\tilde{Q}_2 = \frac{\tilde{\gamma}(I_3 - \tilde{\psi}^2 I_5 \tilde{u}_{ce}^2)}{I_4 \tilde{h}\tilde{u}_{ce}} \tag{12.115}$$

$$\tilde{M}_2 = \alpha_e \tilde{h}\tilde{u}_{ce}\tilde{C}_{af} + \tilde{\gamma}\tilde{\psi}^2(I_2 - I_5)\tilde{b}_j u_{ce}^2 \tilde{C}_{af} - \tilde{\gamma}(1 - I_3)\tilde{b}_j \tilde{C}_{af}$$
$$+ (I_1 - I_4)\tilde{h}\tilde{u}_{ce}\tilde{b}_j \tilde{Q}_2 \tilde{C}_{af} \tag{12.116}$$

The quantities $I_3 = 0.600$, $I_4 = 0.368$ and $I_5 = 0.278$ are numerical constants. The variables $\tilde{r}_j$, $\tilde{b}_j$ and $\tilde{u}_{ce}$ are defined by Eqs. 12.97 through 12.101.

Under the condition $\tilde{\gamma} = 0$, *i.e.*, zero settling velocity, the dimensionless concentration $\tilde{C}(\tilde{x})$ would represent an auto-suspension. In this case Eqs. 12.107 through 12.116 are considerably simplified. Further setting the non-dimensional background concentration $\tilde{C}_{af} = 0$ without loss of generality, $\tilde{C}(\tilde{x})$ is plotted in Figs. 12.51a, b in which the friction parameter $\tilde{f}$ and the slope parameter $\tilde{m}$ are varied. Due to

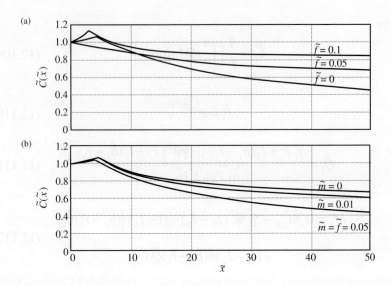

Figure 12.51. Variation of dimensionless jet centerline concentration: (a) constant depth; (b) linearly varying depth with $\tilde{\mu} = 0.05$ (from Mehta and Özsoy [1978]).

the effect of bottom resistance on flow the concentration may reach a higher value within ZFE than C_0 at the entrance.

When $\tilde{\gamma} \neq 0$, $\tilde{C}(\tilde{x})$ is the non-dimensional concentration of depositable suspended matter. The pattern of sedimentation in terms of the deposition flux can now be determined. We will define

$$\tilde{M}_a = \frac{m_a}{C_0 h_0}; \ \tilde{t} = \frac{u_0 t}{b_0} \tag{12.117}$$

The dimensionless deposition flux $\tilde{D}_r$ is then expressed as

$$\tilde{D}_r(\tilde{x}, \tilde{\zeta}) = \frac{d\tilde{M}_a}{d\tilde{t}}$$

$$= \tilde{\gamma}\left\{\tilde{C}_{af}(\tilde{x}) + \left[\tilde{C}(\tilde{x}) - \tilde{C}_{af}(\tilde{x})\right]\tilde{G}(\tilde{\zeta})\right\}\left[1 - \tilde{\psi}^2 \tilde{F}^2(\tilde{\zeta})\tilde{u}_{ce}^2(\tilde{x})\right]$$

$$\tag{12.118}$$

In Fig. 12.52 (where $\tilde{y} = y/b_0$), isopleths of $\tilde{D}_r$ have been plotted with $C_{af} = 0$, $\tilde{f} = 0.05$, $\tilde{m} = 0$, $\tilde{\gamma} = 0.1$ and $\tilde{\psi} = 1$. The output broadly resembles a typical deltaic deposition pattern in current-dominated flows.

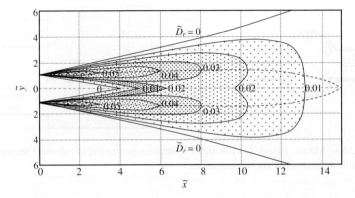

Figure 12.52. Isopleths of dimensionless deposition flux $\tilde{D}_r$ (from Özsoy [1977]).

12.5.4 *Basin with through-flow*

Consider a marina basin as a segment of a narrow waterway (Figs. 12.53a, b). It may be viewed as a basin having two flow entrances. The flow depth h is constant everywhere, B is the basin width and L_b is the length. Although the sketch shows the basin to be wider than the channel, the two widths are equal in the idealization. We will assume that an advective change $U\partial C/\partial x$ (where U is the steady, cross-sectional mean flow velocity) in the suspended sediment concentration C results from the net effect of settling represented by $w_s\partial C/\partial z$, and turbulent diffusion by $\partial(D_{sz}\partial C/dz)/\partial z$, where D_{sz} is the vertical mass diffusivity. Under these conditions, mass balance for suspended sediment in the basin is given by

$$U\frac{\partial C}{\partial x} = \frac{\partial}{\partial z}\left(D_{sz}\frac{\partial C}{\partial z}\right) + w_s\frac{\partial C}{\partial z} \qquad (12.119)$$

Equation 12.119 is readily obtained as the steady-state form of Eq. 6.107 by ignoring longitudinal (x-direction) diffusion and assuming a constant settling velocity. Consider uniformly distributed concentration C_0 at the flow entrance. Thus the boundary condition there is

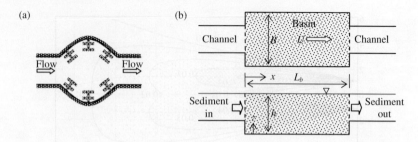

Figure 12.53. Marina basins: (a) Illustrative planform geometry; (b) channel segment as a basin with through-flow via two entrances.

$$C = C_0 \text{ at } x = 0 \tag{12.120}$$

We will select the case of deposition without resuspension of the deposit, for which the bottom boundary condition is

$$D_{sz} \frac{\partial C}{\partial z} = 0 \text{ at } z = 0 \tag{12.121}$$

Furthermore there is no transport across the water surface, *i.e.*,

$$D_{sz} \frac{\partial C}{\partial z} + w_s C = 0 \text{ at } z = h \tag{12.122}$$

For turbulent open channel flows, Eq. 12.119 together with the above conditions can be solved for C and therefore the sediment load provided D_{sz} and the velocity distribution $u(z)$ are known. We will select the parabolic distribution of D_{sz} based on its assumed equivalence to the momentum diffusivity ε_m (Eq. 2.64), and the logarithmic velocity defect law obtained by subtracting Eq. 2.59 from Eq. 2.57 (along with setting $\bar{u} = u$ and $\bar{u}_m = U$)

$$\frac{u - U}{u_*} = \frac{1}{\kappa} \left(\ln \frac{z}{h} + 1 \right) \tag{12.123}$$

where the Karman constant κ is taken as 0.4. The solution is plotted in Fig. 12.54, in which the suspended sediment removal ratio r_s is defined as

$$r_s = \frac{q_{si} - q_{se}}{q_{si}} \qquad (12.124)$$

where q_{si} is the unit sediment load entering the basin, and q_{se} is the load leaving the basin [Sarikaya, 1977].

Example 12.3: Consider the waterway shown in Fig. 12.55. The basin is located between flow cross-sections *A-A* and *B-B*. Although the flow is (semi-diurnal) tidal, sediment enters the basin only through *A-A* during flood flow. Part (or all) of the incoming sediment may deposit in the basin. Any remaining sediment exits *B-B* but does not return during ebb flow. Resuspension of the deposit is negligible under the prevailing current. Relevant parameters are: mean sediment flux through *A-A* during flood = 2×10^{-3} kg m^{-2} s^{-1}, average width of the basin, $B = 180$ m, distance L_b between *A-A* and *B-B* = 1,500 m, mean water depth $h = 4.92$ m, mean flood velocity $U = 0.2$ m s^{-1} and settling velocity $w_s = 2.1 \times 10^{-4}$ m s^{-1}. Assume the Darcy–Weisbach friction factor $f_c = 0.025$ and dry density of the deposit 400 kg m^{-3}. Determine the rate of shoaling in the basin.

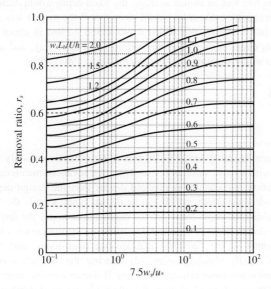

Figure 12.54. Nomogram for estimating the sediment removal ratio (based on Sarikaya [1977]).

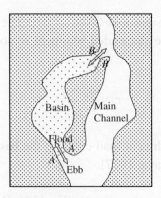

Figure 12.55. Basin with two open connections and selective sediment transport (from Mehta *et al.* [1984]).

We obtain the friction velocity $u_* = \sqrt{f_c / 8}\ U = 0.011$ m s^{-1}. With $w_s/u_* = 0.019$ and $w_s L_b/Uh = 0.32$, from Fig. 12.54 $r_s = 0.24$, which means that 24% of the incoming sediment settles out in the basin. Given the tide-mean sediment influx of 2×10^{-3} kg m^{-2} s^{-1} and a flow cross-section of 885.6 m^2, the sediment load would be 1.77 kg s^{-1}. Over the semi-diurnal flood duration of 6.21 h, 3.1×10^4 kg would be deposited. Since the bed density is 400 kg m^{-3}, the volume of deposit per cycle would become 24 m^3. Given the bed area of 2.7×10^5 m^2, the shoaling rate would be 8.8×10^{-5} m per cycle.

If the shoaling rate was an annual average, the basin depth would decrease by 6.2 cm in a year. Sedimentation of this order would mean that if the marina was excavated with 0.6 m of over-depth dredging[1], this added depth would be lost in about 9 years. After 10–15 years, boats could begin to feel the bottom when underway, and at low tide may rest on the bottom where shoaling is heavy [Mehta *et al.*, 1981; 1984].

---

[1] Navigation channel shoaling depths in US coastal areas have commonly varied between 0.3 m and 2.4 m with a mean value of 0.9 m. Thus, for advanced maintenance of channel against sedimentation, the channel is dredged deeper than the design depth. This extra depth is the over-dredged depth, which can be of two types. One is the "allowable pay over-depth", an additional depth for which the client agrees to pay the dredging contactor to ensure that the design depth is achieved, *i.e.*, allowing for inaccuracies in dredging and surveying. The other is the "advance maintenance over-depth" for: (1) maintaining project depth in rapidly shoaling areas, (2) reducing the frequency of maintenance dredging, or (3) allowing more efficient dredging if deeper cuts are more cost-effective. In most channels that are 9 to 14 m deep, the allowable pay over-depth varies from 0 to 0.9 m with clustering at 0.6 m. Advanced maintenance dredging is typically 0.6 m [U.S. Army Corps of Engineers, 1981; Marine Board, 1983].

12.5.5 *Basin with one entrance*

Consider a small basin with a short entrance channel (Fig. 12.56). As the channel is wide, tidal elevation inside the basin is practically the same as outside. Furthermore, tidal velocities in the basin are negligible. These conditions permit a zero-dimensional approach because it is time-dependent but independent of distance.

Flow continuity relates instantaneous discharge $Q(t)$ through the entrance to instantaneous water surface variation, $\eta(t)$ by

$$Q = A_B \frac{d\eta}{dt} \tag{12.125}$$

where A_B is the basin surface area. Given instantaneous sediment concentration in the basin $C(t)$, settling velocity w_s, mean water depth h and concentration outside the basin C_0, sediment mass balance is

$$A_B \frac{d}{dt}\left[C(h+\eta)\right] = C_{I0}Q - w_s' C A_B \tag{12.126}$$

where

$$C_{I0} = \begin{cases} C_0 & \text{during flood} \\ -C & \text{during ebb} \end{cases} \tag{12.127}$$

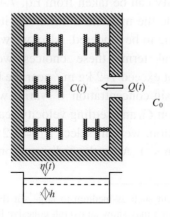

Figure 12.56. Schematic drawing of a small marina basin with one entrance.

The effective settling velocity is $w_s' = \alpha_{ef} w_s$, in which the settling velocity w_s is in quiescent water. The coefficient α_{ef} may be lower or higher than unity depending on the effects of turbulent shear rate and concentration (Eq. 7.114). In addition, α_{ef} depends on sediment composition and the residence time of suspended particles in the basin. The range of variation of α_{ef} is $O(10^{-1}-10^{1})$ [Winterwerp, 2005], which indicates that sedimentation must not be modeled without calibration of α_{ef}.

It is assumed that the basin is depositional without resuspension of the deposit by, for example, waves. Combining Eqs. 12.126 and 12.127 yields

$$(h+\eta)\frac{dC}{dt}+(C-C_{I0})\frac{d\eta}{dt}+\alpha_{ef}w_sC=0 \qquad (12.128)$$

Let the tide be of the form $\eta = -a_0\cos\sigma t$, where a_0 is the amplitude, $\sigma = 2\pi/T$ is the angular frequency and T is the period. Equation 12.128 then becomes

$$(h-a_0\cos\sigma t)\frac{dC}{dt}+(C-C_{I0})a_0\sigma\sin\sigma t+\alpha_{ef}w_sC=0 \qquad (12.129)$$

This equation can be solved numerically with an arbitrary cold-start initial value of concentration $C(0)$.

The settling velocity can be taken from Eq. 7.47 with $n_w = 4/3$. This would mean that while the material is cohesive, its concentration is not high enough for settling to be hindered. Also, we will conveniently select $\alpha_{ef} = 1$. In practical terms these choices would imply that the concentration must not exceed 1–2 kg m^{-3} near the bottom.[m]

The tide–mean basin concentration $\bar{C}$ varies with mean water depth h, outside concentration C_0 and settling velocity scaling coefficient a_w' in Eq. 7.47. For illustration, we will select $a_w' = 1.2\times10^{-5}$ (such that C in mg L^{-1} yields w_s in m s^{-1}). After the sediment concentration $C(t)$ in the

---

[m] For cohesionless sediment such as medium to large silt (but not fine silt, especially when mixed with clay, which may show an overall cohesive behavior), w_s may be taken to be independent of concentration, *i.e.*, $n_w = 0$.

basin is calculated, the incremental shoaling thickness Δh per time-step Δt is given by $w_s C \Delta t / \rho_D$, where ρ_D is the dry density of the deposit. The total depth of shoaling at any given instant is obtained by adding incremental shoaling from the beginning of hot-start computation, *i.e.*, after the effect of the initial condition inherent in cold-starting has become negligible. The mean shoaling rate *SR* is then calculated by dividing the total depth of shoaling by the total elapsed time. In reality the mean water depth will decrease as shoaling progresses. However, over short durations on the order of a few tidal cycles this decrease is usually small in comparison with the total water depth.

Mean shoaling rates in the basin for two tidal amplitudes (1 and 1.5 m) and an assumed mean water depth of 3 m are shown in Fig. 12.57. These results are for a marina in Florida (Camachee Cove Yacht Harbor near St. Augustine) with a basin of area $A_B = 33{,}370 \text{ m}^2$ and semi-diurnal tidal period $T = 12.4$ h. The shoaling rate increases with tidal amplitude and concentration outside the basin. The slightly non-linear increase with concentration results from its influence on the settling velocity.

For use of plots such as Fig. 12.57 at actual basins, long-term variability in C_0 must be accounted for. For Camachee Cove, Fig. 12.58 gives the frequency of occurrence Φ of concentration based on measurements in water over a period of several months.

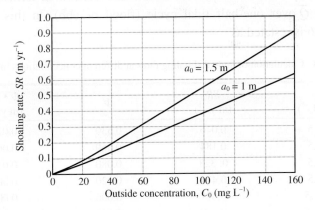

Figure 12.57. Variation of shoaling rate in Camachee Cove Yacht Harbor marina with tidal amplitude and suspended sediment concentration outside the basin (from Mehta and Maa [1985]).

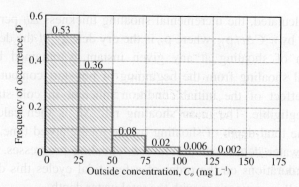

Figure 12.58. Suspended sediment concentration histogram based on measurements at Camachee Cove Yacht Harbor marina (from Mehta and Maa [1985]).

Example 12.4: For Camachee Cove Yacht Harbor marina determine the shoaling rate using information from Figs. 12.57 and 12.58. Assume tidal amplitude $a_0 = 1.5$ m.

Calculations are given in Table 12.14 by taking the concentration frequency into account. The sum of the product $\Phi \cdot SR$ is 0.132 m yr^{-1} = 13.2 cm yr^{-1}. Simulated isopleths of deposited sediment using depth-averaged numerical hydrodynamic modeling are shown in Fig. 12.59 [Hayter, 1983]. Note the loss of water depth near the entrance due to heavy shoaling associated with the local flow pattern (entrance vortex). Sedimentation close to the entrance is typical, and identifies a site where dredging would be required.

Let us now consider the case of small tidal range compared to water depth, *i.e.*, $\eta \ll h$, and represent inflow and outflow in terms of mean discharge $\overline{Q}$ over one-half tidal cycle (flood or ebb). In this instance Eq. 12.126 is restated as

Table 12.14. Calculation of annual shoaling rates at different concentrations in the basin.

C_0 (mg L^{-1})	SR (m yr^{-1})	Φ	$\Phi \cdot SR$ (m yr^{-1})
12.5	0.046	0.530	0.024
37.5	0.184	0.360	0.066
62.5	0. 328	0.080	0.026
87.5	0.474	0.020	0.009
112.5	0.621	0.006	0.004
137.5	0.768	0.002	0.002

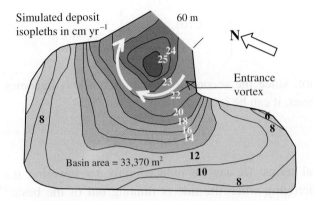

Figure 12.59. Model based simulation of fine sediment deposition isopleths in Camachee Cove Yacht Harbor marina (from Hayter [1983]).

$$A_B h \frac{dC}{dt} = C_0 \bar{Q} - C\bar{Q} - \alpha_w w_s C A_B \qquad (12.130)$$

Given the initial condition $C = C(0)$ at $t = 0$, this equation can be integrated over time to yield

$$C = C_{eq} + \left[C(0) - C_{eq} \right] e^{-\left(\frac{\bar{Q}}{hA_B} + \frac{\alpha_w w_s}{h} \right) t}; \quad C_{eq} = \frac{C_0 \bar{Q}}{\bar{Q} + \alpha_w w_s A_B} \qquad (12.131)$$

where C_{eq} is the equilibrium concentration in the basin [Winterwerp, 2005]. Note that when tidal variation is not ignored, C_{eq} must be defined as a tide-averaged value.

Let $T_h = hA_B / \bar{Q}$ and $T_v = h / \alpha_w w_s$ be the residence time-scales for suspended sediment concentration in the horizontal and the vertical directions, respectively. The shoaling rate SR is given by

$$SR = \left\{ \frac{T_v}{T_v + T_h} + \left[\frac{C(0)}{C_0} - \frac{T_v}{T_v + T_h} \right] e^{-\left(\frac{1}{T_v} + \frac{1}{T_h} \right) t} \right\} \alpha_{ef} \rho_D^{-1} w_s C_0 \qquad (12.132)$$

If the sediment settles very slowly, *i.e.*, $T_v \gg T_h$, the above equation reduces to

$$SR = \left\{ 1 + \left[\frac{C(0)}{C_0} - 1 \right] e^{-\frac{t}{T_h}} \right\} \alpha_{ef} \rho_D^{-1} w_s C_0 \qquad (12.133)$$

Furthermore, since the second term within brackets becomes small as time increases, it can be ignored. Then,

$$SR \approx \alpha_{ef} \rho_D^{-1} w_s C_0 \qquad (12.134)$$

In practical terms SR can be reduced only by lowering α_{ef}, e.g., by agitation-dredging when the tide is running out of the basin. This has been achieved for example by running the propeller of a tugboat where water is sufficiently shallow [Mehta *et al.*, 1981].

12.5.6 *Closed-end canal*

Closed-end narrow canals such as dead-end residential waterways and harbor docks can be havens for deposited fine sediment derived from the body of water to which they are connected. In such areas, particularly those where the tidal range is small or nil, sediment influx can occur by turbidity driven current into the canal basin. The driving head is the density difference between turbid water outside the basin and clearer water inside. Thus, in addition to gravity underflow, turbid density flows can occur over horizontal beds. As the current is usually too weak to resuspend the deposit, heavy sedimentation occurs when turbidity is high.

Figure 12.60 shows early laboratory flume experiments of Migniot [1968] in 1 m water depth and variable initial (and dense) concentration of natural mud at the entrance. At 300 kg m^{-3} the mud had low fluidity and settled out within a distance of 3.5 m. At 200 kg m^{-3} the distance increased to 7 m. These dense flows amount to gravity slides with final slope determined by the yield strength of viscoplastic mud. When the initial concentration was reduced to 88 kg m^{-3} the mud was transported considerably further and may be called a dense turbid current. It was found that deposition occurred when the frontal Reynolds number $Re = u_f h_f / v$ (where u_f is the turbid frontal velocity, h_f is a

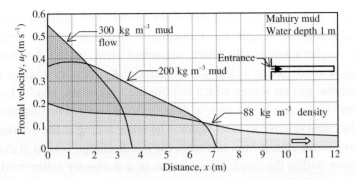

Figure 12.60. Variation of frontal velocity with distance in a laboratory flume at selected initial mud densities (based on Migniot [1968]).

characteristic front height and v is initial frontal viscosity) was less than 2,000.

Typical turbid fronts in closed-end canals tend to be far lower in concentration than 88 kg m^{-3} (which may have been close to the space-filling concentration). If the canal is long enough the finest particles that can deposit will settle out at some distance over which a wedge-like and stationary suspended sediment front can develop. As sediment deposits the concentration decreases with distance from the entrance (mouth) with the result that sedimentation also decreases with distance. Data from the 1.2 km long Rose–Fern Waterway in Marco Island, Florida (Fig. 12.61) seemingly corroborate such a trend.

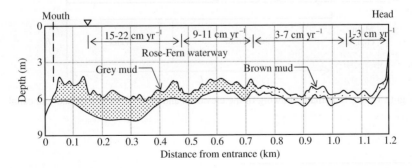

Figure 12.61. Mud deposition in Rose–Fern Waterway at Marco Island, Florida (adapted from Wanless [1975]).

Referring to Fig. 12.62, in a stationary turbid front inside a narrow closed-end canal of length L_b, momentum diffusion in the vertical direction is characteristically more important than in the horizontal direction. A small setup of the water surface $\eta(x)$ occurs against the closed end due to two forces which oppose the hydrostatic head. One is the pressure force due to the excess density $\Delta\rho_0$ of sediment suspension outside the canal relative to clear water. The other is drag at the bottom.

Due to overall equilibrium of forces the net discharge is zero at every cross-section within the front. However, as in a stationary saline wedge, there is force imbalance at different elevations resulting in inward flow at the bottom and outward flow above at an equal rate [McDowell, 1971; Gole *et al.*, 1973; Lin and Mehta, 1997].

Suspended sediment from the main channel is transported into the canal within the turbid layer $[0 < z < \eta(x)]$ and practically sediment-free water exits from the upper layer $[\eta(x) < z < h(x)]$. With increasing distance from the entrance the velocity $u(x,z)$ diminishes in both layers and becomes nil at the head of the front.[n] The outward surface velocity $u_s(x)$ increases from zero at the front head to maximum at the entrance. The velocity profile $u(x, z)$ at any distance from the entrance can be

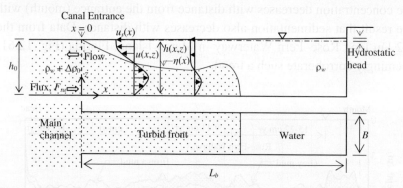

Figure 12.62. Schematic drawing of the elevation and plan views of a closed-end canal.

---

[n] The turbid front head is less well-defined than a saline head because at that distance practically no depositable matter remains in suspension. In laboratory setups the head tends to be only faintly visible when residual non-depositing particles are present.

obtained by solving the equations of motion (and continuity) in both layers. When the flow is non-turbulent the solutions are obtained analytically from linearized governing equations [Lin, 1987]. In this case the velocity profile is

$$
\tilde{u} = \begin{cases} \dfrac{\left(\dfrac{\zeta}{\tilde{\xi}}\right)^3 + \left(-3+1.5\tilde{\xi}^2 - 0.375\tilde{\xi}^3\right)\left(\dfrac{\zeta}{\tilde{\xi}}\right)^2 + \left(3-3\tilde{\xi}+0.75\tilde{\xi}^2\right)\left(\dfrac{\zeta}{\tilde{\xi}}\right)}{0.375\tilde{\xi}-0.5} \, ; 0<\zeta<\tilde{\xi} \\[4mm] \dfrac{\left(1.5-0.375\tilde{\xi}\right)\zeta^2 - \left(3-0.75\tilde{\xi}\right)\zeta+1}{0.375\tilde{\xi}-0.5} \, ; \tilde{\xi}<\zeta<1 \end{cases}
$$

(12.135)

where $\tilde{u}=u/|u_s|$, $\tilde{\xi}=x/h$ and $\zeta = z/h$. In Fig. 12.63, $\tilde{u}(\tilde{\xi},\zeta)$ is plotted against ζ and compared with measurements from a 14.7 m long, 0.1 m wide and 0.2 m high model of a closed-end canal built from Plexiglas. Turbid stationary front was generated using a suspension of kaolinite of 1,005 kg m^{-3} initial concentration. Data scatter is partly due to uncertainty in the measurement of the velocity profile, which was done by injecting in water a vertical streak of rhodamine dye and tracking its movement [Lin, 1987].

The study using kaolinite was extended to test suspensions of a non-cohesive flyash and a natural (cohesive) mud. Based on all test results

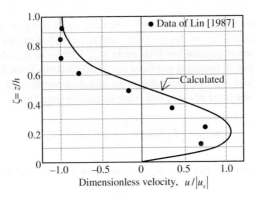

Figure 12.63. Normalized horizontal velocity profile in a stationary turbid front (from Lin and Mehta [1997]).

the steady-state horizontal flux F_{su} of suspended sediment at the entrance was found as follows

$$F_{su} = \alpha_{su} u_\Delta C_0 \tag{12.136}$$

where α_{su} is a proportionality coefficient, C_0 is the depth–mean suspended sediment concentration in the main (connecting) channel and u_Δ is the densimetric velocity defined as

$$u_\Delta = \sqrt{g \frac{\Delta \rho_0}{\rho_w} h_0} = \sqrt{g' h_0} \tag{12.137}$$

where h_0 is the water depth at the entrance and g' is reduced gravity. Note that $\Delta \rho_0 = C_0(s-1)/s$, where s is the specific gravity of particles. Then, from Eqs. 12.136 and 12.137 we obtain

$$F_{su} = \alpha_{su} \sqrt{\frac{g h_0}{\rho_w} \left(\frac{s-1}{s} \right)} C_0^{3/2} = K_{su} C_0^{3/2} \tag{12.138}$$

For a given canal the entire term before $C_0^{3/2}$ can be treated as a characteristic constant K_{su}. As seen in Fig. 12.64, the laboratory data yield a mean value of $K_{su} = 0.015$ (kg$^{-1/2}$ m$^{7/2}$ min^{-1}). Since K_{su} is dependent on basin dimensions, it must be determined from site-specific calibration.

Example 12.5: Consider a 1 km long and 5 m deep canal basin in which the (semi-diurnal) tide amplitude is 0.5 m, $\Delta \rho_0/\rho_w = 10^{-3}$ and $\alpha_{su} = 0.35$. Estimate the contribution of turbid current mass deposition per tidal cycle m_{turb} relative to m_{tide} due to tide-driven transport. Assume that all incoming sediment due to tide deposits in the basin.

The sediment mass m_{turb} deposited due to the turbid current through the lower half of the entrance during the flood phase of tide is approximately given by

$$m_{turb} = \frac{1}{4} F_{su} TBh_0 = \frac{\alpha_{su}}{4} \sqrt{g'} TBh_0^{3/2} C_0 \tag{12.139}$$

where T is the tidal period. The mass m_{tide} is associated with the tidal prism of the basin, i.e.,

$$m_{tide} = 2\alpha_T a_0 L_b BC_0 \tag{12.140}$$

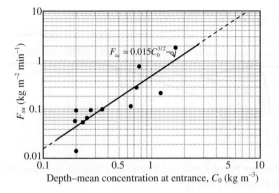

Figure 12.64. Sediment flux into the canal plotted against suspended fine sediment concentration at the entrance (from Lin and Mehta [1997]).

where α_T is the fraction of tide-driven sediment which deposits and a_0 is the tidal amplitude. Thus, dividing Eq. 12.141 by Eq. 12.140 yields

$$\frac{m_{turb}}{m_{tide}} = \frac{1}{8}\frac{\alpha_{su}}{\alpha_T}\sqrt{\frac{\Delta\rho_0}{\rho_w}}\left(\frac{h_0}{a_0}\right)\left(\frac{T\sqrt{gh_0}}{L_b}\right) \tag{12.141}$$

in which $T\sqrt{gh_0}$ is the tidal wave length. We have the following information: $\alpha_{su} = 0.35$, $\alpha_T = 1$ (indicating total deposition of incoming sediment), $\Delta\rho_0/\rho_w = 10^{-3}$, $h_0 = 5$ m, $a_0 = 0.5$ m, $T = 44{,}712$ s and $L_b = 1{,}000$ m. This yields $m_{turb}/m_{tide} = 4.33$, which underscores the dominant role of turbidity-induced deposition in this basin. Prototype applications of this type including details of the sedimentation patterns, as in Fig. 12.61, require that the effects of turbulence be included. For this, numerical simulation of the governing equations is essential [Lin, 1987].

Salinity difference in the longitudinal direction can modify the tidal current and thereby influence turbidity transport in entrance channels. Such an effect was recorded at the Mayport Naval Basin in Florida (Fig. 12.65). A 2.3 km long tidal channel with a nominal depth of 11 m connects the basin (area 5×10^5 m^2) to the St. Johns River close to its entrance (with two jetties) on the Atlantic coast of Florida.

The following method was applied to determine the sediment load. This was done to test the effects of placing a sediment deflecting wall at the junction of the channel with the river to reduce sedimentation in the basin [Headland, 1991; Eysink and Vermaas, 1983; Eysink, 1989].

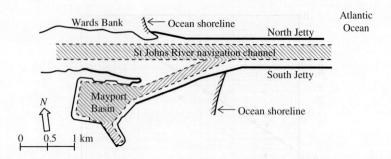

Figure 12.65. Mayport Naval Basin and its entrance through St. Johns River in northern Florida (based on Headland [1991]).

The horizontal, tide-mean suspended dry sediment mass load into the basin is

$$G_{ss} = C_0 \bar{Q}_\Delta \qquad (12.142)$$

where C_0 is a characteristic suspended sediment concentration. The tide-mean densimetric water discharge is obtained from

$$\bar{Q}_\Delta = \xi_1 A_c T \sqrt{\frac{1}{2}\left(\frac{\bar{\rho}_{max} - \bar{\rho}_{min}}{\bar{\rho}}\right)gh} - \xi_2 P \qquad (12.143)$$

where A_c is the channel entrance cross-sectional area, h is the mean water depth, T is the tidal period, $\bar{\rho}_{max}$ and $\bar{\rho}_{min}$ are the maximum and minimum saline water densities during a tidal cycle, $\bar{\rho}$ is the mean water density, P is the tidal prism and ξ_1, ξ_2 are empirical coefficients.

Example 12.6: Characteristic parameters for the Mayport channel and basin are: $h = 11$ m, $A_c = 2,000$ m^2, $T = 12.42$ h, $\bar{\rho} = 1,020$ kg m^{-3}, $\bar{\rho}_{max} - \bar{\rho}_{min} = 2.5$ kg m^{-3} and $P = 3.5 \times 10^6$ m^3, $C_0 = 25$ mg L^{-1}, $\xi_1 = 0.15$ and $\xi_2 = 0.95$. Calculate the sediment load if the basin surface area is 0.7 km^2 and the dry density of deposit $\rho_D = 600$ kg m^{-3}.

From Eq. 12.143, $\bar{Q}_\Delta = 1.55 \times 10^6$ m^3 per tidal cycle. Therefore, the dry sediment mass load $G_{ss} = 25 \times 1.5 \times 10^6 / 1,000 = 37.5 \times 10^3$ kg = 38.8 tons per tidal cycle. Given dry sediment deposit density of 600 kg m^{-3} and 705.3 tidal cycles per year, the annual shoaling rate would be 6.5 cm, which is within the prevailing range of values in this region of Florida.

12.5.7 *Sedimentation trap*

Sedimentation traps or pits dredged in a tidal channel or a stream take advantage of the reduction in flow velocity due to an increase in water depth [Parchure *et al.*, 2000; Bedient and Huber, 2002]. A relevant calculation is the rate at which the pit fills up, also called the volume-of-cut method [PIANC, 2008]. With reference to Fig. 12.66, consider the rate of change of elevation $h_p(t)$ above the dredged depth to be proportional to shoaling thickness $h_f - h_p(t)$, where h_f is the final elevation in the trap. In other words,

$$\frac{dh_p}{dt} = K_P \left[h_f - h_p(t) \right] \tag{12.144}$$

where K_P is a site-specific filling constant. An application of Eq. 12.144 was made for estimating the rate of shoaling in South San Francisco Bay, where a dike was to be breached to convert land area behind it into a wetland [Krone, 1985]. Due to desiccation of the land soil initially created by dike construction, and subsequent rise of mean sea level, the bed level in the bay had become higher than the land surface behind the dike. It was desired to estimate the time for the resubmerged land elevation to rise to its bay level by deposition of sediment from the bay after dike removal (Fig. 12.67).

At the time of dredging when $h_p(0) = 0$, integration of Eq. 12.144 yields

$$h_p(t) = h_f \left(1 - e^{-K_P t} \right) \tag{12.145}$$

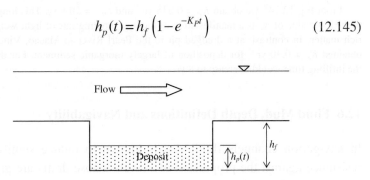

Figure 12.66. Schematic drawing of a partially filled sedimentation trap.

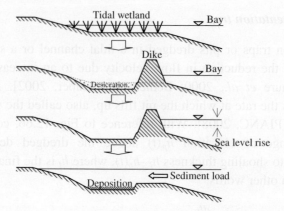

Figure 12.67. Sequence of engineering and natural events in San Francisco Bay (based on Krone [1985]).

Conveniently choosing $h_p(t)/h_f = 0.90$, *i.e.*, considering the shoaling thickness to be equal to 90% of the initial pit depth to represent a filled pit, the corresponding time $t_{90\%} = 2.3/K_P$. Given $h_p(1)$ as the shoaling thickness at the end of the first year, from Eq. 12.145 we obtain

$$K_P = \frac{h_P(1)}{[h_f - h_P(1)]} \ \text{yr}^{-1} \tag{12.146}$$

Example 12.7: A pit was dredged to a depth of 3 m below ambient water depth in a channel. Depth measurements at the end of the first year since dredging indicated that 0.13 m of organic-rich sediment had accumulated in the pit. Determine the time of infilling.

From Eq. 12.146 we obtain $K_P = 0.045$ yr^{-1} and $t_{90\%} = 50.8$ yr. This long period due to the low value of K_P is a manifestation of the low shoaling rate of light-weight organic-rich matter. In contrast, at a dredged pit in the Pearl River in Macao, Vincente [1992] obtained $K_P = 0.30$ yr^{-1} for deposition of largely inorganic sediment. For this K_P value the infilling time would decrease to 8 yr.

12.6 Fluid Mud, Depth Definitions and Navigability

In navigation channels fluid mud is known to induce significant drag resistance against the propulsion of vessels whose drafts are greater than the depth of water. In several large coastal harbors the so-called nautical depth concept is applied to identify the depth to which mud must be dredged to permit easier movement of vessels.

In general the upper level of the fluid mud layer is represented by the floc volume fraction ϕ_{hs} at which hindered settling flux commences. This implies that just below this level the settling flux is lower than above (Chapter 7). At the bottom of fluid mud the floc volume fraction corresponds to its space-filling value ϕ_{sf} at which fluid mud changes to a bed. Since floc volume fractions are not readily measured in the field we may replace them by the respective wet bulk densities ρ_{hs} and ρ_{sf}, which are less accurate but more easily obtained (Fig. 12.68). In Chapter 10 we noted that for cohesive inorganic muds undergoing settling in the absence of flow, representative densities are $\rho_{hs} = 1,050$ kg m^{-3} and $\rho_{sf} = 1,200$ kg m^{-3}.

Multi-frequency fathometers operating simultaneously at the (nominally) standard 200 kHz as well as lower frequencies (typically in the range of 10–33 kHz) are commonly used to identify the fluid mud layer. The higher frequency (acoustic) wave mainly reflects off the upper level of fluid mud while the lower frequency registers the lower level. The reflections actually mark the upper and lower *gradients* in density, the upper gradient being the lutocline and the lower one marking the bed surface [Kirby *et al.*, 1994]. Thus fathometer levels are rough measures of the desired density-based levels. A mismatch may also arise when current or wave-induced oscillations occur, as shearing of mud can modulate hindered settling as well as space-filling. Due to agitation-induced diffusion the hindered settling density ρ_{hs} may differ from its value under quiescent settling. The space-filling density ρ_{sf} also changes

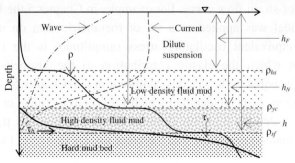

Figure 12.68. Nautical depth definition (based on Parker [1986]).

as agitation retards the formation of permanent inter-particle bonds. For these reasons, and also because the transitory fluid mud levels and thickness depend on the flow, time-series of fathometer soundings based on frequent measurements are helpful even though not ideal.

Lack of detection of fluid mud can impact harbor navigation if the vessel is forced to propel through dense fluid mud. Accordingly, nautical depth is the depth over which the vessel's keel does not experience unacceptably high resistance due to mud [Parker, 1986; Herbich *et al.*, 1989; Parker, 1994]. In Fig. 12.68 three definitions of depth are indicated:

(1) Hydrodynamic depth h at which the (tidal) flow velocity or wave velocity is nil. This is the depth at which erosion and deposition are calculated from the bed shear stress τ_b in typical (numerical) models (Chapters 7 and 9).

(2) Acoustic depth h_F determined by a standard (200 kHz) fathometer. This is the depth of hard-bottom surveys.

(3) Nautical depth h_N defined by mud property that influences the ease of navigation.

A reasonably reliable measure of resistance to propulsion due to mud is the undrained stress τ_y marking the onset of plastic yield. As we noted in Chapter 10, its value is practically nil in low density fluid mud but increases rapidly as density increases with depth (Fig. 12.69). As we noted briefly in Chapter 5, τ_y associated with a current (rotational flow in a rheometer) differs from τ_y under waves (oscillatory flow in a rheometer). Under non-oscillating flow, τ_y and viscosity η are conveniently obtained by fitting a simple constitutive model of the stress versus rate-of-strain flow curve. For example, in Chapter 5 the Herschel–Bulkley model was applied to a set of rheometric data on muds. For waves the equivalent oscillatory stress (amplitude) is the flow-point stress τ_{yf} at which the elastic contribution to the viscoelastic shear modulus G' is equal to the viscous contribution G'' (Chapter 5).

In tidal flow, h_N is the depth at which τ_y exceeds a critical or threshold value τ_{yc} at a density ρ_{yc}. This density, which lies between ρ_{sf} and ρ_{hs}, varies with mud composition, and therefore differs from port to port as in

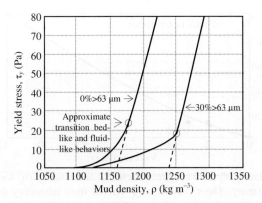

Figure 12.69. Variation of τ_y of a marine mud with density ρ from controlled-shear stress rheometry (adapted from Parker [1994]).

Table 12.15. In Fig. 12.69, using laboratory rheometric data for a sandy marine mud, ρ_{sf} is estimated by extrapolation. The point $\tau_y = \tau_{yc}$ marks the transition from a steep slope when $(\tau_y > \tau_{yc})$ to a gentler slope when $(\tau_y < \tau_{yc})$. The plots illustrate the approximate basis of extrapolation as well as how slope transition changes when the sand fraction is removed (0% > 63 μm). Specifically, $\tau_{yc} = 18$ Pa at $\rho_{sf} = 1{,}240$ kg m^{-3} for the original sediment is lower than 23 Pa at 1,160 Pa for the wholly fine sediment (0% > 63 μm).

Gravity probes have been used to report *in situ* mud viscosity η [Dasch and Wurpts, 2001]. In Fig. 12.70, viscosity data obtained in this way at different depths (and densities) in the soft mud layer in the Emden Outer Harbor, Germany, have been plotted against the yield stress τ_y. The yield stress was obtained from the viscoplastic flow curve in a

Table 12.15. Transition densities for increase in yield stress.

Site	Space-filling density (kg m^{-3})
Wouri River (Cameroon)	1,118
Mahury River (Fr. Guiana)	1,158
Hamiz River (Algeria)	1,274
Sandy mud (30% > 63 μm)	1,240
Fine mud (0% > 63 μm)	1,160

Source: Migniot [1968], Parker [1994].

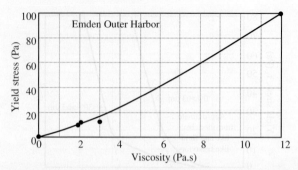

Figure 12.70. Relationship between yield stress (τ_y) and viscoity (η) for mud in Emden Outer Harbor, Germany. The τ_y values were derived from laboratory rheometry and η was obtained *in situ* (from Dasch and Wurpts [2001]).

laboratory rheometer at corresponding field densities. Although the data are sparse, they suggest a well behaved, somewhat non-linear relationship between η and τ_y which in turn permits the use of viscosity alone to identify h_N.

In Fig. 12.71a two sets of data from Emden have been plotted. A dual frequency fathometer was used to obtain two density levels in mud, the upper one by acoustic sounding at 210 kHz frequency and the lower one at 15 kHz. The lateral transect shown was made downstream of the Nesserland Sea-Lock. Fluid mud occurred between the two levels. Also plotted are constant viscosity lines, or isoviscs, for 1, 2, 5 and 10 Pa.s. These measurements, on April 9, 1999 were repeated on July 6, 1999 (Fig. 12.71b). While the upper and lower levels of the fluid mud remained practically unchanged over the nearly two-month intervening duration, the elevations of the isoviscs changed. For instance, the 1 and 2 Pa.s lines rose within the fluid mud layer. These changes suggest the thixotropic nature (Chapter 5) of this sediment.

When viscosity is used to represent mud resistance to vessel movement, *in situ* detection of isoviscs rather than constant density (isopycnal) lines becomes essential. Teeter [1992] proposed 8 Pa.s viscosity as the threshold for nautical depth at the Gulf Port and Calcasieu channels in Louisiana. In other words, the minimum depth for navigation was defined by the depth at which the 8 Pa.s isovisc occurred, with higher values below that level. At Emden 12 Pa.s was proposed [Dasch and Wurpts, 2001].

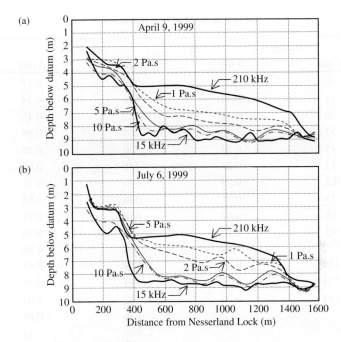

Figure 12.71. Density horizons from a dual frequency fathometer and measured isoviscs in Emden Outer Harbor: (a) April 9, 1999 survey; (b) July 6, 1999 survey (from Dasch and Wurpts [2001]).

12.7 Exercises

12.1 Plot mud beach profiles A and B from Table E12.1 and find their best-fit semi-empirical coefficients for Eq. 12.14.

12.2 A steady underflow of mud (density $\rho_m = 1,180$ kg m^{-3}) is to be maintained in a harbor channel beneath still water of density $\rho_w = 1,005$ kg m^{-3} (Fig. E12.1). The channel bed is inclined at angle $\theta = 3.5°$. (a) Considering the mud constitutive behavior to be Bingham plastic with a yield stress of 45 Pa, determine the minimum mud thickness h_m required to maintain the underflow. (b) If for an underflow depth of 0.5 m its downflow velocity is 1.0 m d^{-1} and the width of the channel is 72 m, find the length of a 3 m deep (below existing channel bed) trap of the same width as the channel to capture underflow sediment. Dredging of the filled trap is to be carried out once every year. Assume that the deposit consolidates rapidly and reaches a density of 1,250 kg m^{-3}. What will be the result if the trap width is only 36 m? Under what conditions would it be feasible to dig such a narrow trap?

Table E12.1. Mud beach profiles.

Profile A		Profile B	
x (m)	h (m)	x (m)	h (m)
9	0.1	50	0.17
14	0.14	158	0.34
26	0.12	237	0.45
33	0.16	265	0.46
45	0.24	330	0.56
53	0.27	438	0.66
63	0.29	510	0.74
75	0.37	546	0.69
85	0.47	618	0.78
95	0.53	632	0.78
122	0.59	661	0.86
148	0.64	704	0.83
174	0.71	797	0.97
198	0.74	833	1.04
		876	0.89
		926	1.15
		1048	1.29
		1127	1.35
		1149	1.39
		1228	1.35
		1260	1.44

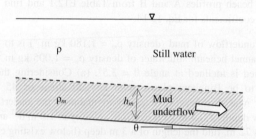

Figure E12.1. Mud underflow definition sketch.

12.3 Given the mud profile parameters in Table 12.10, plot $q_s(x)$ for two additional values of $H_0 = 2$ m and 1.5 m. Select the range of $\bar{x}$ from 0 to 10. Calculate the total load $Q_s(\text{kg s}^{-1})$ in each case.

12.4 In the basin of text Example 12.4, you are required to reduce the annual sedimentation rate to 6 cm. Calculate the change in water depth needed to achieve that goal.

12.5 For the basin in text Example 12.4, show that solely using the weighted-mean value of the frequency of occurrence of the outside concentration yields a significantly lower rate of shoaling. Briefly indicate the reason.

12.6 For waters outside the basin in text Example 12.4, make use of Fig. E12.2 (instead of the histogram of Fig. 12.58) for the cumulative frequency of occurrence of suspended sediment concentration, and recalculate the rate of shoaling. The plot is based on just a few measurements of suspended sediment concentration. When data points are sparse as in this case, they may be conveniently plotted on log-probability coordinates, and the best-fit line used to obtain the required concentration histogram [Ray Krone, personal communication].

12.7 At a marina basin simulation of sedimentation using a two-dimensional, depth-averaged numerical model for cohesive sediment transport led to the plot of basin-mean rate of shoaling as a function of sediment settling velocity in Fig. E12.3. Determine the annual rate of shoaling assuming that the concentration histogram of Fig. 12.58 is applicable *inside* the basin. Select

$$w_s = 2\times10^{-5}C^{1.2} \qquad (\text{E}12.1)$$

where the units of C and w_s are mg L^{-1} and m s^{-1}, respectively. Why is the exponent 1.07 of the shoaling rate curve different from 1.0?

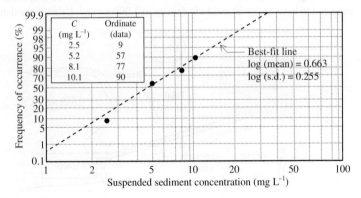

Figure E12.2. Cumulative frequency of occurrenceof suspended sediment concentration outside a marina basin.

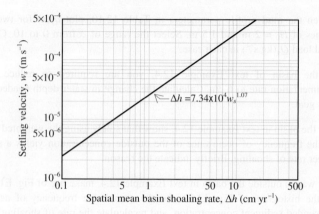

Figure E12.3.   Variation of shoaling rate with settling velocity inside a marina basin (based on Hayter [1983]).

12.8   In order to improve navigation in and out of a marina basin, a 300 m long and 3 m deep approach channel is to be dredged (Fig. E12.4). The width of the channel at the bottom is to be 25 m, and side slopes 1:5. The channel shoals at the rate of 3,000 $m^3 y^{-1}$. This is the net deposition rate representing an annual trapping efficiency of 67%. In other words, one-third of the total sediment deposited in the channel is resuspended and transported out of the channel (due to storm waves). In order to dispose the dredged material, a 150 m by 200 m site at the end of the channel, naturally 3 m deep, has been designated. Assume the depth elsewhere to be 2 m.

Storm winds of 55 km h^{-1} blow from the SW and NW with fetches of 4 and 12 km, respectively. Quasi-steady currents of about 0.1 m s^{-1} occur in the area. The natural wet bulk density of the bed is 1,350 kg m^{-3}, while the disposed slurry is expected to have an initial wet bulk density of 1,200 kg m^{-3}. Using these data decide

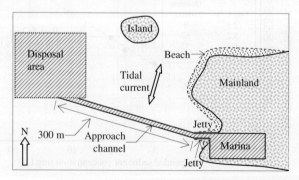

Figure E12.4.   Marina and proposed dredged material disposal site (map not drawn to scale).

whether the dredging disposal practice will increase water turbidity in the region. This is a matter of interest because high turbidity would impact the aquatic habitat and the nearby clean sandy beach. State your assumptions as well as the values of all relevant variables from the text.

12.9 A small port basin is to be constructed in a bay with a fine-sediment bottom (Fig. E12.5). Tidal range in the bay is 0.5 m at neap and 2 m at spring. On a clear day the suspended sediment concentration is 30–50 mg L^{-1} at mid-depth and its settling velocity is 5×10^{-4} m s^{-1}. Typical river discharge is on the order of 10% of the tidal prism. The maximum wave height in the vicinity of the proposed basin is 0.8 m during storms, when the mid-depth concentration rises to about 100 mg L^{-1}. Determine the sedimentation potential of the basin. How would you minimize sedimentation? Hint: Use the following O'Brien–LeConte equation (Eq. E12.2) to obtain the tidal prism P (m^3)$^\circ$ from the flow cross-sectional area A_c (m^2) at the bay mouth [LeConte, 1905; O'Brien, 1933; O'Brien, 1969; Powell *et al.*, 2006].

$$A_c = 9 \times 10^{-4} P^{\,0.85} \qquad \text{(E12.2)}$$

State your assumptions and values of any variables you select from the text.

$^\circ$ The tidal prism of a bay is the volume of sea water that enters the bay during flood flow between low water slack (LWS) and the next high water slack (HWS). Thus,

$$P = \int_{LWS}^{HWS} Q(t)\,dt - \int_{LWS}^{HWS} Q_r(t)\,dt = \int_0^{T/2} (Q - Q_r)\,dt \qquad \text{(F12.1)}$$

where $Q(t)$ is the tidal water discharge, $Q_r(t)$ is the river water discharge (outflow) and T is the tidal period. Assume that Q can be expressed as a simple harmonic function with amplitude *i.e.*, peak discharge, Q_m:

$$Q = Q_m \sin \sigma t \qquad \text{(F12.2)}$$

where $\sigma = 2\pi/T$ is the tidal frequency. Assume also that $Q_r/Q \ll 1$. Substitution of Eq. F.12.2 into Eq. F12.1 gives

$$P = \frac{Q_m T}{\pi} \qquad \text{(F12.3)}$$

In order to account for the contribution of overtide to the actual discharge Q, Eq. F12.3 must be divided by an overtide correction coefficient C_K [Keulegan, 1967]. Its value ranges from 0.81 to 1, with a mean value of 0.86. Therefore, since $Q_m = u_{max}A_c$, where u_{max} is the tidal velocity amplitude and A_c is the mean entrance channel cross-sectional area, Eq. F12.3 becomes

$$P = \frac{u_{max} A_c T}{\pi C_K} \qquad \text{(F12.5)}$$

which permits the determination of P from measurement of u_{max}.

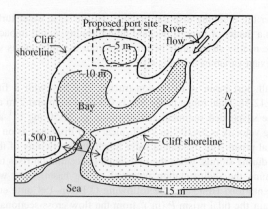

Figure E12.5. Bay in which a small port is to be constructed.

12.10 A port facility is to be constructed on the coast for unloading Liquefied Natural Gas to be supplied by LNG carriers. The dredged channel, turning basin and breakwater are as shown in Fig. E12.6. Waves occur from the southwest during four rainy months (mid-June to mid-September) and from the northwest during non-rainy months. Assume a rainy season wave height of 3.3 m and non-rainy season height of 1 m. The wave period is 10 s throughout. When the (semi-diurnal) tide ebbs out of the estuary, the associated fine suspended sediment moves southward along the length of the dredged channel due to a 0.8 m s^{-1} current. Assume that suspended sediment concentration in ebb flow is 300 mg L^{-1} during rainy months and 40 mg L^{-1} during non-rainy months.

Estimate the annual rate of sedimentation in the proposed ship channel, and determine if the deposited mud will be liquefied by waves. What will be the implication of fluid mud in the channel if it occurs? Estimate the dredging frequency

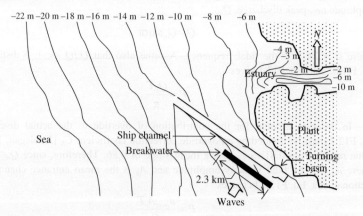

Figure E12.6. LNG port facility. The length of the breakwater is 2.3 km. Assume that the plot is drawn to scale.

and volume of material. State your assumptions and values of variables you select from the text.

12.11 Suspended sediment distribution in a turbid plume issuing into the sea from a coastal entrance is to be modeled.

(1) Show that for a jet over a flat bottom without friction, the centerline velocity and width (plotted in Fig. 12.49) would be

ZFE $(\tilde{x} < \tilde{x}_{fe})$

$$\tilde{u}_{ce} = 1, \quad \tilde{r} = \frac{I_1 - I_2(1 + \alpha_e \tilde{x})}{I_1 - I_2}, \quad \tilde{b} = \frac{(1 - I_2)(1 + \alpha_e \tilde{x}) - (1 - I_1)}{I_1 - I_2} \quad (E12.3)$$

ZEF $(\tilde{x} > \tilde{x}_{fe})$

$$\tilde{u}_{ce} = \frac{1}{\sqrt{1 + \frac{2\alpha_e I_2}{I_1}(\tilde{x} - \tilde{x}_{fe})}}, \quad \tilde{b} = \frac{1 + \frac{2\alpha_e I_2}{I_1}(\tilde{x} - \tilde{x}_{fe})}{I_2} \quad (E12.4)$$

(2) Using the following values plot the jet half-width b, centerline velocity u_{ce} and suspended sediment concentration C as functions of seaward distance x.

h_0	5	m
b_0	200	m
u_0	0.8	m s^{-1}
u_{cd}	0.2	m s^{-1}
w_s	0.000001	m s^{-1}
C_0	2	kg m^{-3}
C_{af}	0.1	kg m^{-3}
I_1	0.45	
I_2	0.316	
I_3	0.6	
I_4	0.368	
I_5	0.278	
α_e	0.036	(ZFE)
α_e	0.05	(ZEF)

12.12 In order to design dry containment areas for disposal of dredged marine material the Bulking Factor (*BF*), *i.e.*, volume of material after dredging divided by volume before dredging, is required. We will define this ratio as

$$BF = \frac{\rho_{Dpr}}{\rho_{Dpo}} = \frac{\rho_{Dpr} - \rho_w}{\rho_{Dpo} - \rho_w} \quad (E12.5)$$

where ρ_{Dpr} is the pre-dredging (dry bulk) density and ρ_{Dpo} is the post-dredging density. From a navigation channel 1,520 m^3 of shoaled sediment (wet bulk density 1,535 kg m^{-3}) is to be dredged by a trailing-suction hopper dredger. The square containment area on dry land for dredged material disposal is to be 1.7 m in height, including free-board (containment wall height above maximum fill elevation) of 0.5 m. Determine the containment area if the discharge from the dredger has a wet bulk density of 1,090 kg m^{-3}. Assume water density of 1,025 mg m^{-3}.

12.13   You are given the density profile in Fig. E12.7 from a turbid bay containing fluid mud. The bottom sediment is entirely fine, *i.e.*, smaller than 63 microns. Determine the nautical depth if the critical yield stress for navigation (as defined by keel resistance) is 40 Pa. What is likely to be the depth reported by a standard fathometer?

12.14   As alternatives to breakwaters using hard materials such as rocks and concrete, a variety of soft solutions to prevent shore erosion have been attempted. They include planting *Spartina* or other saltmarsh vegetation [Kirby, 1995]; clay dikes; non-traditional combination of poles, palm tree leaves and concrete blocks [Augustinus, 1988]; shore-parallel berms of mud to absorb wave energy [Dredging Research Technical Notes, 1992], *etc.* The location of a 2.75 km long and 300 m wide Mobile Berm using mud dredged from the Mobile (Alabama) Ship Channel is shown in Fig. E12.8a. Wave spectra in Fig. 12.8b indicate the modal wave period to be about 5 s and loss of wave energy between the offshore and inshore wave stations about 50%.

   Consider Fig. E12.9. The height of water above the berm crest is 1.8 m, incident wave amplitude is 0.65 m, wave period is 8.7 s, water density is 1,020 kg m^{-3}, and mud dynamic viscosity is 0.1 Pa.s. At the berm crest a 0.3 m thick

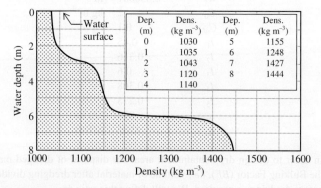

Figure E12.7. Typical mud wet bulk density profile below water surface in a bay laden with fluid mud.

(a)

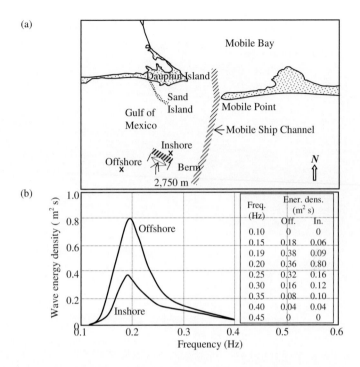

(b)

Figure E12.8. Mobile Berm: (a) Location; (b) wave spectra at the offshore and inshore wave stations in August 1988 (from McClellan *et al.* [1990]).

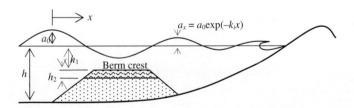

Figure E12.9. Definitions related to mud berm.

layer of mud with a (wet bulk) density of 1,300 kg m^{-3} participates in the wave damping process. Calculate how much wave energy is lost as the wave travels across the 378 m wide crest. Hint: Verify that at the top of the berm the wave is in shallow water. Use Eq. 11.85 to calculate the wave damping coefficient.

Figure E12.8. Mobile Berm. (a) Location. (b) wave spectrum at the offshore and inshore wave sensors in August 1988 (from McClellan et al. [1990]).

Figure E12.9. Definitions related to mud berm.

layer of mud with a wet bulk density of 1,300 kg m⁻³ participates in the wave damping process. Calculate how much wave energy is lost as the wave travels across the 4 km wide creek. Hint: Verify that at the top of the berm the wave is in shallow water. Use Eq. 12.88 to calculate the wave damping coefficient.

Appendix A

Atomic Weights and
Physicochemical Constants

Table A.1 Atomic weights of selected elements.

Name	Symbol	Atomic number	Atomic weight
Aluminum	Al	13	26.98
Arsenic	As	33	74.92
Barium	Ba	56	137.33
Beryllium	Be	4	9.01
Bismuth	Bi	83	208.98
Boron	B	5	10.81
Bromine	Br	35	79.90
Cadmium	Cd	48	112.41
Cesium	Cs	55	132.91
Calcium	Ca	20	40.08
Carbon	C	6	12.01
Chlorine	Cl	17	35.45
Chromium	Cr	24	52.00
Cobalt	Co	27	58.93
Copper	Cu	29	63.55
Fluorine	F	9	19.00
Gold	Au	79	196.97
Hydrogen	H	1	1.01
Iodine	I	53	126.90
Iron	Fe	26	55.85
Lead	Pb	82	207.21
Lithium	Li	3	6.94
Magnesium	Mg	12	24.31
Manganese	Mn	25	54.94
Mercury	Hg	80	200.59
Nickel	Ni	28	58.69
Nitrogen	N	7	14.01
Oxygen	O	8	16.00
Phosphorus	P	15	30.97

Potassium	K	19	39.10
Rubidium	Rb	37	85.47
Silicon	Si	14	28.09
Silver	Ag	47	107.87
Sodium	Na	11	22.99
Strontium	Sr	38	87.62
Sulfur	S	16	32.07
Tin	Sn	50	118.71
Titanium	Ti	22	47.87
Tungsten	W	74	183.84
Zinc	Zn	30	65.41

Table A.2 Physicochemical constants.

Quantity	Symbol	Value
Velocity of light	c_v	2.998×10^8 m s^{-1}
Planck's constant	h_p	6.626×10^{-34} J s
Avogadro number	N_A	6.022×10^{23} mole^{-1}
Faraday constant	F_D	96485 Coulombs mole^{-1}
Absolute temperature of ice point (0°C)	T	273.15 °K ($\approx$ 273 °K)
Electronic charge	$e_c = F_D/N_A$	1.602×10^{-19} Coulomb
Gas constant	R	8.314 J °K^{-1} mole^{-1} 8.314 L kPa °K^{-1} mole^{-1}
Boltzmann constant	$\kappa_B = R/N_A$	1.381×10^{-23} J °K^{-1}

Appendix B

Equations of Flow Continuity and Fluid Motion

B.1 Equation of Continuity

Rectangular coordinates (x, y, z):

$$\frac{\partial \rho}{\partial t} + \frac{\partial}{\partial x}(\rho u_x) + \frac{\partial}{\partial y}(\rho u_y) + \frac{\partial}{\partial z}(\rho u_z) = 0 \tag{B.1}$$

where u_x, u_y and u_z are general representations of u, v and w, respectively.

Cylindrical coordinates (r, θ, z):

$$\frac{\partial \rho}{\partial t} + \frac{1}{r}\frac{\partial}{\partial r}(\rho r u_r) + \frac{1}{r}\frac{\partial}{\partial \theta}(\rho u_\theta) + \frac{\partial}{\partial z}(\rho u_z) = 0 \tag{B.2}$$

Spherical coordinates (r, θ, ϕ):

$$\frac{\partial \rho}{\partial t} + \frac{1}{r^2}\frac{\partial}{\partial r}(\rho r^2 u_r) + \frac{1}{r\sin\theta}\frac{\partial}{\partial \theta}(\rho u_\theta \sin\theta)$$

$$+ \frac{1}{r\sin\theta}\frac{\partial}{\partial \phi}(\rho u_\phi) = 0 \tag{B.3}$$

B.2 Equation of Motion in Rectangular Coordinates

In terms of shear stress τ:

x-component:

$$\rho\left(\frac{\partial u_x}{\partial t}+u_x\frac{\partial u_x}{\partial x}+u_y\frac{\partial u_x}{\partial y}+u_z\frac{\partial u_x}{\partial z}\right)$$

$$=-\frac{\partial p}{\partial x}-\left(\frac{\partial \tau_{xx}}{\partial x}+\frac{\partial \tau_{yx}}{\partial y}+\frac{\partial \tau_{zx}}{\partial z}\right)+\rho g_x \qquad (B.4)$$

y-component:

$$\rho\left(\frac{\partial u_y}{\partial t}+u_x\frac{\partial u_y}{\partial x}+u_y\frac{\partial u_y}{\partial y}+u_z\frac{\partial u_y}{\partial z}\right)$$

$$=-\frac{\partial p}{\partial y}-\left(\frac{\partial \tau_{xy}}{\partial x}+\frac{\partial \tau_{yy}}{\partial y}+\frac{\partial \tau_{zy}}{\partial z}\right)+\rho g_y \qquad (B.5)$$

z-component:

$$\rho\left(\frac{\partial u_z}{\partial t}+u_x\frac{\partial u_z}{\partial x}+u_y\frac{\partial u_z}{\partial y}+u_z\frac{\partial u_z}{\partial z}\right)$$

$$=-\frac{\partial p}{\partial y}-\left(\frac{\partial \tau_{xz}}{\partial x}+\frac{\partial \tau_{yz}}{\partial y}+\frac{\partial \tau_{zz}}{\partial z}\right)+\rho g_z \qquad (B.6)$$

In terms of velocity gradients for a Newtonian fluid with constant ρ and η:

x-component:

$$\rho\left(\frac{\partial u_x}{\partial t}+u_x\frac{\partial u_x}{\partial x}+u_y\frac{\partial u_x}{\partial y}+u_z\frac{\partial u_x}{\partial z}\right)$$

$$=-\frac{\partial p}{\partial x}-\eta\left(\frac{\partial^2 u_x}{\partial x^2}+\frac{\partial^2 u_x}{\partial y^2}+\frac{\partial^2 u_x}{\partial z^2}\right)+\rho g_x$$

(B.7)

y-component:

$$\rho\left(\frac{\partial u_y}{\partial t}+u_x\frac{\partial u_y}{\partial x}+u_y\frac{\partial u_y}{\partial y}+u_z\frac{\partial u_y}{\partial z}\right)$$

$$=-\frac{\partial p}{\partial y}-\eta\left(\frac{\partial^2 u_y}{\partial^2 x}+\frac{\partial^2 u_y}{\partial^2 y}+\frac{\partial^2 u_y}{\partial^2 z}\right)+\rho g_y$$

(B.8)

z-component:

$$\rho\left(\frac{\partial u_z}{\partial t}+u_x\frac{\partial u_z}{\partial x}+u_y\frac{\partial u_z}{\partial y}+u_z\frac{\partial u_z}{\partial z}\right)$$

$$=-\frac{\partial p}{\partial z}-\eta\left(\frac{\partial^2 u_z}{\partial^2 x}+\frac{\partial^2 u_z}{\partial^2 y}+\frac{\partial^2 u_z}{\partial^2 z}\right)+\rho g_z$$

(B.9)

B.3 Equation of Motion in Cylindrical Coordinates

In terms of shear stress τ:

r-component:

$$\rho\left(\frac{\partial u_r}{\partial t}+u_r\frac{\partial u_r}{\partial r}+\frac{u_\theta}{r}\frac{\partial u_r}{\partial \theta}-\frac{u_\theta^2}{r}+u_z\frac{\partial u_r}{\partial z}\right)$$

$$=-\frac{\partial p}{\partial r}-\left(\frac{1}{r}\frac{\partial}{\partial r}\left(r\tau_{rr}\right)+\frac{1}{r}\frac{\partial \tau_{r\theta}}{\partial \theta}-\frac{\tau_{\theta\theta}}{r}+\frac{\partial \tau_{rz}}{\partial z}\right)+\rho g_r \qquad (B.10)$$

θ-component:

$$\rho\left(\frac{\partial u_\theta}{\partial t}+u_r\frac{\partial u_\theta}{\partial r}+\frac{u_\theta}{r}\frac{\partial u_\theta}{\partial \theta}+\frac{u_r u_\theta}{r}+u_z\frac{\partial u_\theta}{\partial z}\right)$$

$$=-\frac{1}{r}\frac{\partial p}{\partial \theta}-\left(\frac{1}{r^2}\frac{\partial}{\partial r}\left(r^2\tau_{r\theta}\right)+\frac{1}{r}\frac{\partial \tau_{\theta\theta}}{\partial \theta}+\frac{\partial \tau_{\theta z}}{\partial z}+\rho g_\theta\right) \qquad (B.11)$$

z-component:

$$\rho\left(\frac{\partial u_z}{\partial t}+u_r\frac{\partial u_z}{\partial r}+\frac{u_\theta}{r}\frac{\partial u_z}{\partial \theta}+u_z\frac{\partial u_z}{\partial z}\right)$$

$$=-\frac{\partial p}{\partial z}-\left(\frac{1}{r}\frac{\partial}{\partial r}\left(r\tau_{rz}\right)+\frac{1}{r}\frac{\partial \tau_{\theta z}}{\partial \theta}+\frac{\partial \tau_{zz}}{\partial z}\right)+\rho g_z \qquad (B.12)$$

In terms of velocity gradients for a Newtonian fluid (with constant ρ and η):

r-component:

$$\rho\left(\frac{\partial u_r}{\partial t}+u_r\frac{\partial u_r}{\partial r}+\frac{u_\theta}{r}\frac{\partial u_r}{\partial \theta}-\frac{u_\theta^2}{r}+u_z\frac{\partial u_r}{\partial z}\right)$$

$$=-\frac{\partial p}{\partial r}+\eta\left[\frac{\partial}{\partial r}\left(\frac{1}{r}\frac{\partial}{\partial r}\left(ru_r\right)\right)+\frac{1}{r^2}\frac{\partial^2 u_r}{\partial \theta^2}-\frac{2}{r^2}\frac{\partial u_\theta}{\partial \theta}+\frac{\partial^2 u_r}{\partial z^2}\right]+\rho g_r \qquad (B.13)$$

θ-*component*:

$$\rho\left(\frac{\partial u_\theta}{\partial t} + u_r \frac{\partial u_\theta}{\partial r} + \frac{u_\theta}{r}\frac{\partial u_\theta}{\partial\theta} + \frac{u_r u_\theta}{r} + u_z \frac{\partial u_\theta}{\partial z}\right)$$

$$= -\frac{1}{r}\frac{\partial p}{\partial\theta} + \eta\left[\frac{\partial}{\partial r}\left(\frac{1}{r}\frac{\partial}{\partial r}(r u_\theta)\right) + \frac{1}{r^2}\frac{\partial^2 u_\theta}{\partial\theta^2} + \frac{2}{r^2}\frac{\partial u_r}{\partial\theta} + \frac{\partial^2 u_\theta}{\partial z^2}\right] + \rho g_\theta \qquad (\text{B.14})$$

z-component:

$$\rho\left(\frac{\partial u_z}{\partial t} + u_r \frac{\partial u_z}{\partial r} + \frac{u_\theta}{r}\frac{\partial u_z}{\partial\theta} + u_z \frac{\partial u_z}{\partial z}\right)$$

$$= -\frac{\partial p}{\partial z} + \eta\left[\frac{1}{r}\frac{\partial}{\partial r}\left(r\frac{\partial u_z}{\partial r}\right) + \frac{1}{r^2}\frac{\partial^2 u_z}{\partial\theta^2} + \frac{\partial^2 u_z}{\partial z^2}\right] + \rho g_z \qquad (\text{B.15})$$

B.4 Equation of Motion in Spherical Coordinates

In terms of shear stress τ:

r-component:

$$\rho\left(\frac{\partial u_r}{\partial t} + u_r \frac{\partial u_r}{\partial r} + \frac{u_\theta}{r}\frac{\partial u_r}{\partial\theta} + \frac{u_\phi}{r\sin\theta}\frac{\partial u_r}{\partial\phi} - \frac{u_\theta^2 + u_\phi^2}{r}\right)$$

$$= -\frac{\partial p}{\partial r} - \left(\frac{1}{r^2}\frac{\partial}{\partial r}\left(r^2 \tau_{rr}\right) + \frac{1}{r\sin\theta}\frac{\partial}{\partial\theta}\left(\tau_{r\theta}\sin\theta\right)\right. \qquad (\text{B.16})$$

$$\left. + \frac{1}{r\sin\theta}\frac{\partial\tau_{r\phi}}{\partial\phi} - \frac{\tau_{\theta\theta} + \tau_{\phi\phi}}{r}\right) + \rho g_r$$

θ-*component*:

$$\rho\left(\frac{\partial u_\theta}{\partial t}+u_r\frac{\partial u_\theta}{\partial r}+\frac{u_\theta}{r}\frac{\partial u_\theta}{\partial\theta}+\frac{u_\phi}{r\sin\theta}\frac{\partial u_\theta}{\partial\phi}+\frac{u_r u_\theta}{r}-\frac{u_\phi^2\cot\theta}{r}\right)$$

$$=-\frac{1}{r}\frac{\partial p}{\partial\theta}-\left(\frac{1}{r^2}\frac{\partial}{\partial r}\left(r^2\tau_{r\theta}\right)+\frac{1}{r\sin\theta}\frac{\partial}{\partial\theta}\left(\tau_{\theta\theta}\sin\theta\right)\right. \tag{B.17}$$

$$\left.+\frac{1}{r\sin\theta}\frac{\partial\tau_{\theta\phi}}{\partial\phi}+\frac{\tau_{r\theta}}{r}-\frac{\cot\theta}{r}\tau_{\phi\phi}\right)+\rho g_\theta$$

ϕ-*component*:

$$\rho\left(\frac{\partial u_\phi}{\partial t}+u_r\frac{\partial u_\phi}{\partial r}+\frac{u_\theta}{r}\frac{\partial u_\phi}{\partial\theta}+\frac{u_\phi}{r\sin\theta}\frac{\partial u_\phi}{\partial\phi}+\frac{u_\phi u_r}{r}+\frac{u_\theta u_\phi}{r}\cot\theta\right)$$

$$=-\frac{1}{r\sin\theta}\frac{\partial p}{\partial\phi}-\left(\frac{1}{r^2}\frac{\partial}{\partial r}\left(r^2\tau_{r\phi}\right)+\frac{1}{r}\frac{\partial\tau_{\theta\phi}}{\partial\theta}+\frac{1}{r\sin\theta}\frac{\partial\tau_{\phi\phi}}{\partial\phi}\right. \tag{B.18}$$

$$\left.+\frac{\tau_{r\phi}}{r}+\frac{2\cot\theta}{r}\tau_{\theta\phi}\right)+\rho g_\phi$$

In terms of velocity gradients for a Newtonian fluid with constant ρ and η:

r-component:

$$\rho\left(\frac{\partial u_r}{\partial t}+u_r\frac{\partial u_r}{\partial r}+\frac{u_\theta}{r}\frac{\partial u_r}{\partial\theta}+\frac{u_\phi}{r\sin\theta}\frac{\partial u_r}{\partial\phi}-\frac{u_\theta^2+u_\phi^2}{r}\right)$$

$$=-\frac{\partial p}{\partial r}+\eta\left(\nabla^2 u_r-\frac{2}{r^2}u_r-\frac{2}{r^2}\frac{\partial u_\theta}{\partial\theta}-\frac{2}{r^2}u_\theta\cot\theta-\frac{2}{r^2\sin\theta}\frac{\partial u_\phi}{\partial\phi}\right)+\rho g_r \tag{B.19}$$

where

$$\nabla^2 = \frac{1}{r^2}\frac{\partial}{\partial r}\left(r^2\frac{\partial}{\partial r}\right) + \frac{1}{r^2\sin\theta}\frac{\partial}{\partial\theta}\left(\sin\theta\frac{\partial}{\partial\theta}\right) + \frac{1}{r^2\sin^2\theta}\left(\frac{\partial^2}{\partial\phi^2}\right) \qquad (B.20)$$

θ-*component*:

$$\rho\left(\frac{\partial u_\theta}{\partial t} + u_r\frac{\partial u_\theta}{\partial r} + \frac{u_\theta}{r}\frac{\partial u_\theta}{\partial\theta} + \frac{u_\phi}{r\sin\theta}\frac{\partial u_\theta}{\partial\phi} + \frac{u_r u_\theta}{r} - \frac{u_\phi^2\cot\theta}{r}\right)$$

$$= -\frac{1}{r}\frac{\partial p}{\partial\theta} + \eta\left(\nabla^2 u_\theta + \frac{2}{r^2}\frac{\partial u_r}{\partial\theta} - \frac{u_\theta}{r^2\sin^2\theta} - \frac{2\cos\theta}{r^2\sin^2\theta}\frac{\partial u_\phi}{\partial\phi}\right) + \rho g_\theta \qquad (B.21)$$

φ-*component*:

$$\rho\left(\frac{\partial u_\phi}{\partial t} + u_r\frac{\partial u_\phi}{\partial r} + \frac{u_\theta}{r}\frac{\partial u_\phi}{\partial\theta} + \frac{u_\phi}{r\sin\theta}\frac{\partial u_\phi}{\partial\phi} + \frac{u_\phi u_r}{r} + \frac{u_\theta u_\phi}{r}\cot\theta\right)$$

$$= -\frac{1}{r\sin\theta}\frac{\partial p}{\partial\phi} + \eta\left(\nabla^2 u_\phi - \frac{u_\phi}{r^2\sin^2\theta} + \frac{2}{r^2\sin\theta}\frac{\partial u_r}{\partial\phi} + \frac{2\cos\theta}{r^2\sin^2\theta}\frac{\partial u_\theta}{\partial\phi}\right) \qquad (B.22)$$

$$+ \rho g_\phi$$

Particle Minerals, Miller Indices and Settling Time

Table C.1 Common minerals and rock types encountered in XRD analysis.

A	**Clastic Rocks** [a]	B-2	*Insoluble residue (after treatment*
A-1	*Coarse fraction (> 2 μm)*		*with HCl)*
	Quartz		Quartz
	Potassium feldspars		Muscovite
	Plagioclase feldspars		Chlorite group
	Muscovite		Smectite group (including
	Zeolites		montmorillonite)
A-2	*Fine fraction (≤ 2 μm)*		Kaolinite group
	Quartz		Illite group
	Smectite group (including		Opal (as disordered cristobalite)
	montmorillonite)	C	**Carbonaceous Organic Rocks,**
	Kaolinite group		**Pore Fillings, etc.**
	Illite group	C-1	*Humic Coal Insoluble Residues*
	Chlorite group		*(after treatment with HF)*
	Vermiculite group		Graphite [b]
	Potassium feldspars	C-2	*Asphaltic Insoluble Residues*
	Plagioclase feldspars		*(after treatment with HF)*
	Goethite		Graphite [b]
	Gibbsite	D	**Evaporates**
	Gypsum	D-1	*Bulk (ground)*
	Pyrite		Halite
	Calcite		Anhydrite
	Siderite		Gypsum
	Opal (as disordered cristobalite)	D-2	*Insoluble Residues (after*
B	**Carbonate Rocks**		*treatment with H_2O)*
B-1	*Bulk (ground)*		Gypsum
	Calcite (including high-Mg		Anhydrite
	variety)		Clay mineral of all types
	Dolomite		
	Aragonite		
	Apatite		
	Quartz		

[a] These are fragments of larger rock. Clastic, or detrital, sedimentary rocks is one common category of clastic rock.
[b] Disordered and expanded; crystallites with relatively large dimensions.
Source: Griffin [1971].

Table C.2 d_M spacing, relative intensities I, and hkl indices of common sedimentary minerals.

Quartz			Orthoclase			Plagioclase, Albite (low)			Analcite		
d_M(Å)	I	hkl	d_M(Å)	I	hkl	d_M(Å)	I	hkl	d_M(Å)	I	hkl
4.26	35	100	6.66	10	110	6.39	20	001	6.87	<10	200
3.34	100	101	6.52	20	020	5.94	2	11$\bar{1}$	5.61	80	211
2.46	12	110	5.82	20	11$\bar{1}$	5.59	2	$\bar{1}$11	4.86	40	220
2.28	12	102	4.24	60	20$\bar{1}$	4.03	16	20$\bar{1}$	3.67	20	321
2.24	6	111	3.94	30	1$\bar{1}$1	3.86	8	1$\bar{1}$1	3.43	100	400
2.13	9	200	3.87	10	200	3.78	25	111	2.93	80	332
1.980	6	201	3.79	100	130	3.68	20	130	2.80	20	422
1.817	17	112	3.62	20	13$\bar{1}$	3.66	16	131	2.69	50	431
1.801	<1	003	3.56	20	22$\bar{1}$			130			510
1.672	7	202	3.46	60	11$\bar{2}$	3.51	10	11$\bar{2}$	2.51	50	521
1.659	3	103	3.33	100	220	3.48	2	22$\bar{1}$	2.43	30	440
1.608	<1	210	3.28	70	202	3.80	8	$\bar{1}$12	2.23	40	611
1.541	15	211	3.26	50	040	3.20	100	002			532
1.453	3	113	3.22	90	002	3.151	10	220	2.17	<10	620
			3.00	60	131	2.96	10	131	2.12	<10	541
			2.93	20	22$\bar{2}$	2.93	16	022	2.02	10	631
			2.91	40	041	2.87	8	131			543
						2.84	2	132	1.94	<10	550
						2.79	2	022			710
						2.64	6	132	1.90	50	640
						2.56	8	24$\bar{1}$			633
						2.54	2	312	1.87	40	721
						2.51	2	1$\bar{1}$2			552
						2.50	6	221	1.83	<10	642
						2.46	6	221	1.74	60	732
						2.44	4	$\bar{2}$41			651
						2.43	2	151	1.72	30	800
						2.41	2	240			741
						2.39	4	310	1.69	40	811
						2.32	4	33$\bar{1}$			554
						2.28	2	113	1.66	10	820
						2.19	4	042			644
						2.13	8	060	1.62	20	822
						2.12	6	151			660
						2.08	2	2$\bar{4}$1			831
						2.04	2	241	1.60	10	743
						2.00	2	202			750
						1.980	4	061	1.498	20	842
						1.927	2	42$\bar{1}$			761
						1.889	8	222	1.480	20	921
											655

(Orthoclase column) See note under 3rd column (Plagioclase, Albite (low))

(Plagioclase, Albite (low) column) Plus 6 lines to 1.785

Note: The ~4.03 line distinguishes plagioclase from K-feldspars, but specific identification of plagioclase species is difficult.

Table C.2 — continued

Chlorite, Monoclinic			Muscovite, 2M₁			Biotite			Vermiculite		
d_M(Å)	I	hkl	d_M(Å)	I^a	hkl	d_M(Å)	I	hkl	d_M(Å)	I^a	hkl
14.2	80	001	9.99	S	002	10.1	100	001	14.4	VVS	002
7.12	100	002	4.98	M	004	4.59	20	110, 020	7.20	VW	004
4.75	80	003	4.47	VS	110, $11\bar{1}$	3.37	100	003	4.79	VW	006
3.56	100	004	4.29	W	111	3.16	20	112	4.60	S	02*l*, 11*l*
2.85	40	005	4.11	W	022	2.92	20	113	3.59	M	008
2.58	30	131, 202	3.95	VW	112	2.66	80	201, 130	2.87	M	00 10, 130
2.55	50	132, 201	3.87	M	113	2.52	40	004, 113	2.66	MW	200, 202
2.44	40	132, 203	3.72	M	023	2.45	80	201	2.60	M	132, 204, 134
2.38	20	133, 202	3.55	VW	113	2.28	20	040, 132	2.55	MW	202, 00 12
2.27	30	133, 204	3.48	M	$11\bar{4}$	2.18	80		2.39	MS	136, 204
2.04	20	007	3.32	VS	024, 006	2.00	80		2.27	VVW	136, 208
2.01	40	135, 204	3.20	MS	114	1.91	20		2.21	VW	138, 206
1.891	20	135, 206	3.1	VW	115	1.75	20		2.08	W	138, 20,10
1.833	20	136, 205	2.98	S	025	1.67	80		2.05	VW	00 14
1.732	10	136, 207	2.86	M	115	1.54	80		2.01	VW	208
1.672	10	137, 206	2.78	M	116	1.47	20		1.835	VVW	1,3,12, 2,0,10
1.577	20	137, 208	2.59	W	131, 200	1.43	20		1.748	W	2,0,14
1.541	60	060, 331	2.56	VS	202, 131	1.36	60		1.677	MW	1,3,14, 2,0,12
1.507	20	062, 331	2.49	W	008	1.33	40		1.574	VVW	1,3,14, 2,0,16, 060
1.429	10	0,0,10	2.46	W	202, 133	1.31	40		1.537	MS	1,3,16, 2,0,14, 330, 332, 334
			2.39	M	204				1.508	VVW	332, 336
			2.38	M	133						
			2.25	W	204, 135						
			2.19	W	223						
			2.14	M	206						
			2.13	M	135						
			2.05	VW	044						
			1.99	S	00 10, 206						
			1.95	W	137						
			1.83	VW							
			1.76	W	138						
			1.65	W	2,0,10						
			1.64	M	139						
			1.504	S	060, 331						

Note: Basal peak intensities vary with the Mg/Fe^{+++} ratio.

[a] VS=strong, S=strong, MS=moderately strong, M=moderate, W=weak, VW=very weak.

Table C.2 — continued

Montmorillonite Wyoming Bent.			Glauconite Well ordered			Talc			Kaolinite		
d_M(Å)	I	hkl	d_M(Å)	I	hkl	d_M(Å)	I	hkl	d_M(Å)	I	hkl
~15	100	00l	10.1	100	001	9.3	48	002	7.16	100+	001
		001	4.98	10	002	4.58	64	020	4.46	40	020
~5	80	003	4.53	80	020			111	4.36	50	$1\bar{1}0$
~3	100	005	4.35	20	$11\bar{1}$	4.11	5	113	4.18	50	$1\bar{1}1$
		hk	4.12	10	021	3.13	40	006	4.13	30	$1\bar{1}\bar{1}$
4.61	100	11	3.63	40	$11\bar{2}$	2.62	32	130	3.85	40	021
		02	3.33	60	003	2.49	100	132	3.74	20	$02\bar{1}$
2.56	80	13			022			204	3.57	100+	002
		20	3.09	40	112	2.22	14	134	3.37	40	111
		22	2.89	5	$11\bar{3}$			206	3.14	30	$11\bar{2}$
2.22	30	04	2.67	10	023	2.10	8	136	3.10	30	$1\bar{1}\bar{2}$
		31			130			204	2.75	30	022
1.692	60	15	2.59	100	131	1.95	3	136			$1\bar{3}0$
		24			200			208	2.56	60	201
1.492	100	33	2.40	60	132	1.87	3	0,0,10			$1\bar{3}0$
		06			201	1.72	11	$\bar{3}11$	2.53	40	$1\bar{3}1$
1.289	60	26	2.26	20	040			$\bar{3}13$			112
		40			$22\bar{1}$	1.68	5	138			$1\bar{3}\bar{1}$
			2.21	10	220			$2,0,1\bar{0}$	2.49	80	200
					041	1.60	2	208			112
			2.15	20	133	1.56	2	0,0,12	2.38	60	003
					202	1.53	64	060			202
			1.994	20B	005			$\bar{3}32$	2.39	90	131
			1.817	5	$22\bar{4}$	1.46	2				131
					311	1.40	5		2.29	80	$1\bar{3}1$
			1.715	10	$24\bar{1}$	1.32	11				131
					240	1.30	13		2.25	20	132
			1.66	30B	312	1.27	3				040
					310				2.19	30	220
					241				2.13	30	$02\bar{3}$
			1.511	60	060				2.06	20	$2\bar{2}\bar{2}$
					$33\bar{1}$				1.989	60	203
			1.495	10	330						132
			1.307	30	260				1.939	40	132
					400				1.896	30	$1\bar{3}3$
					170				1.869	20	042
			1.258	10	350						$1\bar{3}\bar{3}$
					420				1.839	40	202
											$2\bar{2}\bar{3}$
									1.809	20	$11\bar{4}$
									1.781	40	004
									1.707	20	$2\bar{2}\bar{2}$
									1.685	20	$24\bar{1}$
									1.662	70	204
											$1\bar{3}3$
									1.619	60	133
									1.584	40	134
									1.542	50B	$1\bar{3}\bar{4}$
											060
									1.489	80	$33\bar{1}$
											$3\bar{3}\bar{1}$

Notes: 00l Positions vary with moisture and interlayer cation population. Most peaks are broad, suggesting mixed hydration stages. Ethylene glycol solvation must be used for certain identification.

Note: 4.98 Å peak is obtained only on an oriented sample; 10 is a typical intensity value.

Table C.2 — continued

Hematite			Goethite			Gibbsite			Phillipsite		
d_M(Å)	I	hkl	d_M(Å)	I	hkl	d_M(Å)	I	hkl	d_M(Å)	I	hkl
3.67	35	102	4.98	15	020	4.85	100		7.64	100	
2.69	100	104	4.18	100	110	4.37	40		6.91	100	
2.51	75	110	3.38	10	120	4.31	20		6.34	20	
2.20	25	113	2.69	30	130	3.35	6		5.24	50	
2.07	3	202	2.58	8	021	3.31	10		4.91	50	
1.838	30	204	2.52	3	101	3.18	7		4.56	20	
1.692	45	116	2.49	15	040	3.10	4		4.25	70	
1.635	2	121	2.45	25	111	2.45	15		4.07	70	
1.597	15	108	2.25	10	121	2.42	4		3.54	50	
1.484	20	214	2.19	20	140	2.38	25		3.18	100	
1.452	25	300	2.01	2	131	2.29	4		2.94	50	
1.346	6	208	1.920	6	041	2.24	6		2.71	70	
1.310	10	1,0,10	1.799	7	211	2.17	8		2.52	50	
		119	1.770	2	141	2.08	1		2.40	50D	
1.257	8	220	1.721	20	221	2.04	15		2.16	20	
			1.694	10	240	2.02	1		2.07	20	
			1.661	4	060	1.991	8		1.97	50	
			1.606	6	231	1.916	6		1.91	20	
			1.564	15	160	1.801	10		1.84	20	
					151	1.750	9		1.78	50	
			1.509	10	250	1.685	7		1.72	50	
			1.467	4	320	1.655	2		1.67	20D	
			1.453	10	061	1.590	2		1.61	20	
						1.574	1		1.55	20	
						1.555	1		1.49	20	
						1.533	1		1.38	50	
						1.485	1		1.34	50	
						1.477	1		1.28	50D	
						1.457	8				
						1.440	4				
						1.411	5				
						1.402	4				

Note:
D = doublet

Note: Zeolites are usually characterized by sharp, intense peaks in the low angle region.

Table C.2 — continued

Heulandite			Cristobalite, Low			Aragonite			Calcite		
d_M (Å)	I	hkl	d_M (Å)	I	hkl	d_M (Å)	I	hkl	d_M (Å)	I	hkl
8.90	100		4.04	100	101	4.21	2	110	3.86	12	102
7.94	20		3.14	12	111	3.40	100	111	3.04	100	104
6.80	10		2.85	14	102	3.27	52	021	2.85	3	006
6.63	10		2.49	18	200	2.87	4	002	2.50	14	110
5.92	10		2.47	6	112	2.73	9	121	2.29	18	113
5.58	10		2.34	<1	201	2.70	46	012	2.10	18	202
5.24	10		2.12	4	211	2.48	33	200	1.927	5	204
5.09	10		2.02	3	202	2.41	14	031	1.913	17	108
4.89	10		1.932	4	113	2.37	38	112	1.875	17	116
4.69	20		1.874	4	212	2.34	31	130	1.626	4	211
4.45	20		1.756	1	220	2.33	6	022	1.604	8	212
4.36	10		1.736	1	004	2.19	11	211	1.587	2	1,0,10
3.97	20		1.692	3	203	2.11	23	220	1.525	5	214
3.89	30		1.642	1	104	1.977	65	221	1.518	4	208
3.83	10		1.612	5	301	1.882	32	041	1.510	3	119
3.71	10		1.604	2	213	1.877	25	202	1.473	2	215
3.56	10		1.574	1	310	1.814	23	132			
3.47	10				222			230			
3.40	20		1.535	2	311	1.759	4	141			
3.12	10		1.495	3	302	1.742	25	113			
3.07	10		1.432	2	312	1.728	15	231			
3.03	10										
2.97	40		*Note:* Opal produces								
2.80	10		a disordered								
2.72	10		cristobalite pattern,								
2.67	10		otherwise rare in								
2.48	10		sediments.								
2.43	10										
2.35	10										
2.28	20										

Table C.2 — continued

Dolomite			Siderite			Apatite Hydroxy-			Gypsum		
d_M(Å)	I	hkl	d_M(Å)	I	hkl	d_M(Å)	I	hkl	d_M(Å)	I	hkl
4.03	3	101	3.59	25	102	8.17	11	100	7.56	100	020
3.69	5	102	2.79	100	104	5.26	5	101	4.27	51	$12\bar{1}$
2.89	100	104	2.56	2	006	4.72	3	110	3.79	21	031
2.67	10	006	2.34	15	110	4.07	9	200			040
2.54	8	105	2.13	20	113	3.88	9	111	3.16	3	$11\bar{2}$
2.41	10	110	1.962	15	201	3.51	1	201	3.06	57	$14\bar{1}$
2.19	30	113	1.794	10	204	3.44	40	002	2.87	27	002
2.07	5	201	1.736	20	108	3.17	11	102	2.79	5	$21\bar{1}$
2.02	15	202	1.730	20	116	3.08	17	210	2.68	28	022
1.848	5	204	1.526	5	121	2.81	100	211			051
1.804	20	106	1.504	10	212	3.78	60	112	2.59	4	150
1.786	30	116	1.438	3	1,0,10	2.72	60	300			$20\bar{2}$
1.781		009				2.63	25	202	2.53	<1	060
1.567	8	211				2.53	5	301	2.50	6	200
1.545	10	212				2.30	7	212	2.45	4	$22\bar{2}$
1.496	2	1,0,10				2.26	20	310	2.40	4	141
1.465	6	214				2.23	1	221	2.22	6	$15\bar{2}$
1.445	4	208				2.15	9	311	2.14	1	$24\bar{2}$
1.431	10	119				2.13	3	302	2.08	10	$12\bar{3}$
1.413	4	215				2.07	7	113	2.07	8	112
						2.04	1	400			$25\bar{1}$
						2.00	5	203	1.990	4	170
						1.943	30	222	1.953	2	211
						1.890	15	312			080
						1.871	5	320	1.898	16	062
						1.841	40	213			013
						1.806	20	321	1.879	10	$14\bar{3}$
						1.780	11	410	1.864	4	$31\bar{2}$
						1.754	15	402	1.843	1	231
								303	1.812	10	$26\bar{2}$
						1.722	20	004			
								411			
						1.684	3	104			
						1.644	9	322			
								223			
						1.611	7	313			
						1.587	3	501			
								204			

Table C.2 — continued

Anhydrite			Halite			Pyrite		
d_M (Å)	I	hkl	d_M (Å)	I	hkl	d_M (Å)	I	hkl
3.87	6	111	3.26	13	111	3.13	36	111
3.50	100	020	2.82	100	200	2.71	84	200
		002	1.994	88	220	2.42	66	210
3.12	3	200	1.701	2	311	2.21	52	211
2.85	33	210	1.628	15	222	1.92	40	220
2.80	4	121	1.410	6	400	1.633	100	311
2.47	8	022	1.294	1	331	1.564	14	222
2.33	22	202	1.261	11	420	1.503	20	230
		220				1.445	24	321
2.21	20	212				1.243	12	331
2.18	8	103				1.211	14	420
2.09	9	113				1.182	7	421
1.993	6	301				1.155	6	332
1.938	4	222				1.106	6	422
1.869	15	230				1.043	27	511
1.852	4	123						
1.749	11	004						
1.748	10	040						
1.648	14	232						
1.594	3	133						
1.564	5	024						
		042						
1.525	4	204						
		240						
1.515	1	313						
		331						
1.490	5	214						
1.424	3	402						
		420						
1.418	1	323						
1.398	3	242						

Source: Griffin [1971].

Table C.3 Time in minutes required for particles of given diameter and settling velocity (w_s) at standard 20°C to fall 1 m in water at different temperatures (T °C).

T °C	w_s at 20°C (mm s^{-1})												
	8.13	4.90	3.50	1.75	0.873	0.43	0.21	0.108	0.0545	0.0271	0.0136	0.00632	0.00341
5			7.21	14.41	28.9	57.7	115.7	233	463	931	1851	3696	7405
6			6.99	13.98	28.1	55.9	112.3	226	449	903	1795	3584	7181
7			6.78	13.55	27.2	54.2	108.8	219	435	875	1741	3476	6964
8			6.58	13.15	26.4	52.6	105.6	212	422	849	1689	3373	6757
9			6.39	12.77	25.6	51.1	102.5	206	410	825	1641	3276	6563
10	2.60	4.35	6.21	12.41	24.9	49.6	99.6	200	398	801	1594	3182	6373
11	2.54	4.25	6.03	12.06	24.2	48.2	96.8	195	387	779	1549	3093	6197
12	2.48	4.15	5.87	11.73	23.5	46.9	94.2	189	377	758	1507	3008	6026
13	2.42	4.05	5.71	11.41	22.9	45.6	91.6	184	366	737	1466	2926	5862
14	2.36	3.95	5.55	11.11	22.3	44.4	89.2	179	357	717	1427	2849	5706
15	2.30	3.85	5.41	10.82	21.7	43.3	86.8	175	347	699	1389	2774	5557
16	2.25	3.76	5.27	10.54	21.1	42.2	84.6	170	338	681	1354	2703	5414
17	2.20	3.67	5.14	10.27	20.6	41.1	82.4	166	330	663	1319	2633	5276
18	2.15	3.58	5.01	10.01	20.1	40.0	80.4	162	321	647	1286	2568	5144
19	2.10	3.49	4.88	9.76	19.6	39.1	78.4	158	314	631	1254	2504	5017
20	2.05	3.40	4.76	9.53	19.1	38.1	76.5	154	306	615	1224	2443	4895
21	2.01	3.33	4.65	9.30	18.7	37.2	74.6	150	299	601	1194	2385	4777
22	1.97	3.26	4.54	9.08	18.2	36.3	72.9	147	292	586	1166	2328	4664
23	1.93	3.19	4.43	8.87	17.8	35.5	71.2	143	285	573	1139	2274	4556
24	1.89	3.12	4.33	8.66	17.4	34.6	69.5	140	278	559	1113	2221	4450
25	1.85	3.05	4.23	8.47	17.0	33.9	68.0	137	272	547	1087	2171	4350
26	1.82	3.00	4.14	8.28	16.6	33.1	66.4	134	266	534	1063	2122	4252
27	1.79	2.95	4.05	8.09	16.2	32.4	65.0	131	260	523	1039	2075	4158
28	1.76	2.90	3.96	7.92	15.9	31.7	63.5	128	254	511	1017	2030	4067
29	1.73	2.85	3.87	7.74	15.5	31.0	62.2	125	249	500	995	1986	3979
30	1.70	2.80	3.79	7.58	15.2	30.3	60.8	122	243	489	973	1944	3894
d^a	100	75	62.5	44.2	31.2	22.1	15.6	11.0	7.8	5.5	3.9	2.76	1.95

^{a} Particle diameter (μm) for spheres of specific gravity 2.65.
Source: Owen [1970].

Appendix D

Normal Probability Integral Values

$$\frac{1}{\sqrt{(2\pi)}} \int_0^z e^{-\omega^2/2} d\omega \qquad (D.1)$$

Table D.1 Negative z-values and normal probabilities.

z	0.09	0.08	0.07	0.06	0.05	0.04	0.03	0.02	0.01	0.00
-3.9	0.0000	0.0000	0.0000	0.0000	0.0000	0.0000	0.0000	0.0000	0.0000	0.0000
-3.8	0.0001	0.0001	0.0001	0.0001	0.0001	0.0001	0.0001	0.0001	0.0001	0.0001
-3.7	0.0001	0.0001	0.0001	0.0001	0.0001	0.0001	0.0001	0.0001	0.0001	0.0001
-3.6	0.0001	0.0001	0.0001	0.0001	0.0001	0.0001	0.0001	0.0001	0.0002	0.0002
-3.5	0.0002	0.0002	0.0002	0.0002	0.0002	0.0002	0.0002	0.0002	0.0002	0.0002
-3.4	0.0002	0.0003	0.0003	0.0003	0.0003	0.0003	0.0003	0.0003	0.0003	0.0003
-3.3	0.0003	0.0004	0.0004	0.0004	0.0004	0.0004	0.0004	0.0005	0.0005	0.0005
-3.2	0.0005	0.0005	0.0005	0.0006	0.0006	0.0006	0.0006	0.0006	0.0007	0.0007
-3.1	0.0007	0.0007	0.0008	0.0008	0.0008	0.0008	0.0009	0.0009	0.0009	0.0010
-3.0	0.0010	0.0010	0.0011	0.0011	0.0011	0.0012	0.0012	0.0013	0.0013	0.0013
-2.9	0.0014	0.0014	0.0015	0.0015	0.0016	0.0016	0.0017	0.0018	0.0018	0.0019
-2.8	0.0019	0.0020	0.0021	0.0021	0.0022	0.0023	0.0023	0.0024	0.0025	0.0026
-2.7	0.0026	0.0027	0.0028	0.0029	0.0030	0.0031	0.0032	0.0033	0.0034	0.0035
-2.6	0.0036	0.0037	0.0038	0.0039	0.0040	0.0041	0.0043	0.0044	0.0045	0.0047
-2.5	0.0048	0.0049	0.0051	0.0052	0.0054	0.0055	0.0057	0.0059	0.0060	0.0062
-2.4	0.0064	0.0066	0.0068	0.0069	0.0071	0.0073	0.0075	0.0078	0.0080	0.0082
-2.3	0.0084	0.0087	0.0089	0.0091	0.0094	0.0096	0.0099	0.0102	0.0104	0.0107
-2.2	0.0110	0.0113	0.0116	0.0119	0.0122	0.0125	0.0129	0.0132	0.0136	0.0139
-2.1	0.0143	0.0146	0.0150	0.0154	0.0158	0.0162	0.0166	0.0170	0.0174	0.0179
-2.0	0.0183	0.0188	0.0192	0.0197	0.0202	0.0207	0.0212	0.0217	0.0222	0.0228
-1.9	0.0233	0.0239	0.0244	0.0250	0.0256	0.0262	0.0268	0.0274	0.0281	0.0287
-1.8	0.0294	0.0301	0.0307	0.0314	0.0322	0.0329	0.0336	0.0344	0.0351	0.0359
-1.7	0.0367	0.0375	0.0384	0.0392	0.0401	0.0409	0.0418	0.0427	0.0436	0.0446
-1.6	0.0455	0.0465	0.0475	0.0485	0.0495	0.0505	0.0516	0.0526	0.0537	0.0548
-1.5	0.0559	0.0571	0.0582	0.0594	0.0606	0.0618	0.0630	0.0643	0.0655	0.0668
-1.4	0.0681	0.0694	0.0708	0.0721	0.0735	0.0749	0.0764	0.0778	0.0793	0.0808
-1.3	0.0823	0.0838	0.0853	0.0869	0.0901	0.0901	0.0918	0.0934	0.0951	0.0968
-1.2	0.0985	0.1003	0.1020	0.1038	0.1056	0.1075	0.1093	0.1112	0.1131	0.1151
-1.1	0.1170	0.1190	0.1210	0.1230	0.1251	0.1271	0.1292	0.1314	0.1335	0.1357
-1.0	0.1379	0.1401	0.1423	0.1446	0.1469	0.1492	0.1515	0.1539	0.1562	0.1587
-0.9	0.1611	0.1635	0.1660	0.1685	0.1711	0.1736	0.1762	0.1788	0.1814	0.1841
-0.8	0.1867	0.1894	0.1922	0.1949	0.1977	0.2005	0.2033	0.2061	0.2090	0.2119
-0.7	0.2148	0.2177	0.2206	0.2236	0.2266	0.2296	0.2327	0.2358	0.2389	0.2420
-0.6	0.2451	0.2483	0.2514	0.2546	0.2578	0.2611	0.2643	0.2676	0.2709	0.2743
-0.5	0.2776	0.2810	0.2843	0.2877	0.2912	0.2946	0.2981	0.3015	0.3050	0.3085
-0.4	0.3121	0.3156	0.3192	0.3228	0.3264	0.3300	0.3336	0.3372	0.3409	0.3446
-0.3	0.3483	0.3520	0.3557	0.3594	0.3632	0.3669	0.3707	0.3745	0.3783	0.3821
-0.2	0.3859	0.3897	0.3936	0.3974	0.4013	0.4052	0.4090	0.4129	0.4168	0.4207
-0.1	0.4247	0.4286	0.4325	0.4364	0.4404	0.4443	0.4483	0.4522	0.4562	0.4602
-0.0	0.4641	0.4681	0.4721	0.4761	0.4801	0.4840	0.4880	0.4920	0.4960	0.5000

Table D.2 Positive z-values and normal probabilities.

z	0.00	0.01	0.02	0.03	0.04	0.05	0.06	0.07	0.08	0.09
0.0	0.5000	0.5040	0.5080	0.5120	0.5160	0.5199	0.5239	0.5279	0.5319	0.5359
0.1	0.5398	0.5438	0.5478	0.5517	0.5557	0.5596	0.5636	0.5675	0.5714	0.5753
0.2	0.5793	0.5832	0.5871	0.5910	0.5948	0.5987	0.6026	0.6064	0.6103	0.6141
0.3	0.6179	0.6217	0.6255	0.6293	0.6331	0.6368	0.6406	0.6443	0.6480	0.6517
0.4	0.6554	0.6591	0.6628	0.6664	0.6700	0.6736	0.6772	0.6808	0.6844	0.6879
0.5	0.6915	0.6950	0.6985	0.7019	0.7054	0.7088	0.7123	0.7157	0.7190	0.7224
0.6	0.7257	0.7291	0.7324	0.7357	0.7389	0.7422	0.7454	0.7486	0.7517	0.7549
0.7	0.7580	0.7611	0.7642	0.7673	0.7704	0.7734	0.7764	0.7794	0.7823	0.7852
0.8	0.7881	0.7910	0.7939	0.7967	0.7995	0.8023	0.8051	0.8078	0.8106	0.8133
0.9	0.8159	0.8186	0.8212	0.8238	0.8264	0.8289	0.8315	0.8340	0.8365	0.8389
1.0	0.8413	0.8438	0.8461	0.8485	0.8508	0.8531	0.8554	0.8577	0.8599	0.8621
1.1	0.8643	0.8665	0.8686	0.8708	0.8729	0.8749	0.8770	0.8790	0.8810	0.8830
1.2	0.8849	0.8869	0.8888	0.8907	0.8925	0.8944	0.8962	0.8980	0.8997	0.9015
1.3	0.9032	0.9049	0.9066	0.9082	0.9099	0.9115	0.9131	0.9147	0.9162	0.9177
1.4	0.9192	0.9207	0.9222	0.9236	0.9251	0.9265	0.9279	0.9292	0.9306	0.9319
1.5	0.9332	0.9345	0.9357	0.9370	0.9382	0.9394	0.9406	0.9418	0.9429	0.9441
1.6	0.9452	0.9463	0.9474	0.9484	0.9495	0.9505	0.9515	0.9525	0.9535	0.9545
1.7	0.9554	0.9564	0.9573	0.9582	0.9591	0.9599	0.9608	0.9616	0.9625	0.9633
1.8	0.9641	0.9649	0.9656	0.9664	0.9671	0.9678	0.9686	0.9693	0.9699	0.9706
1.9	0.9713	0.9719	0.9726	0.9732	0.9738	0.9744	0.9750	0.9756	0.9761	0.9767
2.0	0.9772	0.9778	0.9783	0.9788	0.9793	0.9798	0.9803	0.9808	0.9812	0.9817
2.1	0.9821	0.9826	0.9830	0.9834	0.9838	0.9842	0.9846	0.9850	0.9854	0.9857
2.2	0.9861	0.9864	0.9868	0.9871	0.9875	0.9878	0.9881	0.9884	0.9887	0.9890
2.3	0.9893	0.9896	0.9898	0.9901	0.9904	0.9906	0.9909	0.9911	0.9913	0.9916
2.4	0.9918	0.9920	0.9922	0.9925	0.9927	0.9929	0.9931	0.9932	0.9934	0.9936
2.5	0.9938	0.9940	0.9941	0.9943	0.9945	0.9946	0.9948	0.9949	0.9951	0.9952
2.6	0.9953	0.9955	0.9956	0.9957	0.9959	0.9960	0.9961	0.9962	0.9963	0.9964
2.7	0.9965	0.9966	0.9967	0.9968	0.9969	0.9970	0.9971	0.9972	0.9973	0.9974
2.8	0.9974	0.9975	0.9976	0.9977	0.9977	0.9978	0.9979	0.9979	0.9980	0.9981
2.9	0.9981	0.9982	0.9982	0.9983	0.9984	0.9984	0.9985	0.9985	0.9986	0.9986
3.0	0.9987	0.9987	0.9987	0.9988	0.9988	0.9989	0.9989	0.9989	0.9990	0.9990
3.1	0.9990	0.9991	0.9991	0.9991	0.9992	0.9992	0.9992	0.9992	0.9993	0.9993
3.2	0.9993	0.9993	0.9994	0.9994	0.9994	0.9994	0.9994	0.9995	0.9995	0.9995
3.3	0.9995	0.9995	0.9995	0.9996	0.9996	0.9996	0.9996	0.9996	0.9996	0.9997
3.4	0.9997	0.9997	0.9997	0.9997	0.9997	0.9997	0.9997	0.9997	0.9997	0.9998
3.5	0.9998	0.9998	0.9998	0.9998	0.9998	0.9998	0.9998	0.9998	0.9998	0.9998
3.6	0.9998	0.9998	0.9999	0.9999	0.9999	0.9999	0.9999	0.9999	0.9999	0.9999
3.7	0.9999	0.9999	0.9999	0.9999	0.9999	0.9999	0.9999	0.9999	0.9999	0.9999
3.8	0.9999	0.9999	0.9999	0.9999	0.9999	0.9999	0.9999	0.9999	0.9999	0.9999
3.9	1.0000	1.0000	1.0000	1.0000	1.0000	1.0000	1.0000	1.0000	1.0000	1.0000

For calculation of erosion probability p_e in Chapter 6 consider

$$p_e = 1 - \frac{1}{\sqrt{2\pi}} \int_{-1.0}^{0.2} e^{-\frac{\omega^2}{2}} d\omega = 1 - \left(\frac{1}{\sqrt{2\pi}} \int_{-\infty}^{0.2} e^{-\frac{\omega^2}{2}} d\omega - \frac{1}{\sqrt{2\pi}} \int_{-\infty}^{-1.0} e^{-\frac{\omega^2}{2}} d\omega \right) \quad \text{(D.2)}$$

From Tables D.2 and D.1 $p_e = 1 - (0.5793 - 0.1379) = 0.5586$.

Appendix E

Vertical Diffusion

For estimation of the rate of vertical mass diffusivity from the eddy diffusivity, recourse is made to k–ε ("kay–epsilon") equations. The standard high-Reynolds number k–ε model considers uniform flow at equilibrium. From Eq. 2.94 the governing equations for the transport of turbulent kinetic energy TKE (Eq. 2.4), simply designated as k following convention, and the turbulent kinetic energy dissipation rate ε_t per unit mass are

$$-\frac{\partial}{\partial z}\left\{\nu+\frac{\varepsilon_m}{\sigma_\varepsilon}\right\}\frac{\partial k}{\partial z}=\varepsilon_m\left(\frac{\partial \overline{u}}{\partial z}\right)^2+\frac{g}{\overline{\rho}}\frac{\varepsilon_m}{Sc}\frac{\partial \overline{\rho}}{\partial z}-\varepsilon_t \qquad (E.1)$$

$$-\frac{\partial}{\partial z}\left\{\nu+\frac{\varepsilon_m}{\sigma_k}\right\}\frac{\partial \varepsilon_t}{\partial z}=c_{1\varepsilon}\varepsilon_m\frac{\varepsilon_t}{k}\left(\frac{\partial \overline{u}}{\partial z}\right)^2+\left(1-c_{3\varepsilon}\right)\frac{\varepsilon_t}{k}\frac{g}{\overline{\rho}}\frac{\varepsilon_m}{Sc}\frac{\partial \overline{\rho}}{\partial z}-c_{2\varepsilon}\frac{\varepsilon_t^2}{k} \qquad (E.2)$$

where $c_{1\varepsilon}$, $c_{2\varepsilon}$, $c_{3\varepsilon}$, σ_ε, σ_k are coefficients and Sc is the Prandtl-Schmidt number (or Schmidt number, Chapter 6). The k–epsilon model enables the eddy diffusivity ε_m to be defined in terms of k and ε_t as

$$\varepsilon_m = C_\mu \frac{k^2}{\varepsilon_t} \qquad (E.3)$$

where C_μ is a proportionality coefficient. Note that the mass diffusivity $D_s = \varepsilon_m / Sc$. Equations E.1, E.2 and E.3 comprise the standard high-Reynolds number form of the model equations. Thus, the coefficients in Table E.1 are applicable to fully rough flows specified by the wall Reynolds number $Re^k = \sqrt{k}z/\varepsilon_m > 70-100$.

An Introduction to Hydraulics of Fine Sediment Transport

Table E.1 High-Reynolds number coefficients for k–epsilon equations.

C_μ	$c_{\varepsilon 1}$	$c_{\varepsilon 2}$	$c_{\varepsilon 3}$	σ_ε	σ_k
0.09	1.44	1.92	1.2	1	1.3

The boundary conditions for solving Eqs. E.1 and E.2 are

$$k\big|_{z=z_B} = \frac{u_*^2}{\sqrt{C_\mu}} \qquad \varepsilon_t\big|_{z=z_B} = \frac{u_*^3}{kz_0}$$

$$k\big|_{z=z_S} = 0 \qquad \varepsilon_t\big|_{z=z_S} = 0 \qquad\qquad\text{(E.4)}$$

where z_B and z_S denote the bottom and surface elevations, respectively, at which these boundary conditions are applied. The equations and the boundary conditions have been modified for sediment in water and low Reynolds numbers. They permit the estimation of mass diffusivity D_s by avoiding the use of the empirical function Φ_s (Chapter 6) to account for sediment-induced damping [Hanjalić, 2004; Hsu *et al.*, 2007; Letter, 2008]. Note that the modified equations continue to depend on empirical coefficients, because mass diffusion is sensitive to the nature of particles and flow. However, empiricism involved in the use of Φ_s is reduced by specifying TKE transport.

Appendix F

Wind Scales and Sea Description

Table F.1 Beaufort wind scale and wave heights.

Beaufort scale	Seaman's description of wind	Wind velocity (knots)	Estimating wind velocities on sea	International scale sea description and wave heights
0	Calm	Less than 1 knot	Calm; sea like a mirror.	Calm glassy
1	Light air	1 to 3 knots	Light air; ripples — no foam crests.	0
2	Light breeze	4 to 6 knots	Light breeze; small wavelets, crests have glassy appearance and do not break.	Rippled 0 to 0.30 m
3	Gentle breeze	7 to 10 knots	Gentle breeze; large wavelets, crests begin to break. Scattered whitecaps.	Smooth 0.30 to 0.60 m
4	Moderate breeze	11 to 16 knots	Moderate breeze; small waves becoming longer. Frequent whitecaps.	Slight 0.60 to 1.2 m
5	Fresh breeze	17 to 21 knots	Fresh breeze; moderate waves taking a more pronounced long form; mainly whitecaps, some spray.	Moderate 1.2 to 2.4 m
6	Strong breeze	22 to 27 knots	Strong breeze; large waves begin to form extensive whitecaps everywhere, some spray.	Rough 2.4 to 4 m

973

7	High wind (Moderate gale)	28 to 33 knots	Moderate gale; sea heaps up and white foam from breaking waves begins to be blown in streaks along the direction of the wind.	-
8	Gale (Fresh gale)	34 to 40 knots	Fresh gale; moderately high waves of greater length; edges of crests break into spindrift. The foam is blown in well-marked streaks along the direction of the wind.	Very rough 4 to 6 m
9	Strong gale	41 to 47 knots	Strong gale; high waves, dense streaks of foam along the direction of the wind. Spray may affect visibility. Sea begins to roll.	-
10	Whole gale	48 to 55 knots	Whole gale; very high waves. The surface of the sea takes on a white appearance. The rolling of sea becomes heavy and shock-like. Visibility affected.	High 6 to 9 m
11	Storm	56 to 63 knots	Storm; exceptionally high waves. Small and medium-sized ships are lost to view long periods.	Very high 9 to 14 m
12	Hurricane	64 and above	Hurricane; the air is filled with foam and spray. Sea completely white with driving spray; visibility very seriously affected.	Phenomenal over 14 m

Source: Adapted from Bascom [1964].

Bibliography

Abraham, G., Diephuis, J. G. H. R. (1959). Stable clay suspensions as substitute for sea water in model tests on density currents, *Proc. 8th Cong. Int. Assoc. Hydr. Res.*, 19 p. Also: (1959). *Delft Hydraulics, Publication 21* (Delft, The Netherlands).

Abramovich, G. N. (1963). *The Theory of Turbulent Jets* (MIT Press).

Ab'Sáber, A. N., Holmquist, C. (2001). *Coast of Brazil* (Metalivros, São Paulo, Brazil).

Agrawal, Y. C., Pottsmith, H. C. (2000). Instruments for particle size and settling velocity observations in sediment transport, *Mar. Geol.*, 168(1–4), pp. 89–114.

Airy, G. B. (1842). *Encyclopedia Metropolitania*, Vol. 5, "Tides and Waves," pp. 241–396.

Albertson, M. (1953). Effect of shape on the fall velocity of gravel particles, *Proc. 5th Iowa Hydraul. Conf.*, Iowa City, Iowa, pp. 243–261.

Albinia, A. (2008). *Empires of the Indus: The Story of a River* (Norton).

Alexander, C. R., Nittrouer, C. A., Demaster, D. J., Park, Y.-A., Park, S.-C. (1991). Macrotidal mudflats of the southwestern Korean coast: A model for interpretation of intertidal deposits, *J. Sediment. Petrol.*, 61(5), pp. 805–824.

Alger, G. R., Simons, D. B. (1968). Fall velocity of irregular shaped particles, *J. Hydraul. Div.*, ASCE, 94(3), pp. 721–737.

Ali, K. H. M., Georgiadis, K. (1991). Laminar motion of fluid mud. *Proc. Inst. Civ. Eng. Res. Theor., Part 2*, London, pp. 795–821.

Alishahi, M. R., Krone, R. B. (1964). Suspension of cohesive sediments by wind-generated waves, *Technical Report HEL-2-9*, Hydraulic Engineering Laboratory, University of California, Berkeley.

Alkhalidi, M. (2004). Sedimentary equilibrium and morphology of inlets and estuaries, PhD Thesis, University of Florida, Gainesville.

Alkhalidi, M., Mehta, A. J. (2005). Discussion of, "Measurements of sediment erosion and transport with the adjustable shear stress erosion and transport flume," by Roberts, J. D. *et al.*, *J. Hydraul. Eng.*, 131(7), pp. 624–625.

Alldredge, A. L., Gotschalk, C. (1988). In-situ settling behavior of marine snow, *Limnol. Oceanogr.*, 33, pp. 339–351.

Allen, J. R. L. (1977). *Physical Processes of Sedimentation,* 4th Impression with revised readings (Allen and Unwin).

Allen, J.R.L. (2004). Annual textural banding in Holocene estuarine silts, Severn Estuary Levels (SW Britain): patterns, cause and implications. *The Holocene,* 14(4), pp. 536–552.

Allen, R. H. (1972). A glossary of coastal engineering terms, *Miscellaneous Papers 2-72,* U. S. Army Corps of Engineers Coastal Engineering Research Center.

Amos, C. L., Brylinsky, M., Sutherland, T. F., O'Brien, D., Lee, S., Cramp, A. (1998). *Sedimentary Processes in the Intertidal Zone,* eds. Black, K. S., Paterson, D. M., and Camp, A., "The Stability of a Mudflat in the Humber Estuary, South Yorkshire, UK," (The Geological Society, London) pp. 25–44.

Amos, C. L., Droppo, I. G., Gomez, E. A., Murphy, T. P. (2003). The stability of a remediated bed in Hamilton Harbour, Lake Ontario, Canada, *Sedimentology,* 50, pp. 149–168.

Amos, C. L., Feeney, T., Sutherland, T. F., Luternauer, J. L. (1997). The stability of fine-grained sediments from the Fraser River Delta, *Estuarine Coastal Shelf Sci.,* 45, pp. 507–524.

Amos, C. L., Sutherland, T. F., Cloutier, D., Patterson, S. (2000). Corrasion of remoulded cohesive bed by saltating littorinid shells, *Cont. Shelf Res.,* 20, pp. 1291–1315.

Annandale, G. W. (1995). Erodibility, *J. Hydraul. Res.,* 33(4), pp. 471–493.

Ariathurai, R. (1974). A finite element model for sediment transport in estuaries, PhD Thesis, University of California, Davis.

Ariathurai, R., Arulanandan, K. (1978). Erosion rate of cohesive soils, *J. Hydraul. Div.,* ASCE, 104(2), pp. 279–283.

Ariathurai, R., MacArthur, R. C., Krone, R. B. (1977). Mathematical model of estuarial sediment transport, *Technical Report D-77-12,* U. S. Army Engineer Waterways Experiment Station, Vicksburg, MS.

Arnot, J. A., Gobas, F. A. P. C. (2004). A food web bioaccumulation model for organic chemicals in aquatic systems, *Environ. Toxicol. Chem.,* 23(10), pp. 2343–2355.

Arulanandan, K., Gillogley, E., Tully, R. (1980). Development of a quantitative method to predict critical shear stress and rate of erosion of natural undisturbed cohesive soils, *Technical report No. GL-80-5,* U. S. Army Engineer Waterways Experiment Station, Vicksburg, MS.

Arulanandan, K., Loganathan, P., Krone, R. B. (1975). Pore and eroding fluid influences on surface erosion of soil, *J. Geotech. Eng. Div.,* ASCE, 101(1), pp. 51–66.

Arulanandan, K., Sargunam, A., Loganathan, P., Krone, R. B. (1973). *Soil Erosion: Causes and Mechanisms; Prevention and Control,* ed. Gray, D. H., Special Report 135, "Application of Chemical and Electrical Parameters to Prediction of Erodibility," (Highway Research Board, Washington, D. C.) pp. 42–51.

Ashley, R. M., Verbanck, M. A. (1996). Mechanics of sewer sediment erosion and transport, *J. Hydraul. Eng.,* 34(6), pp. 753–769.

ASTM (1993a). Standard test method for amount of material in soils finer than the no. 200 (75-μm) sieve, *Annual Book of ASTM Standards,* American Society for Testing and Materials, Sec. 4, Vol. 04.08, pp. 191–193.

ASTM (1993b). Standard test methods for operating performance of particulate cation-exchange materials, *Annual Book of ASTM Standards*, American Society for Testing and Materials, Sec. 11, Vol. 11.02, pp. 772–776.

ASTM (1993c). Standard test methods for moisture, ash, and organic matter of peat and other organic soils, *Annual Book of ASTM Standards*, American Society for Testing and Materials, Sec. 4, Vol. 04.08, pp. 400–402.

Atkinson, J. F. (1988). Note on "Interfacial mixing in stratified flows," *J. Hydraul. Res.*, 26(1), pp. 27–31.

Augustinus, P. G. E. F. (1978). The changing shoreline of Suriname, PhD Thesis, University of Utrecht, The Netherlands.

Augustinus, P. G. E. F. (1987). *International Geomorphology 1986, Part I*, ed. Gardener, V., "The Geomorphologic Development of the Coast of Guyana Between the Corentyne River and the Essequibo River," (Wiley) pp. 1281–1292.

Augustinus, P. G. E. F. (1988). *Artificial Structures and Shorelines*, ed. Walker, H. J., "Surinam," pp. 689–693.

Azam, M. H. (1998). Breaking wave over mud flats, PhD Thesis, Universiti Teknologi Malaysia, Kuala Lumpur.

Babatope, B., Williams, D. J. A., Williams, P. R. (1999). Viscoelastic wave dispersion and rheometry of cohesive sediments, *J. Hydraul. Eng.*, 125(3), pp. 295–298.

Bache, D. H., Rasool, E., Moffatt, D., McGilligan, F. J. (1999). On the strength and character of alumino-humic flocs, *Water Sci. Technol.*, 40(9), pp. 81–88.

Bagnold, R. A. (1956). The flow of cohesionless grains in fluids, *Proc. R. Soc. Philos. Trans.*, London, pp. 235–297.

Bagnold, R. A. (1966). An approach to the sediment transport problem from general physics, *Professional Paper No. 442-I*, U. S. Geological Survey.

Bakker, S. A. (2009). Uncertainty analysis of the mud infill prediction of the Olokola LNG approach channel, M.Sc. Thesis, Faculty of Civil Engineering and Geosciences, Delft University of Technology, The Netherlands.

Baldock, T. E., Tomkins, M. R., Nielsen, P., Hughes, M. G. (2004). Settling velocity of sediments at high concentrations, *Coastal Eng.*, 51, pp. 91–100.

Bardet, J. P. (1997). *Experimental Soil Mechanics* (Prentice–Hall).

Barenblatt, G. I. (1953). On the motion of suspended particles in a turbulent flow, *Prikladnaya Matematika i Mekhanika* (*Applied Mathematics and Mechanics*), 17(3), pp. 261–274 (in Russian).

Barnes, H. A. (1995). A review of the slip (wall depletion) of polymer solutions, emulsions and particle suspensions in viscometers: its cause, character, and cure, *J. Non-Newtonian Fluid Mech.*, 56, pp. 221–251.

Barnes, H. A. (1997). Thixotropy-a review, *J. Non-Newtonian Fluid Mech.*, 70, pp. 1–33.

Barnes, H. A. (2000). *A Handbook of Elementary Rheology* (Institute of Non-Newtonian Fluid Mechanics, University of Wales, UK).

Barnes, H. A., Hutton, J. F., Walters, K. (1989). *An Introduction to Rheology* (Elsevier).

Barry, J. M. (1998). *Rising Tide: The Great Mississippi Flood of 1927 and How it Changed America* (Simon and Schuster).

Barry, K. M. (2003). The effect of clay particles in pore water on the erosion of sand, PhD Thesis, University of Florida, Gainesville.

Barry, K. M., Thieke, R. J., Mehta, A. J. (2006). Quasi-hydrodynamic lubrication effect of clay particles on sand grain erosion, *Estuarine Coastal Shelf Sci.*, 67 (1–2), pp. 161–169.

Bascom, W. (1964). *Waves and Beaches: The Dynamics of the Ocean Surface* (Doubleday).

Bassoullet, P., Le Hir, P., Gouleau, D., Robert, S. (2000). Sediment transport over an intertidal mudflat: field investigations and estimation of fluxes within the "Baie de Marennes–Oleron" (France), *Cont. Shelf Res.*, 20 (12/13), pp. 1635–1653.

Batchelor, G. K. (1977). The effect of Brownian motion on the bulk stress in a suspension of spherical particles, *J. Fluid Mech.*, 83, pp. 97–117.

Battjes, J. A., Janssen, J. P. F. M. (1978). Energy loss and set-up due to breaking of random waves, *Proc. 16th Int. Conf. Coastal Eng.*, ASCE, pp. 569–587.

Bear, J. (1988). *Dynamics of Fluids in Porous Media* (Dover).

Bedford, K. W., Onyx, W., Libicki, C. M., Van Evra, R. (1987). Sediment entrainment and deposition measurements in Long Island Sound, *J. Hydraul. Eng.*, 113(10), pp. 1325–1342.

Bedient, P. B., Huber, W. C. (2002). *Hydrology and Floodplain Analysis*, 3rd Ed. (Prentice–Hall).

Been, K., Sills, G. C. (1981). Self-weight consolidation of soft soils: An experimental and theoretical study, *Geotechnique*, 31(4), pp. 519–535.

Berg, R. C., Collinson, C. (1976). Bluff erosion, recession rates and volumetric losses on the Lake Michigan shore of Illinois, *Environ. Geol. Notes No. 76*, Illinois State Geological Survey, Champaign.

Berlamont, J., Ockenden, M., Toorman, E., Winterwerp, J. (1993). The characterization of cohesive sediment properties, *Coastal Eng.*, 21(1–3), pp. 105–128.

Bernatchez, P., Dubois, J. M. (2008). Seasonal quantification of coastal processes and cliff erosion on fine sediment shorelines in a cold temperate climate, north shore of the St. Lawrence Maritime Estuary, Québec, *J. Coastal Res.*, 24(1A), pp. 169–180.

Berry, W., Rubinstein, N., Melzian, B., Hill, B. (2003). The biological effects of suspended and bedded sediment (SABS) in aquatic systems: A review, Internal report to U. S. EPA Office of Research and Development, Narragansett, RI.

Berthault, G. (1986). Experiments on lamination of sediments, *C.R. Acad. Sci., Ser. II*, Paris, 303(17), pp. 1569–1574.

Berthault, G. (1988). Sédimentation d'un mélange hétérogranulaire, Lamination expérimentale en eau calme et en eau courante, *C.R. Acad. Sci., Ser. II*, Paris, 306, pp. 717–724.

Biggs, C. A., Lant, P. A. (2000). Activated sludge flocculation: on-line determination of floc size and the effect of shear, *Water Res.*, 34(9), pp. 2542–2550.

Biot, M. A. (1941). General theory of three-dimensional consolidation, *J. Appl. Phys.*, 12, pp. 155–164.

Bird, E. C. F. (1993). *Submerging Coasts: The Effects of Rising Sea Level on Coastal Environments* (Wiley).

Bird, R. B., Stewart, W. E., Lightfoot, E. N. (1960). *Transport Phenomena* (Wiley).

Bishop, A. A., Simons, D. B., Richardson, E. V. (1965). Total bed-material transport, *J. Hydraul. Div.*, ASCE, 91(2), pp. 175–191.

Bishop, C. T., Skafel, M. G. (1992). Detailed description of laboratory tests on the erosion by waves of till profiles, *NWRI Contribution No. 92–26*, National Water Research Institute, Burlington, Ontario, Canada.

Bishop, C. T., Skafel, M. G., Nairn, R. (1992). Cohesive profile erosion by waves, *Proc. 23rd Conf. Coastal Eng.*, ed. Edge, B. L., ASCE, pp. 2976–2989.

Black, K. S. (1991). The erosion characteristics of cohesive estuarine sediments: Some in situ experiments and observations, PhD Thesis, University of Wales, Menai Bridge, UK.

Black, K. S., Tolhurst, T. J., Paterson, D. M., Hagerthey, S. E. (2002). Working with natural cohesive sediments, *J. Hydraul. Eng.*, 128(2), pp. 2–9.

Blench, T. (1960). *Regime Behaviour of Canals and Rivers* (Butterworth).

Blom, A. (2008). Different approaches to handling vertical and streamwise sorting in modeling river morphodynamics, *Water Resour. Res.*, 44, W03415, doi:10.1029/2006WR005474.

Bo, M. W. (2008). *Compressibility of Ultra-Soft Soil* (World Scientific).

Bogle, M. G. V., Croucher, A. E., O'Sullivan, M. J., Davis, M. D., Kinley, P., Paterson, G. (2007). The integrated catchment study of Auckland City (New Zealand): Fate of contaminants in coastal receiving environments, *Proc. 2006 World Environ. Water Res. Congr.*, ASCE, dx.doi.org/10.1061/40856 (200) 309.

Boon, J. D., Byrne, R. J. (1981). On basin hypsometry and morphodynamic response of coastal inlet systems, *Mar. Geol.*, 40, pp. 27–48.

Borsje, B. W., de Vries, M. B., Hulscher, S. M. J. H., de Boer, G. J. (2008). Modelling large scale cohesive sediment transport by including small scale biological influences, *Estuarine Coastal Shelf Sci.*, 78, pp. 468–480.

Bosworth, R. C. L. (1956). The kinetics of collective sedimentation, *J. Colloid Sci.*, 11(4–5), pp. 496–500.

Boudreau, B. P. (1994). Is burial velocity a master parameter for bioturbation?, *Geochim. Cosmochim. Acta*, 58(4), pp. 1,243–1,249.

Boudreau, B. P. (1997). *Diagenetic Models and Their Implementation* (Springer).

Bowles, J. E. (1991). *Physical and Geotechnical Properties of Soils* (McGraw–Hill).

Bradley, J. B., McCutcheon, S. C. (1987). *Sediment Transport in Gravel-bed Rivers*, eds. Thorne, C. R., Bathurst, J. C., Hey, R. D., "Influence of Large Suspended-sediment Concentrations in Rivers" (Wiley).

Braswell, R. N., Manders, C. F. (1970). A new finite range probability distribution function (FRPDF) with parameters — nomogram and tables, *Reports of Statistical*

Application Research JUSE, Union of Japanese Scientists and Engineers, 17(2), 11 p.

Bremer, L. G. B., van Vliet, V., Wastra, P. (1989). Theoretical and experimental study of the fractal nature of the structure of casein gels, *J. Chem. Soc., Faraday Trans. 1,* 85, pp. 3359–3372.

Brenner, M., Whitmore, T. J. (1998). Historical sediment and nutrient accumulation rates and past water quality in Newnans Lake, Final Report, Department of Fisheries and Aquatic Sciences, University of Florida, Gainesville.

Briaud, J. L., Ting, F. C. K., Chen, H. C., Cao, Y., Han, S. W., Kwak, K. W. (2001). Erosion function apparatus for scour rate predictions, *J. Geotech. Geoenviron. Eng.,* 127(2), pp. 105–113.

Bristow, R. C. (1938). *History of Mudbanks, Vols. I and II* (Cochin Government Press, Cochin, India).

Brogioli, D., Vailati, A. (2001). Diffusive mass transfer by non equilibrium fluctuations: Fick's law revisited. *Phys. Rev. E,* 63(1), 4 p.

Brown, C. B. (1950). *Engineering Hydraulics,* ed. Rouse, H., Chapter 12, "Sediment Transportation" (Wiley).

Brownlie, W. R. (1981). Prediction of flow depth and sediment discharge in open channels, PhD Thesis, Caltech, California.

Bruens, A. W. (2003). Entraining mud suspensions, PhD Thesis, Delft University of Technology, The Netherlands.

Bruens, A. W., Booij, C., Kranenburg, J. C., Winterwerp, J. C. (2000). Applicability of the rotating flume for entrainment experiments, *Proc. 5th Int. Symp. Stratified Flows,* eds. Lawrence, G. A., Pieters, R., Yonemitsu, N., IAHR, pp. 1173–1178.

Bruun, P. (1954). Coastal erosion and development of beach profiles, *Technical Memorandum No. 44,* Beach Erosion Board, U. S. Army Corps of Engineers.

Bruun, P. (1962). Sea level rise as a cause of shore erosion, *J. Waterw. Harbors Coastal Eng. Div.,* ASCE, 88(1), pp. 117–130.

Bruun, P., Schwartz, M. L. (1985). Analytic predictions of beach profile change in response to sea level rise, *Z. Geomorph.,* Suppl. 57, pp. 33–50.

Burban, P. Y., Lick, W., Lick, J. (1989). The flocculation of fine-grained sediments in estuarine waters, *J. Geophys. Res.,* 94(C6), pp. 8323–8330.

Burt, T. N. (1986). *Estuarine Cohesive Sediment Dynamics,* ed. Mehta, A. J., "Field Settling Velocities of Estuary Muds," (Springer–Verlag) pp. 126–150.

Burt, T. N., Parker, W. R. (1984). Settling and density in beds of natural mud during successive sedimentation, *Report No. IT 262,* HR, Wallingford, UK.

Burt, T. N., Stevenson, J. R. (1983). Field settling velocity of Thames mud, *Report No. IT 251,* HR, Wallingford, UK.

Buscall, R., McGowan, I. J., Mills, P. D. A., Stewart, R. F., Sutton, D., White, L. R., Yates, G. E. (1987). The rheology of strongly-flocculated suspensions, *J. Non-Newtonian Fluid Mech.,* 24, pp. 183–202.

Buscall, R., Mills, P. D. A., Goodwin, J. W., Lawson, D. W. (1988). Scaling behaviour of the rheology of aggregate networks formed from colloidal particles, *J. Chem. Soc., Faraday Trans. 1*, 84(12), pp. 4249–4260.

Caline, B. (1994). The calcale-cherrueix embayment: an intertidal seabed environment, *Senck. Marit.*, 24(1/6), pp. 22–27.

Camenen, B. (2007). Simple and general formula for the settling velocity of particles, *J. Hydraul. Eng.*, 133(2), pp. 229–233.

Cargill, K.W. (1984). Prediction of consolidation of very soft soil, *J. Geotech. Eng. Div.*, ASCE, 110(6), pp. 775–795.

Carpenter, J. R., Timmermans, M.–L. (2012). Temperature steps in salty seas, *Phys. Today*, 65(3), pp. 66–67.

Carreau, P. J. (1972). Rheological equations from molecular network theories, *Trans. Soc. of Rheol.*, 16, pp. 99–127.

Carrier, W. D. (2003). Goodbye Hazen; hello Kozeny–Carman, *J. Geotech. Geoenviron. Eng.*, 129(11), pp. 1054–1056.

Casson, M. (1959). *Rheology of Disperse Systems*, ed. Mills, C. C., "A Flow Equation for the Pigment-Oil Suspensions of the Printing Ink Type," (Pergamon) pp. 84–104.

Cervantes, E. E. (1987). A laboratory study of fine sediment resuspension by waves, MS Thesis, University of Florida, Gainesville.

Cervantes, E., Mehta, A. J., Li, Y. (1995). A laboratory-based examination of "episodic" resuspension of fine-grained sediments by waves and current, *Proc. Int. Conf. Coastal Port Eng. Dev. Countries*, Rio de Janeiro, Brazil, pp. 1–13.

Chapman, D. L. (1913). A contribution to the theory of electrocapillarity, *Philos. Mag.*, 25(6), pp. 129–184.

Cheng, D. C.–H. (1980). Viscosity-concentration equations and flow curves for suspensions, *Chem. Ind.* (London), 17, pp. 403–406.

Cheng, N.–S. (1997a). Simplified settling velocity formula for sediment particles, *J. Hydraul. Eng.*, 123(2), pp. 149–152.

Cheng, N.–S. (1997b). Effect of concentration on settling velocity formula for sediment particles, *J. Hydraul. Eng.*, 123(8), pp. 728–731.

Chien, N., Wan, Z. (1999). *Mechanics of Sediment Transport* (ASCE).

Chiu, T. Y. (1972). Sand transport by water or air, *Technical Report UFL/COEL/TR-040*, Coastal and Oceanographic Engineering Department, University of Florida, Gainesville.

Chou, H. T. (1989). Rheological response of cohesive sediments to water waves, PhD Thesis, University of California, Berkeley.

Christensen, B. A. (1965). Discussion of "Erosion and deposition of cohesive soils" by Partheniades, E., *J. Hydraul. Div.*, ASCE, 91(5), pp. 301–308.

Christensen, B. A. (1972). *Sedimentation Symposium to Honor Professor H. A. Einstein*, ed. Shen, H. W., Chapter 4, "Incipient Motion on Cohesionless Channel Banks" (H. W. Shen Publisher, Fort Collins, CO).

Christensen, B. A. (1975). On the stochastic nature of scour initiation, *Proc. 16th Cong. Int. Assoc. Hydr. Res.*, 2, pp. 65–72.

Christensen, B. A., Chiu, T. Y. (1973). Water and air transport of cohesionless materials, *Proc. 15th Cong. Int. Assoc. Hydr. Res.*, 1, pp. 245–252.

Christensen, R. W., Das, B. M. (1974). *Soil Erosion: Causes and Mechanisms; Prevention and Control*, ed. Gray, D. H., Special Report 135, "Hydraulic Erosion of Remolded Cohesive Soils," (Highway Research Board, Washington, D. C.) pp. 8–19.

Christodoulou, G. G. (1986). Interfacial mixing in stratified flows, *J. Hydraul. Res.*, 24(2), pp. 77–92.

Christoffersen, J. B., Jonsson, I. G. (1985). Bed friction and dissipation in a combined current and wave motion, *Ocean Eng.*, 12(5), pp. 387–423.

Chu, J., Bo, M. W., Chang, M. F., Choa, V. (2004). Discussion of, "Consolidation and permeability properties of Singapore marine clay," *J. Geotech. Geoenviron. Eng.*, 130(3), pp. 339–340.

Clift, R., Grace, J. R., Weber, M. E. (1978). *Bubbles, Drops and Particles* (Academic Press).

Coakley, J. P., Skafel, M. G., Davidson–Arnott, R. G. D., Zeman, A. J., Rukavina, N. A. (1988). Computer simulation of nearshore profile evolution in cohesive material, *Proc. IAHR Symp. Math. Model. Sediment Transp.*, Copenhagen, pp. 290–299.

Coleman, N. L. (1981). Velocity profile with suspended sediment, *J. Hydraul. Res.*, 19(3), pp. 211–229.

Committee on Tidal Hydraulics. (1950). Evaluation of the present state of knowledge of factors affecting tidal hydraulics, *Technical Report 1*, U. S. Army Corps of Engineers Waterways, Experiment Station, Vicksburg, MS.

Committee on Tidal Hydraulics. (1960). Soil as a factor in shoaling processes, *Technical Bulletin 4*, U. S. Army Corps of Engineers Waterways Experiment Station, Vicksburg, MS.

Concha F., Bustos, M. C. (1985). *Flocculation, Sedimentation & Consolidation*, eds. Moudgil, B. M., Somasundaran, P., "Theory of Sedimentation of Flocculated Fine Particles," (American Society of Chemical Engineers) pp. 275–284.

Cornelisse, J. M. (1996). The field pipette withdrawal tube (FIPIWITYU), *J. Sea Res.*, 36(1/2), pp. 37–39.

Costa, R. C. F. G. (1989). Flow-fine sediment hysteresis in sediment-stratified coastal waters, MS Thesis, University of Florida, Gainesville.

Costa, R. G., Mehta, A. J. (1990). Flow-fine sediment hysteresis in sediment stratified coastal waters, *Proc. 22nd Coastal Eng. Con.*, ASCE, pp. 2047–2060.

Cousot, P., Piau, J. M. (1994). On the behavior of fine mud suspensions, *Rheol. Acta*, 33(3), pp. 175–184.

Cramp, A., Lee, S. V., Herniman, J., Hiscott, R. N., Manley, P. L., Piper, D. J. W., Deptuck, M., Johnston, S. K., Black, K. S. (1997). Data report: Interlaboratory comparison of sediment grain-sizing techniques — Data from Amazon Fan upper levee complex sediments, *Proc. Ocean Drill. Program Part B Sci. Results*, eds. Flood, R.D., Piper, D. J. W., Klaus, A., Peterson, L. C., (Texas A & M Publications, College Station), Vol. 155, pp. 217–228.

Cross, M. M. (1965). Rheology of non-Newtonian Fluids: a new flow equation for pseudo-plastic systems, *J. Colloid Sci.*, 20, pp. 417–437.

Culkin, D., Ridout, P. (1989). Salinity: definitions, determinations, & standards, *Sea Technol.*, October, pp. 47–49.

Curran, K. J., Hill, P. S., Milligan, T. G. (2003). Time variation of floc properties in a settling column, *J. Sea Res.*, 49, pp. 1–9.

Dade, W. B., Nowell, A. R. M. (1991). Moving muds in the marine environment, *Proc. Coastal Sediments'91*, eds. Kraus, N. C., Ginerich, K. J., Kriebel, D. L., ASCE, pp. 54–71.

Dade, W. B., Nowell, A. R. M., Jumars, P. A. (1992). Predicting erosion resistance of muds, *Mar. Geol.*, 105, pp. 285–297.

Daily, J. W., Harleman, D. R. F. (1966). *Fluid Dynamics* (Addison–Wesley).

Dallavalle, J. (1948). *The Technology of Fine Particles* (Pitman).

Dalrymple, R. A., Liu, P. L.–F. (1978). Waves over soft muds: a two-layer fluid model, *J. Phys. Oceanogr.*, 8(6), pp. 1121–1131.

Danel, P. (1952). *Gravity Waves*, "On the Limiting Clapotis," (National Bureau of Standards Circular 521, Washington, D. C.) pp. 35–38.

Daniels, F., Alberty, R. A. (1955). *Physical Chemistry* (Wiley).

Dankers, P. (2006). On the hindered settling of suspension of mud and mud-sand mixtures, PhD Thesis, Delft University of Technology, The Netherlands.

Darcy, H. (1856). *Histoire des fontaines publiques de Dijon*, éd., Dalmont, V., Appendice, Note D, "Détermination des lois d'écoulement de l'eau à travers le sable" (Libraire des Corps Impériaux des Ponts et Chaussées et de Mines, Paris).

Dasch, W., Wurpts, R. (2001). Isoviscs as useful parameters for describing sedimen-tation, *Terra et Aqua*, 82, pp. 3–7.

Davidson–Arnott, R. G. D. (1986). Erosion of the nearshore profile in till: rates, controls, and implications for shore protection, *Proc. Symp. Cohesive Shores*, Burlington, Ontario, Canada, pp. 137–149.

Davis, W. R. (1991). Evaluating sediment variables that may predict seabed response to resuspension energy, *Proc. Worksh. Hydraul. Res. Needs*, eds. Krone, R. B., McAnally, W. H., U. S. Army Engineer Waterways Experiment Station, Vicksburg, MS.

Davis, W. R. (1993). The role of bioturbation in sediment resuspension and its interaction with physical shearing, *J. Exp. Mar. Biol. Ecol.*, 171, pp. 187–200.

Day, P. R., Ripple, C. D. (1966). Effect of shear on suction in saturated clays, *Soil Sci. Soc. Am. Proc.*, 30, pp. 675–690.

Dean, R. G. (1973). Heuristic models of sand transport in the surf zone, *Proc. Conf. Eng. Dyn. Surf Zone*, Sydney, Australia, 6 p.

Dean, R. G. (1977). Equilibrium beach profiles: U. S. Atlantic and Gulf Coasts, *Ocean Engineering Report No. 12*, Department of Civil Engineering, University of Delaware, Newark.

Dean, R. G. (1987). Coastal sediment processes: toward engineering solutions, *Proc. of Coastal Sediments'87*, ASCE, pp. 1–24.

Dean, R. G., Dalrymple, R. A. (1991). *Water Wave Mechanics for Engineers and Scientists* (World Scientific).

Dean, R. G., Dalrymple, R. A. (2002). *Coastal Processes: With Engineering Applications* (Cambridge University Press).

De Boer, J., Merckelbach, L. M., Winterwerp, J. C. (2006). *Estuarine and Coastal Fine Sediment Dynamics*, eds. Maa, J. P.–Y., Sanford, L. P., Schoellhamer, D. H., "A Parameterized Consolidation Model for Cohesive Sediment," (Elsevier) pp. 243–262.

Decho, A. W. (1990). Microbial exopolymer secretions in ocean environments: their role(s) in food webs and marine processes, *Oceanogr. Mar. Biol. Annu. Rev.*, 28, pp. 73–153.

de Jonge, V. N., van den Bergs, J. (1987). Experiments on the resuspension of estuarine sediments containing benthic diatoms, *Estuarine Coastal Shelf Sci.*, 24, pp. 725–740.

Delo, E. A., Ockenden, M. C. (1992). Estuarine mud manual, *Report No. SR 309*, HR, Wallingford, Wallingford, UK.

Dennett, K. E. (1990). Coagulation of natural dissolved organic matter using ferric chloride, MS Thesis, Georgia Institute of Technology, Atlanta.

Dennett, K. E., Sturm, T. W., Amirtharajah, A., Mahmood, T. (1998). Effects of adsorbed natural organic matter on the erosion of kaolinite sediments, *Water Environ. Res.*, 70(3), pp. 268–275.

De Philipps, R., Vincenzini, M. (1998). Exocellular polysaccharides from cyanobacteria and their possible applications, *FEMS Microbiol. Rev.*, 22, pp. 151–175.

Devine, P. T., Mehta, A. J. (1999). Modulation of microtidal inlet ebb deltas by severe sea, *Proc. Coastal Sediments'99*, eds. Kraus, N. C., McDougal, W. G., ASCE, pp. 1387–1401.

de Wit, P. J. (1995). Liquefaction of cohesive sediments caused by waves, PhD Thesis, Delft University of Technology, The Netherlands.

Diamond, J. (1999). *Guns, Germs and Steel: the Fates of Human Societies* (Norton).

Dickhudt, P. J., Friedrichs, C. T., Sanford, L. P. (2011). Mud matrix solids fraction and bed erodibility in the York River, USA, and other muddy environments, *Continental Shelf Res.*, 31(10 Supp. 1), pp. s3–s13.

Dickhudt, P. J., Friedrichs, C. T., Schaffner, L. C., Sanford, L. P. (2009). Spatial and temporal variation in cohesive sediment erodibility in the York River estuary, eastern USA: A biologically influenced equilibrium modified by seasonal deposition, *Mar. Geol.*, 267(3–4), pp. 128–140.

Dieckmann, R., Osterthun, M., Partenscky, H. W. (1987). Influence of water-level elevation and tidal range on the sedimentation in a German tidal flat area, *Prog. Oceanogr.*, 18, pp. 151–166.

Dietrich, W. (1982). Settling velocities of natural particles, *Water Resour. Res.*, 18(6), pp. 1615–1626.

Dixit, J. G. (1982). Some laboratory measurements of critical bed shear stress for incipient motion of selected bivalve shells, *Report No. UFL/COEL-82/006*, Department of Coastal and Oceanographic Engineering, University of Florida, Gainesville.

Dixit, J. G., Mehta, A. J., Partheniades, E. (1982). Redepositional properties of cohesive sediments deposited in a long flume, *Report UFL/COEL-82/002*, Coastal and Oceanographic Engineering Department, University of Florida, Gainesville.

Dixon, S., Lanyon, B. (2005). Phase change measurements of ultrasonic shear waves on reflection from a curing epoxy system, *J. Phys. D: Appl. Phys.*, 38, pp. 4115–4125.

Doulah, M. S. (1977). Mechanisms of disintegration of biological cells in ultrasonic cavitation, *Biotechnol. Bioeng.*, 19, pp. 649–660.

Dredging Research Technical Notes. (1992). Monitoring Alabama berms, *Report No. DRP-1-08*, U. S. Army Engineer Waterways Experiment Station, Vicksburg, MS.

Du Boys, P. (1879). Le Rhône et les riviers a lit affouillable, *Ann. Ponts Chaussées*, *Mem. Doc.*, 5.

DuMontelle, P. B., Stoffel, K. L., Brossman, J. J. (1977). Hydrologic, geologic and engineering aspects of surficial materials on the Lake Michigan shore in Illinois, *Illinois Coastal Zone Management Development Program Project Report*, Illinois State Geological Survey, Champaign.

Dyer, K. R. (1986). *Coastal and Estuarine Sediment Dynamics* (Wiley).

Dyer, K. R. (1989). Sediment processes in estuaries: future research requirements, *J. Geophys. Res.*, 94(C10), pp. 14327–14339.

Dyer, K. R., Christie, M. C., Wright, E. W. (2000). The classification of intertidal mudflats, *Cont. Shelf Res.*, 20, pp. 1039–1060.

E, X., Hopfinger, E. J. (1987). Stratification by solid particle suspensions, *Proc. 3rd Int. Symp. Stratified Flows*, IAHR, pp. 488–495.

Edelvang, K., Austen, I. (1997). The temporal variation of flocs and fecal pellets in a tidal channel, *Estuarine Coastal Shelf Sci.*, 44(3), pp. 361–367.

Edil, T. B., Vallejo, L. E. (1977). Shoreline erosion and landslides in the Great Lakes, *Proc. 9th Int. Conf. Soil Mech. Found. Eng.*, ASCE, pp. 51–57.

Edzwald, J. K., Upchurch, J. B., O'Melia, C. R. (1974). Coagulation in estuaries, *Environ. Sci. Technol.*, 8, pp. 58–63.

Eedy, W., Rodgers, B., Saunders, K. E., Akindunni, F. (1994). Application of satellite remote sensing to coastal recession in Nigeria, *Proc. Coast Zone Manage. Conf.*, Halifax, Nova Scotia, Canada, pp. 1–4.

Einstein, A. (1911). Berichtigung zu meiner Arbeit: eine neue bestimmung der molecul dimension, *Ann. Phys. (Leipzig)*, 34, pp. 591–592.

Einstein, H. A. (1941). The viscosity of highly concentrated underflows and its influence on mixing, *Trans. 22nd Ann. Meeting Am. Geophys. Union*, Part I: Hydrology Papers, National Research Council, Washington, D. C., pp. 597–603.

Einstein, H. A. (1950). The bed-load function for sediment transportation in open channel flows, *Technical Bulletin No. 1026*, U. S. Department of Agriculture.

Einstein, H. A., El–Samni, E. A. (1949). Hydrodynamic lift forces on a rough wall, *Rev. Mod. Phys.*, 21(3), pp. 520–524.

Eisma, D. (1986). Flocculation and de-flocculation of suspended matter in estuaries, *Neth. J. Sea Res.*, 20, pp. 183–199.

Eisma, D. (1991). Particle size of suspended matter in estuaries, *Geo-Mar. Lett.*, 11, pp. 147–153.

Eisma, D., Augustinus, P. G. E. F., Alexander, C. (1991). Recent and subrecent changes in the dispersal of Amazon mud, *Neth. J. Sea Res.*, 28, pp. 181–192.

Elgar, S., Raubenheimer, B. (2008). Wave dissipation by muddy seafloors, *Geophys. Res. Lett.*, 35, L07611, doi:10.1029/2008GL033245.

Engelund, F., Fredsøe, J. (1976). A sediment transport model for straight alluvial channels, *Nordic Hydrol.*, 7, pp. 293–306.

Engelund, F., Zhaohui, W. (1984). Instability of hyperconcentrated flow, *J. Hydraul. Eng.*, 110(3), pp. 219–233.

Eskinazi, S. (1968). *Principles of Fluid Mechanics* (Allyn and Bacon).

Espey, W. H. Jr. (1963). A new test to measure the scour of cohesive sediments, *Report No. Hyd 01-6301*, Hydraulic Engineering Laboratory, Department of Civil Engineering, University of Texas, Austin.

Etter, R. J., Hoyer, R. P., Partheniades, E., Kennedy, J. F. (1968). Depositional behavior of kaolinite in turbulent flow, *J. Hydraul. Div.*, ASCE, 94(6), pp. 1439–1452.

Everts, C. H. (1980). Design of enclosed harbors to reduce sedimentation, *Proc. 17th Int. Conf. Coastal Eng.*, ASCE, pp. 1512–1527.

Eyring, H. (1936). Viscosity, plasticity and diffusion as examples of absolute reaction rates, *J. Chem. Phys.*, 4, pp. 283–281.

Eysink, W. D. (1989). Sedimentation in harbor basins: Small density differences may cause serious effects, *Delft Hydraul. Publ. no. 417*, Delft, The Netherlands.

Eysink, W. D., Vermaas, H. (1983). Computational methods to estimate the sedimentation in dredged channels and harbor basins in estuarine environments, *Delft Hydraul. Publ. no. 307*, Delft, The Netherlands.

Faas, R. W. (1986). Mass-physical and geotechnical properties of superficial sediments and dense nearbed sediment suspensions on the Amazon continental shelf, *Cont. Shelf Res.*, 6(1/2), pp. 189–208.

Faas, R. W. (1995). Mudbanks of the southwest coast of India III: Role of non-Newtonian flow properties in the generation and maintenance of mudbanks, *J. Coastal Res.*, 11(3), pp. 911–917.

Farrow, J. B., Warren, L. J. (1989). *Flocculation and Dewatering*, eds. Moudgil, B. M., Scheiner, B. J., "Measurement of Floc Density – Floc Size Distributions," (Engineering Foundation, New York) pp. 153–165.

Feng, J. (1992). Laboratory experiments on cohesive soil bed fluidization by water waves, MS Thesis, University of Florida, Gainesville.

Feng, J., Mehta, A. J. (1992). Laboratory experiments on cohesive soil bed fluidization by water waves, Part I: Relationship between the rate of bed fluidization and the rate of wave energy dissipation, *Report UFL/COEL-92/015*, Department of Coastal and Oceanographic Engineering, University of Florida, Gainesville.

Fennessy, M. J., Dyer, K. R., Huntley, D. A. (1994). INSSEV: An instrument to measure the size and settling velocity of flocs *in situ*, *Mar. Geol.*, 117, pp. 107–117.

Ferguson, R., Church, M. (2004). A simple universal equation for grain settling velocity, *J. Sed. Res.*, 74, pp. 933–937.

Fettweis, M. (2008). Uncertainty of effective density and settling velocity of mud flocs derived from in-situ measurements. *Estuarine Coastal Shelf Sci.*, 78(2), pp. 426–436.

Flemming, B. W. (2000). A revised textural classification of gravel-free muddy sediments on the basis of ternary diagrams, *Cont. Shelf Res.*, 20, pp. 1125–1137.

Flemming, B. W., Delafontaine, M. T. (2000). Mass physical properties of muddy intertidal sediments: some applications, misapplications and non-applications, *Cont. Shelf Res.*, 20, pp. 1179–1197.

Foda, M. A. (1989). Sideband damping of water waves over a soft bed, *J. Fluid Mech.*, 201, pp. 189–201.

Foda, M. A., Hunt, J. R., Chou, H.–T. (1993). A nonlinear model for the fluidization of marine mud by waves, *J. Geophys. Res.*, 98(C4), pp. 7,039–7,047.

Foda, M. A., Tzang, S. Y., Maeno, Y. (1991). Resonant soil liquefaction by water waves, *Proc. Geo-Coast'91*, Port and Harbor Research Institute, Yokohama, 1, pp. 549–583.

Fofonoff, N. P., Millard, R. C. Jr. (1983). Algorithms for computation of fundamental properties of seawater, *Unesco Technical Papers in Marine Science 44*, Division of Marine Sciences, Paris.

Fotyma, M., Mercik, S. (1992). *Chemia Rolna (Agricultural Chemistry)* (Polish Scientific Publishers, Warsaw)(in Polish).

Friedlander, S. K. (2000). *Smoke, Dust and Haze: Fundamentals of Aerosol Dynamics* (Oxford).

Friedrichs, C. T. (1993). Hydrodynamics and morphodynamics of shallow tidal channels and intertidal flats, PhD Thesis, Woods Hole Oceanographic Institution, Woods Hole, MA.

Friedrichs, C. T., Aubrey, D. G. (1996). *Mixing in Estuaries and Coastal Seas*, ed. Pattiaratchi, C., "Uniform Bottom Shear Stress and Equilibrium Hypsometry of Intertidal Flats," (American Geophysical Union) pp. 405–429.

Friedrichs, C. T., Scully, M. E. (2007). Modeling deposition by wave-supported gravity flows on the Po River prodelta: From seasonal floods to prograding clinoforms, *Cont. Shelf Res.*, 27(3–4), pp. 322–337.

Friedrichs, C. T., Wright, L. D., Hepworth, D. A., Kim, S. C. (2000). Bottom-boundary-layer processes associated with fine sediment accumulation in coastal seas and bays, *Cont. Shelf Res.*, 20, pp. 807–841.

Fugate, D. C., Friedrichs, C. T. (2003). Controls on suspended aggregate size in partially mixed estuaries, *Estuarine Coastal Shelf Sci.*, 58, pp. 389–404.

Gade, H. G. (1958). Effects of a non-rigid, impermeable bottom on plane surface waves in shallow water, *J. Mar. Res.*, 16(2), pp. 61–82.

Ganju, N. K. (2001). Trapping organic-rich fine sediment in an estuary, MS Thesis, University of Florida, Gainesville.

Garcia, M., Parker, G. (1993). Experiments on the entrainment of sediment into suspension by dense bottom current, *J. Geophys. Res.*, 98(C3), pp. 4793–4807.

Garcia, M. H., ed., (2008). *Sedimentation Engineering: Processes, Measurements, Modeling and Practice* (ASCE).

Gelinas, P. J., Quigley, R. M. (1973). The influence of geology on erosion rates along the north shore of Lake Erie, *Proc. 16th Conf. Great Lakes Res.*, International Association for Great Lakes Research, Ann Arbor, MI, pp. 421–430.

Gibbs, R. J. (1971). *Procedures in Sedimentary Petrology*, ed. Carver, R. E., Chapter 23, "X-ray Diffraction Mounts" (Wiley).

Gibbs, R. J. (1985). Estuarine flocs: their size, settling velocity and density, *J. Geophys. Res.*, 90(C2), pp. 3249–3251.

Gibson, R. E., England, G. L., Hussey, M. J. L. (1967). The theory of one-dimensional consolidation of saturated clays, I: Finite nonlinear consolidation of thin homogeneous layers, *Geotechnique*, 17, pp. 261–273.

Gilbert, K. G. (1914). The transportation of debris by running water, *Professional Paper 86*, U. S. Geological Survey, Washington, D. C.

Goddard, J. D., Miller, C. (1967). Nonlinear effects in the rheology of dilute suspensions, *J. Fluid Mech.*, 28, pp. 657–673.

Goldstein, S. (1929). The steady flow of viscous fluid past a fixed spherical obstacle at small Reynolds numbers, *Proc. R. Soc. London, Ser. A*, 23(791), pp. 225–235.

Gole, C. V., Tarapore, Z. S., Brahme, S. B. (1971). Prediction of siltation in harbour channels and basins, *Proc. 14th Int. Congr. IAHR*, Vol. 4, Paris, pp. 33–40.

Gole, C. V., Tarapore, Z. S., Gadre, M. R. (1973). Siltation in tidal docks due to density currents, *Proc. 15th Congr. IAHR*, Vol. 1, Istanbul, pp. 335–340.

Goodman, R. W. (1999). *Karl Terzaghi* (ASCE).

Gorham, J., Seidle, P. N., Miller, C. (2008). Observed performance of a no-impact beach nourishment project: Sector 7, Indian River County, Presentation at Florida Shore Beach Preserv. Assoc. Mtg., Captiva Island, FL.

Gornitz, V. (1995). Sea-level rise: A review of recent past and near-future trends, *Earth Surf. Proc. Landforms*, 20(1), pp. 7–20.

Gosselin, M. (1993). Stratified flow over a sill, Unpublished class project report, Department of Coastal and Oceanographic Engineering, University of Florida, Gainesville.

Gouy, G. (1910). Sur la constitution de la charge électrique à la surface d'un electrolyte, *Ann. Phys. (Paris)*, Séries, 4(9), pp. 457–468.

Govindaraju, R. S., Ramireddygari, S. R., Shrestha, P. L., Roig, L. C. (1999). Continuum bed model for estuarine sediments based on nonlinear consolidation theory, *J. Hydraul. Eng.*, 125(3), pp. 300–304.

Gowland, J. E. (2001). Selective bottom sediment properties of Lake Bonnet, Florida, *Report UFL/COEL/MP-2001/01*, Coastal and Oceanographic Engineering Program, Department of Civil and Coastal Engineering, University of Florida, Gainesville.

Gowland, J. E. (2002a). Laboratory experiments on the erosional and settling properties of sediment from the Cedar/Ortega River system, *Report UFL/COEL/CR-2002/01*, Coastal and Oceanographic Engineering Program, Department of Civil and Coastal Engineering, University of Florida, Gainesville.

Gowland, J. E. (2002b). Wind induced wave resuspension and consolidation of cohesive sediment in Newnans Lake, Florida, MS Thesis, University of Florida, Gainesville.

Gowland, J. E., Mehta, A. J., Stuck, J. D., John, C. V., Parchure, T. M. (2007). *Estuarine and Coastal Fine Sediment Dynamics*, eds. Maa, J. P.-Y., Sanford. L. P., Schoellhamer, D. H., "Organic-rich Fine Sediments in Florida, Part II: Resuspension in a Lake," (Elsevier) pp. 167–188.

Grace, R. A. (1978). *Marine Outfall Systems: Planning, Design and Construction* (Prentice–Hall).

Graf, W. H. (1971). *Hydraulics of Sediment Transport* (McGraw–Hill).

Graham, D. I., Jones, T. E. R., Davies, J. M., Delo, E. A. (1992). Measurement and prediction of surface shear stress in annular flume, *J. Hydraul. Eng.*, 118(9), pp. 1270–1286.

Gratiot, N., Michallet, H., Mory, M. (2005). On the determination of the settling flux of cohesive sediments in a turbulent fluid, *J. Geophys. Res.*, 110, C06004, doi:10.1029/2004JC002732.

Gregory, J. (1975). Interaction of unequal double layers at constant charge, *J. Colloid Interface Sci.*, 51(1), pp. 44–51.

Grey, J. R., Gartner, J. W. (2010). *Sedimentology of Aqueous Systems*, eds. Poleto, C., Charlesworth, S., "Surrogate Technologies for Monitoring Suspended-sediment Transport in Rivers" (Blackwell).

Griffin, G. M. (1971). *Procedures in Sedimentary Petrology*, ed. Carver, R.E., Chapter 24, "Interpretation of X-ray Diffraction Data" (Wiley).

Grim, R. E. (1968). *Clay Mineralogy* (McGraw–Hill).

Grinsted, A., Moore, J. C., Jevrejeva, S. (2009). Reconstructing sea level from paleo and projected temperatures 200 to 2100 A.D., *Climate Dynamics*, doi.org/10.1007/s00382-008-0507-2.

Gularte, R. C. (1978). Erosion of cohesive sediment as a rate process, PhD Thesis, University of Rhode Island, Kingston.

Gularte, R. C., Kelly, W. E., Nacci, V. A. (1977). Threshold erosional velocities and rates of erosion for redeposited estuarine dredge materials, *Proc. 2nd Int. Symp. Dredg. Technol.*, Paper H3, BHRA Fluid Engineering, Cranfield, UK.

Guphua, Y., Shadong, B. (1987). Erosion and control of a silty-muddy beach, *Proc. 5th Symp. Coastal Ocean Manage. Coastal Zone'89*, ASCE, pp. 821–835.

Gust, G. (1976). Observations of turbulent drag reduction in a dilute suspension of clay in seawater, *J. Fluid Mech.*, 75, pp. 29–47.

Gust, G., Muller, V. (1997). *Cohesive Sediments*, eds. Burt, N., Parker, R., Watts, J., "Interfacial Hydrodynamics and Entrainment Functions of Currently Used Erosion Devices," (Wiley) pp. 149–174.

Håkanson, L., Janssen, M. (1983). *Principles of Lake Sedimentology* (Springer–Verlag).

Hallermeier, R. J. (1978). Uses for a calculated limit depth to beach erosion, *Proc. 16th Int. Conf. Coastal Eng.*, ASCE, pp. 1493–1512.

Hallermeier, R. J. (1981). Terminal settling velocity of commonly occurring sand grains, *Sedimentology*, 28(6), pp. 859–865.

Han, M. (1989). Mathematical modeling of heterogeneous flocculent sedimentation, PhD Thesis, University of Texas, Austin.

Hanjalić, K. (2004). *Closure models for incompressible turbulent flows*, Lecture series, "Introduction to Turbulence Modeling," von Karman Institute for Fluid Dynamics, Sint–Genesius–Rode, Belgium.

Hanson, G. J. (1990). Surface erodibility of earthen channels at high stresses, Part I: Developing an *in situ* testing device, *Trans. Am. Soc. Agric. Eng.*, 33(1), pp. 132–137.

Hardy, W. B. (1900). A preliminary investigation of the colloids which determine the stability of irreversible sols, *Proc. R. Soc. London*, 66, pp. 110–125.

Hawley, N. (1982). Settling velocity distribution of natural aggregates, *J. Geophys. Res.*, C87, pp. 9489–9498.

Hawley, N., Lesht, B. M. (1992). Sediment resuspension in Lake St. Clair, *Limnol. Oceanogr.*, 37(8), pp. 1720–1737.

Hayter, E. J. (1983). Prediction of cohesive sediment transport in estuarial waters, PhD Thesis, University of Florida, Gainesville.

Hayter, E. J., Mehta, A. J. (1986). Modeling cohesive sediment transport in estuarial waters, *Appl. Math. Modell.*, 10, pp. 294–303.

Hazel, P. (1972). Numerical studies of the stability of inviscid parallel shear flows, *J. Fluid Mech.*, 51, pp. 39–61.

Headland, J. R. (1991). An engineering evaluation of fine sedimentation at the Mayport Naval Basin, Mayport, Florida, *Proc. Coastal Sediments'91*, eds. Kraus, N. C., Ginerich, J. K., Kriebel, D. L., ASCE, pp. 803–816.

Hedges, J. I., Keil, R. G. (1999). Organic geochemical perspectives on estuarine processes: Sorption reactions and consequences, *Mar. Chem.*, 65, pp. 55–65.

Heffler, D. E., Syvitski, J. P. M., Asprey, K. W. (1991). *Principles, Methods, and Application of Particle Size Analysis*, "The Floc Camera Assembly," (Cambridge) pp. 209–221.

Henderson, F. M. (1966). *Open Channel Flow* (Macmillan).

Herbich, J., Darby, R., Gordon, W., Krafft, K., De Hart, D. (1989). Definition of navigable depth in fine-sediments, *CDS Report No. 312*, Center for Dredging Studies, Civil Engineering Department, Texas A & M University, College Station.

Heywood, N. I. (1991). *Slurry Handling Design of Solid-Liquid Systems*, eds. Brown, N. P., Heywood, N. I., "Rheological Characterization of Non-settling Slurries," (Elsevier) pp. 53–87.

Hill, P. S. (1992). Reconciling aggregation theory with observed vertical fluxes following phytoplankton bloom, *J. Geophys. Res.*, 97 (C2), pp. 2295–2308.

Hill, P. S. (1998). Controls on floc size in the coastal ocean, *Oceanography*, 11, pp. 13–18.

Hill, P. S., Milligan, T. G., Geyer, W. R. (2000). Controls on effective settling velocity of suspended sediment in the Eel River flood plume, *Cont. Shelf Res.*, 20, pp. 2095–2111.

Hill, P. S., Nowell, A. R. M. (1990). The potential role of large, fast-sinking particles in clearing nepheloid layers, *Philos. Trans. R. Soc. London*, Ser. A, 331, pp. 103–117.

Hinze, J. O. (1959). *Turbulence: An Introduction to Its Mechanism and Theory* (McGraw–Hill).

Ho, H. W. (1964). Fall velocity of a sphere in a field of oscillating fluid, PhD Thesis, Iowa State University, Ames.

Hollick, M. (1976). Towards a routine assessment of the critical tractive forces cohesive soils, *Trans. Am. Soc. Agric. Eng.*, 19(6), pp. 1076–1081.

Holmboe, J. (1962). On the behavior of symmetric waves in stratified shear layers, *Geofys. Publ.*, 24, pp. 67–113.

Holmes, R. S., Hearn, W. E. (1942). Chemical and physical properties of some of the important alluvial soils of the Mississippi drainage basin, *Technical Bulletin 833*, U. S. Department of Agriculture, Washington, D. C.

Holthuijsen, L. H. (2007). *Waves in Ocean and Coastal Waters* (Cambridge).

Hopfinger, E. J., Linden, P. F. (1982). Formation of thermoclines in zero-mean-shear turbulence subjected to a stabilizing buoyancy flux, *J. Fluid Mech.*, 114, pp. 157–173.

Hopfinger, E. J., Toly, J. A. (1976). Spatially decaying turbulence and its relation to mixing across density interfaces, *J. Fluid Mech.*, 78, pp. 155–175.

Hor, T.–L. (1991). Coastal protection along a mud shore: a pilot study, *Proc. Conf. Coastal Eng. Nat. Dev.*, Institution of Engineers, Malaysia and Institution of Civil Engineers, Kuala Lumpur, pp. 128–137.

Horowitz, A. J. (1991). *A Primer on Sediment-Trace Element Chemistry*, 2nd Ed. (Lewis).

Hough, B. K. (1957). *Basic Soils Engineering* (Ronald Press).

Houwing, E.–J. (1999). Determination of the critical erosion threshold of cohesive sediments on intertidal mudflats along the Dutch Wadden Sea coast, *Estuarine Coastal Shelf Sci.*, 49, pp. 545–555.

Hsu, T.–J. (2002). A two-phase flow approach for sediment transport, PhD Thesis, Cornell University.

Hsu, T.–J, Jenkins, J. T., Liu, P. L.–F. (2003). On two-phase sediment transport: Dilute flow, *J. Geophys. Res.*, 108 (C3), p. 3057, doi: 10.1029/2001JC001276.

Hsu, T.–J., Jenkins, J. T., Liu, P. L.–F. (2004). On two-phase sediment transport: Sheet flow of massive particles, *Proc. R. Soc. London, Ser. A*, 460 (2048), pp. 2,223–2,250.

Hsu, T.–J., Traykovski, P. A., Kineke, G. C. (2007). On modeling boundary layer and gravity-driven fluid mud transport, *J. Geophys. Res.*, 112(C04011), doi:10.1029/2006JC003719.

Huang, S., Han, N., Zhong, X. (1980). Analysis of siltation at mouth bar of the Yangtze River estuary, *Proc. Int. Symp. Riv. Sedimentation*, Paper 6, Chinese Society of Hydraulic Engineering, Beijing (in Chinese).

Huang, X., Garcia, M. H. (1996). A perturbation solution for Bingham-plastic mudflows, *J. Hydraul. Eng.*, 123(11), pp. 986–994.

Huang, X., Garcia, M. H. (1998). A Herschel–Bulkley model for mud flow down a slope, *J. Fluid Mech.*, 374, pp. 305–333.

Huang, X., Garcia, M. H. (1999). Modeling of non-hydroplaning mudflows on continental slopes, *Mar. Geol.*, 154, pp. 131–142.

Huisman, J., Sommeijer, B. (2002). Maximal sustainable sinking velocity of phytoplankton, *Mar. Ecol. Prog. Ser.*, 244, pp. 39–48.

Hunt, J. R. (1982). Particle dynamics in seawater: Implications for predicting the fate of discharged particles, *Environ. Sci. Technol.*, 16(6), pp. 303–309.

Hunt, L. M., Groves, D. G., eds. (1965). A *Glossary of Ocean Science & Undersea Technology Terms* (Compass, Arlington, VA).

Hunt, S. D. (1981). A comparative review of laboratory data on erosion of cohesive sediment beds, *Report no. UFL/COEL/MP-81/007*, Coastal and Oceanographic Engineering Department, University of Florida, Gainesville.

Hunt, S. D., Mehta, A. J. (1985). An evaluation of laboratory data on erosion of fine sediment beds, Paper presented at 16th Ann. Meeting Fine Particle Soc., Miami (unpublished).

Huygens, M., Verhofen, R. (1996). Some fundamental particularities on partly cohesive sediment transport in a circular flume test, *Adv. Fluid Mech.*, 9, pp. 11–20.

Hwang, K.–N. (1989). Erodibility of fine sediment in wave dominated environments, MS Thesis, University of Florida, Gainesville.

Hwang, P. A. (1985). Fall velocity of particles in oscillating flow, *J. Hydraul. Eng.*, 111(3), pp. 485–502.

Ibe, A. C., Awosika, L. F., Ihenyen, A. E., Ibe, C. E., Awani, P. E. (1989). Erosion management strategies for the Mahin mud beach, Ondo State, Nigeria, *Proc. Coastal Zone'89 Conf.*, ASCE, pp. 821–835.

Imai, G. (1981). Settling behavior of clay suspensions, *Soils and Found.*, 20(1), pp. 7–20.

Inglis, C. C., Allen, F. H. (1957). The regimen of the Thames estuary as affected by currents, salinities and river flow, *Proc. Inst. Civil Eng.*, London, 7, pp. 827–868.

Isobe, M., Huynh, T. N., Watanabe, A. (1992). A study on mud mass transport under waves based on an empirical rheology model, *Proc. 22nd Int. Conf. Coastal Engineering*, ASCE, pp. 3093–3106.

Jacinto, R. S., Le Hir, P. (2001). *Coastal and Estuarine Fine Sediment Transport Processes*, eds. McAnally, W. H., Mehta, A. J., "Response of Stratified Muddy Beds to Water Waves," (Elsevier) pp. 95–108.

Jackson, G. A. (1990). A model of the formation of marine algal flocs by physical coagulation processes, *Deep Sea Res.*, 37, pp. 1197–1211.

Jacobs, W. (2011). Sand-mud erosion from a soil mechanical perspective, PhD Thesis, Delft University of Technology, The Netherlands.

Jaeger, J. M., Hart, M. (2001). Sedimentary processes in the Loxahatchee River Estuary: 5000 years ago to the present, Final Report, to Jupiter Inlet District Commission, Jupiter, Florida, by Department of Geology and Geophysics, University of Florida, Gainesville.

Jaeger, J. M., Mehta, A., Faas, R., Grella, M. (2009). Anthropogenic impacts on sedimentary sources and processes in a small urbanized subtropical estuary, Florida, *J. Coastal Res.*, 25(1), pp. 30–47.

Jaeger, J. M., Sun, M. Y., White, J. R., Hendrickson, J. (2004). A seasonal time-series study of sediment resuspension in a high-organic-carbon Blackwater estuary, Lower St. Johns River Florida: Implications for estuarine nutrient fluxes, *Eos Trans. Am. Geophys. Union*, 84(52), Ocean Science Meeting Supplement, Abstract OS21D-02.

Jain, M. (2007). Wave-mud interaction in shallow waters, PhD Thesis, University of Florida, Gainesville.

Jain, M., Mehta, A. J. (2009a). *Handbook of Coastal and Ocean Engineering*, ed. Kim, Y. C., Chapter 27, "Wave-induced Resuspension of Fine Sediment" (ASCE).

Jain, M., Mehta, A. J. (2009b). Role of basic rheological models in determination of wave attenuation over muddy seabeds, *Cont. Shelf Res.*, 29(3), pp. 642–651.

Jain, M., Mehta, A. J., Hayter, E. J., Di, J. (2007). *Sediment and Ecohydraulics: INTERCOH 2005*, eds. Kusuda, T., Yamanishi, H., Spearman, J., Gailani, J. Z., "Fine Sediment Resuspension and Nutrient Transport in Newnans Lake, Florida," (Elsevier) pp. 295–311.

James, A. E., Williams, D. J. A., Williams, P. R. (1987). Direct measurement of static yield properties of cohesive suspensions, *Rheol. Acta*, 26, pp. 437–446.

James, A. E., Williams, D. J. A., Williams, P. R. (1988). *Physical Processes in Estuaries*, eds. Dronkers J., van Leussen, W., "Small Strain, Low Shear Rate Rheometry of Cohesive Sediments," (Springer–Verlag) pp. 488–500.

Jaramillo, S., Sheremet, A., Allison, M. A., Reed, A. H., Holland, K. T. (2009). Wave-mud interactions over the muddy Atchafalaya subaqueous clinoform, Louisiana, United States: Wave-supported sediment transport, *J. Geophys. Res.*, 114, C04002, doi:10.1029/2008JC004821.

Jarvis, P., Jefferson, B., Gregory, J., Parsons, S. A. (2005). A review of floc strength and breakage, *Water Res.*, 39, pp. 3121–3137.

Jeng, D. S. (1997). Wave-induced seabed response in front of a breakwater, PhD Thesis, University of Western Australia, Nedlands.

Jeng, D.-S., Lin, M. (2003). Comparison of existing poroelastic models for wave damping in a porous seabed, *Ocean Eng.*, 30 (11), pp. 1335–1352.

Jepsen, R., Roberts, J., Lick, W. (1997). Effects of bulk density on sediment erosion rates, *Water Air Soil Pollut.*, 99(1–4), pp. 21–31.

Jiang, F. (1993). Bottom mud transport due to water waves, PhD Thesis, University of Florida, Gainesville.

Jiang, F., Mehta, A. J. (1992). *Dynamics and Exchanges in Estuaries and the Coastal Zone*, ed. Prandle, D., "Some Observations on Fluid Mud Response to Water Waves," (American Geophysical Union, Washington, D. C.) pp. 351–376.

Jiang, F., Mehta, A. J. (1995). Mudbanks of the southwest coast of India IV: Mud viscoelastic properties, *J. Coastal Res.*, 11(3), pp. 918–926.

Jiang, F., Mehta, A. J. (1996). Mudbanks of the southwest coast of India V: Wave attenuation, *J. Coastal Res.*, 12(4), pp. 890–897.

Jiang, J. (1999). An examination of estuarine lutocline dynamics, PhD Thesis, University of Florida, Gainesville.

Jiang, J., Ganju, N. K., Mehta, A. J. (2004). Estimation of contraction scour in a riverbed using SERF, *J. Waterw. Port Coastal Ocean Eng.*, 130(3), pp. 215–218.

Jiang, J., Mehta, A. J. (1999). Consolidation modeling for cohesive sediment transport, *Report No. UFL/COEL-99/006*, Coastal and Oceanographic Engineering Program, Department of Civil and Coastal Engineering, University of Florida, Gainesville.

Jiang, J., Mehta, A. J. (2000). Lutocline behavior in a high-concentration estuary, *J. Waterw. Port Coastal Ocean Eng.*, 126(6), pp. 324–328.

Jiang, J., Mehta, A. J. (2002). *Fine Sediment Dynamics in the Marine Environment*, eds. Winterwerp, J. C., Kranenburg, C., "Interfacial Instabilities at the Lutocline in the Jiaojiang Estuary, China," (Elsevier) pp. 125–137.

Jimenez, J., Madsen O. (2003). A simple formula to estimate settling velocity of natural sediments, *J. Waterw. Port Coastal Ocean Eng.*, 129(2), pp. 70–78.

Jinchai, P. (1998). An experimental study of mud slurry flows in pipes, MS Thesis, University of Florida, Gainesville.

Jobson, H. E., Sayre, W. W. (1970). Vertical transfer in open channel flow, *J. Hydraul. Div.*, ASCE, 96(3), pp. 7148–7152.

Johnson, B. D., Barry, M. A., Boudreau, B. P., Jumars, P. A., Dorgan, K. M. (2011). In situ tensile fracture toughness of surficial cohesive marine sediments. *Geo-Mar. Let.*, doi 10.1007/s00367-011-0243-1.

Jones, C., Gailani, J. (2007). Discussion of "Comparison of two techniques to measure sediment erodibility in the Fox River, Wisconsin" by T. Ravens, *J. Hydraul. Eng.*, 133(1), pp. 111–115.

Jonsson, I. G. (1966). Wave boundary layers and friction factors, *Proc. 10th Int. Conf. Coastal Eng.*, ASCE, pp. 127–148.

Julien, P. Y. (1995). *Erosion and Sedimentation* (Cambridge).

Jumars, P. A., Deming, J. W., Hill, P. S., Karp–Boss, L., Yager, P. L., Dade, W. B. (1993). Physical constraints on marine osmotrophy in an optimal foraging context, *Mar. Microb. Food Webs*, 7, pp. 121–159.

Kaihatu, J. M., Sheremet, A., Holland, K. T. (2007). A model for the propagation of nonlinear surface waves over viscous muds, *Coastal Eng.*, 54, pp. 752–764.

Kalinske, A. A. (1947). Movement of sediment as bed load in rivers, *Trans. Am. Geophys. Union*, 28(4), pp. 615–620.

Kamphuis, J. W. (1975). Friction factor under oscillatory waves, *J. Waterw. Harbors Coastal Eng. Div.*, ASCE, 101(1), pp. 135–144.

Kamphuis, J. W. (1986). Erosion of cohesive bluffs, a model and a formula, *Proc. Symp. Cohesive Shores*, ed. Skafel, M. G., National Research Council, Canada, Ottawa, pp. 226–245.

Kandiah, A. (1974). Fundamental aspects of surface erosion of cohesive soils, PhD Thesis, University of California, Davis.

Karthikayan, M., Tan, T. S., Mimura, M., Yoshimura, M., Choon, P. E. (2007). Improvements in nuclear-density cone penetrometer for non-homogeneous soils, *Jap. Geotech. Soc.*, 47(1), pp. 109–117.

Keedwell, M. J. (1984). *Rheology and Soil Mechanics* (Elsevier).

Kelly, W. E., Gularte, R. C. (1981). Erosion resistance of cohesive soils, *J. Hydraul. Div.*, ASCE, 107(10), pp. 1211–1224.

Kelly, W. E., Gularte, R. C., Nacci, V. A. (1979). Erosion of cohesive sediments as a rate process, *J. Geotech. Eng. Div.*, ASCE, 105(5), pp. 673–676.

Kemp, G. P. (1986). Mud deposition at the shoreface: Wave and sediment dynamics on the Chenier Plain of Louisiana, PhD Thesis, Louisiana State University, Baton Rouge.

Kendrick, M. P., Derbyshire, B. V. (1985). Monitoring of a near-bed turbid layer, *Report No. SR44*, HR, Wallingford, UK.

Keulegan, G. H. (1966). *Estuary and Coastline Hydrodynamics*, ed. Ippen, A. T., "The Mechanism of an Arrested Saline Wedge," (McGraw–Hill) pp. 546–574.

Keulegan, G. H. (1967). Tidal flow in entrances: Water level fluctuations of basins in communication with the seas, *Technical Bulletin No. 14*, Committee on Tidal Hydraulics, U. S. Army Engineer Waterways Experiment Station, Vicksburg, MS.

Khelifa, A., Hill, P. S. (2006). Models for effective density and settling velocity of flocs, *J. Hydraul. Res.*, 44(3), pp. 390–401.

Kim, M. H., Koo, W. C., Hong, S. Y. (2000a). Wave interactions with 2D structures on/inside porous seabed by a two-domain boundary element method, *Appl. Ocean Res.*, 22(5), pp. 255–266.

Kim, S.-C., Friedrichs, C. T., Maa, J. P.-Y., Wright, L. D. (2000b). Estimating bottom shear stress in tidal boundary layer from acoustic Doppler velocimeter data, *J. Hydraul. Eng.*, 126(6), pp. 399–406.

Kineke, G. C. (1993). Fluid muds on the Amazon continental shelf, PhD Thesis, University of Washington, Seattle.

Kineke, G. C., Higgins, E. E., Hart, K., Velasco, D. (2007). Fine-sediment transport associated with cold front passages on the shallow shelf, Gulf of Mexico, *Cont. Shelf Res.*, 26, pp. 2073–2091.

Kineke, G. C., Sternberg, R. W. (1995). Distribution of fluid muds on the Amazon Continental Shelf, *Mar. Geol.*, 125, pp. 193–223.

King, I. P. (1988). A finite element model for three dimensional hydrodynamic systems, Report to U. S. Army Engineer Waterways Experiment Station, Vicksburg, MS.

King, I. P., Norton, W. R., Orlob, G. T. (1973). A finite element solution for two-dimensional density stratified flow, Report for Office of Water Resources Research, U. S. Department of the Interior.

King, J. N. (2007). Selective mechanisms for benthic water flux generation in coastal waters, PhD Thesis, University of Florida, Gainesville.

King, J. N., Mehta, A. J., Dean, R. G. (2009). Generalized analytic model for benthic water flux forced by surface gravity waves, *J. Geophys. Res.*, 114, C04004, doi:10.1029/2008JC005116.

King, J. N., Mehta, A. J., Dean, R. G. (2010). Analytic models for the groundwater tidal prism and associated benthic water flux, *Hydrogeol. J.*, 18(1), pp. 203–215, doi:10.1007/s10040-009-0519-y.

Kirby, J. M. (1988). Rheological characteristics of sewage sludge: A granuloviscous model, *Rheol. Acta*, 27, pp. 326–334.

Kirby, R. (1986). Suspended fine cohesive sediment in the Severn estuary and Inner Bristol Channel, *Report No. ESTU-STP-4042*, Department of Atomic Energy, Harwell, UK.

Kirby, R. (1992). *Dynamics and Exchanges in Estuaries and the Coastal Zone*, ed. Prandle, D., "Effect of Sea-level Rise on Muddy Coastal Margins," (American Geophysical Union) pp. 313–334.

Kirby, R. (1995). Tidal flat regeneration – beneficial use of muddy dredged material, *Proc. 14th World Dredg. Congr.*, WODA, Amsterdam, pp. 319–332.

Kirby, R., Hobbs, C. H., Mehta, A. J. (1994). Shallow stratigraphy of Lake Okeechobee, Florida: a preliminary reconnaissance, *J. Coastal Res.*, 10(2), pp. 339–350.

Kirby, R., Parker, W. R. (1974). Seabed density measurements related to echosounder records, *Dock and Harbour Auth.*, 54, pp. 423–424.

Kirkham, D., Powers, W. L. (1972). *Advanced Soil Physics* (Wiley).

Kirkham, M. B. (2005). *Principles of Soil and Plant Water Relations* (Elsevier).

Kironoto, B. A., Graf, W. H. (1994). Turbulence characteristics in rough uniform open-channel flow, *Water Marit. Energy Proc. Inst. Civ. Eng.*, 106(4), pp. 333–344.

Knapp, R. T. (1938). Energy balance in stream-flows carrying suspended load, *Trans. Am. Geophys. Union*, 19, pp. 501–505.

Kolsky, H. (1992). *Mechanics and Mechanisms of Material Damping*, eds. Kinra, V. K., Wolfenden, A., "The Measurement of Material Damping of High Polymers Over Ten Decades of Frequency and Its Interpretation," (ASTM) pp. 4–27.

Komar, P. D. (1998). *Beach Processes and Sedimentation*, 2nd Ed. (Prentice–Hall).

Kranck, K. (1986). *Estuarine Cohesive Sediment Dynamics*, ed. Mehta, A. J., "Settling Behavior of Cohesive Sediment," (Springer–Verlag) pp. 151–169.

Kranck, K., Milligan, T. G. (1980). Macroflocs: Production of marine snow in the laboratory, *Mar. Ecol. Prog. Ser.*, 3(July), pp. 19–24.

Kranck, K., Milligan, T. G. (1992). Characteristics of suspended particles at an 11-hour anchor station in San Francisco Bay, California, *J. Geophys. Res.*, 97(C7), pp. 11,373–11,382.

Kranck, K., Petticrew, E., Milligan, T. G., Droppo, I. G. (1993). *Nearshore and Estuarine Cohesive Sediments Transport*, ed. Mehta, A. J., "*In situ* Particle Size Distributions Resulting from Flocculation of Suspended Sediment," (Springer–Verlag) pp. 60–74.

Kranenburg, C. (1994). The fractal structure of cohesive sediment aggregates, *Estuarine Coastal Shelf Sci.*, 39(5), pp. 451–460.

Kranenburg, C., Winterwerp, J. C. (1997). Erosion of fluid mud layers, I: Entrainment model, *J. Hydraul. Eng.*, 123(6), pp. 504–511.

Kranenburg, W. M. (2008). Modeling wave damping by fluid mud, MSc Thesis, Delft University of Technology, The Netherlands.

Kranenburg, W. M., Winterwerp, J. C., de Boer, G. J., Cornelisse, J. M., Zijlema, M. (2011). SWAN-Mud: Engineering model for mud-induced wave damping, *J. Hydr. Eng.*, DOI: 10.1061/(ASCE)HY.1943-7900.0000370.

Krieger, I. M., Dougherty, T. J. (1959). A mechanism for non-Newtonian flow of suspensions of rigid spheres, *Trans. Soc. Rheol.*, 3, pp. 137–152.

Krishna, G. (2009). A numerical model based examination of conditions for ignitive turbidity currents, MS Thesis, University of Florida, Gainesville.

Krizek, R. J. (1971). Rheological behavior of clay soils subjected to dynamic loads, *Trans. Soc. Rheol.*, 15(3), pp. 433–489.

Krone, R. B. (1957). Silt transport studies utilizing radioisotopes, *First Annual Progress Report*, Hydraulic Engineering Laboratory and Sanitary Engineering Research Laboratory, University of California, Berkeley.

Krone, R. B. (1959). Silt transport studies utilizing radioisotopes, *Second Annual Progress Report*, Hydraulic Engineering Laboratory and Sanitary Engineering Research Laboratory, University of California, Berkeley.

Krone, R. B. (1962). Flume studies of the transport of sediment in estuarial shoaling processes, *Final Report*, Hydraulic Engineering Laboratory and Sanitary Engineering Research Laboratory, University of California, Berkeley.

Krone, R. B. (1963). A study of rheological properties of estuarial sediments, *Technical Bulletin No. 7*, Committee on Tidal Hydraulics, U. S. Army Engineer Waterways Experiment Station, Vicksburg, MS.

Krone, R. B. (1979). *San Francisco Bay: an Urbanized Estuary*, ed. Conomos, T. J., "Sedimentation in the San Francisco Bay System," (American Association for the Advancement of Science, San Francisco) pp. 85–96.

Krone, R. B. (1983). A viscosity-temperature relation for Newtonian liquids, *Chem. Eng. Commun.*, 22(3–4), pp. 161–180.

Krone, R. B. (1985). Simulation of marsh growth under rising sea levels. *Proc. Conf. Hydraul. Hydrol. Small Comput. Age*, ed. Waldrup, W. R., ASCE, pp. 106–115.

Krone, R. B. (1986). *Estuarine Cohesive Sediments Dynamics*, ed. Mehta, A. J., "The Significance of Aggregate Properties to Transport Processes," (Springer–Verlag) pp. 66–84.

Krone, R. B. (1993). *Nearshore and Estuarine Cohesive Sediment Transport*, ed. Mehta, A. J., "Sedimentation Revisited," (American Geophysical Union) pp. 108–125.

Krone, R. B. (1999). Effects of bed structure on erosion of cohesive sediments, *J. Hydraul. Eng.*, 125(12), pp. 1297–1301.

Krotov, M., Mei, C. C. (2007). A generalized viscoelastic model of fluid mud and damping of sea waves, Abstract of paper presented at *Int. Conf. on Cohesive Sediment Transp. Processes*, Brest, France.

Krumbein, W. C. (1936). Application of logarithmic moments to size frequency distribution of sediments, *J. Sediment. Petrol.*, 6(1), pp. 35–37.

Kuijper, C., Cornelisse, J. M., Winterwerp, J. C. (1989). Research on erosive properties of cohesive sediments, *J. Geophys. Res.*, 94(C10), pp. 14341–14350.

Kuo, A. Y., Park, K. (1992). *Dynamics and Exchanges in Estuaries and the Coastal Zone*, ed. Prandle, D., "Transport of Hypoxic Waters: An Estuary-Subestuary Exchange," (American Geophysical Union) pp. 599–615.

Kusuda, T., Umita, T., Koga, T., Futawatari, T., Awaya, Y. (1984). Erosional process of cohesive sediments, *Water Sci. Technol.*, 17(6–7), pp. 891–901.

Kynch, G. J. (1952). A theory of sedimentation, *Trans. Faraday Soc.*, 48, pp. 166–176.

Lamb, H. (1945). *Hydrodynamics*, 6th Ed. (Dover).

Lamb, M. P., D'Asaro, E., Parsons, J. D. (2004). Turbulent structure of high-density suspensions formed under waves, *J. Geophys. Res.*, 109(C12026), doi:10.1029/2004JC002355.

Lambe, T. W., Whitman, R. V. (1979). *Soil Mechanics*, Standard Ed. (Wiley).

Lambrechts, J., Humphrey, C., McKinna, L., Gourge, O., Fabricius, K. E., Mehta, A. J., Lewis, S., Wolanski, E. (2010). Importance of wave-induced bed liquefaction in the fine sediment budget of Cleveland Bay, Great Barrier Reef, *Estuarine Coastal Shelf Sci.*, 89, pp. 154–162.

Lau, Y. L. (1994). Temperature effect on settling velocity and deposition of cohesive sediments, *J. Hydraul. Res.*, 32(1), pp. 41–51.

Lau, Y. L., Krishnappan, B. G. (1994). Does reentrainment occur during cohesive sediment settling?, *J. Hydraul. Eng.*, 120(2), pp. 236–244.

Laufer, J. (1949). Investigation of turbulent flow in a two-dimensional channel, *Report No. 1053*, National Advisory Committee for Aeronautics, Washington, D. C.

Laufer, J. (1954). The structure of turbulence in fully developed pipe flow, *Report No. 1174*, National Advisory Committee for Aeronautics, Washington, D. C.

Law, B. A., Hill, P. S., Milligan, T. J., Curran, K. J., Wiberg, P. L., Wheatcroft, R. A. (2008). Size sorting of fine-grained sediments during erosion: Results from the Gulf of Lions, *Cont. Shelf Res.*, 28, pp. 1935–1946.

Law, D. J., Bale, A. J. (1998). *Sedimentary Processes in the Intertidal Zone*, eds. Black, K. S., Patterson, D. M., Cramp, A., "*In situ* Characterization of Suspended Particles Using Focused-beam, Laser Reflectance Particle Sizing," (Geological Society, London) pp. 57–68.

Law, D. J., Bale, A. J., Jones, S. E. (1997). Adaptation of focused beam reflectance measurement to in-situ particle sizing in estuaries and coastal waters, *Mar. Geol.*, 140, pp. 47–59.

Lawrence, G. A., Ward, P. R. B., MacKinnon, M. D. (1991). Wind-wave-induced suspension of mine tailing in disposal ponds – a case study, *Can. J. Civ. Eng.*, 18, pp. 1047–1053.

LeConte, L. J. (1905). Discussion on the paper, "Notes on the improvement of river and harbor outlets in the United States," by D. A. Watt, Paper No. 1009, *Trans. ASCE*, 55 (December), pp. 306–308.

Lee, B. J., Toorman, E., Molz, F. J., Wang, J. (2011). A two-class population balance equation yielding bimodal flocculation of marine or estuarine sediments, *Water Res.*, 45, pp. 2131–2145.

Lee, C., Wu, C. H., Hoopes, J. A. (2004). Automated sediment erosion testing system using digital imaging, *J. Hydraul. Eng.*, 130(8), pp. 771–782.

Lee, H. J. (1985). *Strength Testing of Marine Sediments and In-situ Measurements*, eds. Chaney, R. C., Demars, K. R., ASTM STP 883, "State of the Art: Laboratory Determination of the Strength of Marine Soils," (American Society of Testing and Materials) pp. 181–250.

Lee, J. (1978). Settling velocity, flow resistance, and the angle of repose of selected bivalve shells, MS Thesis, University of Florida, Gainesville.

Lee, K., Sills, G. C. (1981). The consolidation of a soil stratum including self-weight effects and large strains, *Int. J. Numer. Anal. Methods Geomech.*, 5(4), pp. 405–428.

Lee, S.–C. (1995). Response of mud shore profiles to waves, PhD Thesis, University of Florida, Gainesville.

Lee, S.–C., Mehta, A. J. (1997). Problems in characterizing dynamics of mud shore profiles, *J. Hydraul. Eng.*, 123(4), pp. 351–361.

Lee, T. L., Tsai, J. R. C., Jeng, D.–S. (2002). Ocean waves propagating over a Coulomb-damped poroelastic seabed of finite thickness, *Comput. Geotech.*, 29(2), pp. 119–149.

Le Hir, P. (1997). *Cohesive Sediments*, eds. Burt, N., Parker, R., Watts, J., "Fluid and sediment integrated modeling application to fluid mud flows in estuaries," (Wiley) pp. 417–428.

Le Hir, P., Roberts, W., Cazaillet, O., Christie, M., Bassoullet, P., Bacher, C. (2000). Characterization of intertidal flat hydrodynamics, *Cont. Shelf Res.*, 20, pp. 1433–1459.

Leliavsky, S. (1955). *An Introduction to Fluvial Hydraulics* (Constable).

Lerman, A., Lal, D., Dacey, M. (1974). *Suspended Solids in Water*, ed. Gibbs, R. J., "Stokes Settling and Chemical Reactivity of Suspended Particles in Natural Waters," (Plenum) pp. 17–47.

Letter, J. V. (2009). Significance of probabilistic parameterization in cohesive sediment bed exchange, PhD Thesis, University of Florida, Gainesville.

Letter, J. V., McAnally, W. H. (1981). Physical modeling of harbors: Assessment of predictive capabilities, Report 3, Model study of shoaling, Brunswick Harbor, Georgia, *Research Report H-75-3*, U. S. Army Engineer Waterways Experiment Station, Vicksburg, MS.

Letter, J. V., Mehta, A. J. (2011). A heuristic examination of cohesive sediment bed exchange in turbulent flows, *Coastal Eng.*, 58, pp. 779–789.

Levich, V. G. (1962). *Physicochemical Hydrodynamics* (Prentice Hall).

Lewis, E. L. (1980). The Practical Salinity Scale 1978 and its antecedents, *IEEE J. Oceanic Eng.*, OE-5(1), pp. 3–8.

Li, M. Z., Gust, G. (2000). Boundary layer dynamics and drag reduction in flows of high cohesive sediment suspensions, *Sedimentology*, 47, pp. 71–86.

Li, Y. (1996). Sediment-associated constituent release at the mud-water interface due to monochromatic waves, PhD Thesis, University of Florida, Gainesville.

Li, Y., Mehta, A. J. (1998). Assessment of hindered settling of fluid mudlike suspensions, *J. Hydraul. Eng.*, 124(2), pp. 176–178.

Li, Y., Mehta, A. J. (2001). *Coastal and Estuarine Fine Sediment Transport Processes*, eds. McAnally, W. H., Mehta, A. J., "Fluid Mud in the Wave-dominated Environment Revisited," (Elsevier, Amsterdam) pp. 79–93.

Li, Y., Parchure, T. M. (1998). Mudbanks of the southwest coast of India VI: Suspended sediment profiles, *J. Coastal Res.*, 14(4), pp. 1363–1372.

Li, Y., Wolanski, E., Xie, Q. (1993). Coagulation and settling of suspended sediment in the Jiaojiang River estuary, China, *J. Coastal Res.*, 9(2), pp. 390–402.

Liang, H.–B., Williams, P. (1993). An assessment of the impact of the operation of an additional ferry on shoreline erosion in Corte Madera Bay, *Report No. 847/877*, Philip Williams & Associates, San Francisco.

Lick, W. (1982). Entrainment, deposition and transport of fine-grained sediment in lakes, *Hydrobiologia*, 91(2), pp. 31–40.

Lick, W. (2009). *Sediment and Contaminant Transport is Surface Waters* (CRC Press).

Lick, W., Huang, H. (1993). *Nearshore and Estuarine Cohesive Sediment Transport*, ed. Mehta, A. J., "Flocculation and the Physical Properties of Flocs," (American Geophysical Union) pp. 21–39.

Lick, W., Lick, J., Ziegler, C. K. (1992). *Sediment/Water Interactions, Hydrobiologia*, eds. Hart, B. T., Sly, P. G., "Flocculation and Its Effect on the Vertical Transport of Fine-grained Sediments," 235/236(July), pp. 1–16.

Lide, D. R., ed. (2001). *CRC Handbook of Chemistry and Physics*, 82nd Ed. (CRC).

Liggett, J. A. (1994). *Fluid Mechanics* (McGraw–Hill).

Lin, C.–P. (1987). Turbidity currents and sedimentation in closed-end canals, PhD Thesis, University of Florida, Gainesville.

Lin, M. (2001). The analysis of silt behaviour induced by water waves, *Sci. China Ser. E: Technol. Sci.*, 31(1), pp. 86–96.

Lin, M. Y., Lindsay, H. M., Weitz, D. A., Ball, R. C., Klein, R., Meakin, P. (1990a). Universal reaction-limited colloid aggregation, *Phys. Rev. A: At. Mol. Opt. Phys.*, 41, pp. 2005–2020.

Lin, M. Y., Lindsay, H. M., Weitz, D. A., Ball, R. C., Klein, R., Meakin, P. (1990b). Universal diffusion-limited colloid aggregation, *J. Phys. Condens. Matter*, 2, pp. 3090–3113.

Lin, P. C.–P., Mehta, A. J. (1997). A study of fine sedimentation in an elongated laboratory basin, *J. Coastal Res.*, SI25, pp. 19–30.

Lindenberg, J., van Rijn, L. C., Winterwerp, J. C. (1989). Some experiments on wave-induced liquefaction of soft cohesive soils, *J. Coastal Res.*, SI 5, pp. 127–137.

Lintern, G., Sills, G. (2006). Techniques for automated measurement of floc properties, *J. Sed. Res.*, 76, pp. 1183–1195.

Liu, K., Mei, C. C. (1989). Effects of wave-induced friction on a muddy seabed modelled as a Bingham-plastic fluid, *J. Coastal Res.*, 5(4), pp. 777–789.

Liu, P. L.–F. (1973). Damping of water waves over porous bed, *J. Hydraul. Div.*, ASCE, 99(12), pp. 2263–2271.

Longuet–Higgins, M. S. (1972). *Waves on Beaches and Resulting Sediment Transport*, "Recent Progress in the Study of Longshore Currents," (Academic Press) pp. 203–248.

Longuet–Higgins, M. S., Stewart, R. W. (1962). Radiation stress and mass transport in gravity waves, with application to 'surf beat,' *J. Fluid Mech.*, 13, pp. 481–504.

Lott, J. W. (1986). Laboratory study on the behavior of turbidity current in a closed-end channel, MS Thesis, University of Florida, Gainesville.

Lowe, H. (2007). *Aquamud Scale for Cohesive Sediment Erodibility* (Aquamud Ltd., Dorset, England).

Lowes, C. A. (1993). Soft mud Lagrangian mass transport, Engineering Report, Department of Civil, Construction and Environmental Engineering, Oregon State University, Corvallis.

Luettich, R. A., Harleman, D. R. F., Somlyody, L. (1990). Dynamic behavior of suspended sediment concentrations in a shallow lake perturbed by episodic wind events, *Limnol. Oceanogr.*, 35(5), pp. 1050–1067.

Luettich, R. A. Jr., Wells, J. T., Kim, S. Y. (1993). *Nearshore and Estuarine Cohesive Sediments Transport, Coastal and Estuarine Studies, Vol. 42*, ed. Mehta, A. J., "*In situ* Variability of Large Aggregates: Preliminary Results on the Effects of Shear," (American Geophysical Union) pp. 447–466.

Lyklema, J. (1995). *Fundamentals of Interface and Colloid Science, Volume II: Solid-Liquid Interfaces* (Academic Press).

Lyle, W. M., Smerdon, E. T. (1965). Relation of compaction and other soil properties to erosion and resistance of soils, *Trans. Am. Soc. Agric. Eng.*, 8(3), pp. 419–422.

Maa, J., Lee, P.–Y., Chen, F. J. (1995). Bed shear stress measurement for VIMS Sea Carousel, *Mar. Geol.*, 129, pp. 129–136.

Maa, J. P.–Y., Lee, D.–Y. (2002). *Fine Sediment Dynamics in the Marine Environment*, eds. Winterwerp, J. C., Kranenburg, C., "A preliminary study on the measurement of high resolution marine sediment bed structure," (Elsevier) pp. 469–482.

Maa, J. P.–Y., Wright, L. D., Lee, C.–H., Shannon, T. W. (1993). VIMS Sea Carousel: A field instrument for studying sediment transport, *Mar. Geol.*, 115, pp. 271–287.

Maa, P.–Y. (1986). Erosion of soft mud by waves, PhD Thesis, University of Florida, Gainesville.

Maa, P.–Y., Mehta, A. J. (1987). Mud erosion by waves: a laboratory study, *Cont. Shelf Res.*, 7(11/12), pp. 1269–1284.

Maa, P.–Y., Mehta, A. J. (1990). Soft mud response to water waves, *J. Waterw. Port Coastal Ocean Eng.*, 116(5), pp. 634–650.

MacPherson, H. (1980). The attenuation of water waves over a non-rigid bed, *J. Fluid Mech.*, 97(4), pp. 721–742.

Madsen, O. S. (1974). Stability of sand bed under breaking waves, *Proc. 14th Int. Conf. Coastal Eng.*, ASCE, pp. 777–794.

Madsen, O. S. (1978). Wave-induced pore pressures and effective stresses in a porous bed, *Geotechnique*, 28(4), pp. 377–393.

Madsen, O. S. (1991). Mechanics of cohesionless sediment transport in coastal waters, *Proc. Coastal Sediments '91*, ASCE, pp. 15–27.

Maggi, F., Mietta, F., Winterwerp, J. C. (2007). Effect of variable fractal dimension on the floc size distribution suspended cohesive sediment, *J. Hydrol.*, 343, pp. 43–55.

Malcherek, A., Zielke, W. (1996). The role of macroflocs in estuarine sediment dynamics and its numerical modeling, *Proc. Estuarine Coastal Model. Conf.*, ASCE, pp. 695–706.

Malyasian Economic Planning Unit. (1986). National coastal erosion study, Phase II Report, Stanley Consultants, Jurutera Konsultant (S. E. A.) and Moffatt and Nichols Engineers, Government of Malaysia, Kuala Lumpur.

Manning, A. J. (2004). Observations of the properties of flocculated cohesive sediment in three western European estuaries, *J. Coastal Res.*, SI41, pp. 70–81.

Manning, A. J. (2006). LabSFLOC — A laboratory system to determine the spectral characteristics of flocculating cohesive sediments, *Technical Report, TR 156*, HR, Wallingford, UK.

Manning, A. J., Dyer, K. R. (2002). *Fine Sediment Dynamics in the Marine Environment*, eds. Winterwerp, J. C., Kranenburg, C., "A Comparison of Floc Properties Observed During Neap and Spring Tidal Conditions," (Elsevier, Amsterdam) pp. 233–250.

Mantz, P. A. (1977). Incipient transport of fine grains and flakes by fluids – Extended Shields diagram, *J. Hydraul. Div.*, ASCE, 103(6), pp. 601–615.

Marine Board (1983). Criteria for the depths of dredged navigation channels, Report by the Panel on Criteria for Dredged Depths of Navigation Channels, National Academy Press, Washington, D. C.

Marine Board (1987). *Responses to Changes in Sea Level: Engineering Implications* (National Academy, Washington, D. C.).

Marván, F. G. (2001). A two-dimensional numerical transport model for organic-rich cohesive sediments in estuarine waters, PhD Thesis, Heriot–Watt University, Edinburgh.

Mathew, J. (1992). Wave-mud interaction in mudbanks, PhD Thesis, Cochin University of Science and Technology, Kochi, Kerala, India.

Mathew, J., Baba, M. (1995). Mudbanks of the southwest coast of India, II: Wave mud interactions, *J. Coastal Res.*, 11(1), pp. 179–187.

Mathew, J., Baba, M., Kurian, N. P. (1995). Mudbanks of the southwest coast of India, I: Wave characteristics, *J. Coastal Res.*, 11(1), pp. 168–178.

Maude, A. D., Whitmore, R. L. (1958). A generalized theory of sedimentation, *J. Appl. Phys.*, 9, pp. 477–482.

McAnally, W. H. (1999). Transport of fine sediments in estuarial waters, PhD Thesis, University of Florida, Gainesville.

McAnally, W. H., Mehta, A. J. (2001). *Coastal and Estuarine Fine Sediment Processes*, eds. McAnally, W. H., Mehta, A. J., "Collisional Aggregation of Estuarial Fine Sediment," (Elsevier) pp. 19–39.

McAnally, W. H., Mehta, A. J. (2002). Significance of aggregation of fine sediment particles in their deposition, *Estuarine Coastal Shelf Sci.*, 54, pp. 643–653.

McAnally, W. H., Teeter, A., Schoellhamer, D., Friedrichs, C., Hamilton, D., Hayter, E., Shrestha,, P., Rodriguez, H., Sheremet, A., Kirby R. (2007). Measurement of fluid mud in estuaries, bays, and lakes II: Measurement, modeling and management, *J. Hydr. Eng.*, 133(1), pp. 23–38.

McAnally, W. H., Thomas, W. A., Ariathurai, R. (1983). *Frontiers in Hydraulic Engineering*, ed. Shen, H. T., "Multi-dimensional Modeling of Sediment Transport," (ASCE) pp. 91–95.

McCave, I. N. (1984). Size spectra and aggregation of suspended particles in the deep ocean, *Deep Sea Res.*, 31(4), pp. 329–352.

McCave, I. N., Hall, I. R. (2006). Size sorting in marine muds: Processes, pitfalls, and prospects for paleo-speed proxies, *Geochem. Geophys. Geosyst.*, 7(10), 10.1029/2006CG001284.

McClellan, T. N., Pope, M. K., Burke, C. E. (1990). Benefits of nearshore placement, *Proc. 3rd Ann. Nat. Beach Preserv. Technol. Conf.*, Florida Shore and Beach Preservation Association, Tallahassee, pp. 339–353.

McCowan, J. (1891). On the solitary wave, *Philos. Mag.*, 32, pp. 45–58.

McCutcheon, S. C. (1981). Vertical velocity profiles in stratified flows, *J. Hydraul. Div.*, ASCE, 107(8), pp. 973–988.

McDowell, D. M. (1971). Currents induced in water by settling solids, *Proc. 14th Congr. IAHR*, Vol. 1, Paris, pp. 191–198.

McFetridge, W. F. (1985). Sediment suspension by non-breaking waves over rippled beds, MS Thesis, University of Florida, Gainesville.

McLaughlin, R. T. Jr. (1959). The settling properties of suspensions, *J. Hydraul. Div.*, ASCE, 85(12), pp. 9–41.

McLean, S. R. (1985). Theoretical modeling of deep sediment transport, *Mar. Geol.*, 66, pp. 243–265.

McNeal, C. S., Mehta, A. J. (2002). Some observations on the flushing of pocket marina basins, *Proc. 4th Int. Conf. Role of Eng. towards a Better Environ.*, University of Alexandria, Egypt, pp. 217–232.

McNeil, J., Taylor, C., Lick, W. (1996). Measurements of erosion of undisturbed bottom sediments with depth, *J. Hydraul. Eng.*, 122(6), pp. 316–324.

Meadows, P. S., Tait, J. (1989). Modification of sediment permeability and shear strength by two burrowing invertebrates, *Mar. Biol.*, 101, pp. 75–82.

Mehta, A. J. (1969). Stability of a mixed suspension crystallizer with classified product withdrawal, MS Thesis, University of Florida, Gainesville.

Mehta, A. J. (1973). Depositional behavior of cohesive sediments, PhD Thesis, University of Florida, Gainesville.

Mehta, A. J. (1978). Bed friction characteristics of three tidal entrances, *Coastal Eng.*, 2, pp. 69–83.

Mehta, A. J. (1981). A review of erosion functions for cohesive sediment beds, *Proc. First Indian Conf. Ocean Eng.*, Vol. 1, Indian Institute of Technology, Madras (Chennai), pp. 122–130.

Mehta, A. J. (1991). Review notes on cohesive sedimenterosion, *Proc. Coastal Sed. '91*, ASCE, pp. 40–53.

Mehta, A. J. (2002). *Muddy Coasts of the World: Processes, Deposits and Function*, eds. Healy, T., Wang, Y., Healy, J.–A., "Mudshore Dynamics and Controls," (Elsevier) pp. 19–60.

Mehta, A. J., Ariathurai, R., Maa, P.–Y., Hayter, E. J. (1984). Fine sedimentation in small harbors, *Report UFL/COEL-TR/051*, Coastal and Oceanographic Engineering Department, University of Florida, Gainesville.

Mehta, A. J., Byrne, R. J., DeAlteris, J. T. (1976). Measurement of bed friction in tidal inlets, *Proc. 15th Int. Conf. Coastal Eng.*, ASCE, pp. 1701–1720.

Mehta, A. J., Christensen, B. A. (1977). Hydrodynamic and geometric parameters of shells as dredge material, *Proc. Second Int. Symp. Dredg. Technol.*, BHRA Fluid Engineering, Cranfield, Bedford, UK, pp. B1-1–B1-14.

Mehta A. J., Christensen, B. A. (1983). Initiation of sand transport over coarse beds in tidal entrances, *Coastal Eng.*, 7, pp. 61–75.

Mehta, A. J., Grella, M. J., Ganju, N. K., Paramygin, V. A. (2005). Sediment management in estuaries: The Loxahatchee, Florida, *Port Coastal Eng.*, ed. Bruun, P., *J. of Coastal Res.*, SI 46, pp. 276–303.

Mehta, A. J., Jaeger, J. M., Valle–Levinson, A., So, S., Hayter, E. J., Wolanski, E., Manning, A. J. (2009). Resuspension dynamics in Lake Apopka, Florida, Final Report, St. Johns River Water Management District, Palatka, FL.

Mehta, A. J., Kirby, R. (1996). Mitigation measures for eroding muddy shores, *Proc. N. Am. Water Environ. Congr.*, Session W-17, ASCE, 6 p.

Mehta, A. J., Kirby, R., Hayter, E. J. (2000). Ortega/Cedar River basin, Florida, Restoration: Work plan to assess sediment-contaminant dynamics, *Report UFL/COEL-99/019*, Coastal and Oceanographic Engineering Program, Department of Civil and Coastal Engineering, University of Florida, Gainesville.

Mehta A. J., Lee, J., Christensen, B. A. (1980). Fall velocity of shells as coastal sediment, *J. Hydraul. Div.*, ASCE, 106(11), pp. 1727–1744.

Mehta, A. J., Lee, S.–C. (1994). Problems in linking the threshold condition for the transport of cohesionless and cohesive sediment grains, *J. Coastal Res.*, 10(1), pp. 170–177.

Mehta, A. J., Lee, S.–C., Li, Y., Vinzon, S. B., Abreu, M. G. (1994). Analyses of some sedimentary properties and erodibility characteristics of bottom sediments from the Rodman Reservoir, Florida, *Report UFL/COEL/MP-94/03*, Coastal and Oceanographic Engineering Program, Department of Civil and Coastal Engineering, University of Florida, Gainesville.

Mehta, A. J., Lott, J. W. (1987). Sorting of fine sediment during deposition. *Proc. Coastal Sediments' 87*, ed. Kraus, N. C., ASCE, pp. 348–362.

Mehta, A. J., Maa, P.–Y. (1985). *Sedimentation, Flocculation and Consolidation*, eds. Moudgil, B. M., Somasundaran, P., "Fine Sedimentation in Small Harbor Basins," (Engineering Foundation, New York) pp. 405–414.

Mehta, A. J., Nayak, B. U., Hayter, E. J. (1983). A preliminary investigation of fine sediment dynamics in Cumbarjua Canal, Goa, *Mahasagar: Bull. Nat. Inst. Oceanogr., Goa, India*, 16(2), pp. 95–108.

Mehta, A. J., Özsoy, E. (1978). *Stability of Tidal Inlets: Theory and Engineering*, ed. Bruun, P., "Inlet Hydraulics" (Elsevier).

Mehta, A. J., Parchure, T. M. (2000). *Muddy Coast Dynamics and Resource Management*, eds. Flemming, B. W., Delafontaine, M. T., Liebezeit, G., "Surface Erosion of Fine-grained Sediment Revisited," (Elsevier) pp. 55–74.

Mehta, A. J., Parchure, T. M., Dixit, J. G., Ariathurai, R. (1982). *Estuarine Comparisons*, ed. Kennedy, V. S., "Resuspension Potential of Deposited Cohesive Sediment Beds," (Academic Press) pp. 591–609.

Mehta, A. J., Partheniades, E. (1975). An investigation of the depositional properties of flocculated fine sediments, *J. Hydraul. Res.*, 4, pp. 361–381.

Mehta A. J., Partheniades, E. (1979). Kaolinite resuspension properties, *J. Hydraul. Div.*, ASCE, 105(4), pp. 409–416.

Mehta, A. J., Partheniades, E. (1982). Resuspension of deposited cohesive sediment beds, *Proc. 18th Int. Conf. Coastal Eng.*, ASCE, pp. 1569–1588.

Mehta, A. J., Rao, P. V. (1985). Angle of repose of selected bivalve shell beds, *J. Coastal Res.*, 1(4), pp. 365–374.

Mehta, A. J., Srinivas R. (1993). *Nearshore and Estuarine Cohesive Sediment Transport*, ed. Mehta, A. J., "Observations on the Entrainment of Fluid Mud in Shear Flow," (American Geophysical Union) pp. 224–246.

Mehta, A. J., Weckmann, J., Christensen, B. A. (1981). Sediment management in coastal marinas: A case study, *Proc. Int. Symp. Urban Hydrol. Hydraul. Sediment Control*, University of Kentucky, Lexington, pp. 83–90.

Mehta, A. J, Williams, D. J. A., Williams, P. R., Feng, J. (1995). Tracking dynamic changes in mud bed due to waves, *J. Hydraul. Eng.*, 121(5), pp. 504–506.

Mehta, A. J., Zeh, T. A. (1980). Influence of a small inlet in a large bay, *Coastal Eng.*, 4, pp. 157–176.

Mei, C. C., Liu, K. F. (1987). A Bingham-plastic model for a muddy seabed under long waves, *J. Geophys. Res.*, 92(C13), pp. 14581–14594.

Merckelbach, L. M., Kranenburg, C. (2004a). Equations for effective stress and permeability of soft mud-sand mixtures, *Géotechnique*, 54(4), pp. 235–243.

Merckelbach, L. M., Kranenburg, C. (2004b). Determining effective stress and permeability equations for soft mud from simple laboratory experiments, *Géotechnique*, 54(9), pp. 581–591.

Meriam, J. L. (1959). *Mechanics, Part II: Dynamics*, 2nd Ed. (Wiley).

Merz, B., Thieken, A. (2005). Separating natural and epistemic uncertainty in flood frequency analysis, *J. Hydrol.*, 309, pp. 114–132.

Meyer–Peter, E., Favre, H., Einstein, H. A. (1934). Neuere versuchresultate über den geschiebetrieb, *Schweiz. Bauze.*, 103(4), pp. 89–91.

Meyer–Peter, E., Müller, R. L. (1948). Formulas for bed-load transport, *Int. Assoc. Hydraul. Res.*, 2nd Meeting, Stockholm, Sweden, pp. 39–64.

Mezger, T. G. (2006). *The Rheology Handbook: For Users of Rotational and Oscillatory Rheometers*, 2nd Ed. (Vincentz).

Michaels, A. S., Bolgers J. C. (1962). Settling rates and sediment volumes of flocculated kaolin suspensions, *Ind. Eng. Chem. Fundam.*, 1(1), pp. 24–32.

Michell, J. H. (1893). On the highest waves in water, *Philos. Mag.*, 36(5), pp. 430–435.

Midun, Z. B., Lee, S.–C. (1989). Mud coast protection – The Malaysian experience, *Proc. 6th Symp. Coastal Ocean Manage.*, ASCE, pp. 806–820.

Mietta, F., Chassagne, C., Winterwerp, J.C., 2009. Shear-induced flocculation of a suspension of kaolinite as function of pH and salt concentration, *J. Colloid. Interf. Sci.*, 336, pp. 134–141.

Migniot, C. (1968). A study of the physical properties of different very fine sediments and their behavior under hydrodynamic action, *La Houille Blanche*, 7, pp. 591–620 (in French, with abstract in English).

Migniot, C., Hamm, L. (1990). Consolidation and rheological properties of mud deposits, *Proc. 22nd Coastal Eng. Conf.*, ASCE, pp. 2975–2983.

Mikkelsen, O. A., Hill, P. S., Milligan, T. G. (2006). Single-grain, microfloc and macrofloc volume variations observed with a LISST-100 and a digital floc camera, *J. Sea Res.*, 55, pp. 87–102.

Miller, R. L., Byrne, R. J. (1966). The angle of repose for a single grain on a fixed rough bed, *Sedimentology*, 6, pp. 303–314.

Mimura, N. (1993). *Nearshore and Estuarine Cohesive Sediment Transport*, ed. Mehta, A. J., "Rates of Erosion and Deposition of Cohesive Sediments Under Wave Action," (American Geophysical Union) pp. 247–264.

Mindlin, R. D., Deresiewicz, H. (1953). Elastic spheres in contact under varying oblique forces, *J. Appl. Mech.*, 20, pp. 327–344.

Mirtskhoulava, T. E. (1991). Scouring by flowing water of cohesive and noncohesive beds, *J. Hydraul. Res.*, 29(3), pp. 341–354.

Mitchell, J. K. (1993). *Fundamentals of Soil Behavior*, 2nd Ed. (Wiley).

Mitchell, J. K., Campanella, R. G., Singh, A. (1968). Soil creep as a rate process, *J. Soil Mech. Found. Div.*, ASCE, 94(1), pp. 231–253.

Montague, C. L., Paulic, M., Parchure, T. M. (1993). *Nearshore and Estuarine Cohesive Sediment Transport*, ed. Mehta, A. J., "The Stability of Sediments Containing Microbial Communities: Initial Experiments with Varying Light Intensity," (American Geophysical Union) pp. 348–359.

Mooney, M. (1951). The viscosity of a concentrated suspension of spherical particles, *J. Colloid Sci.*, 6, pp. 162–170.

Moore, D. M., Reynolds, R. C. Jr. (1997). *X-ray Diffraction and the Identification and Analysis of Clay Minerals,* 2nd Ed. (Oxford).

Moore, W. L., Masch, F. D. Jr. (1962). Experiments on the scour resistance of cohesive sediments, *J. Geophys. Res.*, 67(4), pp. 1437–1449.

Morgan, J. P., Van Lopik, J. R., Nichols, L. G. (1953). Occurrence and development of mudflats along the western Louisiana coast, *Technical Report 2*, Coastal Studies Institute, Louisiana State University, Baton Rouge.

Moudgil, B. M., Vasudevan, T. V. (1989). *Flocculation and Dewatering*, eds. Moudgil, B. M., Scheiner, B. J., "Characterization of Flocs," (Engineering Foundation, New York) pp. 167–178.

Müller, G., Förstner, U. (1968). General relationship between suspended sediment concentration and water discharge in the Alpenrhein and other rivers, *Nature,* 217, pp. 244–245.

Munk, W. H. (1950). Origin and generation of waves, Proc. 1st Int. Conf. Coastal Eng., ASCE, pp. 1–4.

Munk, W. H., Anderson, E. A. (1948). Notes on a theory of the thermocline, *J. Mar. Res.*, 7, pp. 276–295.

Nair, A. S. K. (1988). Mudbanks (chakara) of Kerala – a marine environment to be protected, *Proc. Nat. Semin. Environ. Issues*, University of Kerala Golden Jubilee Seminar, Trivandrum, India, pp. 76–93.

Nairn, R. B. (1992). Erosion processes evaluation paper, Final Report for International Joint Commission Great Lakes–St. Lawrence River Reference Study Board, by W.F. Baird and Associates to Water Planning and Management Branch, Environment Canada, Burlington, Ontario.

Nalluri, C., Alvarez, E. M. (1992). The influence of cohesion on sediment behavior, *Water Sci. Technol.*, 25(8), pp. 151–164.

Narayana, A. C., Manojkumar, P., Tatavarti, R. (2001). *Coastal and Estuarine Fine Sediment Processes*, eds. McAnally, W. H., Mehta, A. J., "Beach Dynamics Related to the Ambalapuzha Mudbank Along the Southeast Coast of India," (Elsevier) pp. 495–507.

Needham, J. (collaborators Ling, W., Gwei–Djen, L.). (1974). *Science and Civilization in China, Vol. IV: Physics and Physical Technology*, "Part III: Civil Engineering and Nautics" (Cambridge).

Neiheisel, J. (1966). Significance of clay minerals in shoaling problems, *Technical Bulletin 10*, U. S. Army Corps of Engineers, Waterways Experiment Station, Vicksburg, MS.

Neumeier, U., Lucas, C. H., Collins, M. (2006). Erodibility and erosion patterns of mud-flat sediments investigated using an annular flume, *Aquatic Ecol.*, 40, pp. 543–554.

Ng, C.–O. (2000). Water waves over a muddy bed: a two-layer Stokes' boundary layer model, *Coastal Eng.*, 40(3), pp. 221–242.

Nichols, M. M. (1984/85). Fluid mud accumulation process in an estuary, *Geo-Mar. Lett.*, 4(3–4), pp. 171–176.

Nielsen, P. (1992). *Coastal Bottom Boundary Layers and Sediment Transport* (World Scientific).

Nigam, R., Hashimi, N. H. (2002). Has sea level fluctuation modulated human settlements in Gulf of Khambhat?, *J. Geol. Soc. India*, 59(6), pp. 583–584.

Norton, W. R., King, I. P., Orlob, G. T. (1973). A finite element model for Lower Granite Reservoir, Report to Walla Walla District Office of the U. S. Army Corps of Engineers, Walla Walla, WA.

Obi, S., Inoue, K., Furukawa, T., Masuda, S. (1996). Experimental study on the statistics of wall shear stress in turbulent channel flows, *Int. J. Heat Fluid Flow*, 17(3), pp. 187–192.

O'Brien, M. P. (1933). Review of the theory of turbulent flow and its relation to sediment-transportation, *Trans. 14th Ann. Meeting Am. Geophysic. Union*, NRC, pp. 487–491.

O'Brien, M. P. (1969). Equilibrium flow areas of inlets on sandy coasts, *J. Waterw. Harbors Coastal Eng. Div.*, ASCE, 95(1), pp. 43–52.

Ockenden, M. C. (1993). A model for the settling of non-uniform cohesive sediment in a laboratory flume and a field setting, *J. Coastal Res.*, 9(4), pp. 1094–1105.

O'Connor, B. A., Tuxford, C. (1980). Modeling siltation at dock entrances, *Third Int. Conf. Dredg. Technol.*, BHRA Fluid Engineering, Cranfield, UK, Paper F2, pp. 359–371.

Odd, N. M. V., Bentley, M. A., Waters, C. B. (1993). *Nearshore and Estuarine Cohesive Sediment Transport*, ed. Mehta, A. J., "Observations and Analysis of the Movement of Fluid Mud in an Estuary," (American Geophysical Union) pp. 430–446.

Odd, N. V. M., Rodger, J. G. (1986). An analysis of the behaviour of fluid mud in estuaries, *Report No. SR 84*, HR, Wallingford, Wallingford, UK.

Ohtsubo, K., Muraoka, K. (1986). Re-suspension of cohesive sediments by currents, *Proc. 3rd Int. Symp. River Sedimentation*, University of Mississippi, Oxford, MS, pp. 1680–1689.

Oldroyd, J. G. (1964). *Second-Order Effects in Elasticity, Plasticity and Fluid Dynamics*, ed. Abir, D., "Non-linear Stress, Rate of Strain Relations at Finite Rates of Shear in So-called 'Linear' Elastico-viscous Liquids," (Pergamon) pp. 520–529.

Orlob, G. T., Shubinski, R. P., Feigner, K. D. (1967). Mathematical modeling of water quality in estuarial systems, *Proc. Nat. Symp. Estuarine Pollut.*, Stanford University, Palo Alto, pp. 646–675.

Oseen, C. (1927). *Hydrodynamik*, Ch. 10 (Akademische Verlagsgesellschaft, Lepizig).

Overbeek, J. T. G. (1952). *Colloid Science, Vol. 1*, ed. Kruyt, H. R., "Kinetics of Flocculation," (Elsevier) pp. 278–300.

Owen, M. W. (1970). A detailed study of the settling velocity of an estuary mud, *Report No. IT 78*, HR, Wallingford, Wallingford, UK.

Owen, M. W. (1971). The effect of turbulence on the settling velocities of silt flocs, *Proc. 15th Congr. IAHR*, Vol. 4, Paris, pp. 27–32.

Owen, M. W. (1972). The effect of temperature on the settling velocity of estuary mud. *Report No. INT 106*, Hydraulics Research Station, Wallingford, UK.

Owen, M. W. (1977). *The Sea, Vol. 6: Marine Modeling*, eds. Goldberg, E. D., McCave, I. N., O'Brien, J. J., Steele, J. H., "Problems in the Modeling of Transport, Erosion and Deposition of Cohesive Sediments," (Wiley–Interscience) pp. 515–537.

Ozdemir, C. E. (2010). Numerical investigation of particle laden flow in oscillatory channel and its implication to wave induced fine sediment transport, PhD Thesis, University of Florida, Gainesville.

Özsoy, E. (1977). Flow and mass transport in the vicinity of tidal inlets, PhD Thesis, University of Florida, Gainesville.

Özsoy, E. (1986). Ebb-tidal jets: a model of suspended sediment and mass transport at tidal inlets, *Estuarine Coastal Shelf Sci.*, 22(1), pp. 45–62.

Özsoy, E., Ünlüata, Ü. (1982). Ebb-tidal flow characteristics near inlets, *Estuarine Coastal Shelf Sci.*, 14(3), pp. 251–263.

Ozturgut, E., Lavelle, J. W. (1986). Settling analysis of fine sediment in salt water at concentration low enough to preclude flocculation, *Mar. Geol.*, 69, pp. 353–362.

Paaswell, R. E. (1973). *Soil Erosion: Causes and Mechanisms; Prevention and Control*, ed. Gray, D. H., Special Report 135, "Causes and Mechanisms of Cohesive Soil Erosion: The State of the Art," (Highway Research Board, Washington, D. C.) pp. 52–74.

Pal, R. (2003). Rheology of concentrated suspensions of deformable elastic particles such as human erythrocytes, *J. Biomech.*, 36, pp. 981–989.

Panagiotopoulos, I., Voulgaris, G., Collins, M. B. (1997). The influence of clay on the threshold of movement of fine sandy beds, *Coastal Eng.*, 32, pp. 19–43.

Papanicolaou, A. N. (1999). Discussion of, "Pickup probability for sediment entrainment," by N.–S. Cheng and Y.–M. Chiew, *J. Hydraul. Eng.*, 125(7), pp. 788–789.

Papanicolaou, A. N., Diplas, P. (1999). Numerical solution of a non-linear model for self-weight solids settlement, *Appl. Math. Modell.*, 23, pp. 345–362.

Papanicolaou A. N., Diplas, P., Evaggelopoulos, N., Fotopoulos, S. (2002). Stochastic incipient motion criterion for spheres under various bed packing conditions, *J. Hydraul. Eng.*, 128(4), pp. 369–380.

Papanicolaou A. N., Elhakeem, M., Hilldale, R. (2007). Fluvial erosion of cohesive river banks, *Water Resources Res.*, 43, W12418, doi:1029/2006WR005763.

Paramygin, V. A. (2002). Effect of sediment trapping in an estuary, MS Thesis, University of Florida, Gainesville.

Parchure, T. M. (1984). Erosional behavior of deposited cohesive sediments, PhD Thesis, University of Florida, Gainesville.

Parchure, T. M., Brown, B., McAdory, R. T. (2000). Design of sediment trap at Rollover Pass, Texas, *Report ERDC/CHL TR-00-23*, U. S. Army Engineer Research and Development Center, Vicksburg, MS.

Parchure, T. M., Mehta, A. J. (1985). Erosion of soft cohesive sediment deposits, *J. Hydraul. Eng.*, 111(10), pp. 1308–1326.

Parker, G. (1982). Conditions for ignition of catastrophically erosive turbidity currents, *Mar. Geol.*, 46, pp. 307–327.

Parker, G., Fukushima, Y., Pantin, H. M. (1986). Self-accelerating turbidity currents, *J. Fluid Mech.*, 171, pp. 145–181.

Parker, G., Garcia, M., Fukushima, Y., Yu, W. (1987). Experiments on turbidity currents over an erodible bed, *J. Hydraul. Res.*, 25(1), pp. 123–147.

Parker, W. R. (1986). *Estuarine Cohesive Sediment Dynamics*, ed. Mehta, A. J., "On the Observation of Cohesive Sediment Behavior for Engineering Purposes," (Springer–Verlag) pp. 270–289.

Parker, W. R. (1987). Observations on fine sediment transport phenomena in turbid coastal environments, *Cont. Shelf Res.*, 7(11/12), pp. 1285–1293.

Parker, W. R. (1994). *Coastal, Estuarial and Harbour Engineer's Reference Book*, eds. Abbott, M. B., Price, W. A., "Determining Depth and Navigability in Fine Sediment Areas," (Chapman and Hall) pp. 611–614.

Parker, W. R., Kirby, R. (1982). *Estuarine Comparisons,* ed. Kennedy, V. S., "Time Dependent Properties of Cohesive Sediment Relevant to Sedimentation Management — European Experience," (Academic Press) pp. 573–589.

Partheniades, E. (1962). A study of erosion and deposition of cohesive soils in salt water, PhD Thesis, University of California, Berkeley.

Partheniades, E. (1965). Erosion and deposition of cohesive soils, *J. Hydraul. Div.,* ASCE, 91(1), pp. 105–138.

Partheniades, E. (1977). Unified view of wash load and bed material load, *J. Hydraul. Div.,* ASCE, 103(9), pp. 1037–1057.

Partheniades, E. (1993). *Nearshore and Estuarine Cohesive Sediment Transport,* ed. Mehta, A. J., "Turbulence, Flocculation, and Cohesive Sediment Dynamics," (American Geophysical Union) pp. 40–59.

Partheniades, E. (2009). *Cohesive Sediments in Open Channels* (Elsevier).

Partheniades, E., Cross, R. H., Ayora, A. (1968). Further research on the deposition of cohesive sediments, *Proc. 11th Int. Conf. Coastal Eng.,* ASCE, pp. 723–742.

Partheniades, E., Dermissis, V. (1978). Interfacial friction coefficients in two layered stratified flow, *Proc. 16th Int. Conf. Coastal Eng.,* ASCE, pp. 2778–2797.

Partheniades, E., Dermissis, V., Mehta, A. J. (1975). On the shape and interfacial resistance of arrested saline wedges, *Proc. 16th Congr. IAHR,* São Paulo, Brazil, pp. 157–164.

Partheniades, E., Kennedy, J. F., Etter, R. J., Hoyer, R. P. (1966). Investigations of the depositional behavior of fine cohesive sediments in an annular rotating channel, *Report No. 96,* Hydrodynamics Laboratory, MIT, Cambridge.

Paterson, D. M. (1997). *Cohesive Sediments,* eds. Burt, N., Parker, R., Watts, J., "Biological Mediation of Sediment Erodibility: Ecology and Physical Dynamics," (Wiley) pp. 215–229.

Perlin, M., Dean, R. G. (1983). A numerical model to simulate sediment transport in the vicinity of coastal structures, *Miscellaneous Report 83-10,* Coastal Engineering Research Center, U. S. Army Corps of Engineers.

Pezzeta, J. M. (1974). Sedimentation off the Kewaunee nuclear power plant, *Sea Grant College Technical Report No. WIS-SG-74-221,* University of Wisconsin, Madison.

Phan–Thien, N. (1983). A similarity solution for Rayleigh flow of a Bingham fluid, *J. Appl. Mech.,* 50(1), p. 229.

PIANC. (2008). Minimising harbor siltation, *Maritime Navigation Commission Report No. 102,* PIANC Secretariat General, Brussels.

Piedra–Cueva, I., Mory, M. (2001). *Coastal and Estuarine Fine Sediment Transport Processes,* eds. McAnally, W. H., Mehta, A. J., "Erosion of a Deposited Layer of Cohesive Sediment," (Elsevier) pp. 41–51.

Posmentier, E. S. (1977). The generation of salinity fine structure by vertical diffusion, *J. Phys. Oceanogr.,* 7, pp. 298–300.

Postma, H. (1981). Exchange of materials between the North Sea and the Wadden Sea, *Mar. Geol.,* 40, pp. 199–213.

Potter, P. E., Heling, D., Shimp, N. F., VanWie, W. (1975). Clay mineralogy of modern alluvial muds of the Mississippi River Basin, *Bull. Cent. Rech Pau-SNPA*, 9, pp. 353–389.

Powell, M. A., Thieke, R. J., Mehta, A. J. (2006). Morphodynamic relationships for ebb and flood delta volumes at Florida's tidal entrances, *Ocean Dyn.*, 56, pp. 295–307.

Prandtl, L. (1952). *Essentials of Fluid Dynamics* (Hefner).

Purandare, U. V., Mehta, A. J. (1987). Laboratory study of tidal ebb jet, *Second Nat. Conf. Dock Harbour Eng.*, Indian Institute of Technology, Madras, pp. 60–67.

Quigley, R. M., Gelinas, P. J., Bou, W. T., Packer, R. W. (1977). Cyclic erosion-instability relationships: Lake Erie north shore bluffs, *Can. Geotech. J.*, 14(3), pp. 310–323.

Rao, S. R. (1973). *Lothal and the Indus Civilization* (Asia Publishing House, Bombay).

Rao, S. R. (1994). *The Lost City of Dvārakā* (Aditya Prakashan, New Delhi).

Raudkivi, A. J. (1998). *Loose Boundary Hydraulics*, 3rd Ed. (Telford).

Raudkivi, A. J., Hutchison, D. L. (1974). Erosion of kaolinite by flowing water, *Proc. R. Soc. London*, 337, pp. 537–544.

Raveendran, P., Amritharajah, A. (1995). Role of short-range forces in particle detachment during filter backwashing, *J. Environ. Eng.*, 121(13), pp. 860–868.

Ravens, T. M., Gschwend, P. M. (1999). Flume measurements of sediment erodibility in Houston Harbor, *J. Hydraul. Eng.*, 125(10), pp. 998–1005.

Ravisangar, V., Dennett, K. E., Sturm, T. W., Amrithrajah, A. (2001). Effect of sediment pH on resuspension of kaolinite sediments, *J. Environ. Eng.*, 127(6), pp. 531–538.

Reichardt, H. (1951). VollstandigeDarstellung der turbulenten Geschwindigkeits-verteilung in glatten Leitungen, *Zeitschrift für angewandte Mathematik und Physik* (*J. Appl. Math. Phys.*), 31(7), pp. 208–219.

Reid, R. O., Kajiura, K. (1957). On the damping of gravity waves over a permeable seabed, *Trans. Am. Geophys. Union*, 38(5), pp. 662–666.

Reiner, M. (1956). *Rheology*, ed. Eirich, F. R., Chapter 2, "Phenomenological Macrorheology" (Academic Press).

Reiner, M. (1969). The Deborah number, *Phys. Today*, 17, p. 62.

Rektorik, R. J., Smerdon, E. T. (1964). Critical shear stress in cohesive soils from a rotating shear apparatus, Paper presented at 17th Ann. Meeting Am. Soc. Agric. Eng., Chicago, Paper no. 64-216.

Ren, M. (1992). Human impact on coastal landform and sedimentation—Yellow River example, *GeoJournal*, 28(4), pp. 443–448.

Richards, A. (2003). Sea breeze genesis—A missing factor, MSc Thesis, School of Mathematics Meteorology and Physics, University of Reading, Reading, UK.

Richardson, J. F., Zaki, W. N. (1954). The sedimentation of a suspension of uniform spheres under conditions of viscous flow, *Chem. Eng. Sci.*, 3, pp. 65–73.

Righetti, M., Lucarelli, C. (2007). May the Shields theory be extended to cohesive and adhesive sediments?, *J. Geophys. Res.*, 112, C05039, doi:10.1029/2006JC003669.

Roberts, J., Jepsen, R., Lick, W. (1998). Effects of particle size and bulk density on erosion of quartz particles, *J. Hydraul. Eng.*, 124(12), pp. 1261–1267.

Roberts, J. D., Jepsen, R. A., James, S. C. (2003). Measurements of sediment erosion and transport with the adjustable shear stress erosion and transport flume, *J. Hydraul. Eng.*, 129(3), pp. 862–871.

Roberts, W., Le Hir, P., Whitehouse, R. J. S. (2000). Investigation using mathematical models of the effect of tidal currents and waves on the profile shape of intertidal mudflats, *Cont. Shelf Res.*, 20, pp. 1079–1097.

Robertson, I. D. M., Eggleton, R. A. (1991). Weathering of granitic muscovite to kaolinite and halloysite and of plagioclase-derived kaolinite to halloysite, *Clays Clay Miner.*, 39, pp. 113–126.

Robillard, D. J. (2009). A laboratory investigation of mud seabed thickness contributing to wave attenuation, PhD Thesis, University of Florida, Gainesville.

Rodriguez, H. N. (1997). A mechanism for non-breaking wave-induced transport of fluid mud at open coasts, MS Thesis, University of Florida, Gainesville.

Rodriguez, H. N. (2000). Mud bottom evolution at open coasts, PhD Thesis, University of Florida, Gainesville.

Rodriguez, H. N., Jiang, J., Mehta, A. J. (1997). Determination of selected sedimentary properties and erodibility of bottom sediments from the lower Kissimmee River and Taylor Creek–Nubbin Slough basins, Florida, *Report UFL/COEL/MP-97/09*, Coastal and Oceanographic Engineering Program, Department of Civil and Coastal Engineering, University of Florida, Gainesville.

Rodriguez, H. N., Mehta, A. J. (1998). *Sedimentary Processes in the Intertidal Zone*, eds. Black, K. S., Patterson, D. M., Cramp, A., Special Publication 139, "Considerations on Wave-induced Fluid Mud Streaming at Open Coasts," (Geological Society, London) pp. 177–186.

Rodriguez, H. N., Mehta, A. J. (2000). Longshore transport of fine-grained sediment, *Cont. Shelf Res.*, 20, pp. 1419–1432.

Rodriguez, H. N., Mehta, A. J. (2001). Modeling muddy coast response to waves, *J. Coastal Res.*, SI27, pp. 137–148.

Rohan, K., Lefebvre, G., Douville, S., Millette, J.–P. (1986). A new technique to evaluate erodibility of cohesive material, *Geotech. Test. J.*, 9(2), pp. 87–92.

Rose, C., Thorne, P. (2001). Measurements of suspended sediment transport parameters in a tidal estuary, *Cont. Shelf Res.*, 21, pp. 1551–1575.

Rosenqvist, I. Th. (1953). Considerations on the sensitivity of Norwegian quick clays, *Geotechnique*, 3(5), pp. 195–200.

Rosillon, R., Volkenborn, C. (1964). Sedimentación de material cohesivo en agua salada, Thesis, University of Zulia, Maracaibo, Venezuela (in Spanish).

Ross, M. A. (1988). Vertical structure of estuarine fine sediment suspensions, PhD Thesis, University of Florida, Gainesville.

Ross, M. A., Mehta, A. J. (1989). On the mechanics of lutoclines and fluid mud, *J. Coastal Res.*, SI5, pp. 51–61.

Ross, M. A., Mehta, A. J., Lin, C. P. (1987). On the definition of fluid mud, *Proc. Hydraul. Eng. Conf.*, ASCE, pp. 231–236.

Rossby, C. G. (1932). Conditions of kinematic similarity, *MIT Meteorol. Pap.*, 1(4), pp. 9–14.

Rossby, C. G., Montgomery, R. B. (1935). The layer of frictional influence in wind and ocean currents, *Pap. Phys. Oceanogr. Meteorol.*, 3(3), pp.1–101.

Rouse, H. (1951). *Engineering Hydraulics* (Wiley).

Russell, J. S. (1844). Report on waves, *Proc. 14th Meeting Br. Assoc. Adv. Sci.*, York, UK, pp. 311–390.

Safak, I., Sheremet, A., Allison, M. A., Hsu, T.–J. (2010). Bottom turbulence on the muddy Atchafalaya Shelf, *J. Geophys. Res.*, 115(C12019), doi:10.1029/2010JC006157.

Saffman, P. G., Turner, J. S. (1956). On the collision of drops in turbulent clouds, *J. Fluid Mech.*, 1(1), pp. 16–30.

Sahin, C., Safak, I., Sheremet, A., Mehta, A. J. (2012). Observations on cohesive bed reworking by waves: Atchafalaya Shelf, Louisiana, *J. Geophys. Res.*, 117, C09025, doi:10.1029/2011JC007821.

Sakakiyama, T., Bijker, E. W. (1989). Mass transport velocity in mud layer due to progressive waves, *J. Waterw. Port Coastal Ocean Eng.*, 115(5), pp. 614–633.

Sampath, A. (2009). Sedimentation of Lake Apopka mud and comparison with other sediments, MS Thesis, University of Florida, Gainesville.

Samsami, F., Khare, Y. P., Mehta, A. J. (2012). Rheometric characterization of the fluid mud forming potential of a bay mud, *Proc. 8th Int. Symp. Lowland Tech.*, Bali, Indonesia, pp. 917–921.

Sanchez, M. (2006). Settling velocity of the suspended sediment in three high-energy environments, *Ocean Eng.*, 33, pp. 665–678.

Sanchez, M., Levacher, D. (2007). The influence of particle size of the dispersed mineral fraction on the settlement of marine and estuarine muds, *Geo-Mar Lett.*, 27, pp. 303–313.

Sanford, L. P., Halka, J. P. (1993). Assessing the paradigm of mutually exclusive erosion and deposition of mud, with examples from upper Chesapeake Bay, *Mar. Geol.*, 114(1–2), pp. 35–37.

Sanford, L. P., Maa, J. P.–Y. (2001). A unified erosion formulation for fine sediments, *Mar. Geol.*, 179, pp. 9–23.

Sanford, L. P., Panageotou, W., Halka, J. P. (1991). Tidal resuspension of sediments in northern Chesapeake Bay, *Mar. Geol.*, 97(1/2), pp. 87–103.

Sankalia, H. D. (1987). *Prehistoric and Historic Archeology of Gujarat* (Munshiram Manoharlal Publishers, New Delhi).

Sarikaya, H. Z. (1977). Numerical model for discrete settling, *J. Hydraul. Div.*, ASCE, 103(8), pp. 866–876.

Scarlatos, P. D., Mehta, A. J. (1990). *Residual Currents and Long-Term Transport*, ed. Cheng, R. T., "Some Observations on Erosion and Entrainment of Estuarine Fluid Muds," (Springer–Verlag) pp. 321–332.

Scarlatos, P. D., Mehta A. J. (1993). *Nearshore and Estuarine Cohesive Sediment Transport*, ed. Mehta, A. J., "Instability and Entrainment Mechanisms at the Stratified Fluid Mud-water Interface" (American Geophysical Union), pp. 205–223.

Scarlatos, P. D., Wilder, B. J. (1990). *Hydraulics/Hydrology of Arid Lands (H2AL)*, ed. French, R. H., "Experimental Investigation of Density-driven Hyperconcentrated Flows," (ASCE) pp. 633–638.

Scarlatos, P. D., Zhang, Y. (1991). *Hydraulic Engineering*, ed. Shane, R. M., "Fluid Mud Problems in Lakes, Reservoirs and Detention Ponds," (ASCE) pp. 1127–1132.

Schelske, C. L. (1997). Sediment and phosphorous deposition in Lake Apopka, *Special Publication SJ97-SP21*, St. Johns River Water Management District, Palatka, FL.

Schijf, J. B., Schoenfeld, J. C. (1953). Theoretical considerations on the motion of salt and fresh water, *Proc. Minnesota Int. Hydraul. Convention*, IAHR and ASCE, pp. 321–333.

Schiller, L., Naumann, A. (1933). Über die grundlegenden berechnungen bei der Schwerkraftaufbereitung, *Z. Ver. Dtsch. Ing.*, 77, pp. 318–320.

Schlichting, H. (1968). *Boundary-Layer Theory*, 6th Ed. (McGraw–Hill).

Schlichting, H., Gersten, K. (2000). *Boundary-Layer Theory*, 8th Ed. revised and enlarged (McGraw–Hill).

Schreuder, F. W. A. M., van Dieman, A. J. G., Stein, H. N. (1986). Viscoelastic properties of concentrated suspensions, *J. Colloid Interface Sci.*, 111(1), pp. 35–43.

Schwartzberg, J. E., ed. (1992). *A Historical Atlas of South Asia*, 2nd impression with additional material (Oxford).

Schweim, C., Spork, V., Prochnow, J. V., Köngeter, J., Zhou J. (2000). Large eddy simulation of a lid-driven rotating annular flume flow, *Proc. 4th Int. Conf. Hydroinf.*, Iowa Institute of Hydraulic Research, Iowa City (CD-ROM).

Scully, M. E., Friedrichs, C. T., Wright, L. D. (2002). Application of an analytical model of critically stratified gravity-driven sediment transport and deposition to observations from the Eel River continental shelf, Northern California, *Cont. Shelf Res.*, 22, pp. 1951–1974.

Sharif, A. R. (2002). Critical shear stress and erosion of cohesive soils, PhD Thesis, State University of New York at Buffalo.

Sharma, J. N. (1973). Effect of flow characteristics and boundary roughness on the intensity and statistical distribution of turbulent velocity fluctuations, MS Thesis, University of Florida, Gainesville.

Sharma, J. N., Dean, R. G. (1979). Development and evaluation of a procedure for simulating a random directional second order sea surface and associated wave forces, *Ocean Engineering Report No. 20*, Department of Civil Engineering, University of Delaware, Newark.

Shemdin, O. H. (1970). River-coast interaction: laboratory simulation, *J. Waterw. Harbors Coastal Eng. Div.*, ASCE, 96(4), pp. 755–766.

Sheng, Y. P. (1986). *Estuarine Cohesive Sediment Dynamics*, ed. Mehta, A. J., "Modeling Bottom Boundary Layer and Cohesive Sediment Dynamics in Estuarine and Coastal Waters," (Springer–Verlag) pp. 360–400.

Sheng, Y. P., Chen, X. J. (1992). A three-dimensional numerical model of hydrodynamics, sediment transport and phosphorus dynamics in Lake Okeechobee: Theory, model development, and documentation, Spring 1989, Final Report to the South Florida Water Management District, West Palm Beach.

Sheng, Y. P., Villaret, C. (1989). Modeling the effect of suspended sediment stratification on bottom exchange processes, *J. Geophys. Res.*, 94(C10), pp. 14429–14444.

Shepard, F. P. (1963). *Submarine Geology* (Harper).

Sheremet, A., Mehta, A., Liu, B., Stone, G.W. (2005). Wave-sediment interaction on a muddy inner shelf during Hurricane Claudette, *Estuarine Coastal Shelf Sci.*, 63, pp. 225–233.

Shields, A. (1936). Application of similarity principles and turbulence research to bed-load movement, *Mitt. der Pruss. Versuch. Wasser. und Schiff.*, 26, Berlin (Translated from German by W.P. Ott and J.C. Van Uchlen, Caltech).

Shrestha, P. L., Orlob, G. T. (1996). Multiple distribution of cohesive sediments and heavy metals in estuarine systems, *J. Environ. Eng.*, 122(8), pp. 730–740.

Shubinski, R. P., Krone, R. B. (1970). A proposed model for sediment transport in estuaries, Paper presented at *ASCE Conf. on Water Res. in the Seventies*, Memphis, TN.

Shubinski, R. P., McCarty, J. C., Lindorf, M. R. (1965). Computer simulation of estuarial networks, *J. Hydraul. Div.*, ASCE, 91(5), pp. 33–49.

Sills, G. C., Elder, D. McG. (1986). *Estuarine Cohesive Sediment Dynamics*, ed. Mehta, A. J., "The Transition from Sediment Suspension to Settling Bed," (Springer–Verlag) pp. 192–205.

Simon, D. B., Albertson, M. L. (1960). Uniform water conveyance channels in alluvial material, *J. Hydraul. Div.*, ASCE, 86(5), pp. 33–71.

Sisko, A. W. (1958). The flow of lubricating greases, *Ind. Eng. Chem.*, 50(1), pp. 1789–1792.

Skafel, M. G., Bishop, C. T. (1994). Flume experiments on the erosion of till shores by waves, *Coastal Eng.*, 23(3–4), pp. 329–348.

Sleath, J. F. A. (1991). Velocities and shear stresses in wave-current flows, *J. Geophys. Res.*, 96(C8), pp. 15237–15244.

Smerdon, E. T., Beasley, R. P. (1959). Tractive force theory applied to stability of open channels in cohesive soils, *Research Bulletin No. 715*, Agricultural Experiment Station, University of Missouri, Columbia, MO.

Smith, T. J., Kirby, R., Christiansen, H. (2001). *Coastal and Estuarine Fine Sediment Processes*, eds. McAnally, W. H., Mehta, A. J., "Entrance Flow Control to Reduce Siltation in Tidal Basins," (Elsevier) pp. 459–484.

Smoluchowski, M. Z. von (1917). Versuch einer mathematischen theorie der koagulations-kinetic kolloider losungen, *Z. Phys. Chem.*, 92, pp. 129–168.

So, S. (2009). Fine sediment resuspension in Lake Apopka, Florida, MS Thesis, University of Florida, Gainesville.

Sogreah Consulting Engineers. (1988). Problems posed by dredging muddy beds in access channels to port installations, *Report 5-8500*, Grenoble, France.

Soil Conservation Service. (1992). Ion exchange analysis, *Soil Survey Laboratory Methods Manual*, U. S. Department of Agriculture, 145 p.

Son, M. (2009). Flocculation and transport of cohesive sediment, PhD Thesis, University of Florida, Gainesville.

Son, M., Hsu, T.–J. (2008). Flocculation model of cohesive sediment using variable fractal dimension, *Environ. Fluid Mech.*, 8 (1), pp. 55–71.

Soulsby, R. L. (1983). *Physical Oceanography of Coastal and Shelf Seas*, ed. Johns, B., Chapter 5, "The Bottom Boundary Layer of Shelf Seas," (Elsevier) pp. 189–266.

Soulsby, R. L. (1997). *Dynamics of Marine Sands: A Manual for Practical Applications* (Telford).

Soulsby, R. L., Hamm, L., Klopman, G., Myrhaug, D., Simons, R. R., Thomas, G. P. (1993). Wave-current interaction within and outside the bottom boundary layer, *Coastal Eng.*, 21(1), pp. 41–69.

Spencer, K. L., Manning, A. J., Droppo, I. G., Leppard, G. G., Benson, T. (2010). Dynamic interactions between cohesive sediment tracers and natural mud, *J. Soils Sediment*, 10, 1, pp. 401–1,414.

Spicer, P. T., Pratsinis, S. E., Raper, J., Amal, R., Bushell, G., Meesters, G. (1998). Effect of shear schedule on particle size, density, and structure during flocculation in stirred tanks, *Powder Technol.*, 97, pp. 26–34.

Srinivas. R. (1989). Response of fine sediment-water interface to shear flow, MS Thesis, University of Florida, Gainesville.

Srivastava, M. (1983). Sediment deposition in a coastal marina, *Report no. UFL/COEL/ MP-83/001*, Coastal and Oceanographic Engineering Department, University of Florida, Gainesville.

Stern, O. (1924). Zur theorie de elektrolytischen doppelschicht, *Z. Elektrochem.*, 30, pp. 508–516.

Stevenson, T. (1886). *Design and Construction of Harbors*, 3rd Ed. (Adam and Charles Black, Edinburgh).

Stoddard, D. M. (2002). Evaluation of sediment trap efficiency in an estuarine environment, *Report No. UFL/COEL/MPR-2001/003*, Coastal and Oceanographic Engineering Department, University of Florida, Gainesville.

Stoker, J. J. (1992). *Water Waves: The Mathematical Theory with Applications*, Paperback Ed. (Wiley).

Stokes, G. G. (1847). On the theory of oscillatory waves, *Trans. Br. Philos. Soc.*, 8, pp. 441–455.

Stolzenbach, K. D., Harleman, D. R. F. (1971). An analytic and experimental investigation of surface discharges of heated water, *Report No. 135*, Ralph M. Parsons Laboratory of Water Resources and Hydrodynamics, MIT.

Strickler, A. (1923). Contributions to the question of a velocity formula and roughness data for streams, channels and closed pipelines. (Translation T-10 from German by T. Roesgan and W.R. Brownlie, W. M. Keck Laboratory of Hydraulics and Water Resources, Cal. Inst. Tech., Pasadena, CA, 1981).

Strom, K., Keyvani, A. (2011). An explicit full-range settling velocity equation for mud flocs, *J. Sed. Res.*, 2011, 81, pp. 921–934.

Stuck, J. D. (1996). Particulate phosphorus transport in the water conveyance systems of the Everglades Agricultural Area, PhD Thesis, University of Florida, Gainesville.

Stupples, P., Plater, A. J. (2007). Statistical analysis of the temporal and spatial controls on tidal signal preservation in late—Holocene tidal rhythmites, Romney Marsh, Southeast England, *Int. J. of Earth Sci.*, 96(5), pp. 957–976.

Sumer, M. B., Deigaard, R. (1981). Particle motions near the bottom in turbulent flow in an open channel, Part 2, *J. Fluid Mech.*, 109, pp. 311–337.

Sumer, M. B., Fredsøe, J. (2002). *The Mechanics of Scour in the Marine Environment* (World Scientific).

Sutherland, T. F., Grant, J., Amos, C. L. (1998). The effect of carbohydrate production by the diatom *Nitzschia curvilineata* on the erodibility of sediment, *Limnol. Oceanogr.*, 43(1), pp. 65–72.

Swamee, P. K., Ojha, C. S. P. (1991). Drag coefficient and fall velocity of nonspherical particles, *J. Hydraul. Eng.*, 117(5), pp. 660–667.

Syvitski, J. P. M., Asprey, K. W., Leblanc, K. W. G. (1995). In-situ characteristics of particles settling within a deep-water estuary, *Deep Sea Res. Part II*, 42(1), pp. 223–256.

Syvitski, J. P. M., Lewis, A. G. (1980). Sediment ingestion by *Tigriopus californicus* and other zooplankton: mineral transformation and sedimentological considerations, *J. Sediment. Petrol.*, 50, pp. 869–880.

Szeri, A. Z. (1998). *Fluid Film Lubrication: Theory and Design* (Cambridge).

Taki, K. (2001). *Coastal and Estuarine Fine Sediment Transport Processes*, eds. McAnally, W. H., Mehta, A. J., "Critical Shear Stress for Cohesive Sediment Transport," (Elsevier) pp. 53–61.

Tarigan, A. P. M. (1996). An examination of the dependence of mud shore profiles on the nearshore environment, MS Thesis, University of Florida, Gainesville.

Tarigan, A. P. M. (2002). Modeling of shoreline evolution at an open mud coast, PhD Thesis, University Teknologi Malaysia, Johor Darul Ta'zim.

Tarigan, A. P. M., Lee, S.–C., Mehta, A. J. (1996). Recession of the muddy coast: a departure from the Bruun Rule? *Proc. 10th Cong. Asia Pacific Div. IAHR*, eds. Lee, S.–C., Hiew, K.–L., Ong, S.–H., National Hydraulics Research Institute of Malaysia, Kuala Lumpur, pp. 468–477.

Task Committee on Erosion of Cohesive Materials. (1968). Erosion of cohesive sediments, *J. Hydraul. Div.*, ASCE, 94(4), pp. 1017–1049.

Taylor, G. I. (1935). Statistical theory of turbulence, Parts 1–4, *Proc. R. Soc. London*, A151, pp. 421–478.

Tchen, C. (1947). Mean value and correlation problems connected with the motion of small particles suspended in a turbulent fluid, DSc Thesis, Technische Hogeschool, Delft, The Netherlands.

Teeter, A. M. (1986). *Estuarine Cohesive Sediment Transport*, ed. Mehta, A. J., "Vertical Transport in Fine-grained Suspension and Newly-deposited Sediment," (Springer–Verlag) pp. 170–191.

Teeter, A. M. (1992). Evaluation of new fluid mud survey system at field site, *Dredging Research Technical Note No. DRP-2-05*, U. S. Army Engineer Waterways Experiment Station, Vicksburg, MS.

Teeter, A. M. (2001a). *Coastal and Estuarine Fine Sediment Transport Processes*, eds. McAnally, W. H., Mehta, A. J., "Clay-silt Sediment Modeling Using Multiple Grain Classes, Part I: Settling and Deposition," (Elsevier) pp. 157–171.

Teeter, A. M. (2001b). *Coastal and Estuarine Fine Sediment Transport Processes*, eds. McAnally, W. H., Mehta, A. J., "Clay-silt Sediment Modeling Using Multiple Grain Classes, Part II: Application to Shallow-water Resuspension and Deposition," (Elsevier) pp. 173–187.

Teisson, C. (1997). *Cohesive Sediments*, eds. Burt, N., Parker, R., Watts, J., "A Review of Cohesive Sediment Transport Models," (John Wiley, Chichester, UK) pp. 367–381.

Tennekes, H., Lumley, J. L. (1972). *A First Course in Turbulence* (MIT, Cambridge).

Terzaghi, K. (1943). *Theoretical Soil Mechanics* (Wiley).

Thimakorn, P. (1980). An experiment on clay suspension under water waves, *Proc. 17th Int. Conf. Coastal Eng.*, ASCE, pp. 2894–2406.

Thimakorn, P. (1984). *Seabed Mechanics*, ed. Denness, B., "Resuspension of Clays Under Waves," (Graham and Trotman) pp. 91–196.

Thomas, D.G. (1963). Non-Newtonian suspensions, Part I: Physical properties and laminar transport characteristics, *Ind. Eng. Chem.*, 55(11), pp. 18–29.

Thompson, R. O. R. Y., Imberger, J. (1980). Response of a numerical model of a stratified lake to wind stress, eds. Carstens, T., McClimans, T., *Proc. 2nd Int. Symp. Stratified Flows*, IAHR, pp. 562–570.

Thorn, M. F. C. (1981). Physical processes of siltation in tidal channels, *Proc. Conf. Hydraul. Model. Appl. Marit. Eng. Prob.*, Institution of Civil Engineers, London, pp. 47–55.

Thorn, M. F. C., Parsons, J. G. (1980). Erosion of cohesive sediments in estuaries: An engineering guide, *Proc. 3rd Int. Symp. Dredg. Technol.*, BHRA, Cranfield, UK, pp. 349–358.

Thornton, E. B., Guza, R. T. (1982). Energy saturation and phase speeds measured on a natural beach, *J. Geophys. Res.*, 87, pp. 9499–9508.

Toorman, E. A. (1996). Sedimentation and self-weight consolidation, Part I: General unifying theory, *Géotechnique*, 46(1), pp. 103–113.

Toorman, E. A. (2000a). Parameterisation of turbulence damping in sediment-laden flow, *Report No. HYD/ET/00/COSINUS3*, Hydraulics Laboratory, Civil Engineering Department, Catholic University of Leuven, Haverlee, Belgium.

Toorman, E. A. (2000b). Modelling of turbulence damping in sediment-laden flow, Part 3: Drag reduction in sediment-laden turbulent flow, *Report No. HYD/ET/00/COSINUS5*, Hydraulics Laboratory, Civil Engineering Department, Catholic University of Leuven, Haverlee, Belgium.

Toorman, E. A. (2001). *Coastal and Estuarine Fine Sediment Transport Processes*, eds. McAnally, W. H., Mehta, A. J., "Cohesive Sediment Transport Modeling: European Perspective," (Elsevier) pp. 1–18.

Toorman, E. A. (2002). *Fine Sediment Dynamics in the Marine Environment*, eds. Winterwerp, J. C., Kranenburg, C., "Modelling of Turbulent Flow with Suspended Cohesive Sediment," (Elsevier) pp. 155–169.

Toorman, E. A., Berlamont, J. E. (1993). *Nearshore and Estuarine Cohesive Sediment Transport*, ed. Mehta, A. J., "Mathematical Modeling of Cohesive Sediment Settling and Consolidation," (American Geophysical Union) pp. 148–184.

Torfs, H. (1995). Erosion of mud/sand mixtures, PhD Thesis, Catholic University of Leuven, Belgium.

Torfs, H., Jiang, J., Mehta, A. J. (2001). *Coastal and Estuarine Fine Sediment Transport Processes*, eds. McAnally, W. H., Mehta, A. J., "Assessment of Erodibility of Fine/Coarse Sediment Mixtures," (Elsevier) pp. 109–123.

Trammel, M. A. (2004). Laboratory apparatus and methodology for determining water erosion rates of erodible rock and cohesive sediments, MS Thesis, University of Florida, Gainesville.

Trowbridge, J. H., Kineke, G. C. (1994). Structure and dynamics of fluid muds on the Amazon continental shelf, *J. of Geophys. Res.*, 99(C1), pp. 865–874.

Tsai, C. H., Iacobellis, S., Lick, W. (1987). Flocculation of fine-grained lake sediments due to a uniform shear stress, *J. Great Lakes Res.*, 13(2), pp. 135–146.

Tsai, C. H., Lick, W. (1986). Portable device for measuring sediment resuspension, *J. Great Lakes Res.*, 12(4), pp. 314–321.

Tsuruya, H., Nakano, S. (1987). Interactive effects between surface waves and a muddy bottom, *Proc. Specialty Conf. Advances Understanding Coastal Sediment Processes*, ed. Kraus, N.C., ASCE, pp. 50–62.

Tubman, M. W., Suhayda, J. N. (1976). Wave action and bottom movements in fine sediments, *Proc. 15th Int. Conf. Coastal Eng.*, ASCE, pp. 1168–1183.

Tyler, D. F. (1979). *Ocean and Marine Dictionary* (Cornell Maritime).

Tzang, S. Y., Hunt, J. R., Foda, M. A. (1992). Resuspension of seabed sediments by water waves, *Proc. 23rd Int. Conf. Coastal Eng.*, ASCE, pp. 69–70.

Uittenbogaard, R. E. (1995). The importance of internal waves for mixing in a stratified estuarine tidal flow, PhD Thesis, Technical University of Delft, The Netherlands.

Ulam, S. (1957). Marian Smoluchowski and the theory of probabilities in physics, *Am. J. Phys.*, 25, pp. 475–481.

U. S. Army Corps of Engineers. (1981). Engineering and design: Deep draft navigation project design, *Engineer Regulation 1110-2-1404*, Waterways Experiment Station, Vicksburg, MS.

Vallejo, I. E. (1977). Mechanics of the stability and development of the Great Lakes coastal bluffs, PhD Thesis, University of Wisconsin, Madison.

Vallejo, I. E. (1980). Analysis of wave-induced erosion of cohesive soils forming coastal bluffs, *Proc. Canadian Coastal Conf.*, NRC, Canada, Ottawa, pp. 220–228.

Vand, V. (1948). Viscosity of solutions and suspensions I: Theory, *J. Phys. Colloid Chem.*, 52(2), pp. 277–299.

van de Graaf, E. R., Koomans, R. L., Limburg, J., de Vries, K. (2006). In situ radiometric mapping as a proxy of sediment contamination: Assessment of the underlying geochemical and physical principles, *Appl. Radiat. Isot.*, 65(5), pp. 619–633.

van der Lee, W. T. B. (2000). Temporal variation of floc size and settling velocity in the Dollard estuary, *Cont. Shelf Res.*, 20, pp. 1495–1511.

van Kessel, T., Blom, C. (1998). Rheology of cohesive sediments: comparison between a natural and an artificial mud, *J. Hydraul. Res.*, 36(4), pp. 591–612.

van Kessel, T., Kranenburg, C. (1998). Wave-induced liquefaction and flow of subaqueous mud layers, *Coastal Eng.*, 34, pp. 109–127.

van Ledden, M., van Kesteren, W. G. M., Winterwerp, J. C. (2004). A conceptual framework for the erosion behaviour of sand-mud mixtures, *Continental Shelf Res.*, 24, pp. 1–11.

van Leussen, W. (1994). Estuarine macroflocs and their role in fine-grained sediment transport, PhD Thesis, University of Utrecht, The Netherlands.

van Leussen, W., van Velzen, E. (1989). High concentration suspensions: their origin and importance in Dutch estuaries and coastal waters, *J. Coastal Res.*, SI5, pp. 1–22.

van Maren, D. S., Winterwerp, J. S., Wang, Z. Y., Pu, Q. (2009). Suspended sediment dynamics and morphodynamics in the Yellow River, China, *Sedimentology*, 56, pp. 785–806.

van Olphen, H. (1977). *An Introduction to Clay Colloid Chemistry*, 2nd Ed. (Wiley).

van Olphen, H. (1987). *Chemistry of Clays and Clay Minerals*, ed. Newman, A. C. D., Mineralogical Society Monograph No. 6, "Dispersion and Flocculation," (Longman) pp. 203–224.

Vanoni, V. A. (1941). Velocity distribution in open-channels, *Trans. ASCE*, 11(6), pp. 356–357.

van Prooijen, B. C., Winterwerp, J. C. (2010). A stochastic formulation for erosion of cohesive sediments, *J. Geophys. Res.*, 115, C01005, doi:10.1029/2008JC005189.

van Rijn, L. C. (1984a). Sediment transport, Part I: Bed load transport, *J. Hydraul. Eng.*, 110(10), pp. 1431–1455.

van Rijn, L. C. (1984b). Sediment transport, Part II: Suspended load transport, *J. Hydraul. Eng.*, 110(11), pp. 1613–1641.

van Rijn, L. C. (1993). *Principles of Sediment Transport in Rivers, Estuaries and Coastal Seas* (Aqua Publications).

van Rijn, L. C., Nienhuis, L. E. A. (1985). In-situ determination of fall velocity of suspended sediment, *Proc. 21st Cong., IAHR*, Melbourne, Australia, pp. 145–148.

van Rijn, L.C. (2007a). Unified view of sediment transform by currents and waves. I: Initiation of motion, bed roughness, and bed-load transport, *J. Hydraul.Eng.*, 133(6), pp. 649-667.

van Rijn, L.C.(2007b) . Unified view of sediment transform by currents and waves. II: Suspended transport, *J. Hydraul. Eng.* 133(6), pp. 668-689.

Verbeek, H., Kuijper, C., Cornelisse, J. M., Winterwerp, J. C. (1993). *Nearshore and Estuarine Cohesive Sediment Transport*, ed. Mehta, A. J., "Deposition of Graded Natural Muds in The Netherlands," (American Geophysical Union) pp. 185–204.

Verreet, G., Berlamont, J. (1988). *Encyclopedia of Fluid Mechanics, Vol. VII: Rheology and Non-Newtonian Flows*, ed. Cheremisinoff, N. P., "Rheology and Non-Newtonian Behavior of Sea and Estuarine Mud," (Gulf Publishing) pp. 135–149.

Villaret, C., Paulic, M. (1986). Experiments on the erosion of deposited and placed cohesive sediments in an annular flume and a rocking flume, *Report No. UFL/COEL-86/07*, Coastal and Oceanographic Engineering Department, University of Florida, Gainesville.

Vincente, C. M. (1992). Experimental dredged pit of KA-HO: analysis of shoaling rates, *Proc. Int. Conf. Pearl River Estuary Surrounding Area of Macao*, Vol. 2, Civil Engineering Laboratory of Macao, paper P6.4.

Vinzon, S. B. (1998). A preliminary examination of Amazon shelf sediment dynamics, Engineer Degree Thesis, University of Florida, Gainesville.

Vinzon, S. B., Mehta, A. J. (1998). Mechanism for formation of lutoclines by waves, *J. Waterw. Port Coastal Ocean Eng.*, 124(3), pp. 147–149.

Vinzon, S. B., Mehta, A. J. (2001). *Coastal and Estuarine Fine Sediment Processes*, McAnally, W. H., Mehta, A. J., "Boundary Layer Effects Due to Suspended Sediment in the Amazon Estuary," (Elsevier) pp. 359–372.

Vinzon, S. B., Mehta, A. J. (2003). Lutoclines in high concentration estuaries: some observations at the mouth of the Amazon, *J. Coastal Res.*, 19(2), pp. 243–253.

Vinzon, S. B., Winterwerp, J. C., Nogueira, R., de Boer, G. J. (2009). Mud deposit formation on the open coast of the larger Patos Lagoon–Cassino Beach system, *Cont. Shelf Res.*, 29(3), pp. 572–588.

Visser, P. J., Booij, R., Melis, H. (1992). Flow field observations in a rotating annular flume, Paper presented at Conf. on Rem. of Sediments, Rutgers University, East Brunswick, NJ.

Viswanath, D. S., Ghosh, T. K., Prasad, D. H. L., Dutt, N. V. K., Rani, K. Y. (2007). *Viscosity of Liquids: Theory, Estimation, Experiment, and Data* (Elsevier).

Vollmers, H. (1976). Harbour inlets on tidal estuaries, *Proc. 15th Int. Conf. Coastal Eng.*, ASCE, 1, pp. 854–1,867.

Voyutsky, S. (1978). *Colloid Chemistry* (Central Books Ltd., London; English translation by N. Borov).

Wagner, N. J., Brady, J. F. (2009). Shear thickening in colloidal dispersions, *Phys. Today*, 62(10), pp. 27–32.

Walkley, A., Black, I. A. (1934). An examination of the Degtjareff method for determining soil organic matter and a proposed modification of the chromic acid titration method, *Soil Sci.*, 37, pp. 29–38.

Wan, Z. (1982). Bed material movement in hyperconcentrated flow, *Series Paper No. 31*, Institute of Hydrodynamics and Hydraulic Engineering, Technical University of Denmark, Copenhagen.

Wang, Y., Collins, M. B., Zhu, D. (1988). A comparative study of open coast tidal flats: The wash (U.K.), Bohai Bay and West Huanghai Sea (Chinese mainland), *Proc. ISCZC*, China Ocean Press, Beijing, pp. 120–134.

Wanless, H. R. (1975). Sedimentation in canals, Unpublished report, Division of Marine Geology and Geophysics, Rosenstiel School of Marine and Atmospheric Science, University of Miami, FL.

Wasp, E. J., Kenny, J. P., Gandhi, R. L. (1977). *Solid-Liquid Flow Slurry Pipeline Transportation* (Trans Tech Publications, San Francisco).

Watt, A. (1982). *Longman Illustrated Dictionary of Geology* (Longman).

Weaver, C. E. (1989). *Clays, Mud and Shale* (Elsevier).

Weaver, C. E., Beck, K. C. (1977). Miocene of the southeastern U. S.: A model for chemical sedimentation in a perimarine environment, *Sediment. Geol.*, 17(1/2), pp. 1–234.

Wells, J. T. (1983). Dynamics of coastal fluid muds in low-, moderate-, and high-tide-range environments, *Can. J. Fish. Aquat. Sci.*, 40 (Supplement 1), pp. 130–142.

Wells, J. T., Kemp, G. P. (1986). *Estuarine Cohesive Sediment Dynamics*, ed. Mehta, A. J., "Interaction of Surface Waves and Cohesive Sediments: Field Observations and Geologic Significance," (Springer–Verlag) pp. 43–65.

Wells, M. L., Goldberg, E. D. (1993). Colloid aggregation in seawater, *Mar. Chem.*, 41, pp. 353–358.

Welton, J. E. (1984). *SEM Petrology Atlas* (American Association of Petroleum Geologists, Tulsa, OK).

West, G. B., Brown, J. H. (2004). Life's universal scaling laws, *Phys. Today*, 57(9), pp. 36–42.

West, J. R., Oduyemi, K. O. K. (1989). Turbulence measurements of suspended solids concentration in estuaries, *J. Hydraul. Eng.*, 115(44), pp. 457–474.

Wheatcroft, R. A., Smith, C. R., Jumars, P. A. (1989). Dynamics of surficial trace assemblages in the deep sea, *Deep-Sea Res.*, 36, pp. 71–91.

White, S. J. (1970). Plane bed thresholds of fine grained sediments, *Nature*, 228(Oct.), pp. 152–153.

Whitehouse, R., Soulsby, R., Roberts, W., Mitchener, H. (2000a). *Dynamics of Estuarine Muds: A Manual for Practical Applications* (Thomas Telford).

Whitehouse, R. J. S., Bassoullet, P., Dyer, K. R., Mitchener, H. J., Roberts, W. (2000b). The influence of bedforms on flow and sediment transport over intertidal mudflats, *Cont. Shelf Res.*, 20 (10/11), pp. 1099–1124.

Wiberg, P. L, Smith, J. D. (1987). Calculations of the critical shear stress for motion of uniform and heterogeneous sediments, *Water Resour. Res.*, 23(8), pp. 1471–1479.

Widdows, J., Brown, S., Brinsley, M. D., Salkeld, P. N., Elliott, M. (2000). Temporal changes in intertidal sediment erodibility: influence of biological and climatic factors, *Cont. Shelf Res.*, 20, pp. 1275–1289.

Wilkinson, W. L. (1960). *Non-Newtonian Fluids: Fluid Mechanics, Mixing and Heat Transfer* (Pergamon).

Williams, D. J. A. (1986). *Estuarine Cohesive Sediment Dynamics*, ed. Mehta, A. J., "Rheology of Cohesive Suspensions," (Springer–Verlag) pp. 110–125.

Williams, D. J. A., James, A. E. (1978). Rheology of Brisbane and Rotterdam mud, Report commissioned by Hydraulic Research Station, Department of Chemical Engineering, University College, Swansea, UK.

Williams, P. R., Williams, D. J. A. (1992). The determination of dynamic moduli at high frequencies, *J. Non-Newtonian Fluid Mech.*, 42(3), pp. 267–282.

Winterwerp, J. C. (1998). A simple model for turbulence induced flocculation of cohesive sediment, *J. Hydraul. Res.*, 36(3), pp. 309–326.

Winterwerp, J. C. (1999). On the dynamics of high-concentrated mud suspensions, PhD Thesis, Technical University of Delft, The Netherlands.

Winterwerp, J. C. (2001). Stratification effects by cohesive and noncohesive sediment, *J. Geophys. Res.*, 106(C10), pp. 22,559–22,574.

Winterwerp, J. C. (2002). On the flocculation and settling velocity of estuarine mud, *Cont. Shelf Res.*, 22(9), pp. 1339–1360.

Winterwerp, J. C. (2005). Reducing harbor siltation. I: Methodology, *J. Waterw. Port Coastal Ocean Eng.*, 131(6), pp. 258–266.

Winterwerp, J. C. (2006). Stratification effects by fine suspended sediment at low, medium and very high concentrations, *J. Geophys. Res.*, 111(C05012), doi:10.1029/2005JC003019.

Winterwerp, J. C., Kranenburg, C. (1997). Erosion of fluid mud layers—II: Experiments and model validation, *J. Hydraul. Eng.*, 123(6), pp. 512–519.

Winterwerp, J. C., Manning, A. J., Martens, C., de Mulder, T., Valede, J. (2006). A heuristic formula for turbulence-induced flocculation of cohesive sediment, *Estuarine Coastal Shelf Sci.*, 68, pp. 195–207.

Winterwerp, J. C., Marieke, L., He, Q. (2009). Sediment-induced buoyancy destruction and drag reduction in estuaries, *Ocean Dyn.*, DOI 10.1007/s10236-009-0237-y.

Winterwerp, J. C., van Kesteren, W. G. M. (2004). *Introduction to the Physics of Cohesive Sediment in the Marine Environment* (Elsevier).

Winterwerp, J. C., van Kesteren, W. G. M., van Prooijen, B., Jacobs, W. (2012). A conceptual framework for shear-flow induced erosion of soft cohesive sediment beds, *J. Geophys. Res.*, C10020, doi:10.1029/2012JC008072.

Wolanski, E. (2007). *Estuarine Ecohydrology* (Elsevier).

Wolanski, E., Asaeda, T., Imberger, J. (1989). Mixing across a lutocline, *Limnol. Oceanogr.*, 34(5), pp. 931–938.

Wolanski, E., Brush, L. M., Jr. (1975). Turbulent entrainment across stable density step structures, *Tellus*, 27(3), pp. 259–268.

Wolanski, E., Delesalle, B., Gibbs, R. (1994). Carbonate mud in Matavia Atoll, French Polynesia: Suspension and export, *Mar. Pollut. Bull.*, 29(1–3), pp. 36–41.

Wolanski, E., Gibbs, R., Ridd, P., King, B., Hwang, K. Y., Mehta, A. (1991). Fate of dredge spoil, Cleveland Bay, Townsville, *Proc. 10th Australasian Conf. Coastal Ocean Eng.*, Auckland, New Zealand, pp. 63–66.

Wolanski, E., Gibbs, R., Ridd, P., Mehta, A. (1992). Settling of ocean-dumped dredged material, Townsville, Australia, *Estuarine Coastal Shelf Sci.*, 35, pp. 473–489.

Wolanski, E., Jones, M., Williams, W. T. (1981). Physical properties of Great Barrier Reef Lagoon waters near Townsville. II, Seasonal variations, *Aust. J. Mar. Freshwater Res.*, 32, pp. 321–334.

Wood, W. H., Granquist, W. T., Krieger, I. M. (1955). Viscosity studies on dilute clay suspensions, *Clays Clay Miner.*, 4(1), pp. 240–250.

Wotherspoon, D. J. J., Ashley, R. M. (1992). Rheological measurement of the yield strength of combined sewer deposits, *Water Sci. Technol.*, 25(8), pp. 165–170.

Wright, L. D., Friedrichs, C. T., Kim, S. C., Scully, M. E. (2001). The effects of ambient currents and waves on the behavior of turbid hyperpycnal plumes on continental shelves, *Mar. Geol.*, 175, pp. 25–45.

Wright, L. D., Wiseman, W. J., Bornhold, B. D., Prior, D. B., Suhayda, J. N., Keller, G. H., Yang, Z. S., Fan, Y. B. (1988). Marine dispersal and deposition of Yellow river silts by gravity driven underflows, *Nature*, 332, pp. 629–632.

Wright, S. A., Schoellhamer, D. H. (2005). Estimating sediment budgets at the interface between rivers and estuaries with application to the Sacramento–San Joaquin River Delta, *Water Resour. Res.*, 41, doi: 10.1029/2004WR003753.

Wu, W., Wang, S. S. Y. (2006). Formulas for sediment porosity and settling velocity, *J. Hydraul. Eng.*, 132(8), pp. 858–862.

Yamamoto, T. (1983). On the response of a Coulomb-damped poroelastic bed to water waves. *Mar. Geotech.*, 5(2), pp. 93–130.

Yamamoto, T., Koning, H. L., Sellmeijer, H., van Hijum, E. (1978). On the response of a poro-elastic bed to water waves, *J. Fluid Mech.*, 87(Part 1), pp. 193–206.

Yamamoto, T., Takahashi, S. (1985). Wave damping by soil motion, *J. Waterw. Port Coastal Ocean Eng.*, 111(1), pp. 62–77.

Yamanishi, H., Higashi, O., Kusuda, T., Watanabe, R. (2001). *Coastal and Estuarine Fine Sediment Processes*, eds. McAnally, W. H., Mehta, A. J., "Mud Scour on a Slope Under Breaking Waves," (Elsevier) pp. 63–77.

Yeh, H. Y. (1979). Resuspension properties of flow deposited cohesive sediment beds, MS Thesis, University of Florida, Gainesville.

Yeung, A. K. C., Pelton, R. P. (1996). Micromechanics: a new approach to studying the strength and breakup of flocs, *J. Colloid Interface Sci.*, 184, pp. 579–585.

Young, I. R., Verhagen, L. A. (1996). The growth of fetch limited waves in water of finite depth, Part 1: Total energy and peak frequency, *Coastal Eng.*, 29, pp. 47–78.

Yu, Z., Jin, L., Zhang, Q. (1994). Considerations of hydrodynamic characteristics, sediments & environmental problems of muddy coast in the construction of Lianyungang harbor, *Chin. J. Oceanol. Limnol.*, 12(2), pp. 97–105.

Yu, Z., Jin, L., Zhang, Y. (1987a). Washing and silting of muddy beach due to discharge of dredged material in Lianyungang and analysis of beach profile evolution, *Proc. 2nd Conf. Coastal Port Eng. in Dev. Countries (COPEDEC)*, China Ocean Press, Beijing, pp. 1574–1584.

Yu, Z., Zhang, Y., Chen, D., Jin, L. (1987b). The hydrodynamic characteristics of nearshore waters and the discharging process of mudflats, a case study of the mudflats near Lianyungang Island, China, *Chin. J. Oceanol. Limnol.*, 5(2), pp. 97–108.

Zabawa, C. F. (1978a). Flocculation in the turbidity maximum of northern Chesapeake Bay, PhD Thesis, University of South Carolina, Columbia, SC.

Zabawa, C. F. (1978b). Microstructure of agglomerated suspended sediments in northern Chesapeake Bay estuary, *Science*, 202, pp. 49–51.

Zeller, J. (1963). Einfürung in den sedimenttransport offner Gerinne (Broschiert), *Schweiz. Bauzeitung*, Jgg. 81.

Zienkiewicz, O. C., Chang, C. T., Bettess, P. (1980). Drained, undrained, consolidating and dynamic behaviour assumptions in soils, *Geotechnique*, 30(4), pp. 385–395.

Zilitinkevich, S. S. (1975). Comments on "A model for the dynamics of the inversion above a convective boundary layer," *J. Atmos. Sci.*, 32, pp. 991–992.

Index